Handbook of Lipid Research   **6**

Glycolipids, Phosphoglycolipids, and Sulfoglycolipids

Handbook of Lipid Research

Editor: Donald J. Hanahan
The University of Texas Health Science Center at San Antonio
San Antonio, Texas

Handbook of Lipid Research   *6*

Glycolipids, Phosphoglycolipids, and Sulfoglycolipids

Edited by
Morris Kates
University of Ottawa
Ottawa, Ontario, Canada

Plenum Press · **New York and London**

QP
752
.G56
G59x
1990

Library of Congress Catalog Card Number 88-640002

ISBN 0-306-43355-9

© 1990 Plenum Press, New York
A Division of Plenum Publishing Corporation
233 Spring Street, New York, N.Y. 10013

Printed in the United States of America

Contributors

Werner Fischer, Institut für Biochemie der Medizinischen Fakultät, Universität Erlangen-Nürnberg, D-8520 Erlangen, Federal Republic of Germany

Jill Gigg, Laboratory of Lipid and General Chemistry, National Institute for Medical Research, London NW7 1AA, England

Roy Gigg, Laboratory of Lipid and General Chemistry, National Institute for Medical Research, London NW7 1AA, England

Mayer B. Goren, Division of Molecular and Cellular Biology, Department of Pediatrics, National Jewish Center for Immunology and Respiratory Medicine, Denver, Colorado 80206

Morris Kates, Department of Biochemistry, University of Ottawa, Ottawa, Ontario, Canada K1N 6N5

Robert K. Murray, Departments of Biochemistry and Pathology, University of Toronto, Toronto, Ontario, Canada M5S 1A8

Rajagopalan Narasimhan, Departments of Biochemistry and Pathology, University of Toronto, Toronto, Ontario, Canada M5S 1A8

Alexander P. Tulloch†, Plant Biotechnology Institute, National Research Council of Canada, Saskatoon, Saskatchewan, Canada S7N OW9

† Deceased

Preface

The early history and development of the field of glycolipids was concerned mainly with the predominant glycolipids found in higher animal tissues, namely the glycosphingolipids, as has been extensively documented by J. N. Kanfer and S. Hakomori in Volume 3 of this series. The major glycolipids in organisms of the plant kingdom, however, such as bacteria, yeasts and fungi, algae, and higher plants, are glycoglycerolipids, although glycosphingolipids are also present as minor components in these organisms, except for bacteria.

It is of interest that one of the pioneers in glycosphingolipid research, Herbert E. Carter, also pioneered the discovery and structural elucidation of the plant galactosyldiacylglycerols. This class of glycolipids is present in chloroplast membranes and must surely be one of the most ubiquitous and abundant natural substances in the world, thereby deserving the attention of lipid biochemists. It is therefore surprising to learn that in contrast to the glycosphingolipids, which were discovered in the 1870s, glycoglycerolipids were not discovered until the 1950s. Since that time investigations of the structure and distribution of these glycolipids have proceeded at an exponentially increasing rate, and much information is now available for representatives of many genera of bacteria, yeasts, algae, and higher plants. Glycoglycerolipids have also been identified in animal cells, particularly in the brain, testes, and sperm. At the present time, interest in glycoglycerolipids has shifted towards an understanding of their role in membrane structure and function, and such studies are being actively pursued by several research groups.

The present volume will cover the glycoglycerolipids in bacteria (Chapters 1 and 2), plants (Chapter 3), and animals (Chapter 4). Bacteria also contain phosphoglycoglycerolipids that are associated with lipoteichoic acids, which act as a bridge between the cell wall and the plasma membrane. These anionic glycolipids will also be covered in Chapter 2. Another group of anionic glycolipids, the sulfated glycosylglycerolipids and the plant sulfoquinovosyldiacylglycerols will be covered in Chapters 1 and 3, respectively.

Apart from the glycoglycerolipids and glycosphingolipids, there are three other glycolipid classes, usually present as minor components in non-animal organisms, that deserve attention. These are the plant glycosylsterols, reviewed in Chapter 3; the acylated carbohydrates, reviewed in Chapter 5; and the glycosylated hydroxy fatty acids, covered in Chapter 6.

In contrast to the lack of knowledge concerning the structure–function relationships of sterol glycosides in plant membranes, much effort has been devoted to investigations on the function of one group of acylated carbohydrates, the diacyltrehaloses present in *Mycobacteria*. These acyl trehaloses and their sulfate esters are believed to be responsible for the toxicity of *Mycobacteria*, and the historical development of knowledge in this area is presented in detail in Chapter 5. The glycosylated hydroxy fatty acids, which occur in bacteria and yeasts, are also known to have a specific function: they are excreted into the growth medium where they act as detergents, enhancing the organism's ability to take up and digest lipid material, such as hydrocarbons. This is described in Chapter 6.

Establishment of the structure of these glycolipids, particularly the glycoglycerolipids has depended and still is dependent on chemical synthesis to provide pure material of known structure and stereochemical configuration for comparison with the natural substances. This synthetic material is also useful in investigations of the physical properties and structure–function relationships of the natural glycolipids. Development of procedures for the synthesis of acylated trehaloses and of glycoglycerolipids is reviewed in Chapter 5 and Chapter 7, respectively.

During the preparation of this volume one of our authors, Alexander "Pat" Tulloch, passed away on February 14, 1987, a victim of leukemia. Pat Tulloch was an outstanding carbohydrate chemist and one of the pioneers in the study of the glycosylated hydroxy fatty acids, as is clear from his contribution to this book (Chapter 6). Pat Tulloch will be remembered for his definitive investigations on the chemistry, synthesis, and metabolism of hydroxy fatty acids and their glycosylated derivatives. His premature death represents a great loss to the field of glycolipids and to science generally. For these reasons, I dedicate this volume to the memory of Dr. Alexander P. Tulloch.

There are a number of persons to whom I am indebted for their invaluable assistance in preparing the manuscripts for Chapters 1 and 3: Hélène Amyot for typing the manuscripts of Chapters 1 and 3 and the front papers, Eva Szabo for preparation of many of the figures, Paul Brunon for photography, Clem Kazakoff for the mass spectra, and Raj Capoor for the NMR spectra. The contributors to this volume are grateful to the scientists and publishers who kindly gave permission for citation of published material and for reproduction of published figures.

Morris Kates

Ottawa, Canada

Abbreviations

Abbreviations of glycolipid nomenclature (see IUPAC—IUB Lipid Nomenclature, 1978, *Chem. Phys. Lipids* **21**:159) and other abbreviations used in this volume are as follows:

A	Archaeol
C	Caldarchaeol
DAG	Diacylglycerol
GalAAG	Galactosylacylalkylglycerol
GalCer	Galactoceramide
GalDAG	Galactosyldiacylglycerol
Gal$_2$DAG	Digalactosyldiacylglycerol
Gal$_3$DAG	Trigalactosyldiacylglycerol
Gal$_4$DAG	Tetragalactosyldiacylglycerol
GDAG	Glycosyldiacylglycerol
GGro	Glycosylglycerol
GGroL	Glycoglycerolipid
GlcDAG	Glucosyldiacylglycerol
Gro	Glycerol
GSL	Glycosphingolipid
LacCer	Lactosylceramide
ManDAG	Mannosyldiacylglycerol
PAPS	3'-Phosphoadenosine-5'-phosphosulfate
SulfoGalAAG	Sulfatoxygalactosylacylalkylglycerol
SulfoGalCer	Sulfatoxygalactosylceramide(cerebroside sulfate)
SulfoGalDAG	Sulfatoxygalactosyldiacylglycerol

Contents

Chapter 2

Bacterial Phosphoglycolipids and Lipoteichoic Acids

Werner Fischer

Chapter 3

Glycolipids of Higher Plants, Algae, Yeasts, and Fungi

Morris Kates

Chapter 4

Glycoglycerolipids of Animal Tissues
Robert K. Murray and Rajagopalan Narasimhan

Chapter 5

Mycobacterial Fatty Acid Esters of Sugars and Sulfosugars

Mayer B. Goren

Chapter 6

Glycosides of Hydroxy Fatty Acids

Alexander P. Tulloch

Chapter 7

Synthesis of Glycoglycerolipids

Jill Gigg and Roy Gigg

Chapter 1

Glyco-, Phosphoglyco- and Sulfoglycoglycerolipids of Bacteria

Morris Kates

1.1. Introduction

In view of their ubiquity in bacteria, plants, and animals, it is rather surprising that glycosyldiacylglycerols were discovered relatively recently in plants (during the late 1950s) (see review by Carter *et al.*, 1965) and even more recently in bacteria and animals (during the early 1960s) (see review by Ishizuka and Yamakawa, 1985). The sulfoglycoglycerolipids were also rather late in being discovered, partly because of the lack of specific and sensitive reagents for the detection of sulfate and sulfonate. Thus, the plant sulfonolipid, sulfoquinovosyldiacylglycerol, was discovered in 1959 (Benson, 1963), the sulfated triglycosyldiphytanylglycerol diether of extreme halophiles in 1967 (Kates, 1978), and the sulfogalactoglycerolipid, seminolipid, of testis in 1972 (see review by Murray *et al.*, 1976, 1980).

This chapter reviews the isolation, structure determination, metabolism, and distribution of glycoglycerolipids, phosphoglycoglycerolipids, and sulfoglycoglycerolipids in bacteria, including the ether analogues of these lipids in archaebacteria. Comprehensive general reviews on eubacterial glycoglycerolipids (Kates, 1964, 1966; Lennarz, 1966, 1970; Sastry, 1974; Ishizuka and Yamakawa, 1985) and on glyco-, phosphoglyco-, and sulfoglycoglycerolipids of archaebacteria (Kates, 1972, 1978, 1988; Langworthy, 1985; De Rosa *et al.*, 1986a; Kamekura and Kates, 1988) may be consulted for further background material on this subject. The eubacterial phosphoglycoglycerolipids are not dealt with extensively in this chapter, as they are reviewed in detail in Chapter 2 (*this volume*), along with bacterial lipoteichoic acids.

1.1.1. Classification and Nomenclature of Glycoglycerolipids

Glycoglycerolipids (GGroLs) are defined as glycolipids containing one or more glycerol residues (IUPAC–IUB Nomenclature, 1978) and are divided

---

Morris Kates • Department of Biochemistry, University of Ottawa, Ottawa, Ontario, Canada KIN 6N5.

into neutral and acidic subgroups. The neutral GGroLs include glycosyldia-cylglycerols (Gacyl$_2$Gros or GDAGs), acylated glycosyldiacylglycerols, glycosyl-acylalkylglycerols, and glycosyldialkylglycerols. The glycosyldiacylglycerols also known as glycosyldiglycerides or glycosylglycerides (covered in Chapters 1–4, *this volume*), contain one or more sugar residues attached glycosidically to the free hydroxyl group (usually in the *sn*-3-position) of a diacyl (or monoacyl) glycerol. The acylated glycosyldiacylglycerols (Chapters 1, 2, and 3) contain acyl groups esterified to the glycosylhydroxyls, usually at C-6, but sometimes at either C-2 or C-3, or both. The glycosylalkylacylglycerols (Chapter 4) contain one or more glycosyl groups attached to the free hydroxyl of 1-alkyl-2-acyl glycerol, and the glycosyldialkylglycerols (Chapter 1) contain glycosyl groups attached to the free hydroxyl group of diphytanylglycerol ether.

The acidic glycoglycerolipids include the sulfoglycoglycerolipids (covered in Chapters 1, 3, and 4) and the glycerophosphoglycolipids and lipoteichoic acids (covered in Chapter 2). The term sulfoglycoglycerolipid, like the general term sulfolipid, is defined here as any sulfur-containing glycolipid, following Haines's (1971) definition, and therefore include the plant sul-foglycosyldiacylglycerol, 6-sulfoquinovosyldiacylglycerol (Chapter 3), the sul-fated glycosyldiacyl or alkylacylglycerols of animals (Chapter 4) and the sul-fated glycosyldialkylglycerols of archaebacteria (Chapter 1).

In general, the system of nomenclature recommended by IUPAC–IUB (1978) for glycolipids is followed in this volume. Thus, the stereospecific numbering (*sn*) system is used to designate the position of attachment of substituents on asymmetrically substituted glycerol residues, rather than the numbering of the D,L system. The latter is used, however, for the carbohy-drate groups of glycolipids. The IUPAC–IUB-recommended short symbols and abbreviations are used to designate the detailed structure of glycoglycero-lipids, for example, the bacterial monoglucosyldiacylglycerol, 1,2-di-acyl-3-*O*-α-D-glucopyranosyl-*sn*-glycerol (structure **1**, Fig. 1-1), is abbreviated Glc*p*α 1-3-*sn*-DAG; the bacterial diglucosyldiacylglycerol, 1,2-diacyl-3-O-[α-D-glucopyranosyl-(1–2)-O-α-D-glucopyranosyl]-*sn*-glycerol (structure **13**, Fig. 1-2), is abbreviated Glc*p*α1–2Glc*p*α1–3-*sn*-DAG; and the archaebacterial sul-fated diglycosyldiphytanylglycerol (archaeol), 2,3-di-*O*-phytanyl-1-*O*-[(6-*O*-sulfato)-α-mannopyranosyl-(1–2)-α-glucopyranosyl]-*sn*-glycerol (structure **33'**, Fig. 1-9), is abbreviated [HSO$_3$-*O*-6Man*p*α1–2Glc*p*α1–1]2,3-di-phytanyl-*sn*-Gro.

1.1.2. Historical Background

Although the glycosphingolipids have been known for more than 100 years, having been isolated from brain and identified by Thudicum in 1874–1876 [see Kanfer and Hakomori (1983) in Vol. 3 in this series], the existence of glycoglycerolipids was unknown until 1956, when Carter *et al.* (1956) re-ported the isolation from wheat flour of a "crude mixture of glycolipids" derived from mono- and digalactosylglycerols. These glycolipids were later identified as mono- and digalactosyldiacylglycerols (Carter *et al.*, 1961*a,b*), and are discussed in detail in Chapter 3 (*this volume*).

Following the discovery of the plant galactolipids, bacterial glycoglycerolipids were detected by Macfarlane first in *Micrococcus lysodeikticus* and identified as mannosyldiacylglycerol (Macfarlane, 1961) and then in *Staphylococcus aureus* as a glycosyldiacylglycerol (Macfarlane, 1962a). The latter glycolipid was tentatively identified as a diglucosyldiacylglycerol by Polonovski *et al.* (1962). These discoveries stimulated a wide-ranging search for glycolipids in bacteria resulting in the finding of a large variety of glycosyldiacylglycerols, particularly in gram-positive bacteria, but also in gram-negative bacteria, including photosynthetic bacteria, and the glycosyldiphytanylglycerol diethers (archaeols) and dibiphytanyldiglycerol tetraethers (caldarchaeols) of archaebacteria (see Tables 1-7 and 1-13–1-16, and reviews by Kates, 1964, 1978; Sastry, 1974; Shaw, 1970, 1975; Gigg, 1980; Fischer, 1976, 1981; Ward, 1981; Asselineau, 1981, 1983; Ishizuka and Yamakawa, 1985; Langworthy, 1985; De Rosa *et al.*, 1986a; and Kamekura and Kates, 1988).

At the same time, considerable interest was aroused in the biosynthesis and metabolism of glycoglycerolipids in bacteria, resulting in the elucidation of the biosynthetic pathways for many of these glycolipids (see reviews by Kates, 1966; Lennarz, 1970; Goldfine, 1972; and Ambron and Pieringer, 1973). The glycoglycerolipid structure has been found to be correlated with bacterial taxonomic classification, with respect to both the carbohydrate residue and the fatty acid moieties (Shaw, 1974); similar correlations have also been found for the glyco- and sulfoglycoglycerolipids of extremely halophilic bacteria (Torreblanca *et al.*, 1986; Kamekura and Kates, 1988).

1.2. Glycosyldiacylglycerols in Eubacteria

The main types of glycosyldiacylglycerols are the mono-, di-, and tri-glycosyldiacylglycerols, the tetra- and polyglycosyldiacylglycerols being minor but important components (see Figs. 1-1 to 1-4, for structures). This section deals with their isolation, distribution in eubacteria, and structure determination.

1.2.1. Extraction, Isolation, and Analysis

1.2.1.1. Extraction Procedures

Glycosyldiacylglycerols, being membrane components, are extracted, along with neutral lipid and phospholipid components, by alcohol-containing organic solvent systems that ensure disruption of lipid–protein complexes (see Kates, 1986a). Bacteria are best extracted in the wet state, directly after harvesting, by a modification (Kates, 1986a) of the procedure of Bligh and Dyer (1959). In this modified procedure, the cells are not freeze-dried or homogenized before extraction, as this may result in enzymatic degradation or chemical change in the lipids. Instead, the wet cell paste is suspended in water (0.5–1 g of wet cells per 1 ml) and extracted directly by stirring with

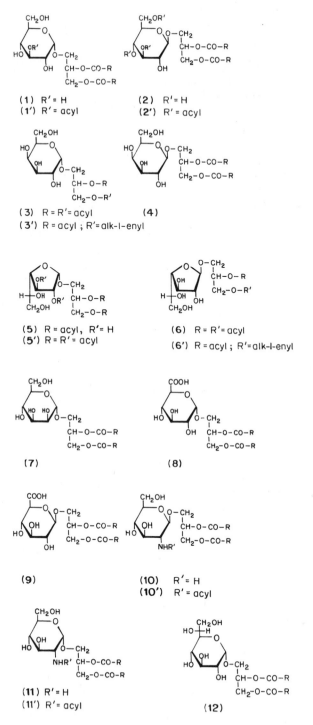

Figure 1-1. Structures of eubacterial monoglycosyldiacyglycerols. (1) Glcpα1–3DAG; (1′) 3-acyl-Glcpα1–3DAG; (2) Glcpβ1–3DAG; (2′) 3,4,6-acyl-Glcpβ1–3DAG; (3) Galpα1–3DAG; (3′) 1-alkenyl-2-acyl-3-Galpα-Gro; (4) Galpβ1–3DAG; (5) Galfα1–3DAG; (5′) 2,3-diacyl-Galfα1–3DAG; (6) Galfβ1–3DAG; (6′) 1-alkenyl-2-acyl-3-Galfβ-Gro; (7) Manpα1–3DAG; (8) GlcUpα1–3DAG; (9) GlcUpβ1–3DAG; (10) GlcNpβ1-3DAG; (10′) N-acyl-GlcNpβ1-3DAG; (11) GlcNpα1–3DAG; (11′)N-acylGlcNpα1–3DAG; (12) *glycero-gluco*-Heptpα1–3DAG.

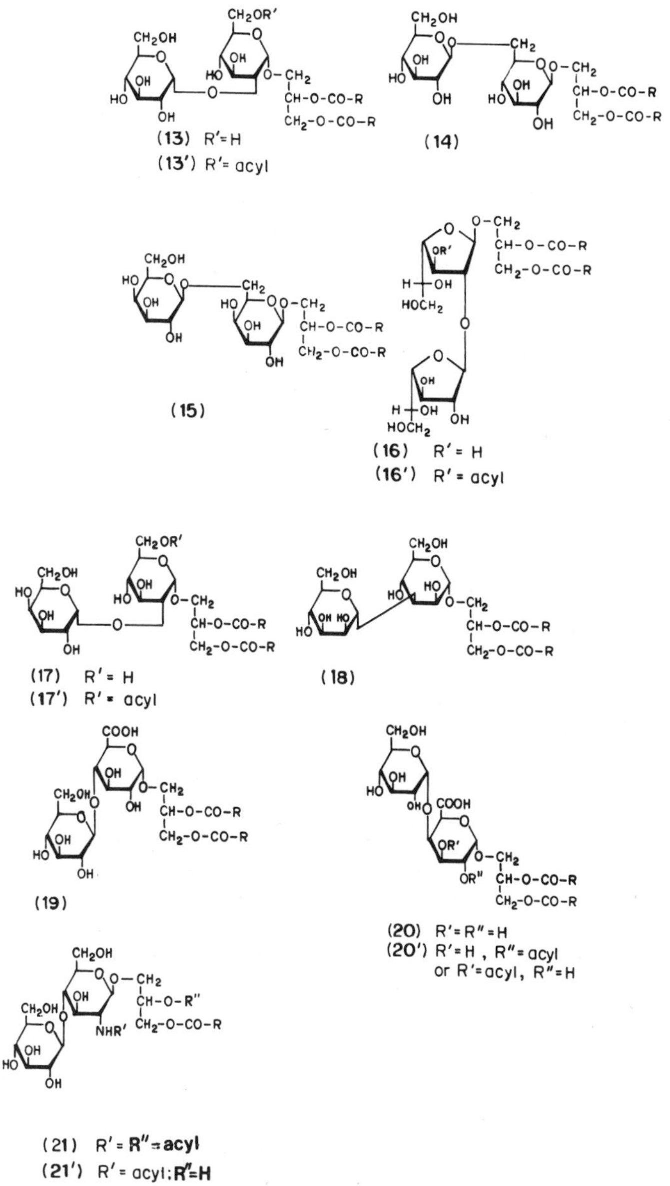

Figure 1-2. Structures of eubacterial diglycosyldiacylglycerols. (**13**) Glc*p*α1–2Glc*p*α1–3DAG; (**13'**) Glc*p*α1–2(6-acyl)Glc*p*α1–3DAG; (**14**) Glc*p*β1–6Glc*p*β1–3DAG; (**15**) Gal*p*β1–6 Gal*p*β1–3DAG; (**16**) Gal*f*β1–2Gal*f*β1–3DAG; (**16'**) Gal*f*β1–2(3-acyl)Gal*f*β1–3DAG; (**17**) Gal*p*α1–2Glc*p*α1–3DAG; (**17'**) Gal*p*α1–2(6-acyl)Glc*p*α1–3DAG; (**18**) Man*p*α1–3Man*p*α1–3DAG; (**19**) Glc*p*β1–4 GlcU*p*α1–3DAG; (**20**) Glc*p*α1–4GalU*p*α1–3DAG; (**20'**) Glc*p*α1–4(2- or 3-acyl)GalU*p*α1–3DAG; (**21**) *N*-acyl-Glc*p*β1–4GlcN*p*β1–3DAG; (**21'**) *N*-acyl-Glc*p*β1–4GlcN*p*β1–3-monoacylGro.

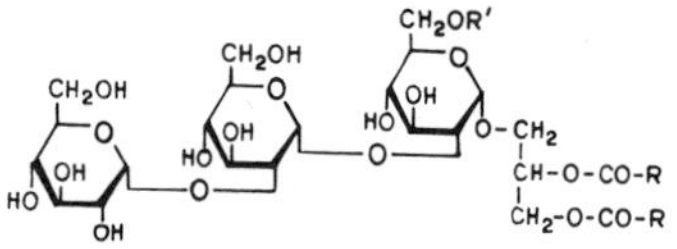

(22) R' = H
(22') R' = acyl

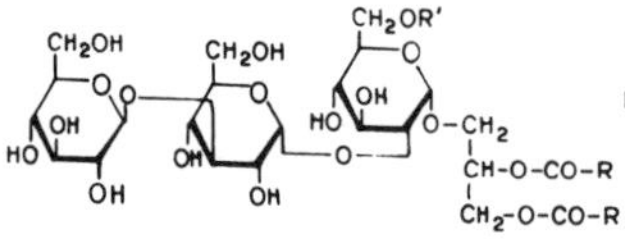

(23)

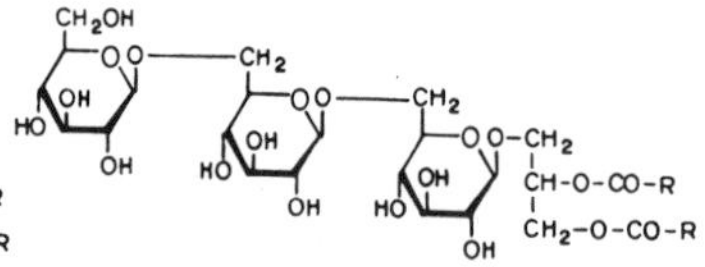

(24)

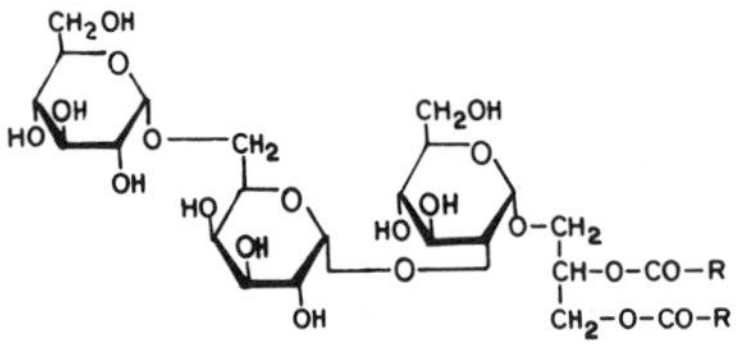

(25)

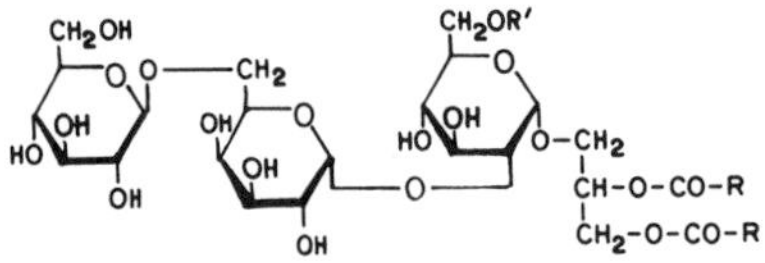

(26) R' = H
(26') R' = acyl

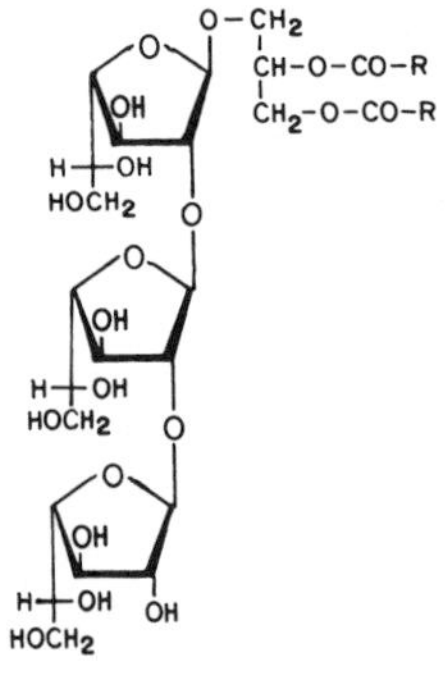

(27)

Figure 1-3. Structures of eubacterial triglycosyldiacylglycerols. (**22**) (Glc*p*α1–2)$_2$ Glc*p*α1–3DAG; (**22'**) (Glc*p*α1–2)$_2$(6-acyl)Glc*p*α1–3DAG; (**23**) Glc*p*β1–3 Glc*p*α1–2 Glc*p*α1–3DAG (R'=H or acyl); (**24**) (Glc*p*β1–6)$_2$ Glc*p*β1–3DAG; (**25**) Glc*p*α1–6 Gal*p*α1–2 Glc*p*α1–3DAG; (**26**) Glc*p*β1–6Gal*p*α1–2Glc*p*α1–3DAG; (**26'**) Glc*p*β1–6 Gal*p*α1–2(6-acyl)Glc*p*α1–3DAG; (**27**) (Gal*f*β1–2)$_2$Gal*f*β1–3DAG.

methanol–chloroform (2 : 1, v/v; 3.75 volumes per volume of cell suspension) for 0.5–1 hr at room temperature. The mixture is then centrifuged, the one-phase extract, methanol–chloroform–water (2 : 1 : 0.8, v/v), is decanted, and the cell pellet is suspended in water and extracted as before. The combined extracts are then diluted with equal volumes of chloroform and water to form

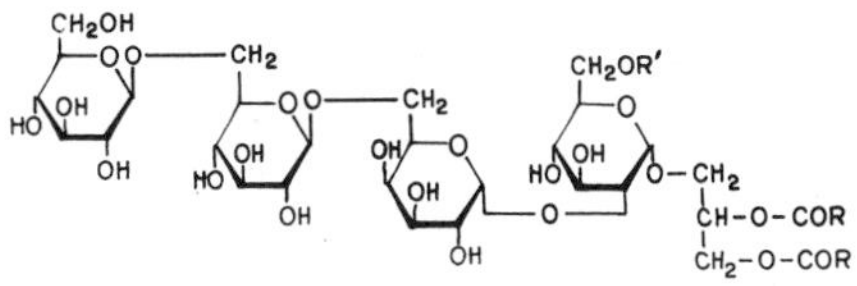

(28) R′ = H
(28′) R′ = acyl

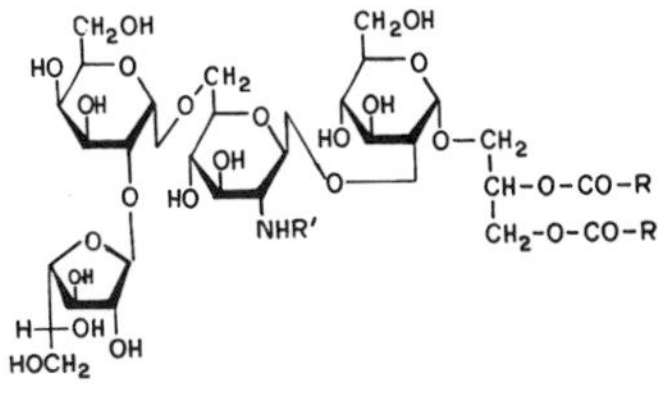

(29) R′ = iso – 17 : 0

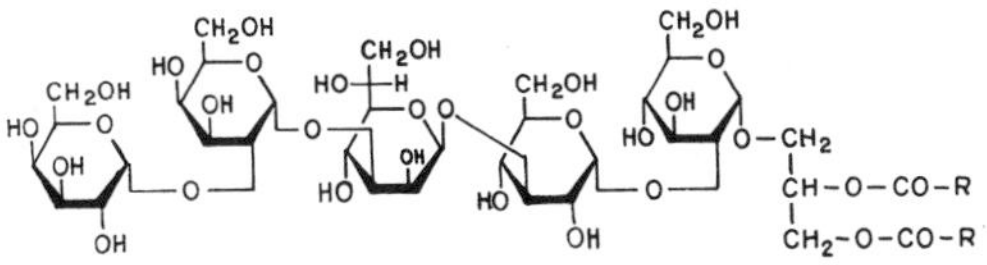

(30)

Figure 1-4. Structures of eubacterial polyglycosyldiacylglycerols. (**28**) (Glc*p*β1–6)$_2$Gal*p*α1–2Glc*p*α1–3DAG; (**28′**) (Glc*p*β1–6)$_2$Gal*p*α1–2(6-acyl)Glc*p*α1–3DAG; (**29**) Gal*f*β1–2Gal*p*α1–6GlcN*p*(*N*-iso-17 : 0)β1–2Glc*p*α1–3DAG; (**30**) Gal*p*α1–2Gal*p*α1–3D-*glycero*-D-*manno*Hept*p*β1–3Glc*p*α1–2Glc*p*αDAG.

a two-phased system (methanol–chloroform –water, 1:1:0.9, v/) in which the lower phase is chloroform, containing the total lipids and the upper phase is methanol–water (10 : 9, v/v).

This partitioning step is sufficient to remove water-soluble, nonlipid contaminants such as sugars and amino acids, in the methanol–water phase, and no further treatment, such as washing with water or filtration through cellulose or Sephadex, is required. The methanol–water phase should be saved to check for the presence of sulfated polyglycosylglycerolipids and other highly polar lipids that may partition into this phase. The chloroform phase is then withdrawn, diluted with benzene (to aid in removal of traces of water), and brought to dryness *in vacuo* (rotary evaporator). The residual lipid is immediately dissolved in chloroform–methanol (2 : 1) and stored at −20°C or subjected immediately to fractionation procedures for separation of the glycolipids.

Other extraction procedures using chloroform–methanol (2 : 1, v/v) at room temperature (Brundish *et al.,* 1965; Reeves *et al.,* 1964), or brief extraction with hot (60–65°C) methanol followed by chloroform–methanol (2 : 1

v/v) (Vorbeck and Marinetti, 1965*a;* Exterkate and Veerkamp, 1969), have also been used.

1.2.1.2. Separation, Isolation, and Analytical Procedures

Several types of procedures have been used to isolate glycoglycerolipids from the total lipid extracts, such as solvent (e.g., acetone) precipitation, countercurrent distribution, column chromatography on silicic acid, DEAE cellulose, etc. (see reviews by Sastry, 1974; Kates, 1986*a*). However, the most useful and widely used procedure involves column chromatography on silicic acid and final purification by preparative thin-layer chromatography (TLC). Thus, chromatography of total lipids of bacteria on a column of silicic acid eluted first with chloroform to remove neutral lipids and pigments, and then with chloroform–acetone (1 : 1) and acetone, yields the glycolipid fraction containing mono- and diglycosyldiacylglycerols, well separated from the phospholipids, the latter being eluted with increasing amounts of methanol in chloroform (Vorbeck and Marinetti, 1965*a;* Kates, 1986*a*).

Separation and purification of these glycolipids is best achieved by preparative TLC on silica gel H or G in the solvent system chloroform–methanol–28% ammonia (70 : 20 : 2, or 65 : 25 : 5, v/v) (Kates, 1986*a*). Other solvent systems that are effective have been reviewed by Sastry (1974). For preparative purposes, glycolipids are best visualized with iodine or Rhodamine 6G and eluted from the plates with chloroform–methanol (1 : 1, v/v).

A final partition into chloroform by the Bligh–Dyer procedure is used to remove any water-soluble contaminants, but it should be cautioned that glycolipids with more than three glycosyl groups may partition partly into the methanol–water phase (see Smallbone and Kates, 1981; Kates, 1986*a*). In such cases, passage of the glycolipids through a column of Sephadex G-25 (Wells and Dittmer 1963) would be preferable.

For analytical purposes, glycolipids are best separated by two-dimensional TLC on silica gel H first in chloroform–methanol–28% ammonia (65 : 35 : 5, v/v) and then in chloroform–acetone–methanol–acetic acid–water (10 : 4 : 2 : 2 : 1, v/v) and visualized by spraying the plate with 50% sulfuric acid and heating on a hot plate (glycolipids appear first as red spots), or with α-naphthol and heating in an oven at 120°C (glycolipids give blue-purple spots; other polar lipids give yellow spots) (Siakotos and Rouser 1965; Kates, 1986*a*). The periodic acid–Schiff stain for vicinal hydroxyls (Shaw 1968*b*), although not specific for sugars, is also very useful for detection of glycolipids.

Mono- and diglycosyldiacylglycerols can also be separated from phospholipids by high-performance liquid chromatography (HPLC) on silica columns (Marion *et al.,* 1984; Demandre *et al.,* 1985). Each glycolipid can be further resolved into molecular species by reverse-phase HPLC on two coupled columns of Biophase–ODS using isocratic elution with methanol-1 mM phosphate buffer (pH 7.4) (95 : 5, v/v) at a flow rate of 1 ml/min (Lynch et al., 1983); or on a C_{18} Bondapak column eluted with methanol–water–acetonitrile (90.5 : 7 : 7.5, v/v) at a flow rate of 1.5 ml/min (Demandre *et al.,* 1985). Argentation TLC (Nichols and Moorehouse, 1969) can also be used for separation of molecular species of glycosylDAG.

Quantitative analysis of glycosylglycerolipids is carried out by estimation of the sugar content in the intact lipid or the water-soluble deacylated glycolipid by the anthrone procedure (Radin *et al.*, 1955) or by the phenol–sulfuric acid procedure (Dubois *et al.*, 1956) modified for use with lipids (Kushwaha and Kates, 1981). The latter method may also be used for quantitation of glycolipid spots on TLC plates (Kates, 1986a). Quantitation may also be carried out on the glycosylglycerols, obtained by deacylation of the glycosyldiacylglycerols, using gas–liquid chromatography (GLC) of the *O*-trimethylsilyl (TMS) derivatives on 3% SE-30 with *O*-trimethylsilyl maltose as internal standard (Inoue *et al.*, 1971), or on 3% SE-52 with *O*-TMS-Glc*p*αGro as standard (Brundish and Baddiley, 1968; see Table 1-3c). Another useful quantitative procedure, especially for glycolipids separated on TLC plates, involves methanolysis with methanolic-HCl and GLC analysis of the fatty acid methyl esters using a suitable fatty acid methyl ester (e.g., 19 : 0) as internal standard (Kates, 1986a); analysis of the individual sugars liberated as methyl glycosides can also be done by GLC or GC/MS of the *O*-trimethylsilyl (TMS) ether derivatives on a column of 5% SE-30 (Sweeley and Vance, 1967) (see Section 1.2.2.2).

1.2.2. Structure Determination (see Hakomori, 1983)

1.2.2.1. Characterization of Glycosyldiacylglycerols

In approaching the structure determination of a glycosyldiacylglycerol, it is useful to characterize it by determining its physicochemical properties, such as TLC mobilities in several solvent systems (Table 1-1) and staining behavior with diagnostic reagents such as α-naphthol, periodic acid–Schiff, ninhydrin, and carbazole reagent; mobilities on HPLC; and spectral behavior, e.g., infrared (IR), nuclear magnetic resonance (NMR), and mass spectra (see Hakomori, 1983). Finally, elemental analysis and analysis of sugar, fatty acid, glycerol, and other constituents should be carried out to determine the mole ratios of these constituents (see Table 1-2).

1.2.2.1a. Thin-Layer Chromatography. TLC R_f values in a few solvent systems of several bacterial glycoglycerolipids are given in Table 1-1; and a two-dimensional TLC plate of *Staphylococcus epidermidis* lipids containing a mono- and diglucosyl DAG is shown in Fig. 1-5. Mono-, di-, tri-, and polyglycosylglycerolipids can readily be distinguished by their TLC mobilities, which decrease in the order mono >di >tri >, etc. (Table 1-1). Staining of the TLC plates with specific reagents such as α-naphthol–H_2SO_4, diphenylamine, or orcinol–H_2SO_4 (Kates, 1986a) will detect the presence of all sugar-containing lipids; subsequent ninhydrin staining of the sugar positive spots will detect the presence of free amino sugars, and if negative, then hypochlorite (Chlorox)/benzidine staining will detect the presence of *N*-acylated amino sugars in the glycolipid (Kates 1986a); finally, carbazole reagent (Galambos 1967) will detect the presence of uronic acids. The periodic acid–Schiff spray (Shaw, 1968b), which is specific for vicinal hydroxyl groups, will also detect the presence of glycolipids, provided any pyranose sugars are unsubstituted in the 3-

Table 1-1. TLC Mobilities of Glycosyldiacylglycerols in Various Solvents

Glycosyl group	Solvents[a] ($R_f \times 100$)								References
	A	B	C	D	E	F	G	G'	
Glcpα (1)		75				67		82	Shaw et al. (1968b); Wilkinson and Galbraith (1979); Nakano and Fisher (1977)
Glcpβ (2)	80	60	75						Komaratat and Kates (1975)
Galpα (3)	77								Matthews et al. (1980)
Galpβ (4)		66							Shaw and Stead (1971)
Galfα (5)				49	15				Mayberry and Smith (1983)
2,3-diacyl-Galfα (5')				100	55				Mayberry and Smith (1983)
Galfβ (6)		81							Plackett (1967)
GlcUpα (8)						24			Wilkinson and Galbraith (1979)
GlcNpβ (10)		73							Phizackerley et al. (1972)
Glcpα1–2Glcpα (13)		65							Shaw et al. (1968b)
Glcpα1–2(6-acyl)Glcpα (13')	49						21		Fischer et al. (1978a)
Glcpβ1–6Glcpβ (14)		50	46				60		Fischer et al. (1978a)
Galpβ1–6Galpβ (15)		43							Komaratat and Kates (1975)
Galpα1–2Glcpα (17)		48							Shaw and Stead (1971)
Galpα1–2(6-acyl)Glcpα (17')		72						38	Nakano and Fischer (1977)
Manpα1–3Manpα (18)		51						78	Nakano and Fischer (1977); Shaw and Stead (1971)
Glcpβ1–6Galpα1–2Glcpα (26)		32						17	Nakano and Fischer (1977)
Glcpβ1–6Galpα1–2(6-acyl)Glcpα (26')		55						40	Nakano and Fischer (1977)
(Glcpβ1–6)₂Galpα1–2Glcpα (28)		16						4	Nakano and Fischer (1977)

[a]Solvents: A: chloroform–methanol–conc. ammonia (65 : 35 : 5, v/v) (Rouser et al., 1970); B: chloroform–methanol–90% acetic acid (30 : 20 : 4, v/v) (Komaratat and Kates, 1975); C: chloroform–methanol–water (65 : 25 : 4, v/v) (Lepage, 1964); D: chloroform–methanol–water (60 : 10 : 1, v/v) (Mayberry and Smith, 1983); E: chloroform–methanol (20 : 1, v/v) (Mayberry and Smith, 1983); F: chloroform–methanol–7 M ammonia (65 : 25 : 4, v/v) (Wilkinson and Galbraith, 1979); G or G': chloroform–acetone–methanol–acetic acid–water (80 : 20 : 10 : 10 : 4, or 50 : 20 : 10 : 10 : 5, respectively, v/v) (Fischer et al., 1978a).

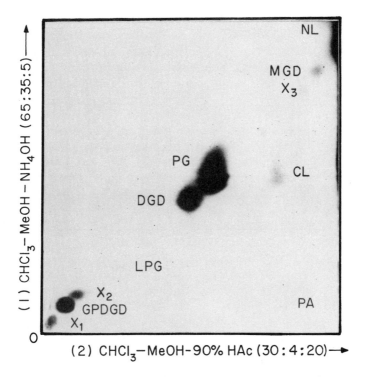

Figure 1-5. Two-dimensional thin-layer chromatography (TLC) plate of *Staphylococcus epidermidis* lipids (halotolerant strain). Identity of components: PG, phosphatidylglycerol; LPG, lysophosphatidylglycerol; MGD, monoglucosyldiacylglycerol; DGD, diglucosyldiacylglycerol; GPDGD, glycerophosphoryldiglucosyldiacylglycerol; CL, cadiolipin; PA, phosphatidic acid; NL, neutral lipids; X_1, X_2, and X_3, unidentified. (From Komaratat and Kates, 1975.)

position, and any furanose sugars are unsubstituted in the 3- and 5- or 6-positions.

1.2.2.1b. Spectrometric Analysis. Infrared spectra can be taken in solution (in CCL_4 or $CHCl_3$) or in a KBr pellet (see Kates, 1986*a*), or by Fourier transform infrared (FTIR) spectroscopy. Glycosyldiacylglycerols show strong absorption bands for OH groups (3400 cm^{-1}), hydrocarbon CH_2 and CH_3 groups (2940, 2860, 1460, and 710 cm^{-1}), ester C=O (1730 cm^{-1}) and C-O-(1165 cm^{-1}) groups, alcohol C-O- groups (1070 cm^{-1}) and, if present, for *cis* double bonds (3000, 1650, and 690 cm^{-1}) (see representative spectra in Fig. 1-6). IR spectra are helpful in confirming the presence of the main structural features of the glycolipid, such as the presence of ester- or ether-linked alkyl chains, sugar groups, and sulfate and phosphate groups.

More detailed structural information is obtainable from NMR (proton and ^{13}C) spectra, in particular, the configuration of the anomeric carbons and protons (see Fig. 1-6, Table 1-6, and Section 1.2.2.6, for further discussion).

Newer techniques of mass spectrometry, such as chemical ionization (CI) and fast atom bombardment mass spectrometry (FAB–MS) are particularly

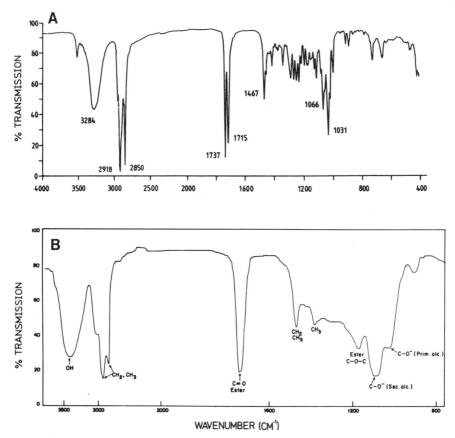

Figure 1-6. Infrared spectra, in chloroform. (A) 1,2-Dipalmitoyl-3-*O*-Glc*p*βGro (**2**) (synthetic; FTIR spectrum. (From Mannock *et al.*, 1987.) (B) Glc*p*β1–6Glc*p*β1–3DAG (**14**). (From Koma-ratat, 1974.) (C) one- and two-dimensional [¹H]-NMR spectra in C²HCl₃ of 1,2-di-*O*-tri-decanoyl-3-*O*-Glc*p*β-Gro (**2**) (synthetic). (From Mannock *et al.*, 1987.)

useful in structure determination, since they do not require derivatization of the sample and give the parent molecular ion as well as a few important diagnostic ions (see Asselineau and Asselineau, 1984). A mass spectrum (CI) of a glycosyldiacylglycerol is shown in Fig. 1-7. Fragment ions derived from successive cleavage of the glycosyl groups at their glycosidic linkages establish the sequence of sugars in the carbohydrate moiety (Hakomori, 1983; Asselineau and Asselineau, 1984); for further discussion, see Section 1.2.2.7).

1.2.2.1c. Analysis of Molecular Constituents. For determination of mole ratios of the glycolipid constituents, quantitative analysis of neutral sugars is determined on the intact lipid or on the water-soluble portion of acid hydroly-sates by the phenol-H_2SO_4 procedure (Kushwaha and Kates, 1981); ester groups are determined by the hydroxamic acid procedure (Snyder and Ste-phens, 1959; Renkonen, 1961); fatty acids are quantitated by GLC of their

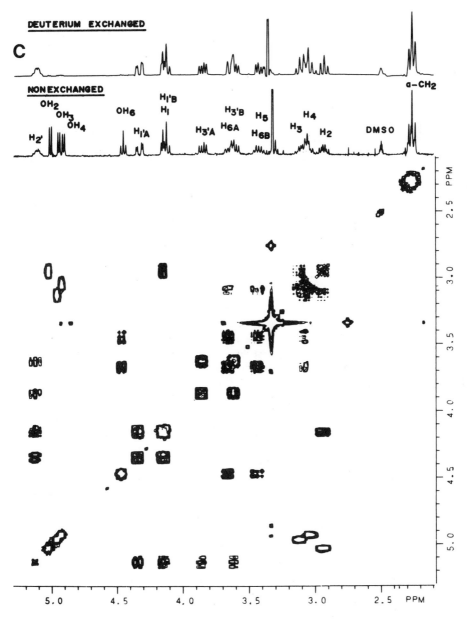

Figure 1-6. (*continued*)

methyl esters with an internal standard (Kates, 1986*a*); and glycerol is determined by the chromotropic acid procedure after acid hydrolysis to release the
glycerol residue (Renkonen, 1962). Acid groups in acidic glycolipids are quantitated by titration (Kates, 1986*a*) and uronic acids are analyzed, after release
by acid hydrolysis, using the carbazole procedure (Galambos, 1967); amino
sugars are quantitated by the Elson–Morgan procedure (Dittmer and Wells,

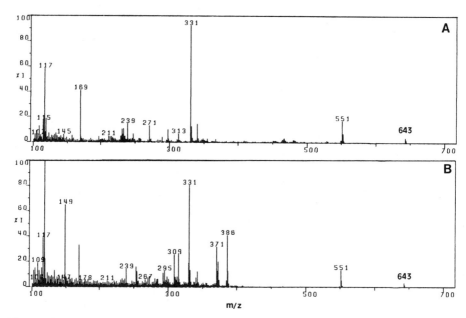

Figure 1-7. Chemical ionization (ethyl ether) mass spectrum. (A) 1,2-Dipalmitoyl-3-*O*-Glc*p*α-Gro (**1**) (peracetylated). (B) 1,2-Dipalmitoyl-3-*O*-Glc*p*β-Gro (**2**) (peracetylated) (synthetic compounds, gift of Dr. D. Mannock). Major peaks: [M-palmitate]$^+$, m/z=643; [M−(Ac)$_4$Glc]$^+$, or [di-16:0-Gro]$^+$, m/z=551; [M−di-16:0-Gro]$^+$ or [(Ac)$_4$Glc]$^+$, m/z=331 (M$^+$ calc. = 898). (From Kates and Mannock, unpublished data.)

1969). Mole ratios of these constituents for various glycosyldiacylglycerols are given in Table 1-2.

1.2.2.2. Identification of Sugar Groups

To identify the sugar groups present in a glycosyldiacylglycerol, the sample is first subjected to acid methanolysis to release the carbohydrate residues (and the glycerol and fatty acid groups) (see Hakomori, 1983). This is most conveniently carried out (Hakomori, 1983; Kates, 1986a) by acid methanolysis in 0.6 N HCl in methanol (methanol–conc. HCl, 20:1, v/v) in screw-capped, Teflon-lined tubes in a heating block at 80°C for 5 hr; after dilution with 10% of water and extraction of the fatty acid methyl esters with low-boiling petroleum ether or hexane, the methanol–water phase is concentrated to dryness under a stream of nitrogen or in a rotary evaporator and dried in a desiccator over KOH, and the residual methyl glycosides of the sugars are converted to TMS derivatives by reaction with pyridine–hexamethylsilazane–trimethylchlorosilane (5:2:1, v/v) and identified by their GLC retention times relative to TMS–mannitol (internal standard) (Sweeley and Vance, 1967; Laine *et al.*, 1972). If amino sugars are present, the sample is subjected to *N*-acetylation (Perry, 1964) before TMS derivatization. GLC is carried out on 2% SE-30 at 160°C or on 3% SE-30 with temperature program-

Table 1-2. Mole Ratios of Constituents of Various Glycosyldiacylglycerols

Glycolipid	Mole ratios[a]						
	Fatty acid	Glycerol	Hexose	Hexosamine	Hexosuronic acid	NH$_2$	Acid groups
MonohexosylDAG	2	1	1	—	—	—	—
AcylmonohexosylDAG	3	1	1	—	—	—	—
DihexosylDAG	2	1	2	—	—	—	—
AcyldihexosylDAG	3	1	2	—	—	—	—
TrihexosylDAG	2	1	3	—	—	—	—
AcyltrihexosylDAG	3	1	3	—	—	—	—
MonohexosaminylDAG	2	1	—	1	—	1	—
N-AcylhexosaminylDAG	3	1	—	1	—	1	—
MonohexosylmonohexosaminylDAG	2	1	1	1	—	—	—
HexosuronylDAG	2	1	—	—	1	—	1
MonohexosylmonohexosuronylDAG	2	1	1	—	1	—	1

[a]See Kates (1986a) for analytical procedures.

Table 1-3a. Relative Retentions of Monosaccharide and Partially Methylated Derivatives on GLC

Monosaccharide	Relative retentions					
	TMS-methyl glycosides[a]			TMS-2-acetamido-2-deoxyglycoside[b]	Alditol acetates[c]	TMS-aldono-1,4-lactones[d]
	β	α	γ			
Free						
Glucose	0.76	0.67	—		1.00	
Galactose	0.61	0.52	0.46		0.92	
Mannose	—	—	—		0.68	
Galactosamine	0.42	0.39	0.38	1.78	7.05	
Glucosamine	—	—	—	2.17	6.25	
Fucose	0.19	0.17	0.15		0.16	
Rhamnose	—	—	0.15		—	
Glucuronic acid						0.274
Galacturonic acid						0.213
Mannuronic acid						0.512

Partially methylated	Methyl glycosides[e]		0-TMS-methylglycosides[e]	
	β	α	β	α
Glucose, 2,3-Me$_2$	—	8.14	—	0.76
2,4-Me$_2$	4.68	6.84	0.64	0.76
2,6-Me$_2$	6.01	8.14	1.00	1.31
3,4-Me$_2$	4.56	5.40	0.53	0.61
3,6-Me$_2$	5.28	6.68	0.76	0.91
4,6-Me$_2$	—	4.90	—	1.05
3,4,6-Me$_3$	1.88	2.22	0.68	0.82
2,4,6-Me$_3$	1.95	3.86	0.88	0.98
2,3,6-Me$_3$	2.10	2.84	0.70	1.09
2,3,4-Me$_3$	1.56	2.19	0.56	0.78

[a]Relative to TMS–mannitol, on 2% SE-30 at 160°C (Sweeley and Vance, 1967).

[b]Relative to TMS-2-acetamido-2-deoxyglucitol on 10% neopentylglycol sebacate polyester/chromosorb W at 190°C (Perry, 1964).

[c]Relative to glucitol hexaacetate, on 3% ECNSS-M/Gas Chrom Q at 170°C (Hakomori, 1983).

[d]Relative to pentacetylarabinitol on 10% NPGS at 176°C (Perry and Hulyalkar, 1965).

[e]Relative to 2,3,4,6-tetramethyl-α-methylglucoside, on 3% neopentylglycol succinate polyester (NPGS)/Chromosorb W, at 160°C for methyl glucosides and 140°C for TMS ethers (Matsubara and Hayashi, 1974).

ming over the range 150–250°C (Sweeley and Vance, 1967). Each sugar methyl glycoside gives two to three peaks, corresponding to different anomeric and ring forms, which are usually well resolved by this GLC procedure (Table 1-3a).

To avoid any interference by hexosamines, it is advisable to remove them from the methanolysate by passage through a column of Dowex 50 (H$^+$); they can then be eluted with HCl and analyzed separately as TMS derivatives (Karkkainen *et al.*, 1965; Penick and McCluer, 1966; Langworthy *et al.*, 1976). The presence of hexuronic acid methyl esters in the methanolysate may result in several peaks arising from mixtures of various lactones and methyl esters on GLC analysis of the TMS ether derivatives (Perry and Hulyalkar, 1965).

Table 1-3b. Relative Retentions of Glycosylglycerol TMS Derivatives on GLC

Glycosylglycerol[a]	Structure of glycosylDAG	Relative retentions[b] on 3% SE-52	
		185°C	225°C
Glc*p*αGro	**(1)**	1.00	
Glc*p*βGro	· **(2)**	1.36	
Gal*p*αGro	**(3)**	0.97	
Gal*p*βGro	**(4)**	1.08	
Gal*f*βGro	**(6)**		
Gal*p*α1–2Glc*p*αGro	**(17)**		1.00
Glc*p*α1–2Glc*p*αGro	**(13)**		1.06
Glc*p*β1–2Glc*p*αGro			1.06
Glc*p*β1–2Glc*p*βGro			0.93
Glc*p*β1–3Glc*p*Gro			0.91
Glc*p*β1–4Glc*p*βGro			1.10
Glc*p*β1–6Glc*p*βGro	**(14)**		1.33
Gal*f*β1–2Gal*f*βGro	**(16)**		1.06

[a]Derivatives of *sn*-3-glycerol.
[b]Retention times relative to Glc*p*αGro at 185°C, or to Gal*p*α1–2Glc*p*αGro at 225°C on a 150-cm column of 3% SE-52 (Brundish and Baddiley, 1968).

Table 1-3c. Relative Retentions of Partially Methylated Alditol Acetates[a]

Location of methyl group	Ara	Rib	Xyl	Gal	Glc	Man	Fuc	Rha
2	—	—	2.92	8.1	7.9	7.9	1.67	1.52
3	—	—	2.92	11.1	9.6	8.8	2.05	1.94
4	—	—	2.92	11.1	11.5	8.8	—	1.72
6	—	—	—	5.10	5.62	4.48	—	—
2,3	—	—	1.54	5.68	5.39	4.38	1.18	0.98
2,4	1.40	—	1.34	6.35	5.10	5.44	1.12	0.99
2,5	1.10	—	—	3.70	—	—	—	—
2,6	—	—	—	3.65	3.83	3.35	—	—
3,4	1.38	—	1.54	6.93	5.27	5.37	—	0.92
3,5	0.91	0.77	1.08	6.35	—	5.44	—	—
3,6	—	—	—	4.35	4.40	4.15	—	—
4,6	—	—	—	3.64	4.02	3.29	—	—
2,3,4	0.73	—	0.68	3.41	2.49	2.48	0.65	0.46
2,3,5	0.48	0.40	—	3.28	—	—	0.62	—
2,3,6	—	—	—	2.42	2.50	2.20	—	—
2,4,6	—	—	—	2.28	1.95	2.09	—	—
2,5,6	—	—	—	2.25	—	—	—	—
3,4,6	—	—	—	2.50	1.98	1.95	—	—
1,3,4,6	—	—	—	—	0.72	0.72	—	—
2,3,4,6	—	—	—	1.25	1.00	1.00	—	—
2,3,5,6	—	—	—	1.15	—	—	—	—

[a]Relative to 2,3,4,6-tetramethyl glucitol-1,5-diacetate on a 2-m × 4-mm column of 1% ECNSS-M on Gas Chrom Q (100 to 120 mesh) at 170°C or 180°C (Björndal *et al.*, 1967; M. Perry, unpublished results).

This difficulty can be avoided by reductive hydrolysis of the sample, which converts the hexuronic acids to aldonic acids and aldoses to alditols, followed by TMS derivation, which converts the aldonic acids to tetra-O-TMS–aldono-1,4-lactones and the alditols to TMS–alditols. Such a mixture is well resolved by GLC on 10% neopentylsebacate polyester (Perry and Hulyalkar, 1965) (Table 1-3a).

The sugar methyl glycosides may be converted to the free sugars by hydrolysis in 1 N HCl at 100°C for 3 hr. The free sugars can be identified by their mobilities and staining behavior [AgNO$_3$–NaOH (Trevelyan *et al.*, 1950), HIO$_4$–Schiff (Shaw, 1968*b*), *p*-anisidine, etc., reagents] on paper chromatography in various solvent systems (see Table 1-4) (Hakomori, 1983; Kates, 1986*a*), or by GLC of their TMS derivatives (with TMS mannitol as internal standard) on 2% SE-30 at 160°C or 3% SE-30 at 200°C (Sweeley and Vance, 1967; Hakomori, 1983) (Table 1-3). However, since the α-glucopyranosyl and β-galactopyranosyl forms are often incompletely resolved, it is recommended that free sugars be analyzed as alditol acetates (Table 1-3a), which are prepared by borohyride reduction followed by acetylation with acetic anhydride at 100°C for 2 hr; GLC analysis is then carried out on 1% or 3% ECNSS-M at 180°C or 170°C, respectively (Hakomori, 1983; Kushwaha *et al.*, 1981). The distinct advantage of this procedure is that only one peak is obtained for each sugar (Table 1-3a). The mole ratios of each of the sugar components is thus readily calculated from the areas of the peaks of the alditol acetates relative to that of the internal standard, glucitol hexaacetate, or mannitol hexaacetate.

1.2.2.3. Characterization and Identification of Glycosylglycerols

The glycosylglycerol moieties of glycosyldiacylglycerols can be useful degradation products for identification of the intact glycolipids. Glycosylglycerols are obtained by mild alkaline deacylation of the glycosyldiacylglycerols, togther with the methyl esters of the constituent fatty acids. The deacylation procedure is conveniently carried out by a modification (Kates, 1986*a*) of the method of Dawson (1967), as follows.

To a solution of the sample of glycosyldiacylglycerol (1–16 mg) in 0.5 ml of chloroform and 0.75 ml of methanol in a 15-ml glass stoppered or screw-capped centrifuge tube is added 1.25 ml of 0.2 N methanolic NaOH; the mixture is vortexed and kept at room temperature for 30 min. A biphasic mixture is then made by addition of 1.5 ml of chloroform and 1.8 ml of water the mixture is centrifuged briefly and the aqueous–methanol (upper) phase containing the glycosylglycerols is removed by Pasteur pipet and treated with 1–2 ml of cation-exchange resin (Dowex-50 H$^+$) until neutral or acidic, if uronic acids and sulfate or phosphate groups are present. The methanol–water phase is then brought to neutrality with 0.2 N methanolic ammonia and concentrated to a small volume under a stream of nitrogen. If amino sugars (unacetylated) are present, the aqueous–methanol phase is treated with ethyl formate (0.2 ml) for 5–10 min, instead of cation-exchange resin, to neutralize the NaOH. The fatty acid components released as methyl esters in the lower chloroform phase may be analyzed by GLC (see Kates 1986a). If amino sugars acylated with long chain fatty acids are present, the N-acylglycosylglycerol will appear in the chloroform phase and can be completely deacylated by hydrolysis in 1 N KOH at 125°C for 2 hr (Oshima and Yamakawa, 1974).

The glycosylglycerols are then identified tentatively by paper chromatography in a suitable solvent (Table 1-4) (Sastry, 1974; Komaratat and Kates, 1975; Langworthy *et al.*, 1976; Kates, 1986a), together with GLC of the TMS ether derivatives (Brundish and Baddiley, 1968; Langworthy *et al.*, 1976) (see Table 1-3b). These chromatographic procedures can readily distinguish between monoglycosyl- and diglycosylglycerols and their respective anomers and positional isomers, but not their diastereomers (Tables 1-3c, 1-4) (Brundish *et al.*, 1966; Steim, 1967; Brundish and Baddiley, 1968).

For further characterization and identification, the glycosylglycerols may be separated and isolated by preparative paper chromatography on Whatman 3MM paper in the solvent system pyridine–ethyl acetate–water (2:5:5, upper phase) (Sastry and Kates, 1964) or butanol–pyridine–water (6:4:3, v/v) (Carter *et al.*, 1961a). Bands corresponding to the separated glycosylglycerols are eluted with water, the eluates are concentrated to dryness, and the residual glycosides crystallized from a suitable solvent, such as methanol–ethyl ether (Carter *et al.*, 1956; Sastry and Kates, 1964; Heinz, 1967; Brundish and Baddiley, 1968). The purified glycosylglycerols are then characterized by melting point (Wickberg, 1958a,b), optical rotation (Table 1-5), GLC of the TMS derivative (Inoue *et al.*, 1971), and IR (Wickberg, 1958b), NMR, and mass spectrometry (Asselineau and Asselineau, 1984).

The glycosylglycerols may also be hydrolyzed in 2 N HCl at 100°C for 3 hr and the free sugars and glycerol identified by paper chromatography (Table 1-4) or by GLC of the TMS derivatives or alditolacetates (Table 1-3a), as described in Section 1.2.2.2. The glycosylglycerols may also be analyzed for sugar sequences, sugar linkage positions, and anomeric configuration (see Sections 1.2.2.4, 1.2.2.5, and 1.2.2.6, respectively).

1.2.2.4. *Determination of Sugar Sequences*

Sugar sequence determination of di-, tri-, and higher glycosyldiacylglycerols containing more than one sugar may be carried out by mild acid methanolysis, for example, in chloroform–0.25 N methanolic HCl (3:4, v/v) at 25°C for 96 hr (Kates and Deroo, 1973), to hydrolyze the sugar glycosidic linkages sequentially. While this method is ideally suited for glycolipids containing hydrocarbon chains joined to glycerol by acid-stable linkages, such as in the diphytanylglycerol ether lipids of archaebacteria (see Section 1.3) or in cerebrosides, it can also be applied to glycosylDAGs, but concomitant methanolysis of the fatty acid ester linkages may occur, resulting in low yields of the partially deglycosylated glycolipid. The reaction is monitored by TLC on silica gel H plates in chloroform–90% acetic acid–methanol (30:20:4, v/v) to determine the optimum time for the formation of adequate amounts of each of the intermediate partially deglycosylated diacylglycerols. The latter are isolated from the methanolysate by addition of chloroform and water to form a

Table 1-4. Mobilities of Glycosylglycerols and Their Constituents on Paper Chromatography (Whatman No. 1) in Various Solvents

Glycosylglycerol[a] or constituent	Structure of glycosylDAG	Solvents[b] ($R_f \times 100$)							References
		A	B	C R_{Glc}	D	E	F R_{Man}	G	
Constituent									
Glycerol		93	50		70	75			Kates (1986a)
Galactose		32							Kates (1986a)
Glucose		38	20			28			Kates (1986a)
Mannose		43							Kates (1986a)
Arabinose		52							Kates (1986a)
Glucosamine		19							Kates (1986a)
Inositol		12							Kates (1986a)
Glycosylglycerols									
Glc*p*αGro	**(1)**	(95)[c]		100				⎱ 63	Brundish and Baddiley (1968)
									Nakano and Fischer (1977)
Glc*p*βGro	**(2)**	27(70)[c]		102		58		⎰	Brundish and Baddiley (1968); Sastry (1974)
Gal*p*αGro	**(3)**			86					Brundish and Baddiley (1968)
Gal*p*βGro	**(4)**	26	25	85	30	62			Sastry (1974); Kates (1986a)

Glycolipid					Reference
GalfβGro	(6)	(76)[d]	(32)[e]	62	Exterkate and Veerkamp (1969)
Glcpα1–2GlcpαGro	(13)	(82)[c]	53(20)[e]		Brundish and Baddiley (1968); Pieringer (1968)
Glcpβ1–2GlcpαGro			57		Brundish and Baddiley (1968)
Glcpβ1–2GlcpβGro			65		Brundish and Baddiley (1968)
Glcpβ1–3GlcpGro			79		Brundish and Baddiley (1968)
Glcpβ1–4GlcpβGro			43		Brundish and Baddiley (1968)
Glcpβ1–6GlcpβGro	(14)		49		Brundish and Baddiley (1968)
Galfβ1–2GalfβGro	(16)	(17)[d]	(13)[e]	51	Exterkate and Veerkamp (1969)
Galpα1–2GlcpαGro	(17)	(40)[c]	43		Brundish and Baddiley (1968); Brundish et al. (1965)
Manpα1–3ManpαGro	(18)			77	Lennarz and Talamo (1966)
Glcpα1–6Galpα1–2GlcαGro	(25)		23		Shaw et al. (1968a)
Glcpβ1–6Galpα1–2GlcαGro	(26)			25	Nakano and Fischer (1977)
(Glcpβ1–6)₂Galpα1–2GlcpαGro	(28)			11	Nakano and Fischer (1977)

[a]Glycosyl derivatives of *sn*-3-glycerol.
[b]A, pyridine–ethyl acetate–water (2 : 5 : 5 v/v, upper phase) (Sastry and Kates, 1964); B, butanol–propionic acid–water (15 : 7 : 10, v/v) (Dawson, 1967); C, butanol–pyridine–water (6 : 4 : 3, v/v) (Carter *et al.*, 1961a); D, butanol–acetic acid–water (5 : 3 : 1, v/v) (Kates, 1986a); E, phenol–water (saturated) (100 : 38, w/v) (Dawson, 1967); F, 2-propanol–pyridine–water (12 : 4 : 1, v/v) (Lennarz and Talamo, 1966); G, 1-propanol–pyridine–water (7 : 4 : 2, v/v) (Nakano and Fischer, 1977).
[c]Values in parentheses are relative to glucose.
[d]Values in parentheses are relative to galactose.
[e]R_f value.

Table 1-5. Molecular Rotations for Various Methyl Glycosides and Glycosylglycerols

Glycoside	$[\alpha]_D{}^a$ (degrees)	M_D (degrees) Found[b]	Calc[c]	References
Methyl glycosides				
α-D-Glcp	+159	+309	—	Stanek et al. (1963)
β-D-Glcp	− 34	− 66	—	Stanek et al. (1963)
α-D-Galp	+179	+381	—	Stanek et al. (1963)
β-D-Galp	0	0	—	Stanek et al. (1963)
α-D-Galf	+104	+202	—	Green (1966)
β-D-Galf	−112	−217	—	Green (1966)
α-D-Manp	+ 79	+154	—	Stanek et al. (1963)
β-D-Manp	− 53	−136	—	Stanek et al. (1963)
α-D-GlcNp(HCl)	+127	+239	—	Stanek et al. (1963)
β-D-GlcNp(HCl)	− 23	− 52	—	Stanek et al. (1963)
α-D-GlcNAcp	+131	+308	—	Stanek et al. (1963)
β-D-GlcNAcp	− 47	−110	—	Stanek et al. (1963)
β-D-GlcNp(free)	− 48	− 93	—	Oshima and Ariga (1976)
α-6-deoxy-D-Glc(quinovose)	+153	+272	—	Stanek et al. (1963)
β-6-deoxy-D-Glc(quinovose)	− 55	− 98	—	Stanek et al. (1963)
α-L-Rhap	− 63	−112	—	Stanek et al. (1963)
β-L-Rhap	+ 95	+169	—	Stanek et al. (1963)
α-L-Fucp	−197	−351	—	Stanek et al. (1963)
β-L-Fucp	+ 14	+ 25	—	Stanek et al. (1963)

Glycosylglycerols[d]

	[α]_D	M_D	M_D calc.	Reference
D-Glcpα1–3Gro	+ 93	+236	+291	Brundish and Baddiley (1968)
D-Glcpβ1–3Gro	− 32	− 81	− 84	Brundish and Baddiley (1968)
D-Galpα1–3Gro	+158	+401	+363	Brundish and Baddiley (1968)
D-Galpβ1–3Gro	± 1	± 2	− 18	Brundish and Baddiley (1968)
D-Galpβ1–3Gro	− 7	− 17	− 18	Wickberg (1958b)
D-Gal/β1–3Gro	−101	−259	−235	Veerkamp and Van Schaik (1974)
D-Manpα1–3Gro	+ 43.9	+112	+154	Lennarz and Talamo (1966)
D-Glcpα1–2Glcpα1–3Gro (α-kojibiosylglycerol)	+165	+687	+600	Fischer and Seyferth (1968)
D-Glcpβ1–2Glcpα1–3Gro	+ 66.5	+277	+225	Brundish and Baddiley (1968)
D-Glcpβ1–2Glcpβ1–3Gro	− 20	− 83	−150	Brundish and Baddiley (1968)
D-Glcpβ1–3Glcpα,β1–3Gro	+ 10	+ 42	+ 38	Brundish and Baddiley (1968)
D-Glcpβ1–4Glcpα,β1–3Gro (cellobiosylglycerol)	− 8.7	− 36	− 84	Brundish and Baddiley (1968)
D-Glcpβ1–6Glcpβ1–3Gro (β-gentiobiosylglycerol)	− 38.2	−159	−150	Komaratat and Kates (1975)
D-Galpα1–2Glcpα1–3Gro	+150	+624	+672	Shaw et al. (1968a)
D-Galpα1–6Galpβ1–3Gro	+ 86.4	+359	+363	Wickberg (1958a)
D-Manpα1–3-D-Manpα1–3Gro	+ 91.4	+380	+290	Lennarz and Talamo (1966)

[a]Specific rotation in degrees, in aqueous solution.

[b]$M_D = [\alpha]_D \times$ mol. wt./100 (aqueous solution).

[c]Calculated by Hudson's (1909) isorotation rule: M_D calc. $= M_{Dx} + M_{Dy} \ldots + M_{DGro}$, where M_{Dx}, $M_{Dy} \ldots$ are molecular rotation contributions of the constituent glycosyl groups, taken from the values of the corresponding methyl glycosides (given in this table) and M_{DGro} is the molecular rotation contribution of sn-3-glycerol taken from the M_D of sn-3-propanoylglycerol (−18.3°, in water; Baer and Fischer, 1945).

[d]Derivatives of sn-3-glycerol.

Bligh–Dyer biphasic system (Kates, 1986a). The lower chloroform phase contains the glycosyldiacylglycerol mixture which is separated by preparative TLC in the above solvent system into mono-, di-, tri-, and higher glycosyldiacylglycerol fractions; sugar analysis of each of these (see Section 1.2.2.2) is usually sufficient to determine the sequence of sugars in the intact glycolipid. The sequence may also be checked by monitoring the methanol–water phases periodically for methylglycosides of free sugars or disaccharides by TLC on kieselguhr G-silica gel G in ethyl acetate–methanol–water (68 : 23 : 9, v/v) (Kates and Deroo, 1973).

It should be noted that the sugar glycosidic linkages are not all hydrolyzed at the same rate; for example, galactofuranosides are hydrolyzed much faster than hexopyranosides, while free glycosaminopyranosides (unacylated) are very resistant to acid hydrolysis (Oshima and Yamakawa, 1974). Of the hexopyranosides, mannopyranoside appears to be hydrolyzed more readily than either galacto- or glucopyranosides (Kates and Deroo, 1973).

These variations in rates of hydrolysis have been used to advantage in designing hydrolysis conditions for glycolipids with unusual sugar sequences. For example: (1) the galactofuranose- and glucosamine-containing tetraglycosylDAG in *Thermis thermophilus* (structure **29**, Fig. 1-4; Table 1-7), or the fully deacylated glycolipid, were subjected to hydrolysis in 0.1 N HCl in 50% methanol at 80°C for 30 min to release the galactofuranosyl group and the fully deacylated glycolipid was hydrolyzed with 1 N HCl at 100°C for 3 hr to hydrolyze all glycosidic linkages but not that of the glucosaminyl group (Oshima and Yamakawa, 1974); (2) the glucuronosyl-containing diglycosylDAG (structure **19**, Fig. 1-2) in *P. diminuta* and *P. vesicularis* (Table 1-7), after deacylation, was subjected to acid hydrolysis in 0.1 N HCl at 105°C for 40 min, which released the glucopyranosyl group but not the glucuronosyl group (Wilkinson, 1969); (3) the glucose and galactose-containing di- and triglycosylDAGs (structures **17** and **25**, Figs. 1-2 and 1-3, respectively) in *Lactobacilli*, after deacylation, were hydrolyzed in 0.025 N HCl at 100°C for 5 hr, releasing galactose from the deacylated diglycosylDAG (**17**) and the disaccharide Glc–Gal from the deacylated triglycosylDAG (**25**), in both cases leaving glucose glycosidically bound to glycerol (Shaw *et al.,* 1968a).

Sugar sequences (and also the anomeric configuration of the glycosidic linkage hydrolyzed) may also be determined with the aid of stereospecific "exo" glycosidases, which specifically hydrolyze only a terminal nonreducing sugar residue. The following enzymes are available commercially (Sigma or Boehringer–Mannheim) in pure form for this purpose: α-glucosidase from yeast (EC 3.2.1.20), β-glucosidase from almonds (EC 3.2.1.21), α-galactosidase of almonds (EC 3.2.1.22), the β-galactosidase of *Escherichia coli* (EC 3.2.1.23), and the α-mannosidase of almonds (EC 3.2.1.24). Other specific glycosidases that may be useful with glycosylDAGs are: the β-*N*-acetylhexosaminidase of jack bean, the α-*N*-acetylgalactosaminidase of hog liver, the α-L-fucosidase of *Turbo cornutus,* and sialidase from various sources (see Hakomori, 1983).

This procedure, although limited by the availability of pure specific glycosidases, is most useful in distinguishing terminal glucose and galactose groups, since the pure glucosidases and the galactosidases are the most readily available. The enzyme reactions are carried out on the intact glycolipid in the presence of a suitable detergent such as sodium taurocholate or deoxy-taurocholate (see Hakomori, 1983). Incubation is carried out at 37°C in a suitable buffer containing detergent at the appropriate pH optimum (α-glucosidase, 0.1 M phosphate, pH 7.2; β-glucosidase, 0.1 M acetate, pH 5.0) and appropriate incubation times (see Hakomori, 1983, for details). The reaction mixture is then converted to a 2-phase Bligh–Dyer system by the addition of appropriate amounts of methanol and chloroform (see Kates, 1986a) and the free sugar liberated in the methanol–water phase is identified by paper chromatography or GLC, as described in Section 1.2.2.2. The corresponding glycosylDAG in the chloroform phase, containing one sugar residue less, may be isolated by preparative TLC and checked for the identity of the remaining bound sugars. It may then be subjected to further hydrolysis by the appropriate glycosidase as before, and this procedure is repeated if necessary, until all the sugar components have been accounted for.

This enzymatic procedure may also be applied to the deacylated glycolipid (glycosylGros), in which case the detergent is omitted and the hydrolysis products are examined by paper chromatography or TLC on kieselguhr G–silica gel G (Kates and Deroo, 1973; Komaratat and Kates, 1975), or by GLC (Langworthy *et al.*, 1976). It should be noted that each glycolipid will require an individual and perhaps unique approach to the determination of its sugar sequence, depending on the availability of specific enzymes and the particular sequence involved.

1.2.2.5. Determination of Sugar Linkage Positions

The positions of the glycosyl linkages in a glycosylDAG are most reliably determined by permethylation analysis (see Hakomori, 1983, for a detailed description of these procedures). For both glycosylDAGs and the derived glycosylGros, permethylation may be carried out in dimethylformamide with methyl iodide and silver oxide at 30°C with shaking for 30–36 hr (Heinz, 1967; Brundish *et al.*, 1967). A more effective procedure is that described by Hakomori (1983), involving reaction of the glycolipid or deacylated glycolipid with methylsulfinyl carbanion in anhydrous dimethylsulfoxide (DMSO) in a Teflon-lined screw-cap tube with stirring for 3 hr at room temperature, followed by reaction with methyl iodide for 2 hr. The permethylated product is isolated and purified on a small column of Sephadex LH20 eluted with chloroform–methanol (1 : 1, v/v).

Another procedure that works well for glycosylDAGs, glycosyldiphytanylglycerol ethers, and glycosyldibiphytanyldiglycerol tetraethers involves heating (under reflux, or in a screw-capped tube) a solution of the glycolipid in methyl iodide or methyl iodide–benzene (1 : 1) in the presence of silver oxide at 40°C with magnetic stirring for 24–26 hr (Brundish *et al.*, 1966; Kushwaha *et al.*, 1981; Kates, 1986a). The reaction is monitored by TLC in ethyl ether or ethyl ether–methanol (99 : 1, v/v) and continued if necessary for 16 hr with further addition of methyl iodide and silver oxide, until methylation is complete as indicated by the presence only of a fast-moving spot on TLC in the solvent mentioned. The reaction mixture is diluted with benzene and freed of silver salts by centrifugation; the supernatant is brought to dryness *in vacuo.* The residue is purified by preparative TLC in ethyl ether or in ethyl ether–methanol (99 : 1, v/v) and checked for completeness of methylation by the absence of OH absorption at 3500–3200 cm^{-1} in the IR region. Note that permethylation of acidic glycosylDAGs (containing uronic acids or sulfate or phosphate groups) by this procedure is best carried out on the free acid prepared by the acid Bligh–Dyer partition procedure (see Kates, 1986a); the corresponding methyl esters of uronic acid, sulfate or phosphate groups in the permethylated products will be obtained by this permethylation procedure. Note that permethylation of glycosylDAGs containing amino sugars or acetylated amino sugars by the Hakomori (1983) or Brundish *et al.* (1966) procedures yields the partially methylated *N*-methyl acetamidohexoses.

The purified permethylated glycolipid or glycosylglycerol is then subjected to methanolysis in 0.6 N methanolic–HCl at 80°C for 2–3 hr in a screw-capped tube. After dilution with 10% water and extraction of fatty acid methyl esters and other lipid material with petroleum ether, the methanol–water phase is taken to dryness and the residual methyl glycosides of the partially methylated sugars are identified as such by GLC or GC–MS on 3% neopentylglycolsuccinate polyester at 160°C, or at 140°C for the TMS ethers (Matsubara and Hayashi, 1974) (Table 1-3a). The free partially methylated methyl glycosides may also be analyzed by GLC on Carbowax 6000 at 175°C (Kates *et al.*, 1967) or on butanediolsuccinate polyester at 175°C (Brundish *et al.*, 1967). Analysis of methyl glycosides may be complicated by the presence of α- and β-anomeric forms (Table 1-3a) (see Section 1.2.2.2). A more definitive analysis of partially methylated sugars is achieved by GLC of the alditol acetate derivatives prepared by acid hydrolysis of the methyl glycosides, borohydride reduction of the free partially methylated sugars, and acetylation with acetic anhydride and pyridine (Yang and Hakomori, 1971; Kates and Deroo, 1973; Kushwaha *et al.*, 1981). GLC or GC–MS is carried out on a column of 3% ECNSS at 170°C (Björndahl *et al.*, 1967; Kates and Deroo, 1973), on 1% ECNSS at 180°C (Björndahl *et al.*, 1967; Kushwaha *et al.*, 1981) or on 3% OV-225 at 175°C (Hakomori, 1983). Relative retentions of partially methylated alditol acetates are given in Table 1-3c.

Application of periodate oxidation followed by borohydride reduction and acid hydrolysis (Smith degradation) to the intact or deacylated glycosyl-DAGs followed by examination of the water soluble products by paper chromatography (Komaratat and Kates, 1975) or by GLC (Langworthy *et al.*, 1976) can provide useful data on the linkage positions of the sugars. When applied to the intact glycolipid in aqueous suspension, only the terminal nonreducing sugar residue can be oxidized with periodate, so that repetition of the procedure on the resulting isolated glycolipid will give the sequence of the sugars as well as their linkage positions (Langworthy *et al.*, 1976; see Hakomori, 1983).

1.2.2.6. Determination of Anomeric Configuration

Configuration of the glycosidic linkages in glycosylDAGs or glycosyl diphytanylglycerols may be determined by hydrolysis of the glycolipid or the deacylated glycolipid (glycosylGros) with specific α- and β-glycosidases and chromatographic identification of the free sugars liberated, as described in Section 1.2.2.4 (Carter *et al.,* 1956; Brundish *et al.,* 1966).

Another procedure that can be used is chromium trioxide oxidation of the acetylated glycosylDAGs or glycosylGros (Laine and Renkonen, 1975). Acetylated β-linked hexopyranosides (equatorial glycosidic linkages) are oxidized to 5-ketohexulosonates, but α-linked (axial) acetylated hexopyranosides are completely inert to oxidation and are converted to alditol acetates and analyzed by GLC on OV-225 at 200°C (see Section 1.2.2.5). However, chromium trioxide oxidation does not give unambiguous results with furanosides, since both α- and β-linked furanosides are oxidized (Smallbone and Kates, 1981), although the α-isomer is more readily oxidized than β-isomer, which is the opposite to the action of chromium trioxide on hexopyranosides (Oshima and Ariga, 1976). This method is therefore somewhat limited but may be very useful, particularly in conjunction with the use of specific glycosidases.

The anomeric configuration of each of the glycosidic linkages in a glycosylDAG can, in principle, be reliably assigned by ^{1}H- and ^{13}C-NMR spectrometry (Usui *et al.,* 1973; Heinz *et al.,* 1974; Ritchie *et al.,* 1975; Falk *et al.,* 1979), as described in detail by Hakomori (1983). Some NMR stereochemical assignments of anomeric protons and carbons are given in Table 1-6.

Another approach to the determination of anomeric configuration is by means of molecular rotation data of the deacylated glycosylDAGs, whereby the molecular rotation is compared to values calculated for all possible glycoside configurations using Hudson's (1909) isorotation rule (Stanek *et al.,* 1963; Greene, 1966). Molecular rotations calculated for various glycosylGros are given in Table 1-5.

1.2.2.7. Mass Spectrometry of GlycosylDAGs

Mass spectrometry is a useful tool in the diagnosis of glycolipid structure (see Hakomori, 1983, for a detailed account of this procedure). It is capable of providing molecular species analyses of natural mixtures of a given glycosylDAG, along with sugar sequence and linkage positional data, but not the anomeric configuration of the glycosidic linkages. Analyses may be carried out on the acetylated or TMS derivatives (Heinz *et al.,* 1974; Zepke *et al.,* 1978), but the permethylated derivatives are more stable and volatile and give more easily interpreted spectra (see Hakomori, 1983). Also, permethylation methylates *N*-acetamido groups of hexosamines to *N*-methylacetamido groups, as well as uronic acids to methyl esters; subsequent reduction with LiAlH$_4$ converts *N*-methylacetamido groups to *N*-methyl-*N*-ethyl groups and carboxymethyl esters to carbinols that can be trimethylsilated. These derivatives are thus well suited for detection and analysis of *N*-acetyl amino sugar and uronic acid containing glycolipids by mass spectrometry.

Both electron impact (EI) and chemical ionization (CI) mass spectrometry (MS) have been used, but EI–MS gives more extensive fragmentation, which is more useful for sugar sequence determination. Fast atom bombardment

Table 1-6. NMR Assignments of Anomeric Protons and Carbons of Glycosides

Glycosides	NMR chemical shift (ppm)				References
	Anomeric proton		Anomeric carbon		
	δ_H	J_{Hz}	δ_c	Linkage	
Me-αGlcp	4.68	3.5		—	Komaratat and Kates (1975)
			100.5	—	Usui et al. (1973)
Me-βGlcp	4.35	7.5		—	Komaratat and Kates (1975)
			104.5	—	Usui et al. (1973)
Me-αGlcf			104.0	—	Ritchie et al. (1975)
Me-βGlcf			110.0	—	Ritchie et al. (1975)
Me-αGalp			93.0	—	Perlin et al. (1970)
Me-βGalp			88.7	—	Perlin et al. (1970)
Me-αGalf			103.8	—	Ritchie et al. (1975)
Me-βGalf			109.9	—	Ritchie et al. (1975)
Me-αManp	4.78	2.4		—	Lennarz and Talamo (1966)
			91.8	—	Perlin et al. (1970)
Me-βManp	4.55	2.4		—	Lennarz and Talamo (1966)
			91.5	—	Perlin et al. (1970)
Glcpα1–2Glcpβ (β-kojibiose)			99.0	α1 → 2	Usui et al. (1973)
Glcpβ1–2Glcpβ (β-sophorose)			103.9	β1 → 2	Usui et al. (1973)
Glcpα1–4Glcpβ (β-maltose)			101.0	α1 → 4	Usui et al. (1973)
Glcpβ1–6Glcpβ (β-gentiobiose)			103.8	β1 → 6	Usui et al. (1973)
Glcpβ1–6GlcpβGro (gentiobiosylglycerol)	4.25, 4.31	7.0, 7.0			Komaratat and Kates (1975)
ManpαGro	4.88	4.0			Lennarz and Talamo (1966)
Manpα1–3ManpαGro	4.82, 5.12	4.0, 4.0			Lennarz and Talamo (1966)

Compound	δ H-1	J (Hz)	δ C-1	Linkage	Reference
Glcpα1–4GalUpαDAG (**20**)	4.74, 4.59	3.6, 3.6			Belgelson *et al.* (1970)
Glcpβarchaeol (MGA-2) (**37**)	4.38	7.8	103.9	β1 → 1	Ferrante *et al.* (1986)
Glcpβ1–6Glcpβarchaeol (DGA-3) (**38**)		7.8, 7.6	103.8, 104.1	β1 → 1, β1 → 6	Ferrante *et al.* (1986)
Manpα1–3Galpβarchaeol (DGA-4) (**39**)	4.93, 4.43		97.0, 104.6	α1 → 3, β1 → 1	Ferrante *et al.* (1988)
Galpβ1–6Galpβhydroxyarchaeol (DGA-4) (**40**)	4.39, 4.37		104.4, 104.6	β1 → 6, β1 → 1	Ferrante *et al.* (1988)
Glcpα1–2Galpβarchaeol (DGA-I) (**41**)	4.90, 4.94	4.0	98.5, 106.1	α1 → 2, β1 → 1	Kushwaha *et al.* (1981)
Galpβ1–6Manpα1–2Glcpαarchaeol[a] (TGA-1) (**34**)	4.25, 4.96, 5.0	7.4, 3.1		β1 → 6, α1 → 1, { α1 → 2, α1 → 1 }	Falk *et al.* (1980)
Glcpβ1–6Manpα1–2Glcpαarchaeol (TGA-2) (**35**)	4.47, 4.95				Evans *et al.* (1980)
Glcpα1–2Galpβcaldarchaeol-P-Gro (PGC-I) (**43′**)	4.90, 5.02	4.0	99.1, 106.7	α1 → 2, β1 → 1	Kushwaha *et al.* (1981)
Galpβ1–6Galpβcaldarchaeol-P-Gro (PGC-II) (**44′**)			108.4, 109.0	β1 → 6, β1 → 1	Kushwaha *et al.* (1981)
Glcpβcaldarchaeol (**47**)			103.6	β1 → 1	Thurl and Schäfer (1988)
Glcpβcaldarchaeol-P-inos. (**47′**)			103.6	β1 → 1	De Rosa *et al.* (1980a)
GlcpβGalpβcaldarchaeol (**48**)			103.4, 103.9	1 → ?, 1 → ?	De Rosa *et al.* (1980a)
Glcpβnonitolcaldarchaeol (**50**)			103.4	1 → 3′	De Rosa *et al.* (1980a)
Glcpα1–4Galpβcaldarchaeol (**48a**)	4.80, 4.20	1, 7	98.6, 104.3	1 → 4, 1 → 1	Lanzotti *et al.* (1987)
Glcpα1–4Galpβcaldarcaheol-P-inos. (**48a′**)	4.80, 4.20	1, 7	98.8, 104.3	1 → 4, 1 → 1	Lanzotti *et al.* (1987)

[a] Permethylated and desulfated derivative.

mass spectrometry (FAB–MS) gives molecular ion peaks with relatively little fragmentation and is thus well suited for molecular species analysis of a given glycosylDAG class, in underivatized form (Rinehart, 1982). Negative-ion FAB–MS in particular is well suited for analysis of underivatized anionic glycosylglycerolipids containing carboxyl (from uronic acids), sulfate, or phosphate as anionic groups, e.g., the sulfated triglycosyldiphytanyl glycerol from extremely halophilic bacteria (Falk *et al.*, 1980) (see Fig. 1-17).

1.2.3. Distribution in Eubacteria

Since the discovery of glycosyldiacylglycerols in *M. lysodeikticus* by Mac-farlane (1961), a wide variety of these glycolipids have been identified in eubacteria, including mono-, di-, tri-, and polyglycosyl diacylglycerols, containing neutral, basic, and acidic carbohydrates (see structures shown in Figs. 1-1–1-4). The range of sugar residues in bacterial glycoglycerolipids is much more diverse than in plants (see Chapter 3, *this volume*), or animals (see Chapter 5, *this volume*) and includes the neutral sugars glucose, galactose, mannose, quinovose and mannoheptose, as well as the basic sugars, glucosamine and fucosamine, and the acidic sugars glucosuronic and galactosuronic acids.

The distribution of these glycolipids is presented for gram-negative and gram-positive eubacteria along taxonomic lines in order to highlight the correlations that may exist between glycosyldiacylglycerol structure and bacterial classification (see Shaw, 1970; Sastry, 1974; Goldfine, 1972, 1982; Asselineau, 1983; Ishizuka and Yamakawa, 1985; see also Table 1-7). Bacterial classification and nomenclature follow that given in Bergey's Manual (1974, 1984). It will be obvious that only a very small proportion of bacterial families and genera have been examined for their glycolipid composition; most of the available data deal with the most common gram-positive and gram-negative species.

1.2.3.1. Gram-Negative Eubacteria

GlycosylDAGs have a relatively limited distribution among gram-negative bacteria in comparison with gram-positive bacteria (Table 1-7). They have so far been identified in photosynthetic (anoxy) bacteria, and in some species of *Spirochaetes, Pseudomonads, Bacteroides,* and *Mycoplasma.*

Anoxyphotobacteria (see Kenyon, 1978). Phototrophic purple bacteria generally contain small to trace quantities of glycolipids, but some species (*Ectothiorhodospira*) are apparently devoid of glycolipids (Imhoff *et al.*, 1982; Asselineau and Trüper, 1982). The plant sulfolipid, sulfoquinovosylDAG (see Chapter 3 for structure) occurs in most *Rhodopseudomonas* species, but not the plant mono- or digalactosylDAGs (see Chapter 3 for structures) (Table 1-7). All the *Chromatiaceae* examined, except the genus *Ectothiorhodospira* (Imhof *et al.*, 1982; Asselineau and Trüper, 1982), however, contain several glycolipids, including the plant mono- and digalactosylDAGs and sulfoquinovosylDAG (Table 1-7).

Spirochaetaceae. Five species of the genus *Treponoma* have so far been examined for the presence of glycolipids (Table 1-7) and two of these [*T.*

Table 1-7. Distribution of Glycosyldiacylglycerols among Eubacteria

Family/Genus/Species	Glycosyl group[a]	% Total lipids	References
Gram-negative bacteria			
Anoxyphotobacteria			
Rhodospirillaceae			
Rhodopseudomonas	HSO₃Qp		Kenyon (1978)
R. sphaeroides	HSO₃Qp		Marinetti and Cattieu (1981)
R. sulfidophila	HSO₃Qp		Marinetti and Cattieu (1981)
Chromatiaceae			Imhoff *et al.* (1982)
Thiocapsa roseopersicina	Gal (**4**); Gal–Gal,		
Thiocystis gelatinosa	HSO₃Qp		Imhoff *et al.* (1982)
Chromatium vinosum			
C. minus			
Chromatium strain D	Glc;Man–Glc;Man–Man–Glc		Steiner *et al.* (1969)
Chlorobiaceae			
Chlorobium limicola	Gal–Gal		Nichols and James (1965)
Chloropseudomonas ethylicum	Galβ (**4**);Gal–Rha–unknown		Constantopoulos and Bloch (1967)
Spirochaetaceae			
Treponema			
T. hyodysenteriae	Galα[b,c] (**3,3′**)	30	Matthews *et al.* (1980)
T. pallidum (Kazan 5)	Galα(**3**)	49	Livermore and Johnson (1970, 1974)
T. pallidum (Reiter)	Gal (**3?**)	25	Meyer and Meyer (1971)
T. reiteri	Galβ (**4**)		Coulon-Morelec *et al.* (1969)
T. zuelzerae	Glc	37	Meyer and Meyer (1971)
Pseudomonadaceae			
Pseudomonas			
P. diminuta	Glcpα (**1**)	7–11	Wilkinson (1969)
	GlcUpα (**8**)	3.5	
	Glcpβ1–4GlcUpα (**19**)	4.1	
P. vesicularis	Glcpα (**1**), GlcUpα (**8**)		Wilkinson and Galbraith (1979)
	Glcpβ1–4GlcUpα (**19**)		
	glycero-gluco-Hept (**12**)		

(continued)

Table 1-7. (continued)

Family/Genus/Species	Glycosyl group[a]	% Total lipids	References
Moderate halophile (unidentified)	GlcU (8?)		Peleg and Tietz (1973)
	Glcpα-PG[d]		Peleg and Tietz (1973)
Alteromonas (Pseudomonas) rubescens	Glcpβ (2)	12	Wilkinson (1968)
	GlcUpβ (9)	3	Wilkinson (1968)
Thermus thermophilus	Galfβ1–2Galp1–6Glc p		Oshima and Yamakawa (1974); Oshima and Ariga (1976)
	(acyl)β1–2Glcpα (29)		
Vibrionaceae			
Vibrio			
V. cholerae	Gal		Raziuddin (1976)
Campylobacter (Vibrio) fetus strain	Gal–Gal		Tornabene and Ogg (1971)
V. costicola	oligoglycosyl (unident.)		Hanna et al. (1984)
Enterobacteriaceae			
Yersinia (Pasteurella) pseudotuberculosis	Glc; Glc–Glc	0.5–1.5	Tornabene (1973)
Salmonella typhimurium	—	0	Macfarlane (1962a)
Klebsiella pneumoniae	Glc	9	Wassef (1976)
	Glc–Glc	16	
Bacteroidaceae			
Bacteroides symbiosus	Galfβ (6)		Reeves et al. (1964)
Butyrivibrio fibrisolvens	Gal; Gal–Gal		Kunsman (1970)
Butyrivibrio spp.	Galfβ[b,c] (6')		Clarke et al. (1976)
Mycoplasmataceae			
Mycoplasma			
M. mycoides	Galfβ (6)		Plackett (1967)
M. neurolyticum	Galpβ (2);		Smith (1972)
	Glcpβ1–6Glcpβ (14)		
M. pneumoniae	Glc–Glc (14?); Gal–Gal (16?); Gal–Gal–Gal (27?)		Plackett et al. (1969)
Acholeplasmataceae			
Acholeplasma			
A. axanthum	Galfα[c] (5,5')		Mayberry and Smith (1983)

A. laidlawii strains B and A-EF22	Glcpᶜ (**1,1'**) Glcpα1–2Glcpα (**13**)	45	Shaw et al. (1968b); Smith (1986)
A. granularum	Glcpα (**1**); Glcpβ1–3-Glcpα1–2Glcpα (**23**); Glcpα1–2Glcpα (**13**)		Smith et al. (1980)
A. modicum	Galpα1–2Galp1–3glycero-manno- Heptpβ1–3-Glcpα1–2Glcpα (**30**)		Mayberry et al. (1976)
Gram-positive bacteria Lactobacillaceae *Lactobacillus* L. acidophilum	Galpα1–2Glcpα (**17**); Glcpα1–6Galpα1–2Glcpα (**25**); Glcpβ1–6Glcpβ1–6Galpα1–2Glcpα (**28?**)		Shaw (1970)
L. buchneri L. casei ATCC 7469 L. casei DSM 20021	Galpα1–2Glcpα (**17**) Galpα1–2Galpᶜ (**17,17'**) Glcpα (**1**); Galpα1–2Galpα (**17**) Glcpα1–6Galpα1–2Glcpα (**25**); Glcpβ1–6Galpα1–2Glcpαᶜ (**26,26'**) (Glcpβ1–6)₂Galpα1–2Glcpα (**28**)	17–20	Shaw (1970) Shaw et al. (1968a) Nakano and Fisher (1977) Fischer et al. (1978b)
L. fermenti L. helvticus	Gal–Glc (**17?**) Galpα1–2Glcpα (**17**); Glcpα1–6Galpα1–2Glcpα (**25**); Glc–Glc–Gal–Glc (**28?**)		Winken and Knox (1970) Shaw (1970)
L. plantarum	Galpα1–2Glcpα (**17**) Glc–Gal–Glc (**25?**)		Shaw (1970)
Listeria monocytogenes Bacillaceae *Bacillus*	Gal–Glc (**17?**)		Carroll et al. (1968)
B. cereus	Glcpβ1–6Glcpβ (**14**)		Lang and Lundgren (1970); Saito and Mukoyama (1971)
B. licheniformis	Glcpβ1–6Glcpβ (**14**) (Glcpβ1–6)₂Glcpβ (**24**)		Button and Hemmings (1976) Fischer et al. (1978b)
B. megaterium	GlcNpβᶜ (**10,10'**)	5	Phizackerley et al. (1972)

(continued)

Table 1-7. (Continued)

Family/Genus/Species	Glycosyl group[a]	% Total lipids	References
B. subtilis	GlcNp-PG[e]		Op den Kamp and Van Deenen (1966); MacDougall and Phizackerley (1969)
	Glc$p$β1–6Glc$p$β (**14**) (Glc$p$1–6)$_2$Glc$p$β (**24**)	1–2	Brundish and Baddiley (1968); Bishop *et al.* (1967); Fischer *et al.* (1978b)
B. acidocaldarius	Glc$p$β1–4GlcNp(acyl)4GB (**21,21′**) HSO$_3$Qp	16	Langworthy *et al.* (1976)
Clostridium perfringens	Man	10	Macfarlane (1962a)
Micrococcaceae			
Micrococcus			
M. luteus (lysodeikticus)	Man$p$α1–3Man$p$α (**18**)	7–15	Macfarlane (1961) Lennarz and Talamo (1966)
Staphylococcus			
S. aureus	Glc$p$β1–6Glc$p$β (**14**)	1–2	Macfarlane (1962b); Polonovski *et al.* (1962); Fischer *et al.* (1978c)
			White and Frerman (1967)
S. epidermidis	Glc$p$ (**22?**); Glc–Glc (**14?**) Glc$p$β (**2**)	4	Komaratat and Kates (1975)
S. lactis I3	Glc$p$β1–6Glc$p$β (**14**) Glc$p$β1–6Glc$p$β (**14**)	16 3	Brundish *et al.* (1967); Quinn and Sherman (1971)
S. lactis 7944	Glc$p$β1–6Glc$p$β (**14**)	1–2	Brundish and Baddiley (1968)
S. saprophyticus	Glc$p$β1–6Glc$p$β (**14**)	1–2	Brundish and Baddiley (1968)
Planococcus citreus	Glc (**2?**)		Thirkell and Summerfield (1977)
Streptococcaceae			
Streptococcus			
S. faecalis ATCC9790	Glc$p$α (**1**); Glc$p$α1–2Glc$p$α (**13**) Glc$p$α1–2(6-acyl)Glc$p$α (**13′**)		Vorbeck and Marinetti (1965b); Pieringer (1968) Ambron and Pieringer (1971)
S. faecalis strain 9	Glc$p$α1–2Glc$p$α (**13**)	1–2	Brundish *et al.* (1966) Brundish and Baddiley (1968)

Organism	Structure	No.	References
S. faecium	Glc*p*α1-2Glc*p*α (13) (Glc*p*α1-2)$_2$Glc*p*α (22)		Pieringer (1968) Fischer *et al.* (1973)
S. hemolyticus	Glc*p*α1-2Glc*p*α (13) (Glc*p*α1-2)$_2$Glc*p*α (22) Glc*p*α1-2Glc*p*αᶜ (13,13')	44 1	Ishizuka and Yamakawa (1968); Fischer (1976) Fischer and Seyferth (1968); Fischer *et al.* (1978*a*, 1979)
S. (Diplococcus) pneumoniae	Glc*p*α (1)	11	Ishizuka and Yamakawa (1968)
S. pyogenes	Glc–Glc (13?)		Cohen and Panos (1966)
S. MG	Glc*p*α (1)		Plackett and Shaw (1967)
Pneumococcus types I and XIV	Glc*p*α1-2Glc*p*α (13) Glc*p*α (1) Gal*p*α1-2Glc*p*α (17)	34	Brundish *et al.* (1965) Kaufman *et al.* (1965)
Corynebacteriaceae *Corynebacterium* *C. aquaticum* *C. diphtheriae* *Arthrobacter* *A. crystallopoites,* *A. globiformis,* and *A. pascens*	Man (7?); Man–Man (18?); PI–Man*f* PI-di-Man(di- and triacyl)*f* Gal*p*β (4) Gal*p*β1-6Gal*p*β (15) Man*p*α1-3Man*p*α (18)	13	Khuller and Brennan (1972*a*) Brennan and Lehane (1971) Walker and Bastl (1967); Shaw (1970); Shaw and Stead (1971)
Microbacterium *M. lacticum* *M. thermosphactum*	Man*p*1-3Man*p* (18) Man–Man (18?)		Shaw (1968*a*) Shaw and Stead (1970)
Actinomycetaceae *Bifidobacterium* *B. bifidum*	Gal*p*β (4); Gal*p*β (6,6') Gal*p*β1-2Gal*p*βᶜ (16,16') (Gal*p*β1-2)$_2$Gal*p*β (27) Gal		Exterkate and Veerkamp (1969, 1971); Veerkamp (1972); Veerkamp and van Shaik (1974)
Actinomyces viscosa	Hex*p*-Hex*p* Hex*p*-Hex*p*-Hex*p*		Yribarren *et al.* (1974); Pandhi and Hammond (1978)
Actinopolyspora halophila		12	Gochnauer *et al.* (1975)
Streptomycetaceae *Streptomyces* sp. LA7017	Glc*p*α1-4GalU*p*αᶜ (20,20')	6	Bergleson *et al.* (1970); Batrakov and Bergelson (1978)

(continued)

Table 1-7. (Continued)

Family/Genus/Species	Glycosyl group[a]	% Total lipids	References
Mycobacteriaceae			
Mycobacterium			
M. tuberculosis and M. phlei	PI-2,6-di-Manpα(diacyl)[f]		Lee and Ballou (1965); Brennan and Ballou (1968)
M. smegmatis	Glcp;Glcp–Glcp; Galp; Manp; Gal–Glc		Schultz and Elbein (1974)
Propionibacterium			
P. shermanii	PI-2-Manpα(diacyl)[f]		Prottey and Ballou (1968)

[a]Glycosyl groups are linked to 1,2-di-*O*-acyl-*sn*-glycerol(DAG) unless otherwise stated.
[b]Glycosyl groups are linked to 1-*O*-alkenyl-2-*O*-acyl-*sn*-glycerol.
[c]Glycosyl groups are also acylated.
[d]Glcpα-PG: α-glucopyranosylphosphatidylglycerol.
[e]GlcNp-PG: glucosaminylphosphatidylglycerol.
[f]PI-Man: phosphatidylinositol monomannoside; PI-2,6-di-Manpα : 1-phosphatidyl-L-*myo*-inositol-2,6-di-α-D-mannopyranoside.

hydodysenteriae and *T. pallidum* (Kazan)] and probably also *T. pallidum* (Reiter) have been found to contain a major monogalactolipid, α-galactopyranosyl-DAG (**3**), which is the anomer of the plant monogalactolipid (**4**) (see Chapter 3, *this volume*). However, the monogalactosylDAG in *T. hydodysenteriae* is present largely in plasmalogen form (88%) and also contains a third fatty acyl group esterified to the galactose residue (Matthews *et al.*, 1980). By contrast, *T. reiteri* does contain the plant monogalactosylDAG (**4**) (Coulon-Morelec *et al.*, 1969). However, another *Treponoma* species, *T. zuelzerae*, contains a glucosylDAG, the detailed structure of which has not yet been reported (Meyer and Meyer, 1971).

Pseudomonadaceae. The major glycolipid in species of the *Pseudomonas* genus of this family of gram-negative anaerobic rods and cocci so far studied, such as *P. diminuta* and *P. vesicularis,* is the α-glucopyranosyldiacylglycerol (**1**), which may account for up to 10% or more of the total polar lipids in some species (Table 1-7). Minor glycolipids are the α-glucuronosylDAG (**8**) and glucosyl-α-glucuronosylDAG (**19**) in *P. diminuta* (Wilkinson, 1969); a novel -D-*glycero*-D-*gluco*-heptopyranosylDAG (**12**) has also been identified in *P. vesicularis* (Wilkinson and Galbraith 1979) (Table 1-7). It is interesting that the anomeric β-glucosylDAG (**2**) and β-glucuronosylDAG (**9**) have been found in *Alteromonas rubescens,* formerly classified as a pseudomonad (Wilkinson, 1968). A monoglucuronosylDAG, different from the β-GlcU*p*DAG of *A. rubescens* and thus probably identical with the α-GlcU*p*DAG (**8**) of *P. diminuta*, has been found in an unidentified moderately halophilic gram-negative bacterium along with an α-glucosylphosphatidylglycerol (Peleg and Tietz, 1971, 1973). An unusual tetraosylDAG has been identified in an extreme thermophile, *Thermus thermophilus,* as Gal*f*β1–2Gal*p*α1–6GlcN*p*β-1–2Glc*p*αDAG, the glucosaminyl group being *N*-acetylated with isoheptadecanoic acid (Oshima and Yamakawa, 1974; Oshima and Ariga, 1976) (Table 1-7).

Vibrionaceae. Glycolipid data (incomplete) on only two species in this family are available. *Vibrio cholerae* has been reported to contain a galactosyl glycolipid (Raziuddin, 1976), and the moderately halophilic *Vibrio costicola* has been reported to contain a glycolipid with very low mobility on TLC (Hanna *et al.*, 1984). An unidentified digalctosylDAG has been reported in a *Campylobacter (Vibrio) fetus* strain (Tornabene and Ogg, 1971).

Enterobacteriaceae. Glycoglycerolipids (structures not determined) have so far been detected only in two species of this family of facultative anaerobic rods, *Yersinia pseudotuberculosis* (Tornabene, 1973) and *Klebsiella pneumoniae* (Wassef, 1976), but not in the common Enterobacteria *E. coli* and *Salmonella typhimurium* (Macfarlane, 1962*a*). However, a wider range of species must be examined and the glycolipid structures determined before any meaningful conclusions can be drawn.

Bacteroidaceae. The one species in this family so far examined, *Bacteroides symbiosus,* contains a β-galactofuranosyl DAG (**6**) (Reeves *et al.*, 1964). Galactofuranosyl DAGs (plasmalogen form) have also been found in a *Butyrivibrio* species (uncertain affiliation) (Clark *et al.*, 1976) and possibly also in *Butyrivibrio fibrisolvens* (Kunsman, 1970), which suggest that this genus may be related to the Bacteroidaceae (Table 1-7).

Mycoplasmataceae. Some species of *Mycoplasma* (wall-less gram-negative bacteria requiring cholesterol for growth), e.g., *M. mycoides* and *M. pneumoniae,* contain mono-, di-, and trigalactofuranosylDAGs (structures **6** and probably **16** and **27,** respectively). Another species, *M. neurolyticum,* contains only the mono- and diglucopyranosylDAGs (structures **2** and **14,** respectively), but *M. pneumoniae* also contains a diglucosylDAG (**14**?) in addition to the galactosyl DAGs (Table 1-7).

Acholeplasma species (not requiring cholesterol) contain mono- and di-glucopyranosylDAGs (structures **1** and **13,** respectively) different from those present in the Mycoplasmas. One species (*A.axanthum*) contains the α-galac-tofuranosylDAGs (structures **5, 5′**) anomeric to that (structure **6**) present in *Mycoplasma* (Mayberry and Smith, 1983) (Table 1-7). A triglycosylDAG, Glc*p*β1–3Glc*p*α1–2Glc*p*αDAG (**23**) (Fig. 1.3), together with mono- and di-glycosylDAGs (**1** and **13**) are present in *A. granularum* (Table 1-7).

Some *Acholeplama* contain polyglycosylDAGs, e.g., *A. modicum* contains the pentaglycosylDAG, Gal*p*α1–2Gal*p*α1-3-D-*glycero*-D-*manno*hept*p*β1–3-Glc*p*α1–2 Glc*p*αDAG (**30**) (Fig. 1-4; Mayberry *et al.,* 1976); and *A. granularum* contains a lipoglycan with 27 sugar residues, including hexoses (Glc, Gal) and hex-osamines (FucNac, GlcNac) (Smith, 1981).

1.2.3.2. Gram-Positive Bacteria

Lactobacillaceae. The genus *Lactobacillus* in this bacterial family is charac-terized by the presence of high contents of the diglycosylDAG, Gal*p*α1-2 Glc*p*αDAG (**17**) [plus small amounts of its 6-*O*-acyl derivative (**17′**)], and in some species the related tri- and tetraglycosylDAGs, Glc*p*α1–6Gal*p*α1–2Glc*p*αDAG (**25**), Glc*p*β1–6Gal*p*α1–2Glc*p*αDAG (**26**) [plus small amounts of its 6-*O*-acyl derivative (**26′**)] and (Glc*p*β1–6)$_2$Gal*p*α1–2Glc*p*αDAG (**28**), re-spectively. Glc*p*αDAG (**1**) is also found in some species (Table 1-7). Some species also contain the triglucosylDAG, (Glc*p*β1–2)$_2$Glc*p*αDAG (**22**) and its 6-*O*-acyl derivative (**22′**) (Nakano and Fischer, 1977). It is noteworthy that some of these glycosyl sequences are also found in the phosphoglycolipids in this bacterial family (see Chapter 2, *this volume*).

Bacillaceae. The one genus in this family that has been studied in detail is the *Bacillus* genus; the species that have been examined (*B. cereus, B. lichenifor-mis and B. subtilis*) are characterized by the presence of the diglucosylDAG, Glc*p*β1-6Glc*p*βDAG (**14**) and the homologous triglucosylDAG, (Glc*p*β1–6)$_3$DAG (**24**) (Fig. 1-3; Table 1-7). However, a glucosaminylglycolipid, GlcN*p*βDAG (**10**) and its *N*-acyl derivative (**10′**) (Phizackerley *et al.,* 1972) are present in *B. megaterium,* which also contains a glucosaminylphosphatidylglyc-erol (Op den Kamp and van Deenen, 1966). The derived Glc*p*β1–4GlcN*p*β-DAG as its *N*-acyl derivative (**21**) and lyso derivative (**21′**) are found in the nonarchaebacterial thermoacidophile *B. acidocaldarius* (Langworthy *et al.,* 1976). This eubacterial thermoacidophile also contains a sulfonoglycosylDAG, probably identical to the plant sulfoquinovosyl DAG, and a *N*-acylglucosami-nylpentacyclic (hopane) tetrol (Langworthy *et al.,* 1976). It is interesting that of four strains of alkaliphilic *Bacillus* spp. investigated (Koga *et al.,* 1982) none contained any glycolipids or phosphoglycolipids. Only one species of another genus in this family, *Clostridium perfringens* (formerly *C. welchii*) has been reported to contain a high proportion of a mannosyl DAG (Macfarlane, 1962a), but no further studies on its structure have been reported.

Micrococcaceae. The only *Micrococcus* species studied so far is *M. lysodeikticus* (*luteus*) which has been shown to contain the mono- and dimannosideDAGs, Man$p\alpha$1–3DAG (**7**) and Man$p\alpha$1–3Man$p\alpha$DAG (**18**) (Macfarlane, 1961; Lennarz and Talamo, 1966). By contrast, several species of *Staphylococcus* have been found to contain a single diglucosylDAG, Glc$p\beta$1–6Glc$p\beta$DAG (**14**) (Fig. 1-2; Table 1-7), suggesting that this glycolipid is characteristic of the *Staphylococcus* genus. However, the diglucosylDAG (**14**) is also present in some bacilli and in at least one species of *Mycoplasma* (Table 1-7). It is of interest that a halotolerant species of *S. epidermidis* has also been found to contain Glc$p\beta$1–6Glc$p\beta$DAG (**14**) along with the monoglucosyl homologue, Glc$p\beta$DAG (**2**); the diglucosylDAG (**14**) is also a constituent of the phosphoglycolipid in this organism (Komaratat and Kates, 1975).

Streptococcaceae. In this family, all the streptococci examined for the presence of glycolipids (e.g., *S. faecalis*, *S. hemolyticus*, *S. lactis*) have been found to contain the diglucosylDAG, Glc$p\alpha$1–2Glc$p\alpha$DAG (**13**); some species also contain the monoglucosyl homologue, Glc$p\alpha$DAG (**1**) or the triglucosyl homologue, (Glc$p\alpha$1–2)$_3$DAG (**22**) (Fig. 1-3). However, *Acholeplasma laidlawii* also contains the same mono- and diglucosylDAGs (**1** and **13,** respectively) as the streptococci (Table 1-7).

Corynebacteria (Coryneform Group). This group has not been extensively studied but the same dimannosylDAG (structure **18**) present in *Micrococcus luteus* has been found in a *Corynebacterium* species (Khuller and Brennan, 1972a) and in two species of *Microbacterium* (Shaw, 1968a; Shaw and Stead, 1970). The dimannosylDAG (structure **18**), together with a monogalactosylDAG (structure **4**) and digalactosylDAG (structure **15**), are also present in several species of *Arthrobacter* (Fig. 1-2; Table 1-7). The two unrelated diglycosylDAGs (**15** and **18**) in *Arthrobacter* spp. may be representative of the two morphological forms of these species.

Actinomycetaceae. Of the two species studied in this family, *Bifidobacterium bifidum* is characterized by the presence of a homologous series of mono-, di-, and tri-β-galactofuranosylDAGs (structures **6, 16,** and **27,** respectively); the digalactosylDAG (**16**) is also present in a 3-*O*-acylated form (**16'**) (Fig. 1-2; Table 1-7). Note that the monogalactofuranosylDAG (**6**) is also present in species of *Bacteroides*, *Mycoplasma*, and *Acholeplasma* (Table 1-7). A β-galactopyranosylDAG (structure **4**) has also been found in *B. bifidum* (Exterkate and Veerkamp, 1971). An unidentified galactosylglycerolipid has also been found in *Actinomyces viscosa* (Yribarren *et al.*, 1974; Pandhi and Hammond 1978). An extremely halophilic *Actinomycete*, *Actinopolyspora halophila* (Gochnauer *et al.*, 1975) contains a dihexosyl and a trihexosylDAG of unidentified structure.

Streptomycetaceae. The one species of *Streptomyces* examined contains a glucosylgalactosuronylDAG (structure **20**; Fig. 1-2), which may be acylated at position-2 or -3 of the GalUp group (Bergelson *et al.*, 1970; Batrakov and Bergelson, 1978) (Table 1-7).

Mycobacteriaceae. Apart from the trehalose acyl esters (see Chapter 4, *this volume*), little is known concerning glycosylDAGs in *Mycobacteria,* although their presence in *Nocardia* spp. has been reported (Khuller and Brennan, 1972*b*). However, mono- and diglucosylDAGs, monogalactosylDAG, and monomannosylDAG have been identified in *Mycobacterium smegmatis* (Schultz and Elbein, 1974). Phosphatidylinositol-2,6-dimannosides have long been known in *M. tuberculosis* and *M. phlei* (Lee and Ballou, 1965) (Table 1-7), and a phosphatidylinositol monomannoside (diacylated) has been identified in *Propionibacterium shermanii* (Prottey and Ballou, 1968) (see review by Asselineau, 1981).

Summary: Eubacteria. The available data on the glycolipid composition of eubacteria show some interesting trends that can be correlated with taxonomic classification on the level of the genus, but there is insufficient information to permit any definitive conclusions to be drawn on the level of a bacterial family (Table 1-8); such a correlation has been suggested previously for bacterial fatty acids and phospholipids (Kates, 1964; Asselineau, 1983). On the genus level, correlation with glycolipid composition is clearer for the diglycosylDAGs and triglycosylDAGs than for the monoglycosylDAGs (see Table 1-8). Also, the presence in some genera of uronic acid or amino sugar-containing glycolipids is helpful in discerning correlations of glycolipid composition and structure with taxonomic classification. The following generalizations may be made on the basis of the data available (see Table 1-8):

Among the gram-negative bacteria, the genus *Treponoma* is clearly distinguished by the α-monogalactosylDAG (**3**), the genus *Pseudomonas* by the presence of mono- and diglycosylDAGs containing glucuronic acid (structures **8** and **19,** respectively), and both the *Bacteroides* and *Butyrivibrio* genera by the presence of the β-galactofuranosylDAG (**6**). The latter glycolipid is also present in a *Mycoplasma* species, and the anomeric α-galactofuranosylDAG (**5**) is present in one *Acholeplasma* species. These two genera are further distinguished from each other and from other gram-negative genera by the di- and triglycosylDAGs present: structures **14, 16,** and **27** in *Mycoplasma* and structures **13** and **23** in *Acholeplasma* (see Table 1-8). The presence of these gram-positive-associated glycolipids (structures **13, 14, 16,** see below) in *Mycoplasma* and *Acholeplasma* species raises the question whether the Mycoplasmatales should be classified with the gram-positive rather than with the gram-negative bacteria (Shaw, 1975).

Among gram-positive bacteria, the *Lactobacilli* are characterized by the presence of both the galactosylglucosylDAG (**17**) and the related triglycosylDAG (structure **25;** Fig. 1-3), distinguishing this genus from *Pneumococci,* which contain the diglycosylDAG (**17**) but not the triglycosylDAG (**25**); *Bacilli* by the presence of both the β-diglucosylDAG (**14**) and the β-triglycosylDAG (**24**) and also the amino-sugar glycolipid (**10**); *Staphylococci* by the presence of only the β-diglucosylDAG (**14**); *Streptococci* by the presence of the α-diglucosylDAG (**13**) and the triglycosylDAG (**22**); *Arthrobacter* species by the presence of both the β-digalactosylDAG (**15**) and the dimannosylDAG (**18**) (the latter is also present, as the sole glycolipid, in one species of *Micrococcus* and two species of *Microbacterium*); and *Bifidobacteria* by the presence of the β-di- and trigalactofuranosylDAGs (**16** and **27,** respectively); *Mycobacteria* are

Table 1-8. Summary of Distribution of GlycosylDAGs among Eubacterial Genera[a]

Genus/Family	Structure of glycosylDAG[b]				
	Mono-	Di-	Tri-	Uronic	Amino
Gram-negative					
Anoxyphotobacteria					
Rhodospirillaceae					
Rhodopseudomonas	HSO₃Q				
Chromatiaceae					
Spirochaetaceae					
Trepanoma	(3), (4)				
Pseudomonadaceae					
Pseudomonas	(1)			(8) (19)	
Bacteroidaceae					
Bacteroides	(6)				
Butyivibrio	(6,6')				
Mycoplasmataceae					
Mycoplasma	(6) (2)	(14) (16?)	(27?)		
Acholeplasma	(1)	(13)	(23)		
Gram-positive					
Lactobacillaceae					
Lactobacillus		(17,17')	(25)		
Bacillaceae					
Bacillus		(14)	(24)		(10)
Micrococcaceae					
Micrococcus		(18)			
Staphylococcus		(14)			
Streptococcaceae					
Streptococcus	(1)	(13)	(22)		
Pneumococcus	(1)	(17)			
Corynebacteriaceae					
Corynebacterium		(18), PI-diMan			
Arthrobacter		(15) (18)			
Microbacterium		(18)			
Actinomycetaceae					
Bifidobacterium	(4) (6)	(16, 16')	(27)		
Streptomyces				(20, 20')	
Mycobacteriaceae					
Mycobacterium		PI-diMan			

[a]See Table 1-7.
[b]Structures of the glycosylDAGs are given in Figs. 1-1–1-3.

characterized by the presence of phosphatidyl-*myo*-inositol mannosides (see review by Asselineau, 1981), but only one species (*M. smegmatis*) is so far known to contain mono- or diglycosylDAGs (see Table 1-7). Clearly, further information on the distribution of glycolipids in a wider variety of bacteria will be needed to verify the preceding tentative correlations.

1.2.4. Fatty Acid and Alkyl Chain Composition

In general, the fatty acid composition of bacterial glycosylDAGs reflects that of the total lipids in both gram-negative and -positive bacteria, which

Table 1-9. Fatty Acid Components of Eubacterial GlycosylDAGs

Species	Glycosyl group	Major fatty acids[a]	References
Gram-negative			
Treponoma hyodysenteriae	Galpα (**3,3'**)	alkenyl : 14 : 0, *i*-15 : 0,16 : 0 acyl : 16 : 0, *i*-15 : 0 *a*-15 : 0	Matthews *et al.* (1980)
Chloropseudomonas ethylicum	Galpβ (**4**)	16 : 0,16 : 1 (18 : 1)	Constantopoulos and Bloch (1967)
Pseudomonas diminuta	Glcpα (**1**) GlcUpα (**8**) GlcpβGlcUpα (**19**)	16 : 0,18 : 1 16 : 0,18 : 1 16 : 0,18 : 1, 19 : *cy*	Wilkinson (1969)
Alteromonas rubescens	Glcpβ (**2**) GlcUpβ (**9**)	16 : 0,16 : 1,18 : 1 17 : *cy*,15 : *br*	Wilkinson (1968)
Moderate halophile	GlcU	16 : 0,17 : *cy*, 19 : *cy*	Peleg and Tietz (1971)
Mycoplasma mycoides	Galfpβ (**6**)	16 : 0,18 : 0,18 : 1	Plackett (1967)
Acholeplasma axanthum	Galfα (**5**) acyl-Galfα (**5'**)	14 : 0,16 : 0,18 : 0 14 : 0,16 : 0,18 : 0	Mayberry and Smith (1983)
Gram-positive			
Lactobacillus casei	Glcpα (**1**)	16 : 0,18 : 1, 19 : *cy*	Shaw *et al.* (1968a)
	Galpα–Glcpα (**17**)	16 : 0,18 : 1, 19 : *cy*	
	GlcpβGalpαGlcpα (**26**)	16 : 0,18 : 1, 19 : *cy*	Nakano and Fischer (1977)

Organism	Structure	Fatty acids	Reference
Listeria monocytogenes	Gal–Glc	16 : 0,15 : *br,* 17 : *br*	Carroll *et al.* (1968)
Bacillus cereus	Glc*p*β–Glc*p*β (**14**)	16 : 0,15 : *br,* 17 : *br*	Saito and Mukoyama (1971)
Bacillus acidocaldarius	Glc*p*β–GlcN*p*β (**21**)	17 : *br,*17 : *cy,* 19 : *cy*	Langworthy *et al.* (1976)
Micrococcus luteus	Man*p*αMan*p*α (**18**)	15 : *br,*17 : *br,* 16 : 0,18 : 0	Macfarlane (1962)
Staphylococcus lactis	Glc*p*βGlc*p*β (**14**)	16 : 0,16 : 1 15 : *br,*17 : *br*	Brundish *et al.* (1967)
Staphylococcus epidermidis	Glc*p*β (**2**)	15 : *br,*17 : *br*	Komaratat and Kates (1975)
Streptococcus hemolyticus	Glc*p*βGlc*p*β (**14**) Glc*p*αGlc*p*α (**13**) Glc*p*αGlc*p*αGlc*p*α (**22**)	15 : *br,*17 : *br* 16 : 0,16 : 1,18 : 1 16 : 0,16 : 1,18 : 1	Fischer (1976)
Pneumococcus types I and XIV	Glc*p*α (**1**) Gal*p*αGlc*p*α (**17**) PI-di-Man	12 : 0+14 : 0,16 : 0 16 : 1,18 : 1	Kaufman *et al.* (1965)
Corynebacterium diphtheriae	Gal*p*β (**4**) Man*p*αMan*p*α (**18**)	14 : 0,15 : *br,*15 : 0 16 : *br,*16 : 0,16 : 1	Brennan and Lehane (1971)
Arthroacter crystallopoietes	Gal*p*β (**6**)	15 : *br,*17 : *br* 15 : *br,*17 : *br*	Shaw and Stead (1971)
Bifidobacterium bifidum	Glc*p*αGlcU*p*α (**20**)	16 : 0,18 : 0,18 : 1, 19 : *cy*	Veerkamp and van Schaik (1974)
Streptomyces LA 7017	Glc	15 : 0,15 : 1	Bergelson *et al.* (1970)
Mycobacterium smegmatis	Glc–Glc Gal–Glc	16 : 0,18 : 1 16 : 0,18 : 0,18 : 1 16 : 0,18 : 0,18 : 1	Schultz and Elbein (1974)

*a*Abbreviations: The first number represents the fatty acid chain length, the second represents the number of double bonds in the chain; *a*, anteiso; *br*, branched; *cy*, cyclopropyl; *i*, iso.

consist mainly of saturated, monounsaturated, branched, and/or cyclopropane fatty acids. The major fatty acids in glycosylDAGs of some representative
eubacteria of both types are given in Table 1-9. Thus, in the anoxyphotobac-
terium *C. ethylicum*, the galactosylDAG contains mostly palmitic and pal-
mitoleic acid and is completely lacking any polyunsaturated fatty acids characteristic of the galactosylDAGs of higher plants (see Chapter 3, *this volume*).
Other gram-negative bacteria, such as *Pseudomonads* and *Mycoplasmas*, contain
mainly saturated acids (16:0, 18:0) and monoenes (16:1, 18:1) in their
glycosylDAGs, and some (e.g., *Alteromonas rubescens*) also contain branched-
chain (15:br, 17:br) acids, reflecting the total fatty acids of these organisms.
However, an unidentified moderate halophile contains unusually high proportions of cyclopropane acids (17:cy, 19:cy) in its glycolipid as well as its
phospholipid components (Peleg and Tietz, 1971).

Some gram-positive bacteria, such as *Streptococcus* and *Pneumococcus*, contain saturated (16:0) and monoenoic (16:1, 18:1) acids in their glycolipids;
some, such as *Lactobacillus*, contain in addition cyclopropane acids (17:cy
19:cy) derived from the corresponding monoenoic acids (16:1 and 18:1,
respectively); and others, including *Bacillus*, *Micrococcus*, *Staphylococcus*, and
Listeria, contain branched-chain acids (15:br, 17:br) as major fatty acid components of their glycolipids (Table 1-9). A eubacterial acidothermophile, *B.
acidocaldarius*, contains both branched and cyclopropane fatty acids in its
glycosylDAGs (Langworthy *et al.*, 1976); another eubacterial thermophile (anaerobic), *Thermodesulfotobacterium commune*, contains both straight- and
branched-chain fatty acids but also normal-, and iso- and anteiso-branched
glycerol diethers (Langworthy *et al.*, 1983).

Plasmalogen forms of glycerophospholipids, although generally rare in
bacteria, are known to occur in strictly anaerobic bacteria (Goldfine and
Hagen, 1972; Clarke *et al.*, 1976). Plasmalogenic glycosylDAGs have so far
only been reported present in *T. hyodysenteriae* (Matthews *et al.*, 1980) and in
anaerobic rumen bacteria of the genus *Butyrivibrio* (Clarke *et al.*, 1976), occurring almost exclusively as the 1-alkenyl-2-acyl-*sn*-Gro (plasmalogen) form, in
which the aldehyde chains were mainly 14:0, *i*-15:0 and 16:0, and the acyl
chains were mainly 16:, *i*-15:0, and *a*-15:0 (Table 1-9). Little is known concerning the structure of the plasmalogen glycolipid in *Butyrivibrio* species,
other than that it contains a galactofuranosyl group and has a fatty aldehyde-
to-fatty acid ratio of 0.75:1.25 (Clarke *et al.*, 1976).

1.3. Glyco-, Sulfoglyco-, and Phosphoglycoglycerolipids of Archaebacteria

The archaebacteria are a newly delineated kingdom of organisms (Woese
and Wolfe, 1985) that have evolved under the conditions of the primitive
earth, such as high temperature, anaerobic atmosphere, and high salinity.
Three groups of archaebacteria are known: the extreme halophiles, the methanogens, and the thermoacidophiles. They are clearly distinguished from all
other organisms, particularly the eubacteria, by three major criteria: (1) their

ribosomal RNA(16S) sequences differ distinctly from those of eubacteria and eukaryotes; (2) their cell walls consist of glycosylated proteins rather than the peptidoglycan structure in eubacteria; and (3) their membrane lipids are unique in consisting entirely of derivatives of a dialkylglycerol, diphytanyl-glycerol diether (structure **31**, Fig. 1-8A) (Kates, 1978), or its dimer, di-biphytanyldiglyceroltetraether (structure **32**, Fig. 1-8) (Langworthy, 1977*a*).* Furthermore, the stereochemistry of the glycerol residue of the di- or tetraethers is opposite to that in the diacylglycerols of all other organisms: the diphytanylglycerol (**31**) has the configuration 2,3-diphytanyl-*sn*-glycerol (Kates, 1978), and the configuration of the two glycerols in the tetraether (**32**) is the same as in the diether (Kushwaha *et al.*, 1981; Langworthy, 1985). Also, the phytanyl groups have the configuration $3R,7R,11R$ (Kates 1978), and the biphytanyl groups in the glycerol tetraether have the configuration $3R,7R,11R,15R,15'R,\ 11'R,7'R,3'R$ (Heathcock *et al.*, 1985).

Variants of the diphytanylglyceroldiether and the dibiphytanyldiglycerol-tetraether structures (**31** and **32,** respectively) are also found in some arch-aebacteria. For example, lipids of alkaliphilic halophiles (*Natronobacterium* and *Natronococcus*) contain 2-*O*-sesterterpanyl-3-*O*-phytanyl-*sn*-glycerol (C_{20}-C_{25}-diether; structure **31′**, Fig. 1-8) and 2,3-di*O*-sesterterpanyl-*sn*-glycerol (C_{25}-C_{25}-diether; structure **31″**, Fig. 1-8) (De Rosa *et al.*, 1982*a*, 1983*a*). Lipids of *Methanococcus jannaschii* contain a macrocyclic C_{40}-diether (structure **31a,** Fig. 1-8A) (Comita and Gagosian, 1983) and *Methanosarcina barkeri* contains a tetritol–diphytanyldiether (structure **31b,** Fig. 1-8A) (De Rosa *et al.*, 1986*b*). Species of the thermoacidophilic *Sulfolobus* genus contain lipids derived from a dibiphytanylglycerolnonitoltetraether (structure **32′**, Fig. 1-8A) (De Rosa *et al.*, 1986*a*). *Methanosarcina* spp. and *Sulfolobus* spp. also contain lipids derived from dibiphytanyldiglyceroltetraethers or dibiphytanylglycerolnonitoltetra-ether containing one to four cyclopentane rings in the C_{40} biphytanyl groups (structures **32a–d,** Fig. 1-8B) (De Rosa *et al.*, 1986*a,b*).

Present nomenclature of these complex lipids is cumbersome, and trivial names appear to be required. Nishihara *et al.* (1987) suggested that di-phytanylglyceroldiether (**31**) and its variants (**31′**, **31″**, **31a,** and **31b**) be called archaeol (modified by adding the appropriate alkyl group designations, e.g., C_{20}, C_{25}) and that dibiphytanyldiglyceroltetraether (**32**) be called cal-darchaeol and its variant nonitolcaldarchaeol (**32′**). This forms the basis of a useful and practical system of nomenclature for archaebacterial lipids, fol-lowed in this chapter.

This section deals with the glycolipids, the sulfated glycolipids, and the phosophoglycolipids of archaebacteria, in which they play a prominant mem-brane structural role. Archaebacterial lipids, including the di- and tetraether-

---

*Although only archaebacterial lipids are derived from *sn*-2,3-diphytanylglycerol diether and/or the dibiphytanyldiglycerol tetraether, a eubacterial thermophilic anaerobe has been found to contain lipids derived from *sn*-1,2-dialkylglycerol diether (containing *a*-17 : 0, *i*-16 : 0, etc., chains) (Langworthy *et al.*, 1983). However, this finding does not negate the criteria that arch-aebacterial lipids are distinguished by having *sn*-2,3-diphytanylglycerol or its dimer as a basic lipid core.

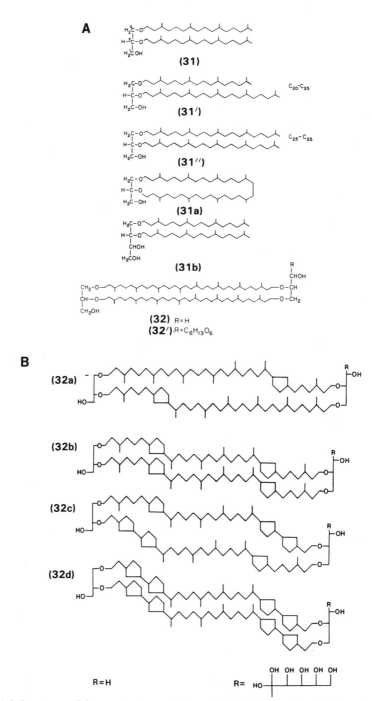

Figure 1-8. Structures of the parent isopranyl glycerol ether lipid core of archaebacteria. (A) (31) 2,3-Di-O-phytanyl-sn-glycerol (phytanyl diether; C_{20}-diether; archaeol); (31′) 2-O-sesterterpanyl-3-O-phytanyl-sn-glycerol (C_{20},C_{25}-diether; C_{20},C_{25}-archaeol); (31″) 2,3-di-O-sesterterpanyl-sn-glycerol (C_{25}, C_{25}-diether; C_{25}, C_{25}-archaeol); (31a) 2,3-cyc-biphytanyl-sn-glycerol (C_{40}-macrocyclic diether; cyc-archaeol); (31b) di-O-phytanyl tetritol diether (tetritolarchaeol); (32) di-biphytanyldiglyceroltetraether (R = H; C_{40}-tetraether; caldarchaeol); (32′) di-biphytanylglycerolnonitoltetraether (R = $C_6H_{13}O_6$; nonitolcaldarchaeol). (B) (32a–32d) variant caldarchaeols (R = H) or nonitolcaldarchaeols (R = $C_6H_{13}O_6$) with one, two, three, and four cyclopentane rings in the biphytanyl chains, respectively.

derived glycolipids have been the subject of several previous reviews (Kates, 1972, 1978, 1988; Kates *et al.*, 1982; Kates and Kushwaha, 1978; Ishizuka and Yamakawa, 1985; Langworthy, 1982*a,b*, 1985; De Rosa *et al.*, 1986*a;* Kamekura and Kates, 1988).

1.3.1. Extraction, Isolation, and Quantitation of Archaebacterial Lipids

Extraction of lipids from extremely halophilic archaebacteria is carried out essentially as described in Section 1.2.1 for eubacteria, by the modified procedure (Kates, 1986*a*) of Bligh and Dyer (1959), except that the wet cell paste is suspended in 4 M NaCl instead of water, to avoid cell lysis before extraction of lipids. Methanogenic bacteria may also be extracted by the modified Bligh–Dyer procedure (see Kushwaha *et al.*, 1981; Morii *et al.*, 1986; Ferrante *et al.*, 1986, 1987), but extraction with chloroform–methanol acidified with trichloroacetic acid (TCA) has also been used for cells with a rigid pseudomurein cell wall (e.g., *M. thermoautotrophicum*) that are difficult to extract with neutral solvents (Nishihara and Koga, 1987). Lipid extraction from thermoacidophilic bacteria has been carried out with chloroform–methanol (2 : 1, v/v) at room temperature, using Sephadex G25 to remove nonlipid contaminants (Wells and Dittmer, 1963; Langworthy *et al.*, 1974), or by the Bligh–Dyer (1959) procedure (Langworthy *et al.*, 1974; De Rosa *et al.*, 1983*b*); more drastic extraction conditions (continuous extraction with chloroform–methanol, 1 : 1, under reflux in a Soxhlet for 12 hr) have been used (De Rosa *et al.*, 1982*b*), but such conditions do not appear to be necessary and may result in partial degradation of the complex lipids.

Lipoglycan or lipopolysaccharide, e.g., that present in *Thermoplasm acidophilum*, may be extracted, after prior solvent extraction of the total lipids, by stirring the cells with 45% aqueous phenol at 70°C for 15 min, collecting the aqueous phase by centrifugation, dialyzing it against deionized water and lyophilizing it (Mayberry-Carson *et al.*, 1974; Smith, 1980).

1.3.1.1. Glycolipid Isolation from Extreme Halophiles

The major glycolipid in the genus *Halobacterium* is the sulfated triglycosylarchaeol, S-TGA-1 (structure **34'**, Fig. 1-9) (Kates *et al.*, 1967; Kates and Deroo, 1973), but minor amounts of the sulfated tetraglycosylarchaeol, S-TeGA (structure **36'**, Fig. 1-9), the unsulfated triglycosylarchaeol, TGA-1 (structure **34**, Fig. 1-9) and the unsulfated tetraglycosylarchaeol, TeGA (structure **36**, Fig. 1-9) are also present (Smallbone and Kates, 1981). Separation and isolation of these glycolipids from extreme halophiles of the *Halobacterium* genus (e.g., *H. cutirubrum*) requires a rather complex procedure to separate them from the phospholipid components. The most convenient procedure (Smallbone and Kates, 1981) uses prior precipitation of the major phospholipid, phosphatidylglycerophosphate (PGP) in form of its barium salt (Kates *et al.*, 1965), followed by fractionation of the glycolipids plus minor phospholipids in the supernatant by preparative TLC on silica gel H (for details, see Section 1.2.1.2), using the following solvent systems: (1) chloroform–methanol–90% acetic acid (30 : 4 : 20, v/v) yielding a mixture of

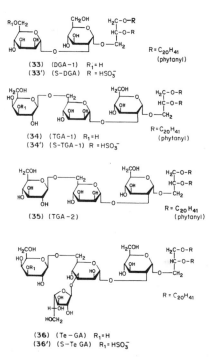

Figure 1-9. Structures of glycolipids and sulfated glycolipids of extremely halophilic archaebacteria. (**33**) Man*p*α1–2Glc*p*α1–1-archaeol (DGA-1); (**33'**) 6-HSO₃-Man*p*α1–2Glc*p*α1–1-archaeol (S-DGA-1); (**34**) Gal*p*β1–6Man*p*α1–2Glc*p*α1–1-archaeol (TGA-1); (**34'**) 3-HSO₃-Gal*p*β1–6Man*p*α1–2Glc*p*α1–1-archeol (S-TGA-1); (**35**) Glc*p*β1–6Man*p*α1–2Glc*p*α1–1-archaeol (TGA-2); (**36**) Gal*p*β1–6Man*p*(3–1Gal*f*β)α1–2Glc*p*α1–1-archaeol (TeGA); (**36'**) 3-HSO₃-Gal*p*β1–6Man*p*(3–1Gal*f*β)α1–2Glc*p*α1–1-archaeol (S-TeGA).

S-TeGA (**36'**), TeGA (**36**), and S-TGA-1 (**34'**), and a mixture of TGA-1 (**34**) and phosphatidylglycerosulfate (PGS); (2) chloroform–methanol–diethylamine–water (110 : 50 : 8 : 5) to separate the mixture of TeGA (**36**), S-TeGA (**36'**), and S-TGA-1 (**34'**); and (3) chloroform–methanol–28% ammonium hydroxide (65 : 35 : 5) to separate TGA-1 from PGS (see Table 1-10); all glycolipids were detected by iodine vapor and eluted from the silica with chloroform–methanol–water (1 : 1 : 0.1, v/v); they were chromatographically pure (see Fig. 1-12).

A more direct isolation of the major glycolipid S-TGA-1 (**34'**) may be achieved by preparative TLC on silica gel H of the total polar lipids in the solvent system chloroform–methanol–90% acetic acid (30 : 4 : 20, v/v), and elution from the silica as the free acid with chloroform-methanol-0.1 N HCl (1 : 2 : 0.8, v/v) followed by conversion to the ammonium salt (Kates and Deroo, 1973).

In the genus *Haloarcula* the major glycolipid is a triglycosyl archaeol, TGA-2 (structure **35**, Fig. 1-9), which is the analog of TGA-1 (**34**) in *Halobacteria*, containing a terminal β-glucopyranosyl group instead of a β-galactopyranosyl group (Evans *et al.*, 1980). An unidentified diglycosylarchaeol

Table 1-10. TLC Mobilities of Archaebacterial Glycolipids

Glycosyl group of glycolipid	Solvents[a] ($R_f \times 100$)						References
	A	B	C	D	E	F	
Glycosylarchaeols (Figs. 1-9 and 1-10)							
Glcpα (MGA-1)	98						Kates and Deroo (1973)
Glcpβ (MGA-2, **37**)		86		95			Ferrante et al. (1986)
Manpα1-2Glcpα (DGA-1, **33**)	77	42					Kates and Deroo (1973)
				61			Kushwaha et al. (1982a)
Man–Glc (DGA-2)	57	22	86				Evans et al. (1980)
Glcpβ1-6Glcpβ (DGA-3, **38**)				57			Nishihara et al. (1987b); Ferrante et al. (1986)
Manpα1-3Galpβ (DGA-4, **39**)				57			Ferrante et al. (1988)
Galpβ1-6Galpβ (DGA-5, **40**)[b]				50			Ferrante et al. (1988)
Glcpα1-2Galpβ (DGA-I, **41**)			68	46			Ferrante et al. (1988)
Galfβ1-6Galpβ (DGA-II, **42**)			78	52			Kushwaha et al. (1981)
6-HSO$_3$–Manpα1-2Glcpα (S-DGA-1, **33'**)				62			Kushwaha et al. (1981); Kushawa et al. (1982a)
Galpβ1-6Manpα1-2Glcpα (TGA-1, **34**)	34		38	36			Smallbone and Kates (1981)
3-HSO$_3$-Galpβ1-6Manpα1-2Glcpα (S-TGA-1, **34'**)	15	11	63	16			Kates and Deroo (1973); Smallbone and Kates (1981)
Glcpβ1-6Manpα1-2Glcpα (TGA-2, **35**)	43	8	48				Evans et al. (1980)
Galpβ1-6Manp(3-1βGalf)α1-2Glcpα (TeGA, **36**)	17		20				Smallbone and Kates (1981)
3-HSO$_3$-Galpβ1-6Manp(3-1βGalf)α1-2Glcpα (S-TeGA, **36'**)	5		46	9			Smallbone and Kates (1981)
Glycosylcaldarchaeols (Figs. 1-10 and 1-11)							
Glcpα1-2Galpβ (DGC-I, **43**)			73	57			Kushwaha et al. (1981)
Galfβ1-6Galpβ (DGC-II, **44**)			83	66			Kushwaha et al. (1981)
Glcpβ1-6Glcpβ (DGC-3, **45**)					75	69	Nishihara et al. (1987b)
GlcpβGalpβ (DGC-4, **48**)					80		Thurl and Schäfer (1988)
Glcpα1-4Galpβ (DGC-5, **48a**)					70		Langworthy (1977b); Lanzotti et al. (1987)
(Glcpβ1-6)$_2$Glcpβ (DGC-6, **49**)					50		Thurl and Schäfer (1988)

(continued)

Table 1-10. (Continued)

Glycosyl group of glycolipid	Solvents[a] ($R_f \times 100$)						References
	A	B	C	D	E	F	
Phosphoglycosylcaldarchaeols (Figs. 1-10 and 1-11)							
Glcpα1–2Galfβ (PGC-I, **43'**)			34	8			Kushwaha *et al.* (1981)
Galfβ1–6Galfβ (PGC-II, **44'**)			46	11			Kushwaha *et al.* (1981)
Glcpβ1–6Glcpβ (PGC-3, **45'**)							Nishihara *et al.* (1987b)
GlcpβGalpβ (PGC-4, **48'**)					20		Langworthy (1977b)
Glcpβ (PGC-5, **47'**)					20	19	Thurl and Schäfer (1988)
Glycosylnomitolcaldarchaeols (Fig. 1-11)							
Glcpβ1–3' (GNC, **50**)					80		De Rosa *et al.* (1986a); Langworthy (1977b)

[a]Solvents: A: chloroform–methanol–90% acetic acid (30 : 4 : 20, v/v); B: chloroform–methanol–28% ammonium hydroxide (65 : 35 : 5, v/v); C: chloroform–methanol–diethylamine–water (110 : 50 : 8 : 5, v/v); D: chloroform–methanol–acetic acid–water (85 : 22.5 : 10 : 4, v/v); E: chloroform–methanol–water (65 : 25 : 4, v/v); F: chloroform–acetone–methanol–acetic acid–water (20 : 15 : 10 : 1.5 : 1, v/v).
[b]Glycosyl groups are linked to 2-phytanyl-3-(3'-hydroxyphytanyl)-*sn*-glycerol.

(DGD-2), present in minor amounts, is the only other glycolipid in species of this genus. These two glycolipids were isolated from a representative *Haloarcula* species, *H. marismortui* (Evans et al., 1980), by TLC of the total polar lipids on silica gel G in solvent system chloroform–methanol–diethylamine–water (110 : 50 : 8 : 5, v/v) for TGA-2 and chloroform–methanol 28% ammonium hydroxide (65 : 35 : 5, v/v) for DGA-2.

Species of the genus *Haloferax* contain a major sulfated glycolipid, the sulfated diglycosyl archaeol, S-DGA-1 (structure **33′**, Fig. 1-9), and the corresponding unsulfated glycosylarchaeol, DGA-1 (structure **33**, Fig. 1-9). These two glycolipids have been isolated from the total polar lipids of a typical *Haloferax*, *H. mediterranei*, by preparative TLC on silica gel G in the solvent system chloroform–methanol–acetic acid–water (85 : 20 : 10 : 4, v/v) (Kushwaha *et al.*, 1982*a*).

1.3.1.2. Glycolipid Isolation from Methanogens

The glycolipids of only a few species of methanogens have so far been studied; glycosyl as well as phosphoglycosyl derivatives of both archaeol (**31**) and caldarchaeol (**32**) have been identified in methanogens. Some of these glycolipids (structures **41,42,43,43′,44,44′**, Fig. 1-10A) have been separated and isolated from *Methanospirillum hungatei* by fractionation of the total lipids on a column of silicic acid (Bio-sil A) eluted with chloroform (neutral lipids), acetone (glycolipids and phosphatidylglycerol), chloroform–methanol (3 : 2), and methanol (phosphoglycolipids) (Kushwaha *et al.*, 1981). Final separation and purification of the individual glycolipids (diglycosylarchaeols, **41** and **42**, Fig. 1-10A; and diglycosylcaldarchaeols, **43** and **44**, Fig. 1-10A) and phosphoglycolipids (diglycosylglycerophosphorylcaldarchaeols, **43′** and **44′**, Fig. 1-10A) was achieved by preparative TLC on silica gel G plates using chloroform–methanol–acetic acid–water (85 : 15 : 10 : 3, v/v) for the glycolipids and chloroform–methanol–diethylamine–water (110 : 50 : 8 : 3.5, v/v) for separation of the phosphoglycolipids. Lipid components were detected by exposure to iodine vapor and were eluted from the silica gel with a one-phase Bligh–Dyer system (chloroform–methanol–water, 1 : 2 : 0.8) followed by a two-phase partition (see Kates, 1986*a*).

A more detailed fractionation of the lipids of *M. hungatei* has been carried out by Ferrante *et al.* (1987) by silicic acid (Bio-sil A) column chromatography using the following elution sequence: chloroform (neutral lipids), chloroform–methanol (9 : 1) (unidentified glycolipids and diglycosylcaldarchaeol, structure **44**, Fig. 1-10A); chloroform–methanol (6 : 1) (diglycosylarchaeols, structures **41** and **42**, Fig. 1-10A; diglycosylcaldarchaeols structures **43** and **44**, Fig. 1-10A); chloroform–methanol (3 : 2) (glycolipid **41**, phosphoglycolipids **43′** and **44′**, unidentified lipids); chloroform–methanol (1.25 : 1.1 : 1 and 1 : 2) (phosphoglycolipids **43′** and **44′**, phosphopentitol dimethylaminodiether, and unidentified lipid); chloroform–methanol (1 : 6) (phosphoglycolipids **43′** and **44′**, phosphopentitoltrimethylaminodiether). Final purification of the glyco- and phosphoglycolipids was carried out (Ferrante *et al.*, 1986, 1987) by preparative TLC on silica gel G (1-mm-thick layers) in the solvent system chlo-

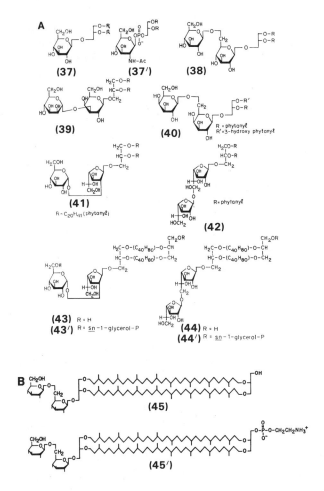

Figure 1-10. Structures of the glycolipids and phosphoglycolipids of methanogenic archaebacteria. (A) (37) Glc*p*β1–1archaeol (MGA-2); (37′) GlcN*p*Ac1-P-archaeol; (38) Glc*p*β1–6Glc*p*β1–1archaeol (DGA-3); (39) Man*p*α1–3Gal*p*β1–1archaeol (DGA-4); (40) Gal*p*β1–6Gal*p*β1–1-hydroxyarchaeol (DGA-5; R′ = 3-hydroxyphytanyl); (41) Glc*p*α1–2Gal*f*β1–1archaeol (DGA-I); (42) Gal*f*β1–6Gal*f*β1–1archaeol (DGA-II); (43) Glc*p*α1–2Gal*f*β1–1caldarchaeol (DGC-I; R = H); (43′) Glc*p*α1–2Gal*f*β1–1caldarchaeol-P-Gro (PGC-I; R = *sn*-3-glycerol-P); (44) Gal*f*β1–6Gal*f*β1–1caldarchaeol (DGC-II; R = H); (44′) Gal*f*β1–6Gal*f*β1–1caldarchaeol-P-Gro (PGC-II; R = *sn*-3-glycerol-P). (B) (45) Glc*p*β1–6Glc*p*β1–1caldarchaeol (DGC-3); (45′) Glc*p*β1–6Glc*p*β1–1caldarchaeol-P-ethanolamine (PGC-3).

roform–methanol–acetic acid–water (85 : 22.5 : 10 : 4, v/v). Lipid components were detected and eluted as described above.

The lipids of *Methanococcus voltae* have been found to be derived entirely from the diether archaeol (31) and not from the tetraether caldarchaeol (32) (Ferrante *et al.*, 1986). These lipids, including a monoglucosylarchaeol (structure 37, Fig. 1-10A), a diglucosylarchaeol (structure 38, Fig. 1-10A) and a novel *N*-acetylglucosamine-1-phosphate-archaeol (structure 37′, Fig. 1-10A) were fractionated on a column of silicic acid (Bio-sil A) as described above for

M. hungatei (Kushwaha *et al.*, 1981) and finally purified by preparative TLC as described above (Ferrante *et al.*, 1986). Another methanogen, *Methanothrix concilii*, was also found to contain two glycosylarchaeols (structures **39** and **40,** Fig. 1-10A), which were isolated and purified as described for *M. voltae* (Ferrante *et al.*, 1988a).

The glycolipids of the methanogen *Methanobacterium thermoautotrophicum* are derived both from the diether archaeol and the tetraether caldarchaeol (Nishihara *et al.*, 1987; Koga *et al.*, 1987). These are isolated from the total lipids by fractionation on a column of DEAE-cellulose (acetate form) using the following elution scheme: chloroform (neutral lipids); chloroform–methanol (9 : 1) (diglucosylcaldarchaeol, structure **45**, Fig. 1.10B); chloroform–methanol (4 : 1) (aminoethylphosphocaldarchaeol); and chloroform–methanol (1 : 1) (aminoethylphosphodiglycosylcaldarchaeol, structure **45'**, Fig. 1-10B) (Morii *et al.*, 1986; Nishihara *et al.*, 1987). Final purification of these glycolipids was achieved by preparative TLC on silica gel 60 (Merck) plates in the solvent system chloroform–acetone–methanol–acetic acid–water (200 : 150 : 100 : 15 : 10, v/v) (Nishihara *et al.*, 1987).

1.3.1.3. Isolation of Glycolipids from Thermoacidophiles

Only a few species of thermoacidophiles (e.g., *Thermoplasma* and *Sulfolobus* spp.) have been examined for glyco- and phosphoglycolipids. Lipids of *Thermoplasma acidophilum* were fractionated on a column of silicic acid (Unisil) eluted with chloroform (neutral lipids), acetone (glycolipids), and methanol (phospholipids + phosphoglycolipids) (Langworthy *et al.*, 1972). Final separation and purification of individual components were carried out by preparative TLC on silica gel H using the solvent systems chloroform–methanol (9 : 1, v/v) for glycolipids and chloroform–methanol–acetic acid–water (100 : 20 : 12 : 5, v/v) for phosphoglycolipids. Final purification of the lipoglycan (**51,** Fig. 1-11C) from *T. acidophilum* was achieved by passage of the aqueous solution of the lipoglycan through a column of AG-3-X4a anion-exchange resin to remove traces of nucleic acids and proteins (Smith, 1980).

Lipids of *Sulfolobus acidocaldarius* have been fractionated (Langworthy *et al.*, 1974) by column chromatography on silicic acid into neutral lipid, glycolipid, and polar lipid fractions as described above for *T. acidophilum* followed by further fractionation of the glycolipid and polar lipid fractions on a DEAE-cellulose column (acetate form) and final purification by preparative TLC using chloroform–methanol (9 : 1, v/v) for polar lipids and glycolipids, respectively. Isolated bands of lipids, detected by iodine vapour, were scraped and eluted through sintered glass filters with chloroform–methanol–water (60 : 30 : 4.5, v/v).

Lipids of *Sulfolobus sulfataricus* (formerly *Caldariella acidophila*) were fractionated (De Rosa *et al.*, 1980a) on a silica gel (Merck Kieselgel) column eluted with chloroform (neutral lipids), chloroform–methanol (6 : 1, v/v) (glycolipids) and chloroform–methanol–water (65 : 24 : 4, v/v) (polar lipids). The glycolipid fraction was further separated into glycolipid A and B by preparative TLC on Kieselgel–boric acid plates in solvent system chloroform–

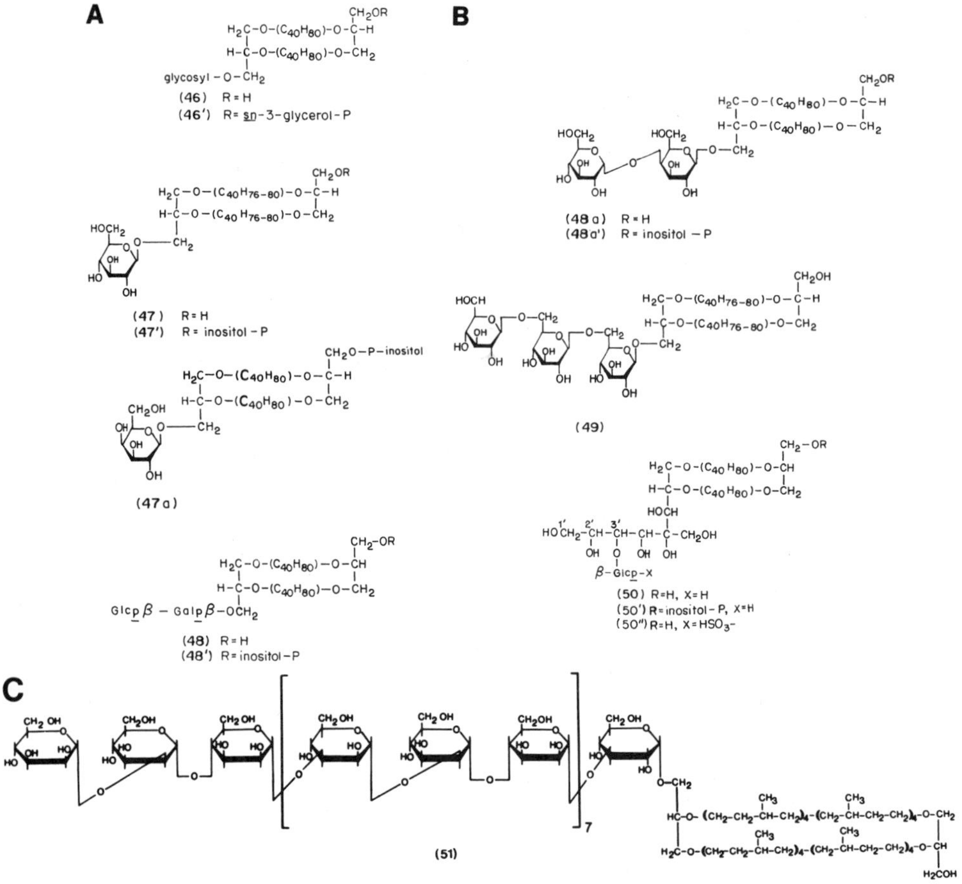

Figure 1-11. Structures of glycolipids, phosphoglycolipids, and sulfoglycolipids of thermoacidophilic archaebacteria. (A) (46) glycosyl–caldarchaeol (R = H); (46′) glycosyl–caldarchaeol-P-Gro (R = sn-3-glycerol-P); (47) Glc$p$β1–1caldarchaeol (MGC-1) (R = H); (47′) Glc$p$β1–1caldarchaeol-P-inositol (PGC-5) (R = P-inositol); (47a) Gal$p$β1–1caldarchaeol-P-inositol (PGC-6); (48) Glc$p$βGal$p$β1–1caldarchaeol (DGC-4; R = H); (48′) Glc$p$βGal$p$β1–1caldarchaeol-P-inositol (PGC-4; R = inositol-P). (B) (48a) Glc$p$α1–4Gal$p$β1–1caldarchaeol (DGC-5) (R = H); (48a′) Glc$p$α1–4Gal$p$β1–1cardarchaeol-P-inositol (PGC-7) (R = inositol-P); (49) (Glc$p$β1–6)₂Glc$p$β1–1caldarchaeol; (50) Glc$p$β1–3′ nonitolcaldarchaeol (GNC; R = H, X = H); (50′) Glc$p$β1–3′-nonitolcaldarchaeol-P-inositol (R = inositol-P, X = H); (50″) HSO₃-Glc$p$β1–3′-nonitolcaldarchaeol (R = H, X = HSO₃). (C) (51) [Man$p$α1–2Man$p$α1–4Man$p$α1–3]₈ Glc$p$α1–1caldarchaeol. (From Smith, 1980.)

methanol–water (65 : 25 : 4, v/v), and the polar lipids were resolved by preparative TLC on silica gel in solvent system chloroform–methanol–water (65 : 54 : 4, v/v) into phospholipid A, sulfoglycolipid, and phosphoglycolipids B and C (De Rosa *et al.*, 1980a).

1.3.1.4. Quantitative Procedures

Quantitative analysis of archaebacterial glycolipids can be carried out essentially by the same methods described for eubacterial lipids (see Section

1.2.1.2): (1) sugar analysis of the intact lipid in solution or as TLC spots, by the modified phenol–sulfuric acid procedure (Kushwaha and Kates, 1981); (2) analysis of individual sugar methyl glycosides, liberated from the glycolipid by methanolysis, by GLC of the TMS ether derivatives with TMS–mannitol as internal standard (Sweeley and Vance, 1967). Because of the stability of the alkyl ether linkages, it is not feasible to cleave the alkyl ether groups (e.g., with BCl_3 or HI) for analysis of the resulting glycosylglycerols, since the glycosidic linkages of the sugars will also be cleaved.

1.3.2. Characterization of Archaebacterial Lipids

1.3.2.1. Thin-Layer Chromatography

Thin-layer chromatographic mobilities of glycosylarchaeols from halophiles and methanogens and glycosylcaldarchaeols from methanogens and thermoacidophiles are given in Table 1-10. A TLC plate showing the separation of the glycolipids of a typical *Halobacterium* species, *H. cutirubrum*, is shown in Fig. 1-12 (Smallbone and Kates, 1981); a two-dimensional TLC plate of the lipids of a *Haloarcula* species, *H. marismortui*, is shown in Fig. 1-13 (Evans *et al.*, 1980); a one-dimensional TLC plate of the lipids of a *Haloferax* species, *H. mediterranei* (Kushwaha *et al.*, 1982a), is shown in Fig. 1-14; and a TLC plate showing the separation of both glycosylarchaeol, glycosylcaldarchaeol and phosphoglycosylcaldaracheols from *M. hungatei* is shown in Fig.

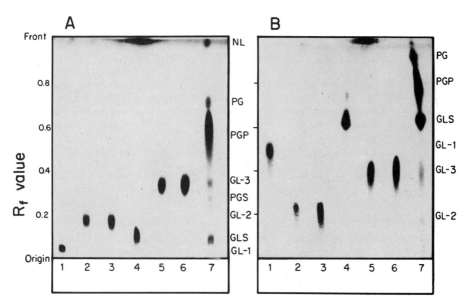

Figure 1-12. Thin-layer chromatograms (TLC) of glycolipids of *Halobacterium cutirubrum*, on silica gel G. (A) Chloroform–methanol–90% acetic acid (30 : 4 : 20, v/v). (B) Chloroform–methanol–diethylamine–water (110 : 50 : 8 : 5, v/v). 1, S-TeGA (**36′**); 2, desulfated S-TeGA (GL-2) (**36**); 3, natural TeGA (GL-2) (**36**); 4, S-TGA-1 (GLS) (**34′**); 5, desulfated S-TGA-1 (GL-3) (**34**); 6, natural TGA-1 (GL-3) (**34**); 7, total polar lipids. (From Smallbone and Kates, 1981.)

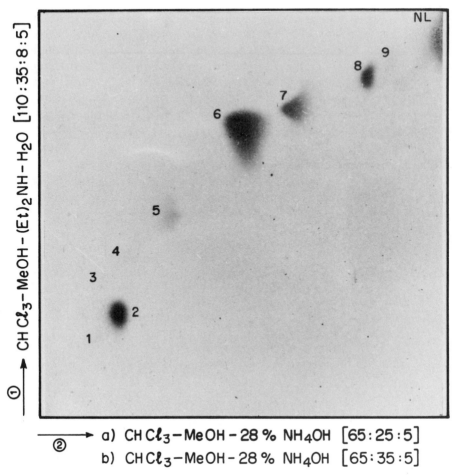

Figure 1-13. Two-dimensional thin-layer chromatogram (TLC) of the lipids of *Haloarcula marismortui* on silica gel G in solvent systems indicated. Spots 1,3,4,9, unidentified; spot 2, TGA-2 (**35**); spot 5, unidentified diglycosylarchaeol (DGA-2); spots 6–8, PGP (archaeol-P-Gro-P), PGS (archaeol-P-Gro-S), and PG (archaeol-P-Gro), respectively. (From Evans *et al.,* 1980.)

1-15 (Kushwaha *et al.,* 1981; Ferrante *et al.,* 1987). It is of interest that not only are the mono-, di-, and triglycosylarchaeols distinguished by TLC (as was also found for the glycosyldiacylDAGs, Table 1-1), but isomeric triglycosylarchaeols (TGA-1 and TGA-2, structures **34** and **35,** respectively) and diglycosylaracheols [DGA-1(**33**), DGA-2, DGA-3(**38**), DGA-I(**41**) and DGA-II(**42**)] are also separated by TLC in an appropriate solvent (Table 1-10).

Separation of the glycosylcaldarchaeols (**46,48**), the glycosylnonitolcardarchaeols (**50**) and phosphoglycosylcaldarchaeols (**46′,48′,50′**) of thermoacidophiles has been achieved by one-dimensional TLC (Langworthy *et al.,* 1972, 1974; De Rosa *et al.,* 1980a) (see Table 1-10). Specific stains for neutral and aminosugars and uronic acids can be used to characterize these glycolipids on TLC plates (see Section 1.2.2.1a).

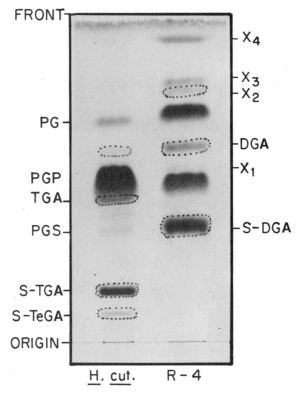

Figure 1-14. Thin-layer chromatogram (TLC) of the lipids of *Halobacterium cutirubrum* (*H. cut.*) and *Haloferax mediterranei* (R-4), in solvent system chloroform–methanol–acetic acid–water (85 : 22.5 : 10 : 4, v/v; double development), stained with α-naphthol sugar stain then charred (dotted spots are sugar positive). S-DGA-1 (**33′**); DGA (**33**); S-TGA-1 (**34′**); TGA-1 (**34**); S-TeGA (**36′**); X_1–X_4, unidentified; PG, phosphatidylglycerol (archaeol-P-Gro); PGP, phosphatidylglyc-erolphosphate (archaeol-P-Gro-P); PGS, phosphatidylglycerosulfate (archaeol-P-Gro-S). (From Kushwaha *et al.*, 1982*a*.)

1.3.2.2. Spectrometric Analyses

Infrared spectra of archaebacterial glycolipids and phosphoglycolipids are clearly distinguished from those of analogous diacylglycolipids (glycosyl-DAGs) by the absence of any absorption bands for ester C═O or C—O groups. Instead strong absorption for alkyl ether C—O—C groups (1100 cm^{-1}) is characteristic of the spectra of all archebacterial glycolipids, both derivatives of archaeols and caldarchaeols (Fig. 1-16A,B). Another diagnostic absorption band common to all glycosylarchaeols is that for the isopropyl group [1385–1375 cm^{-1} (doublet)] arising from the phytanyl chains. This doublet is absent in the spectra of glycocaldarchaeols in which the biphytanyl groups lack a terminal isopropyl group; a singlet absorption band is present, however, at 1375 cm^{-1} due to the branched methyl groups (Fig. 1-16). The presence of sugar groups in these glycolipids of course gives rise to charac-teristically strong OH (3400 cm^{-1}) and C—OH (1070–1050 cm^{-1}) absorp-

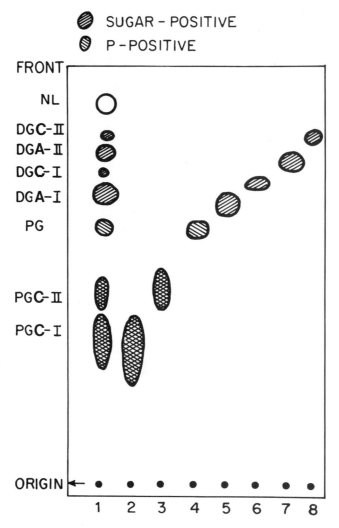

Figure 1-15. Thin-layer chromatogram (TLC) of the lipids of *Methanospirillum hungatei*, in solvent system chloroform–methanol–diethylamine–water (110 : 50 : 8 : 3.5, v/v) stained with the phosphate stain and α-naphthol sugar stain. 1, total lipids; 2, PGC-I (**43′**); 3, PGC-II (**44′**); 4, PG, phosphatidylglycerol (archaeol-P-Gro); 5, DGA-I (**41**); 6, DGC-I (**43**); 7, DGA-II (**42**); 8, DGC-II (**44**). (From Kushwaha *et al.*, 1981.)

tion bands. Fourier transform IR spectrometry (FT-IR) after self-deconvolution is useful in characterizing the various caldarchaeols containing cyclopentane rings (Mancuso *et al.*, 1986). More detailed structural information can be obtained from ^{1}H- or ^{13}C-NMR spectra, particularly concerning the configuration of the anomeric protons or carbons (see Table 1-6 and discussion in Section 1.2.2.6).

FAB–mass spectrometry is useful in supplying the parent molecular ion of the glycolipid, and hence its molecular weight. In particular, negative ion

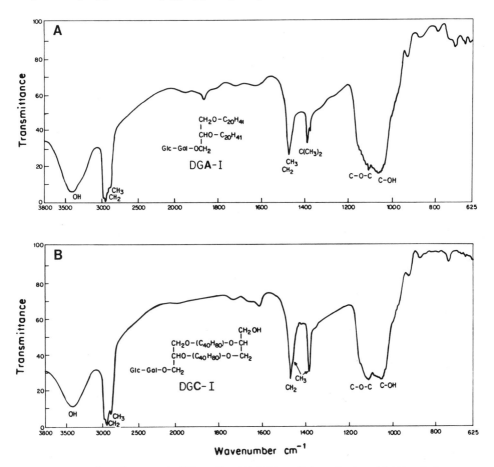

Figure 1-16. Infrared spectra in KBr pellet. (A) DGA-I, diglycosylarchaeol-I (**41**). (B) DGC-I, diglycosylcaldarchaeol-I (**43**) isolated from *Methanospirillum hungatei*. (From Kushwaha *et al.*, 1981.)

FAB–MS, when applied to the analysis of anionic glycolipids, such as the sulfated glycosylarchaeols (e.g., S-TGA-1 and S-TeGA, see Fig. 1-17A,B) or phosphoglycosylcaldarchaeols (Nishihara *et al.*, 1987), gives the molecular ion peak (M-H)⁻.

1.3.2.3. Analysis of Molecular Constituents

Determination of the mole ratios of the glycolipid constituents such as sugars, sulfate, phosphate, diether (archaeol), or tetraether (caldarchaeol) is an important step in characterizing the archaebacterial glycolipid and determination of its structure (see Section 1.2.2.1), as follows: sugars are determined on the intact lipid by the phenol–sulfuric acid procedure (Kushwaha and Kates, 1981) or by GLC of the methylglycosides derived by acid methanalysis with mannitol as internal standard (see Section 1.2.2.2); sulfate and

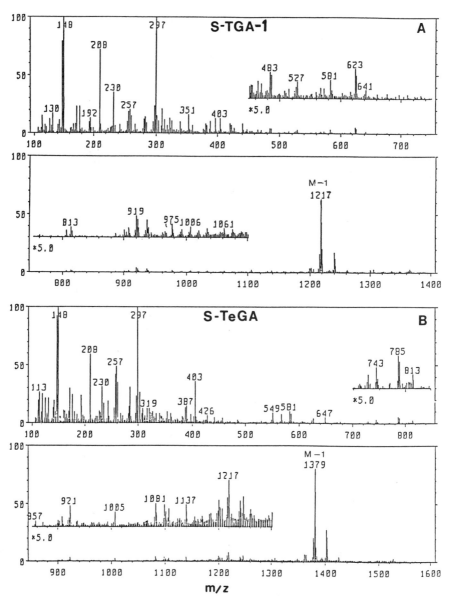

Figure 1-17. Negative-ion fast atom bombardment (FAB) mass spectrum. (A) Sulfated tri-glycosylarchaeol (S-TGA-1) (**34′**) (free acid, 1218M_r). (B) Sulfated tetraglycosylarchaeol (S-TeGA) (**36′**) (free acid, 1380M_r). (From M. Kates and A. Hayashi, unpublished data.)

phosphate are assayed as described elsewhere (Kates, 1986*a*); and archaeol or caldarchaeol by gravimetric analysis (Kates, 1986*a*) or HPLC analysis (Man-cuso *et al.*, 1986; Thurl and Schäfer, 1988) of the nonpolar constituent re-leased on acid methanolysis (see Section 1.3.3.2).

1.3.3. Structure Determination

The particular procedure to be used for determination of the structure of a given archaebacterial glycolipid will depend on the structure in question. However, any approach used must deal with (1) the structure and configuration of the carbohydrate moiety, (2) the structure of the nonpolar backbone, and (3) the location of any sulfate or phosphate groups. These three aspects of glycolipid structure determination will now be discussed in detail.

1.3.3.1. Identification of Carbohydrate Structure and Configuration

Procedures for structure determination of the carbohydrate moiety of archaebacterial glycolipids are essentially the same as described for eubacterial glycolipids (see Section 1.2.2): identification of sugar groups (see Section 1.2.2.2), and determination of sugar sequences (see Section 1.2.2.4), sugar linkage positions (see Section 1.2.2.5), and anomeric configuration (see Section 1.2.2.6). The latter can also be determined from the molecular rotation of the intact glycolipid (see Table 1-11; see also Section 1.2.2.6). Procedures for determining the carbohydrate structure and configuration in lipoglycans have been detailed by Smith (1980).

1.3.3.2. Identification of the Nonpolar (Alkyl Ether) Backbone

The characteristic difference between eubacterial and archaebacterial glycoglycerolipids is the presence of archaeol or caldarchaeol backbones in the latter glycolipids. These alkyl ether backbone residues are obtained as the nonpolar moieties (chloroform or hexane soluble) after the standard acid methanolysis procedure (Kates, 1986a) (see Section 1.2.2.2), or by acetolysis of the intact complex lipids followed by acid methanolysis (Nishihara and Koga, 1987). Examination of the nonpolar material by TLC in petroleum ether–ethyl ether–acetic acid (50 : 50 : 1, v/v) (Kushwaha *et al.*, 1981); Nishihara *et al.*, 1987) will distinguish (Fig. 1-18) between the diphytanylglycerol diether (archaeol) (R_f, 0.65) and the dibiphytanylglycerol tetraether (cardarchaeol) (R_f, 0.35).

If variant diethers are present, such as those from haloalkaliphiles (Ross *et al.*, 1981; De Rosa *et al.*, 1982a, 1983a) or nonalkaliphilic extreme halophiles (Onishi *et al.*, 1985), they may be distinguished from the usual di-C_{20}-diether and the tetraether by TLC first in petroleum ether–acetone (95 : 5, v/v), then in the same direction in toluene/acetone (97 : 3, v/v) (Ross *et al.*, 1981); the di-C_{20}-, C_{20}-C_{25}-, and di-C_{25}-diethers have R_f values of 0.60, 0.63, and 0.66, respectively (Ross *et al.*, 1981; De Rosa *et al.*, 1983a), and the tetraether has an R_f value of 0.1 (Ferrante *et al.*, 1986).

For further identification, the nonpolar material is fractionated by preparative TLC in chloroform–ethyl ether (9 : 1, v/v) or by silicic acid column chromatography (Joo *et al.*, 1968) for the diether or its variants, and further resolved into individual di-C_{20}-, C_{20}-, C_{25}-, and di-C_{25}-archaeols by pre-

Table 1-11. Molecular Rotations of Diphytanylglycerol Diether (Archaeol), Dibiphytanyldiglycerol Tetraether (Caldarchaeol), and Their Glycosyl or Phosphoglycosyl Derivatives

Compound	$[\alpha]_D{}^a$ (degrees)	$M_D{}^a$ (degrees)		References
		Found	Calc.[b]	
Natural archaeol (**31**)	+8.4	+55		Kates (1978)
Synthetic archaeols				
sn-2,3 (R,R,R,)	+8.5	+55		Kates (1978)
sn-2,3 (R,R,R-S,R,R)	+7.6	+50		Kates (1978)
sn-1,2 (R,R,R-S,R,R)	−7.0	−46		Kates (1978)
sn-1,2 (R,R,R)	+1.1	+7		Kates (1978)
Natural caldarchaeol (**32**)				
M. hungatei	+8.7	+113		Kushwaha et al. (1981)
S. sulfataricus	+8.3	+108		De Rosa et al. (1980a)
M. thermoautotrophicum	+8.5	+110		Nishihara et al. (1987b)
Glcpαarchaeol (MGA-1)	+43.6	+355	+365	Kushwaha et al. (1982a)
Glcpβarchaeol (MGA-2, **37**)	−11.3	−92	−93	Ferrante et al. (1986)
Manpα1–2Glcpαarchaeol (DGA-1, **33**)	+61.1	+597	+519	Kushwaha et al. (1982a)
6-HSO₃Manpα1–2Glcpαarchaeol (S-DGA-1, **33′**)	+42.8	+467	+519	Kushwaha et al. (1982a)
Glcpβ1–6Glcpβarchaeol (DGA-3, **38**)	−25.1	−245	−243	Ferrante et al. (1986)

Manpα1–3Galpβarchaeol (DGA-4, **39**)	+23.7	+228	+210	Ferrante *et al.* (1988*a*)
Galpβ1–6Galpβhydroxyarchaeol (DGA-5, **40**)	−12.3	−122	+40	Ferrante *et al.* (1988*a*)
Glcpα1–2Galpβarchaeol (DGA-I, **41**)	+15.3	+150	+148	Kushwaha *et al.* (1981)
Galpβ1–6Galpβarchaeol (DGA-II, **42**)	−37.8	−369	−379	Kushwaha *et al.* (1981)
Galpβ1–6Manpα1–2Glcpαarchaeol (TGA-1, **34**)	+46.8	+533	+518	Smallbone and Kates (1981) Kates and Deroo (1973)
3HSO₃Galpβ1–6Manpα1–2Glcpαarchaeol (S-TGA-1, **34′**)	+46.8	+579	+518	Kates and Deroo (1973)
3–HSO₃Galpβ1–6Manpβ(3–1βGalfα1–2Glcpαarchaeol (S-TeGA, **36′**)	+52.4	+733	+720	Smallbone and Kates (1981)
Glcpβ1–6Manpα1–2Glcpαarchaeol (TGA-2, **35**)	+34.6	+394	+448	Evans *et al.* (1980)
Glcpβcaldarchaeol (MGC-1, **47**)	+1.6	+24	+45	Thurl and Schäfer (1988)
Glcpα1–2Galpβcaldarchaeol (DGC-I, **43**)	+13.6	+221	+205	Kushwaha *et al.* (1981)
Galfβ1–6Galpβcaldarchaeol (DGC-II, **44**)	−14.0	−227	−322	Kushwaha *et al.* (1981)
Glcpα1–2Galpβcaldarchaeol-P-Gro (PGL-I, **43′**)	+7.6	+137	+202	Kushwaha *et al.* (1981)
Galfβ1–6Galpβcaldarchaeol-P-Gro (PGL-I, **44′**)	−19.4	−352	−325	Kushwaha *et al.* (1981)

[a]Specific rotation or molecular rotation in chloroform solution.

[b]Calculated by Hudson's (1909) isorotation rule using the sugar methyl glycoside M_D values given in Table 1-5, and the M_D values for achaeol and caldarchaeol given in this table.

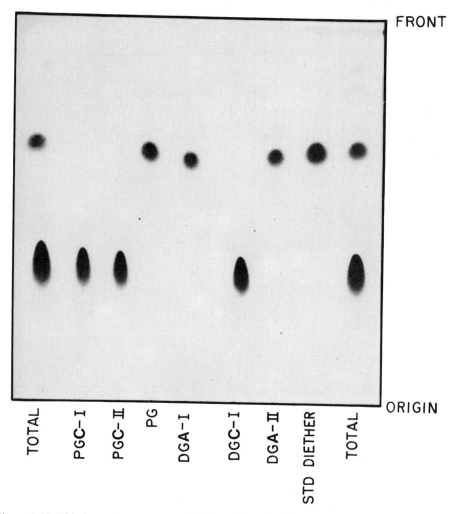

Figure 1-18. Thin-layer chromatogram (TLC) on silica gel G in petroleum ether–ethyl ether–acetic acid (50 : 40 : 1, v/v) of archaeol (**31**) or caldarchaeol (**32**) derived from (left to right): total polar lipids, PGC-I (**43'**), PGC-II (**44'**), PG, DGA-I (**41**), DGC-I (**40**), DGA-II (**42**) of *Methanospirillum hungatei*. (From Kushwaha *et al.*, 1981.)

parative TLC or by HPLC (De Rosa *et al.*, 1982*a*, 1983*a*). For mixtures of di- and tetraethers, fractionation is done by preparative TLC in petroleum ether–ethyl ether–acetic acid (50 : 50 : 1, v/v) (Kushwaha *et al.*, 1981; Langworthy, 1982*b*; Nishihara *et al.*, 1987). Separation of archaeol and acyclic, mono-cyclic, and bicyclic caldarchaeols can be achieved by HPLC of the underivatized components (Mancuso *et al.*, 1986) or of the *p*-nitrobenzoyl derivatives (Thurl and Schäfer, 1988). The purified arachaeols or caldarchaeols are then identified by IR (Fig. 1-19), NMR (Fig. 1-20), and mass spectrometry (Fig. 1-21) (Kates, 1978; Langworthy, 1977*a*; De Rosa *et al.*,

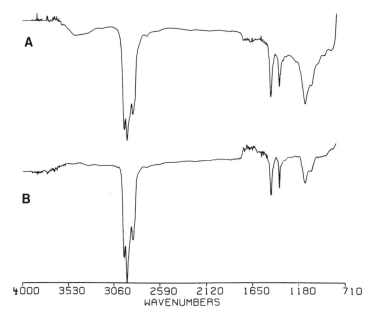

Figure 1-19. Infrared (FTIR) spectrum in CCl₄. (A) Archaeol (**31**) from *Methanosarcina barkeri.* (B) Caldarchaeol (**32**) from *Methanobacterium thermoautotrophicum.* (From Mancuso *et al.*, 1986.)

1977*a,b*, 1982*a*, 1983*a;* Kushwaha *et al.*, 1981; Ferrante *et al.*, 1987; Mancuso *et al.*, 1986; Thurl and Schäfer, 1988).

Final identification of the actual hydrocarbon type and chain length is done by cleavage of the ether linkages with BCl_3 or HI (Kates, 1978; Langworthy, 1982*a,b*) and analysis of the derived alkyl chlorides, iodides, acetates, or hydrocarbons (Fig. 1-22) by GLC or GLC–MS (Kates, 1978; Langworthy, 1977*a*, 1982*b;* Langworthy *et al.*, 1983; De Rosa *et al.*, 1977*b*, 1982*a;* Thurl and Schäfer, 1988). [^{13}C]-NMR is particularly useful in identifying and locating the cyclopentane rings in the hydrocarbons derived from caldarchaeols (De Rosa *et al.*, 1977*a,b;* Thurl and Schäfer, 1988). Quantitation of glycerol released by BCl_3 treatment of the nonpolar material and determination of the mole ratio alkyl groups/glycerol is useful in distinguishing between linear diethers and macrocyclic diethers, and linear diethers and tetraethers (Langworthy, 1977*a*). The distribution of archaeols, caldarchaeols, and their variants among the extreme halophiles, methanogens, and thermacidophiles is given in Table 1-12.

An important characteristic of archaebacterial lipids, distinguishing them from the lipids of all other organisms, is the *sn*-2,3-stereochemical configuration of the glycerols in the diphytanylglyceroldiethers (Kates, 1978; De Rosa *et al.*, 1982*a*, 1983*a*) and the dibiphytanyldiglyceroltetraethers (Langworthy, 1977*a;* Kushwaha *et al.*, 1981; De Rosa *et al.*, 1986*a*). This is determined by measurement of the optical rotation of the purified diethers and tetraethers and comparison with those of known standards (see Table 1-11). A complicat-

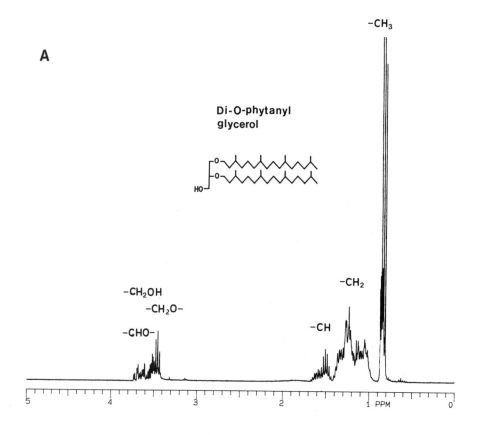

A

Di-O-phytanyl glycerol

−CH₃

−CH₂OH
−CH₂O−
−CHO−

−CH₂

−CH

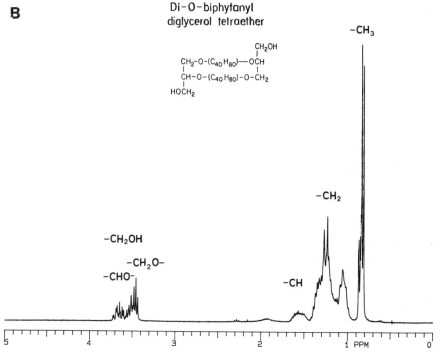

B

Di-O-biphytanyl diglycerol tetraether

−CH₃

−CH₂OH
−CH₂O−
−CHO−

−CH₂

−CH

Figure 1-20. [^{1}H]-NMR spectra in C^2HCl_3. (A) Archaeol (**31**). (B) Caldarchaeol (**32**). (From Kates, unpublished data.)

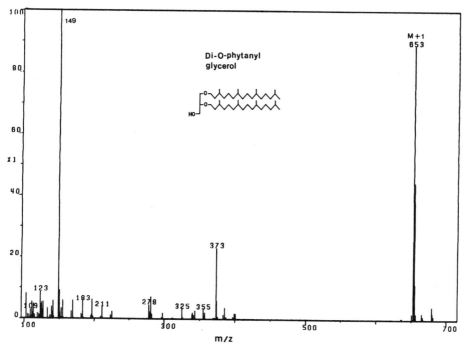

Figure 1-21. Chemical ionization mass spectrometry (CI–MS) of archaeol (**31**); M⁺ = 652. (From Kates, unpublished data.)

ing factor in determining the configuration of the diether or tetraether is the presence of chiral centers in the alkyl substituents in addition to that of the glycerol group(s). Thus, the C_{20} phytanyl group has the $3R,7R,11R$ configuration (Kates, 1978), and the C_{40} biphytanyl group has the $3R,7R,11R,15R,15'R,11'R,7'R,3'R$ configuration (Heathcock *et al.*, 1985). The contribution of these chiral centers to the optical rotation of the diether, however, amounts to about 13% and is presumably the same for the tetraether (see Table 1-11).

The configuration of the chiral centers in the C_{25} sesterterpanyl group found in lipids of haloalkaliphiles (De Rosa *et al.*, 1982a, 1983a) has not yet been determined but is presumed to be $3R,7R,11R,15R$. It would be of interest to determine the configuration of the new chiral centers present in the cyclized C_{40} biphytanyl chains containing cyclopentane rings found in lipids of *Sulfolobus* spp. (De Rosa *et al.*, 1986a), as well as the configuration of the chiral centers in the nonitol group of some of the tetraethers in these species.

1.3.3.3. Location of Sulfate or Phosphate Groups

Sulfated glycolipids (e.g., structures **33'** and **34'**) represent major lipid components in extreme halophiles, particularly the *Halobacteria*, but are only minor components in other archaebacteria (see Tables 1-13–1-16). Phos-

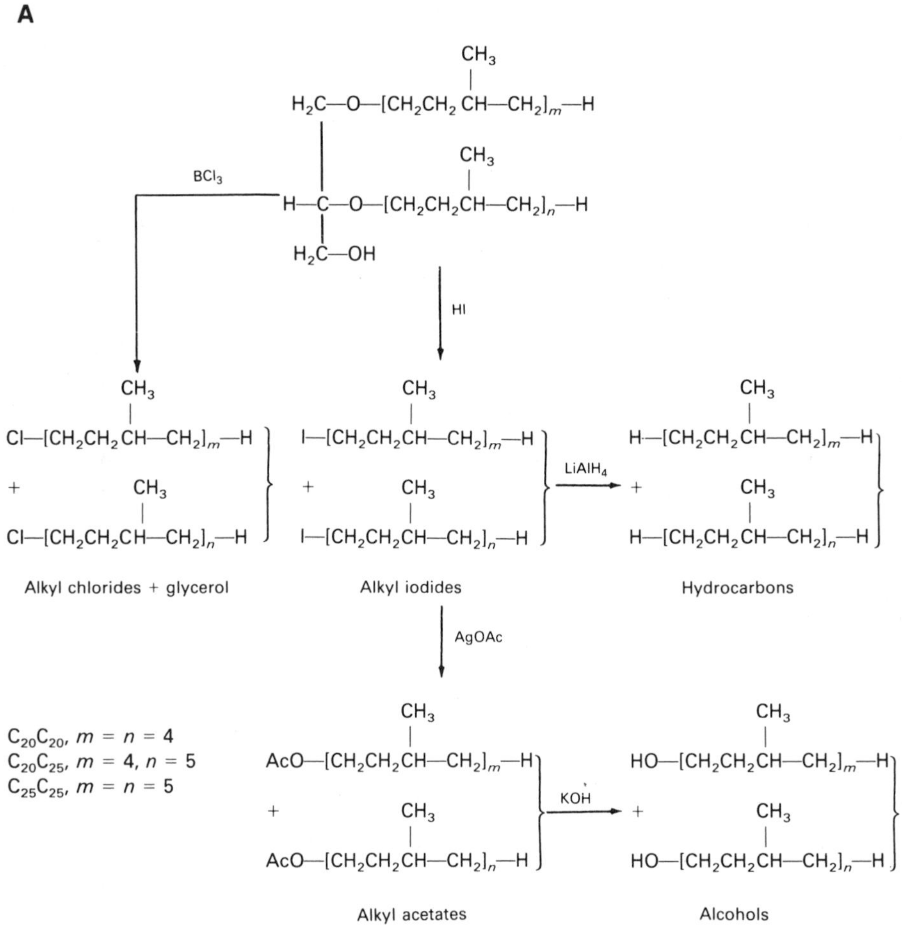

A

CH_3
$H_2C-O-[CH_2CH_2\,CH-CH_2]_m-H$

CH_3
$H-C-O-[CH_2CH_2CH-CH_2]_n-H$
H_2C-OH

BCl$_3$

HI

CH_3
$Cl-[CH_2CH_2CH-CH_2]_m-H$
+ CH_3
$Cl-[CH_2CH_2CH-CH_2]_n-H$

Alkyl chlorides + glycerol

CH_3
$I-[CH_2CH_2CH-CH_2]_m-H$
+ CH_3
$I-[CH_2CH_2CH-CH_2]_n-H$

Alkyl iodides

LiAlH$_4$

CH_3
$H-[CH_2CH_2CH-CH_2]_m-H$
+ CH_3
$H-[CH_2CH_2CH-CH_2]_n-H$

Hydrocarbons

AgOAc

$C_{20}C_{20}$, $m = n = 4$
$C_{20}C_{25}$, $m = 4$, $n = 5$
$C_{25}C_{25}$, $m = n = 5$

CH_3
$AcO-[CH_2CH_2CH-CH_2]_m-H$
+ CH_3
$AcO-[CH_2CH_2CH-CH_2]_n-H$

Alkyl acetates

KOH

CH_3
$HO-[CH_2CH_2CH-CH_2]_m-H$
+ CH_3
$HO-[CH_2CH_2CH-CH_2]_n-H$

Alcohols

Figure 1-22. Reaction scheme for conversion of (A) archaeols (**31**), and (B) caldarchaeols (**32**) or nonitolcaldarchaeols (**32′**), to alkyl chlorides, iodides, acetates, alcohols, and hydrocarbons. (From Kates, 1978; Langworthy, 1977.)

phoglycolipids, however, have not been detected in extreme halophiles, but are major lipid components (e.g., the glycosylglycerophosphorylcald-archaeols, structures **43′**, **44′** and **46′**) in some methanogens and thermoacidophiles (see Tables 1-13–1-16).

Detection of sulfated glycolipids is difficult, since there is no reliable stain for sulfate on TLC plates; they can be detected, however, by elemental analysis for sulfur in the purified sulfolipid, by sulfate analysis after acid hydrolysis (Kates, 1986a), or by the change in TLC mobility of the sulfolipid after subjection to solvolysis in tetrahydrofuran–0.008 N HCl (Kates and Deroo, 1973; Smallbone and Kates, 1981). Location of the sulfate group in the carbohydrate moiety can be determined by permethylation analysis carried out on the sulfolipid and on the desulfated glycolipid, and comparison of their respec-

B

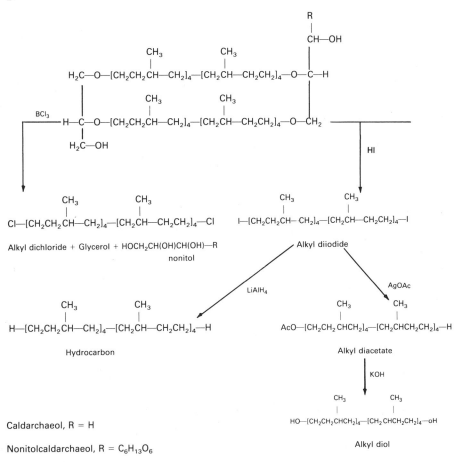

Figure 1-22. (*continued*)

tive methylation pattern (Kates and Deroo, 1973; Smallbone and Kates, 1981; Kushwaha *et al.*, 1982*a*).

By contrast, phosphoglycolipids are readily detectable on TLC plates by positive staining with the phosphate and sugar stains (see Fig. 1-15), and the location of the phosphate group is relatively easy to establish. For example, the presence of a glycerophosphate group or an inositol phosphate group on one of the hydroxyls of a tetraether (e.g., structure **43′** or **48′**, Figs. 1-10 or 1-11, respectively) can be established by strong alkaline hydrolysis of the phosphoglycolipid to yield the water-soluble phosphate ester moiety and the glycosylcaldarchaeol (structures **43** and **48,** respectively). Identification of the water-soluble phosphate ester is achieved by paper chromatography (Kates, 1986*a*; Kushwaha *et al.*, 1981) or by hydrolysis in 6 N HCl at 100°C for 72 hr and GLC of the TMS derivative of the released glycerol or inositol (Langwor-

Table 1-12. Distribution of Archaeol, Caldarchaeol, and Their Variants among Archaebacteria[a]

Genus	Archaeols			Caldarchaeols					
	$C_{20}C_{20}$ (31)	$C_{20}C_{25}$ (31')	$C_{25}C_{25}$ (31'')	(32)	(32')	(32a)	(32b)	(32c)	(32d)
Extreme halophiles[b]									
Halobacterium	+	(+)[h]							
Haloarcula	+								
Haloferax	+								
Holococcus[b']	+	+							
Haloalkaliphiles[c]									
Natronobacterium	+	+	+						
Natromococcus	+	+	+						
Methanogens[d]									
Methanococcus	+			trace					
Methanothrix	+								
Methanolobus	+	+		trace					
Methanosarcina	+	+		trace					
Methanoplanus	+			+					
Methanobacterium	++			+					

Methanobrevibacter	+					+
Methanospirillum	+					+
Methanogenium	+					+
Methanothermus	+					++
Thermoacidophiles[c]						
Thermoplasma[e]	+			+	+	+
Sulfolobus[e]	trace	+	+	+	+	++
Thermoproteus[f]	+	+		+	+	++
Thermophilic anaerobes[g]						
Desulfurococcus	+					+

[a]Data taken from Langworthy (1985), Langworthy *et al.* (1982), and Langworthy (unpublished data); Tornabene and Langworthy (1979). [b]Kates, 1978; Kates *et al.* (1966); Kamekura and Kates (1988). [b']Montero *et al.*, 1989. [c]De Rosa *et al.* (1982a, 1986a). [d]Tornabene and Langworthy (1979); Koga *et al.* (1987); Grant *et al.* (1985). Unidentified novel core lipids are present in some methanogen species (Grant *et al.*, 1985; Langworthy, 1985). [e]Langworthy (1985); De Rosa *et al.* (1986a). [f]Thurl and Schäfer (1988). [g]Zillig *et al.* (1981, 1982); Lanzotti *et al.* (1987). [h]An unidentified extreme halophile (nonpigmented, nonalkalophilic *Halobacterium*, strain 172) contains both $C_{20}C_{20}$ and $C_{20}C_{25}$ core lipids (Onishi *et al.* 1985; Kamekura and Kates, 1988).

thy *et al.*, 1974). Permethylation analysis of the phosphoglycosylcardarchaeol before and after alkaline hydrolysis will determine the location of the phosphate moiety unambiguously as being attached to one of the tetraether hydroxyls, which will be monomethylated only after alkaline hydrolysis (Kushwaha *et al.*, 1981; Thurl and Schäfer, 1988). If the phosphate moiety is glycerophosphate, its configuration can be determined enzymatically with the stereospecific glycerophosphate dehydrogenase (Kates, 1986a).

When a sugar is linked glycosidically through a phosphate group attached to a diether, as in structure **37′** (Fig. 1-10), mild acid methanolysis results in the liberation of the sugar glycoside and the diether phosphate; the sugar glycoside is converted to the free sugar and analyzed by paper chromatography or by GC–MS of the alditol acetate derivative, while the diether phosphate is identified by TLC and negative FAB–MS (Ferrante *et al.*, 1986).

1.3.4. Distribution in Archaebacteria

The range of glycolipid types found in archaebacteria up to the present is fairly extensive and may equal or even surpass that in eubacteria as more species are examined. The glycolipids found vary not only in the carbohydrate portion but in the nonpolar portion as well, the latter being characteristic of the archaebacterial family (see Table 1-12). Thus, the extreme halophiles and haloalkaliphiles contain only archaeol (diether)-derived lipids, the methanogens contain either archaeols or mixtures of archaeols and caldarchaeols (tetraethers), and the thermoacidophiles contain largely caldarchaeols and nonitolcaldarchaeols with only small to trace amounts of archaeols. The thermoacidophiles are further distinguished by the presence of one to four cyclopentane rings in the biphytanyl chains of the caldarchaeol group (structures **32a–d,** Fig. 1-8), the number of rings increasing with increasing growth temperature (De Rosa *et al.*, 1980*b;* Langworthy, 1982*a*); the presence of the nonitolcaldarchaeol appears to be characteristic of the *Sulfolobus* species (De Rosa *et al.*, 1986*a*).

In regard to the carbohydrate portion, there would appear to be a correlation between glycolipid structure and archaebacterial classification on the level of the genus, at least for the extreme halophiles (Torreblanca *et al.*, 1986). The distribution of glycolipids among halophiles, methanogens and thermoacidophiles will be presented along taxonomic lines (Tables 1-13–1-16), following the general classification and nomenclature in Bergy's Manual (1988; see Grant and Larsen, 1988).

1.3.4.1. Extreme Halophiles (Order Halobacteriales)

The Halobacteriales are extremely halophilic aerobic archaebacteria which include two families, Halobacteriaceae and Halococcaciae; the Halobacteriaceae have traditionally been classified in the one genus, *Halobacterium* and the Halococcaciae in the genus *Halococcus* (Bergy, 1974). Recently, two new genera, *Haloarcula* and *Haloferax*, in addition to the well-known genus *Halobacterium* have been proposed (Torreblanca *et al.*, 1986; Grant and Larsen, 1988) to cover the Halobacteriaceae, on the basis of numerical taxonomy and lipid composition, particularly their glycolipid composition (Kushwaha *et al.*, 1982*b*). In addition, the alkaliphilic halobacteria, which require a medium with pH 9–10 and a low magnesium concentration, have been classified into two genera, *Natronobacterium* and *Natronococcus* (Tindall *et al.*, 1984). The glycolipid composition of the above-described genera is as follows (see Table 1-13).

1.3.4.1a. Halobacteriaceae.

Halobacterium. Well-known species of this genus, such as *H. salinarium, H. cutirubrum, H. halobium* and *H. saccharovorum* are found in salted foods, such as salted fish, in salt flats, and in common salt obtained from these flats, and were the first archaebacteria to be studied for lipids (Kates, 1972, 1978). Subsequently, several species of *Halobacterium* have been isolated, including *H. sodomense* and undesignated species from Spain (Y12, 3.5 rp4, 3.1 palp 4, Gca-19), and their lipids have been examined (Kushwaha *et al.*, 1982*b*; Torreblanca *et al.*, 1986).

Most of the halobacteria contain one major glycolipid, a sulfated triglycosylarchaeol (S-TGA-1, **34'**), which constitutes about 25% by weight of the total polar lipids (Kates and Deroo, 1973), and three minor glycolipids: a sulfated tetraglycosylarchaeol (S-TeGA, **36'**), the corresponding desulfated tetraglycosylarchaeol (TeGA, **36**), and the desulfated triglycosylarchaeol (TGA-1, **34**) (Smallbone and Kates, 1981) (see Table 1-13). This pattern of glycolipids, together with the presence of three archaeol-based phospholipids (PGP, PG, and PGS), is characteristic of all the halobacteria examined to date, except for *H. sodomense,* and two halophiles from Spain, 3.5 rp4 and 3.1 palp4, which contain an unidentified glycolipid GL-2 replacing S-TGA-1 (**34'**) and S-TeGA (**36'**) (see Table 1-13.) (Kushwaha *et al.*, 1982*b*; Torreblanca *et al.*, 1986). The latter three species may form a separate subgroup of halobacteria.

Haloarcula. The species of this newly designated genus constitute a less homogeneous group than the halobacteria and have a more diverse distribution, covering media of a wide range of salt concentrations and geographical sites. Thus, *H. marismortui* is found in the Dead Sea, *H. vallismortis* in Death Valley, *H. californiae* in Californian salterns, *H. hispanica,* and strains Gaa-3, Gaa-6, Ma 2.25, and I 165 in Spanish salt flats (Torreblanca *et al.*, 1986; Juez *et al.*, 1986). *Haloarcula* species are characterized by the presence of a single major triglycosylarchaeol, TGA-2 (**35**), and by the absence of any sulfated glycosylarchaeols (Kushwaha *et al.*, 1982*b*; Torreblanca *et al.*, 1986) (see Table 1-13). To compensate for this lack of sulfated glycolipids, the *Haloarcula* contain high concentrations of the phosphosulfolipid PGS (see Table 1-14) (Evans *et al.*, 1980; Kushwaha *et al.*, 1982*a,b*; Torreblanca *et al.*, 1986). In addition, the species of this genus contain a minor diglycosylarchaeol, DGA-2, of unidentified structure but containing glucose and mannose (Evans *et al.*, 1980). The major glycolipid, TGA-2 (**35**) has a structure similar to that of TGA-1 (**34**) found in *Halobacteria,* except that the terminal sugar is glucose instead of galactose.

One species that appears in this group, *H. trapanicum,* is unusual, since it

Table 1-13. Distribution of Glycolipids among Various Species of Extreme Halophiles

Genus/Species	DGA-1 (33)	DGA-2	TGA-1 (34)	TGA-2 (35)	S-DGA-1 (33')	S-TGA-1 (34')	S-TeGA (36')	GL-1	GL-2
Halobacterium[a]									
H. salinarium	+	–	+	–	–	+++	++	–	–
H. cutirubrum	+	–	+	–	–	+++	+	–	–
H. halobium	–	–	+	–	–	+++	+	–	–
H. saccharovorum	–	–	–	–	?	++	++	–	–
Y 12	–	–	–	–	?	++	++	–	–
Gca-19	+	–	+	–	–	+	+	–	–
3.5 rp4	–	–	–	–	–	–	–	–	+++
3.1 palp 4	–	–	–	–	–	–	–	–	+++
H. sodomense	–	–	–	–	–	–	–	–	+++
Haloarcula[a]									
H. hispanica	–	+	–	+++	–	–	–	–	–
Gaa-3 and Gaa-6	–	–	+	+	–	–	–	–	–
H. trapanicum	–	?	+	++	–	–	–	++	–
H. marismortui (Volcani)	–	+	–	+++	–	–	–	–	–
H. vallismortis	–	+	–	+++	–	–	–	–	–
Ma 2.25	–	+	–	+++	–	–	–	–	–
I 165	–	+	–	+++	–	–	–	–	–

	1	2	3	4	5	6	7	8	9
H. californiae	−	−	−	−	−	+++	−	+	−
H. sinaiiensis	−	−	−	−	−	+++	−	+	−
Amoebobacter morrhuae	−	−	−	−	−	−	+++	−	−
Ma 2.20	−	?	−	−	−	−	+++	−	−
Haloferax[a]									
H. volcanii	−	−	−	−	+++	−	−	−	+
H. gibbonsii (Ma 2.38)	−	−	−	−	+++	−	−	−	+
H. mediterranei (R-4)	−	−	−	−	+++	−	−	−	+
M 2.5	−	−	−	−	+++	−	−	−	+
Halococcus[b]									
H. morrhuae	−	−	++	+	−	−	++	−	−
Sarcina sp.	+	−	−	+++	+	−	+	−	−
H. saccharolyticus[b']	−	−	−	−	+	−	−	−	−
Natronobacterium[c]									
N. pharaonis	−	−	−	−	−	−	−	−	−
Natronococcus occultus[c]	−	−	−	−	−	−	−	−	−

[a] Data from Torreblanca et al. (1986); Kushwaha et al. (1982b).
[b] Data from Kates et al. (1966); Kocur and Hodgkiss (1973). [b'] Data from Montero et al. (1989); a phosphoglycolipid and two minor unidentified glycolipids are also present.
[c] Data from Ross and Grant (1985); Tindall et al. (1984); De Rosa et al. (1988).

Table 1-14. Polar Lipid Composition of Three Major Genera of Extreme Halophiles[a]

Lipid component	Halobacterium cutirubrum (mole %)	Haloarcula marismortui (mole %)	Haloferax mediterranei (mole %)
Phosphatidylglycerol (PG)	4	11	44
Phosphatidylglycerophosphate (PGP)	70	62	30
Phosphatidylglycerosulfate (PGS)	4	17	—
Triglycosylarchaeol-1 (TGA-1)	trace	—	—
Sulfated triglycosylarchaeol-1 (S-TGA-1)	21	—	—
Triglycosylarchaeol-2 (TGA-2)	—	11	—
Sulfated diglycosylarchaeol (S-DGA-1)	—	—	25
Number of negative charges/mole polar lipid[b]	2.4–1.7	2.3–1.7	1.6–1.3

[a]Data from Kushwaha *et al.* (1982a).

[b]The first and second values given are calculated on the basis of 3 and 2 negative charges per molecule of PGP, respectively.

contains in addition to TGA-2, the isomeric TGA-1, as well as another unidentified glycolipid, GL-1 (see Table 1-13). Two other species that appear to belong in the *Haloarcula* genus, *Amoebobacter morrhuae* and Ma 2.20, contain only the TGA-1 (**34**) and not the TGA-2 (**35**) characteristic of the *Haloarcula* genus (see Table 1-13). Perhaps these latter species may form a separate subgroup of *Haloarcula*, but further study is necessary to resolve these discrepancies.

Haloferax. Species of the third genus, *Haloferax*, are clearly distinguished from *Halobacteria* both genotypically and by their distinctive glycolipid composition. So far there are only three designated species, *H. volcanii* (Mullakhanbhai and Larsen, 1975), *H. mediterranei* (formerly strain R-4, Rodriguez-Valera *et al.*, 1983), and *H. gibbonsii* (formerly strain Ma 2.38, Juez *et al.*, 1986) and some undesignated strains, e.g., M 2.5 and Gla 2.2 (Kushwaha *et al.*, 1982b; Torreblanca *et al.*, 1986). All contain a single major glycolipid, the sulfated diglycosylarchaeol S-DGA-1 (**33′**), together with minor amounts of the unsulfated diglycosylarchaeol DGA-1 (**33**). No other sulfated glycolipid or sulfophospholipid (PGS) is present (see Table 1-13).

Summary: Halobacteriaceae. It is of interest that the glycolipids of the above three genera of Halobacteriaceae, in spite of their apparent diversity, are actually all derivatives of the diglycosylarchaeol DGA-1 (**33**), as is shown in Fig. 1-23. This observation suggests that there is a genetic interrelation between these genera on the level of the genes involved in glycolipid biosynthesis. Studies on the comparative pathways of glycolipid biosynthesis in these three genera would therefore be rewarding (for further discussion of this topic, see Section 1.4.1). Another point that should be noted is that in spite of the different molar compositions of polar anionic lipids in these three genera, the number of negative charges per mole ionic lipid appears to be maintained at ~1.5–2.0 (see Table 1-14; see further discussion in Section 1.6.2.1).

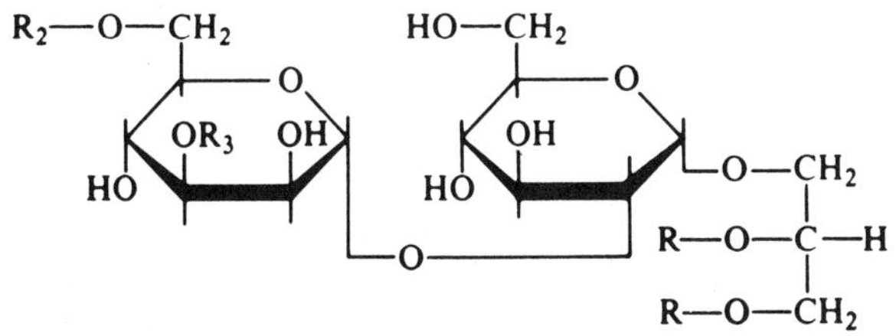

$$R_2-O-CH_2$$

$$HO-CH_2$$

DGD, $R_2 = H$, $R_3 = H$
TGD-1, $R_2 = \beta\text{-gal}p$, $R_3 = H$
TGD-2, $R_2 = \beta\text{-glc}p$ $R_3 = H$
S-DGD, $R_2 = -SO_2-OH$ $R_3 = H$
S-TGD-1, $R_2 = 3\text{-}SO_3^- \text{-}\beta\text{-gal}p$ $R_3 = H$
S-TeGD, $R_2 = 3\text{-}SO_3^- \text{-}\beta\text{-gal}p$ $R_3 = \alpha\text{-gal}f$

$$\text{R = phytanyl group: } CH_3CH(CH_2)_3CH(CH_2)_3CH(CH_2)_3CH(CH_2)_2-$$

Figure 1-23. Structural relationships between glycolipids of *Halobacteriacea*. (From Kushwaha *et al.*, 1982*b*.)

1.3.4.1b. Halococcaciae.

Halococcus. The lipids of only two species of *Halococcus* have so far been examined, *Halococcus morrhuae* (formerly *Sarcina literalis*) (Kocur and Hodgkiss, 1973) and an unidentified *Sarcina* sp. (Kates *et al.*, 1966). On the basis only of chromatographic mobilities and staining with rhodamine 6G, both species appear to contain the sulfated triglycosylarchaeol S-TGA-1 (**34'**) and the unsulfated TGA-1 (**34**). *H. morrhuae* also appears to contain the sulfated tetraglycosylarchaeol S–TeGA (**36'**), and the *Sarcina* sp. contains another unidentified glycolipid. Both species contain the major phospholipid PGP but not the minor PGS; only *H. morrhuae* also contains the minor phospholipid PG (see Table 1-13). Although the lipids of these halococci appear to be similar to those of the halobacteria, on the basis of TLC motilities, further detailed studies of the lipids of halococci are required before any meaningful comparisons can be made with those of halobacteria. That the lipids of halococci may differ from those of halobacteria was recently shown in a study of *Halococcus saccharolyticus*, a newly designated *Halococcus* species from Spanish salt flats (Montero *et al.*, 1989). This *Halococcus* was found to contain a sulfated diglycosylarchaeol (probably S-DGA-1), but not S-TGA-1, as well as a minor phosphoglycolipid; its phospholipids are archaeol analogues of PGP and PG containing $C_{20}-C_{20}$ and $C_{20}-C_{25}$ isopranyl chains, as well as small amounts of PA and an unidentified phospholipid.

Natronobacterium and Natronococcus. The complex lipids of these two genera of haloalkaliphiles have only recently been studied, particularly the species *Natronobacterium pharaonis*, *Nb. gregoryi*, *Nb. magadii*, and *Natronococcus occultus* (Tindall *et al.*, 1984; Morth and Tindall, 1985; Ross and Grant, 1985; Grant and Ross, 1986; Ross *et al.*, 1985; De Rosa *et al.*, 1988). These haloalkaliphiles

Table 1-15. *Distribution of Glycolipids and Phosphoglycolipids among Methanogenic Bacteria*

Genus/Species	Glycosyl group (lipid core)[a]	References
Methanobacterium		
M. thermoautotrophicum	Glc*p*β1–6Glc*p*β (archaeol) (DGA-3, **38**)	Koga *et al.* (1987)
	Glc*p*β1–6Glc*p*β (caldarchaeol) (**45**)	Nishihara *et al.* (1987)
	Glc*p*β1–6Glc*p*β (caldarchaeol)-P-ethanolamine (**45′**)	Nishihara *et al.* (1987)
	Unidentified phosphoglycolipid	Nishihara *et al.* (1987)
M. formicum	Unidentified phosphoglycolipid	Koga *et al.* (1987)
M. bryanti	Unidentified phosphoglycolipid	Koga *et al.* (1987)
Methanobrevibacter		
M. arboriphilus	Unidentified phosphoglycolipid	Koga *et al.* (1987)
(strains DH1, Az, Dc, A₂)		
M. smithii (strain AL1, PS)	Unidentified phosphoglycolipid	Koga *et al.* (1987)
M. ruminantium	Unidentified phosphoglycolipid	Koga *et al.* (1987)
Methanococcus		
M. voltae	Glc*p*β (archaeol) (MGA-2, **37**)	Ferrante *et al.* (1986)
	GlcN*p*Acβ-P-(archaeol) (**37′**)	Ferrante *et al.* (1986)
	Glc*p*β1–6Glc*p*β (archaeol) (DGA-3, **38**)	Ferrante *et al.* (1986)
M. jannaschii	Glc*p* (*cyc*-archaeol)	Ferrante *et al.* (1990)
	Glc*p* 1–6Glc*p* (*cyc*-archaeol)	
	6-(P-ethanolamine)Glc*p* (*cyc*-archaeol)	
Methanospirillum		
M. hungatei	Glc*p*α1–2Gal*f*β (archaeol) (DGA-I, **41**)	Kushwaha *et al.* (1981)
	Gal*f*β1–6Gal*f*β (archaeol) (DGA-II, **42**)	Kushwaha *et al.* (1981)
	Glc*p*α1–2Gal*f*β (caldarchaeol) (DGC-I, **43**)	Kushwaha *et al.* (1981)
	Gal*f*β1–6Gal*f*β (caldarchaeol) (DGC-II, **44**)	Kushwaha *et al.* (1981)
	Glc*p*α1–2Gal*f*β (caldarchaeol)-P-Gro (PGC-I, **43′**)	Kushwaha *et al.* (1981)
	Gal*f*β1–6Gal*f*β (caldarchaeol)-P-Gro (PGC-II, **44′**)	Kushwaha *et al.* (1981)
Methanothrix		
M. concilii	Man*p*α1–3Gal*p*β (archaeol) (DGA-4, **39**)	Ferrante *et al.* (1988*a,b*)
	Gal*f*β1–6Gal*p*β (hydroxyarchaeol) (**40**)	Ferrante *et al.* (1988*a,b*)

[a]Numbers in parentheses refer to the structures in Fig. 1-10.

have a rather simple polar lipid composition, the main components being the C_{20},C_{20} and C_{25},C_{20} archaeol analogues of phosphatidylglycerophosphate (PGP) and phosphatidylglycerol (PG), and one to three unidentified minor phospholipids; no PGS or glycolipids have been detected. The haloalkaliphiles are thus clearly distinguished from all other species in the Halobacteriaceae, although the presence of PGP and PG is consistent with their being extremely halophilic archaebacteria.

1.3.4.2. Methanogenic Bacteria

The taxonomy of the methanogens is much more complex than that of the extreme halophiles, encompassing three orders, five families, and ten or more genera (Whitman, 1985). Recent investigations of Kushwaha *et al.* (1981), Koga *et al.* (1987), Nishihara *et al.* (1987), and Ferrante *et al.* (1986, 1987, 1988*a,b*) have provided detailed structural data on the glycolipids of a few species belonging to three of the four known methanogen families (see Table 1-15). Preliminary data based on two-dimensional TLC and specific staining have been reported on the glycolipids and other lipids of several other strains of methanogens covering 10 genera (Grant *et al.*, 1985; Nishihara and Koga, 1987; Koga *et al.*, 1987). The methanogen lipids are based on a variety of lipid cores ranging from the C_{20},C_{20}-, C_{20},C_{25}-archaeols (**31, 31′, 31″**) to caldarchaeols (**32**) with and without cyclopentane rings (**32a–d**) (Table 1-12) (see Section 1.3.3.2.). The available information is given in Table 1-15 and is summarized as follows.

1.3.4.2a. Methanobacteriaceae.

Methanobacterium. The only species in this genus in which the glycolipids have been structurally characterized is *Mba. thermoautotrophicum* (Nishihara *et al.*, 1987, 1989; Koga *et al.*, 1987). Three glycolipids have been identified: a diglucosylarchaeol (gentiobiosylarchaeol), Glcpβ1–6G1cpβ1–1 archaeol (DGA-3, structure **38,** Fig. 1-10); the corresponding diglucosylcaldarchaeol (DGC-3, structure **45,** Fig. 1-10); and the corresponding phosphoglycosylcaldarchaeol (PGC-3, structure **45′,** Fig. 1-10), in which the diglucosyl group is on one hydroxyl of the tetraether and a phosphoethanolamine is on the other. In addition, an unidentified phosphoglycolipid (PGC-4) containing a diglucosylated caldarchaeol attached to an acid-labile phosphate group was detected in this species and in several other methanobacteria (Table 1-15) (Koga *et al.*, 1987). More recently, gentiobiosylcaldarchaeol-P-serine and -P-inositol have been identified (Nishihara *et al.*, 1989).

Methanobrevibacter. No detailed glycolipid analyses have been carried out on the lipids of several species in this genus examined by two-dimensional TLC (Koga *et al.*, 1987; Grant *et al.*, 1985). Only the unidentified phosphoglycolipid (PGC-4) found in the methanobacteria was reported present (Table 1-15).

1.3.4.2b. Methanococcaceae.

Methanococcus. The lipids of one species in this genus, *Methanococcus voltae,* have been characterized (Ferrante *et al.,* 1986) and found to be derivatives only of archaeol. The major glycolipid was found to be the diglucosylarchaeol (DGA-3, structure **38**), also present in *Mba. thermoautotrophicum* (Koga *et al.,* 1987); minor glycolipids, monoglucosylarchaeol, Glc$p\beta$1–1archaeol (MGA, structure **37**, Fig. 1-10), presumably the precursor of DGA-3 (**38**), and a novel GlcNAc-1-P-archaeol (structure **37'**, Fig. 1-10) were also identified (Table 1-15). Another *Methanococcus* species that has been studied recently (Ferrante *et al.,* 1990) is *M. jannaschii,* a deep-sea thermophilic methanogen. The glycolipids of this unusual bacterium consist of the C_{40}-macrocyclic diether analogues of the monoglucosyl-archaeol (MGA, structure **37**, Fig. 1-10) and the diglucosyl-archaeol (DGA-3, structure **38**) found in *M. voltae* (see Table 1-15). In addition, a major novel aminophosphoglycolipid, 6-(phosphoethanolamine)-monoglucosyl-*cyc*-archaeol was also identified (Table 1-15).

1.3.4.2c. Methanomicrobiaceae.

Methanospirillum. The lipids of one species in this genus, *Msp. hungatei* have been characterized in detail (Kushwaha *et al.,* 1981; Ferrante *et al.,* 1987). Two phosphoglycolipids derived from caldarchaeol, PGC-I (structure **43'**, Fig. 1-10) and PGC-II (structure **44'**, Fig. 1-10), the corresponding diglycosylcaldarchaeols, DGC-I (**43**) and DGC-II (**44**), respectively, and the diglycosylarchaeols DGA-I (**41**) and DGA-II (**42**) comprise the bulk of the total lipids, with PGC-I (50% by weight), PGC-II (14%) and DGA-I (17%) being the major components (Kushwaha *et al.,* 1981). These glycolipids form an interesting set of six structures based on the same two diglycosyl groups: I, Glc$p\alpha$1–2Gal$f\beta$, and II, Gal$f\beta$1–6Gal$f\beta$, and on the archaeol and caldarchaeol lipid core types. These six glycolipids are most likely related biosynthetically (see section 1.4.2).

1.3.4.2d. Methanosarcinaceae.

Methanothrix. The one species in this genus, *Methanothrix concilii,* that has been studied was found to contain two glycolipids of different structure, both archaeol-based: glycolipid DGA-4 (structure **39**, Fig. 1-10) has a Man$p\alpha$1–3Gal$p\beta$disaccharide linked to archaeol, and DGA-5 (structure **40**, Fig. 1-10) has a Gal$p\beta$1–6Gal$p\beta$disaccharide linked to an archaeol that has a 3-hydroxy group on the *sn*-1-phytanyl group (Ferrante *et al.,* 1988*a,b*) (Table 1-15).

Other Species. In addition to the above *Methanogen* spp., several other species covering genera *Methanobacterium, Methanobrevibacter, Methanococcus, Methanomicrobium, Methanogenium, Methanoplasma, Methanosarcina,* and *Methanolobus* have been surveyed for glycolipids by two-dimensional TLC (Grant *et al.,* 1985; Koga *et al.,* 1987). All have been found to contain several glycolipids, phosphoglycolipids, and aminoglycolipids that were not further characterized.

Summary: Methanogenic Bacteria. Clearly, the methanogens offer an extensive array of novel glycolipid types that will occupy the attention of microbiologists–biochemists for years to come. The present detailed structural data (Table 1-15), covering only four species, is insufficient to permit any conclusions concerning correlations with taxonomic classification. However, it is interesting that both DGA-3 (**38**) and the aminophosphoglycolipid PGC-3 (**45'**) are found in *Mba. thermoautotrophicum* and *Mco. voltae,* species belonging to two different genera, *Methanobacterium* and *Methanococcus.* Further structural studies of glycolipids from this fascinating group of archaebacteria will surely be very rewarding.

1.3.4.3. Thermoacidophiles

The thermoacidophiles are characterized as aerobic or anaerobic archaebacteria that live at pH values well below 7 and at optimal temperatures in the range 60–105°C (Stetter and Zillig, 1985). The taxonomical classification of thermoacidophiles is not as complex as that of the methanogens: eight genera are known so far, distributed among four orders, Thermoplasmatales, Sulfolobales, Thermoproteales, and Thermococcales. Only a relatively small number of thermoacidophiles have been examined for their lipids, mainly *Sulfolubus, Thermoplasma, Thermoproteus,* and *Desulfurococcus* species, but their core lipids are more complex and varied than any of the other archaebacteria (De Rosa *et al.,* 1986a; Langworthy, 1985; Thurl and Schäfer, 1988; Lanzotti *et al.,* 1987). The available data on lipid composition of thermoacidophiles are listed in Table 1-16.

1.3.4.3a. Thermoplasma. The one species in this genus that has been examined is *Thermoplasma acidophilum,* a wall-less *Mycoplasma*-like aerobe. Its polar lipids consist of at least six different glycolipids and at least seven phosphorus-containing lipids, one of which is the major component (80% of the polar lipids), a phosphoglycolipid of unassigned structure. However, the latter is known to be a caldarchaeol with a single unidentified sugar attached to one hydroxyl of the tetraether and a *sn*-3-glycerophosphate group attached to the other (structure **46'**, Fig. 1-11) (Langworthy *et al.,* 1972; Langworthy, 1985). The corresponding monoglycosylcaldarchaeol containing the unidentified sugar (structure **46,** Fig. 1-11) is also present, as well as monoglucosyl and diglucosylcaldarachaeol derivatives of unidentified structure (perhaps **47** and **48,** respectively) (Langworthy *et al.,* 1972; Langworthy, 1982a). The rest of the polar lipids are mostly derived from caldarchaeol but a few are derived from archaeol (Langworthy, 1985). Detailed structural studies on the glycolipids of *Thermoplasma* have still to be done. By contrast, the structure of the lipoglycan of this organism has been fully established in detail as having a glycan of 24 mannose residues and 1 glucose linked as shown in structure **51** (see Fig. 1-11) to one of the free hydroxyls in caldarachaeol (Mayberry-Carson *et al.,* 1974; Smith, 1980). Similar lipoglycans have been identified in the eubacterial *Acholesplasmas* (Smith, 1981). Lipoglycans may help stabilize the membrane of these wall-less organisms.

Table 1-16. Distribution of Glycolipids and Phosphoglycolipids among Thermoacidophilic Bacteria

Genus/Species	Glycosyl group (lipid core)[a]	References
Thermoplasma		
T. acidophilum	Glycosyl (caldarchaeol) (**46**)	Langworthy *et al.* (1972, 1982)
	Glycosyl(caldarchaeol)-P-Gro (**46**')	Langworthy (1985)
Sulfolobus		
S. acidocaldarius	Glc–Gal(cardarchaeol) (**48**?)	Langworthy (1977b)
	Glc (nonitolcaldarchaeol) (**50**)	Langworthy (1977b)
	Glc–Gal(caldarchaeol)-P-inositol (**48**'?)	Langworthy (1977b)
	Glc(nonitolcardarchaeol)-P-inositol (**50**'?)	Langworthy (1977b)
	HSO₃Glc(nonitolcaldarchaeol) (**50**''?)	Langworthy (1977b)
S. sulfataricus	Glc*p*βGal*p*β(caldarchaeol) (**48**)	De Rosa *et al.* (1980a, 1986a)
	Glc*p*βGal*p*β(caldarchaeol)-P-inositol (**48**')	Langworthy (1985)
	Glc*p*β(nonitolcaldarchaeol) (**50**)	Langworthy (1985)
	Glc*p*β(nonitolcaldarchaeol)-P-inositol (**50**')	Langworthy (1985)
	HSO₃Glc*p*β(nonitolcaldarchaeol)-P-inositol (**50**'')	Langworthy (1985)
Thermoproteus		
T. tenax	Glc*p* (caldarchaeol) (**47**)	Thurl and Schäfer (1988)
	Glc*p* 1–6Glc*p* (caldarchaeol) (**45**)	Thurl and Schäfer (1988)
	Glc*p*β1–6Glc*p*β1–6Glc*p*β(caldarchaeol) (**49**)	Thurl and Schäfer (1988)
	Glc*p*β(caldarchaeol)-P-inositol (**47**')	Thurl and Schäfer (1988)
Desulfurococcus		
D. mobilis	Glc*p*α1–4Gal*p*β(caldarchaeol) (**48a**)	Lanzotti *et al.* (1987)
	Glc*p*α1–4Gal*p*β(caldarchaeol)-P-inositol (**48a**')	Lanzotti *et al.* (1987)
	Gal*p*β(caldarchaeol)-P-inositol (**47**')	Lanzotti *et al.* (1987)

[a]Numbers in parentheses refer to the structures in Fig. 1-11.

1.3.4.3b. Sulfolobus. Species of this genus live in sulfataric hot springs, such as those in Yellowstone National Park, Italy, New Zealand, and Japan, at temperatures between 60° and 100°C and pH values of 1–5, and are sulfur and iron oxidizing (Stetten and Zillig, 1985). The polar lipids of only three species, *Sulflobus acidocaldarius*, *Sulfolobus solfataricus* (formerly *Caldariella acidophila*), and *Sulfolobus brierleyi* (also called *ferrolobus*) have been investigated in detail (Langworthy *et al.*, 1974, 1982; Langworthy, 1977*b*; De Rosa *et al.*, 1980*a*, 1983*b*, 1986*a*). These lipids are derived almost exclusively from either caldarchaeol (**32**) or nonitolcaldarchaeol (**32'**). Polar lipid derivatives of these tetraethers occur in about equal amounts in cells grown heterotrophically in yeast extract, while the nonitolcaldarchaeol-based lipids occur predominantly in cells grown autotrophically on sulfur (Langworthy, 1977*b*). Both types of tetraether also may contain one to four cyclopentane rings in their biphytanyl chains, the extent of cyclization increasing with increasing growth temperature (De Rosa *et al.*, 1980*b*).

The polar lipids of *S. acidocaldarius* are predominantly glycolipids (35%) and phosphoglycolipids (45%), in almost the same proportions for heterotrophically or autotrophically grown cells. The strict autotroph, *ferrolobus*, also has the same proportions of glycolipids and phosphoglycolipids (Langworthy *et al.*, 1974; Langworthy, 1977*b*). The major glycolipids in *S. acidocaldarius* (Table 1-16) and *ferrolobus* were tentatively identified as glycolipid DGC-4, a glucosylgalactosylcaldarchaeol (structure **48**, Fig. 1-11), and glycolipid GNC, a glucosylnonitolcaldarchaeol (structure **50**, Fig. 1-11), the glucosyl group being linked to one of the OH groups of the nonitol group. These two glycolipids occur in almost equal proportion in *S. acidocaldarius* cells grown heterotrophically, while the caldarchaeol-derived glycolipid (DGC-4) predominates (75%) in *S. acidocaldarius* cells grown autotrophically, and in *ferrolobus*. In addition, small amounts of an unidentified diglucosylnonitolcaldarchaeol and of another unidentified glycolipid were also detected in autotrophically grown cells (Langworthy, 1977*b*).

The phosphoglycolipids in both *S. acidocaldarius* and *ferrolobus* (Table 1-16) consisted predominantly (72%) of the phosphoinositol derivatives of glycolipids DGC-4 (**48**) and GNC (**50**): Glc–Gal–caldarchaeol-P-inositol (PGC-4; structure **48'**, Fig. 1-11), in which the diglycosyl group is attached to one OH group of the tetraether and a phosphoinositol group is attached to the other OH group; and Glc-nonitolcaldarchaeol-P-inositol (structure **50'**, Fig 1-11), in which the glucosyl group is linked to one OH in the nonitol group and the phosphoinositol is linked to the free OH in the glycerol residue of caldarchaeol. A minor acidic glycolipid detected in both heterotrophically and autotrophically grown *S. acidocaldarius* cells, but not in *ferrolobus*, was partially characterized as a monosulfate derivative of glycolipid GNC (structure **50''**, Fig. 1-11); the sulfate group is presumably attached to the glucosyl group but its precise location is not known. Another minor acidic glycolipid, found only in autotrophically grown *S. acidocaldarius* cells and in *ferrolobus*, was tentatively characterized as a mixture of a disulfate and a phosphate derivative of glycolipid GNC (**50**) (Langworthy, 1977*b*). The main glycolipids of *S. acidocaldarius* and *ferrolobus* are probably similar if not identical to those in *S. sulfataricus,* but further structural studies are necessary to establish this unambiguously (Langworthy, 1985; De Rosa, 1986*a*).

The glycolipids of *S. sulfotaricus* (De Rosa *et al.,* 1980*a,* 1983*b,* 1986*a*) have been characterized in more detail than those of *S. acidocaldarius,* but their structures are still not unambiguously defined. The major glycolipids are glycolipid DGC-4 (6%), Glc*p*β-Gal*p*β-caldarchaeol (structure **48,** Fig. 1-11), and glycolipid GNC (14%), Glc*p*β-nonitolcaldarchaeol (structure **50,** Fig. 1-11); the linkage positions of the disaccharide in (**48**) and the position of attachment of the glucosyl group to the nonitol group in (**50**) are still to be determined. The phosphoglycolipids consisted mainly of the phosphoinositol derivatives of glycolipids (**48**) (DGC-4) and (**50**) (GNC): Glc*p*βGal*p*βcaldarchaeol-P-inositol (8%, structure **48′,** Fig. 1-11), and Glc*p*β-nonitolcardarcheol-P-inositol (55%, structure **50′,** Fig. 1-11) respectively; the sulfated derivative of glycolipid (**50**), HSO₃-Glc*p*β-nonitolcaldarchaeol (10%, structure **50″,** Fig. 1-11) was also identified.

*1.3.4.3c. Thermoproteus.*One species of this anaerobic, sulfur-dependent genus, *T. tenax,* has been examined for its lipids (Thurl and Schäfer, 1988). Cells grown autotrophically from hydrogen and sulfur, or heterotrophically, contain 5% neutral lipids, 20% glycolipids, and 75% phospholipids and phosphoglycolipids. The ether core lipids consist mainly (95%) of caldarchaeols with minor proportions (5%) of archaeols. The caladarchaeols are acyclic (0 + 0, **32**), or contain one (1 + 1, **32a**), or two (2 + 2, **32b**) cyclopentane rings in each chain, as well as mixed numbers cf rings (0 + 1, 0 + 2, 1 + 2, and 2 + 3), similar to those found in *Sulfolobus* (De Rosa *et al., 1986a*). The polar lipid composition resembles that of *Sulfolobus* but is characteristically different with respect to the glycosyl groups of the glycolipids (Table 1-16). The glycolipids consist of mono-, di-, and triglucosylcaldarchaeols, the sugar groups being β1–6-linked (structures **47, 45,** and **49,** respectively, Figs. 1-10 and 1-11) and the caldarchaeol cores contain 0 to 4 rings (Table 1-16). It is of interest that, except for the presence of rings in the caldarchaeol core, the diglucosylcaldarchaeol (**45**) is identical to that found in the methanogen *M. thermoautotrophicum* (Table 1-15) and the monoglucosyl caldarchaeol (structure **47,** Fig. 1-11) is similar to structure (**46**) in *Thermoplasma,* but the triglucosylcaldarchaeol has not been reported in any other archaebacterium (Thurl and Schäfer, 1988). The major phosphoglucolipid is the monoglucosylcaldarchaeol (**47**) with a phosphoinositol attached to the free hydroxyl (structure **47′,** Fig. 1-11), which is a variation of structure (**46′**) in *Thermoplasma* or (**50′**) in *Sulfolobus* (Table 1-16). No nonitolcaldarchaeol glyco- or phosphoglycolipids appear to be present in *T. tenax.*

1.3.4.3d. Thermococcales. The lipids of one species of this anaerobic sulfur reducing group of thermoacidophiles, *Desulfurococcus mobile,* are derived from caldarchaeol and resemble those in *Sulfolobus* (Table 1-16), except that the lipid cores do not contain any nonitolcaldarchaeols. The major glycolipid is: Glc*p*α1–4Gal*p*βcaldarcheol (structure **48a,** Fig. 1-11), and the phosphoglycolipids are Gal*p*βcardarchaeol-P-inositol (**47a,** Fig. 1-11) and Glc*p*α1–4Gal*p*βcardarchaeol-P-inositol (**48a′,** Fig. 1-11) (Lanzotti *et al.,* 1987).

Summary: Thermoacidophiles. In spite of the limited number of species of thermoacidophiles that have been studied, it appears that at least two genera, *Thermoplasma* and *Sulfolobus*, and probably also *Thermoproteus* and *Desulfurococcus*, can be distinguished by their glycolipid compositions. *Thermoplasma* contains a glycosylcaldarchaeol (**46**) and its glycerophosphoryl derivative (**46'**); *Sulfolobus* contains both diglycosyl caldarchaeol (**48**) and glycosylnonitolcardarchaeol (**50**) and the corresponding inositolphosphoryl derivatives (**48'** and **50'**, respectively); *Thermoproteus* contains mono-, di-, and triglucosylcaldarchaeols (**47, 45,** and **49,** respectively) and the phosphoinositol derivative of the monoglucosylcaldarchaeol (**47'**), and *Desulfurococcus* contains a diglycosylcaldarchaeol (**48a**) and its P-inositol derivative (**48a'**) and a monogalactosylcaldarchaeol-P-inositol (**47a**). It is of interest, however, that the *Thermoplasma* glycolipid (**46**) and its glycerophosphoryl derivative (**46'**) have similar structures to those in the methanogen *M. hungatei*, and *Thermoproteus* has a diglucosylcaldarchaeol (**45**) found in a methanogen (see Tables 1-15 and 1-16), but *Sulfolobus* glycolipids can be distinguished by the presence of phosphoinositol derivatives of the glycosylnonitolcaldarchaeols, which do not occur in other thermoacidophiles or in the methanogens. Further studies of the glycolipids of a wide variety of thermoacidophiles should be of interest.

1.4. Metabolism of Bacterial Glycolipids

1.4.1. Eubacteria

1.4.1.1. Biosynthesis

Despite the wide variety of glycolipids that have been isolated and identified in eubacteria, biosynthetic pathways for these lipids have been studied in relatively few species (see reviews by Sastry, 1974; Ishizuka and Yamakawa, 1985). In general, the mechanisms for biosynthesis of glycosylDAGs in bacteria are analogous to those in algae or higher plants (see Chapter 3, *this volume*), and involve glycosyl transfer from a suitable nucleoside diphosphate sugar to the free hydroxyl group of a 1,2-diacyl-*sn*-glycerol (DAG) to form a monoglycosylDAG, which can accept another glycosyl group to form a diglycosyl-DAG, etc.

The 1,2-diacyl-*sn*-glycerol acceptor is presumably synthesized by stepwise acylation of *sn*-3-glycerophosphate to 1,2-diacyl-*sn*-3-glycerophosphate, followed by dephosphorylation of the diacylglycerophosphate by phosphatidic acid phosphatase, as has been shown in *Escherichia coli* (Lennarz, 1966; van den Bosch and Vagelos, 1970; Raetz, 1978). Another route to the formation of DAG would be the action of phospholipase C or of phospholipase D plus phosphatidate phosphatase on a phospholipid such as phosphatidylethanolamine (PE), phosphatidylglycerol (PG), or cardiolipin (CL) (Cole and Proulx, 1977; Proulx and van Deenen, 1967; Raetz, 1978). Evidence suggesting the formation of DAG from PG in *Staphylococcus epidermidis* has been obtained in pulse studies with ^{14}C glycerol in which an increase in ^{14}C in DAG was observed at the expense of ^{14}C in PG (Komaratat, 1974).

The first demonstration in bacteria of the glycosyltransferase reaction was by Kaufman *et al.* (1965). They were able to show that in *Pneumococcus* type XIV the diglycosylDAG, Gal*p*α1–2Glc*p*α1–3DAG, was synthesized by transfer of a galactosyl group from UDP–Gal to glucosylDAG. Synthesis of the glucosylDAG presumably occurred by transfer of a glucosyl group from UDP–Glc to a diacylglycerol:

(1) 1,2-Diacyl-*sn*-glycerol + UDP-Glc → Glc*p*α1–3DAG (**1**)

(2) Glc*p*α1–3DAG + UDP–Gal → Gal*p*α1–2Glc*p*α1–3DAG (**17**)

In subsequent studies with cell-free systems (Lennarz and Talamo, 1966), the biosynthesis of the mannosylDAGs in *M. lysodeikticus* (*luteus*) was shown to involve a particulate-bound enzyme for transfer of the first mannose from GDP-mannose to a DAG:

(3) 1,2-Diacyl-*sn*-glycerol + GDP–Man → Man*p*α1–3DAG (**7**); and another enzyme present in the soluble fraction for transfer of the second mannose group to the ManDAG:

(4) Man*p*α1–3DAG + GDP–Man → Man*p*α1–3Man*p*α1–3DAG (**18**)

The enzyme catalyzing reaction 3 was specific and optimal for a *sn*-1,2-DAG containing branched-chain fatty acids that are present in *M. lysodeikticus* (*luteus*).

Similar cell-free studies have demonstrated the biosynthesis of glycosyl-DAGs using particulate enzyme preparations from various eubacteria, such as: (*i*) *Streptococcus faecalis* (*faecium*) (Pieringer, 1968) or *Acholeplasma laidlawii* (Smith, 1969):

(1) 1,2-DiacylGro + UDP–Glc → Glc*p*DAG (**1**);

(5) Glc*p* DAG + UDP–Glc → Glc*p*α1–2Glc*p*αDAG (**13**); (*ii*) *Bifidobacterium bifidum* (Veerkamp, 1972):

(6) 1,2-DiacylGro + UDP–Gal → Gal*f*DAG (**6**); (*iii*) a moderate halophile (Stern and Tietz, 1973) or *P. diminuta* (Shaw and Pieringer, 1972):

(7) 1,2-DiacylGro + UDP–GlcU → GlcU*p*αDAG(**8**). (*iv*) the halotolerant *Staphylococcus epidermidis* (Komaratat, 1974):

(8) 1,2-DiacylGro + UDP–Glc → Glc*p*βDAG (**2**) → Glc*p*β1–6Glc*p*βDAG (**14**); and (*v*) *Mycobacterium smegmatis* (Schultz and Elbein, 1974):

(9) 1,2-DiacylGro + UDP–Glc → GlcDAG → Glc–GlcDAG;

(10) 1,2-DiacylGro + UDP–Gal → GalDAG → Gal–GalDAG.

Most of these glycosyl transfer enzyme systems required Mg^{2+} (Pieringer, 1968; Stern and Tietz, 1973) and a suitable DAG, usually containing endogenous fatty acids (Lennarz and Talamo, 1966; Stern and Tietz, 1973), for optimal activity; in some cases anionic or nonionic detergents were also required (Veerkamp and Van Schaik, 1974), but not in others (Pieringer, 1968; Shaw and Pieringer, 1972). The glycosyl transferases are stereospecific with respect to the configuration of the DAG, a *sn*-1,2-diacylglycerol, but not the *sn*-2,3-isomer, being required (Pieringer, 1968; Lennarz and Talamo, 1966). Each glycosyl transferase is also specific for the nucleoside diphosphate sugar involved, the configuration of the glycosidic linkage formed and, in the case of di-, tri-, etc., glycosylDAGs, for the position of attachment of the next sugar residue (Pieringer, 1968).

Biosynthesis of monoacylated Glc*p*αDAG in *Acholeplasma laidlawii* has been shown by pulse-chase studies and by *in vitro* studies with cell-free systems to occur by direct acyl transfer to the monoglucosylDAG (Smith, 1986). Biosynthesis of phosphoglycosylDAGs is not covered in this chapter, since it is dealt with in detail in Chapter 2 (*this volume*).

Biosynthesis of phosphatidylinositol-α-mannosides (PI-mannosides) of *Mycobacterium* (Lee and Ballou, 1965) has been demonstrated with a particulate fraction of *M. phlei* (Brennan and Ballou, 1968) which catalyzed the transfer of mannose from GDP-mannose to PI yielding first PI-monomannoside then PI–dimannoside. However, PI–monomannoside was later found (Takayama and Goldman, 1969) not to serve as acceptor for the second mannose group and it was suggested that the monomannoside has to be modified, perhaps by acylation before transfer of the second mannose.

1.4.1.2. Degradation

Very few studies have been reported on the degradation of glycolipids in eubacteria. In a study of the metabolic turnover of polar lipids in *Acholeplasma laidlawii* strain B (McElhaney and Tourtellotte, 1970), it was found that the glycolipids and the phospholipids showed no loss of ^{14}C or ^{32}P after several hours of chase of cells pulse-labeled either with ^{14}C-fatty acids, [^{14}C]glucose or [^{32}P]phosphate. It was concluded that independent turnover of the fatty acyl groups or polar head groups of glycolipids or phospholipids did not occur during the period of study and that the membrane polar lipids are metabolically stable in actively growing cells.

By contrast, chase studies with whole cells of *Staphylococcus aureus* prelabeled with [^{14}C]glucose showed a rapid loss of [^{14}C]glucose from mono- and diglucosylDAGs (half-life, about 1 generation time), but no loss of ^{14}C-fatty acids (Short and White, 1970). These findings indicate that glycosidases in *S. aureus* are capable of removing glycosyl groups from glycosylDAGs without prior removal of acyl groups.

1.4.2. Archaebacteria

1.4.2.1. Biosynthesis

Available information on glycolipid biosynthesis in archaebacteria is based on results of labeling studies with whole cells (see previous reviews by Kates and Kushwaha, 1978; Bu'Lock *et al.*, 1983; de Rosa *et al.*, 1986a; Kamekura and Kates, 1988). These studies are described for each of the archaebacterial classes.

1.4.2.1a. Extreme Halophiles. Pulse-labeling studies with whole cells of *H. cutirubrum* grown in the presence of [^{35}S]sulfate or [^{14}C]glycerol (Deroo, 1974) have shown the following product-precursor relationship between the glycolipids in this bacterium:

$$(\text{unidentified precursor}) \rightarrow \text{MGA-1} \rightarrow \text{DGA-1} \rightarrow \text{TGA-1} \rightarrow \text{S-TGA-1} \rightarrow$$
$$\text{S-TeGA}$$

These results suggest that S-TGA-1 (**34′**) may be synthesized by stepwise glycosylation of an unidentified diphytanylglycerol ether precursor with

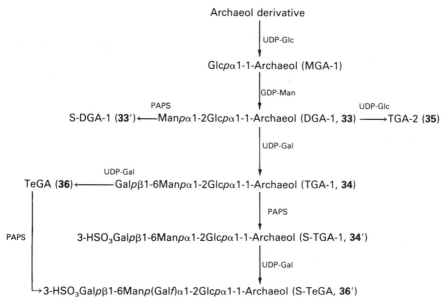

Figure 1-24. Hypothetical pathways for biosynthesis of archaeol glycolipids in *Halobacteria*. (From Kamekura and Kates, 1988.)

glucose, mannose, and galactose followed by sulfation with PAPS (Fig. 1-24). Synthesis of the minor glycolipid S-TeGA (**36′**) might occur by galac-tofuranosylation of S-TGA-1 or by galactofuranosylation of TGA-1 followed by sulfation with PAPs (Fig. 1-24). The minor non-sulfated glycolipids TGA-1 and TeGA may be formed by interruption of the pathway at the appropriate sulfation steps (Fig. 1-24) or by action of sulfatases on S-TGA-1 and S-TeGA, respectively. The sulfated diglycosylarchaeol S-DGA-1 (**33′**) found in *Haloferax* species could be formed by interruption of the pathway at DGA followed by sulfation with PAPS (Fig 1-24). Finally, the unsulfated TGA-2 (**35**) found in *Haloarcula* might be formed by glycosylation of DGA with UDP–Glc instead of UDP–Gal. It should be noted that these proposed pathways are only tentative and have not yet been investigated directly in cell-free systems.

Before discussing glycolipid biosynthesis further, mechanisms for bio-synthesis of the lipid core, diphytanylglycerol (archaeol, **31**), will be consid-ered (see Kamekura and Kates, 1988; Moldoveanu and Kates, 1988). Growth of whole cells of *H. cutirubrum* in presence of [^{14}C]acetate or 2-[^{14}C]mevalonate showed that >98% of the label was incorporated into the phytanyl groups of the polar lipids and <0.3% was found in long-chain fatty acids (Kates *et al.*, 1968). Cell-free studies subsequently demonstrated the presence of a mevalonate pathway for synthesis of isoprenoids that was abso-lutely dependent on high concentrations (4 M) of NaCl (Kates and Kushwaha, 1978), as well as a fatty acid synthetase, which, by contrast, was almost com-pletely inhibited in 4 M NaCl (Pugh *et al.*, 1971). These findings therefore account for the presence of only traces of fatty acids and the predominance of isoprenoid/isopranoid chains in *H. cutirubrum;* the latter are synthesized by the mevalonate pathway, starting from acetate (and involving lysine, Ekiel *et* *al.*, 1986), proceeding to isopentenyl–PP and then to geranylgeranyl–PP (Kushwaha *et al.*, 1976) and perhaps partial reduction to phytyl-PP:

lysine

Acetate $\xrightarrow{\quad\quad}$ mevalonate $\rightarrow$ isopentenyl-PP $\rightarrow$ dimethyally-PP $\rightarrow$

geranyl-PP $\rightarrow$ farnesyl-PP $\rightarrow$ geranylgeranyl-PP $\dashrightarrow$ phytyl-PP

Final reduction of geranylgeranyl-PP (or phytyl-PP) to phytanyl groups apparently takes place only after linkage to the glycerol backbone in form of an unidentified isoprenyl ether derivative, which is the precursor of both phospholipids and glycolipids (Fig. 1-25), as has been shown in recent biosynthetic studies with intact cells of *H. cutirubrum* (Moldoveanu and Kates, 1988, 1989) and labeling studies with *M. hungatei* (Poulter *et al.*, 1988). The unidentified precursor (isoprenyl diether derivative) is probably formed by alkylation of a suitable glycerol derivative, most likely dihydroxyacetone (see review by Kamekura and Kates, 1988), with an isoprenyl pyrophosphate, which could be either geranylgeranyl-PP or phytyl-PP (Fig. 1-25). The choice of dihydroxyacetone (DHA) or a substituted DHA as acceptor of the isoprenyl group is based on the finding that the glycerol moiety of diphytanylglycerol undergoes dehydrogenation at C-2 but not at C-1 (or C-3) (Kates *et al.*, 1970), thus eliminating triose phosphate (DHA-P and glyceraldehyde-P) and glycerol-P as possible acceptors, since these would exchange hydrogen at C-1 due to aldo–keto or keto–enol isomerizations (Fig. 1-25). If glycerol (or a substituted glycerol) were the acceptor, as has been suggested for thermoacidophiles (De Rosa *et al.*, 1982*b*, 1986*a*), it would have to undergo dehydrogenation at C-2 at a later stage to account for the loss of ^{3}H at C-2 observed in *H. cutirubrum* grown in presence of 2-[^{3}H]glycerol (Kates *et al.*, 1970). The involvement of a C_{20}-isoprenyl-PP in the alkylation reaction has been established by inhibition studies with bacitracin (Basinger and Oliver, 1979; Moldoveanu and Kates, 1989. Note that nothing is known about the detailed mechanism of this alkylation reaction for formation of the isoprenyl ether linkage.

The "unidentified isoprenyl ether precursor" is then glycosylated stepwise with glucose, mannose, and galactose to a glycolipid intermediate that is sulfated to give the isoprenyl ether form of S-TGA (pre-S-TGA) and finally reduced to the saturated phytanyl S-TGA (Fig. 1-25). These findings indicate that the pathways for biosynthesis of glycosylarchaeols are much more complex than is indicated in Fig. 1-25. Further studies to identify the key "isoprenyl ether precursor" of both glycolipids and phospholipids are needed to clarify the pathways for glycolipid synthesis in extreme halophiles.

1.4.2.1b. Methanogens. Few studies have been reported on the biosynthesis of archaeol or caldarchaeol glycolipids or phosphoglycolipids in methanogens. Nevertheless, examination of the known structures of methanogen lipids (Fig. 1-10, Table 1-15) suggests possible biosynthetic relationships between archaeol (**31**) and caldarchaeol (**32**) lipids that could be tested experimentally.

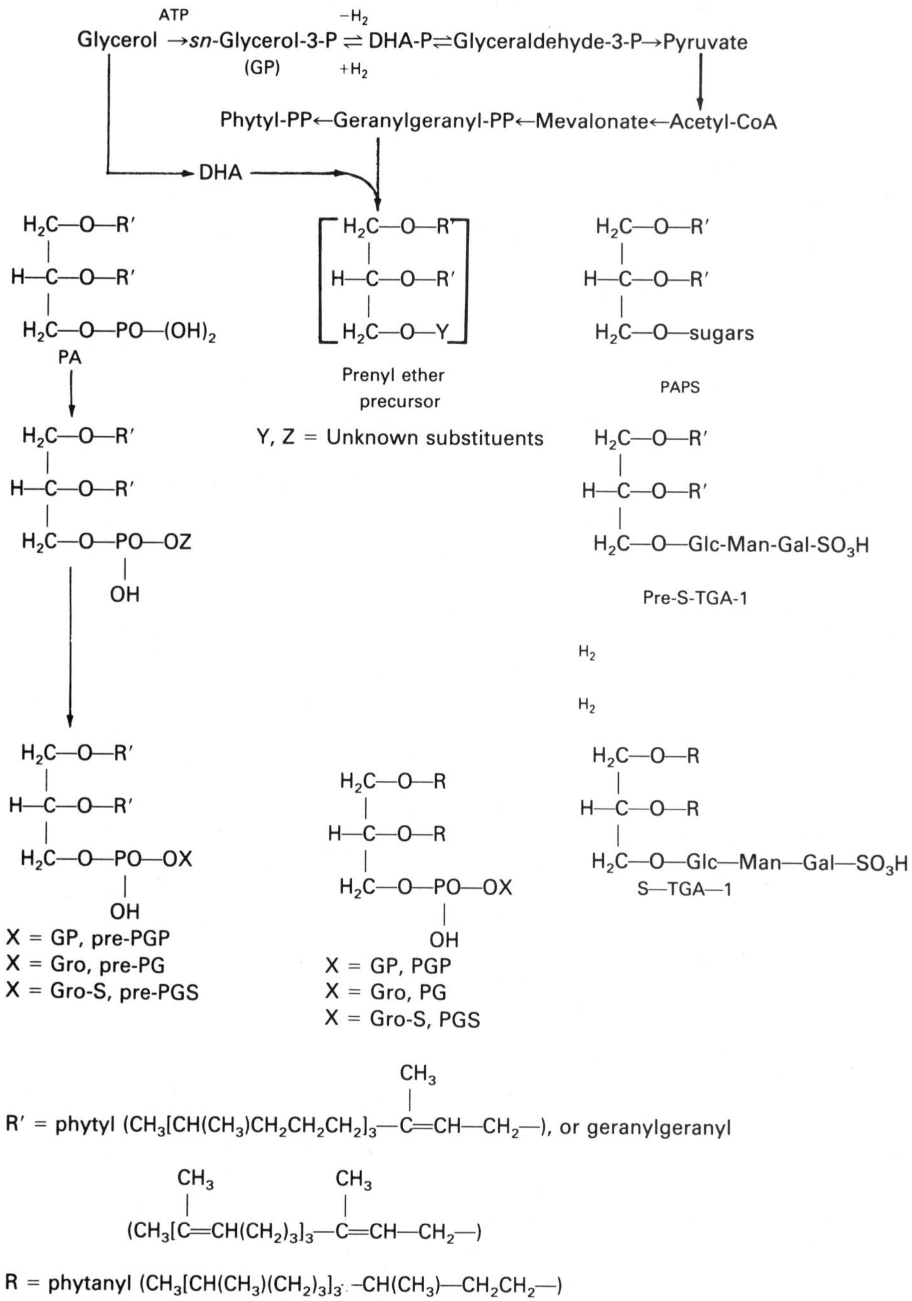

ATP $-H_2$
Glycerol $\rightarrow$ *sn*-Glycerol-3-P $\rightleftharpoons$ DHA-P $\rightleftharpoons$ Glyceraldehyde-3-P $\rightarrow$ Pyruvate
          (GP) $+H_2$

Phytyl-PP $\leftarrow$ Geranylgeranyl-PP $\leftarrow$ Mevalonate $\leftarrow$ Acetyl-CoA

DHA

H_2C—O—R'
|
H—C—O—R'
|
H_2C—O—PO—$(OH)_2$
PA

$\begin{bmatrix} H_2C-O-R' \\ | \\ H-C-O-R' \\ | \\ H_2C-O-Y \end{bmatrix}$

Prenyl ether
precursor

Y, Z = Unknown substituents

H_2C—O—R'
|
H—C—O—R'
|
H_2C—O—sugars

PAPS

H_2C—O—R'
|
H—C—O—R'
|
H_2C—O—PO—OZ
|
OH

H_2C—O—R'
|
H—C—O—R'
|
H_2C—O—Glc-Man-Gal-SO_3H

Pre-S-TGA-1

H_2

H_2

H_2C—O—R'
|
H—C—O—R'
|
H_2C—O—PO—OX
|
OH
X = GP, pre-PGP
X = Gro, pre-PG
X = Gro-S, pre-PGS

H_2C—O—R
|
H—C—O—R
|
H_2C—O—PO—OX
|
OH
X = GP, PGP
X = Gro, PG
X = Gro-S, PGS

H_2C—O—R
|
H—C—O—R
|
H_2C—O—Glc—Man—Gal—SO_3H
S—TGA—1

                                          CH_3
                                           |
R' = phytyl ($CH_3[CH(CH_3)CH_2CH_2CH_2]_3$—C=CH—$CH_2$—), or geranylgeranyl

        CH_3 CH_3
         | |
($CH_3[C$=$CH(CH_2)_3]_3$—C=CH—CH_2—)

R = phytanyl ($CH_3[CH(CH_3)(CH_2)_3]_3$ –$CH(CH_3)$—CH_2CH_2—)

Figure 1-25. Proposed mechanisms for biosynthesis of polar lipids in *Halobacterium cutirubrum*. (From Moldoveanu and Kates, 1988.)

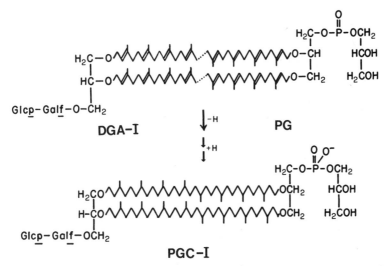

Figure 1-26. Hypothetical mechanism for biosynthesis of caldarchaeol phosphoglycolipids in methanogens, e.g., Glc*p*1–2Gal*f*–caldarchaeol-P-Gro (PGC-I, **43′**) (See Kushwaha *et al.*, 1981). Note that the head-to-head condensations occur with the isoprenyl–ether derivatives of DGA-I and PG (see Poulter *et al.*, 1988).

Langworthy (1977*a*) suggested that caldarchaeol might be formed by head-to-head condensation of two archaeol molecules, which do occur in the free state in methanogens (Langworthy, 1985). It is assumed that the archaeol molecule would be synthesized by the same or similar route as in extreme halophiles (see Fig. 1-25). Glycosyl groups would then be transferred to one of the free hydroxyls of caldarchaeol and a *sn*-glycero-3-phospho group would be attached to the other OH group. Alternatively, Kushwaha *et al.* (1981) proposed, on the basis of the presence in *M. hungatei* of analogous archaeol and caldarchaeol glyco- and phosphoglycolipids, that the biosynthesis of the phosphoglycosylcaldarchaeols (PGC-I (**43′**) and PGC-II (**44′**) might occur by head to head condensation of one molecule of *sn*-3-glycerophosphoarchaeol with DGA-I (**41**) or DGA-II (**42**), respectively (Fig. 1-26). Similarly, biosynthesis of glycocaldarchaeols such as DGC-I (**43**) and DGC-II (**44**) might occur by condensation of archaeol with DGA-I or DGA-II, respectively. This mechanism would be consistent with the presence of archaeol glycolipids in most methanogens (Table 1-15) and would suggest that these glycolipids (or more likely their isoprenyl ether derivatives) are the precursors of the analogous glycocaldarchaeols and phosphoglycocardarchaeols. However, it should be noted that the required *sn*-3-glycerophosphoarchaeol (or its isoprenyl ether derivative) has not been identified in *M. hungatei* (Kushwaha *et al.*, 1981; Ferrante *et al.*, 1987). In recent [32P]phosphate pulse-chase studies with *M. thermoautotrophicum*, a pronounced lag was observed between labeling of the diether and tetraether polar lipids, consistent with the mechanism shown in Fig. 1-26 (Nishihara *et al.*, 1989).

A similar mechanism may also apply to the caldarchaeol lipids of *M. thermoautotrophicum* (Table 1-15, Fig. 1-10) (Koga *et al.*, 1987). In this case, the diglucosylcaldarchaeol-P-ethanolamine (PGC-3, **45′**, Fig. 1-10) could be bio-synthesized by head-to-head condensation of the isoprenyl analogue of the diglucosylarchaeol (DGA-3, **38,** Fig. 1-10) with the isoprenyl analogue of the archaeol-P-ethanolamine idéntified in this methanogen (Nishihara *et al.,* 1987). The alternative mechanism of the condensation of two archaeols to form a caldarchaeol to which the glycosyl and/or phospho groups are added is again also possible but less likely.

However, it should be noted that no direct experimental evidence is available concerning the mechanism of this unique head-to-head condensa-tion, other than that the two carbons involved in the coupling are derived from C-2 of mevalonate (De Rosa *et al.,* 1986a). Also, it is plausible that the head-to-head condensation would involve coupling of terminal carbons on terminal prenyl groups, which would suggest that the intermediates involved in the condensation are isoprenyl ether derivatives perhaps similar to those detected in extreme halophiles (see Fig. 1-26; see also Section 1.4.2.1a). Re-cent labeling studies with *M. hungatei* have eliminated the possibility of cou-pling of saturated archaeol terminals, thus favoring the coupling of terminals in the prenyl analogues (Poulter *et al.,* 1988). Pulse-chase studies with whole cells as well as studies with cell-free systems would help elucidate the biosyn-thetic pathways for caldarchaeol lipids in methanogens.

1.4.2.1c. Thermoacidophiles. Biosynthesis of caldarchaeol and nonitolcald-archaeol lipids in thermoacidophiles has been studied in whole cells in *S. sulfataricus* using U-^{14}C, 1(3)-[^{3}H]glycerol and U-^{14}C, 2-[^{3}H]glycerol as pre-cursors (De Rosa *et al.,* 1982b, 1986a). The biphytanyl moieties of both cald-archaeol and nonitolcaldarcheol were found to undergo complete loss of ^{3}H from 2-[^{3}H]glycerol and 50% loss from 1(3)-[^{3}H]glycerol, as expected for biosynthesis of acetate from glycerol via the glycolytic pathway, and consistent with the findings for phytanyl groups in *H. cutirubrum* (Kates *et al.,* 1970). The glycerol moieties of caldarchaeol and nonitolcaldarchaeol, however, do not undergo loss of hydrogen either from C-1(3) or from C-2, nor does the glycerol of glycerophosphate pools undergo any dehydrogenation. This is in contrast to the situation in extreme halophiles in which the archaeol glycerol moiety loses hydrogen at C-2 but not at C-1(3).

It would appear, then, that the glycerol precursors of caldarchaeol or nonitolcaldarcheol cannot undergo dehydrogenation at C-2, nor aldo–keto or keto–enol isomerizations, thus eliminating the involvement of DHA, DHA-P, and glyceraldehyde-P. It was concluded (De Rosa *et al.,* 1982b, 1986a) that the ether-forming step could occur in thermoacidophiles by direct condensation of geranylgeranyl-PP or similar allylic pyrophosphates with glycerol as accep-tor. However, an activated or derivatized glycerol with the correct chirality at C-2 would perhaps be a more appropriate acceptor.

Presumably, the products of this alkylation reaction would be isoprenyl ether intermediates similar to those detected in the biosynthesis of archaeol lipids in *H. cutirubrum* (Moldoveanu and Kates, 1988; see Fig. 1-25) and in *M. hungatei* (Poulter *et al.,* 1988). However, in thermoacidophiles, these prenyl ether intermediates would probably not be reduced directly but would first participate in head-to-head condensation to form the cardarchaeol polar

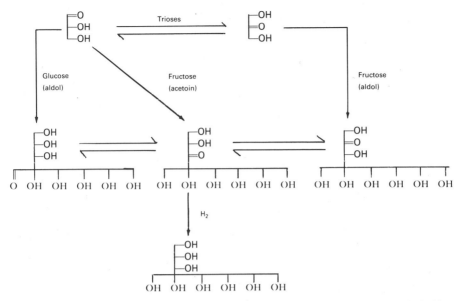

Figure 1-27. Hypothetical mechanism for biosynthesis of the nonitol constituent of nonitolcaldarchaeol. (From De Rosa *et al.*, 1986*a*.)

lipids (see Fig. 1-26); introduction of cyclopentane rings would also occur before final reduction of the chains.

Biosynthesis of the nonitol group in nonitolcaldarchaeol could occur by aldol or acetoin-type condensations between a triose and a hexose precursor, followed by reduction (Fig. 1-27) (De Rosa *et al.*, 1982*b*, 1986*a*). Evidence favouring this mechanism is the complete loss of ^{3}H from 2-[^{3}H]glycerol and 70% retention of ^{3}H from 1(3)-[^{3}H]glycerol at C-1 to C-3 and the incorporation of labeled glucose or fructose in C-4 to C-9 in the nonitol skeleton (De Rosa *et al.*, 1982*b*, 1986*a*) (see Fig. 1-27). Further studies on the biosynthesis of these complex archaebacterial lipids would be of great interest.

1.5. Physical Properties

1.5.1. Glycosyldiacylglycerols of Eubacteria

Relatively few studies have been made on the physical properties of eubacterial glycosylDAGs, largely because of the difficulties of obtaining individual glycolipids with sufficiently homogeneous acyl groups. Recently, synthetic saturated monoglucosylDAGs (Mannock *et al.*, 1987, 1988) and synthetic monoglucosyldialkylglycerols (Hinz *et al.*, 1985) and diglucosylalkylglycerols (Iwamoto *et al.*, 1982) have been made available and their physical properties have been studied. Physical studies have also been done with native glucosylDAGs from *Acholeplasma laidlawii* B, enriched with a single exogenous fatty acid (Silvius *et al.*, 1980; Wieslander *et al.*, 1981). The subject has been reviewed by Quinn and Williams (1983), Boggs (1984, 1987), McElhaney (1984), Rilfors *et al.* (1984), and Curatolo (1987*a*).

Table 1-17. Gel-to-Liquid Crystalline Phase Transition Temperatures for Glycosyldiacylglycerols, Glycosyldiakylglycerol Ethers, and Dipalmitoylphospholipid Standards

Lipid[a,b]	T_m (°C)	ΔH (kJ/mole)	References
GlycosylDAG			
Glc$p\beta$1–3DAG (16 : 0) (2)	61	38	Mannock et al. (1987; 1988)
Gal$p\beta$1–3DAG (18 : 0) (4)	82	67	Quinn and Williams (1983)
Gal$p\alpha$1–6Gal$p\beta$1–3DAG (18 : 0)	51	48	Quinn and Williams (1983)
Glycosyldiethers			
Glc$p\beta$1–3DGE (16 : 0)	64	40	Hinz et al. (1985)
Glc$p\beta$1–3DGE (18 : 0)	72	47	Hinz et al. (1985)
Glc$p\alpha$1–4Glc$p\beta$DGE (16 : 0)	52		Iwamoto et al. (1982)
Glc$p\beta$1–4Glc$p\beta$DGE (16 : 0)	45		Iwamoto et al. (1982)
Phospholipid standards			
Phosphatidylcholine (16 : 0) (DPPC)	41	36	Boggs (1987)
Phosphatidylglycerol (16 : 0) (DPPG)	41	35	Boggs (1987)
Cardiolipin (16 : 0) (di-DPPG)	40		Boggs (1987)
Phosphatidylethanolamine (16 : 0) (DPPE)	63	36	Boggs (1987)
Phosphatidic acid (16 : 0) (DPPA)	65	33	Boggs (1987)

[a]Abbreviations: DAG, 1,2-diacyl-sn-glycerol; DGE, 1,2-dialkyl-sn-glycerol.
[b]Numbers in parentheses refer to acyl or alkyl chain length and number of double bonds.

1.5.1.1. Thermotropic Phase Behavior

Differential scanning calorimetry (DSC) of pure synthetic Glc$p$β1–3-di-palmitoylglycerol (**2**) dispersed in water showed a major transition (T_m 61°C, ΔH 38 kJ/mole) assigned to the lamellar gel to liquid–crystalline phase transition and a minor transition (T_m 75°C, ΔH 6 kJ/mole) assigned to a liquid–crystalline to inverted hexagonal phase transition (Mannock *et al.*, 1987, 1988). More complex behavior was observed for the β-galactosyldistearoylglycerol (**4**) prepared by hydrogenation of chloroplast β-monogalactosylDAGs (Sen *et al.*, 1981; Quinn and Williams, 1983). When compared with the phase transitions of dipalmitoylphospholipids, such as DPPC and DPPE, the transition temperatures of the β-monoglucosyldipalmitoylglycerol (**2**) (Mannock *et al.*, 1987, 1988), the β-monoglucosyldihexadecylglycerol (Hinz *et al.*, 1985) or β-monogalactosyldistearoylglycerol (**4**) (Quinn and Williams, 1983) are much higher than those of DPPC, DPPG, and cardiolipin, but similar to that for DPPE or DPPA (see Table 1-17) (Boggs, 1987). The high transition temperature of monoglycosylDAGs may be attributed to intermolecular hydrogen bonding between the sugar hydroxyls of neighboring molecules (Wieslander *et al.*, 1978; Sen *et al.*, 1981). Synthetic diglycosyldihexadecylglycerols (maltose or cellobiose headgroups) also showed higher T_m values than did the corresponding phosphatidylcholines (Iwamoto *et al.*, 1982), but these were lower than the T_m of monoglycosylDAGs (Table 1-17).

Differential scanning calorimetry studies on acyl homogeneous glycolipids of *A. laidlawii* enriched with isopalmitate showed higher T_m values for both the monoglucosylDAG (25°C) and the diglucosylDAG (35°C) as compared with the total lipids (21°C) or total phospholipids (20.5°C) (Silvius *et al.*, 1980).

1.5.1.2. Lamellar or Hexagonal Phase Formation

Unsaturated monoglycosylDAGs have been shown by X-ray diffraction (XRD) and freeze-etch electron microscopy (EM) to form the hexagonal II phase in aqueous dispersion, while saturated monoglycosylDAGs (e.g., distearoyl forms) form lamellar phases (Quinn and Williams, 1983; Boggs, 1984; Rilfors *et al.*, 1984). By contrast, diglycosylDAGs form lamellar phases in aqueous dispersion, regardless of whether the acyl groups are saturated or unsaturated, and do not enter the hexagonal-II phase even if polyunsaturated acyl groups are present. Formation of hexagonal-II phase by monoglycosylDAGs may be explained by hydrogen bonding interactions between adjacent sugar groups facilitated by the wedge shape of the molecule, while lamellar phase formation by diglycosylDAGs is explained by the larger molecular volume of the hydrocarbon chains in the liquid–crystalline phase, arising from the large head group, being better accommodated in a bilayer and facilitated by the cylindrical shape of the molecule (Boggs, 1984; Rilfors *et al.*, 1984).

In eubacteria containing both mono- and diglycosylDAGs (e.g., *A. laidlawii*), the problem of maintaining a stable bilayer membrane structure over a wide range of environmental conditions must be solved by the organism for its continuing survival. *A. laidlawii* has solved this problem by regulating various parameters, such as the ratio of mono- to diglycosylDAGs, ratio of saturated to unsaturated fatty acids and cholesterol content (Rilfors *et al.*, 1984). Thus, a decrease in growth temperature results in an increase in the mono- to diglycosylDAG ratio and in a decrease in the proportion of charged lipid, which compensates for the increase in proportion of unsaturated fatty acids occuring at lower temperatures. An increase in cholesterol content at a given temperature results in a decrease in mono- to diglycosylDAG ratio, an increase in the proportion of charged lipids, and an increase in unsaturated fatty acids. All these changes have been shown to be correlated with stabilization of the bilayer membrane structure (Rilfors *et al.*, 1984).

1.5.1.3. Orientational Order of Acyl Chains and Conformation of Glycosyl Groups

Acyl chain order in *A. laidlawii* membrane lipids enriched in specifically deuterated or fluorinated fatty acids has been studied by deuterium NMR and ^{19}F-NMR, respectively (McElhany, 1984). It was found that the acyl chain order remained constant from C-2 to C-10, decreasing thereafter toward the chain terminus. This acyl chain order profile is presumed to be characteristic of the glycolipid components, since they are the major lipid components, and has also been found for specifically deuterated DPPC (Curatolo, 1987*a*). Thus acyl chain ordering appears to be similar to that in phospholipids, but further studies on a wider range of glycolipids from different eubacteria should be carried out.

Studies of glycolipid head group conformation by [^{2}H]-NMR of oriented multibilayers of synthetic 1,2-ditetradecyl-3-*O*-β-glycopyranosylglycerol show that the sugar ring is fully extended away from the bilayer surface and that the direction about which motion and order are axially symmetric is the bilayer normal for all the head groups, the glycerol backbone, and hydrophobic core (Jarrel *et al.*, 1987).

1.5.2. Glycosyl Archaeols or Caldarchaeols of Archaebacteria

1.5.2.1. Thermotropic Phase Behavior

Very few physical studies have been made on the glycosylarchaeols or caldarchaeols, or indeed on any of the archaebacterial lipids. However, DSC on the lipids of the extreme halophile, *H. cutirubrum*, has shown only broad and weak endothermic transitions centered at −49 or −45°C for 1 : 1 (w/w) dispersions in water of total polar lipids, the minor phospholipid PG, or the sulfated triglycosylarchaeol (S-TGA-1; **34′**), respectively, and none was detected in the range −80° to +80°C for the major phospholipid PGP or the minor phospholipid PGS (Chen *et al.*, 1974). Also, no lipid phase change was detected in the purple membrane of *H. cutirubrum* or in PGP aqueous dispersions by [^{31}P]-NMR over the temperature range 5–60°C (Ekiel *et al.*, 1981). However, aqueous dispersions of total polar lipids or pure PGP showed a polar head group transition at +8°C which was shifted to +23°C in presence of 3 M NaCl, as measured by spin labels (Plachy *et al.*, 1974). Phase transitions in the range 5–60°C associated with changes in crystal lattice structure of lipid and protein have been detected for intact membranes of *Halobacteria* by spin-label, calorimetric, and [^{13}C]-NMR techniques (Chignell and Chignell, 1975; Jackson and Sturtevant, 1978; Hiraki *et al.*, 1981; Degani *et al.*, 1980).

Differential scanning calorimetry studies of aqueous dispersions of cald-archaeol glycolipids and phosphoglycolipids from *Thermoplasma acidophilium* grown at 50°C and pH 2 showed the presence of a small thermal transition (ΔH 5–9 J/g) at ~ 14–18°C for the total lipids, the phosphoglycolipid fraction (nine components), the main phosphoglycolipid (**47'**), and the glycolipid fraction (10 components, including **47**); the glycolipids also exhibit a metastable polymorphism between −30 and −60°C (Blöcher *et al.*, 1984). In intact membranes of *T. acidophilum*, thermal transitions were observed at around 15 and 40°C (Yang and Haug, 1979), which may be attributable to lipid–protein interaction.

Studies on hydrated free caldarchaeol (**32**) and nonitolcaldarchaeols (**32'**, **32a–d**) derived from lipids of *S. sulfataricus* showed a single broad transition at 30°C (ΔH 20, J/g) for the symmetrical caldarchaeols (**32**) and a very broad multipeaked transition in the range 0–60°C (ΔH, 4 J/g) for the various un-symmetrical nonitolcaldarchaeols (**32'**, **32a–d**) (Gliozzi *et al.*, 1983; De Rosa *et al.*, 1986a). The low enthalphies of the caldarchaeol or nonitolcaldarchaeol transitions are consistent with the fact that both ends of the hydrocarbon chains are constrained by covalent bonding to the two glycerols or to the glycerol and nonitol groups, respectively. Further constraint is offered by the presence of cyclopentane groups, the transition temperatures being directly correlated with the number of such rings in the chains (Gliozzi *et al.*, 1983).

Thus, the thermal behavior of both archaeol and caldarchaeol glycolipids is dominated by the highly branched apolar chains (phytanyl or biphytanyl), which prevent the formation of an ordered gel state except at very low temperatures. This is in marked contrast to the thermal behavior of acyl-linked glycolipids from eubacteria and plants, which is dominated by the glycosyl head groups (see Section 1.5.1.1). It may be concluded that the archaeol or caldarchaeol glycolipids are most likely in a liquid–crystalline state at all temperatures above 0°C, and certainly at the ambient temperatures at which the archaebacteria grow.

1.5.2.2. Lamellar, Hexagonal, and Cubic Phase Formation

Archaeol Lipids. Early physical studies by Chen *et al.* (1974) on the lipids of the extreme halophile *H. cutirubrum* demonstrated the formation of stable multibilayer birefringent liposomes in aqueous dispersions of total polar lipids, as seen by negative-staining electron microscopy, and which acted as ideal osmometers in 5–200 mM NaCl or KCl. At higher concentrations (0.2–4 M), the liposomes shrank to spherical particles that (1) no longer were strongly birefringent, (2) did not show distinct bilayer structure by electron microscopy and (3) no longer behaved as ideal osmometers. This behavior of the liposomes at the high salt concentrations (4–5 M) necessary for growth of the organism could be due to charge shielding, leading to a decrease in the surface area of the polar head groups relative to the phytanyl chains and resulting in tight packing of the lipid chains not consistent with ideal osmometry. To

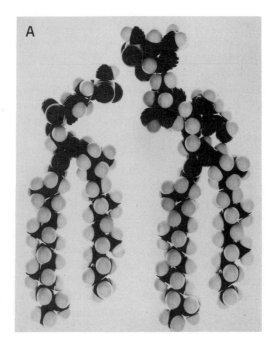

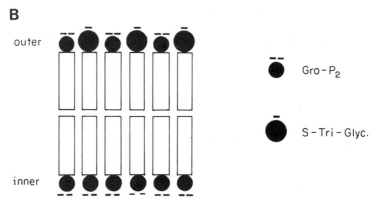

Figure 1-28. (A) CPK model of PGP (left) and S-TGA-1 (right) showing packing of their polar head groups. (B) Hypothetical model of arrangement of PGP and S-TGA-1 in the purple membrane of *Halobacteria*. (From Kates, unpublished data.)

explain further the stability of the membranes of *H. cutirubrum* in 4 M salt, it was suggested that direct interaction of the membrane proteins with the polar lipid components, both hydrophobically with the phytanyl chains and hydrophilically with the polar head groups, was required (Chen *et al.*, 1974).

In regard to the individual polar lipid components, only the sulfated triglycosylarchaeol (S-TGA-1; **34′**), alone or in mixtures with the major phospholipid PGP, formed stable multibilayer liposomes showing ideal osmometry in 5–200 mM KCl or NaCl. These findings suggested that the polar head groups of the phospholipids are too small relative to the effective cross-sectional area of the phytanyl chains for stable bilayer formation and that the S-TGA with its relatively large polar head group may be needed for proper packing into a stable bilayer structure (Chen *et al.*, 1974) (see Fig. 1-28A,B). Recent studies (Stewart, 1988; R. Bittman, L. C. Stewart, and M. Kates, unpublished results) demonstrated marked decreases in urea permeability of PGP liposomes when mixed with S-TGA-1, thus confirming the stabilizing effect of S-TGA-1.

Subsequent freeze fracture studies (Quinn *et al.*, 1986) have shown that the S-TGA or PGP dispersed in water form lamellar structures that are highly hydrated, while dispersion in 1 M NaCl causes the bilayers to stack more tightly (Fig. 1-29A,B); in 5 M NaCl, both lamellar and nonlamellar structures are formed at 20°C. Mixtures of these two lipid components dispersed in 5 M NaCl in mole ratios of 1 : 2, 2 : 1, or 3.5 : 1 showed that phase separation takes place, the sulfated glycolipid tending to form bilayers at 20°C while PGP preferentially formed nonlamellar structures including cubic as well as hexagonal-II phases. Similar behavior was shown by dispersions of total polar lipids in 5 M NaCl. With increasing PGP : S-TGA mole ratio, an increase in the proportion of nonlamellar phase at the expense of the lamellar phase was observed. Thermotropic studies in the range −30 to +70°C showed that bilayer structures were formed at low temperatures in both pure lipids and in mixtures thereof, whereas nonlamellar inverted cubic phases were preferentially formed at high temperatures. These observations are consistent with the idea that nonbilayer structures of the lipids are needed to package the intrinsic membrane proteins into a lipid bilayer matrix (Quinn *et al.*, 1986). Recent XRD studies (Quinn *et al.*, unpublished results) have confirmed the formation of lamellar and nonlamellar (cubic and hexagonal-II) phases in aqueous dispersions of polar lipids of *H. cutirubrum*.

In regard to lipid organization in the membranes of extreme halophiles, XRD studies of the purple membrane (PM) (Blaurock and Stoeckenius, 1971) have confirmed that the lipids are in a bilayer structure; no such studies have been made, however, on the red membrane (RM) of the halophiles. The question then arises as to the asymmetric distribution of the phospholipids and glycolipids between the inner and outer layers of the bilayer. Analytical studies have shown that the major glycolipid S-TGA-1 (**34′**) occurs only in the PM (Table 1-18), in mole ratio PGP:S-TGA-1, 2.2 : 1 (Kushwaha *et al.*, 1975; Kates *et al.*, 1982). Electron microscopy of ferritin/avidin/biotin-labeled PM has shown that S-TGA-1 is located entirely in the outer layer of the PM (Henderson *et al.*, 1978), consistent with an asymmetric structure of the PM as revealed by XRD (Blaurock and King, 1977; Blaurock, 1982). It follows that the mole ratio PGP : S-TGA-1 in the outer layer should then be close to 1 : 1 (Table 1-18). A model showing the packing of PGP and S-TGA-1 as it might occur in the outer layer of the PM is shown in Fig. 1-28B.

Caldarchaeol and Nonitolcaldarchaeol Lipids. The unusual dimeric structures of the caldarchaeol and nonitolcalarcheol core lipids (**32, 32′**) would suggest that the methanogen or thermoacidophile membranes consist of monolayers of glyco- and phosphoglycolipids (**43,43′,44,44′** in methanogens, **47,47′**

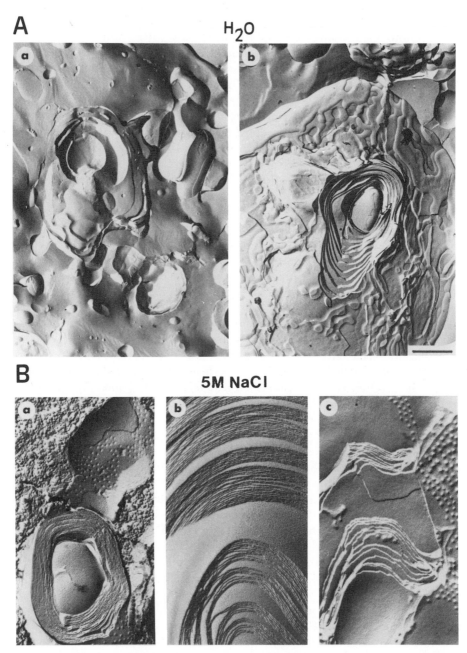

Figure 1-29. Freeze-fracture electron microscopy of aqueous dispersions of polar lipids of *H. cutirubrum*: (A) PGP (a) and S-TGA-1 (b) in water; bar represents 500 nm. (B) PGP (a), S-TGA-1 (b), and PGP/S-TGA-1 (2 : 1) (c) in 5 M NaCl; bar represents 200 nm. (From Quinn *et al.*, 1986.)

Table 1-18. Lipid Composition of H. cutirubrum Membrane Fractions

Lipid component	Whole cells[a]		Purple membrane (PM)[b]		Red membrane (RM)[c]	
	Total lipids (mole %)	Polar lipids (mole %)	Total lipids (mole %)	Polar lipids (mole %)	Total lipids (mole %)	Polar lipids (mole %)
Neutral lipids						
Squalenes	9	—	10	—	12	—
Vit MK-8	1	—	2	—	2	—
Retinal	0.3	—	8	—	0	—
C50 bacterioruberins	2	—	0	—	0.5	—
Polar lipids						
PG	4	5	5	6	3	3
PGP	58	70	49	61	56	62
PGS	4	4	5	5	—	—
GLS	17	21	7 ⎫ 22	⎫ 28	trace	—
TGD	trace	trace	15 ⎭	⎭	trace	—
$(X_1 + X_2)$[d]	—	—	—	—	+	+

[a]From Kates (1978); Kates *et al.* (1982).
[b]From Kushwaha *et al.* (1975).
[c]Discrepancies among whole cell, PM, and RM lipid compositions probably arise from different growth conditions for cells grown under normal conditions and those grown specifically for the production of PM (Kushwaha *et al.* 1975).
[d]Unidentified glycolipids (32-35% by wt.).

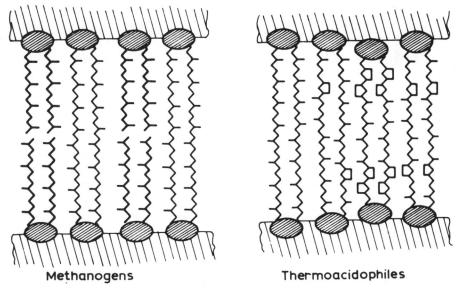

Methanogens   **Thermoacidophiles**

Figure 1-30. Hypothetical arrangement of caldarchaeol lipids as monolayers in methanogens and thermoacidophiles. (From Ishizuka and Yamakawa, 1985; De Rosa *et al.*, 1983a.)

48,48′,50,50′ in thermoacidophiles), which would span the entire thickness of the membrane (Fig. 1-30) (see Ishizuka and Yamakawa, 1985; Langworthy, 1985; De Rosa *et al.*, 1986*a*). Evidence for such a monolayer structure in membranes of *Thermoplasma* and *Sulfolubus* and some methanogens, which are composed entirely of C_{40} biphytanyl chains, is as follows:

1. The dimensions of the caldarchaeol and nonitolcaldaracheol lipids are comparable to the thickness of the methanogen or thermoacidophile membranes (Langworthy 1977*a;* Ourisson and Rhomer, 1982).
2. Methanogen or thermoacidophile membranes do not freeze-fracture tangentially (as do membranes of extreme halophiles and eubacteria) but cross-fracture perpendicularly through the plane of the membrane (Langworthy, 1982*a*, 1985).
3. Electron-spin resonance (ESR) studies show that the *Thermoplasma* membrane is extremely rigid (Smith *et al.*, 1974).
4. Current–voltage measurements on black lipid films of nonitolcaldarchaeol (the symmetrical caldarchaeol does not form stable black films) are consistent with a monolayer structure (De Rosa *et al.*, 1986*a*).
5. XRD studies of dispersions of total lipids and total polar lipids of *S. solfataricus* show the formation of lamellar phases together with hexagonal-II and cubic phases, depending on the temperature and water content (Gulik *et al.*, 1985).

The foregoing evidence thus strongly supports the concept of a monolayer of caldarchaeol and nonitolcaldarchaeol glyco- and phosphoglycolipids serving as the basis of the membrane structure in thermoacidophiles and some methanogens. The question then arises as to whether all the glycosyl groups are oriented on one side of the membrane, and if so, which side, the external or the cytoplasmic? Experiments by De Rosa *et al.* (1983*b*) in which intact cells of *S. solfataricus* were treated with a nonpenetrating mixture of glycosidases from a marine gastropod showed that 82% of the total lipids were hydrolyzed, compared with an untreated control in which 92% of the total lipids were glycosylated on one of the polar head groups. These results are consistent with the hypothesis that the glycosyl groups are located essentially on the exterior surface of the membranes, as was found for the purple membrane of *H. halobium* (Henderson *et al.*, 1978). Presumably, a similar orientation of sugar groups may also occur in the membranes of methanogens containing glycosyl or phosphoglycosylcaldarchaeols, but experiments bearing directly on this point have yet to be carried out. It would follow that the anionic groups (phosphates) would be largely on the interior surface of the membranes of thermoacidophiles and methanogens, in contrast to the situation in membranes of extreme halophiles (see Fig. 1-28B).

1.5.2.3. Conformational Order of the Phytanyl and
*   Biphytanyl Chains and Carbohydrate Groups*

Early ESR studies with 5-doxylstearic acid spin labels indicated that the lipids in the PM of *H. halobium* were relatively highly ordered and tightly bound to the protein (Chignell and Chignell, 1975). A more detailed ESR study of cell envelope vesicles of *H. cutirubrum* using a series of stearic acid spin labels showed that the first 10 carbon atoms in the phytanyl chains of the lipids were highly ordered, but the order decreased rapidly to low values for the terminal carbon atoms (Esser and Lanyi, 1973). For the total lipid dispersions in 3.4 M NaCl, the chain order for the first few carbons was high but less than that for the membrane vesicles, then decreased linearly with carbon number for the first 10 carbons and then decreased markedly for the last carbons. In a later ESR study (Plachy *et al.*, 1974), the results for dispersions of total and individual polar lipids were interpreted in terms of *trans–gauche* cooperative kinking of the phytanyl chains, preferentially at the methyl-branched carbons, resulting in clusters of lipid chains with single kinks at the same positions. More recent deuterium NMR studies on synthetic deuterated diphytanyl ether analogues of PA and PG (Stewart, 1988; Stewart *et al.*, 1990*a*) have shown that the presence of the branched methyl groups causes the phytanyl chains to be much less ordered than normal acyl chains, with respect to segmental dynamics.

Electron-spin resonance studies of the membrane of *T. acidophilum* containing caldarchaeol glyco- and phosphoglycolipids (Table 1-16) have shown that this is the most rigid membrane known so far (Smith *et al.*, 1974). However, ESR spin-label studies of aqueous dispersions of total caldarchaeol and nonotolcaldarchaeol lipids from *S. solfataricus* have indicated that the correlation times of asymmetric nonitolcaldarchaeol lipids (**50,50′,50″**; Fig. 1-11) are 1000-fold greater than those for caldarchaeol lipids (**48,48′**) (Bruno *et al.*, 1986). These studies have also shown that at the growth temperature of this bacterium, the fluidity of the central portion of the hydrocarbon core is similar to that of an egg lecithin bilayer.

The role of the cyclopentane rings in the conformation of the caldarchaeol and nonitolcaldarcheol lipid cores has not been extensively studied, but presumably, these rings can serve to control the rigidity or fluidity of the archaebacterial (thermoacidophile) membrane in response to the environmental temperature in two ways (see Langworthy, 1985): (1) by reducing the rotational freedom and hence the fluidity within the biphytanyl chains; and (2) by reducing the effective biphytanyl chain length (see structures **32a–d**, Fig. 1-8 and Fig. 1-30, hence the thickness of the membrane. Thus, the rigidity or order of the hydrophobic region of the thermoacidophile membrane would be directly proportional to the number of cyclopentane rings in the biphytanyl chains, as has been shown by DSC studies of caldarchaeol in the dry state (Gliozzi *et al.*, 1983). This conclusion would be consistent with the observed increase in cyclopentane rings with increasing growth temperature of *Sulfolobus* (De Rosa *et al.*, 1980*b*) and *Thermoplasma* (Langworthy *et al.*, 1982).

In regard to the conformation of the glycosyl head groups in archaebacterial glycolipids, few studies have so far been carried out on this point, but presumably the sugar groups in the monoglycosylarachaeols would be fully extended away from the bilayer surface, as was shown for the mono-glucosylditetradecylglycerol (Jarrell *et al.*, 1987). Recent monolayer studies of the sulfated triglycosylarchaeol (S-TGA-1,**34′**) at the air–water interface, however, have shown that the polar head group ($3\text{-}SO_3^-$-Gal-Man-Glc) penetrates deeply into the aqueous subphase when S-TGA is spread at low electrolyte concentrations (< 0.01 M), but at high electrolyte concentrations (≥ 1 M) the polar head group is oriented into the interface presumably due to a salting out effect (Tomoaia-Cotisel *et al.*, 1988).

1.6. Membrane Function of Glycolipids

The presence in both eubacterial and archaebacterial membranes of an incredibly varied array of glycolipids raises the question as to their function in the cell membrane. In the following section, this question is discussed separately for the eubacteria and the archaebacteria.

1.6.1. Eubacteria

While glycosylDAGs are relatively minor lipid components in most eubacterial membranes, certain species, such as *Micrococcus lysodeikticus*, *Pneumococcus* species, *Arthrobacter*, *Treponema*, *Microbacteria*, *Mycoplasma*, and *Acholeplasma* spp., contain glycosylDAGs as major components of their membrane lipids (see Table 1-7). Relatively few studies have been made on the role of glycosylDAGs in bacterial membranes, with the exception of *Acholeplasma* species. In the latter organisms, e.g., *Acholeplasma laidlawii*, such studies are made possible by the fact that these wall-less organisms can incorporate fatty acids and cholesterol supplied in the medium into their plasma membrane (see reviews by Razin, 1978; Rottem, 1980; and McElhaney, 1984; Curatolo, 1987*b*).

Thus, it is possible to study directly the effect of changes in fatty acid composition of the membrane lipids on glycolipid composition and functioning of the membranes. Studies by Wieslander and colleagues (Wieslander *et al.*, 1978; Rilfors *et al.*, 1984) have shown that *A. laidlawii* cells enriched in oleate to the extent of 95% exhibited a marked decrease in the ratio of mono- to diglucosylDAG. Examination of model membranes containing mono- and diglucosylDAGs showed that unsaturated monoglucosylDAGs are more disruptive of lamellar phase structure than are saturated monoglucosylDAGs. Thus, these workers concluded that the cells overcome the disruptive effect of the wedge-shaped monoglucosylDAG (which tends to form hexogonal-II phases) by decreasing the mono- to diglucosylDAG ratio. Similarly, increased incorporation of cholesterol also decreased the mono- to diglucosylDAG ratio presumably to counter the tendency of cholesterol to form destabilizing hexagonal-II phases.

Nevertheless, since localized nonbilayer membrane structures may have some cellular function, the cells appear to maintain a balance of bilayer and nonbilayer membrane structure under a variety of environmental conditions, such as temperature and fatty acid composition and cholesterol content available in the growth medium, by variation of the mono- to diglucosylDAG ratio. However, as pointed out by McElhaney (1984), this ratio appears to be unaffected in *A. laidlawii* strain B by any of the above variations in environmental conditions. Thus, it must be concluded that variation in the ratio of mono- to diglycosylDAGs may not have a general bearing on the function of glycolipids in membrane stabilization. However, glycolipids may have an extra membrane-stabilizing effect in *Mycoplasma* (which lack a cell wall) due to interlipid hydrogen bonding of the glycosyl head groups (Boggs, 1987), as supported by the studies of Silvius *et al.* (1980) (see Section 1.5.1.1). Consistent with this proposed function of glycolipids is the finding that wall-less L-forms of gram-positive bacteria have high glycolipid contents, comparable to those of *Mycoplasma,* but much greater than those of their gram-positive precursors (Shaw, 1970).

The question then arises concerning the distribution of the glycolipids between the inside and the outside layers of the *Mycoplasma* membrane. Localization of carbohydrate groups of glycolipids or phosphoglycolipids has been investigated by immunological techniques, agglutination with lectins, and electron microscopic visualization of concanavalin A (Con A)–surface carbohydrate complexes, all of which have indicated the presence of sugar groups at the external surfaces of several species of *Mycoplasma* (see reviews by Rottem, 1980, 1982). This conclusion was supported by the finding that, in all *Mycoplasma and Acholeplasma* species tested, the amount of labeled lectins bound to intact cells was almost the same as that bound to isolated membranes (Kahane and Tully, 1976).

McElhaney (1984) criticized such a distribution of glycolipids all on the outer surface, since more than two thirds of the membrane lipids would be in the outer layer and the inner layer would consist almost entirely of anionic phospholipids and phosphoglucolipids, which would make the inner layer very unstable due to charge repulsion. However, the latter objection might be overcome by postulating negative charge shielding by protonated amino groups in the membrane proteins. Further studies of the transbilayer lipid asymmetry in *Mycoplasma* and other bacteria are needed to settle this point.

Another function served by glycosylDAGs in some eubacteria is a metabolic one, in which these glycolipids may act as precursors of phosphoglycolipids and these in turn as precursors of lipoteichoic acids. For further discussion and details of this subject, see Chapter 2 (*this volume*).

1.6.2. Archaebacteria

The fact that all archaebacterial membranes contain lipids derived from diphytanylglycerol diether (**31**) or its dimer dibiphytanyldiglycerol tetraether (**32**) suggests that these "peculiar" lipids have functions common to all archaebacteria (Langworthy, 1985; Kamekura and Kates, 1988). Indeed, in contrast to the usual acyl ester structure, the alkyl ether structure would impart stability to the lipids over the wide range of pH values encountered by these bacteria, e.g., high pH for haloalkaliphiles and some extreme halophiles, and low pH for the thermoacidophiles. Furthermore, the saturated alkyl chains would impart stability toward oxidative degradation, particularly in the case of the extreme halophiles exposed to air and sunlight; and the unnatural *sn*-1 configuration of the backbone glycerol would impart resistance to attack by phospholipases released by other organisms, providing a survival value for

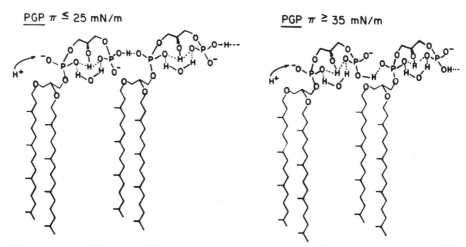

$\underline{PGP} \ \pi \leq 25 \ mN/m$

$\underline{PGP} \ \pi \geq 35 \ mN/m$

Figure 1-31. Hypothetical model of a proton conductance pathway in the red membrane outer layer involving the polar headgroup of phosphatidylglycerolphosphate (PGP). The model shows part of a monolayer of PGP at a water–air interface at low (<25 mN/m) and high (>35 mN/m) surface pressure. (From Teissié *et al.*, 1989.)

the extreme halophiles. The branched isopranyl structure of the archaeol and caldarchaeol would ensure that the membrane lipids are in the liquid crystalline state at all ambient temperatures, and the covalent linking of the terminal ends of the phytanyl chains to form the macrocyclic dimeric structure of caldarchaeol together with the introduction of pentacyclic rings would prevent the lipid monolayers from becoming too fluid as the ambient temperature increased. More specific functions of archaebacterial lipids are considered below for each of the archaebacterial groups.

1.6.2.1. Extreme Halophiles

Glycosylarchaeols, particularly the sulfated derivatives, such as S-TGA-1 (**34′**) in *Halobactaeria* species and S-DGA (**33′**) in *Haloferax* species, are major membrane lipid components in all extremely halophilic bacteria, except for the haloalkaliphiles (Table 1-13). The question concerning the membrane function of these glycolipids has not yet been answered unambiguously on the basis of direct experimentation. However, some interesting suggested functions deserve attention.

The sulfated triglycosylarchaeol (S-TGA-1, **34′**) is known to be associated exclusively with the purple membrane of *Halobacteria* (Kushwaha *et al.*, 1975; Kates *et al.*, 1982), in which it is located entirely in the exterior surface layer of the membrane (Henderson *et al.*, 1978). It is of interest that extreme halophiles that lack the purple membrane, such as *Haloarcula*, also lack the S-TGA or any other sulfated glycolipid and have instead a nonsulfated glycosylarchaeol, TGA-2 (**35**). Apart from the bilayer-stabilizing function of S-TGA-1 (Fig. 1-28A) (see Section 1.5.2.2), its presence in the outer surface of the PM would suggest that it might participate together with the major phospholipid PGP in proton conductance pathways Haines (1983) demonstrated in model systems (Teissie *et al.*, 1985, 1990; Quinn *et al.*, 1989). Such pathways, involving the polar head sulfate group of S-TGA and phosphate groups of PGP, would serve to transport the protons transduced by light-activated bacteriorhodopsin across the outer surface of the PM to the red membrane where the PGP head group phosphates (and possibly also the PG phosphate) would conduct the protons to the sites of the H^+–ATPases in the red membrane, where they can drive ATP synthesis and also participate in Na^+ extrusion through the H^+/Na^+ antiport system (Falk *et al.*, 1980; Lanyi, 1979). A hypothetical model of a proton conductance pathway involving the polar head group of PGP in the red membrane is shown in Fig. 1-31. The involvement of the free OH group of the glycerol head group in the hydrogen-bonding network is particularly noteworthy; this has recently been demonstrated by potentiometric titration and other studies of PGP and its deoxy analogue dPGP (Stewart *et al.*, 1988, 1990*b*; Teissie *et al.*, 1990).

Involvement of S-TGA-1 in K^+ transport in *Halobacteria* has been suggested (Falk *et al.*, 1980): this sulfated glycolipid would function as a selective K^+ receptor in the K^+ uniport (Wagner *et al.*, 1978) for transport of K^+ from the external high-Na^+ and low-K^+ medium into the cytoplasm against a high K^+ gradient. In this connection, it may be hypothesized that in *Haloferax* species the sulfated diglycosylarchaeol (S-DGA-1 **33'**), which replaces S-TGA-1, might play a similar role in K^+ transport. Furthermore, in the *Haloarcula*, which are devoid of sulfated glycolipids, such a role might be undertaken by PGS, which is a major polar lipid component in this genus (Table 1-14).

Another point that should be noted is that the number of negative charges per mole anionic lipid in the membranes of *Halobacteria, Haloarcula*, and *Haloferax* is maintained at ~ 1.5–2.0, despite differences in polar lipid composition (Table 1-14). Such a high negative charge surface density would most likely be shielded by the high Na^+ ion concentration (4 M), preventing disruption of the lipid bilayers due to charge repulsion. A highly negatively charged membrane surface would thus appear to be required for the survival of extreme halophiles in media of high salt concentration.

The sulfated triglycosylarchaeol may have another function in the PM, related to the proton pumping action of bacteriorhodopsin (BR), since it has been demonstrated that reconstitution of BR in vesicles containing S-TGA-1 results in increased rates of proton pumping (Lind *et al.*, 1982). Similar functions might also be envisaged for the sulfated diglycosylarchaeol (S-DGA, **33'**) in *Haloferax* sp., which do contain the PM; however, no experimental evidence is available concerning the function of this glycolipid.

Nothing is known in regard to the functions of the analogous triglycosylarchaeol (TGA-2, **35**) or the unidentified diglycosylarchaeol (DGA-2) in *Haloarcula*, the sulfated tetraglycosylarchaeol (S-TeGA,**36'**) in *Halobacteria*, or the diglycosylarchaeol (DGA-1, **33**) in *Haloferax*. Clearly, further studies relating to the possible functions of these glycolipids would be of great interest.

1.6.2.2. Methanogens and Thermoacidophiles

The membranes of caldarchaeol-containing methanogens and thermoacidophiles consist essentially of a bipolar monolayer structure (Fig. 1-30) (see Section 1.5.2). Such a structure would impart stability to the membrane at the high growth temperatures of thermophilic methanogens and of thermoacidophiles. In the methanogens, possession of caldarchaeol-based membrane lipids would also be an advantage under conditions of high methane concentration that must be present in these cells and might lead to membrane disruption. However, it should be noted that some species of methanogens, e.g., *M. voltae* manage to survive with only the archaeol-type lipids in their membranes.

Another methanogen, *Methanococcus jannaschii*, isolated from a deep-sea hydrothermal vent at temperatures of 85°C or greater, has membrane lipids based on *cyc*-archaeol (**31a**), which was not found in any other species of methanogens surveyed (including *Methanococcales, Methanobacteriales,* and *Methanomicrobiales* and three thermophilic methanogens) and may be unique to methanogens from deep-sea hydrothermal vents (Comita *et al.,* 1984). Thus, it is possible that the presence of *cyc*-archaeol-based lipids may be related to the high pressures under which these deep-sea methanogens live. Further studies on the detailed structures of the polar lipids of these organisms will be of great interest.

The high proportions of glycosylated caldarchaeols present in membranes of both methanogens and thermoacidophiles may further stabilize the membrane structure by interglycosyl head group hydrogen bonding. In addition, the presence of cyclopentane rings in the biphytanyl chains of caldarchaeol may serve to fine-tune the rigidity of the membrane monolayer in direct response to the growth temperature of the thermoacidophile (see Section 1.5.2.3). By contrast, there does not appear to be any specific structural feature of the polar lipids of thermoacidophiles that would be related to the low pH of their growth environment; e.g., no amino or other basic groups are present that might serve to neutralize the high H^+ concentration. However, further study of the lipids of thermoacidophiles growing over a range of pH values may provide new insights concerning this point.

Another puzzling point concerns the asymmetrical orientation at the exterior membrane surface of the glycosyl groups in the caldarchaeol lipids of thermoacidophiles and perhaps of methanogens as well. Of necessity, the anionic groups (phosphates in phosphoglycocaldarchaeols) would all have to be oriented in the inner membrane layer, placing a high negatively charged surface density on one side of the membrane. It is difficult to understand how the membrane would remain stable under such an arrangement, unless the negative charges on the interior side were neutralized or shielded by protonated amino groups in the membrane proteins. Clearly, further studies in this area are needed to answer this question.

1.7. Summary and Conclusions

The foregoing review of the structure, physical properties, and function of membrane lipids of eubacteria and archaebacteria has revealed a remarkable variety of polar lipid classes, particularly glycosylDAGs, in these organisms. The wide diversity of glycosyl groups in eubacteria appears to be genetically controlled on the level of the bacterial genus, but further data are need to corroborate this correlation. In archaebacteria, the polar lipids, including phospholipids, glycolipids, phosphoglycolipids, and sulfolipids, are all derived from the one basic core structure, diphytanylglycerol (**31**). A much clearer genetic control, on the genus level, of the glycosyl structures is evident, particularly in the extreme halophiles.

Even with the relatively limited knowledge we have of the physical properties of glycolipids in eubacteria and archaebacteria, it is clear that they serve an important role in stabilizing the membrane structure under a variety of environmental conditions (e.g., temperature, salt concentration). The archaebacterial lipids are particularly well adapted as membrane components to the extreme environmental conditions of the three groups of archaebacteria, extreme halophiles, methanogens, and thermoacidophiles.

However, much remains to be learned concerning the precise asymmetrical arrangement of the lipids in the membrane bilayers or monolayers, as well as the function of this membrane lipid asymmetry with respect to ion transport, permeability to nutrients, proton transport and conductance, and energy transduction. An understanding of these functions will require knowledge concerning the interaction of the lipids with the membrane proteins. In the light of such knowledge, perhaps the unusual archaebacterial lipids will not appear so strange, and our knowledge of them will help us understand the function of the more familiar lipids in eubacteria and eukaryotes.

ACKNOWLEDGMENTS. Support by the Natural Sciences and Engineering Research Council of Canada and the Medical Research Council of Canada is acknowledged.

References

Ambron, R. T., and Pieringer, R. A., 1971, The metabolism of glyceride glycolipids. V. Identification of the membrane lipid formed from diglucosyl diglyceride in *Streptococcus faecalis* ATCC 9790 as an acylated derivative of glycerylphosphoryl diglucosylglycerol, *J. Biol. Chem.* **246:**4216.

Ambron, R. T., and Pieringer, R. A., 1973, in *Form and Function of Phospholipids* (B. B. Ansell, R. M. C. Dawson, and J. N. Hawthorne, eds.), pp. 289–331, BBA Library 3, Elsevier, Amsterdam.

Asselineau, J., 1981, Complex lipids of the cell envelope of Actinomycetales, in *Actinomycetes* (Schaal and Pulverer, eds.), Zentralblatt Bakteriol. Mikrobiol. Hyg. Abt. 1, Supplement 11, pp. 391–400, Gustav Fischer Verlag, Stuttgart.

Asselineau, J., 1983, Vers l'utilisation des lipides pour l'identification d'une souche bactérienne, *Bull. Inst. Pasteur* **81:**367.

Asselineau, C., and Asselineau, J., 1984, Fatty acids and complex lipids, in *Gas Chromatography/Mass Spectrometry: Applications in Microbiology* (G. Odham, L. Larsson, and P.-A. Mardh, eds.), pp. 57–103, Plenum, New York.

Asselineau, J., and Trüper, H. G., 1982, Lipid composition of six species of the phototrophic bacterial genus Ectothiorhodospira, *Biochim. Biophys. Acta* **712:**111.

Baer, E., and Fischer, H. O. L., 1945, Synthesis of a homologous series of optically active normal aliphatic monoglycerides (L-series), *J. Am. Chem. Soc.* **67:**2031.

Basinger, G. W., and Oliver, J. D., 1979, Inhibition of *Halobacterium cutirubrum* lipid biosynthesis by bacitracin, *J. Gen. Microbiol.* **111:**423.

Batrakov, S. G., and Bergelson, L. D., 1978, Lipids of *Streptomyces*. Structural investigation and biological interrelation, *Chem. Phys. Lipids* **21**:1.

Benson, A. A., 1963, *The plant sulfolipid, Adv. Lipid Res.* **1**:387.

Bergelson, L. D., Batrakov, S. G., and Pilipenko, T. V., 1970, A new glycolipid from *Streptomyces, Chem. Phys. Lipids,* **4**:181.

Bergy, M. G., 1974, *Manual of Determinative Bacteriology,* 8th ed. (R. E. Buchanan and N. E. Gibbons, eds.), Williams & Wilkins, Baltimore.

Bergy, M. G., 1984–1988, *Manual of Systematic Bacteriology,* 1st ed. (J. G. Holt, ed. in chief) Williams & Wilkins, Baltimore.

Bishop, D. G., Rutberg, L., and Samuelson, B., 1967, The chemical composition of the cytoplasmic membrane of *Bacillus subtilis, Eur. J. Biochem.* **2**:448.

Björndal, H., Lindberg, B., and Svensson, S., 1967, Gas–liquid chromatography of partially methylated alditols as their acetates, *Acta Chem. Scand.* **21**:1801.

Blaurock, A. E., 1982, Analysis of bacteriorhodopsin structure by X-ray diffraction, *Methods Enzymol.* **88**:124.

Blaurock, A. E., and Stoeckenius, W., 1971, Structure of the purple membrane, *Nature New Biol.* **233**:152.

Blaurock, A. E., and King, G. I., 1977, Asymmetric structure of purple membrane, *Science* **196**:1101.

Bligh, E. G., and Dyer, W. J., 1959, A rapid method of total lipid extraction and purification, *Can. J. Biochem. Physiol.* **37**:911.

Blöcher, D., Guterman, R., Henkel, B., and Ring, K., 1984, Physico-chemical characterization of tetraether lipids from *Thermoplasma acidophilum, Biochim. Biophys. Acta* **778**:74.

Boggs, J. M., 1984, Intermolecular hydrogen bonding between membrane lipids, in *Membrane Fluidity* (M. Kates and L. A. Manson, eds.), pp. 3–53, Plenum, New York.

Boggs, J. M., 1987, Lipid intermolecular hydrogen bonding: Influence on structural organization and membrane function, *Biochim. Biophys. Acta* **906**:353.

Brennan, P. J., and Ballou, C. E., 1968, Biosynthesis of mannophosphoinositides by *Mycobacterium phlei*. Enzymatic acylation of dimannophosphoinositides, *J. Biol. Chem.* **243**:2975.

Brennan, P. J., and Lehane, D. P., 1971, The phospholipids of corynebacteria, *Lipids* **6**:357.

Brundish, D. E., and Baddiley, J., 1968, Synthesis of glucosylglycerols and diglucosylglycerol and their identification in small amounts, *Carbohyd. Res.* **8**:308.

Brundish, D. E., Shaw, N., and Baddiley, J., 1965, The glycolipids from the non-capsulated strain of *Pneumococcus* I, *Biochem. J.* **97**:158.

Brundish, D. E., Shaw, N., and Baddiley, J. 1966, Bacterial glycolipids. Glycosyl diglycerides in gram-positive bacteria, *Biochem. J.* **99**:546.

Brundish, D. E., Shaw, N., and Baddiley, J., 1967, The structure and possible function of the glycolipid from *Staphylococcus lactis, Biochem. J.* **105**:885.

Bruno, S., Canristraro, S., Gliozzi, A., De Rosa, M., and Gambacorta, A., 1986, A spin label ESR and saturation transfer-ESR study of archaebacteria bipolar lipids, *Eur. Biophys. J.* **13**:67.

Bu'Lock, J. D., de Rosa, M., and Gambacorta, A., 1983, Isoprenoid biosynthesis in Archaebacteria, in *Biosynthesis of Isoprenoid Compounds* (J. W. Porter and S. L. Spurgeon, eds.), John Wiley, New York, pp. 159–189.

Button, D., and Hemmings, N. L., 1976, Lipoteichoic acid from *Bacillus licheniformis* 6346 MH-1. Comparative studies on the lipid portion of the lipoteichoic acid and the membrane glycolipid, *Biochemistry* **15**:989.

Carroll, K. K., Cutts, J. H., and Murray, E. G. D., 1968, The lipids of *Listeria monocytogenes, Can. J. Biochem.* **46**:899.

Carter, H. E., McCluer, R. H., and Slifer, E., 1956, Wheat flour lipids, *J. Am. Chem. Soc.* **78**:3735.

Carter, H. E., Ohno, K., Nojima, S., Tipton, C. L., and Stanacev, N. Z., 1961a, Wheat flour lipids. II. Isolation and characterization of glycolipids of wheat flour and other plant sources, *J. Lipid Res.* **2**:215.

Carter, H. E., Hendry, R. A., and Stanacev, N. Z., 1961b, Wheat flour lipids. III. Structure determination of mono- and digalactosyl diglycerides, *J. Lipid Res.* **2**:223.

Carter, H. E., Johnson, P., and Weber, E. J., 1965, Glycolipids, *Annu. Rev. Biochem.* **34**:109.

Chen, J. S., Barton, P. G., Brown, D., and Kates, M., 1974, Osmometric and microscopic studies on bilayers of polar lipids from extreme halophile, *Halobacterium cutirubrum, Biochim. Biophys. Acta* **352**:202.

Chignell, C. F., and Chignell, D. A., 1975, A spin label study of purple membranes from *Halobacterium halobium, Biochem. Biophys. Res. Commun.* **62**:136.

Clarke, N. G., Hazlewood, G. P., and Dawson, R. M. C., 1976, Novel lipids of *Butyrivibrio* spp., *Chem. Phys. Lipids* **17**:222.

Cohen, M., and Panos, C., 1966, Membrane lipid composition of *Streptococcus pyogenes* and derived L-form, *Biochemistry* **5**:2385.

Cole, R., and Proulx, P., 1977, Further studies on the cardiolipin phosphodiesterase of *Escherichia coli, Can. J. Biochem.* **55**:1228.

Comita, P. B., and Gogosian, R. B., 1983, Membrane lipid from deep-sea hydrothermal vent methanogen: A new macrocyclic glycerol diether, *Science* **222**:1329.

Comita, P. B., Gagosian, R. B., Pang, H., and Costello, C. E., 1984, Structural elucidation of a unique macrocyclic membrane lipid from a new extremely thermophilic deep-sea by hydrothermal vent *Archaebacterium, Methanococcus jannaschii, J. Biol. Chem.* **259**:15234.

Constantopoulos, G., and Bloch, K., 1967, Isolation and characterization of glycolipids from some photosynthetic bacteria, *J. Bacteriol.* **93**:1178.

Coulon-Morelec, M. J., Dupouey, P., and Marechal, J., 1969, *C. R. Acad. Sci. Paris* **269**:854.

Curatolo, W. 1987a, The physical properties of glycolipids, *Biochim. Biophys. Acta* **906**:111.

Curatolo, W., 1987b, Glycolipid function, *Biochim. Biophys. Acta* **906**:137.

Dawson, R. M. C., 1967, Analysis of phosphatides and glycolipids by chromatography of their partial hydrolysis of alcoholysis products, in *Lipid Chromatographic Analysis*, Vol. 1 (G. V. Marinetti, ed.), pp. 163–189, Marcel Dekker, New York.

Degani, H., Danon, A., and Caplan, S. R., 1980, Proton and carbon-13 NMR studies of polar lipids of *Halobacterium halobium, Biochemistry* **19**:1626.

Demandre, C., Tremolieres, A., Justin, A.-M., and Mazliak, P., 1985, Analysis of molecular species of plant polar lipids by high performance and gas–liquid chromatography, *Phytochemistry* **24**:481.

Deroo, P. W., 1974, Studies on the structure and metabolism of glycolipids and glycolipid sulfate in the extreme halophile *Halobacterium cutirubrum*, Ph.D. thesis, University of Ottawa, Canada.

De Rosa, M., De Rosa, S., and Gambacorta, A., 1977a, ^{13}C-NMR assignments and biosynthetic data for the ether lipids of *Caldariella, Phytochemistry* **16**:1909.

De Rosa, M., De Rosa, S., Gambacorta, A., Minale, L., and Bu'Lock, J. D., 1977b, Chemical structure of ether lipids of thermophilic acidophilic bacteria of the *Caldariella* group, *Phytochemistry* **16**:1961.

De Rosa, M., Gambacorta, A., Nicolaus, B., and Bu'Lock, J. D., 1980a, Complex lipids of *Caldariella acidophila*, a thermoacidophile archaebacterium, *Phytochemistry* **19**:821.

De Rosa, M., Eposisto, E., Gambacorta, A., Nicolaus, B., and Bu'Lock, J. D., 1980b, Effects of temperature on ether lipid composition of *Caldariella acidophila, Phytochemistry* **19**:827.

De Rosa, M., Gambacorta, A., Nicolaus, B., Ross, H. N. M., Grant, W. D., and Bu'Lock, J. D., 1982a, An asymmetric archaebacterial diether lipids from alkaliphilic halophiles, *J. Gen. Microbiol.* **128**:343.

De Rosa, M., Gambarcorta, A., Nicolaus, B., and Sodano, S., 1982b, Incorporation of labelled glycerols into ether lipids in *Caldariella acidophila, Phytochemistry* **21**:595.

De Rosa, M., Gambacorta, A., Nicolaus, B., and Grant, W. D., 1983a, A C_{25}, C_{25} diether core lipid from archaebacterial haloalkaliphiles, *J. Gen. Microbiol.* **129**:2333.

De Rosa, M., Gambacorta, A., and Nicolaus, B., 1983b, A new type of cell membrane in thermophilic archaebacteria based on bipolar ether lipids, *J. Membr. Sci.* **16**:287.

De Rosa, M., Gambacorta, A., and Gliozzi, A., 1986a, Structure, biosynthesis, and physicochemical properties of archaebacterial lipids, *Microbiol. Rev.* **50**:70.

De Rosa, M., Gambacorta, A., Lanzotti, V., Trincone, A., Harris, J. E., and Grant, W. D., 1986b, A range of ether core lipids from the methanogenic archaebacterium *Methanosarcina barkeri, Biochim. Biophys. Acta* **875**:487.

De Rosa, M., Gambacorta, A., Grant, W. D., Lanzotti, V., and Nicolaus, B., 1988, Polar lipids and glycine betaine from haloalkaliphilic archaebacteria, *J. Gen. Microbiol.* **134**: 205.

Dittmer, J. C., and Wells, M. A., 1969, Quantitative and qualitative analysis of lipids and lipid components, *Methods Enzymol.* **14:**482.

Dubois, M., Gilles, K. A., Hamilton, J. K., Rebers, P. A., and Smith, F., 1956, Colorimetric method for determination of sugars and related substances, *Anal. Chem.* **28:**350.

Ekiel, I., Marsh, D., Smallbone, B. W., Kates, M., and Smith, I. C. P., 1981, The state of the lipids in the purple membrane of *Halobacterium cutirubrum* as seen by ^{31}P-NMR, *Biochem. Biophys. Res. Commun.* **100:**105.

Ekiel, I., Sprott, G. D., and Smith, I. C. P., 1986, Mevalonic acid is partially synthesized from amino acids in *Halobacterium cutirubrum:* A ^{13}C-NMR study, *J. Bacteriol.* **166:**559.

Esser, A. F., and Lanyi, J. K., 1973, Structure of the lipid phase in cell envelope vesicles of *Halobacterium cutirubrum*, *Biochemistry* **12:**1933.

Evans, R. W., Kushwaha, S. C., and Kates, M., 1980, The lipids of *Halobacterium marismortui*, an extremely halophilic bacterium from the Dead Sea, *Biochim. Biophys. Acta* **619:**533.

Exterkate, F. A., and Veerkamp, J. H., 1969, Biochemical changes in *Bifidobacterium bifidum* var. *pennsylvanicus* after cell wall inhibition. I. Composition of lipids, *Biochim. Biophys. Acta* **176:**65.

Exterkate, F. A., and Veerkamp, J. H., 1971, Biochemical changes in *Bifidobacterium bifidum* var. *pennsylvanicus* after cell wall inhibition. IV. Galactolipid composition, *Biochim. Biophys. Acta* **231:**545.

Falk, K.-E., Karlsson, K.-A., and Samuelsson, B. E., 1979, Proton nuclear magnetic resonance analysis of anomeric structure of glycosphingolipids, *Arch. Biochem. Biophys.* **192:**177.

Falk, K.-E., Karlsson, K.-A., and Samuelsson, B. E., 1980, Structural analysis of mass spectrometry and NMR spectroscopy of the glycolipid sulfate from *Halobacterium salinarium* and a note on its possible function, *Chem. Phys. Lipids* **27:**9.

Ferrante, G., Ekiel, I., and Sprott, G. D., 1986, Structural characterization of the lipids of *Methanococcus voltae* including a novel N-acetylglucosamine-1-phosphate diether, *J. Biol. Chem.* **261:**17062.

Ferrante, G., Ekiel, I., and Sprott, G. D., 1987, Structures of diether lipids of *Methanospirillum hungatei* containing novel head groups N,N-dimethylamino- and N,N,N-trimethylaminopentanetetrol, *Biochim. Biophys. Acta* **921:**281.

Ferrante, G., Ekiel, I., Patel, G. B., and Sprott, G. D., 1988a, Structure of major polar lipids isolated from the aceticlastic methanogen, *Methanothrix concilii*, *Biochim. Biophys. Acta* **963:**162.

Ferrante, G., Ekiel, I., Patel, G. B. and Sprott, G. D., 1988b, A novel core lipid isolated from the aceticlastic methanogen, *Methanothrix concilii* GP6, *Biochim. Biophys. Acta* **963:**173.

Ferrante, G., Richards, J. C., and Sprott, G. D., 1990, Structures of polar lipids from the thermophilic deep sea archaebacterium *Methanococcus jannaschii*, *Biochem. Cell Biol.*, **68:** 274.

Fischer, W., 1976, in *Lipids*, Vol. 1 (R. Paoletti, G. Porcellati, and G. Jacini, eds.), pp. 255–266, Raven, New York.

Fischer, W., 1981, in *Chemistry and Biological Activities of Bacterial Surface Amphiphiles* (G. D. Shockman and A. J. Wicken, eds.), pp. 209–228, Academic, Orlando, Florida.

Fischer, W., and Seyferth, W., 1968, 1-(O-α-D-glucopyranosyl-(1-2)-O-α-D-glycopyranosyl)-glycerin aus den Glycolipiden von *Streptococcus faecalis*, *Streptococcus lactis*, *Hoppe-Seyler's Z. Physiol. Chem.* **349:**1662.

Fischer, W., Ishizuka, I., Landgraf, H. R., and Herrmann, J., 1973, Glycerylphosphoryl diglucosyl diglyceride, a new phosphoglycolipid from *Streptococci*, *Biochim. Biophys. Acta* **296:**527.

Fischer, W., Laine, R. A., Nakano, M., Schuster, D., and Egge, H., 1978a, The structure of acyl-α-kojibiosyldiacylglycerol from *Streptococcus lactis*, *Chem. Phys. Lipids* **21:**103.

Fischer, W., Nakano, M., Laine, R. A., and Bohrer, W., 1978b, On the relationship between glycerophosphoglycolipids and lipoteichoic acids in gram-positive bacteria. I. The occurrence of phosphoglycolipids, *Biochim. Biophys. Acta* **528:**288.

Fischer, W., Laine, A. R., and Nakano, M., 1978c, On the relation between glycerophosphoglycolipids and lipoteichoic acids in gram-positive bacteria. II. Structures of glycerophosphoglycolipids, *Biochim. Biophys. Acta* **528:**298.

Fischer, W., Schuster, D., and Laine, A. R., 1979, Studies on the relationship between glycerophosphoglycolipids and lipoteichoic acids. IV. Trigalactosylglycerophospho-acylkojibiosyldiacylglycerol and related compounds from *Streptococcus lactis* Kiel 42172, *Biochem. Biophys. Acta* **575:**389.

Galambos, J. T., 1967, The reaction of carbazole with carbohydrates. I. Effect of borate and sulfanate on carbazole colour of sugars, *Anal. Biochem.* **19**:119.

Gigg, R., 1980, Synthesis of glycolipids, *Chem. Phys. Lipids* **16**:287.

Gliozzi, A., Paoli, G., Rolandi, R., De Rosa, M., and Gambacorta, A., 1983, Effect of isoprenoid cyclization on the transition temperature of lipids in thermophilic archaebacteria, *Biochim. Biophys. Acta* **735**:234.

Gochnauer, M. B., Leppard, G. G., Komaratat, P., Kates, M., Novitsky, T., and Kushner, D. J., 1975, Isolation and characterization of *Actinopolyspora halophila*, an extremely halophilic *Actinomycete, Can. J. Microbiol.* **21**:1500.

Goldfine, H., 1972, Comparative aspects of bacterial lipids, *Adv. Microbiol. Physiol.* **8**:1.

Goldfine, H., 1982, Lipids of prokaryotes—Structure and distribution, *Curr. Top. Membr. Transp.* **17**:1.

Goldfine, H., and Hagen, P. O., 1972, Bacterial plasmalogens, in *Ether Lipids, Chemistry and Biology* (F. Snyder, ed.), pp. 329–350, Academic, New York.

Grant, W. D., and Larsen, H., 1988, Extremely halophilic archaeobacteria, in *Bergey's Manual of Systematic Bacteriology,* Vol. 3 (J. T. Stalley, ed.), pp. 2216–2233, Williams & Wilkins, Baltimore.

Grant, W. D., and Ross, H. N. M., 1986, The ecology and taxonomy of halobacteria, *FEMS Microbiol. Rev.* **39**:9.

Grant, W. D., Pinch, G., Harris, J. E., De Rosa, M., and Gambacorta, A., 1985, Polar lipids in methanogen taxonomy, *J. Gen. Microbiol.* **131**:3277.

Green, J. N., 1966, The glycofuranosides, *Adv. Carbohydr. Chem.* **21**:95.

Gulik, A., Luzzati, V., De Rosa, M., and Gambacorta, A., 1985, Structure and polymorphism of bipolar isopranyl ether lipids from Archaebacteria, *J. Mol. Biol.* **182**:131.

Haines, T. H., 1971, The chemistry of the sulfolipids, *Prog. Chem. Fats Other Lipids* **11**:297.

Haines, T. H., 1983, Anionic lipid headgroups as a proton-conducting pathway along the surface of membranes: A hypothesis, *Proc. Nat. Acad. Sci. USA* **80**:160.

Hakomori, S., 1983, Chemistry of glycosphingolipids, in *Sphingolipid Biochemistry,* Vol. 3 (J. N. Kanfer and S. Hakomori, eds.), pp. 1–165, Plenum, New York.

Hanna, K., Bengis-Garber, C., Kushner, D. J., Kogut, M., and Kates, M., 1984, The effect of salt concentration on the phospholipid and fatty acid composition of the moderate halophile *Vibrio costicola, Can. J. Microbiol.* **30**:669.

Heathcock, C. H., Finkelstein, H., Aoki, B. L., and Poulter, C. D., 1985, Stereostructure of the archaebacterial C_{40} diol, *Science* **229**:862.

Heinz, E., 1967, Acygalactosyldiglycerid aus Blatthomogenaten, *Biochim. Biophys. Acta* **144**:321.

Heinz, E., Rullkotter, J., and Budzikiewicz, H., 1974, Acyl digalactosyl diglyceride from leaf homogenates, *Hoppe-Seyler's Z. Physiol. Chem.* **355**:612.

Henderson, R., Jubb, J. S., and Whytock, S., 1978, Specific labeling of the protein and lipid on the extracellular surface of purple membrane, *J. Mol. Biol.* **123**:259.

Hinz, H.-J., Six, L., Ruess, K.-P., and Lieflander, M., 1985, Head-group contributions to bilayer stability: Monolayer and calorimetric studies on synthetic, stereochemically uniform glucolipids, *Biochemistry* **24**:806.

Hiraki, K., Hamanaka, T., Mitsui, T., Kito, Y., 1981, Phase transitions of the purple membrane and brown holomembrane, X-ray diffraction, circular dichroism spectrum and adsorption spectrum studies, *Biochim. Biophys. Acta* **647**:18.

Hudson, C. S., 1909, The significance of certain numerical relations in the sugar group, *J. Am. Chem. Soc.* **31**:66.

Imhoff, J. F., Kushner, D. J., Kushwaha, S. C., and Kates, M., 1982, Polar lipids in phototrophic bacteria of the *Rhodospirillaceae* and *Chromatiaceae* families, *J. Bacteriol.* **150**:1192.

Inoue, T., Deshmukh, D. S., and Pieringer, R. A., 1971, The association of the galactosyldiglycerides of brain with myelination. I. Changes in the concentration of monogalactosyldiglycerol in the microsomal and myelin fractions of brain of rats during development, *J. Biol. Chem.* **246**:5688.

Ishizuka, I., and Yamakawa, T., 1968, Glycosyl glycerides from *Streptococcus hemolyticus* strain D-58, *J. Biochem.* **64**:13.

Ishizuka, I., and Yamakawa, T., 1985, Glycoglycerolipids, in *Glycolipids* (H. Wiegandt, ed.), pp. 101–197, Elsevier, Amsterdam.

IUPAC–IUB Commission on Biochemical Nomenclature, 1978, The nomenclature of lipids, *Chem. Phys. Lipids* **21**:159.

Iwamoto, K., Sunamoto, J., Inoue, K., Endo, T., and Nojima, S., 1982, Liposomal membranes. XV. Importance of surface structure in liposomal membranes of glyceroglycolipids, *Biochim. Biophys. Acta* **691**:44.

Jackson, M. B., and Sturtevant, J. M., 1978, Phase transitions of the purple membranes of *Halobacterium halobium*, *Biochemistry* **17**:911.

Jarrell, H. C., Jovall, P. A., Giziewicz, J. B., Turner, L. A., and Smith, I. C. P., 1987, Determination of conformational properties of glycolipid head groups by ^{2}H-NMR of oriented multi-bilayers, *Biochemistry* **26**:1805.

Joo, C. N., Shier, T., and Kates, M., 1968, Characterization and synthesis of mono- and diphytanyl ethers of glycerol, *J. Lipid Res.* **9**:782.

Juez, G., Rodriguez-Valera, F., Ventosa, A., and Kushner, D. J., 1986, *Haloarcula hispanica*, spec. nov. and *Haloferax gibbonsii* spec. nov., two new species of extremely halophilic archaebacteria, *Syst. Appl. Microbiol.* **8**:75.

Kahane, I., and Tully, J. G., 1976, Binding of lectins to mycoplasma cells and membranes, *J. Bacteriol.* **128**:1.

Kamekura, M., and Kates, M., 1988, Lipids of halophilic archaebacteria, in *Halophilic Bacteria*, Vol. II (F. Rodriguez-Valera, ed.), pp. 25–54, CRC Press, Boca Raton, Florida.

Kanfer, J. N., and Hakomori, S. (eds.), 1983, *Sphingolipid Biochemistry*, Vol. 3, Plenum, New York.

Karkkainen, J., Lehtonen, A., and Nikkari, T., 1965, *J. Chromatogr.* **20**:457.

Kates, M., 1964, Bacterial lipids, *Adv. Lipid Res.* **2**:17.

Kates, M., 1966, Biosynthesis of lipids in microorganisms, *Annu. Rev. Microbiol.* **20**:13.

Kates, M., 1972, Ether-linked lipids in extremely halophilic bacteria, in *Ether Lipids, Chemistry and Biology* (F. Snyder, ed.), pp. 351–398, Academic, New York.

Kates, M., 1978, The phytanyl ether-linked polar lipids and isoprenoid neutral lipids of extremely halophilic bacteria, *Prog. Chem. Fats Other Lipids* **15**:301.

Kates, M., 1986a, *Techniques of Lipidology*, 2nd rev. ed., Elsevier, Amsterdam.

Kates, M., 1986b, Influence of salt concentration of membrane lipids on halophilic bacteria, *FEMS Microbiol. Rev.* **39**:95.

Kates, M., 1988, Structure, physical properties and function of archaebacterial lipids, in *Biological Membranes: Aberrations in Membrane Structure and Function, Proceedings of the Twelfth International Conference on Biological Membranes* (M. L. Karnovsky, A. Leaf, and L. C. Bolis, eds.), pp. 357–384, Alan R. Liss, New York.

Kates, M., and Deroo, P. W., 1973, Structure determination of the glycolipid sulfate from the extreme halophile *Halobacterium cutirubrum*, *Biochemistry* **14**:438.

Kates, M., and Kushwaha, S. C., 1978, Biochemistry of the lipids of extremely halophilic bacteria, in *Energetics and Structure of Halophilic Microorganisms* (S. R. Caplan and M. Ginzburg, eds.), pp. 461–480, Elsevier, Amsterdam.

Kates, M., and Volcani, B. E., 1966, Lipids of diatoms, *Biochim. Biophys. Acta* **116**:264.

Kates, M., Yengoyan, L. S., and Sastry, P. S., 1965, A diether analog of phosphatidylglycerophosphate in *Halobacterium cutirubrum*, *Biochim. Biophys. Acta* **98**:252.

Kates, M., Palameta, B., Joo, C. N., Kushner, D. J., and Gibbons, N. E., 1966, Aliphatic diether analogues of glyceride-derived lipids. IV. Occurrence of di-O-dihydrophytylglycerol ether containing lipids in extremely halophilic bacteria, *Biochemistry* **5**:4092.

Kates, M., Palameta, B., Perry, M. P., and Adams, G. A., 1967, A new glycolipid sulfate in *Halobacterium cutirubrum*, *Biochim. Biophys. Acta* **137**:213.

Kates, M., Wassef, M. K., and Kushner, D. J., 1968, Radioisotopic studies on the biosynthesis of the glycerol diether lipids of *Halobacterium cutirubrum*, *Can. J. Biochem.* **46**:971.

Kates, M., Wassef, M. K., and Pugh, E. L., 1970, Origin of the glycerol moieties in the glycerol diether lipids of *Halobacterium cutirubrum*, *Biochim. Biophys. Acta* **202**:206.

Kates, M., Kushwaha, S. C., and Sprott, G. D., 1982, Lipids of purple membrane from extreme halophiles and of methanogenic bacteria, *Methods Enzymol.* **88**:98.

Kaufmann, B., Kunding, F. D., Distler, J., and Roseman, S., 1965, Enzymatic synthesis and structure of two glycolipids from type XIV *Pneumococcus*, *Biochem. Biophys. Res. Commun.* **18**:312.

Kenyon, C. N., 1978, Complex lipids and fatty acids of photosynthetic bacteria, in *The Photosynthetic Bacteria* (R. K. Clayton and W. R. Sistrom, eds.) pp. 281–313, Plenum, New York.

Khuller, G. K., and Brennan, P. J., 1972a, Further studies on the lipids of corynebacteria. The mannolipids of *Corynebacterium aquaticum, Biochem. J.* **127**:369.

Khuller, G. K., and Brennan, P. J., 1972b, The polar lipids of some species of *Nocardia, J. Gen. Microbiol.* **73**:409.

Kocur, M., and Hodgkiss, W., 1973, Taxonomic status of the genus *Halococcus* Schoop, *Int. J. Systm. Bacteriol.* **23**:151.

Koga, Y., Nishihara, M., and Morii, H., 1982, Lipids of alkaliphilic bacteria: identification, composition and metabolism, *J. Univ. Occupat. Environ. Health* **4**:227.

Koga, Y., Ohga, M., Nishihara, M., and Morii, H., 1987, Distribution of diphytanyl ether analog of phosphatidylserine and ethanolamine-containing tetraether lipid in methanogenic bacteria, *Syst. Appl. Microbiol.* **9**:176.

Komaratat, P., 1974, Structural and metabolic studies of the cellular lipids of halotolerant *Staphylococcus epidermidis*, Ph.D. thesis, University of Ottawa, Canada.

Komaratat, P., and Kates, M., 1975, The lipids of a halotolerant species of *Staphylococcus epidermidis, Biochim. Biophys. Acta* **398**:464.

Kunsman, J. E., 1970, Characterization of the lipids of *Butyrivibrio fibrisolvens, J. Bacteriol.* **103**:104.

Kushwaha, S. C., and Kates, M., 1981, Modification of phenol-sulfuric acid method for estimation of sugars in lipids, *Lipids* **16**:372.

Kushwaha, S. C., Kates, M., and Martin, W. G., 1975, Characterization and composition of purple and red membranes from *Halobacterium cutirubrum, Can. J. Biochem.* **53**:284.

Kushwaha, S. C., Kates, M., and Porter, J. W., 1976, Enzymatic synthesis of C_{40} carotenes by cell-free preparations from *Halobacterium cutirubrum, Can. J. Biochem.* **54**:816.

Kushwaha, S. C., Kates, M., Sprott, G. D., and Smith, I. C. P., 1981, Novel polar lipids from methanogen *Methanospirillum hungatei, Biochim. Biophys. Acta* **664**:156.

Kushwaha, S. C., Kates, M., Juez, G., Rodriguez-Valera, F., and Kushner, D. J., 1982a, Polar lipids of an extremely halophilic bacterial strain (R-4) isolated from salt ponds in Spain, *Biochim. Biophys. Acta* **711**:19.

Kushwaha, S. C., Juez-Perez, G., Rodriguez-Valera, F., Kates, M., and Kushner, D. J., 1982b, Survey of lipids of a new group of extremely halophilic bacteria from salt ponds in Spain, *Can. J. Microbiol.* **28**:1365.

Laine, R. L., and Renkonen, O., 1975, Analysis of anomeric configuration in glyceroglycolipids and glycosphingolipids by chromium trioxide oxidation, *J. Lipid Res.* **16**:102.

Laine, R. A., Esselman, W. J., and Sweeley, C. C., 1972, Gas–liquid chromatography of carbohydrates, *Methods Enzymol.* **28**:159.

Lang, D. R., and Lundgren, D. G., 1970, Lipid components of *Bacillus cereus* during growth and sporulation, *J. Bacteriol.* **101**:483.

Langworthy, T. A., 1977a, Long-chain diglycerol tetraethers from *Thermoplasma acidophilum, Biochim. Biophys. Acta* **487**:37.

Langworthy, T. A., 1977b, Comparative lipid composition of heterotrophically and autotrophically grown *Sulfolobus acidocaldarius, J. Bacteriol.* **130**:1326.

Langworthy, T. A., 1982a, Lipids of bacteria living in extreme environments, *Curr. Top. Membr. Transp.* **17**:45.

Langworthy, T. A., 1982b, Lipids of thermoplasma, *Methods Enzymol.* **88**:396.

Langworthy, T. A., 1985, Lipids of archaebacteria, in *The Bacteria*, Vol. 8 (C. R. Woese and R. S. Wolfe, eds.), pp. 459–497, Academic, Orlando, Florida.

Langworthy, T. A., Smith, P. F., and Mayberry, W. R., 1972, Lipids of *Thermoplasma acidophilum, J. Bacteriol.* **112**:1193.

Langworthy, T. A., Mayberry, W. R., and Smith, P. F., 1974, Long-chain glycerol diether and polyol dialkyl glycerol triether lipids of *Sulfolobus acidocaldarius, J. Bacteriol.* **119**:106.

Langworthy, T. A., Mayberry, W. R., and Smith, P. F., 1976, A sulfonolipid and novel glucosamidyl glycolipids from the extreme thermoacidophile *Bacillus acidocaldarius, Biochim. Biophys. Acta* **431**:550.

Langworthy, T. A., Tornabene, T. G., and Holzer, G., 1982, Lipids of archaebacteria, *Zentralbl. Bakteriol. Mikrobiol. Hyg. Abt. 1 Orig. C* **3**:228.

Langworthy, T. A., Holzer, G., Zeikus, J. G., and Tornabene, T. G., 1983, Iso and anteiso-branched glycerol diethers in the thermophilic anaerobe, *Thermodesulfatobacterium commune, Syst. Appl. Microbiol.* **4**:1.

Lanyi, J. K., 1979, The role of Na⁺ in transport processes of bacterial membranes, *Biochim. Biophys. Acta* **559:**377.

Lanzotti, V., De Rosa, M., Trincone, A., Basso, A. L., Gambacorta, A., and Zillig, W., 1987, Complex lipids from *Desulfurococcus mobilis*, a sulfur-reducing archaebacterium, *Biochim. Biophys. Acta* **922:**95.

Lee, Y. C., and Ballou, C. E., 1965, Complete structures of the glycophospholipids of mycobacteria, *Biochemistry* **4:**1395.

Lennarz, W. J., 1966, Lipid metabolism in bacteria, *Adv. Lipid Res.* **4:**175.

Lennarz, W. J., 1970, in *Lipid Metabolism* (S. J. Wakil, ed.), pp. 155–184, Academic, New York.

Lennarz, W. J., and Talamo, B., 1966, The chemical characterization and enzymatic synthesis of mannolipids in *Micrococcus lysodeikticus*, *J. Biol. Chem.* **241:**2707.

Lepage, M., 1964, Isolation and characterization of an esterified form of steryl glucoside, *J. Lipid Res.* **5:**587.

Lind, C., Hojeberg, B., and Khorana, H. G., 1981, Reconstitution of delipidated bacteriorhodopsin with endogenous polar lipids, *J. Biol. Chem.* **256:**8298.

Livermore, B. P., and Johnson, R. C., 1970, Isolation and characterization of glycolipid from *Treponema pallidum* Kazan, *Biochim. Biophys. Acta* **210:**315.

Livermore, B. P., and Johnson, R. C., 1974, Lipids of the Spirochaetales: Comparison of the lipids of several members of the genera *Spirochaeta*, *Trepanoma*, *J. Bacteriol.* **120:**1268.

Lynch, D. V., Gundersen, R. E., and Thompson, G. A., Jr., 1983, Separation of galactolipid molecular species by high performance liquid chromatography, *Plant Physiol.* **72:**903.

MacDougall, J. C., and Phizackerley, P. J. R., 1969, Isomers of glucosaminylphosphatidylglycerol in *Bacillus megaterium*, *Biochem. J.* **114:**361.

Macfarlane, M. G., 1961, Isolation of a phosphatidylglycerol and glycolipid from *Micrococcus lysodeikticus*, *Biochem. J.* **80:**45P.

Macfarlane, M. G., 1962*a*, Lipid components of *Staphylococcus aureus* and *Salmonella typhimurium*, *Biochem. J.* **82:**40P.

Macfarlane, M.G., 1962*b*, Characterization of lipoamino acids as *O*-amino acid esters of phosphatidylglycerol, *Nature (Lond.)* **196:**136.

Mancuso, C. A., Nichols, P. D., and White, D. C., 1986, A method for the separation and characterization of archaebacterial signature ether lipids, *J. Lipid Res.* **27:**49.

Mannock, D. A., Lewis, R. N. A. H., and McElhaney, R. N., 1987, An improved procedure for the preparation of 1,2-di-*O*-acyl-3-*O*-(β-D-glucopyranosyl)-*sn*-glycerols, *Chem. Phys. Lipids* **43:**113.

Mannock, D. A., Lewis, R. N. A. H., Sen, A., and McElhaney, R. N., 1988, The physical properties of glycosyldiacylglycerols. Calorimetric studies of a homologous series of 1,2-diacyl-3-*O*-(β-D-glucopyranosyl)-*sn*-glycerols, *Biochemistry* **27:**6852.

Marinetti, G. V., and Cattieu, K., 1981, Lipid analysis of cells and chromatosphores of *Rhodopseudomonas sphaeroides*, *Chem. Phys. Lipids* **28:**241.

Marion, D., Gandemer, G., and Douillard, R., 1984, Separation of plant phosphoglycerides and galactosylglycerides by high performance liquid chromatography, in *Structure, Function, and Metabolism of Plant Lipids* (P. A. Siegenthaller and W. Eichenberger, eds.), pp. 139–143, Elsevier, Amsterdam.

Matsubara, T., and Hayashi, A., 1974, Determination of the structure of partially methylated sugars as *O*-trimethylsilyl ethers by gas chromatography–mass spectrometry, *Biomed. Mass Spect.* **1:**62.

Matthews, H. M., Yang, T.-K., and Jenkin, H. M., 1980, Alk-1-enyl ether phospholipids (plasmalogens) and glycolipids of *Treponoma hyodysenteriae*, *Biochim. Biophys. Acta* **618:**273.

Mayberry, W. R., and Smith, P. F., 1983, Structures and properties of acyldiglucosyl cholesterol and galactofuranosyl diacylglycerol from *Acholeplasma axanthum*, *Biochim. Biophys. Acta* **752:**434.

Mayberry-Carson, K. J., Langworthy, T. A., Mayberry, W. R., and Smith, P. F., 1974, A new class of lipopolysaccharide from *Thermoplasma acidophilum*, *Biochim. Biophys. Acta* **360:**217.

Mayberry, W. R., Langworthy, T. A., and Smith, P. F., 1976, Structure of the mannoheptose containing pentaglycosyldiacylglycerol from *Acholeplasma modicum*, *Biochim. Biophys. Acta* **441:**115.

McElhaney, R. N., 1984, Structure and function of the *Acholeplasma laidlawii* plasma membrane, *Biochim. Biophys. Acta* **779:**1.

McElhaney, R. N., and Tourtellotte, M. E., 1970, Metabolic turnover of the polar lipids of *Mycoplasma laidlawii* Strain B, *J. Bacteriol.* **101**:72.

Meyer, H., and Meyer, F., 1971, Lipid metabolism in the parasitic and free-living spirochetes *Trepanoma pallidum* (Reiter) and *Trepanoma zuelzerae, Biochim. Biophys. Acta* **231**:93.

Moldoveanu, N., and Kates, M., 1988, Biosynthetic studies of the polar lipids of *Halobacterium cutirubrum.* Formation of isoprenyl ether intermediates, *Biochim. Biophys. Acta* **960**:164.

Moldoveanu, N., and Kates, M., 1989, Effect of bacitracin on growth and phospholipid, glycolipid and bacterioruberin biosynthesis in *Halobacterium cutirubrum, J. Gen. Microbiol.* **135**: 2503.

Montero, C. G., Ventosa, A., Rodriguez-Valera, F., Kates, M., Moldoveanu, N., and Ruiz-Berra-quero, F., 1989, *Halococcus saccharolyticus* sp. nov., a new species of extremely halophilic non-alkaliphilic cocci, *Syst. Appl. Microbiol.* **12**:167.

Morii, H., Nishihara, M., Ohga, M., and Koga, Y., 1986, A diphytanyl ether analog of phos-phatidylserine from a methanogenic bacterium, *Methanobrevibacter arboriphilus, J. Lipid Res.* **27**:724.

Morth, S., and Tindall, B. J., 1985, Variation in polar lipid composition within *Haloalkaliphilic archaebacteria, Syst. Appl. Microbiol.* **6**:247.

Mullakhanbhai, M. F., and Larsen, H., 1975, *Halobacterium volcanii,* new species, a Dead Sea *Halobacterium* with a moderate salt requirement, *Arch. Microbiol.* **104**:207.

Murray, R. K., Levine, M., and Kornblatt, M. J., 1976, Sulfatides: Principal glycolipids of the testes and spermatozoa of chordates, in *Glycolipid Methodology* (L. A. Witting, ed.), pp. 305–327. American Oil Chemists' Society, Champaign, Illinois.

Murray, R. K., Narasimhan, R., Levine, M., and Pinteric, L., 1980, Galactoglycerolipids of mam-malian testis, spermatozoa, and nervous tissue, in *Cell Surface Glycolipids* (C. C. Sweeley, ed.), pp. 105–125, ACS Symposium Series, Vol. 128, American Chemical Society, Washington, D.C.

Nakano, M., and Fischer, W., 1977, The glycolipids of *Lactobacillus casei* DSM 20021, *Hoppe-Seyler's Z. Physiol. Chem.* **358**:1439.

Nichols, B. W., and James, A. T., 1965, Lipids of photosynthetic tissue, *Biochem. J.* **94**:22P.

Nichols, B. W., and Moorehouse, R., 1969, The separation, structure and metabolism of mono-galactosyl diglyceride species in *Chlorella vulgaris, Lipids* **4**:311.

Nishihara, M., and Koga, Y., 1987a, Extraction and composition of polar lipids from the Arch-aebacterium, *Methanobacterium thermoautotrophicum:* Effective extraction of tetraether lipids by an acidified solvent, *J. Biochem.* **101**:997.

Nishihara, M., Morii, H., and Koga, Y., 1987b, Structure determination of a quartet of novel tetraether lipids from *Methanobacterium thermoautotrophicum, J. Biochem.* **101**:1007.

Nishihara, M., Morii, H., and Koga, Y., 1989, Heptads of polar ether lipids of an archaebac-terium, *Methanobacterium thermoautotrophicum:* Structure and biosynthesis relation, *Biochemis-try* **28**:95.

Onishi, H., Kobayashi, T., Iwao, S., and Kamekura, M., 1985, Archaebacterial diether lipids in a non-alkalophilic, non-pigmented extremely halophilic bacterium, *Agric. Biol. Chem.* **49**:3053.

Op den Kamp, J. A. F., and van Deenen, L. L. M., 1966, On the structure of glucosaminyl phosphatidylglycerol in *Bacillus megaterium, Chem. Phys. Lipids* **1**:86.

Oshima, M., and Ariga, T., 1976, Analysis of the anomeric configuration of galactofuranose-containing glycolipid from an extreme halophile, *FEBS Lett.* **64**:440.

Oshima, M., and Yamakawa, T., 1974, Chemical structure of a novel glycolipid from an extreme thermophile, *Flavobacterium thermophilum, Biochemistry* **13**:1140.

Ourisson, G., and Rohmer, M., 1982, Prokaryotic polyterpenes: Phylogenetic precursors of sterols, *Curr. Top. Membr. Transp.* **17**:153.

Pandhi, P. N., and Hammond, F., 1978, The polar lipids of *Actinomyces viscosus, Arch. Oral Biol.* **23**:17.

Peleg, E., and Tietz, A., 1971, Glycolipids of a halotolerant moderately halophilic bacterium, *FEBS Lett.* **15**:309.

Peleg, E., and Tietz, A., 1973, Phospholipids of a moderately halophilic halotolerant bac-terium. Isolation and identification of glycosylphosphatidylglycerol, *Biochim. Biophys. Acta* **306**:368.

Pennick, R. J., and McCluer, R. H., 1966, Quantitative determination of glucose and galactose in ganglioside by gas-liquid chromatography, *Biochim. Biophys. Acta* **116**:288.

Perlin, A. S., Casu, B., and Koch, H. J., 1970, Configurational and conformational influences on carbon-13 chemical shifts of some carbohydrates, *Can. J. Chem.* **48:**2596.

Perry, M. B., 1964, The separation, determination and characterization of 2-amino-2-deoxy-D-glucose (D-glucosamine) and 2-amino-2-deoxy-D-galactose (D-galactosamine) in biological materials by gas-liquid parition chromatography, *Can. J. Biochem.* **42:**451.

Perry, M. B., and Hulyalka, R. K., 1965, The analysis of hexuronic acids in biological materials by gas-liquid partition chromatography, *Can. J. Biochem.* **43:**573.

Phizackerley, P. J. R., MacDougall, J. C., and Moore, R. A., 1972, 1-(*O*-α-Glucosaminyl)-2,3-diglyceride in *Bacillus megaterium, Biochem. J.* **126:**499.

Pieringer, R. A., 1968, The metabolism of glyceride glycolipids, *J. Biol. Chem.* **243:**4894.

Plachy, W. Z., Lanyi, J. K., and Kates, M., 1974, Lipid interactions in membranes of extremely halophilic bacteria. I. Electron spin resonance and dilatometric studies of bilayer structure, *Biochemistry* **13:**4906.

Plackett, P., 1967, The glycerolipids of *Mycoplasma mycoides, Biochemistry* **6:**2746.

Plackett, P., and Shaw, E. J., 1967, Glycolipids from *Mycoplasma laidlawii* and *Streptococcus* MG, *Biochem. J.* **104:**61C.

Plackett, P., Marmion, B. P., Shaw, E. J., and Lemcke, R. M., 1969, Immunochemical analysis of *Mycoplasma pneumoniae, Aust. J. Exp. Biol. Med. Sci.* **47:**171.

Polonovski, J., Wald, R. and Paysant Diamant, 1962, Les lipides de *Staphylococcus aureus, Ann. Inst. Pasteur Paris* **103:**32.

Poulter, C. D., Aoki, T., and Daniels, L., 1988, Biosynthesis of isoprenoid membranes in the methanogenic archaebacterium *Methanospirillum hungatei, J. Am. Chem. Soc.* **110:**2620.

Prottey, C., and Ballou, C. E., 1968, Diacylmyoinositol monomannoside of *Propionibacterium shermanii, J. Biol. Chem.* **243:**6196.

Proulx, P., and van Deenen, L. L. M., 1967, Phospholipase activity of *Escherichia coli, Biochim. Biophys. Acta* **144:**171.

Pugh, E. L., Wassef, M. K., and Kates, M., 1971, Inhibition of fatty acid synthesis in *Halobacterium cutirubrum* and *Escherichia coli* by high salt concentrations, *Can. J. Biochem.,* **49:**953.

Pugh, E. L., Bittman, R., Fugler, L., and Kates, M., 1989, Comparison of steady-state fluorescence polarization and urea permeability of phosphatidylcholine and phosphatidylsulfocholine liposomes as a function of sterol structure, *Chem. Phys. Lipids* **50:**43.

Quinn, P. J., and Sherman, W. R., 1971, Monolayer characteristics and calcium absorption to cerebroside and cerebroside oriented at the air–water interface. *Biochim. Biophys. Acta* **233:**734.

Quinn, P. J., and Williams, W. P., 1983, The structural role of lipids in photosynthetic membranes, *Biochim. Biophys. Acta* **737:**223.

Quinn, P. J., Brain, A. P. R., Stewart, L. C., and Kates, M., 1986, The structure of membrane lipids of the extreme halophile, *Halobacterium cutirubrum* in aqueous systems studied by freeze-fracture, *Biochim. Biophys. Acta* **863:**213.

Quinn, P. J., Kates, M., Tocanne, J.-F., and Tomaoia-Cotisel, M., 1989, Surface characteristics of phosphatidylglycerophosphate from *Halobacterium cutirubrum,* compared with its deoxy analogue at the air–water interface, 1989, *Biochem. J.* **261:**377.

Radin, N. S., Lavin, F. B., and Brown, J. R., 1955, Determination of cerebrosides, *J. Biol. Chem.* **217:**789.

Raetz, C. R. H., 1978, Enzymology, genetics and regulation of membrane phospholipid synthesis in *Escherichia coli, Microbiol. Rev.* **42:**614.

Razin, S., 1978, The *Mycoplasmas, Microbiol. Rev,* **42:**414.

Raziuddin, S., 1976, Effect of growth temperature and culture age on the lipid composition of *Vibrio cholerae* 569B, *J. Gen. Microbiol.* **94:**367.

Reeves, R. E., Latour, N. G., and Lousteau, R. J., 1964, A glycerol galactofuranoside from the lipid of an anaerobe provisionally designated as *Bacteroides symbiosus, Biochemistry* **3:**1248.

Renkonen, O., 1961, A note on spectrophotometric determination of acyl ester groups in lipids, *Biochim. Biophys. Acta* **54:**361.

Renkonen, O., 1962, Determination of glycerol in phosphatides, *Biochim. Biophys. Acta* **56:**367.

Rilfors, L., Lindblom, G., Wieslander, A., and Christiansson, A., 1984, Lipid bilayer stability in biological membranes, in *Membrane Fluidity* (M. Kates and L. A. Manson, eds.), pp. 205–245, Plenum, New York.

Rinehart, K. L., 1982, Fast atom bombardment mass spectrometry, *Science* **218**:254.

Ritchie, R. G. S., Cyr, N., Korsch, B., Koch, H. J., and Perlin, A. S., 1975, Carbon-13 chemical shifts of furanosides and cyclopentanols. Configurational and conformational influence, *Can. J. Chem.* **53**:1424.

Rodriguez-Valera, F., Juez, G., and Kushner, D. J., 1983, *Halobacterium mediterranei*, spec. nov., a new carbohydrate-utilizing extreme halophile, *Syst. Appl. Microbiol.* **4**:369.

Ross, H. N. M., and Grant, W. D., 1985, Nucleic acid studies of halophile archaebacteria, *J. Gen. Microbiol.* **131**:165.

Ross, H. N. M., Collins, M. D., Tindall, B. J., and Grant, W. D., 1981, A rapid procedure for detection of archaebacterial lipids in halophilic bacteria, *J. Gen. Microbiol.* **123**:75.

Ross, H. N. M., Grant, W. D., and Harris, J. E., 1985, Lipids in archaebacterial taxonomy, in *Chemical Methods in Bacterial Systematics* (M. Goodfellow and D. E. Minnikin, eds.), pp. 289–299, Academic, Orlando, Florida.

Rottem, S., 1980, Membrane lipids of *Mycoplasmas, Biochim. Biophys. Acta* **604**:65.

Rottem, S., 1982, Transbilayer distribution of lipids in microbial membranes, *Curr. Top. Membr. Transp.* **17**:235.

Rouser, S., Fleischer, S., and Yamomoto, A., 1970, Two dimensional TLC separation of polar lipids and determination of phospholipids by phosphorus analysis of spots, *Lipids* **5**:494.

Saito, K., and Mukoyama, K., 1971, Diglucosyl diglyceride from *B. cereus, J. Biochem. (Tokyo)* **69**:83.

Sastry, P. S., 1974, Glycosyl glycerides, *Adv. Lipid Res.* **12**:251.

Sastry, P. S., and Kates, M., 1964, Lipid components of leaves. V. Galactolipids, cerebrosides and lecithin of runner-bean leaves, *Biochemistry* **3**:1271.

Schultz, J. C., and Elbein, A. D., 1974, Biosynthesis of glycosyldiglycerides in *Mycobacterium smegmatis, J. Bacteriol.* **117**:107.

Sen, A., Williams, W. P., and Quinn, P. J., 1981, The structure and thermotropic properties of pure 1,2-diacylgalactosylglycerols in aqueous systems, *Biochim. Biophys. Acta* **663**:380.

Shaw, N., 1968a, The lipid composition of *Microbacterium lacticum, Biochim. Biophys. Acta* **152**:427.

Shaw, N., 1968b, The detection of lipids on thin-layer chromatograms with the periodate–Schiff reagents, *Biochim. Biophys. Acta* **164**:435.

Shaw, N., 1970, Bacterial glycolipids, *Bacteriol. Rev.* **34**:365.

Shaw, N., 1974, Lipid composition as a guide to the classification of bacteria, *Adv. Appl. Microbiol.* **17**:63.

Shaw, N., 1975, Bacterial glycolipids and glycophospholipids, *Adv. Microbiol. Physiol.* **12**:141.

Shaw, N., and Pieringer, R. A., 1972, Biosynthesis of glucuronosyl diglyceride by particulate fractions of *Pseudomonas diminuta, Biochem. Biophys. Res. Commun.* **46**:1201.

Shaw, N., and Stead, D., 1970, A study of the lipid composition of *Microbacterium thermosphactum* as a guide to its taxonomy, *J. Appl. Bacteriol.* **33**:470.

Shaw, N., and Stead, D., 1971, Lipid composition of some species of *Arthrobacter, J. Bacteriol.* **107**:130.

Shaw, N., Heatherington, K., and Baddiley, J., 1968a, The glycolipids of *Lactobacillus casei, Biochem. J.* **107**:491.

Shaw, N., Smith, P. F., and Koostra, W. L., 1968b, The lipid composition of *Mycoplasma laidlawii*, strain B, *Biochem. J.* **107**:329.

Short, S., and White, D. C., 1970, Metabolism of glucosyldiglycerides and phosphatidylglucose of *Staphylococcus aureus, J. Bacteriol.* **104**:126.

Siakotos, A. N., and Rouser, G., 1965, Analytical separation of nonlipid water soluble substances and gangliosides from other lipids by dextran gel column chromatography, *J. Am. Oil Chemists Soc.* **42**:913.

Silvius, J. R., Mak, N., and McElhaney, R. N., 1980, Lipid and protein composition and thermotropic phase transitions in fatty acid-homogeneous membrane of *Acholeplasma laidlawii* B, *Biochim. Biophys. Acta* **597**:199.

Smallbone, B. W., and Kates, M., 1981, Structural identification of minor glycolipids in *Halobacterium cutirubrum, Biochim. Biophys. Acta* **665**:551.

Smith, P. F., 1969, Biosynthesis of glucosyldiglycerides by *Mycoplasma laidlawii* Strain B, *J. Bacteriol.* **99**:480.

Smith, P. F., 1972, Lipid composition of *Mycoplasma neurolyticum, J. Bacteriol.* **112**:554.

Smith, P. F., 1980, Sequence and glucoside bond arrangement of sugars in lipopolysaccharide from *Thermoplasma acidophilum*, *Biochim. Biophys. Acta* **619**:367.

Smith, P. F., 1981, Structure of the oligosaccharide chain of lipoglycan from *Acholeplasma*, *Biochim. Biophys. Acta* **665**:92.

Smith, P. F., 1986, Structures of unidentified lipids in *Acholeplasma laidlawii* A-EE22, *Biochim. Biophys. Acta* **879**:107.

Smith, P. F., Patel, K. R., and Al-Shamari, A. J. N., 1980, An aldehydophosphoglycolipid from *Acholeplasma granularum*, *Biochim. Biophys. Acta* **617**:419.

Smith, G. G., Ruwart, M. J., and Haug, A., 1974, Lipid phase transitions in membrane vesicles from *Thermoplasma acidophila*, *FEBS Lett.* **45**:96.

Snyder, F., and Stephens, N., 1959, A simplified spectrophotometric determination of ester groups in lipids, *Biochim. Biophys. Acta* **34**:244.

Stanek, J., Cerny, M., Kocourek, J., and Pacak, J., 1963, in *The Monosaccharides* (K. Mayer, transl.), pp. 50–60, Academic, New York.

Steim, J. M., 1967, Mnogalactosyldiglyceride: A new neurolipid, *Biochim. Biophys. Acta* **144**:118.

Steiner, S., Conti, S. F., and Lester, R. L., 1969, Separation and identification of the polar lipids of *Chromatium* strain D, *J. Bacteriol.* **98**:10.

Stern, N., and Tietz, A., 1973, Glycolipids of a halotolerant, moderate halophilic bacterium I. Biosynthesis of glucuronosyldiglyceride by cell-free particles, *Biochim. Biophys. Acta* **296**:136.

Stetter, K. O., and Zillig, W., 1985, *Thermoplasma* and the thermophilic sulfur-dependent archaebacteria, in *The Bacteria*, Vol. 8 (C. R. Woese and R. S. Wolfe, eds.) pp. 85–170, Academic Press, New York.

Stewart, L. C., 1988, Physical studies of diphytanylglycerol phospholipids, Ph.D. thesis, University of Ottawa, Ottawa, Canada, pp. 241–250.

Stewart, L. C., Kates, M., and Smith, I. C. P., 1988, Synthesis and characterization of deoxy analogs of diphytanylglycerol ether phospholipids, *Chem. Phys. Lipids* **48**:177.

Stewart, L. C., Ekiel, I., Kates, M., and Smith, I. C. P., 1990a, Biophysical studies of deuterated diphytanylglycerol phospholipids by ^{2}H-NMR spectrometry, *Chem. Phys. Lipids*, in press.

Stewart, L. C., Yang, P. W., Mantsch, H. H., and Kates, M., 1990b, Inter- and intramolecular hydrogen bonding in diphytanylglycerol phosphoryl glycerophosphate. An infrared spectroscopic investigation, *Biochem. Cell Biol.* **68**:266.

Sweeley, C. C., and Vance, D. E., 1967, Gas chromatographic estimation of carbohydrates in glycolipids, in *Lipid Chromatographic Analysis*, Vol. 1 (G. V. Marinetti, ed.), pp. 465–495, Marcel Dekker, New York.

Takayama, K., and Goldman, D. S., 1969, Pathway for synthesis of mannophospholipids in *Mycobacterium tuberculosis*, *Biochim. Biophys. Acta* **176**:196.

Teissie, J., Prats, M., Souicaille, P., and Tocanne, J. F., 1985, Evidence for conduction of protons along the interface between water and a polar lipid monolayer, *Proc. Natl. Acad. Sci. USA* **82**:3217.

Teissie, J., Prats, M., Lemassu, A., Stewart, L. C., and Kates, M., 1990, Structural control of lateral proton conduction in monolayers of phospholipids from extreme halophiles. *Biochemistry*, **29**:59.

Thirkell, D., and Summerfield, M., 1977, The membrane lipids of *Planococcus citreus* from cells grown in the presence of three difference concentrations of sea salt added to a basic medium, *Antonie von Leewenhoek* **43**:43.

Thurl, S., and Schäfer, W., 1988, Lipids from the sulfur-dependent archaebacterium *Thermoproteus tenax*, *Biochim. Biophys. Acta* **961**:253.

Tindall, B. J., Ross, H. N. M., and Grant, W. D., 1984, *Natronobacterium* gen. nov. and *Natronococcus* gen. nov. Two new genera of Haloalkaliphilic Archaebacteria, *Syst. Appl. Microbiol.* **5**:41.

Tomoaia-Cotisel, M., Zsako, J., Mocanu, A., Chifu, E., and Quinn, P. J., 1988, Monolayer properties of membrane lipids of the extreme halophile *Halobacterium cutirubrum*, at the air/water interface, *Biochim. Biophys. Acta* **942**:295.

Tornabene, T. G., 1973, Lipid composition of selected strains of *Yersinia pestis* and *Yersinia pseudotuberculosis*, *Biochim. Biophys. Acta* **306**:173.

Tornabene, T. G., and Langworthy, T. A., 1979, Diphytanyl and dibiphytanyl glycerol ether lipids of methanogenic archaebacteria, *Science* **203**:51.

Tornabene, T. G., and Ogg, J. E., 1971, Chromatographic studies of the lipid components of *Vibrio fetus, Biochim. Biophys. Acta* **239**:133.

Torreblanca, M. F., Rodriguez-Valera, F., Juez, G., Ventosa, A., Kamekura, M., and Kates, M., 1986, Classification of non-alkaliphilic *Halobacteria* based on numerical taxonomy and polar lipid composition, and description of *Haloarcula* gen. nov. and *Haloferax* gen. nov., *Syst. Appl. Microbiol.* **8**:89.

Trevelyan, W. E., Proctor, D. P., and Harrison, J. S., 1950, Detection of sugars on paper chromatograms, *Nature (Lond.)* **166**:444.

Usui, T., Yamoaka, N., Matsuda, K., Tuzimura, K., Sugiyama, H., and Seto, S., 1973, ^{13}C Nuclear magnetic resonance spectra of glucobioses, glucotrioses and glucans, *J. Chem. Soc. Perkin* **I**:2425.

van den Bosch, H., and Vagelos, P. R., 1970, Fatty acyl-CoA and fatty acylacyl carrier protein as acyl donors in the synthesis of lysophosphatidate and phosphatidate in *Escherichia coli, Biochim. Biophys. Acta* **218**:233.

Veerkamp, J. H., 1972, Biochemical changes in *Bifodobacterium bifidum* var. pennsylvanicum after cell wall inhibition. V. Structure of the galactosyl diglycerides, *Biochim. Biophys. Acta* **273**:359.

Veerkamp, J. H., and van Shaik, F. W., 1974, Biochemical changes in *Bifidobacterium bifidum* var. pennsylvanicum after cell wall inhibition. VII. Structure of the phosphogalactolipids, *Biochim. Biophys. Acta* **348**:370.

Vorbeck, M. L., and Marinetti, G. V., 1965a, Separation of glycosyl diglycerides from phosphatides using silicic acid column chromatography, *J. Lipid Res.* **6**, 3.

Vorbeck, M. L., and Marinetti, G. V., 1965b, Intracellular distribution and characterization of the lipids of *Streptococcus faecalis* ATCC 9790, *Biochemistry* **4**:296.

Wagner, G., Hartman, R., and Oesterhelt, D., 1978, Potassium uniport and ATP synthesis in *Halobacterium halobium, Eur. J. Biochem.* **89**:169.

Walker, R. W., and Bastl, C. P., 1967, The glycolipids of *Arthrobacter globiformis, Carbohydr. Res.* **4**:49.

Ward, J. B., 1981, Teichoic and teichuronic acids: Biosynthesis, assembly and location, *Microbiol. Rev.* **45**:211.

Wassef, M. K., 1976, Lipids of *Klebsiella pneumoniae:* The presence of phosphatidylcholine in succinate-grown cells, *Lipids* **11**:364.

Wells, M. A., and Dittmer, J. C., 1963, The use of Sephadex for removal of nonlipid contaminants from lipid extracts, *Biochemistry* **2**:1259.

White, D. C., and Frerman, F. E., 1967, Extraction, characterization and cellular localization of lipids of *Staphylococcus aureus, J. Bacteriol.* **94**:1854.

Whitman, W. B., 1985, Methanogenic bacteria, in *The Bacteria*, Vol. 8 (C. R. Woese and R. S. Wolfe, eds.), pp. 3–84, Academic, Orlando, Florida.

Wickberg, B., 1958a, Structure of a glyceritol glycoside from *Polysyphonia fastigiata* and *Corallina officinalis, Acta Chem. Scand.* **12**:1183.

Wickberg, B., 1958b, Synthesis of 1-glyceritol-D-galactopyranosides, *Acta Chem. Scand.* **12**:1187.

Wieslander, A., Ulmius, J., Lindblom, G., and Fontell, K., 1978, Water binding and phase structures for different *Acholeplasma laidlawii* membrane lipids studied by deuteron nuclear magnetic resonance and X-ray diffraction, *Biochim. Biophys. Acta* **512**:241.

Wieslander, A., Rilfors, L., Johansson, L. B.-Å., and Lindblom, G., 1981, Reversed cubic phase with glucolipids from *Acholeplasma laidlawii*, Biochemistry **20**: 730.

Wilkinson, S. G., 1968, Glycosyl diglycerides from *Pseudomonas rubescens, Biochim. Biophys. Acta* **164**:148.

Wilkinson, S. G., 1969, Lipids of *Pseudomonas diminuta, Biochim. Biophys. Acta* **187**:492.

Wilkinson, S. G., and Galbraith, L., 1979, Polar lipids of *Pseudomonas vesicularis*. Presence of a heptosyldiacylglycerol, *Biochim. Biophys. Acta* **575**:244.

Winken, A. J., and Knox, K. W., 1970, Studies on the group F antigen of *lactobacilli:* isolation of a teichoic acid–lipid complex from *Lactobacillus fermenti* NCTC 6991, *J. Gen. Microbiol.* **60**:293.

Woese, C. R., and Wolfe, R. S. (eds.), 1985, *The Bacteria*, Vol. 8: *Archaebacteria*, Academic, Orlando, Florida.

Yang, H. J., and Hakomori, S., 1971, A sphingolipid having a novel type of ceramide and lacto-*N*-fucopentaose III, *J. Biol. Chem.* **246**:1192.

Yang, L. L., and Haug, A., 1979, Structure of membrane lipids and physicobiochemical properties of the plasma membrane from *Thermoplasma acidophilum* adapted to growth at 37°C, *Biochim. Biophys. Acta* **573**:308.

Yribarren, M., Vilkas, E., and Rozanis, J., 1974, Galactosyl-diglycerides from *Actinomyces viscosus*, *Chem. Phys. Lipids* **12**:172.

Zepke, H. D., Heinz, E., Radunz, A., Linscheid, M., and Pesch, R., 1978, Combination and positional distribution of fatty acids in lipids of blue-green algae, *Arch. Microbiol.* **119**:157.

Zillig, W., Stetter, K. O., Schäfer, W., Janekovic, D., Wunderl, S., Holz, I., and Palm, P., 1981, *Thermoproteales*—a novel type of extremely thermoacidophilic anaerobic archaebacterium isolated from Icelandic solfataras, *Zentralbl. Bakteriol. Mikrobiol. Hyg., Abt I, Orig. C* **2**:205.

Zillig, W., Stetter, K. O., Prangishvilli, D., Schäfer, W., Wunderl, S., Janekovic, D., Holz, I., and Palm, P., 1982, *Desulfurococcaceae*, the second family of extremely thermophilic, anaerobic, sulfur-respiring *Thermoproteales, Zentralbl. Bakteriol. Mikrobiol. Hyg., Abt. I, Orig. C* **3**:304.

Bacterial Phosphoglycolipids and Lipoteichoic Acids

Werner Fischer

2.1. History and Nomenclature

Glycoglycerolipids, phosphoglycolipids, and lipoteichoic acids are characteristic constituents of the cytoplasmic membrane of a large number of gram-positive bacteria (Fischer, 1981; Ishizuka and Yamakawa, 1985). These lipid amphiphiles are structurally and biosynthetically related to each other. Glycolipids are the lipid moiety in phosphoglycolipids and lipoteichoic acids. Phosphoglycolipids carry a glycerophosphate, a di(glycerophosphate) or a phosphatidyl residue; the common types of lipoteichoic acid carry a 1,3-phosphodiester-linked poly(glycerophosphate) chain. By contrast, glycophospholipids are derived from phospholipids containing phosphatidylglycerol, cardiolipin, or phosphatidylinositol to which monosaccharides or oligosaccharides are glycosidically linked (for a review, see Ishizuka and Yamakawa, 1985).

2.1.1. Lipoteichoic Acids and Related Compounds

Teichoic acids were detected in 1958 by Baddiley and co-workers (Armstrong *et al.*, 1958, 1959) in a search for the role of CDP-glycerol and CDP-ribitol, which they had earlier discovered in gram-positive bacteria. The first representatives were isolated from cell walls and shown to be polymers of ribitol phosphate or glycerol phosphate usually bearing glycosyl residues or alanine ester or both. In view of the structural diversity recognized in subsequent work, the term teichoic acids was redefined to include all bacterial wall, membrane, or capsular polymers containing glycerophosphate or ribitol phosphate residues (Baddiley, 1972).

From teichoic (Greek: *teichos*, "wall") acids in a stricter sense a group of poly(glycerophosphates) was differentiated and named intracellular teichoic acids because they were extracted from whole cells but, unlike teichoic acids,

---

Werner Fischer • Institut für Biochemie der Medizinischen Fakultät, Universität Erlangen-Nürnberg, D-8520 Erlangen, Federal Republic of Germany.

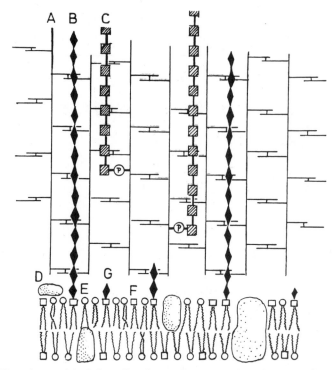

Figure 2-1. Tentative model of the cell wall–membrane complex of gram-positive bacteria. A, peptidoglycan; B, lipoteichoic acid; C, teichoic acid; D, protein; E, phospholipid; F, glycolipid; G, phosphoglycolipid. The orientation of lipoteichoic acid perpendicular to the membrane is arbitrary. An orientation parallel to the membrane might also be possible, as well as fluctuations between these positions. (Redrawn from Driel *et al.*, 1973.)

were absent from cell walls (McCarty, 1959; Critchley *et al.*, 1962). Later, intracellular teichoic acids were located in the cytoplasmic membrane and were therefore renamed membrane teichoic acids (Shockman and Slade, 1964). A third change of name, to lipoteichoic acids, was finally suggested when, 8 years later, teichoic acid lipid complexes were detected (Wicken and Knox, 1970) and subsequently recognized as amphiphiles in which the poly(glycerophosphate) is covalently linked to a membrane glycolipid by a phosphodiester bond (Toon *et al.*, 1972). Thus, as shown in Fig. 2-1, lipoteichoic acids are amphiphiles anchored in the cytoplasmic membrane by hydrophobic interaction, whereas teichoic acids are covalently linked through a particular linkage unit to the peptidoglycan of the cell wall (Coley *et al.*, 1978; Ward, 1981; Hancock and Baddiley, 1985). Many gram-positive bacteria contain both polymers, but lipoteichoic acids are more widespread, and their synthesis seems to be less dependent on growth conditions (Ellwood and Tempest, 1972; Wicken and Knox, 1975*a*).

In addition to the different anchoring in the wall–membrane complex, lipoteichoic acids differ from glycerophosphate-containing wall teichoic acids with respect to the stereochemical configuration and therefore the metabolic origin of their glycerophosphate residues (Fig. 2-2).

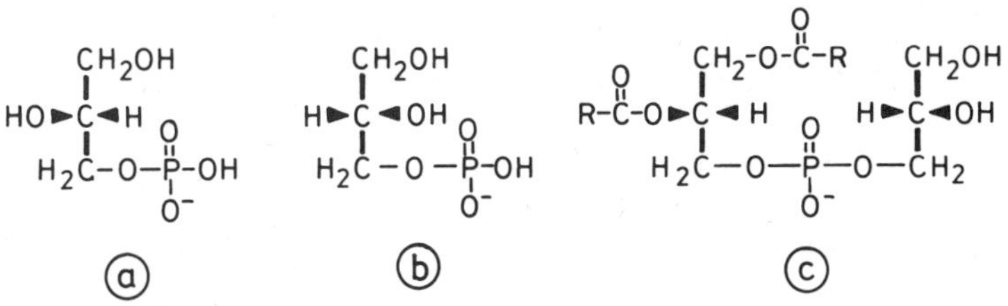

Figure 2-2. Stereochemical configuration of glycerophosphates. *sn*-glycero-3-phosphate (a) is present in wall teichoic acids and phosphatidylglycolipids, *sn*-glycero-1-phosphate (b) in lipoteichoic acids and glycerophosphoglycolipids. Phosphatidylglycerol (c) contains both enantiomers and serves as the biosynthetic donor of phosphatidyl and *sn*-glycero-1-phosphate residues.

In contrast to the structural diversity of wall teichoic acids (Archibald, 1974), all the lipoteichoic acids initially isolated had poly(glycerophosphate) structures. The resulting view that lipoteichoic acids are structurally homogeneous (Lambert *et al.*, 1977; Wicken and Knox, 1975a) is no longer tenable, however, since several exceptions are now known: e.g., in a number of *Streptococcus lactis* strains a poly(digalactosylglycerophosphate) lipoteichoic acid has been detected (Koch and Fischer, 1978; Schleifer *et al.*, 1985); the lipoteichoic acid of *Bifidobacterium bifidum* (Op den Camp *et al.*, 1984) was recently shown to be a lipopolysaccharide, carrying monoglycerophosphate branches (Fischer, 1987; Fischer *et al.*, 1987); and the lipoteichoic acid of *Streptococcus pneumoniae* seems to have a complex carbohydrate structure and to contain ribitol phosphate and choline phosphate in place of glycerophosphate residues (see Section 2.2.2).

A number of glycerophosphate-containing capsular polysaccharides in gram-negative bacteria seem to be lipid-linked amphiphiles. Although differing from lipoteichoic acids in the lipid moiety and the configuration of the glycerophosphate, their structures are briefly summarized in Section 2.2.5.

Several reviews on teichoic and lipoteichoic acids have appeared (Archibald and Baddiley, 1966; Archibald *et al.*, 1968; Baddiley, 1972; Knox and Wicken, 1973; Archibald, 1974; Wicken and Knox, 1975a; Lambert *et al.*, 1977; Ward, 1981; Fischer, 1988). Methodological aspects, in particular, have been considered by Archibald and Baddiley (1966).

2.1.2. Phosphoglycolipids

Several years after the discovery of glycolipids in gram-positive bacteria (Macfarlane, 1961), glycerophosphate-containing glycolipids were detected in these organisms (Ishizuka and Yamakawa, 1968; Fischer and Seyferth, 1968). Subsequent studies differentiated between phosphatidylglycolipids (Fischer, 1970; Wilkinson and Bell, 1971; Fischer *et al.*, 1973b) and glycerophosphoglycolipids (Shaw *et al.*, 1972; Fischer *et al.*, 1973a).

An important development was the finding that the glycerophosphate in glycerophosphoglycolipids had the *sn*-1 configuration (Fischer *et al.*, 1973a). This contrasted with the fact that CDP-glycerol, the common biosynthetic glycerophosphate donor, used, for example, in wall teichoic acid synthesis

(Ward, 1981), had been shown to have the *sn*-3 configuration (Baddiley *et al.*, 1956, 1958). The *sn*-glycero-1-phosphate moiety was therefore suggested to originate from the nonacylated glycerophosphate moiety of phosphatidylglycerol (Fischer *et al.*, 1973*a*). When 1 year later phosphatidylglycerol appeared to be the glycerophosphate donor in the biosynthesis of lipoteichoic acids (Emdur and Chiu, 1974; Glaser and Lindsay, 1974), a metabolic relationship between glycerophosphoglycolipids and lipoteichoic acids could be envisaged (Fischer, 1976). This, together with the subsequent demonstration that in a variety of gram-positive bacteria glycerophosphoglycolipids are the initial part of the respective lipoteichoic acid, led to the idea that glycerophosphoglycolipids might be intermediates or by-products of lipoteichoic acid biosynthesis or enzymatic breakdown products. The functioning of glycerophosphoglycolipids as biosynthetic intermediates was recently established in *Staphylococcus aureus* (Koch *et al.*, 1984).

In contrast to the widespread occurrence of glycerophosphoglycolipids, phosphatidylglycolipids are restricted to a small number of gram-positive bacteria (Fischer *et al.*, 1978*b*). The di-*O*-acyl-α-glycerophosphate residue of phosphatidylglycolipids has the *sn*-3 configuration and is biosynthetically derived from the acylated glycerophosphate moiety of phosphatidylglycerol (Fig. 2-2). Phosphatidylglycolipids, in addition to glycolipids, may serve as lipid anchors of lipoteichoic acids. Phosphoglycolipids have also been detected in a few Mycoplasmatales and gram-negative bacteria, in which they appear to function as membrane lipids (see Sections 2.2.4, 2.2.5).

2.2. Occurrence and Structure

2.2.1. Phosphoglycolipids and Poly(Glycerophosphate) Lipoteichoic Acids

The pioneer work of McCarty (1959) using antibodies against readily extractable poly(glycerophosphate) antigens from gram-positive bacteria has, in retrospect, provided a comprehensive survey of the distribution of poly(glycerophosphate) lipoteichoic acids (Table 2-1). Since then, their occurrence has been verified for many species by isolation and characterization (Knox and Wicken, 1973; Wicken and Knox, 1975*a*; Coley *et al.*, 1975; Fischer *et al.*, 1980*a,b*). It is noteworthy, however, that there are a number of gram-positive bacteria that do not contain poly(glycerophosphate) lipoteichoic acids (Table 2-1). They are also absent from gram-negative bacteria and yeasts (McCarty, 1959; Hamada *et al.*, 1976) but were recently found together with teichoic acids in *Streptomycetes* (Naumova *et al.*, 1980; Potekhina *et al.*, 1983).

An investigation into 33 strains of gram-positive bacteria revealed that glycerophosphoglycolipids are as widespread as poly(glycerophosphate) lipoteichoic acids and parallel their occurrence (Fischer *et al.*, 1978*b*). This investigation also revealed a structural variety of glycerophosphoglycolipids, which in subsequent studies was found to be due to differences in the glycolipid moiety and to the fact that particular glycerophosphoglycolipids reflect the chain elongation principle and the substitution of lipoteichoic acids as shown in Figs. 2-3–2-10.

Table 2-1. Distribution of Poly(Glycerophosphate)
Antigen in Various Gram-Positive Bacteria[a]

Bacteria containing the antigen
  Streptococcus[b]
    Group A, all types
    Groups B, C, D, E, F, G, K, L, N, R
    mutans, salivarius, milleri, sanguis (biotype A)
  Lactobacillus
    casei, plantarum, fermentum
  Leuconostoc mesenteroides
  Staphylococcus
    aureus, albus
  Bacillus
    subtilis, cereus, anthracis, megatherium

Bacteria with positive and negative strains
  Streptococcus[b]
    group H, *viridans, mitis*

Bacteria lacking the antigen
  Streptococcus[b]
    group O, *sanguis* (biotype B)
  Streptococcus pneumoniae
  Staphylococcus citreus
  Micrococcus
    luteus, citreus
  Corynebacterium
    diphtheriae, zerosis, pyogenes, ulcerans, ovis
  Clostridium
    histolyticum, septicum, tetani, sporogenes
  Actinomyces
    naeslundii, viscosus

---

[a]Based on the work of McCarty (1959), completed by data of
Hamada *et al.* (1976, 1980), and Rosan (1978).
[b]For novel classification of streptococci see Collins *et al.*
(1983), Schleifer and Kilpper-Bälz (1984), Kilpper-Bälz and
Schleifer (1984), Schleifer *et al.* (1984), Schleifer *et al.* (1985),
and references cited therein.

In various gram-positive bacteria, the glycolipid moiety of both lipoteichoic acid and glycerophosphoglycolipid proved identical (Figs. 2-3–2-5, 2-7–2-10). There is also strong evidence that it is the same position of the glycolipid to which the poly(glycerophosphate) chain or the glycerophosphate residue is linked (see Section 2.4.2.7). This, together with the common *sn*-1 configuration of the glycerophosphate residues, suggests that glycerophosphoglycolipids of most bacteria represent short-chain homologues of the respective lipoteichoic acid.

In addition to variations in the glycolipid structure, species and genus variations in lipoteichoic acids occur with respect to the length of the poly(glycerophosphate) chain and to the nature and degree of substitution at position 2 of their glycerophosphate residues (Table 2-2). With a single unbranched chain on the glycolipid moiety, the average length of the chain varies between

Table 2-2. Substitution and Chain Length of Poly(Glycerophosphate) Lipoteichoic Acids

Lipoteichoic acid source	Molar ratios to phosphorus		(GroP)$_n$	References[e]
	D-Alanine	Glycosyl substituents[a]		
Micrococcus varians, AICC 29750	None	None	32 ± 4	1,5
Bacillus megaterium, various strains	None	α-Gal 0.05	20–25	2
Lactobacillus casei, 7469, DSM 200021	0.56–0.63	None	40	3,4,5
Lactobacillus helveticus, DSM 20075	0.67	None	24	6
Streptococci group A	0.47–0.56	None	25	7,8
Lactococcus plantarum, NCDO 1869	0.45	None	29	9
Lactococcus raffinolactis, NCDO 617	0.52	None	27	9
Staphylococcus xylosus, DSM 20266	0.29–0.40	None	28	5
Staphylococcus aureus H, DSM 20233	0.30–0.87	α-GlcNAc[b] 0.03–0.10	25–30	7,10,11
Bacillus coagulans, various strains	None	α-Gal 0.40–0.43	17	2
Enterococcus faecalis, ATCC 9790	None	α-Glc$_{1-4}$[c] 0.82	20	1,12
Enterococcus faecalis Kiel 27738	0.48	α-Glc$_2$[c] 0.47	19 ± 1	1,5
Enterococcus faecalis, DSM 20478	0.23	α-Glc$_2$[c] 0.45	19 ± 1	5
Enterococcus faecalis, NCIB 8191	0.88	α-Glc$_{2,3}$[c] >0.90	n.d.	13
Enterococcus faecalis, NCIB 39	0.41	α-Glc$_{1,2}$[c] >0.90	n.d.	13
Lactococcus lactis, various strains	0.10–0.50	α-Gal 0.17–0.60	19–26	1,6,9
Listeria monozytogenes, AICC 15313	0.31	α-Gal 0.17	23	14,15
Listeria, various strains	0.21–0.36	α-Gal 0.11–0.34	16–33	14
Bacillus licheniformis, DSM 13	0.51	α-Gal, α-GlcNAc 0.02, 0.18	27	5
Bacillus licheniformis, AHU 1371	0.69	α-GlcNAc 0.15	28	2
Bacillus subtilis, W 23	0.40	α-Glc, α-GlcNAc 0.20, 0.21	24	16
Bacillus subtilis, various strains	0.35–0.55	α,β-Glc, α-GlcNAc[d] 0.04, 0.21–0.43	25–33	2

[a]Glycosyl residues, as far as studied, belong to the D-series and are in the pyranose form.

[b]Absent from Staphylococcus aureus H gol⁻¹ ΦR (W. Fischer, unpublished data), a mutant lacking the usual N-acetylglucosaminyl substituents on wall teichoic acid (Heckels et al., 1975).

[c]Structures: α-Glc(1— ; α-Glc(1—2)α-Glc(1— ; α-Glc(1—2)α-Glc(1—2)α-Glc(1— ; α-Glc(1—2)α-Glc(1—2)α-Glc(1— .

[d]Not present in all strains, anomeric form strain specific.

[e]References: 1, Fischer et al. (1981); 2, Iwasaki et al. (1986); 3, Kelemen and Baddiley (1961); 4, Nakano and Fischer (1978); 5, W. Fischer (unpublished data); 6, Fischer et al. (1980b); 7, Fischer et al. (1980a); 8, McCarty (1964); 9, Schleifer et al. (1985); 10, RajBhandari and Baddiley (1963); 11, Fischer and Rösel (1980); 12, Cabacungan and Pieringer (1985); 13, Wicken and Baddiley (1963); 14, Ruhland and Fiedler (1987); 15, Uchikawa et al. (1986); 16, Fischer and Koch (1981).

Table 2-3. Chain Composition of Substituted Poly(Glycerophosphate) Lipoteichoic Acids[a,b]

Bacillus subtilis			Lactococcus lactis		Enterococcus faecalis	
	W-23	Marburg	NCDO 712		ssp. *zymogenes*	
Gro[c]	0.23	0.28	Gro[c]	0.31	Gro[c]	0.23
AlaGro	0.42	0.38	AlaGro	0.21	AlaGro	0.29
GlcGro	0.20	0.17	GalGro	0.39	Glc_2Gro	0.29
GlcNAcGro	0.21	0.18	AlaGalGro	0.09	$AlaGlc_2Gro$	0.19

[a]From Fischer and Koch (1981).
[b]Measurements after hydrolysis with HF as described in Section 2.4.2.2*b*; values are molar ratios to phosphorus.
[c]Nonsubstituted glycerol.

18 and 40 glycerophosphate residues, being fairly constant for different strains of a given species or genus. Substituents of the glycerophosphate residues are either D-alanine ester (other amino acids have not been detected) or glycosyl residues, or both (Table 2-2). A lipoteichoic acid lacking any substituent was found in *Micrococcus varians.* If D-alanyl and glycosyl substituents occur at the same polymer, they may be linked to separate glycerol residues, as in the lipoteichoic acid of *Bacillus subtilis* (Table 2-3), or part of the D-alanine ester may be attached to the glycosyl residues, as in *Lactococcus lactis** and *Enterococcus faecalis** (Table 2-3).

Glycerophosphoglycolipids usually constitute a minor fraction of membrane lipids (Fischer *et al.*, 1978*b*). The content of glycerophosphoglycolipid VIII, shown as an example in Table 2-4, varies in a species-specific manner between 0.1 and 10 mole% of the total polar lipids. The content of the other glycerophosphoglycolipids in all bacteria studied lies in a similar range (Fischer *et al.*, 1978*b*). These low concentrations favor the idea of a metabolic role rather than a membrane structural role for glycerophosphoglycolipids.

The lipoteichoic acid content of various gram-positive bacteria was reported to range between 0.4 and 1.6% of the dry weight (Huff, 1982). Expressed as phosphorus or glycerol, approximately 120 µmole per g dry weight was found in *Lactobacillus fermentum* and *S. aureus* (Wicken *et al.*, 1973; Fischer *et al.*, 1983). The contribution of lipoteichoic acid phosphorus to the phosphate-containing polymers of *S. aureus* is shown in Table 2-5. The data suggest that lipoteichoic acid and teichoic acid are present in approximately equimolar amounts if one considers that the former contains 25 (Fischer *et al.*, 1980*a*) and the latter 45 phosphates residues per chain (Ghuysen *et al.*, 1965). A comparison of the content of lipoteichoic acid and polar lipids indicates that in the membrane of *Lactococcus lactis* or *S. aureus* about every tenth and twentieth lipid amphiphile, respectively, is lipoteichoic acid (Fischer, 1981; Koch *et al.*, 1984). If located exclusively in the outer layer of the membrane, the actual concentration may be twice as high. The long-chain lipoteichoic acids contribute considerably to the glycerol content of the membrane lipid amphiphiles,

*Novel genera, previously *Streptococcus lactis* (Schleifer *et al.*, 1985) and *Streptococcus faecalis* (Schleifer and Kilpper-Bälz, 1984), respectively.

Table 2-4. Proportion of Glycerophosphoglycolipid
VIII[a] to Total Membrane Lipids in Bacilli and
Staphylococcus aureus[b]

Organism	Mole %
B. licheniformis	10.5
B. subtilis W-23	3.8
B. subtilis Marburg	1.8
S. aureus	0.1–0.5

[a]For structure, see Fig. 2-5.
[b]From Fischer et al. (1978b).

amounting to 50% in *S. aureus* (Koch *et al.*, 1984) and to 25–30% in *E. faecalis* (Kessler and Shockman, 1979a; Carson *et al.*, 1979).

The structures of phosphoglycolipids and lipoteichoic acids are now presented according to their natural occurrence. Phosphoglycolipids are designated by roman numerals given to them in the original publications and used in recent reviews (Fischer, 1981; Ishizuka and Yamakawa, 1985).

Streptococcus hemolyticus D-58. This group A streptococcus, as shown in Fig. 2-3, contains four phosphatidylglycolipids (I–IV) and one glycerophosphoglycolipid (V) (Fischer *et al.*, 1973b; Fischer, 1976; Landgraf, 1976). These are derived from the membrane glycolipids: Glc(α1–3)DAG, Glc(α1—2)Glc(α1–3)DAG, and Glc(α1–2)Glc(α1–2)Glc(α1–3)DAG (Ishizuka and Yamakawa, 1968). The phosphatidyl residues contain *sn*-glycero-3-phosphate, while the glycerophosphoglycolipid contains the enantiomeric *sn*-glycero-1-phosphate (Fischer *et al.*, 1973b; Fischer, 1976). The structural relationship between the glycerophosphoglycolipid V and the lipoteichoic acid of this organism (Fischer *et al.*, 1980a) (Fig. 2-3) requires proof that the polyglycerophosphate is attached to the nonreducing glucosyl residue of the glycolipid moiety.

Enterococcus faecalis (previously Streptococcus faecalis). *E. faecalis* contains glycerophosphoglycolipid V (Fischer *et al.*, 1973a) and phosphatidylglycolipid III (Ambron and Pieringer, 1971; Fischer *et al.*, 1973b), but there is also a glycerophosphophosphatidylglycolipid, VI (Fischer and Landgraf, 1975), that carries both the phosphatidyl substituents of III and the glycerophosphate

Table 2-5. Phosphate-Containing Polymers of
Staphylococcus aureus[a]

Polymer	μmole Phosphorus / g dry bacteria
Lipoteichoic acid	125
Teichoic acid	275
Nucleic acids	550

[a]From Fischer et al. (1983).

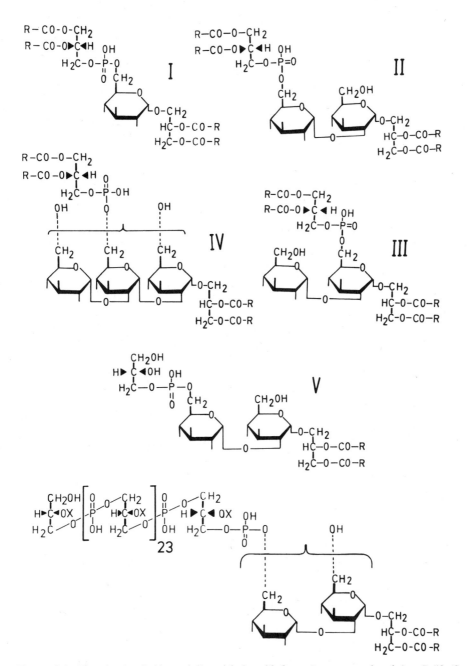

Figure 2-3. Phosphoglycolipids and lipoteichoic acid from *Streptococcus hemolyticus* D-58 (S. *pyogenes*, serological group A, type 3). −X = H or D-alanyl (see Table 2-2).

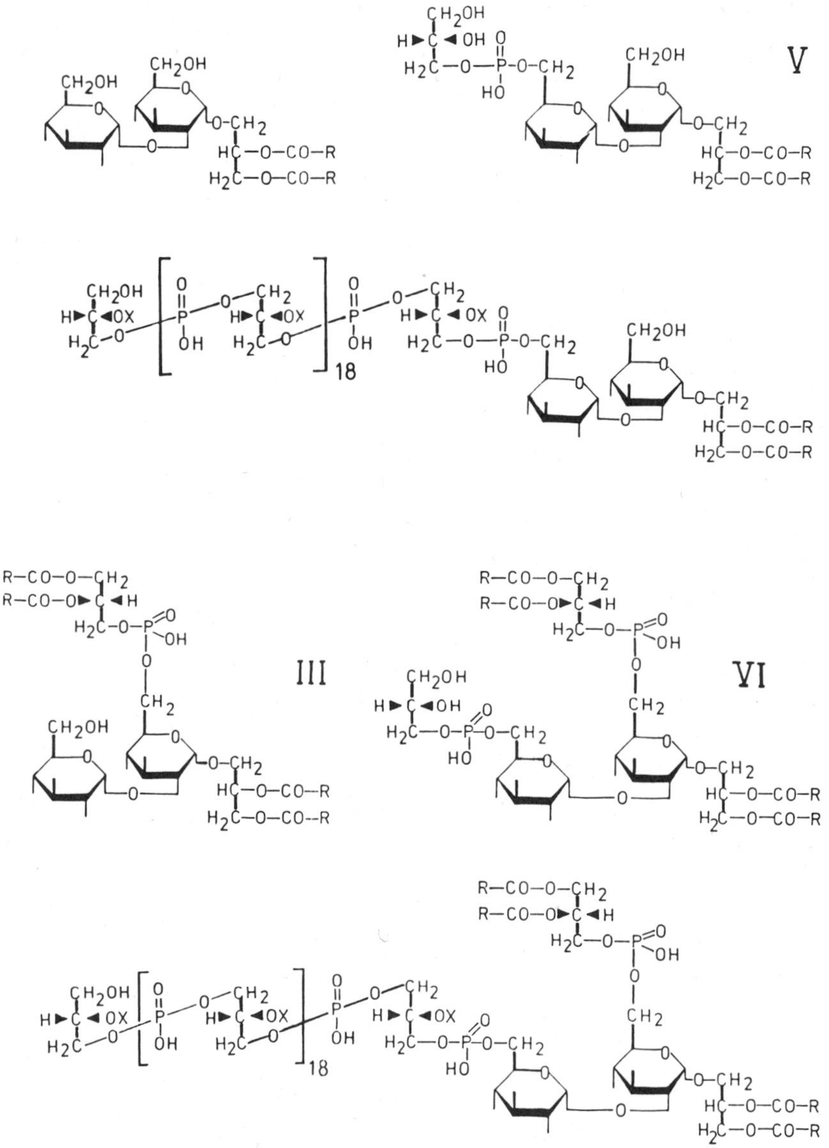

Figure 2-4. Glycolipid, phosphoglycolipids, and lipoteichoic acid species from *Enterococcus faecalis* (formerly *Streptococcus faecalis*). For substitution in lipoteichoic acids (−X), see Table 2-2 and 2-3. Small amounts of Glc(α1–3)acyl$_2$Gro and Ptd→6Glc(α1–3)acyl$_2$Gro are also present. (From Fischer *et al.*, 1978*b*.)

substituent of V at the respective positions of α-kojibiosyldiacylglycerol (Fig. 2-4). In one of the lipoteichoic acids of *E. faecalis*, the poly(glycerophosphate) is linked to phosphatidylglycolipid III, while in the other it is linked to di-glucosyldiacylglycerol (Toon *et al.*, 1972; Ganfield and Pieringer, 1975; Fischer, 1981, Fischer *et al.*, 1981, 1989), suggesting that lipids V and VI are the respective precursors of these two lipoteichoic acid species (Fig. 2-4).

Identical phosphoglycolipid and lipoteichoic acid species have been found in various strains and subspecies of *E. faecalis* (Fischer *et al.*, 1978*b*, 1981). The content of phosphatidylglycolipid III ranged between 8 and 28 mole% of total polar lipids (Fischer *et al.*, 1978*b*), which suggests a function for III as a membrane component in addition to that as a lipid anchor.

Listeria. In *Listeria monocytogenes*, a glycerophosphate-containing Gal(α1–2)Glc(α1–3)DAG and more highly acylated derivatives were observed (Shaw and Stead, 1972). The same glycolipid Gal(α1-2)Glc(α1–3)DAG was released by HF from the lipoteichoic acid together with diacylglycerol, which suggests that, as in *E. faecalis,* the lipid anchor is a mixture of glycolipid and the phosphatidyl derivative thereof (Hether and Jackson, 1983). A mixture of glycolipid and diacylglycerol was also obtained from the lipoteichoic acid of eight *Listeria* species by hydrolysis with HF (Ruhland and Fiedler, 1987). The presence of Gal(α1–2),Ptd→6Glc(α1–3)DAG in addition to the simple glycolipid as lipid anchor has recently been established: the lipoteichoic acid species containing the phosphatidylglycolipid amounts to 10–75% of the total of lipoteichoic acid (Uchikawa *et al.*, 1986; Fischer *et al.*, 1989). In contrast to *E. faecalis*, phosphatidylglycolipid is not detected among membrane lipids of *Listeria* species (W. Fischer and A. Gründler, unpublished data).

Bacillus licheniformis, Bacillus subtilis, and Staphylococcus aureus. The glycerophosphoglycolipid (VIII) and the lipoteichoic acid of these bacteria are derived from β-gentiobiosyldiacyl glycerol (Fig. 2-5), the main membrane glycolipid in these organisms (Fischer *et al.*, 1978*b*). In *Bacillus licheniformis*, the glycerophosphoglycolipid is accompanied by its D-alanyl derivative (Fig. 2-5, VIIIa), which reflects the chain substitution of the lipoteichoic acid (cf. Table 2-2). The content of lipid VIII remained constant through all stages of growth

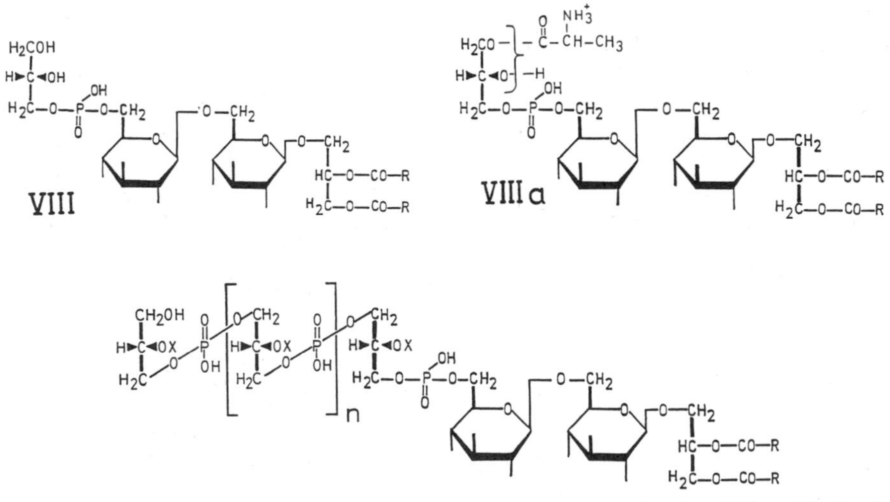

Figure 2-5. Glycerophosphoglycolipids (Fischer *et al.*, 1978*c*; Fischer, 1982) and lipoteichoic acid from *Staphylococcus aureus* (Duckworth *et al.*, 1975; Fischer *et al.*, 1980*a*), *Bacillus licheniformis* (Button and Hemmings, 1976a), and *Bacillus subtilis* (Iwasaki *et al.*, 1986). For chain length and substitution (−X) of lipoteichoic acids, see Table 2-2.

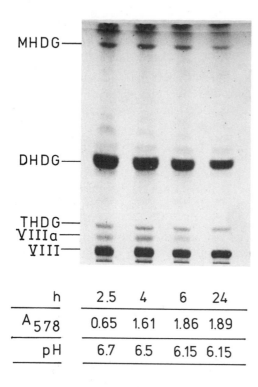

MHDG—

DHDG—

THDG—
VIIIa—
VIII—

h	2.5	4	6	24
A$_{578}$	0.65	1.61	1.86	1.89
pH	6.7	6.5	6.15	6.15

Figure 2-6. Glycerophosphoglycolipids during the growth of *Bacillus licheniformis*. Thin-layer chromatogram of crude lipid extracts on a silica gel plate, developed twice with chloroform, acetone, methanol, acetic acid, water (50:20:10:10:5, by vol.) and stained with α-naphthol/H$_2$SO$_4$. Growth stages are characterized by growth time (h), absorbance (A, 578 nm), and pH of the culture. *Abbreviations:* MHDG, Glc(β1–3)DAG; DHDG, Glc(β1–6)Glc(β1–3)DAG; THDG, Glc(β1–6)Glc(β1–6)Glc(β1–3)DAG. For glycerophosphoglycolipids VIII and VIIIa, see Fig. 2-5. (Adapted from Fischer *et al.*, 1978*b*.)

(Fig. 2-6), which may reflect a steady state between biosynthesis and conversion of the glycerophosphoglycolipid to lipoteichoic acid (see Section 2.5.3.1*c*).

Lipid VIII was also isolated and characterized from a halo-tolerant strain of *Staphylococcus epidermidis* (Komaratat and Kates, 1975). This strain responded to high NaCl concentrations in the growth medium with an increase in the percentage of glycerophosphoglycolipid VIII and cardiolipin (Table 2-6). This change suggested a role for these lipids in controlling the ion permeability of the cytoplasmic membrane (Komaratat and Kates, 1975).

Lactobacilli. As shown in Fig. 2-7, the glycerophosphoglycolipids and the

Table 2-6. Polar Lipid Composition[a] of S. epidermidis in Media Containing Increasing NaCl Concentrations[b]

Lipid component	NaCl concentration (g/dl)				
	0	5	10	15	25
Phosphatidylglycerol	67.1	69.7	69.0	62.4	49.6
Bisphosphatidylglycerol	0.5	0.7	0.9	1.6	10.8
GroP-6Glc(β1–6)Glc(β1–3)acyl$_2$Gro	5.3	6.0	7.2	9.7	14.5
Glc(β1–3)acyl$_2$Gro	6.6	5.9	4.0	3.5	5.2
Glc(β1–6)Glc(β1–3)acyl$_2$Gro	20.5	17.7	18.9	22.8	19.8

[a]Polar lipid composition in mole %.
[b]From Komaratat and Kates (1975).

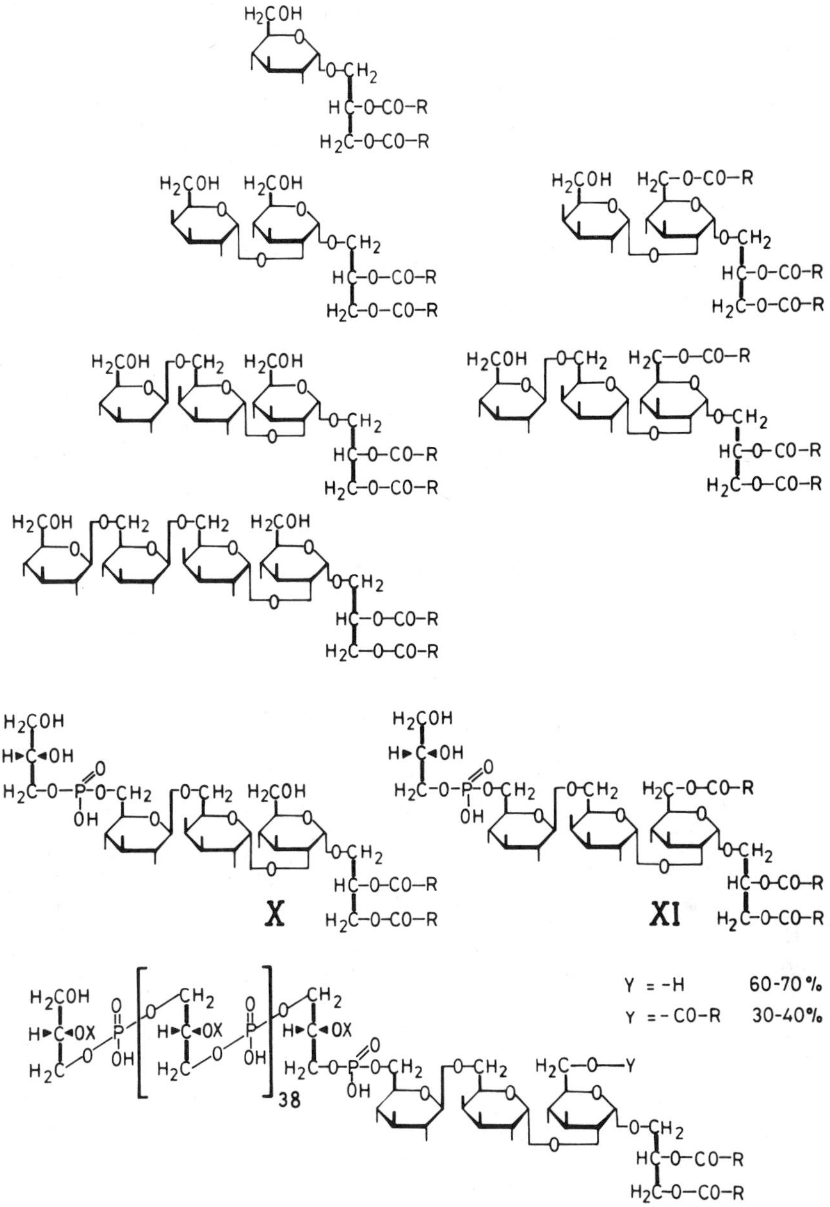

Figure 2-7. Membrane glycolipids, glycerophosphoglycolipids, and lipoteichoic acid from *Lactobacillus casei*. −X = H or D-alanyl (Table 2-2).

lipoteichoic acid of *Lactobacillus casei* contain trihexosyldiacylglycerol and a derivative thereof, which carries a third fatty acid ester on position 6 of the diacylglycerol-linked glucosyl moiety (Fischer *et al.*, 1978c; Nakano and Fischer, 1978). Of the six glycolipids present in *L. casei* (Nakano and Fischer, 1977), only two are involved in the synthesis of glycerophosphoglycolipids (X and XI) and lipoteichoic acid (Fig 2-7).

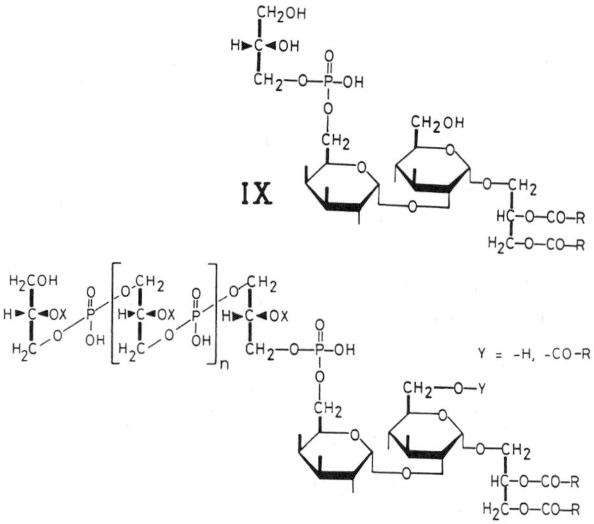

Figure 2-8. Glycerophosphoglycolipid and lipoteichoic acid from *Lactobacillus fermentum* (Fischer, W. and Fleischmann-Sperber, T., unpublished data) and *Streptococcus lactis motile* (Arneth, 1984; Fischer, W. and Arneth, D., unpublished data). For chain length, see Table 2-2; −X = H or D-alanyl (Table 2-2).

Glycerophosphoglycolipids X and XI are also found in *Lactobacillus helveticus* and *Lactobacillus plantarum* (Fischer *et al.*, 1978*b,c*) and represent, as in *L. casei*, the lowest homologues of the two lipoteichoic acid species in those bacilli (Fischer *et al.*, 1980*a,b*; Fleischmann-Sperber, 1982). In *Lactobacillus fermentum* which is lacking Glc(β1−6)Gal(α1−2)Glc(α1−3)DAG (Fischer *et al.*, 1978*b*), glycerophosphoglycolipid (IX) and the corresponding lipoteichoic acid contain Gal(α1−2)Glc(α1−3)DAG (Fig. 2-8). Gal(α1−2), acyl→6Glc(α1−3)DAG is present in the lipoteichoic acid but absent from the glycerophosphoglycolipid (W. Fischer and T. Fleischmann-Sperber, unpublished work).

Streptococcus lactis motile. Four strains of this taxonomically ill-defined species (Schleifer *et al.*, 1985) contained Gal(α1−2)Glc(α1−3)DAG, and Gal(α1−2),acyl→6Glc(α1−3)DAG as membrane lipids and the lipid anchor of lipoteichoic acid (Arneth, 1984; W. Fischer, unpublished data). As in *L. fermentum*, the triacylglycolipid was not detected in the glycerophosphoglycolipid fraction (Arneth, 1984).

Lactococcus lactis (previously Streptococcus lactis). As shown in Fig. 2-9, *Lactococcus lactis* NCDO 712 contains Glc(α1−2)Glc(α1−3)DAG and Glc(α1−2),acyl→6Glc(α1−3)DAG in both glycerophosphoglycolipids (V and VII) and lipoteichoic acid (Fischer *et al.*, 1978*c*, 1981). In addition to lipids V and VII, three unusually polar glycerophosphoglycolipids, XII, XIII, and XIV, were detected (Fischer *et al.*, 1978*b*). They were shown to carry two *sn*-glycero-1-phosphate residues linked to each other by a 1→3) phosphodiester bond (Laine and Fischer, 1978), in accord with the chain-elongation principle of lipoteichoic acids (Fig. 2-9).

Also reflected is the chain substitution (Fischer *et al.*, 1980*a,b*). One of the

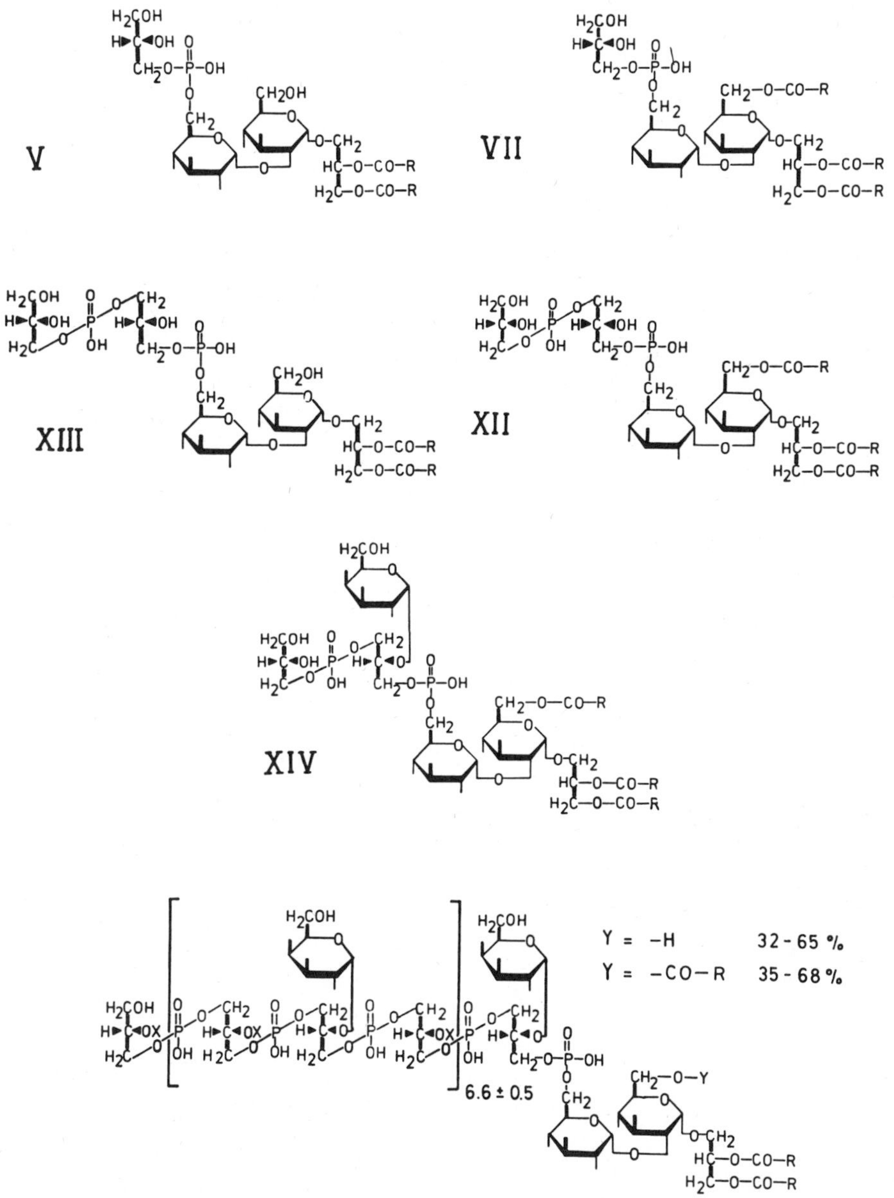

Figure 2-9. Glycerophosphoglycolipids and lipoteichoic acid from *Lactococcus lactis* NCDO 712 (formerly *Streptococcus lactis* NCDO 712). The galactosyl substituents in lipoteichoic acid are arbitrarily inserted; $-X = H$ or D-alanyl (Table 2-2). Lipids XIII and XIV were previously designated as XIVa and XIVb because of their similar chromatographic behavior (*cf.* Fig. 2-18).

glycerophosphate moieties of lipid XIV was found to carry an α-D-galactopyranosyl residue (Laine and Fischer, 1978) and, under appropriate conditions, derivatives of lipids V, XIII, and/or XIV could be isolated with a D-alanine ester on the glycerophosphate moiety (Fischer, 1982).

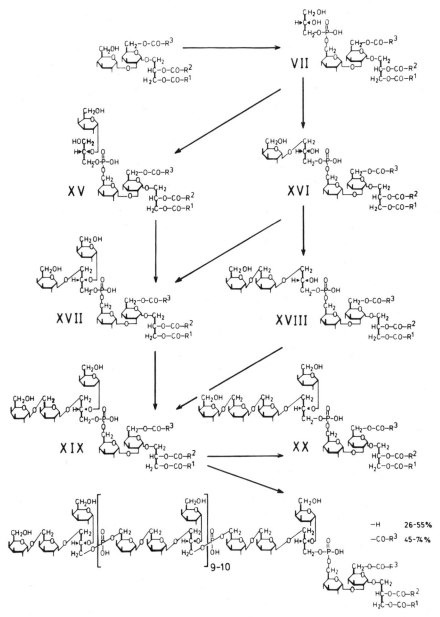

Figure 2-10. Glycerophosphoglycolipids and lipoteichoic acid from *Lactococcus garvieae* (formerly *Streptococcus lactis* Kiel 42172). Arrows indicate possible biosynthetic reactions.

2.2.2. Unusual Lipoteichoic Acid Structures and Related Macroamphiphiles in Gram-Positive Bacteria

Lactococcus garvieae. *Streptococcus lactis* Kiel 4217, now classified as *Lactococcus garvieae* (Schleifer *et al.,* 1985), possesses a set of unusual polar glycerophosphoglycolipids (Fischer *et al.,* 1987*b*). As shown in Fig. 2-10, they all carry a single glycerophosphate residue on the same glycolipid moiety but differ in the number of α-D-galactopyranosyl residues linked to the glycerophosphate moiety (Fischer *et al.*, 1979). The lipoteichoic acid of this bacterium was shown to have a poly(3-digalactosyl-2-galactosylglycero-1-phosphate) structure (Koch and Fischer, 1978) whose major repeating unit is identical to the trigalactosylglycerophosphate moiety of lipid XIX (Fig. 2-10). Minor repeating units in the lipoteichoic acid are the di- and tetragalactosylglycerophosphate, which are present in lipid XVIII and XX, respectively (Koch and Fischer, 1978; W. Fischer, unpublished data). The same poly(digalactosylgalactosylglycerophosphate) lipoteichoic acid was isolated from five strains of *Lactococcus garvieae* (Schleifer *et al.*, 1985). The number of repeating units per lipid anchor varied in the narrow range of 10.8–12.1 (W. Fischer, unpublished data).

As shown in Fig. 2-10, a biosynthetic sequence may be constructed which leads by three routes from lipid VII to lipid XIX, the direct precursor of the lipoteichoic acid. During exponential growth only the highly galactosylated compounds (XIX, XX) were detectable (Fig. 2-11). The lower homologues (XV–XVIII) and the unsubstituted glycerophosphoglycolipids (V, VI) did not

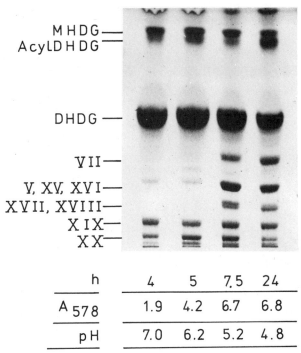

h	4	5	7.5	24
A 578	1.9	4.2	6.7	6.8
pH	7.0	6.2	5.2	4.8

Figure 2-11. Glycerophosphoglycolipids during the growth of *Lactococcus garvieae*. Thin-layer chromatogram of crude lipid extracts on a silica gel plate, developed twice with chloroform, methanol, ammonia (ρ = 0.91 g·ml^{-1}), water (60 : 35 : 4 : 4, v/v) and stained with α-naphthol/ H_2SO_4. Growth stages are characterized by growth time (h), absorbance (A, 578 nm), and pH of the culture. *Abbreviations*: MHDG, Glc(α1–3)DAG; AcylDHDG, Glc(α1–2),acyl–6Glc(α1–3)DAG; DHDG, Glc(α1–2)Glc(α1–3)DAG. For structures of glycerophosphoglycolipids V, VII, XV–XX, see Figs. 2-9 and 2-10. (Adapted from Fischer *et al.*, 1978*b*.)

$$—[\text{-}6\text{-}\beta\text{-}D\text{-}Glc}p\,(1\text{-}3)\text{-}\alpha\text{-}AATGal}p\,(1\text{-}4)\text{-}\alpha\text{-}D\text{-}Gal\,NAc}p\text{-}$$
$$(1\text{-}3)\text{-}\beta\text{-}D\text{-}Gal\,NH_2p(1\text{-}1')\text{ribitol-5- phosphoryl-}]—$$
$$\underset{\underset{\text{phosphorylcholine}}{6|}}{}$$

Figure 2-12. Repeating unit of the wall teichoic acid from *Streptococcus pneumoniae* (Jennings *et al.*, 1980). AATGal, 2-acetamido-4-amino-2,4,6-trideoxy galactose.

appear earlier than the stationary phase of growth, suggesting that they were kept at a low level during lipoteichoic acid biosynthesis.

Streptococcus pneumoniae. Although known since 1943 as lipocarbohydrate (Goebel *et al.*, 1943) and intensively studied for biological activity (Höltje and Tomasz, 1975*a,b;* Tomasz *et al.*, 1975; Briese and Hakenbeck, 1985), the structure of the lipoteichoic acid of *S. pneumoniae* has not yet been established. It is an amphiphile that contains fatty acid ester, ribitol, glucose, *N*-acetylgalactosamine, choline, and phosphorus (Fujiwara, 1967; Brundish and Baddiley, 1968; Briles and Tomasz, 1973). With respect to the hydrophilic components, it is similar to the wall teichoic acid of *S. pneumoniae* whose structure is shown in Fig. 2-12. In spite of the similarity in composition, lipoteichoic acid and wall teichoic acid are considered biosynthetically independent because the distribution of [^{3}H]choline pulse label between lipoteichoic acid and wall teichoic acid remained constant over a subsequent chase period (Briles and Tomasz, 1975).

Bifidobacterium bifidum. *Bifidobacterium bifidum* contains a set of glycolipids with β-D-galactosyl residues in pyranose and furanose form (Veerkamp, 1972). The major glycerophosphoglycolipid (XXI), shown in Fig. 2-13, is unique in so far as it is derived from a monohexofuranosyldiacylglycerol (Veerkamp and van Schaik, 1947). Uncommon also are the two polymeric structures, which were reported for the lipoteichoic acid containing a 1,2-linked poly(glycerophosphate) chain with either a β-D-glucan or a β-D-galactofuranan glycosidically linked to the glycerol terminus (Op den Camp *et al.*, 1984). Recent studies led to the proposal of the different structure depicted in Fig. 2-13, in which the constituents found by Op den Camp *et al.* (1984) are contained in a single lipoglycan (Fischer, 1987; Fischer *et al.*, 1987). The galactofuranan is linked to the glucan and this is joined to D-galactopyranosyldiacylglycerol. The glycerophosphate residues, having the *sn*-1 configuration, are attached in the form of monomeric side branches through phosphodiester bonds to C-6 of the D-galactofuranosyl residues; 20–50% of the glycerophosphate residues is substituted with alanine ester which, in contrast to D-alanine in common type lipoteichoic acids, is the L-isomer (Fischer, 1987). In the light of these results, *B. bifidum* seems to be the first grampositive bacterium in which the glycerophosphoglycolipid is structurally unrelated to lipoteichoic acid, and may therefore be a by-product rather than an intermediate in the synthesis of the polymer (cf. Fig. 2-13).

Glycerophosphate side branches were also found in the poly(ribitolphosphate) teichoic acid of *Streptomyces griseus* (Bews, 1967) and in the capsular

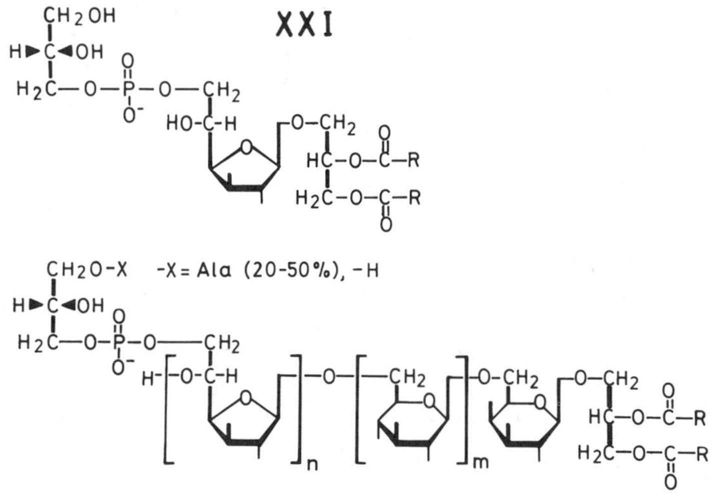

Figure 2-13. Phosphoglycolipid and lipoteichoic acid from *Bifidobacterium bifidum* spp. *pennsylvanicum*. n = 8–15, m = 6–9.

polysaccharide of type 18 (Estrada-Parra *et al.*, 1962) and type 11a (Kennedy, 1964) *Streptococcus pneumoniae.*

Actinomyces and Streptococcus sanguis biotype B. Species of both taxa lack serologically detectable poly(glycerophosphate) lipoteichoic acid (Hamada *et al.*, 1976, 1980). They contain amphiphilic heteropolysaccharides of different composition with small amounts of possibly monomeric glycerophosphate residues (Wicken *et al.*, 1978; Yamamoto *et al.*, 1985).

Mycobacteria. From *Mycobacterium leprae* and *Mycobacterium tuberculosis*, a lipoarabinomannan was isolated that, besides arabinose and mannose, contains glycerol, fatty acids, and myoinositol-1-phosphate and carries lactate and succinate in ester linkage (Hunter *et al.*, 1986). The presence of phosphatidylinositol as lipid anchor is unlikely because after alkaline release of inositolphosphate glycerophosphates were not detected and the glycerol was found glycosidically substituted at C-1 or C-3 in the polysaccharide backbone. The lipoarabinomannan is one of the dominant immunogens of *M. leprae* and immunologically cross-reactive with a like product from *M. tuberculosis*. The authors suggest that lipoarabinomannan may be the long-sought lipoteichoic acid-like polymer of mycobacteria.

Propionibacterium freudenreichii. Cells of this bacterium possess a lipoglycan that contains mannose, inositol, glycerol, fatty acids, and phosphate (Sutcliffe and Shaw, 1989). The presence of the components of phosphatidylinositol enabled the authors to suggest that the lipoglycan might be anchored to the membrane by covalently linked phosphatidylinositol, but other structures could not be excluded.

Micrococci. Micrococcus luteus, M. flavus, and *M. sodonensis* lack lipoteichoic acid (Powell, *et al.*, 1974, 1975) but possess instead an acidic lipomannan (Owen and Salton, 1975a; Pless *et al.*, 1975; Powell *et al.*, 1975). The polymer consists of 50–70 (1–2)-, (1–3)-, and (1–6)-linked D-mannopyranosyl residues

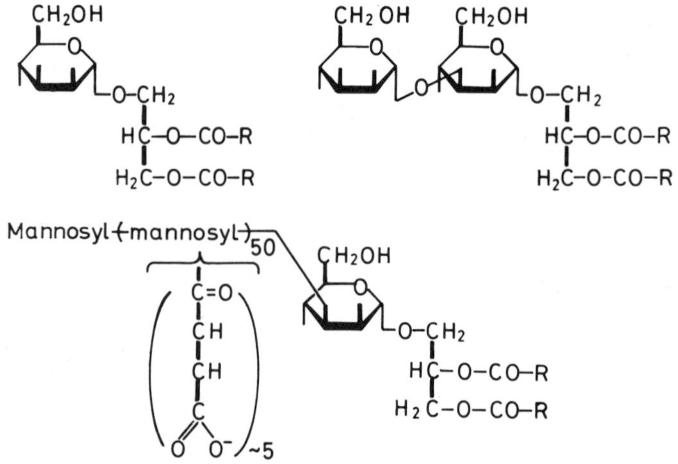

Figure 2-14. Glycolipids and succinylated lipomannan of *Micrococcus luteus*. Most, if not all mannosyl residues in the polymer are α-glycosidically linked.

with two branch points containing 2,4-substituted mannopyranosyl residues (Fig. 2-14); 10–25% of the mannosyl residues are substituted with ester-linked succinate. A neutral lipomannan, lacking succinyl residues, has been isolated from *M. agilis* (Lim and Salton, 1985). The lipid moiety of lipomannan is diacylglycerol (Pless *et al.*, 1975; Powell *et al.*, 1975), which makes it possible that the biosynthesis of the polymer starts from dimannosyldiayclglycerol (Fig. 2-14), the major glycolipid of *M. luteus* (Lennarz and Talamo, 1966).

A functional analogy of lipomannans to lipoteichoic acids has been proposed in view of their amphiphilic nature, membrane localization, net negative charge, and a similar Mg^{2+}-binding capacity (Powell *et al.*, 1975; Wicken and Knox, 1980). The glycerophosphate-containing lipoglycan of *B. bifidum* may now be considered a structural link between lipomannan and lipoteichoic acid.

2.2.3. Taxonomic Note

Instead of polyglycerophosphate lipoteichoic acids, members of the genera *Bifidobacterium, Actinomyces, Mycobacterium, Propionibacterium*, and *Micrococcus* contain lipoglycans that may carry glycerophosphate or possibly inositolphosphate, lactate, and succinate as substituents on the polysaccharide backbone. Interestingly, all these bacteria belong to one subdivision of gram-positive eubacteria whose DNAs contain more than 55% guanine plus cytosine (G + C), whereas the organisms containing poly(glycerophosphate) lipoteichoic acids all belong to the other subdivision in which DNAs have less than 50% G + C (Fox *et al.*, 1980; Woese, 1987). It has been suggested that phosphatidylinositol mannosides that are ubiquitous in actinomycetales may be lower homologues of the lipomannans and play a role as their lipid anchors as do glycerophosphoglycolipids in poly(glycerophosphate) lipoteichoic acids

(Brennan, 1988). This hypothesis, however, requires further evidence by more rigorous structural analysis of the lipomannans and in particular their lipid moieties.

2.2.4. Phosphoglycolipids in Mycoplasmatales

Mycoplasmatales are gram-positive eubacteria of the low G + C group that are unable to synthesize cell walls (Woese, 1987). A limited number of these organisms possess glycolipids; phosphoglycolipids were detected in three species (Langworthy, 1983).

Acholeplasma laidlawii contains glycerophosphodiglucosyldiacylglycerol and phosphatidyldiglucosyldiacylglycerol (Shaw *et al.*, 1972; Smith, 1972) whose structures (Fig. 2-15) are similar to bacterial phosphoglycolipids (see Fig. 2-3). The glycerophosphate residue of lipid XXII, however, has the enantiomeric *sn*-3-configuration so that, in contrast to the situation in bacteria, the glycerophospho- and phosphatidylglycolipid may be metabolically related.

A novel type of glycerophosphorylglycolipid was isolated from *Acholeplas-*

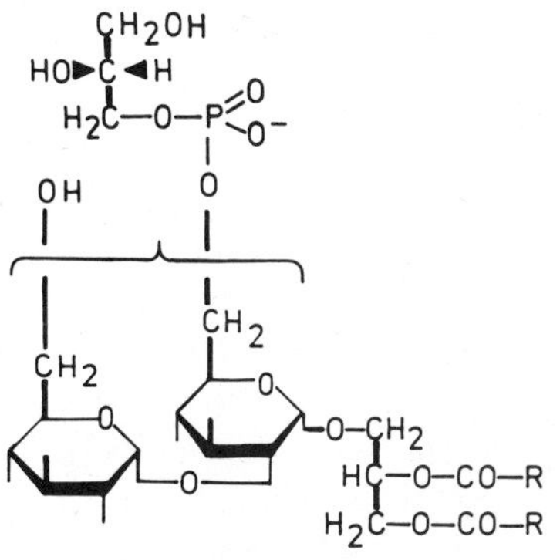

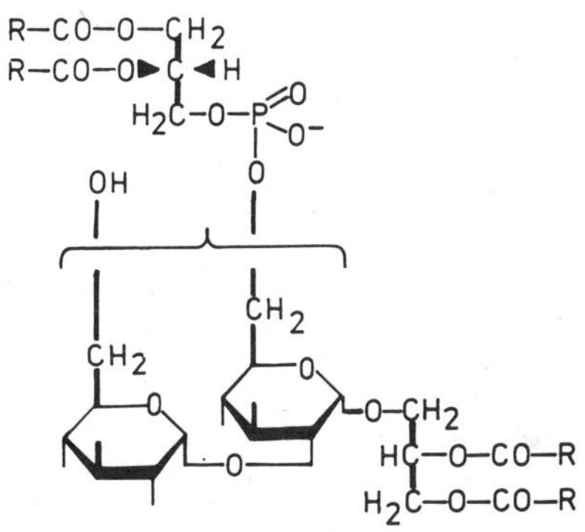

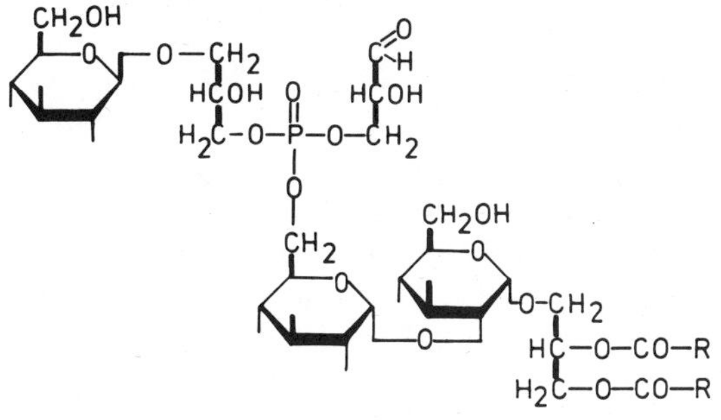

Figure 2-15. Phosphoglycolipids from *Acholeplasma laidlawii* (XXII, XXIII and *Acholeplasma granularum* (XXIV).

ma granularum (Smith *et al.*, 1980). It was proposed to be a glucosylglycero-phosphoryldiglucosyldiacylglycerol to which glyceraldehyde is attached through a phosphotriester bond (Fig. 2-15, lipid XXIV).

An isopranyl diether analogue of glycerophosphoglycosyldiacylglycerol was reported to occur as the major membrane lipid of the archaebacterium *Thermoplasma acidophilum* (Langworthy *et al.*, 1972). Recent investigations, however, suggest a diglycerol tetraether compound with glycerophosphate and unidentified carbohydrate separately attached to the opposite glycerol moieties (Langworthy *et al.*, 1982). This revised structure does not conform to the definition of phosphoglycolipids reviewed in this chapter. Phosphoglycolipids of archaebacteria are reviewed in Chapter 1 (*this volume*).

The high concentration of phosphoglycolipids in mycoplasmatales led to the speculation that these compounds accumulate because mycoplasmatales, unlike their proposed bacterial ancestors, lost the ability to synthesize lipoteichoic acid (Smith *et al.*, 1980; Langworthy, 1983). This hypothesis requires the presence of *sn*-glycero-1-phosphate residues in glycerophosphoglycolipids, which so far have not been detected.

2.2.5. Phosphoglycolipids, Lipoteichoic Acid, and Glycerophosphate-Containing Capsular Polysaccharides in Gram-Negative Bacteria

Pseudomonas diminuta, *Pseudomonas vesicularis*, and the rumen anaerobe *Butyrivibrio fibrisolvens* are so far unique among gram-negative bacteria, as they produce phosphoglycolipids, as does *B. fibrisolvens* lipoteichoic acid. The phosphoglycolipid of *P. diminuta* was identified as Ptd→6Glc(α1–3)DAG (Wilkinson and Bell, 1971), and the compound from *P. vesicularis* seems to possess the same structure (Wilkinson and Galbraith, 1979). The phosphoglycolipids of *B. fibrisolvens* were tentatively characterized as plasmalogen glycerophospho-D-galactofuranosyldiacylglycerol and a derivative containing four acyl chains (Clarke *et al.*, 1976). The lipoteichoic acid contains a poly(glycerophosphate) chain and a glucolipid moiety, the structure of which has not yet been definitely characterized (Hewett *et al.*, 1976). Although *B. fibrisolvens* is gram-negative when stained by the standard technique, this classification has been questioned on the basis of ultrastructural studies (Cheng and Costerton, 1977).

Certain species of various gram-negative genera produce capsular polysaccharides that contain glycerophosphate residues (Table 2-7). They are either linked to a polysaccharide as monomeric side chains or intercalated into the chain of poly(hexosylglycerophosphate) structures. Poly(glycerophosphate) chains have not been reported. Since, as far as has been studied, the glycerophosphate residues have the *sn*-3 configuration (Table 2-7), the capsular polysaccharides are reminiscent of wall teichoic acids rather than of lipoteichoic acids, and CDP–glycerol rather than phosphatidylglycerol may serve as the glycerophosphate donor in their biosynthesis (cf. Ward, 1981). There is increasing evidence that the capsular polysaccharides of gram-negative bacteria are amphiphiles, being anchored in the outer membrane by hydrophobic interaction (Gotschlich *et al.*, 1981; Schmidt and Jann, 1982;

Table 2-7. Glycerophosphate-Containing Capsular Polysaccharides of Gram-Negative Bacteria

Polysaccharide from	Structure	References

Escherichia coli 0100 : K(B):H2[a]	→2Rha(1–3)Rha(1–3)Gal(1–4)GlcNAcl→
	 3↑
	   O
	   ‖
	 O—P—O-Gro
	   |
	   O⁻
		Jann *et al.* (1970; unpublished results)

Escherichia coli K2[a]
 Escherichia coli K62[b]
	O
	               ‖
	 →5-α-D-Gal*f*(1–2)Gro-3-O—P—O→
	               |
	 alternating with O⁻
	               O
	               ‖
	 →4-α-D-Gal*p*(1–2)Gro-3-O—P—O→
	               |
	               O⁻
		Jann *et al.* (1980); Jann and Schmidt (1980); Fischer *et al.* (1982)

Neisseria meningitidis serogroup Z	O
	             ‖
	 →3-α-D-GalNAc(1–1)Gro-3-O—P—O→
	             |
	             O⁻
		Jennings *et al.* (1979)

Neisseria meningitidis serogroup H	O
	          ‖
	 →4-βD-Gal*p*(1–2)Gro-3-O—P—O→
	  /‾\ |
	 3-OAc 2-OAc O⁻
		van der Kaaden *et al.* (1984)

Pasteurella hemolytica serotype T15	O
	          ‖
	 →4-α-D-Gal*p*(1–2)Gro-3-O—P—O→
	  /‾\ |
	 3-OAc 2-OAc O⁻
		Adlam *et al.* (1985)

[a]The glycerophosphate residues have the *sn*-3 configuration (Fischer *et al.*, 1982; W. Fischer, unpublished data).
[b]Acetylated derivative of *Escherichia coli* K2 polysaccharide.

Kuo *et al.*, 1985). In the examples investigated, the lipid moiety was a di-acylglycerol linked by a phosphodiester bond to the reducing end of the polysaccharide chain.

2.2.6. Fatty Acid Composition of Phosphoglycolipids and Lipoteichoic Acids

The constituent fatty acids of membrane lipids in gram-positive bacteria and their metabolism have been reviewed by Shaw (1974), Lechevalier (1977), and Fulco (1983). Enterococci, lactococci, lactobacilli, and group A and B streptococci contain straight-chain saturated and monounsaturated fatty acids, and often a C_{19} cyclopropane fatty acid (*cis*-11,12-methylenoc-

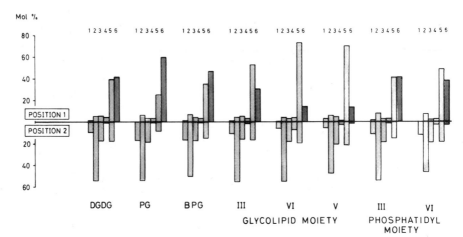

Figure 2-16. Positional fatty acid distribution in the glycolipid, phospholipids, and the glycolipid and phosphatidyl moiety of phosphoglycolipids in *Enterococcus faecalis* (Fischer *et al.*, 1973*b*; Fischer and Landgraf, 1975; Fischer, 1976). Fatty acids: (1) tetradecanoic, (2) hexadecanoic, (3) hexadecenoic, (4) octadecanoic, (5) *cis*-Δ^{11}-octadecenoic, (6) *cis*-11,12-methylenoctadecanoic acid. DGDG, Glc(α1–2)Glc(α1–3)DAG; PG, phosphatidylglycerol; BPG, bisphosphatidylglycerol. For phosphoglucolipid structures (III, V, VI), see Fig. 2-4.

tadecanoic acid) as well, derived from *cis*-Δ^{11}-octadecenoic acid. Iso- and ante-iso-branched fatty acids are the predominant acyl groups in the lipids of *Bacillus, Micrococcus,* and *Staphylococcus* species.

In a variety of gram-positive bacteria, glycolipids, phospholipids, phosphoglycolipids, and lipoteichoic acids have a very similar fatty acid composition, apparently reflecting the connection of the biosynthetic pathways of these amphiphiles through the membrane diacylglycerol pool (see Fig. 2-46) (see Section 2.5.3.3.). In some cases, quantitative differences in fatty acid composition between the lipid classes suggest preference of certain molecular species at biosynthetic branch points or postsynthetic modifications by de- and reacylation. These two types of fatty acid distribution among lipid amphiphiles are illustrated in Figs. 2-16 and 2-17.

In *Enterococcus faecalis*, the diacylglycerol moieties of glyco-, phospho-, and phosphoglycolipids display not only a similar fatty acid composition but also the same distribution pattern: shorter-chain fatty acids (C_{14}, C_{16}) are accumulated at position 2, longer-chain fatty acids (C_{18}, C_{19}) at position 1 (Fig. 2-16). Only the contents of C_{19}–cyclopropane fatty acid and of octadecenoic acid, its biosynthetic precursor are variable. Since the sum of these fatty acids is fairly constant, and cyclopropanization occurs on completed lipids (Thomas and Law, 1966), the difference may reflect different accessibility to cyclopropane synthetase. The lipoteichoic acid of *Enterococcus faecalis* shares the low content of cyclopropane fatty acid with the phosphoglycolipids (Table 2-8) its potential biosynthetic precursors (Fig. 2-4). In *Lactococcus lactis* and *Lactococcus garvieae*, the lipoteichoic acids also have a distinctly lower content of C_{19}–cyclopropane fatty acid than that of the total

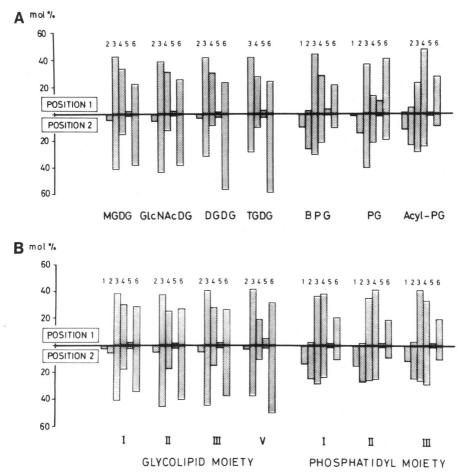

Figure 2-17. Positional fatty acid distribution in the lipids of *Streptococcus hemolyticus* D-58 (Landgraf, 1976). The upper panel shows the positional distribution of fatty acids in glycolipids and phospholipids, the lower panel the positional distribution of fatty acids in the glycolipid and phosphatidyl moiety of phosphatidylglycolipids: I, Ptd-6Glc(α1–2)DAG; II, Ptd-6Glc(α1–2)Glc(α1–3)DAG; III, Glc(α1–2),Ptd-6Glc(α1–3)DAG, and V, Gro*P*-6Glc(α1–2)Glc(α1–3)DAG. (For analysis of fatty acid distribution in phosphoglycolipids, see Fig. 2-24). Fatty acids: (1) Dodecanoic, (2) tetradecanoic, (3) hexadecanoic, (4) hexadecenoic, (5) octadecanoic, (6) *cis*-Δ^{11}-octadecenoic acid. MGDG, Glc(α1–3)DAG; GlcNAcDG, GlcNAc(α1–3)DAG; DGDG, Glc(α1–2)Glc(α1–3)DAG; TGDG, Glc(α1–2)Glc(α1–2)Glc(α1–3)DAG; BPG, bisphosphatidylglycerol; PG, phosphatidylglycerol; AcylPG, acylphosphatidylglycerol (diacylglycerol moiety shown).

membrane lipid (Table 2-8), which suggests that their location, and possibly their synthesis in the outer layer of the cytoplasmic membrane (see Section 2.7), renders them and their precursors less accessible to cyclopropane synthetase.

In *Streptococcus hemolyticus* D-58, glycolipids and phospholipids form two different groups insofar as phospholipids contain a higher proportion of shorter-chain fatty acids (C_{12}, C_{14}) at the expense of octadecenoic acid (Fig.

Table 2-8. Percentage Fatty Acid Composition of Lipoteichoic Acids and Membrane Lipids in Various Gram-Positive Bacteria

	14:0	14:1	16:0	16:1	18:0	18:1	19cy[n]
Enteroccus faecals[a,f]							
Lipoteichoic acid	5.7	—	37.7	8.7	—	40.0[k]	7.9
Total lipds	3.4	—	37.7	7.5	—	35.8[k]	15.5
Lactococcus lactis[b,g]							
Lipoteichoic acid	12.6	—	30.3	3.3	1.9	35.1[k]	16.9
Total lipids	7.7	0.4	21.7	3.9	2.3	24.0[k]	40.9
Lactococcus garvieae[c,g]							
Lipoteichoic acid	11.4	0.5	33.5	4.5	2.9	43.2[k]	2.1
Total lipids	· 4.5	0.5	28.5	2.8	4.7	48.5[k]	10.6
Micrococcus varians[d,g]							
Lipoteichoic acid	26.9	1.6	15.7	12.5	—	43.8[l]	—
Total lipids	25.6	1.9	7.1	22.4	—	42.5[l]	—
Bifidobacterium bifidum[e,h]							
Macroamphiphile[i]	6.5	—	33.5	0.9	13.6	45.6[m]	—
Total lipids	4.6	—	38.7	1.2	12.7	42.7[m]	—

[a]NCIB 8191.
[b]NCDO 712.
[c]Kiel 42172.
[d]ATTCC 29750.
[e]DMS 20239.
[f]Toon *et al.* (1972).
[g]Fischer (unpublished data).
[h]Fischer *et al.* (1987).
[i]For structure of the macroamphiphile, see Fig. 2-12.
[k]Predominantly cis-Δ11-octadecenoic acid.
[l]Octadecenoic acid of unknown structure.
[m]Predominantly cis-Δ9-octadecenoic acid.
[n]Cis-11,12-methylenoctadecanoic acid.

2-17A). This naturally affects the positional distribution, so that in the glycolipids position 2 contains more octadecenoic acid in place of the shorter-chain fatty acids, which are present at position 2 of the phospholipids. In the phosphoglycolipid series (Fig. 2-17B), the glycolipid moieties have the distribution pattern of their biosynthetic glycolipid precursors, whereas the phosphatidyl moieties reflect the pattern found in bisphosphatidylglycerol from which they may be derived by biosynthesis (see Section 2.5.1).

Fatty acid composition similar to that of the respective membrane lipids was also seen in the lipoteichoic acid of *Bacillus licheniformis* and *Staphylococcus aureus* (Table 2-9) and the macroamphiphile of *Bifidobacterium bifidum* (Table 2-8). This similarity in fatty acid composition between macroamphiphiles and membrane lipids is in contrast to the situation in gram-negative bacteria, in which lipopolysaccharides contain hydroxy fatty acids, which are absent from membrane lipids (Lüderitz *et al.*, 1982), and in the enterobacterial common antigen (Mayer and Schmidt, 1979) and capsular polysaccharides (Gotschlich *et al.*, 1981; Schmidt and Jann, 1982) in which palmitic acid is highly concentrated, compared with membrane lipids.

Table 2-9. *Percentage Fatty Acid Composition of Lipoteichoic Acids and Membrane Lipids in Bacillus licheniformis and Staphyloccus aureus*

	Fatty Acids[a]														
	14i	14	15i	15ai	16i	16	17i	17ai	17	18i	18	19i	19ai	19	20
Bacillus licheniformis[b]															
Lipoteichoic acid	6.0	Tr[d]	45.0[e]		28.0	6.0	5.6	13.0[e]	—	—	—	—	—	—	—
Glycolipid	2.0	Tr	41.0[e]		31.0	12.0	5.5	14.0[e]	—	—	—	—	—	—	—
Staphylococcus aureus[c]															
Lipoteichoic acid	0.4	Tr	7.4	30.9	0.8	3.6	5.6	12.2	0.6	1.0	16.5	2.9	4.7	1.4	13.4
Total lipids	0.8	0.5	9.8	36.6	1.2	2.1	5.5	10.9	0.5	1.0	8.8	4.1	5.0	1.7	11.5

[a]Abbreviations: *i* and *ai*, iso- and anteiso-branched fatty acids.
[b]Strain 6346MH-1 (Button and Hemmings, 1976a).
[c]Strain DSM 20233 (W. Fischer, unpublished data).
[d]Trace amount.
[e]Incompletely resolved mixture of iso- and anteiso-branched fatty acid.

Table 2-10. Behavior of Phosphoglycolipids on Thin-Layer Chromatography (TLC) and Column Chromatography (LC) on DEAE Cellulose (Fischer, 1984)

No.	Compound	TLC[a] ($R_F \times 100$)			LC[b]
		A	B	C	NH$_4$OAc (g/liter)
I	Ptd-6Glc(α1–3)acyl$_2$Gro	65	—	—	0.1
II	Ptd-6Glc(α1–2)Glc(α1–3)acyl$_2$Gro	41	—	—	0.1
III	Glc(α1–2),Ptd-6Glc(α1–3)acyl$_2$Gro	34	—	—	0.5
IV	Ptd-[Glc(α1–2)Glc(α1–2)Glc(α1–3)acyl$_2$Gro]	25	—	—	1
V	GroP-6Glc(α1–2)Glc(α1–3)acyl$_2$Gro	5	30	21	2–4
VI	GroP-6Glc(α1–),Ptd-6Glc(α1–3)acyl$_2$Gro	7	—	—	—
VII	GroP-6Glc(α1–2),acyl-6Glc(α1–3)acyl$_2$Gro	—	43	36	1–2
VIII	GroP-6Glc(β1–6)Glc(β1–3)acyl$_2$Gro	—	23	—	—
IX	GroP-6Gal(α1–2)Glc(α1–3)acyl$_2$Gro	—	30	—	—
X	GroP-6Glc(β1–6)Gal(α1–2)Glc(α1–3)acyl$_2$Gro	—	21	—	10
XI	GroP-6Glc(β1–6)Gal(α1–2),acyl-6Glc(α1–2)acyl$_2$Gro	—	32	—	4
XII	(GroP)$_2$-6Glc(α1–2),acyl-6Glc(α1–3)acyl$_2$Gro	—	—	13	10
XIII	(GroP)$_2$-6Glc(α1–2)Glc(α1–3)acyl$_2$Gro	—	—	5	20
XIV	Gal(GroP)$_2$-6Glc(α1–2),acyl-6Glc(α1–3)acyl$_2$Gro	—	—	5	20
XV, XVI	GalGroP-6Glc(α1–2),acyl-6Glc(α1–3)acyl$_2$Gro	—	30	—	2
XVII, XVIII	Gal$_2$GroP-6Glc(α1–2,acyl-6Glc)α1–3)acyl$_2$Gro	—	22	—	4
XIX	Gal$_3$GroP-6Glc(α1–2,acyl-6Glc)α1–3)acyl$_2$Gro	—	10	—	10
XX	Gal$_4$GroP-6Glc(α1–2,acyl-6Glc)α1–3)acyl$_2$Gro	—	5	—	25
	Glc(α1–2)Glc(α1–3)acyl$_2$Gro[c]	39	60	53	0
	Glc(α1–2)Glc(α1–2)Glc(α1–3)acyl2Gro[c]	19	32	24	0
	PtdGro[c]	37	—	54	0.5–1
	Ptd$_2$Gro[c]	43	—	56	2

[a] Chromatography on silica gel plates (Merck 60). Solvents: A, Chloroform, acetone, methanol, acetic acid, water 50:20:10:10:5 (v/v); B, chloroform, methanol, 25% ammonia, water 65:35:4:4 (by vol); C, chloroform, methanol, 25% ammonia, water 65:35:3:3 (v/).

[b] The concentration of ammonium acetate (NH$_4$OAc) in chloroform/methanol (2:1, v/v) is given at which the main fraction of the respective lipid was eluted.

[c] Included for comparison.

2.3. Detection, Isolation, and Purification

On extraction of gram-positive bacteria with appropriate mixtures of organic solvents and water, lipoteichoic acids separate into the aqueous layer and phosphoglycolipids together with the other membrane lipids into the organic phase. The critical size of the hydrophilic moiety that renders amphiphiles water soluble has not yet been determined. However, even the most polar phosphoglycolipids listed in Table 2-10 may be recovered as salts of divalent cations from the organic layer.

2.3.1. Phosphoglycolipids

Phosphoglycolipids are extracted by the procedure of Bligh and Dyer (1959) as modified by Kates (1986). Before phase partition NaCl or $CaCl_2$ is added (Folch *et al.*, 1957), in order to avoid losses of phosphoglycolipids into the aqueous layer (M. Nakano, unpublished data). Lipids that contain amino acids in ester linkage require pH 4–5 at all steps in their isolation (Houtsmuller and van Deenen, 1965; Fischer, 1982); otherwise, the labile linkage hydrolyzes spontaneously. If the lipid composition is to be determined quantitatively, the extraction should be carried out with mechanically disintegrated cells (Fischer *et al.*, 1983; Koch *et al.*, 1984) in order to avoid incomplete extraction of cardiolipin from whole bacterial cells (Bertsch *et al.*, 1969; Filgueiras and Op den Kamp, 1980; Koch *et al.*, 1984).

Phosphoglycolipids in the crude lipid extract are detected by two-dimensional thin-layer chromatography (TLC) (Fig. 2-18) and staining for sugars with α-naphthol/H_2SO_4 (Jacin and Mischkin, 1965) and for phosphate with the reagent of Dittmer and Lester (1964). Glycerophosphoglycolipids, which react very faintly with the latter stain, may be tentatively recognized by their high polarity (low mobility on TLC) (Table 2-10) and by their characteristic reaction with periodate–Schiff reagent (Shaw, 1968).

For isolation of phosphoglycolipids, column chromatography on DEAE cellulose (Rouser *et al.*, 1967) is the procedure of choice (Landgraf, 1976; Fischer, 1984). Uncharged and zwitterionic lipids are eluted with chloroform–methanol mixtures, and phospholipids and phosphoglycolipids by a stepwise gradient of ammonium acetate in chloroform–methanol (2 : 1, v/v). As summarized in Table 2-10 the anionic lipids elute in the order of increasing polarity (cf. phosphatidylglycerol and lipid V). Of great influence, too, are the number and distribution of acyl groups whose hydrocarbon chains exert a shielding effect against hydrogen bonding and electrostatic interaction between lipid and column material (cf. lipids II and V or lipids II and III, respectively) (Table 2-10). The same factors also determine the behavior of the lipids on adsorption TLC (Table 2-10).

Final purification of phosphoglycolipids is achieved by column chromatography on silicic acid (Landgraf, 1976) or, more readily, by preparative TLC on silica gel plates. Appropriate solvent systems are given in Table 2-10. Purified phosphoglycolipids stored as sodium or calcium salts in chloroform–methanol (2 : 1, v/v) at −20°C have remained stable for more than 10 years.

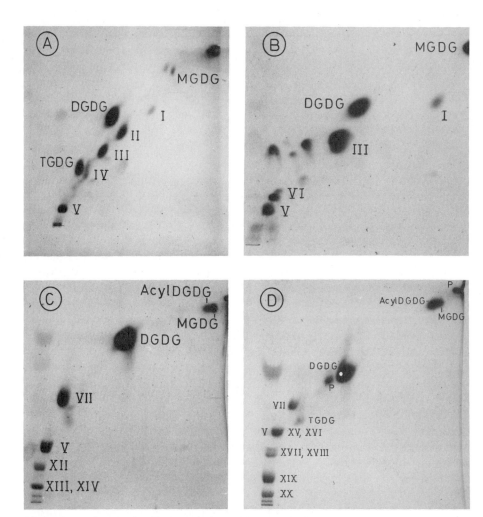

Figure 2-18. Phosphoglycolipids and glycolipids in crude lipid extracts from (A) *Streptococcus hemolyticus* D-58, (B) *Enterococcus faecalis* Kiel 7064, (C) *Lactococcus lactis* NCDO 712, and (D) *Lactococcus garvieae* Kiel 42172. Phosphoglycolipids are designated by their roman numerals (for structures, see Figs. 2-3, 2-4, 2-9, and 2-10). MGDG, Glc(α1–3)DAG; Acyl-DGDG, Glc(α1–2),acyl-6Glc(α1–3)DAG; DGDG, Glc(α1–2)Glc(α1–3)DAG; TGDG, Glc(α1–2)Glc(α1–2)Glc(α1–3)DAG; P, pigment. Two-dimensional thin-layer chromatograms of crude lipid extracts on silica gel plates. Development in the first dimension: plates A and B, once with chloroform–methanol–water (65 : 25 : 4, v/v); plate C, three times with chloroform–methanol–ammonia (sp.gr. = 0.91 g·ml^{-1})–water (60 : 30 : 3 : 3, v/v); plate D, twice with chloroform–methanol–ammonia (sp.gr. = 0.91 g·ml^{-1})–water (60 : 35 : 4 : 4, v/v). Development in the second dimension: plates A–D, once with chloroform–acetone–methanol–acetic acid–water (50 : 20 : 10 : 10 : 5, v/v). Glycolipids and phosphoglycolipids were stained with α-naphthol/H$_2$SO$_4$. [From Fischer and Landgraf, 1975 (B); Landgraf, 1976 (A); Laine and Fischer, 1978 (C); Fischer *et al.*, 1979 (D).]

2.3.2. Lipoteichoic Acids

In culture fluids and crude cellular extracts, lipoteichoic acids are readily detected by serological methods (Knox and Wicken, 1981a) employing antisera usually specific for the poly(glycerophosphate) chain rather than for carbohydrate substituents (for reviews, see Knox and Wicken, 1973; Wicken and Knox, 1975a). For semiquantitative estimation, the hemagglutination reaction has been used, in which the greatest dilution that will fully sensitize erythrocytes to antibody, is determined (Markham *et al.*, 1975). A more recent development is crossed or rocket immunoelectrophoresis by which, under appropriate conditions, lipoteichoic acid and its deacylation product can be separated and quantitatively determined (Kessler *et al.*, 1979). Enzyme-linked immunosorbent assay (ELISA) has also been reported (Kessler and Thivierge, 1983). In this study, it was shown that glycosyl and alanyl substituents may substantially interfere with the binding of lipoteichoic acid to antipolyglycerophosphate antibodies and cause errors in quantitating lipoteichoic acids of unknown composition by immunological techniques. In phenol–water extracts from bacteria, lipoteichoic acids can be detected specifically and quantitated by phosphate determination using a rapid scaled-down version of hydrophobic interaction chromatography (Koch *et al.*, 1984). In defatted cells of bacteria that do not contain poly(glycerophosphate) wall teichoic acids, lipoteichoic acid can be measured as the glycerol released by hydrolysis with HF (Hurst *et al.*, 1975; Fischer *et al.*, 1983), only a few percent of the glycerol being derived from the linkage unit of wall teichoic acid (Fischer *et al.*, 1983).

Because of the neighboring phosphate groups (Kelemen and Baddiley, 1961; Shabarova *et al.*, 1962) alanine ester substituents of the glycerol residues of lipoteichoic acids are, as shown in Table 2-11, susceptible to spontaneous base-catalyzed hydrolysis (Childs and Neuhaus, 1980; Fischer *et al.*, 1980b). For the isolation of alanyl lipoteichoic acids, all operational steps must therefore be done at low pH and, as far as possible, at low temperature (Fig. 2-19). Under these conditions, most of the original D-alanine ester substituents are preserved (Fischer *et al.*, 1980b, 1983; Fischer and Rösel, 1980).

Table 2-11. Base-Catalyzed Hydrolysis of Alanine Ester of Lipoteichoic Acids[a]

Buffer (0.1 M)	pH	Temp. (°C)	$t_{1/2}$[b] (hr)
Tris–HCl	8.0	25	2.2
PIPES	7.0	25	28
MES	6.0	37	155
Sodium acetate	5.0	37	415
Sodium formate	4.0	37	590

[a]From Fischer and Koch (1981).
[b]Calculated from first-order velocity plots.

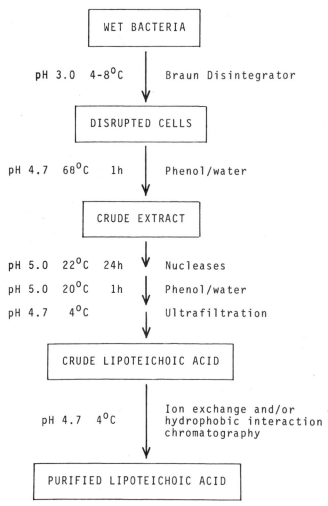

Figure 2-19. Flow sheet for extraction and purification of lipoteichoic acid. When hydrophobic interaction chromatography is used treatment with nucleases and subsequent phenol/water extraction may be omitted (see text). (From Fischer *et al.*, 1983.)

The standard extraction procedure for lipoteichoic acids uses hot* or cold phenol–water. Originally introduced for the extraction of lipopolysaccharides from gram-negative bacteria (Westphal *et al.*, 1952), it was in retrospect first adapted to lipoteichoic acids by Moskowitz (1966) for the extraction of a "red cell sensitizing substance" from streptococci. Referring to earlier work of Burger and Glaser (1964), Wicken and Knox used phenol–water extraction in their discovery of lipoteichoic acid (1970) and compared it with other extraction procedures (Wicken *et al.*, 1973).

Phenol–water extraction yields lipids and some proteins in the lower

*At 68°C, phenol and water are miscible in all proportions. Mixtures containing equal volumes of cell suspension in buffer and 80–95% (w/v) aqueous phenol, as used for extraction of amphiphiles, yield two phases after cooling.

phenol-rich layer, denatured proteins and cell walls at the interface, and nucleic acids and lipoteichoic acid in the upper aqueous layer. The complete partition of lipids into the phenol layer (Fischer *et al.*, 1983) makes the frequently performed additional extraction with chloroform–methanol unnecessary.

With most bacteria, complete extraction of lipoteichoic acid is only achieved when the cell walls have previously been disintegrated (Fischer and Koch, 1981; Huff, 1982; Fischer *et al.*, 1983; Card and Finn, 1983). If whole cells are extracted the recovery may be as low as 10%. Mechanical disintegration at low pH and low temperature (Fig. 2-19) is preferable to digestion with lysozyme because during incubation lipoteichoic acid may be deacylated by endogenous lipases (Kessler and Shockman, 1979*b;* Kessler and van Rijn, 1979; Kessler *et al.*, 1979). The often reported extraction of deacylated lipoteichoic acid together with the native form may be an artifact, being avoidable if the bacteria are harvested at low temperature and are disintegrated immediately at pH 3. Before phenol–water extraction, the pH is adjusted to 5 because at lower values the extractability is greatly decreased (Moskowitz, 1966).

Crude extracts usually contain large amounts of nucleic acids (cf. Fischer *et al.*, 1983), which, however, in most cases can be effectively removed by treatment with nucleases and subsequent dialysis or ultrafiltration (Wicken *et al.*, 1973; Coley *et al.*, 1975; Fischer *et al.*, 1980*b*). Co-extracted polysaccharides, wall teichoic acids, and proteins have presented more serious problems (Coley *et al.*, 1975). Gel permeation chromatography, the standard purification procedure since the first isolation of lipoteichoic acid (Wicken and Knox, 1970), has frequently proved ineffective (Coley *et al.*, 1975; Nakano and Fischer, 1978; Silvestri *et al.*, 1978). Consequently, at the expense of reduced recoveries, efforts were made to avoid contamination by using membrane fractions and ʟ forms or protoplasts (Coley *et al.*, 1975). Lectin affinity chromatography has been used in the purification of a galactosyl-containing lipoteichoic acid (Wicken and Knox, 1975*b*) and of lipomannan from *Micrococcus* species (Powell *et al.*, 1975; Lim and Salton, 1985). Purification by incorporating lipoteichoic acid into lecithin liposomes has also been tried (Silvestri *et al.*, 1978).

These problems have been overcome by the introduction of anion exchange and hydrophobic interaction chromatography (Fiedler and Glaser, 1974*a;* Nakano and Fischer, 1978; Fischer *et al.*, 1983). Anion-exchange chromatography is done on columns of DEAE Sephacel from which, in the presence of Triton X-100, lipoteichoic acid is completely eluted by a buffered salt gradient (Fig. 2-20). Omission of the detergent results in retarded and incomplete elution of lipoteichoic acid. Hydrophobic interaction chromatography has been successfully performed on columns of octyl Sepharose (Fischer *et al.*, 1983). In the standard procedure the sample is applied to the column in buffer, containing 10–15% (by vol.) propanol, and lipoteichoic acid is eluted by a buffered propanol gradient (Fig. 2-21). If the sample is added in the absence of propanol, the binding capacity is diminished, and hydrophobic complexation with protein may completely prevent the interaction of lipotei-

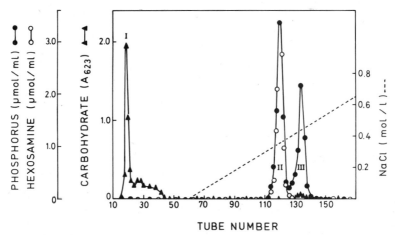

TUBE NUMBER

Figure 2-20. Separation of *Micrococcus varians* lipoteichoic acid (III) from contaminating polysaccharide (I) and poly(N-acetylglucosaminylphosphate) teichoic acid (II) by ion-exchange chromatography on a column of DEAE-Sephacel. The sample was applied to the column in 0.1 M sodium acetate buffer of pH 4.7, containing 0.05% (w/v) Triton X-100. The column was eluted with this buffer–detergent solution containing the NaCl concentrations indicated. (From Fischer *et al.*, 1983.)

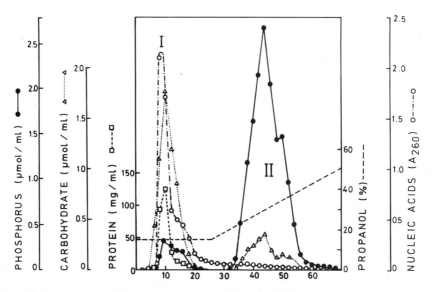

Figure 2-21. Separation of *Lactobacillus casei* lipoteichoic acid (II) from polysaccharide, protein, and nucleic acids (I) by hydrophobic interaction chromatography on a column of octyl Sepharose. The sample was applied to the column in 0.1 M sodium acetate buffer of pH 4.7 containing 15% (v/v) propan-1-ol. The column was eluted with the same buffer containing the propanol concentrations indicated. (From Fischer *et al.*, 1983.)

choic acid with the hydrophobic matrix (Fischer *et al.*, 1983). Hydrophobic binding of lipoteichoic acid is also absolutely dependent on the presence of salts. NaCl and sodium acetate buffer of pH 4.7 at a concentration of 50 mM are equally effective (W. Fischer, unpublished data).

In view of the following advantages, hydrophobic interaction chromatography has become the procedure of choice for isolation of lipoteichoic acids (Fischer *et al.*, 1983; W. Fischer, unpublished data). Large amounts of nucleic acids are separated so effectively from lipoteichoic acid that the digestive step (Fig. 2-19) can be omitted. Negatively charged wall teichoic acids, which cause problems on DEAE Sephacel, are clearly separated from lipoteichoic acid by their lack of fatty acids. Contamination of native acylated lipoteichoic acid with lyso-derivatives and deacylation products is recognized and at least the latter may be effectively removed (Fig. 2-22). Clearly distinguishable are also molecular species which contain two and three or two and four fatty acid ester groups (Fischer *et al.*, 1983; Fischer, 1987). An important advantage is finally the ready removal of propanol as compared with Triton X-100. Scaled-down versions of the described extraction and purification procedures have been worked out for use in labeling experiments (Koch *et al.*, 1984).

2.3.3. Behavior of Isolated Lipoteichoic Acids in Water

Hydrophobic aggregation of crude and purified lipoteichoic acids in aqueous solution is indicated by the behavior on ultracentrifugation and gel

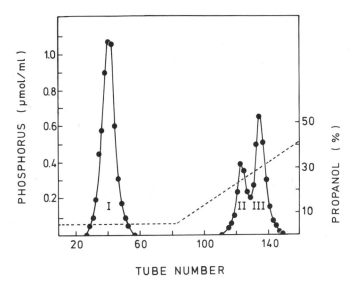

Figure 2-22. Separation of native *Staphylococcus aureus* lipoteichoic acid (III), containing two fatty acids, from the lyso-compound (II) and the fully deacylated derivative (I). Column chromatography on octyl Sepharose. The sample was applied to the column in 0.1 M sodium acetate buffer of pH 4.7 containing 5% (v/v) propan-1-ol. The column was eluted with the same buffer containing the propanol concentrations indicated. (From Fischer *et al.*, 1983.)

permeation chromatography (Wicken and Knox, 1970; Toon *et al.*, 1972; Briles and Tomasz, 1973). After separation from membrane lipids, lipoteichoic acids, because of their large head groups, may form micelles rather than liposomes, which has recently been confirmed by X-ray scattering studies (H. Labischinski, D. Naumann, and W. Fischer, unpublished data). Similar micellar ultrastructure was observed for the diverse lipoteichoic acid structures shown in Figs. 2-5, 2-9, and 2-12. Over a temperature range of 5–53°C, fatty acids were in the disordered, molten α-type. An average micelle of *Staphylococcus aureus* lipoteichoic acid was estimated to contain about 150 molecules arranged in a spherical assembly with a diameter of 18 nm and a thickness of the hydrophilic shell of approximately 6.5 nm. Since the chain contains on the average 25 glycerophosphate residues, each glycerophosphate unit may contribute about 0.26 nm to the thickness of the hydrophobic shell. If compared with a theoretical value of approximately 0.7 nm for the fully extended conformation of the chain, the value observed indicates a highly coiled conformation. The size of micelles appeared to be nearly independent of the degree of alanine substitution of lipoteichoic acid and the ionic strength of the solution. These results indicate that the ultrastructure of lipoteichoic acid in water differs from that of lipopolysaccharides which form highly ordered bilayer structures and show several phase transition temperatures (Labischinski *et al.*, 1985; Naumann *et al.*, 1987).

Determined with dyes, the critical micellar concentration (CMC) of various lipoteichoic acids were in the range of 5×10^{-6} M (Courtney *et al.*, 1986) and 2.8×10^{-7} M to 6.9×10^{-7} M (Wicken *et al.*, 1986), respectively. These values are close to 7×10^{-6} M, the CMC of palmitoylglycerophosphorylcholine (Haberland and Reynolds, 1975), and three to four orders of magnitude higher than the value for dipalmitoylglycerophosphorylcholine (4.6×10^{-10} M) (Smith and Tanford, 1972). Since lipoteichoic acids contain two to four fatty acid residues per molecule, each having 14–18 carbon atoms (Table 2-8), their CMC seems to be considerably increased by electrostatic repulsion between the anionic glycerophosphate chains. For ionic amphiphiles, increased CMC and decreased micellar size, as compared with uncharged analogues, have been well established (Tanford, 1980). As expected for ionic amphiphiles, increasing salt concentration causes a decrease in the CMC of lipoteichoic acids (Courtney *et al.*, 1986). It will be of interest to see whether the CMC of lipoteichoic acids approach that of phospholipids if they are incorporated, presumably remote from each other, in the lipid layer of membranes.

2.4. Structure Determination

2.4.1. Phosphoglycolipids

The degradative steps used in the structure determination of phosphoglycolipids are shown in Figs. 2-23 and 2-24. Lipid VI was chosen as an example because it contains both a phosphatidyl and a glycerophosphate residue.

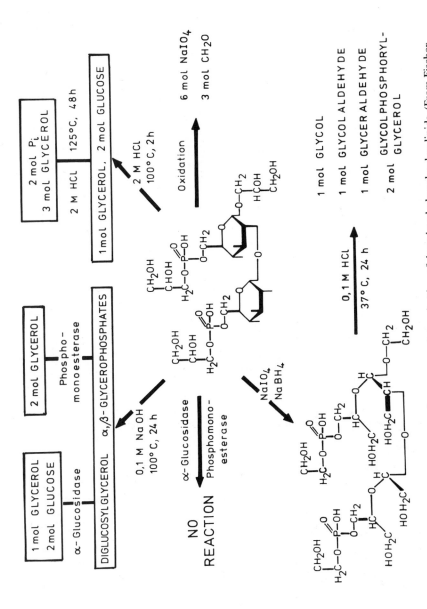

Figure 2-23. Degradative sequences for structure determination of deacylated phosphoglycolipids. (From Fischer and Landgraf, 1975.)

2.4.1.1. Molecular Composition and Basic Structure

Fatty acids are measured by a hydroxamate procedure (Snyder and Stephens, 1959) or for identification of individual components by gas–liquid chromatography (GLC) after conversion to methyl esters by methanolysis. For studies of the deacylated core (Fig. 2-23), fatty acids are removed by mild alkaline treatment according to Prottey and Ballou (1968) or by the procedure of Dawson (1967) as modified by Kates (1986). The components of the deacylation products are released, as shown in Fig. 2-23, by a two-step acid hydrolysis procedure and identified and quantitated by standard procedures (Wilkinson and Bell, 1971; Shaw *et al.*, 1972; Fischer *et al.*, 1973*a*).

The basic structure is elucidated by phosphodiester cleavage with alkali and analysis of the products (Fig. 2-23). Alkaline hydrolysis requires an α-hydroxyl group adjacent to the phosphodiester that participates in cleavage through formation of a cyclic five-membered phosphate. This intermediate is readily hydrolyzed to a mixture of isomeric phosphomonoesters (Baer and Kates, 1948, 1950; Brown and Todd, 1952; Ukita *et al.*, 1955). Since in phosphoglycolipids glycerophosphate or phosphatidyl residues are usually linked to C-6 of a hexosyl moiety (see Section 2.2.1), phosphate cyclization occurs preferentially at the glycerol, resulting essentially in a phosphate-free glycoside and a mixture of approximately 45% α-glycerophosphate and 55% β-glycerophosphate (Fischer, 1970; Shaw *et al.*, 1972; Fischer *et al.*, 1973*a,b*). The additional appearance of glycerobisphosphate and glycerol suggests a di(glycerophosphate) or a higher homologue as the substituent (Laine and Fischer, 1978). The glycoside formed is characterized by standard procedures (Komaratat and Kates, 1975; Fischer *et al.*, 1978*c;* Nakano and Fischer, 1977; Fischer, 1984), and the glycerophosphates are subjected to stereochemical analysis (Fischer, 1970).

2.4.1.2. Stereochemical Configuration of the Glycerophosphates

Whereas the glycerophosphates released by acid hydrolysis undergo rapid phosphoryl migration, this does not occur under alkaline conditions (Baer and Kates, 1950; Ukita *et al.*, 1957; Fischer and Landgraf, 1975). Therefore, the α-glycerophosphate formed by alkaline hydrolysis preserves the stereochemical configuration of the glycerophosphate residue in the parent lipid (Fischer *et al.*, 1973*a,b;* Brotherus *et al.*, 1974; Fischer and Landgraf, 1975).

The configuration is determined by reaction with the stereospecific enzyme *sn*-glycero-3-phosphate dehydrogenase (Fischer, 1970). By this procedure, the glycerophosphate and phosphatidyl residues of all bacterial phosphoglycolipids have been shown to be exclusively *sn*-glycero-1-phosphate and *sn*-glycero-3-phosphate, respectively (see references to the individual phosphoglycolipids). Accordingly, lipid VI yielded an equimolar mixture of both enantiomers. The configuration of each of the two substituents was, however, established by removal first of the glycerophosphate residue by hydrolysis with moist acetic acid (for principle see below) and analysis of the resulting phosphatidyl compound by the alkaline hydrolysis procedure (Fig. 2-24, lower left).

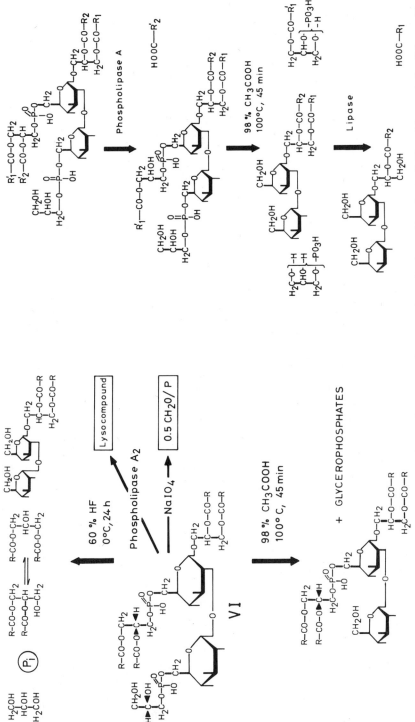

Figure 2-24. Degradative sequences for the location of the acyl groups and the enantiomeric glycerophosphates (left) and for analysis of the positional fatty acid distribution (right). Lipase, 1-position-specific lipase from *Rhizopus arrhizus*. (From Fischer and Landgraf, 1975, and Landgraf, 1976.)

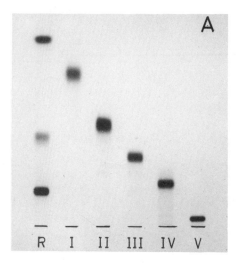

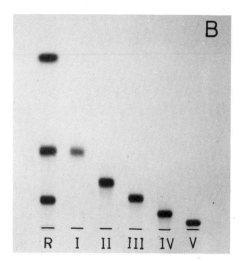

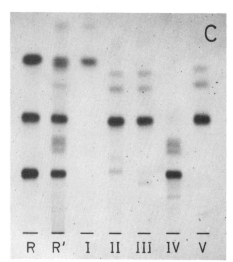

Figure 2-25. Intermediates after sequential treatment of phosphoglycolipids I-V with phospholipase A$_2$ and moist acetic acid (*cf.* Fig. 2-24). Phosphoglycolipids (for structures see Fig. 2-3) before (A) and after treatment with phospholipase A$_2$ (B), and after subsequent hydrolysis with moist acetic acid (C). R, Reference lipids (from top to bottom): Glc(α1–3)DAG; Glc(α1–2)Glc(α1–3)DAG; Glc(α1–2)Glc(α1–2)Glc(α1–3)DAG; R', reference lipids, treated with moist acetic acid. Thin-layer chromatography on silica gel with chloroform–aceton–methanol–acetic acid–water (50 : 20 : 10 : 10 : 5, v/v) as the solvent for A and B, and chloroform–methanol–water (65 : 25 : 4, v/v) as the solvent for C. Visualization with α-naphthol/H$_2$SO$_4$. (From Landgraf, 1976.)

The alkaline hydrolysis procedure can also be applied to the stereochemical analysis of 1,3-linked di(glycerophosphate) residues in glycerophosphoglycolipids (Laine and Fischer, 1978). It is not appropriate, however, for the analysis of higher homologues and lipoteichoic acids because they yield a racemic mixture of α-glycerophosphates due to random phosphodiester cleavage.

In view of the stereospecificity of phospholipase A$_2$ toward *sn*-3-phospholipids (De Haas *et al.*, 1968), the *sn*-glycero-3-phosphate configuration of phosphatidyl residues can also be inferred from their susceptibility to this enzyme (Fischer and Landgraf, 1975; Landgraf, 1976) (Fig. 2-25).

A glycerophosphoglycolipid (lipid VIIIa) in which an alanine ester substituent protected the α-glycerophosphate moiety from oxidation with periodate made another approach possible (Fischer, 1982). The lipid was oxidized with periodate and then incubated at pH 10.5, which hydrolyzed the alanine ester and liberated the α-glycerophosphate for stereochemical analysis. In addition, this release of α-glycerophosphate by β-elimination locates the phosphodiester of C-6 of one of the hexopyranosyl moieties.

2.4.1.3. Linkage Analysis by Smith Degradation Procedure*

On Smith degradation, depicted in Fig. 2-23, the deacylated phosphoglycolipid is oxidized with periodate and the product reduced with borohydride. Then the acetal bonds are hydrolyzed by a mild acid procedure that leaves the phosphodiester bonds intact (Fischer *et al.*, 1973*a,b*, 1982). Identification and quantitation of the products, as shown in Fig. 2-23, are used to locate the linkages between all components of the core. The point of attachment of the glycerophosphates is indicated by the released glycolphosphoglycerol whose glycerol moiety includes C-4 through C-6 of the respective hexopyranosyl residue. If, as in lipids XII and XIV, two glycerophosphates are linked to each other, glycolphosphoglycerophosphoglycerol is identified by mass spectrometry as the phosphate-containing hydrolysis product (Fig. 2-26).

The hexosyl residue in a di- or trihexosyl moiety, to which the glycerophosphate is linked at C-6, can be determined by use of the appropriate glycosidase, which does not release the respective hexosyl residue unless the glycerophosphate is removed (Fischer and Landgraf, 1975; Fischer *et al.*, 1978*c*).

2.4.1.4. Linkage Analysis by Permethylation Procedure*

The location of the linkages of lipid VIII has also been achieved by permethylation studies (Komaratat and Kates, 1975). Permethylation was done under apparently mild conditions (silver oxide as the catalyst), which left the phosphodiester linkage intact. The products obtained from the permethylated lipid, shown in Fig. 2-27, indicate in the parent compound an α-glycerophosphate linked to C-6 of the nonreducing end of a 1→6-linked diglucosyl moiety attached to C-1 of glycerol.

2.4.1.5. Location of the Fatty Acids

The most useful procedure in locating the fatty acids of phosphoglycolipids is hydrolysis with HF,† followed by phase partition and analysis of the

---

*[14]C- and ^{1}H-NMR spectroscopy has not yet been used in structural studies of phosphoglycolipids. With deacylation products, the anomeric conformations and points of attachment of the sugar residues can be established (see Section 2.4.2.9), and two-dimensional techniques such as COSY (correlation spectroscopy) may permit identification of the hexosyl residue that carries the glycerophosphate moiety (Erbing *et al.*, 1986; Fischer *et al.*, 1987).

†After hydrolysis, removal of HF by evaporation *in vacuo* at 2°C (Koch and Fischer, 1978) is preferable to neutralization of HF with $NaHCO_3$ or LiOH (Shaw *et al.*, 1972; Fischer *et al.*, 1973*a*).

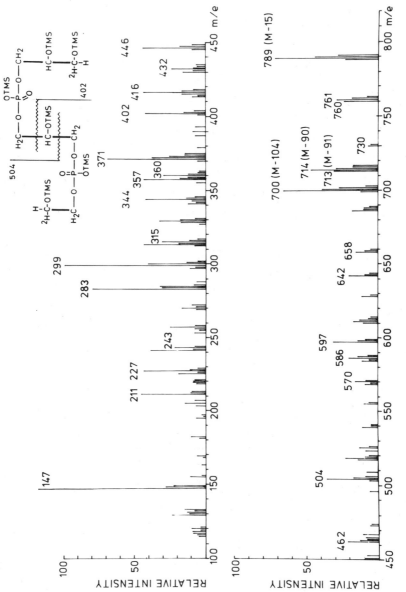

Figure 2-26. Electron impact mass spectrum of hexakis(trimethylsilyl)glycolphosphoglycerophosphoglycerol obtained by Smith degradation of lipid XII (Fig. 2-9). (From Laine and Fischer, 1978.)

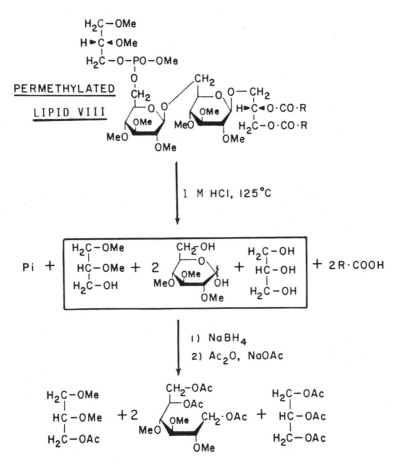

Figure 2-27. Linkage analysis of lipid VIII by permethylation procedure. (From Komaratat and Kates, 1975.)

lipid- and water-soluble products (Shaw *et al.,* 1972; Fischer *et al.,* 1973*a,b*). Under appropriate conditions, HF selectively hydrolyzes phosphodiester and phosphomonoester linkages leaving *O*-glycopyranosidic and fatty acid ester bonds essentially intact (Glaser and Burger, 1964; Lipkin *et al.,* 1969; Toon *et al.,* 1972; Fischer *et al.,* 1973*a,b*; Shaw and Stead, 1974). As shown with lipid VI in Fig. 2-24, glycerol derived from the glycerophosphate residue appears together with the phosphate in the aqueous layer; diacylglycerol, indicative of a phosphatidyl residue, and the basic glycolipid are recovered from the organic layer.

A particular problem with lipid VI was the location of the glycerophosphate and phosphatidyl residues (Fig. 2-24). This was achieved by hydrolysis with moist acetic acid, which selectively hydrolyzes phosphodiester bonds adjacent to a hydroxyl group (Coulon-Morelec *et al.,* 1960; Coulon-Morelec and Douce, 1968; Fischer and Landgraf, 1975). The lipid product was, after deacylation, susceptible to α-glucosidase, which locates the phosphatidyl residue on the diacylglycerol-linked glucosyl moiety.

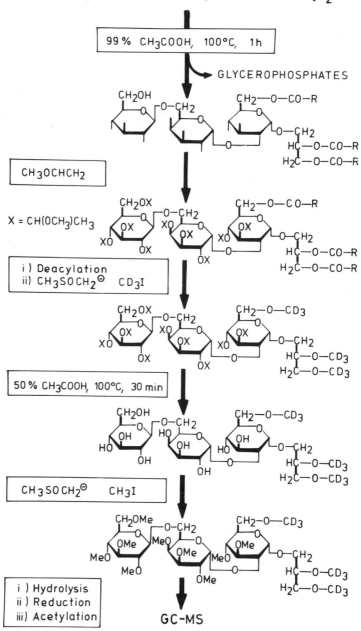

Figure 2-28. Location of fatty acids by two-step methylation procedure. CD₃, trideuteromethyl; GC–MS, gas–liquid chromatography–mass spectrometry (Nakano and Fischer, 1977).

If the glycolipid moiety contains a third fatty acid (lipids VII, XI, XII, XV–XX), the latter is located by a two-step methylation procedure in which the esterified and free hydroxyl groups of the parent compound are methylated with C^2H_3I and CH_3I, respectively (Nakano and Fischer, 1977). Using lipid XI as an example, the sequence of reactions is depicted in Fig. 2-28.

In studying the positional distribution of fatty acids in phosphatidylglycolipids, hydrolysis with HF is inadequate because the fatty acids of the released diacylglycerol randomize by acyl migration (Landgraf, 1976). Reliable results are obtained with the degradative sequence involving phospholipase A_2 and moist acetic acid as shown in Figs. 2-24 and 2-25.

2.4.2. Lipoteichoic Acids

2.4.2.1. Molecular Composition

As with phosphoglycolipids, the components of lipoteichoic acids are released by a two-step acid hydrolysis procedure (cf. Fig. 2-23) and are analyzed and quantitated by standard procedures (Wicken and Knox, 1970; Toon *et al.*, 1972; Nakano and Fischer, 1978; Fischer *et al.*, 1980*b*). Fatty acids can be extracted for analysis after the first step. In view of the potential contamination with other polymers, monosaccharides and amino acids should not be considered as chain substituents unless their linkage to glycerol has been proved.

2.4.2.2. Lipid Anchor and Repeating Units of the Chain

Most informative with respect to the overall structure of lipoteichoic acids is hydrolysis with HF, the specificity of which has been outlined above. The lipid anchor is released and the chain is hydrolyzed into glycerol and inorganic phosphate. Glycosyl substituents and D-alanine ester remain attached to the liberated glycerol units (Toon *et al.*, 1972; Koch and Fischer, 1978; Fischer *et al.*, 1980*b*). Susceptible to HF are, however, acid labile galactofuranosidic bonds (Op den Camp *et al.*, 1984; Fischer, 1987) and a few glycosides with a polyol as aglycon (Bennet and Glaudemans, 1979; Jennings *et al.*, 1980). The released lipid and the water-soluble chain fragments are separated by solvent phase partition.

2.4.2.2a. Lipid Anchor. In most cases, the organic layer contains one or two glycolipids, which, on the same core structure, contain two and three fatty acyl residues, respectively (see Figs. 2-7 and 2-9) (Koch and Fischer, 1978; Nakano and Fischer, 1978). The two glycolipid species are separated by TLC and characterized as outlined above. Small amounts of lysoglycolipids and fatty acids are usually observed (Toon *et al.*, 1972; Koch and Fischer, 1978; Nakano and Fischer, 1978). Lysolipids, if formed by hydrolysis with HF, are not seen when the poly(glycerophosphate) chain is hydrolyzed with moist acetic acid (Nakano and Fischer, 1978). Disadvantages of this procedure are that the released glycolipids are partially acetylated (see Fig. 2-25) (Fischer

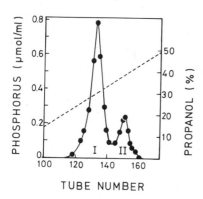

Figure 2-29. Separation of lipoteichoic acid from *Enterococcus faecalis* into molecular species containing two (I) and four fatty acids (II). For lipoteichoic acid structure see Fig. 2-4. Column chromatography on octyl Sepharose, as described in Fig. 2-21. (From Fischer *et al.*, 1983.)

and Landgraf, 1975) and that, according to the mode of cleavage (Coulon-Morelec *et al.*, 1960; Coulon-Morelec and Douce, 1968), glycosylated lipoteichoic acids are incompletely hydrolyzed.

Formation of diacylglycerol in addition to glycolipid, as was found with the lipoteichoic acids of *E. faecalis* (Toon *et al.*, 1972), *Listeria monocytogenes* (Hether and Jackson, 1983), and other *Listeria* species (Ruhland and Fiedler, 1987), indicates the presence of a phosphatidyl residue linked either to the glycolipid or to separate poly(glycerophosphate) chains. Mild acid hydrolysis in organic solvent was reported to release Glc(α1–2),Ptd→6Glc(α1–3)DAG from the lipoteichoic acid of *E. faecalis* (Ganfield and Pieringer, 1975). In another approach, phosphatidic acid, corresponding to the amount of diacylglycerol in the HF hydrolysate, was released from *E. faecalis* and *Listeria* lipoteichoic acids by periodate oxidation and subsequent β-elimination at pH 9 (Fischer *et al.*, 1989). This result establishes the phosphoglycolipid structure and locates the phosphatidyl residue on C-6 of one of the glycosyl moieties. Using the degradability of lipoteichoic acids by moist acetic acid (Nakano and Fischer, 1978), a phosphoglycolipid was recently released from the lipoteichoic acids of *E. faecalis* and *Listeria* strains and characterized as Glc(α1–2)Ptd→6Glc(α1–3)DAG and Gal(α1–2),Ptd→6Glc(α1–3)DAG, respectively (Uchikawa *et al.*, 1986; Fischer *et al.*, 1990). Quantitation of diacylglycerol and glycolipid indicates that in *E. faecalis* lipoteichoic acid chains are linked to both phosphatidylglycolipid and diacylglycerol glycolipid (see Fig. 2-4) (Fischer, 1981; Fischer *et al.*, 1981). These two lipoteichoic acid species separate on hydrophobic interaction chromatography (Fig. 2-29), as do lipoteichoic acids containing two and three fatty acid esters (Fischer *et al.*, 1983).

2.4.2.2b. Pattern of Glycerol Substitution. The water-soluble HF hydrolysis products of poly(glycerophosphate) lipoteichoic acids are analyzed as described by Fischer *et al.* (1980*b*, 1981). A schematic illustration is given in Fig. 2-30. The completeness of the hydrolysis is monitored by measurement of the inorganic phosphate release.* Glycerol is measured (A) directly, (B) after mild

---

*Inorganic phosphate is measured after acid hydrolysis of phosphofluoridate, which is formed during hydrolysis with HF and amounts to approximately 10% of the phosphate released.

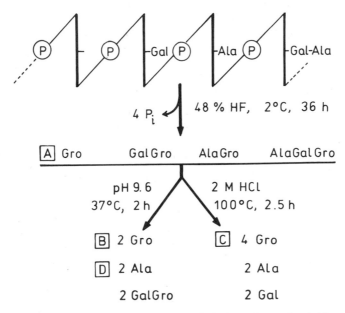

Figure 2-30. Analyses of the water-soluble HF hydrolysis products of poly(glycerophosphate) lipoteichoic acids. For explanation, see text.

alkaline treatment (pH 9.6, 37°C, 2 hr), and (C) after acid hydrolysis (2 M HCl, 100°C, 2 hr). After mild alkaline hydrolysis, total alanine is also determined (D). In relation to the phosphate or to total glycerol, these values give the proportion of

(A)	Nonsubstituted glycerol
(B)–(A)	Alanylglycerol
(C)–(B)	Glycosylglycerol
(D)–(B–A)	Alanylglycosylglycerol

The interpretation would have to be modified if the glycosylglycerol contained more than one alanine residue or in the presence of amino acids other than alanine.

To illustrate these analyses, the results obtained with *E. faecalis* lipoteichoic acids are shown in Table 2-12. The variously substituted glycerols may be directly identified by chromatographic procedures (Toon *et al.*, 1972; Fischer *et al.*, 1980*b*; Cabacungan and Pieringer, 1985), which is particularly important in the case of glycosylation. On the basis of the data in Table 2-12, the lipoteichoic acids of *E. faecalis* ATCC 9790 and *E. faecalis* ssp. *zymogenes* Kiel 27738 contain two glucosyl residues on the average per glycosylated glycerol. As shown in Fig. 2-31, however, GLC reveals a homologous series from mono- to tetraglucosylglycerol for the former, diglucosylglycerol and, as expected from the data in Table 2-12, alanylglycerol and alanyldiglucosylglycerol for the latter.

Table 2-12. Pattern of Glycerol Substitution of the Lipoteichoic Acids from Three Strains of Enterococcus faecalis[a]

| Lipoteichoic acid | Glucose | Alanine | Molar ratio to phosphorus[b] | | | | Chain length[c] |
			Nonsubstituted glycerol	Alanylglycerol	Glycosylglycerol	Alanylglycosyl-glycerol	
E. faecalis ATCC 9790	1.6	0	0.16	0	0.84	0	20
E. faecalis DSM 20478	0.9	0.16	0.41	0.16	0.43	0	19 ± 2.3[d]
E. faecalis subsp. zyogenes Kiel 27738	0.94	0.47	0.23	0.29	0.48	0.19	24

[a]From Fischer et al., 1981 and unpublished data.
[b]Analyses of the water-soluble products in the HF hydrolysate as described in the text.
[c]Molar ratios of phosphorus to glycolipid, measured in the water and organic layer, respectively.
[d]$\bar{x} \pm \sigma_{n-1}$; $n = 5$.

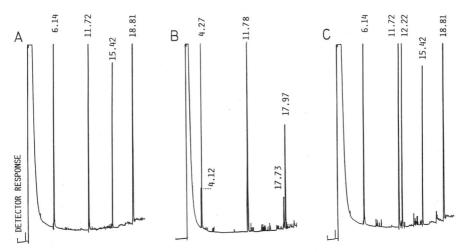

Figure 2-31. Identification of substituted glycerols from the lipoteichoic acids of *Enterococcus faecalis* ATCC 9790 (A, C) and *E. faecalis* ssp. *zymogenes* Kiel 27738 (B) by gas–liquid chromatography (GLC). The water-soluble products of hydrolysis with HF were trifluoroacetylated and analyzed by GLC on a fused silica capillary column (Durabond 5, 20 m, internal diameter 2.37 mm, film thickness 0.25 µm) at a programmed temperature rise of 5°C/min from 150°C to 240°C. Sample A contains (in order of increasing retention times) a homologous series from Glc(α1–2)Gro to Glc(α1–2)Glc(α1–2)Glc(α1–2)Glc(α1–2)Gro with the individual components in fractions of total glycosylated glycerol of 0.18, 0.27, 0.06, and 0.49. Sample B contains AlaGro isomers (t_R = 4.12, 4.27), Glc(α1–2)Glc(α1–2)Gro (t_R = 11.78), and Ala–Glc(α1–2)Glc(α1–2)Gro isomers (t_R = 17.73, 17.97). Sample C was derived, like sample A, from the lipoteichoic acid of *E. faecalis* ATCC 9790, but before hydrolysis with HF the polymer was deacylated. This permits measurement of the ratio of glycosylated chain glycerol [see (A)] to deacylated glycolipid, Glc(α1–2)Glc(α1–3)Gro (t_R = 12.22) for chain length determination (see Section 2.4.2.3). (J. Schurek and W. Fischer, unpublished data.)

2.4.2.2c. Poly(glycosylglycerophosphate) Lipoteichoic Acid from Lactococcus garvieae Kiel 42172. The structure of the galabiosylgalactosylglycerol that was released from this lipoteichoic acid by HF suggested a poly(glycosylglycerophosphate) chain. The degradative sequences used for linkage analysis are summarized in Fig. 2-32. Treatment of the native compound with α-galactosidase and subsequent hydrolysis with HF indicated the galabiosyl residue to be integrated into the chain backbone. The point of attachment of the glycerophosphate on the galabiosyl residues was elucidated by Smith degradation. The di(glycerol)-1,1′-phosphate formed contained the original glycerophosphate and C-4 through C-6 of a D-hexopyranosyl residue that was attached by a phosphodiester linkage at C-6. Fission of the phosphodiester bonds of native lipoteichoic acid by β-elimination after periodate oxidation confirmed this finding.

2.4.2.2d. Monoglycerophosphate Residues Bearing Lipoglycan from Bifidobacterium bifidum. The degradative steps leading to the structure of this compound (Fischer, 1987) are summarized in Fig. 2-33. Because of the acid lability of galactofuranosyl residues, hydrolysis with HF liberated the basic

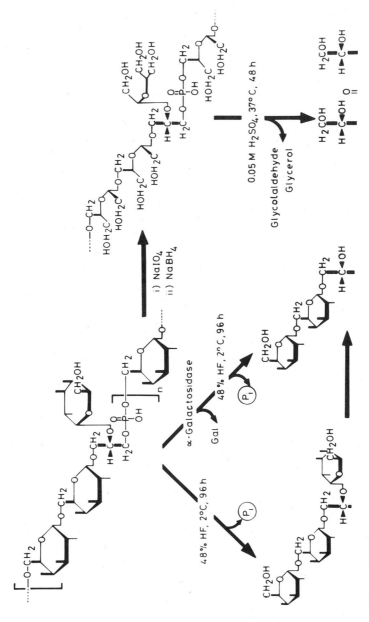

Figure 2-32. Degradative sequences for structure determination of the poly(glycosylglycerophosphate) lipoteichoic acid from *Lactococcus garvieae* Kiel 42172. The structure of the tri- and digalactosyldiacylglycerol was established by permethylation analysis and oxidation with CrO₃. (From Koch and Fischer, 1978.)

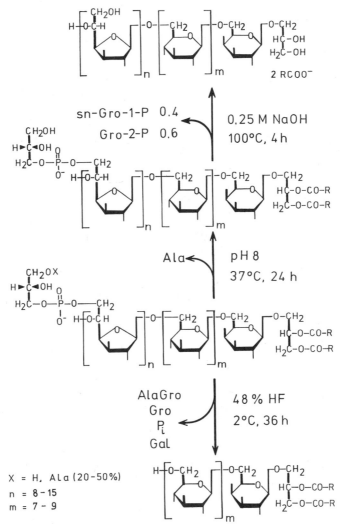

Figure 2-33. Degradative steps for structural analysis of the macroamphiphile from *Bifidobacterium bifidum*. (From Fischer, 1987.)

lipogalactoglucan whose structure was established by permethylation analysis. Mild alkaline hydrolysis selectively removed alanine ester from the parent compound, which made it possible to show that the glycerophosphate residues are susceptible to oxidation with periodate. Smith degradation of the alanine-free derivative cleaved the phosphodiester linkage to C-5 or C-6 of the galactofuranosyl residues. Analysis of the glycerophosphates released by strong alkaline hydrolysis (0.2 M NaOH, 100°C, 6 hr) established the *sn*-1 configuration for the glycerophosphate residues (cf. Section 2.4.2.5). Permethylation study of the polymer obtained by alkali hydrolysis established a linear (1→5)-linked galactofuranan and its attachment to the nonreducing terminus of the glucan moiety. Anomeric configurations were determined by optical rotation; the whole structure was confirmed by nuclear magnetic resonance (NMR) analysis (Fig. 2-39; see also Table 2-16).

2.4.2.3. Average Chain Length of Poly(glycerophosphate) Lipoteichoic Acids

Under the assumption of a single unbranched poly(glycerophosphate) chain on the lipid anchor, the number of repeating units per chain is given by the molar ratio of glycerol to glycolipid. The values may be measured after hydrolysis with HF or moist acetic acid followed by phase partition (Fischer *et al.*, 1980*b*) (Table 2-12.) Alternatively, the ratio may be calculated from the monosaccharides released by hydrolysis with HCl (Button and Hemmings, 1976*a;* Koch and Fischer, 1978), provided the monosaccharides of the lipid anchor and of chain substituents are different and the preparation is free of polysaccharide contamination (Nakano and Fischer, 1978). If lipid anchor and chain substituents contain the same monosaccharide, e.g., in the lipoteichoic acid of *E. faecalis,* an alternative procedure for determination of the chain length involves mild alkaline deacylation of lipoteichoic acid, followed by hydrolysis with HF and GLC of the trifluoroacetylated hydrolysis products. Chain length is calculated by

$$\frac{\text{Sum of glycosylated chain glycerol}}{\text{deacylated glycolipid}} \times \frac{\text{total glycerol}}{\text{glycosylglycerol}}$$

whereby the first term is determined by GLC, as shown in Fig. 2-31C, the second term by direct analysis of the hydrolysate (Table 2-12).

2.4.2.4. Differentiation between Chain Structures

Glycosylated poly(glycerophosphates) and poly(glycosylglycerophosphates) are readily distinguished by their products of acid hydrolysis (1 M HCl, 100°C, 2 hr): poly(glycerophosphates) characteristically form glycerobisphosphate, glycerol, and glyceromonophosphates in molar ratios of approximately 1 : 1 : 3, whereas poly(glycosylglycerophosphates) form glyceromonophosphates exclusively along with a small amount of glycerol that is derived from the glycolipid moiety rather than from the chain (Koch and Fischer, 1978).

Monoglycerophosphate branches, as found in the lipopolysaccharide of *Bifidobacterium bifidum* (see Fig. 2-12), are also released on acid hydrolysis as glyceromonophosphates. They are, however, readily distinguished from structures in which glycerophosphates are part of the chain by destruction of their glycerol moiety with periodate and concomitant formation of formaldehyde or on Smith degradation by conversion into glycolphosphate (Fischer, 1987).

The suggested presence of a 1,2-linked poly(glycerophosphate) structure in the lipoteichoic acid of *Bifidobacterium bifidum* (Op den Camp *et al.*, 1984) underlines the need for procedures that distinguish between this and the more common 1,3-phosphodiester linkage. A product formed on alkaline hydrolysis that is characteristic of a poly(glycerophosphate) containing 1,3-phosphodiester linkages has been shown to be *bis*(glycero-2-phosphate)-1,3′-phosphate (Archibald and Baddiley, 1966):

$$
\begin{array}{ccccccc}
\text{O} & \text{H}_2\text{COH} & & \text{H}_2\text{COH} & \text{O} & & \\
\parallel & | & & | & \parallel & & \\
\text{O}^- \!\!-\!\!\text{P}\!\!-\!\!\text{O} -\!\!\text{CH} & & \text{O} & \text{HC}\!\!-\!\!\text{O}\!\!-\!\!\text{P}\!\!-\!\!\text{O}^- & & \\
| & | & \parallel & | & | & \\
\text{OH} & \text{H}_2\text{C}\!\!-\!\!\text{O}\!\!-\!\!\text{P}\!\!-\!\!\text{O}-\!\!\text{CH}_2 & \text{OH} & \\
& & | & & \\
& & \text{O}^- & &
\end{array}
$$

This compound arises because in 1,3-linked chains a hydroxyl group is adjacent to the phosphodiester linkage at each side, so that hydrolysis can proceed in both directions. In 1,2-linked poly(glycerophosphate) chains, however, where a single hydroxyl-group is adjacent to the phosphodiester, initial hydrolysis of the chain can occur in one direction only. Based on these different modes of alkaline hydrolysis and the fact that lipoteichoic acids are made of *sn*-glycero-1-phosphate residues, another means for distinguishing 1,3- and 1,2-linked poly(glycerophosphate) structures should be the stereochemical configuration of the released α-glycerophosphates. As shown in Table 2-13, the random fission of 1,3-linked chains results in an almost equimolar mixture of *sn*-glycero-3-phosphate and *sn*-glycero-1-phosphate, with the latter slightly predominating because the glycerophosphates at both termini yield only *sn*-glycero-1-phosphate. By contrast, 1,2-linked chains being hydrolyzed through a 2,3-cyclic phosphate intermediate are expected to predominantly yield *sn*-glycero-3-phosphate, the mirror image of the original stereochemical configuration. Finally, exclusively *sn*-glycero-1-phosphate is formed when the parent compound contains monoglycerophosphate side chains as the macroamphiphile of *Bifidobacterium bifidum* (see Fig. 2-33).

2.4.2.5. Stereochemical Configuration of the Glycerophosphate Residues

The biosynthetic origin of the glycerophosphates of lipoteichoic acids from phosphatidyl glycerol (see Section 2.5.3.1a) suggested that they possess the *sn*-glycero-1-phosphate configuration. This was chemically proved by Smith degradation of the poly(galabiosyl,galactosylglycerophosphate) chain, as shown in Fig. 2-32. The di(α-glycerol)phosphate obtained was subjected to alkaline hydrolysis and the resulting α-glycerophosphate tested with *sn*-glycero-3-phosphate dehydrogenase. As expected, it was not oxidized. In another approach, using in succession oxidation with periodate, β-elimination, and hydrazinolysis, the glycerophosphate is completely released as the α-isomer, which can be used for stereochemical analysis (Fischer *et al.*, 1982).

The stereochemical configuration of the glycerophosphate of 1,3-linked poly(glycerophosphate) chains has been more difficult to assess because of the random fission of the phosphodiester bonds on alkaline hydrolysis, as out-

Table 2-13. Analysis of the Alkali Hydrolysis Products of Lipoteichoic Acids for Locating the Phosphodiester Linkages on the Glycerol Residues

Lipoteichoic acid source[a]	Molar ratio to phosphorus[b]			
	Glycerobisphosphate	Glyceromonophosphate	α-Glycerophosphate	sn-Glycero-e-phosphate / α-Glycerophosphate
Lactobacillus casei DSM 20021	0.238	0.538	0.207	0.44
Lactococcus plantarum NCDO 1869	0.232	0.566	0.216	0.46
Lactococcus raffinolactis NCDO 617	0.207	0.567	0.205	0.45
'*Streptococcus lactis*' motile Kiel 48809	0.230	0.540	0.209	0.46
Streptoccus pyogenes IID 698	0.221	0.547	0.221	0.43

[a]Nonglycosylated poly(glycerophosphate) lipoteichoic acids were used.
[b]After alkali hydrolysis (0.25 MNaOH, 100°C, 6 hr), glycerol, glyceromonophosphate, and glycerobisphosphate were separated on a small column of DEAE–Sephadex (Laine and Fischer, 1978). α-Glycerophosphate was measured as described by Fischer *et al.* (1982), sn-glycero-3-phosphate according to Hohorst (1970).

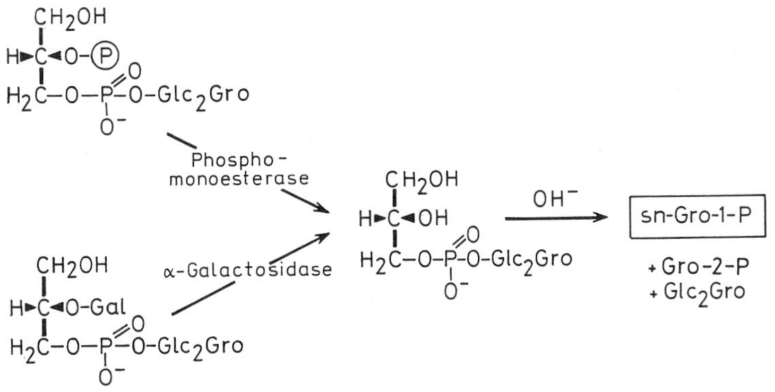

Figure 2-34. Stereochemical analysis of the glycerophosphate of lipoteichoic acids with poly(glycerophosphate) chains. The galactosyl- and phosphomonoester-containing compounds were formed on alkaline hydrolysis of the lipoteichoic acid from *Lactococcus lactis* NCDO 712, and isolated by chromatography on DEAE–Sephacel. (Based on H. U. Koch and W. Fischer, unpublished data.)

lined above. As a consequence of glycosyl substitution, however, part of the glycerophosphate is retained on the deacylated lipid anchor in the original orientation. This made it possible also to establish the *sn*-glycero-1-phosphate stereochemical configuration for 1,3-linked poly(glycerophosphate) chains, as shown in Fig. 2-34.

2.4.2.6. Proof of Single Unbranched Poly(glycerophosphate) Chains

When poly(glycerophosphate) lipoteichoic acids, freed of chain substituents, are treated with periodate, the glycerophosphate terminus is oxidized and by hydrazinolysis converted into a phosphomonoester (Fig. 2-35). The difference of inorganic phosphate measured before and after treatment with phosphomonoesterase serves as a measure of the chain terminus. With most lipoteichoic acids studied (Table 2-14), the molar ratio of inorganic phosphate to total phosphorus (B) was close to the molar ratio of lipid to phosphorus (A), which suggests one glycerol terminus per lipid moiety and by that a single unbranched poly(glycerophosphate) chain. Occasional measurements of formaldehyde led to the same conclusion. With some lipoteichoic acids, the B/A ratios were between 1.2 and 1.7. The additional phosphomonoester separated, however, from these lipoteichoic aids on hydrophobic interaction chromatography (Table 2-14, values marked by asterisks) and can therefore not be derived from chain branches.

2.4.2.7. Linkage between Chain and Lipid Anchor

The fission of the bond between both poly(glycerophosphate) and poly(glycosylglycerophosphate) chains and their glycolipid moieties by HF or moist acetic acid has been taken, in view of the specificity of these hydrolyses, as evidence for the presence of a phosphodiester linkage. The release of the

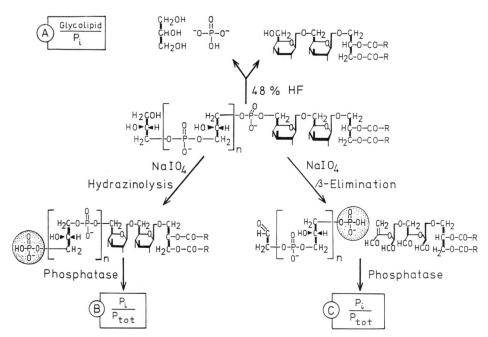

Figure 2-35. Degradative sequences for establishing a single unbranched poly(glycerophosphate) chain and its point of attachment on the glycolipid moiety. Controlled oxidation with $NaIO_4$ (0.05 M, 4°C, 30 min), followed by hydrazinolysis, converts the terminal glycerophosphate into a phosphomonoester (Fischer *et al.*, 1980*a*; Koch *et al.*, 1982). Exhaustive oxidation with $NaIO_4$ (0.2 M, 37°C, 48 hr), followed by β-elimination (pH 10, 37°C, 36 hr), eliminates the inner glycerophosphate terminus if it is linked to C-6 of a hexopyranosyl moiety that is unsubstituted at C-4 and C-3 (Fischer *et al.*, 1982). In both products, the ratio of phosphomonoester to total phosphorus is determined and compared with the molar ratio of glycolipid to phosphorus measured after hydrolysis with HF. The ratios of A, B, and C are unity if a single unbranched poly(glycerophosphate) is linked by a phosphodiester bond to C-6 of a periodate-oxidable hexopyranosyl moiety.

poly(glycerophosphate) chain terminating in a phosphomonoester by periodate oxidation and subsequent incubation at pH 10 (Fischer *et al.*, 1982) confirms the phosphodiester linkage and identifies C-6 of one of the hexosyl moieties as the point of attachment (Fig. 2-35). All lipoteichoic acids listed in Table 2-14 produced phosphomonoester on this treatment and in most cases the molar ratio of phosphomonoester to total phosphorus (C) approached that of glycolipid to phosphorus (A). It is evident that the nonreducing terminus is involved in the phosphodiester linkage for the lipoteichoic acids from *Bacillus licheniformis*, *Bacillus subtilis*, *Micrococcus varians*, *S. aureus*, and *S. xylosus* because the diacylglycerol-linked hexosyl moiety of these lipoteichoic acids is glycosidically substituted at C-6 (see Table 2-14, lipid structure). Location on the nonreducing hexosyl terminus may also be suggested for the lipoteichoic acids of the lactobacilli and *Lactococcus* strains because one of the molecular species of these lipoteichoic acids carries an acyl residue at C-6 of the di-acylglycerol-linked hexosyl moiety, and in the trihexosyldiacylglycerol-containing lipoteichoic acids from lactobacilli the middle hexosyl residue is glycosidically substituted at C-6. The same argument is applicable to the

Table 2-14. *Testing for Single Unbranched Poly(glycerophosphate) Chains and Location of the Chains on the Glycolipid Moieties*[a]

Lipoteichoic acid Provenance	Lipid structure[b]	A Lipid/ phosphorus	B Phosphomonoester/ phosphorus	D CH_2O/ phosphorus	C Phosphomonoester/ phosphorus
Bacillus licheniformis DSM 13	D	0.040	0.037	0.038	0.060
		0.040	0.042	n.d.	0.051
		0.041	0.040	n.d.	0.049
Bacillus megaterium ATCC 14581	I	0.026	0.026	n.d.	0.049
Bacillus subtilis W 23	D	0.049	0.046	0.050	0.047
Enterococcus faecalis DSM 20748	A,C	0.054	n.d.[c]	n.d.	0.053
Enterococcus faecalis Kiel 27738	A,C	0.053	n.d.[c]	n.d.	0.022
Lactobacillus casei DSM 20021	G,H	0.023	0.024	0.024	0.041
Lactobacillus plantarum DSM 20314	G,H	0.043	0.049	0.046	0.046
		0.042	0.041	n.d.	
Lactococcus lactis NCDO 712	A,B	0.045	0.042	0.048	0.042
Lactococcus lactis NCDO 2727	A,B	0.052	0.054	n.d.	0.049
Lactococcus raffinolactis NCDO 617	A,B	0.040	0.046 0.041*	n.d.	0.039
Lactococcus plantarum NCDO 1869	A,B	0.037	0.043 0.040*	0.046	0.044
Leuconostoc mesenteroides DSM 20343	A,B	0.025	0.029	0.031	0.027
Listeria monocytogenes NCTC 7973	E,F	0.044	0.051 0.041*	0.041	0.042
Listeria innocua NCTC 11288	E,F	0.038	0.044 0.040*	0.042	0.037
Listeria welshimeri SLCC 5334	E,F	0.038	0.049 0.044*	0.051	0.038
Micrococcus varians ATCC 29750	D	0.035	0.035	0.036	0.037

(continued)

Table 2-14. (Continued)

Lipoteichoic acid Provenance	Lipid structure[b]	A Lipid/ phosphorus	B Phosphomonoester/ phosphorus		C Phosphomonoester/ phosphorus	D CH$_2$O/ phorphorus
Staphylococcus aureus DSM 20233	D	0.034	0.034		0.035	0.034
		0.039	0.038		0.041	n.d.
Staphylococcus xylosus DSM 20266	D	0.039	0.068[d]	0.045*	0.093[d]	n.d.
		0.041	0.061[d]	0.042*	0.081[d]	0.035
Streptococcus mutans NCTC 10449	A	0.034	0.029		0.035	n.d.
Streptococcus pyogenes II D 698	A	0.035	0.035		0.036	n.d.
Streptococcus sanguis DSM 20567	A	0.033	0.035		0.036	n.d.

[a]For determination of the molar ratio of lipid to phosphorus (A) see text. 2.4.2.3. Molar ratios of phosphomonoester to phosphorus were measured after periodate oxidation followed by (B) hydrazinolysis or (C) β-elimination (Fig. 2-35). The molar ratio of formaldehyde to phosphorus (D) was measured before hydrazinolysis. Prior to analysis galactosyl residues and alanine ester substituents were removed from the chains. Values of preparations from different bacterial cultures are given on separate lines. For values marked by an asterisk see text. n.d., Not determined. From Fischer *et al.* (1989).

[b]Lipid strutures: *A*, Glc(α1–2)Glc(α1–3)acyl$_2$Gro; *B*, Glc(α1–2),acyl→6Glc(α1–3)acyl$_2$Gro; *C*, Glc(α1–2)Ptd→6Glc(α1–3)acyl$_2$Gro; *D*, Glc(β1–6)Glc(β1–3)acyl$_2$Gro; *E*, Gal(α1–2)Glc(α1–3)acyl$_2$Gro; *F*, Gal(α1–2)Ptd→6Glc(α1–3)acyl$_2$Gro; *G*, Glc(β1–6)Gal(α1–2)Glc(α1–3)acyl$_2$Gro; *H*, Glc(β1–6)Gal(α1–2),acyl→6Glc(α1–3)acyl$_2$Gro, *I*, acyl$_2$Gro.

[c]Not determined because the molar ratio of glycosylglycerol/total glycerol was greater than 0.5.

[d]Material generating additional phosphomonoester could be removed by treatment of the lipoteichoic acid at pH 4 (100° C, 25 min) and subsequent purification by chromatography on octyl Sepharose.

lipoteichoic acids from enterococci and *Listeria* strains one molecular species of which carries a phosphatidyl residue at C-6 of the diacylglycerol-linked hexosyl moiety (Uchikawa *et al.*, 1986; Fischer *et al.*, 1990). From these observations it becomes evident that the poly(glycerophosphate) lipoteichoic acid and the glycerophosphoglycolipid of a given bacteria do not only share the same glycolipid moiety but also the point of attachment of the phosphodiester linkage.

2.4.2.8. Distribution of Substituents in Poly(glycerophosphate) Lipoteichoic Acids

The degree of substitution of poly(glycerophosphate) lipoteichoic acids varies between 0.3 and 0.9 (see Table 2-2), raising the question as to whether these lipoteichoic acids contain mixtures of fully substituted and unsubstituted chains or whether all chains are incompletely substituted. In the latter case, the substituents may be randomly or regularily distributed along the chain or accumulated on a certain glycerophosphate sequence.

Incomplete substitution with D-alanine ester of all chains was demonstrated by anion exchange chromatography of *S. aureus* lipoteichoic acid on columns of DEAE–Sephacel (Fischer and Rösel, 1980). As a consequence of the positively charged alanine ester compensating the negative charge of the adjacent phosphate group, alanyl lipoteichoic acid could be separated from the artificial alanine-free derivative (Fig. 2-36). Moreover, molecular species were eluted in the order of decreasing alanine content, which showed further that all chains are substituted within a narrow range independent of the average alanine ester content.

The mode of alanine ester distribution along the poly(glycerophosphate) chain was elucidated by stepwise hydrolysis of the chain by the concerted action of phosphodiesterase and phosphomonoesterase from *Aspergillus niger* (Fischer *et al.*, 1980*b*; Koch *et al.*, 1985):

$$\text{GroP-(GroP)}_n\text{-glycolipid} \xrightarrow{\text{H}_2\text{O}} \text{Gro} + P\text{-(GroP)}_n\text{-glycolipid} \qquad (1)$$

$$P\text{-(GroP)}_n\text{-glycolipid} \xrightarrow{\text{H}_2\text{O}} P_i + \text{(GroP)}_n\text{-glycolipid} \qquad (2)$$

With continuously decreasing chain length, the ratio of alanine to phosphorus or glycerol remained constant, which indicates an average homogeneous distribution of alanine ester along the chain.

Wicken and Knox (1975*b*) used the binding of galactosyl residues by *Ricinus* lectin in order to show that all species of *L. lactis* lipoteichoic acid are partially substituted with α-D-galactosyl residues. Isolation from *L. lactis* under appropriate conditions provides α-D-galactosyl lipoteichoic acid that retains the native substitution with α-D-alanine ester (Fischer *et al.*, 1980*b*). Chromatography of this lipoteichoic acid on DEAE–Sephacel confirmed the absence of unsubstituted chains and indicated a narrow range of substitution for both galactosyl residues and alanine ester (Fig. 2-37). In order to answer the

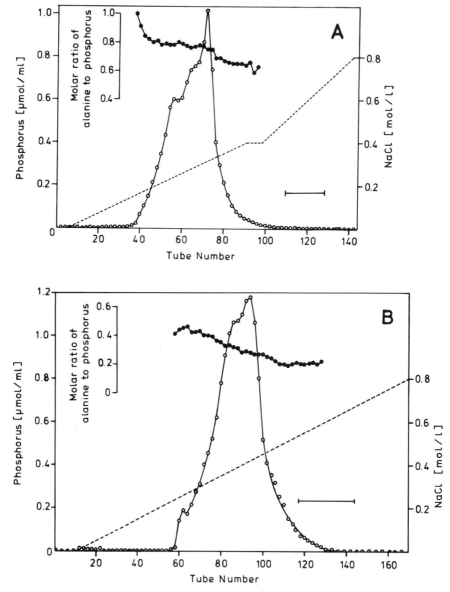

Figure 2-36. Range of alanine ester substitution of the lipoteichoic acid of *Staphylococcus aureus*. Cells were grown at low (A) and high (B) salt concentration. Column chromatography on DEAE–Sephacel, essentially as described in Fig. 2-20. The bars indicate the location of artificial alanine-free lipoteichoic acid determined in a separate run. (From Fischer and Rösel, 1980.)

question as to whether the two substituents occur together on the same chain or separately on different chains, alanine ester was hydrolyzed and the resulting galactosyl lipoteichoic acid chromatographed on DEAE cellulose. The obvious absence of nonsubstituted species clearly favors the first possibility (Fig. 2-37B).

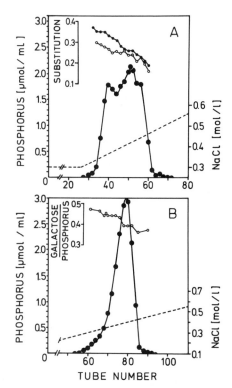

Figure 2-37. Range of substitution with D-alanyl ester (●) and α-D-galactosyl residues (○) of the lipoteichoic acid from *Lactococcus lactis* NCDO 712. The phosphate elution profile is given by the large solid circles. (A, B) Different lipoteichoic acid preparations. (A) Native LTA. (B) Dealanylated LTA. Chromatography on DEAE–Sephacel as described in Fig. 2-20. (Based on H. U. Koch and W. Fischer, unpublished data.)

For studies of the galactosyl distribution along the chain, *L. lactis* lipoteichoic acid with a Gal/Gro ratio of 0.42 was hydrolyzed in alkali (W. Fischer and H. U. Koch, unpublished data). Since sequences of adjacent galactosylglycerophosphates are resistant to alkaline hydrolysis, a regular distribution of the substituents would result in complete hydrolysis; random distribution, however, would yield fragments with the composition of $Gal_n Gro_n P_{n-1}$. Treatment with phosphomonoesterase as well as a second alkaline hydrolysis were necessary in order to remove 2-phosphoglycero-1(3)-phosphoryl terminals from primary fragments. Separation of the final products on DEAE-Sephacel revealed GalGro, $Gal_2 Gro_2 P$, $Gal_3 Gro_3 P_2$, $Gal_4 Gro_4 P_3$, and $Gal_5 P_4$ in molar ratios of $10:4.5:1.5:0.9:0.3$. These results indicate a random rather than a regular distribution of the galactosyl substituents along the chain. Accumulation of galactosyl residues anywhere on the chain is precluded because in this case the higher homologues of the $Gal_n Gro_n P_{n-1}$ series would predominate.

2.4.2.9. Structural Studies by NMR Spectroscopy

Much progress has been made during the last decade in polysaccharide structural analysis by ^{13}C-NMR spectroscopy (Gorin, 1981; Jennings, 1983). In spite of the early use of NMR for analysis of phosphodiester-containing

Table 2-15. ^{13}C Chemical Shifts of Poly(1,3-Glycerophosphate), Poly(2,3-Glycerophosphate), and Poly(Glycosylglycerophosphate) Wall Teichoic Acids and Relevant Monomers[a]

	Chemical shifts[b]								
	Monosaccharide residues						Glycerol residues		
Compound	C-1	C-2	C-3	C-4	C-5	C-6	C-1	C-2	C-3
Glycerol							63.8	73.3	63.8
Glycero-2-phosphate							62.5 (3.8)[f]	75.1 (5.0)[g]	62.5 (3.8)[f]
Glycero-3-phosphate							62.8	71.6 (6.4)[f]	65.2 (4.8)[g]
Poly(1,3-glycerophosphate)							66.3 (5.6)[g]	69.8 (8.1)[f]	66.3 (5.6)[g]
Methyl-α-D-glucopyranoside	99.5	71.8	73.5	70.0	71.5	61.1			
Methyl-β-D-glucopyranoside	103.4	73.4	75.2	70.0	75.2	61.4			
Glcp(β1–1)Gro[c]	103.1	73.6	76.1	70.3	76.3	61.2	71.4	71.1	62.8
Glcp(β1–1)Gro-3-P[c]	103.1	73.7	76.2	70.2	76.3	61.3	70.9	69.5 (7.9)[f]	67.5 (5.1)[g]
+2Gro-3-P+$_n$[d] 1 βGlcp	103.3	73.6	76.1	70.1	76.3	61.2	69.5 (4.0)[f,h]	75.3[i]	66.1[i]
+6Glcp(α1–1)Gro-3-P+$_n$[e]	98.6	71.7	73.2	69.2	70.0 (8.0)[f]	64.3 (5.0)[g]	68.6	69.3 (8.0)[f]	66.3 (5.6)[g]

[a] de Boer et al., (1976; 1978).
[b] In parts per million from external tetramethylsilane; solvent ^2H_2O.
[c] Prepared from teichoic acid of Bacillus subtilis subsp. niger WM by alkali hydrolysis and dephosphorylation, respectively.
[d] Teichoic acid from Bacillus subtilis ssp. niger WM.
[e] Teichoic acid from Bacillus stearothermophilus B65.
[f] $^3J_{^{13}C,^{31}P}$ (Hz).
[g] $^2J_{^{13}C,^{31}P}$ (Hz).
[h] Poorly resolved.
[i] Coupled signals, no coupling constants determined.

polymers (Bundle *et al.*, 1974) and for studies of wall teichoic acids (de Boer *et al.*, 1976; 1978; Shaskov *et al.*, 1979; Jennings *et al.*, 1980; Anderton and Wilkinson, 1985) it is only recently that lipoteichoic acids were studied by this technique.

The usefulness for differentiating chain structures is illustrated in Table 2-15 by the ^{13}C-chemical shifts of three wall teichoic acids. The downfield shifts of resonances due to carbon atoms attached to the phosphodiester group are used to determine the position of the linkages. This effect, caused by deshielding through substituents on hydroxyl groups, has been well documented for phosphate groups (Lapper *et al.*, 1972, 1973; Bundle *et al.*, 1973; de Boer *et al.*, 1976), acetate (Bundle *et al.*, 1973), alanine ester (Batley *et al.*, 1981), and glycosyl residues (cf. Gorin, 1981). The position of a phosphodiester linkage can, however, be determined independently by the coupling between ^{13}C and ^{31}P, which occurs on the α-carbon atom to the phosphate group and extends only to the β-carbon atom (Lapper *et al.*, 1972, 1973). This is illustrated in Table 2-15 for glycero-2-phosphate and glycero-3-phosphate in comparison with glycerol. NMR analysis does not discriminate between C-1 and C-3 of the glycerol moieties, which can, however, be differentiated by stereochemical analysis (see Section 2.4.2.6) or may be assigned on the basis of biosynthetic considerations (cf. Fig. 2-2). Table 2-15. shows further that poly(1,3-glycerophosphate) teichoic acid exhibits, relative to glycerol, a downfield shift (2.5 ppm) of the signals of C-1 and C-3 combined with two-bond ^{13}C–^{31}P coupling (5.6 Hz) and an upfield shift of the signal of C-2 (−3.5 ppm) combined with three-bond ^{13}C–^{31}P coupling (8.1 Hz). Assignments in the ^{13}C-NMR spectrum of the teichoic acid of *Bacillus subtilis* spp. *niger* were made on the basis of the spectra of the glucosylglycerol and glucosylglycerophosphate which were prepared from the teichoic acid by hydrolysis (Table 2-15). The β-glycosidic bond of the glucosyl residue is deduced from the signal at 103.1 ppm. The shifted signals of C-1 (7.6 ppm) and C-2 (−2.2 ppm) of the glycerol moiety locate the glycosidic bond at C-1. The resonances of the glucosylglycerophosphate are similar except for the displaced signals of C-2 and C-3 of the glycerol moiety, which, together with three- and two-bond ^{13}C–^{31}P coupling, are consistent with a 3-phosphate group (Table 2-15.). Compared with the 3-phosphate, the teichoic acid shows the signal of C-2 of the glycerol moiety shifted by 5.8 ppm and those of C-1 and C-3 by −1.4 ppm, indicating a poly(2,3-glycerophosphate) structure. The nondisplaced signals of the glucosyl moiety confirm that it is not an integral part of the chain. This is the case, however, in the teichoic acid of *Bacillus stearothermophilus* B65 (Table 2-15). The phosphate bond is shown here to be on C-6 of the glucosyl residue by two-bond ^{13}C–^{31}P coupling (5.0 Hz) and a downfield shift of 3.2 ppm relative to the C-6 signal of methyl-α-glucoside. As expected, the C-5 resonance shows three-bond ^{13}C–^{31}P coupling (8.0 Hz) and an upfield shift of −1.5 ppm. ^{13}C–^{31}P coupling is present on the C-2 and C-3 signals and not on the C-1 signal of the glycerol moiety, which locates the phosphodiester at C-3.

Figure 2-38 shows the ^{13}C-NMR spectrum of a lipoteichoic acid partially substituted with alanine ester, which gives rise to resonances at 16.35 and 50.0

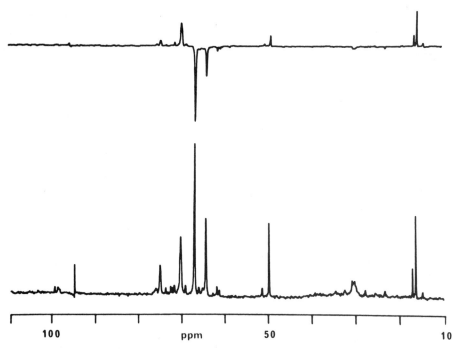

Figure 2-38. ^{13}C-NMR spectra of the alanyl lipoteichoic acid from *Lactobacillus fermentum*. Lower trace: broad-band proton-decoupled spectrum; upper trace: spectrum obtained using INEPT (acronym for *i*nsensitive *n*uclei *e*nhanced by *p*olarization *t*ransfer) pulse sequence, which permits separation of lines for carbon atoms carrying different numbers of protons. Signals of primary and tertiary carbon atoms are directed upward, those of secondary carbon atoms downward. (From Batley *et al.*, 1981.)

ppm. The small resonances slightly downfield from those of ester-linked al-anine are due to free alanine, resulting from hydrolysis in the NMR tube at pH 8. The resonance at 70.6 ppm can be assigned to C-2, that at 67.4 ppm to C-1 and C-3 of the glycerol moiety, indicating a 1,3-linked poly(glycerophos-phate) (cf. Table 2-15). The alanine ester gives rise to another two ^{13}C signals of the glycerol moiety. One is shifted downfield (4.8 ppm) from the signal of C-2, the other upfield (-2.6 ppm) from the signal of C-1 and C-3, consistent with the alanine ester linked to C-2 of the glycerol moiety. The resonances of the alanyl residues give rise to sharper lines than those for glycerol carbons indicating increased mobility of the lateral substituents over that of the main chain. Evidence by ^{13}C-NMR spectroscopy for a 1,3-linked poly(glycerophos-phate) chain was also obtained with the lipoteichoic acid of *Listeria mono-cytogenes* (Hether and Jackson, 1983). A detailed analysis of poly(glycerophos-phate) lipoteichoic acids by ^{13}C-, ^{1}H-, and ^{31}P-NMR spectroscopy was recently published (Batley *et al.*, 1987).

The ^{13}C-NMR analysis of the macroamphiphile from *Bifidobacterium bi-fidum* is summarized in Table 2-16. A detail of the DEPT* spectrum of the

*DEPT, acronym for *d*istorsionless *e*nhancement by *p*olarization *t*ransfer.

Table 2-16. ^{13}C Chemical Shifts of Reference Compounds and the Alkali-Treated and Native Macroamphiphile from Bifidobacterium bifidum[a]

	Chemical shifts[b]								
	Monosaccharide residues						Glycerol residues[c]		
Compound	C-1	C-2	C-3	C-4	C-5	C-6	C-1	C-2	C-3
Methyl-β-D-glucopyranoside[d]	104.0	74.1	76.6	70.6	76.8	61.8	—	—	—
β-D-(1→6)-glucan[e]	104.2	74.2	70.7		76.1	70.0	—	—	—
Methyl-β-D-galactofuranoside[d]	109.9	81.3	78.4	84.7	71.7	63.4	—	—	—
β-D-(1→)-galactofuranan[f]	108.0	83.0	78.3	83.0	77.5	62.5	—	—	—
Methyl-β-D-galactopyranoside[d]	104.5	71.7	73.8	69.7	76.0	62.0	—	—	—
Macroamphiphile, alkalihydrolyzed									
-6βGlc*p*1-	103.7	73.8	76.3	70.2	75.6	69.6	—	—	—
-5βTal*f*1-	107.7	82.1	77.7	82.0	76.3	61.8	—	—	—
βGal*f*1-	108.7	—	—	—	—	—	—	—	—
-6βGal*p*1-, -3Gro	103.6	71.4	73.2	69.4	74.5	61.5	63.0	71.2	71.7
Macroamphiphile, native[j]									
-6βGlc*p*1-, Gro*P*	103.6	73.7	76.2	70.2	75.5	69.6	67.2 (4.5)[g]	71.3 (7.7)[h]	62.8
-5βGal*f*1-, AlaGro*P*	108.3	82.2	77.3	82.2	74.5[i]	66.0[i]	66.7 (4.5)[g]	68.5 (7.7)[h]	67.2

[a]From Fischer et al. (1987).
[b]In parts per million (ppm) from external trimethylsilane; solvent ^2H_2O.
[c]For chemical shifts of glycerol and glycerophosphates see Table 2-15.
[d]Bock and Pedersen (1983).
[e]Saito et al. (1977).
[f]Ogura et al. (1974).
[g]$^2J_{^{13}C,^{31}P}$ (Hz).
[h]$^3J_{^{13}C,^{31}P}$ (Hz).
[i]Split signals, poorly resolved.
[j]Additional signals: alanine CO 171.7, CH 49.5, CH_3 15.9; fatty acids CH_2 30.5, CH_3 24.5.

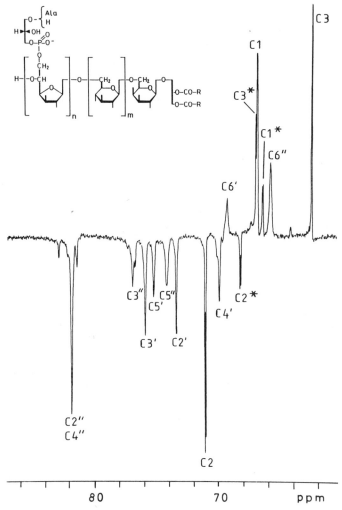

Figure 2-39. Partial ^{13}C DEPT spectrum of the native macroamphiphile from *Bifidobacterium bifidum*. Signals of primary and tertiary carbon atoms are directed downward, those of secondary carbon atoms upward. C1–C3 and C1*–C3*, signals of carbon atoms of glycerophosphate and alanylglycerophosphate moieties; C1′–C6′ and C1″–C6″, signals of the carbon atoms of glucopyranosyl and galactofuranosyl residues, respectively. (From Fischer *et al.*, 1987.)

native polymer is shown in Fig. 2-39. The alkaline hydrolysis product from which fatty acids, alanine, and glycerophosphate were removed showed 12 resonances (Table 2-16) consistent with the β-(1–5)galactofuranan and β-(1–6)glucan moiety previously established by chemical procedures (cf. structure depicted in Fig. 2-39). Smaller resonances could be assigned to the nonreducing galactofuranosyl terminus and the galactopyranosylglycerol moiety. In the native compound the carbohydrate signals were unchanged except for the resonances of C-6 and C-5 of the galactofuranosyl moiety, being displaced by 4.2 and −1.8 ppm, respectively. This locates the phosphodiester bond at C-6 of the galactofuranosyl moiety. The DEPT spectrum of the native amphiphile, shown in Fig. 2-39, differentiates carbon atoms with one or three and two protons, respectively, which facilitates the assignment of the resonances of the glycerol moiety. The signals of C-1 and C-2 showed a splitting of 4.5 and 7.7 Hz, respectively, resulting from two-bond and three-bond $^{13}C-^{31}P$ coupling. Together with the nondisplaced signal of C-3, these observations locate the phosphodiester at C-1 of the glycerol moiety and indicate monoglycerophosphate side chains rather than a poly(glycerophosphate) structure. The carbon atoms of the glycerol moieties displayed a second set of signals resulting from the substitution with alanine ester. The displacement of the C-3 signal by 4.4 ppm and of the C-2 and C-1 signal by -2.8 and -0.5 ppm, respectively, locates the alanyl residues at C-3 of the glycerol moiety.

The ^{31}P-NMR spectra provide considerable information on chain substitution (Batley *et al.*, 1987): Unsubstituted lipoteichoic acid shows a single large resonance at 0.96 ppm. As shown in Fig. 2-40, the spectra of substituted lipoteichoic acids consist of three groups of signals. The group at the lowest frequency, A, represents phosphate residues with both adjacent glycerol residues unsubstituted, group B and C are assigned to phosphate residues with substituents at one and at either side, respectively. Closer inspection of the

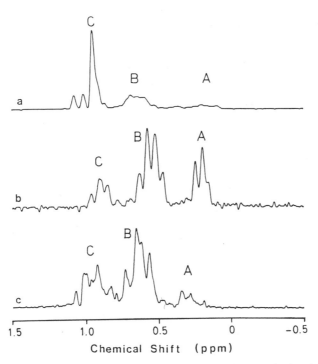

Figure 2-40. ^{31}P-NMR spectra of lipoteichoic acid from *Enterococcus faecium* (a), *Lactobacillus fermentum* (b), and *Lactococcus lactis* (c) (^2H_2O, containing 10 mM EDTA, 30°C). Chain substitution: (a) α-kojibiosyl (Glc/P = 1.02); (b) alanyl (Ala/P = 0.46) and D-galactosyl (Gal/P = 0.06); (c) alanyl (Ala/P = 0.14) and D-galactosyl (Gal/P = 0.46). (From Batley *et al.*, 1987.)

spectrum of the lipoteichoic acid from *Lactobacillus fermentum* reveals four peaks in group B and two in the other two groups. The relative intensity of the lines within each group varied with the degree of substitution, which suggests that the chemical shifts are affected by more distant alanyl residues as well. From low-substituted lipoteichoic acids it became further apparent that an alanyl or glycosyl substituent at one side of the phosphate residue produces two signals (0.57 and 0.62 ppm, and 0.66 and 0.85 ppm, respectively) possibly reflecting separate effects of the chiral substituent on the phosphate residues at C-1 and C-3 of the substituted glycerol moiety.

From the area of the groups, the mean degree of substitution (α) is calculated by

$$\alpha = (A + 0.5\ B)/(A + B + C) \tag{3}$$

The mode of substitution may also be elucidated because for random arrangement the peak areas should be in the ratios

$$A : B : C = \alpha^2 : 2\alpha(1 - \alpha) : (1 - \alpha)^2 \tag{4}$$

According to Eq. (4), the distribution of substituents of lipoteichoic acid from *E. faecium* and *L. fermentum* (Fig. 2-40) proved to be random. The spectra of lipoteichoic acids that possess D-alanine ester and high levels of hexosyl substituents contain overlapping signals and are therefore not amenable to simple interpretation. Recently, ^{31}P-NMR spectroscopy revealed a random distribution also for the D-alanyl substituents of the lipoteichoic acid from *S. aureus* and *L. casei* and the α-kojibiosyl substituents of the lipoteichoic acid of *E. faecalis* (Fischer, 1989; W. Bauer and W. Fischer, unpublished results).

High-resolution ^{31}P-NMR spectroscopy was further used for detection of lipoteichoic acid in intact bacterial cells (Roberts *et al.*, 1985).

2.5. Biosynthesis

The biosynthetic pathways for phosphatidylglycolipids, glycerophospho-glycolipids, and lipoteichoic acids have in common that the required phosphatidyl and glycerophosphate residues are transferred from phosphatidylglycerol or cardiolipin rather than from nucleotide-activated precursors. Since the synthesis of one phosphodiester bond involves breakage of another, the equilibrium constant would be near 1, unless the products were continuously removed.

2.5.1. Phosphoglycolipids

Phosphatidyldiglucosyl diacylglycerol was discovered in *E. faecalis* as a product derived from diglucosyldiacylglycerol and an unknown substance present in the membrane-bound enzyme preparation (Pieringer, 1968). Subsequent studies showed that the phosphatidyl residue was transferred from phosphatidylglycerol and bisphosphatidylglycerol rather than from CDP–diacylglycerol (Pieringer, 1972). Since the reaction rate was twice as fast with bisphosphatidylglycerol and the membrane preparation also catalyzed the formation of bisphosphatidylglycerol from phosphatidylglycerol, the former seems to be the direct phosphatidyl donor and the following sequence of reactions has been suggested to occur:

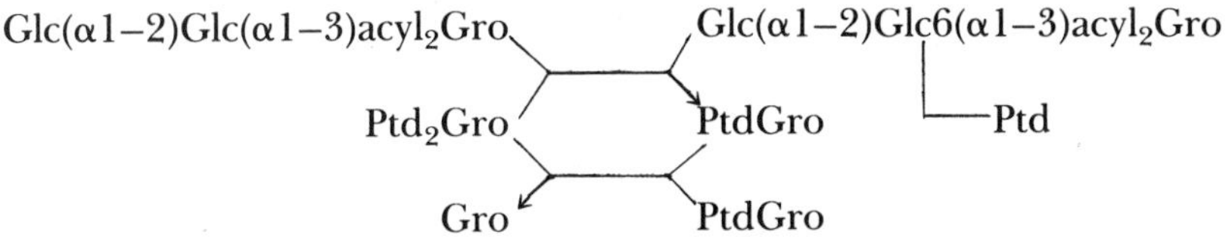

$$\text{Glc}(\alpha 1{-}2)\text{Glc}(\alpha 1{-}3)\text{acyl}_2\text{Gro} \qquad \text{Glc}(\alpha 1{-}2)\text{Glc}6(\alpha 1{-}3)\text{acyl}_2\text{Gro} \qquad (5)$$

Ptd₂Gro / PtdGro — Ptd

Gro / PtdGro

[Reaction scheme 5: Glc(α1–2)Glc(α1–3)acyl₂Gro + Ptd₂Gro ⇌ Gro + PtdGro, and PtdGro → Glc(α1–2)Glc6(α1–3)acyl₂Gro + Ptd]

In the formation of phosphatidylglucosyldiacylglycerol by membrane preparations of *Pseudomonas diminuta*, phosphatidylglycerol served as the phosphatidyl donor (Shaw and Pieringer, 1977):

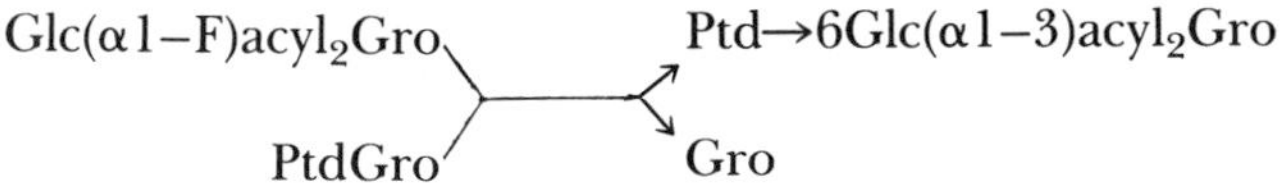

$$\text{Glc}(\alpha 1{-}\text{F})\text{acyl}_2\text{Gro} \qquad \text{Ptd}{\rightarrow}6\text{Glc}(\alpha 1{-}3)\text{acyl}_2\text{Gro} \qquad (6)$$

PtdGro / Gro

[Reaction scheme 6: Glc(α1–F)acyl₂Gro + PtdGro ⇌ Ptd→6Glc(α1–3)acyl₂Gro + Gro]

CDP–diacylglycerol and bisphosphatidylglycerol proved to be inactive.

2.5.2. Glycerophosphoglycolipids

The identification of *sn*-glycero-1-phosphate in the glycerophosphoglycolipids of gram-positive bacteria (see Section 2.2.1) suggested that the glycerophosphate moiety may be derived from the *sn*-glycero-1-phosphate residue in phosphatidylglycerol. With *Bifidobacterium bifidum*, pulse-chase experiments *in vivo* (van Schaik and Veerkamp, 1976) and *in vitro* investigations (Veerkamp, 1976) confirmed the donor function of phosphatidylglycerol

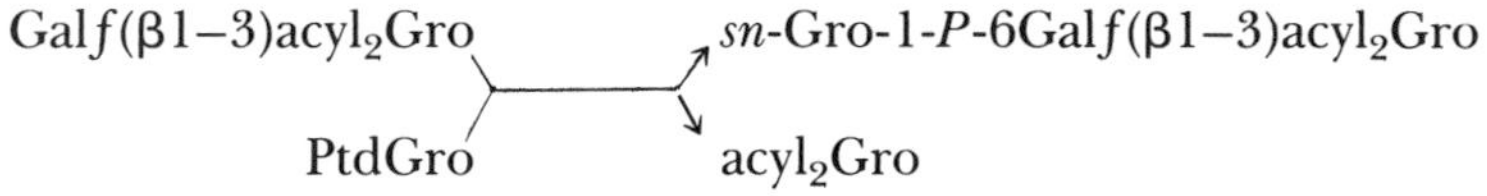

$$\text{Gal}f(\beta 1{-}3)\text{acyl}_2\text{Gro} \qquad sn\text{-Gro-1-}P\text{-6Gal}f(\beta 1{-}3)\text{acyl}_2\text{Gro} \qquad (7)$$

PtdGro / acyl₂Gro

[Reaction scheme 7: Galf(β1–3)acyl₂Gro + PtdGro ⇌ sn-Gro-1-P-6Galf(β1–3)acyl₂Gro + acyl₂Gro]

In vitro, CDP–glycerol, another potential donor of glycerophosphate but of the enantiomeric *sn*-3 configuration, was inactive.

A precursor product relationship between the nonacylated glycerol of phosphatidylglycerol and the glycerophosphate moiety of the glycerophosphoglycolipid was demonstrated in *S. aureus* (Koch *et al.*, 1984):

$$\text{Glc}(\beta 1{-}6)\text{Glc}(\beta 1{-}3)\text{acyl}_2\text{Gro} \qquad sn\text{-Gro-1-}P\text{-6Glc}(\beta 1{-}6)\text{Glc}(\beta 1{-}3)\text{acyl}_2\text{Gro}$$

PtdGro / acyl₂Gro

$$(8)$$

[Reaction scheme 8: Glc(β1–6)Glc(β1–3)acyl₂Gro + PtdGro ⇌ sn-Gro-1-P-6Glc(β1–6)Glc(β1–3)acyl₂Gro + acyl₂Gro]

The reaction to the right is favored by recycling of the diacylglycerol to phosphatidic acid and by elongation of the glycerophosphoglycolipid to lipoteichoic acid (see Section 2.5.3.1*c*).

2.5.3. Lipoteichoic Acid

2.5.3.1. Biosynthesis of the Basic Polymeric Structure

2.5.3.1a. Synthesis of the Poly(glycerophosphate) Chain. Before the glycero-phosphate of lipoteichoic acid was shown chemically to be the *sn*-1 isomer (see Section 2.4.2.5), pulse-chase *in vivo* experiments with *Staphylococcus aureus* and *Streptococcus sanguis* provided suggestive evidence that the glycerophosphate chain is directly derived from phosphatidylglycerol (Glaser and Lindsay, 1974; Emdur and Chiu, 1974). This conclusion was amply verified by subsequent *in vitro* experiments with membrane preparations of *Streptococcus sanguis* (Emdur and Chiu, 1975; Mancuso *et al.*, 1979) and *Enterococcus faecalis* (Ganfield and Pieringer, 1980) and with toluene-treated cells of *Lactobacillus casei* (Taron *et al.*, 1983). As expected, CDP–glycerol was a poor substitute for phosphatidylglycerol (Pieringer *et al.*, 1981). In *in vivo* experiments with *Bacillus subtilis*, 3,4-dihydroxybutyl-1-phosphonate, an analogue of *sn*-glycero-3-phosphate, inhibited phosphatidylglycerol synthesis, hence the biosynthesis of lipoteichoic acid (Deutsch *et al.*, 1980). The synthesis of both phosphatidylglycerol and lipoteichoic acid was also inhibited by growth inhibitory concentrations of racemic 1(3)-dodecylglycerol in *Streptococcus mutans* BHT, possibly through impairment in the synthesis of CDP-diacylglycerol (Brisette *et al.*, 1986).

Moreover, recent pulse-chase experiments with *S. aureus* (Koch *et al.*, 1984) and *B. subtilis* (Koga *et al.*, 1984) *in vivo* revealed a rapid and almost complete turnover of the nonacylated glycerol of phosphatidylglycerol into lipoteichoic acid (Fig. 2-41). Similar results were obtained with *Bacillus stearothermophilus* (Card and Finn, 1983) although in this case the nonacylated glycerol of phosphatidylglycerol also took part in reactions other than lipoteichoic acid synthesis.

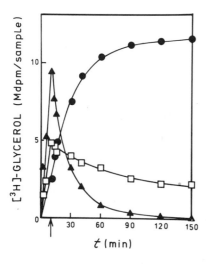

Figure 2-41. Turnover of the nonacylated glycerophosphate moiety of phosphatidylglycerol into lipoteichoic acid of growing *Staphylococcus aureus*. Pulse chase experiment, using [2-³H]glycerol; the arrow indicates the onset of the chase. (●) Lipoteichoic acid; (▲) glycerophosphate; (□) diacylglycerol moiety of phosphatidyl glycerol. (From Koch *et al.*, 1984.)

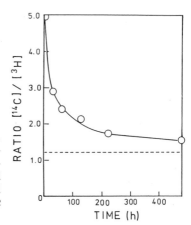

Figure 2-42. Mode of lipoteichoic acid chain extension. In cells of *Lactobacillus casei*, the lipoteichoic acid was labeled in succession with [2-³H]glycerol and [¹⁴C]glycerol, isolated, and stepwise degraded from the glycerol terminus by the joint action of phosphodiesterase and phosphomonoesterase from *Aspergillus*. (From Taron *et al.*, 1983.)

Whether, as was recently suggested (Chiu *et al.*, 1985), phosphatidylglycerophosphate, in addition to phosphatidylglycerol, might serve as a glycerophosphate donor in the lipoteichoic acid synthesis of *Streptococcus sanguis*, requires further experimental evidence. One would expect that glycerobisphosphate rather than glycerophosphate is transferred, which would require subsequent cleavage of the phosphomonoester at the terminus of the chain before elongation can occur.

2.5.3.1b. Mode of Chain Extension. Extension of the poly(glycerophosphate) chain could occur in one of two directions: either by the sequential addition of glycerophosphate residues to the end of the growing chain distal to the lipid anchor or in the direction of the lipid anchor by transfer of the growing chain to the next monomeric unit. The first mode is similar to that of teichoic acid synthesis (Burger and Glaser, 1964; Kennedy and Shaw, 1968), the second one is found in peptidoglycan and lipopolysaccharide synthesis (Robbins *et al.*, 1967; Ward and Perkins, 1973). These possibilities were studied *in vivo* and *in vitro* by differential radioisotope labeling techniques (Cabacungan and Pieringer, 1981; Taron *et al.*, 1983). Specific degradation of the differentially labeled lipoteichoic acid with periodate or by the concerted action of the phosphodiesterase and phosphomonoesterase from *Aspergillus* (see Section 2.4.2.8) indicated chain extension distal to the lipid anchor (Fig. 2-42). This mode of chain extension requires the growing terminus to stay in contact with the membrane until the chain is completed.

2.5.3.1c. Chain Synthesis in Linkage to the Lipid Anchor. When radiolabeled Glc(α1–2),Ptd-6Glc(α1–3)acyl₂Gro, one of the lipid anchors of the lipoteichoic acid of *E. faecalis* (see Fig. 2-4), was added to membrane preparations of this organism, it appeared first in an intermediate that was soluble in chloroform–methanol–water and later in water-soluble lipoteichoic acid (Ganfield and Pieringer, 1980). In a pulse-chase experiment with *S. aureus*, [¹⁴C]acetate-labeled diacylglycerol was, as shown by precursor product relationships (Fig. 2-43), successively incorporated into Glc(β1–3)acyl₂Gro, Glc(β1–6)Glc(β1–

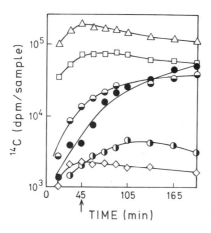

Figure 2-43. Pulse-chase kinetics of lipid amphi-philes in growing *S. aureus* on labeling of fatty acids with [14C]acetate. The arrow indicates the onset of the chase. ($\triangle$) Phosphatidylglycerol; ($\square$) di-acylglycerol; ($\bullet$) lipoteichoic acid; ($\ominus$) di-glucosyldiacylglycerol; ($\mathbb{O}$) glucosyldiacylglycerol; ($\diamond$) phosphatidic acid. (From Koch *et al.*, 1984.)

3)acyl$_2$Gro and lipoteichoic acid. The glycerophosphoglycolipid was found to be an intermediate in this sequence:

$$\text{Glc}_2\text{acyl}_2\text{Gro} + \text{PtdGro} \longrightarrow \text{Gro}P\text{-Glc}_2\text{acyl}_2\text{Gro} \qquad (9)$$
$$+ \text{acyl}_2\text{Gro}$$

$$\text{Gro}P\text{-Glc}_2\text{acyl}_2\text{Gro} + 24 \text{ PtdGro} \rightarrow (\text{Gro}P)_{25}\text{-Glc}_2\text{acyl}_2\text{Gro} \qquad (10)$$
$$+ 24 \text{ acyl}_2\text{Gro}$$

As shown in Fig. 2-44, in the chase after labeling with [2-3H]glycerol, the glycerophosphate moiety of Gro$P$-Glc$_2$acyl$_2$Gro lost most of its radioactivity rapidly and synchronously to the nonacylated glycerol of phosphatidylglycer-

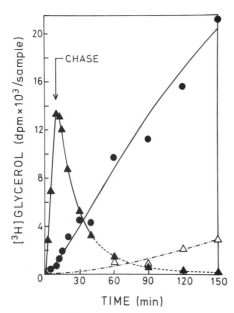

Figure 2-44. Pulse-chase kinetics of the glycerol moieties of the glycerophosphodi-glucosyldiacylglycerol on labeling of growing *Staphylococcus aureus* with [2-3H]glycerol. ($\bullet$) Diglucosyldiacylglycerol moiety; ($\blacktriangle$) glycero-phosphate moiety; ($\triangle$) glycerophosphate–glycolipid, which escaped conversion into lipoteichoic acid. (From Koch *et al.*, 1984.)

ol, whereas the glycerol residue of the glycolipid moiety continuously gained radioactivity from the constantly labeled diacylglycerol pool (cf. Fig. 2-43).

These observations, taken together, indicate that the poly(glycerophosphate) is synthesized in linkage to the definitive glycolipid anchor. The selection of a particular glycolipid in this process (cf. Fig. 2-7) and the frequent appearance of monoglycerophosphoglycolipids as the sole intermediates strongly suggests the involvement of two glycerophosphate transferases: an initiating enzyme with high glycolipid acceptor specificity and a polymerase to synthesize the chain.

In a mutant strain of *Bacillus licheniformis* lacking phosphoglucomutase, hence being unable to synthesize UDP glucose (Button and Hemmings, 1976*b*), the glycolipid of lipoteichoic acid was replaced by diacylglycerol. In *Bacillus coagulans* and *Bacillus megaterium,* in which the poly(glycerophosphate) seems to be naturally linked to diacylglycerol (Iwasaki *et al.,* 1986), the biosynthesis of lipoteichoic acid may normally start from diacylglycerol or phosphatidylglycerol.

2.5.3.1d. Synthesis of Non-Poly(glycerophosphate) Lipoteichoic Acids. The *sn*-glycero-1-phosphate branches of the lipoglycan in *Bifidobacterium bifidum* (see Fig. 2-12) were derived from phosphatidylglycerol, as demonstrated by incubation of membrane preparations with the radiolabeled precursor (Op den Camp *et al.,* 1985*a*). Radioactivity from UDP-[^{14}C]glucose and UDP-[^{14}C]galactose was not incorporated into the polymer in these experiments, which suggests that the glycerophosphate was transferred to preformed lipoglycan (see Fig. 2-12). Possibly the synthesis of the latter requires hexosyl-1-phosphoryl-undecaprenol as intermediates, like the biosynthesis of lipomannan in *Micrococcus luteus* (Scher *et al.,* 1968; Scher and Lennarz, 1969). In the lipoglycan of *Bifidobacterium bifidum,* molecular species were detected that suggest that the lipoglucan moiety is synthesized first and subsequently single galactofuranosyl and glycerophosphate residues are added separately (Fischer, 1987).

Although the biosynthesis of the poly(digalactosyl, galactosylglycerophosphate) lipoteichoic acid of *Lactococcus garvieae* (see Fig. 2-10) has not been investigated, the established *sn*-1 configuration of the glycerophosphate residues (Koch and Fischer, 1978; Fischer *et al.,* 1982) again suggests a phosphatidylglycerol origin. The isolated hypothetical biosynthetic intermediates, shown in Fig. 2-10, point to an assembly of the repeating unit on the growing chain rather than on an undecaprenylphosphate intermediate.

2.5.3.2. Addition of Chain Substituents

2.5.3.2a. Glycosylation. Although the poly(glycerophosphate) chain of *E. faecalis* lipoteichoic acid is highly substituted with mono-, di-, tri-, and tetra-glucosyl residues (Toon *et al.,* 1972; Cabacungan and Pieringer, 1985) the polymer synthesized *in vitro* in the absence of activated glucose is unsubstituted (Ganfield and Pieringer, 1980). Conversely, glycosylation *in vitro* does not require polymer synthesis (Cabacungan and Pieringer, 1985). Studying the glycosyl transfer from UDP-[^{14}C]glucose to lipoteichoic acid (LTA) in membrane preparations of *S. sanguis,* Mancuso and Chiu (1982) found a short-lived [^{14}C]glucosyl lipid intermediate. This was identified as glucosyl-1-phosphorylundecaprenol and the glycosylation was formulated as a two-step reaction:

$$UDP\text{-}Glc + undecaprenylphosphate \rightleftharpoons Glc\text{-}l\text{-}phosphorylundeca-prenol + UDP \qquad (11)$$

$$Glc\text{-}l\text{-}phosphorylundecaprenol + LTA \rightarrow Glc\text{-}LTA + undecaprenylphosphate \qquad (12)$$

The requirement of glucosyl-1-phosphorylundecaprenol as a carrier is noteworthy because glycosylation of wall teichoic acids, being also a membrane-associated process, proceeds directly from hexosyl-1-diphosphorylnucleosides (for a review, see Ward, 1981). Confirmatory evidence comes from recent studies into polymer synthesis by membrane preparations of *Bacillus coagulans:* UDP–galactose is directly used in the synthesis of poly(galactosylglycerophosphate) wall teichoic acid (Yokoyama *et al.,* 1987), whereas a galactosyl phosphorylpolyprenol apparently serves as the donor for galactosylation of lipoteichoic acid (Yokoyama *et al.,* 1988). Formation of galactosyl phosphorylpolyprenol and transfer from this donor of galactosyl to lipoteichoic acid might occur at separate sites of the cell as is suggested by the different pH optima of the two enzymes involved.

2.5.3.2b. Incorporation of D-*Alanine.* The *in vitro* incorporation of D-[^{14}C]alanine into membranes of *Lactobacillus casei* requires, in addition to ATP and Mg^{2+}, the supernatant fraction, which catalyzes the following two-step reaction sequence (Reusch and Neuhaus, 1971; Linzer and Neuhaus, 1973; Neuhaus *et al.,* 1974):

$$enzyme + \text{D-alanine} + ATP \rightleftharpoons enzyme \cdot \text{D-alanyl-AMP} + PP_i \qquad (13)$$

$$enzyme \cdot \text{D-alanyl-AMP} + membrane\ acceptor \xrightarrow{\text{ligase}}$$

$$\text{D-alanyl-membrane acceptor} + enzyme + AMP \qquad (14)$$

The ultimate membrane acceptor has been established to be the endogenous poly(glycerophosphate) lipoteichoic acid which becomes alanylated at position 2 of the glycerol residues (Childs and Neuhaus, 1980). It is not yet known whether the ligase functions as alanyl transferase or is required for binding the enzyme · D-alanyl–AMP complex to a membrane site, prior to alanine incorporation (Linzer and Neuhaus, 1973). Mutants of *L. casei* deficient in the alanyl content of lipoteichoic acid suggest that an additional membrane-associated factor is required (Ntamere *et al.,* 1987): In spite of a normal or increased lipoteichoic acid content, membrane fragments of two of the mutant strains were unable to incorporate D-alanine when incubated with supernatant proteins from the parent strain, whereas the reverse combination showed activity. In *L. casei*, the same system that inserts alanyl residues into lipoteichoic acid also incorporated D-alanine into several lipophilic compounds which have chromatographic mobilities as expected for glycerophosphoglycolipids (Brautigan *et al.*, 1981). Interesting in this context is the isolation and characterization of 3(2)-*O*-D-alanyl-*sn*-glycerol-1-phospho-6Glc(β1–6)Glc(β1–3)DAG from the lipid extract of *Bacillus licheniformis* (Fischer, 1982). Since the alanine-containing lipophilic compounds seem to be elongated to lipoteichoic acid in *L. casei* (Brautigan *et al.*, 1981), poly(glycerophosphate) synthesis and incorporation of D-alanine may occur concomitantly but incorporation into completed lipoteichoic acid is also possible (Childs and Neuhaus, 1980).

2.5.3.2c. Turnover of D-*Alanyl Substituents and Replacement by Re-esterification.* Pulse-chase experiments with growing *S. aureus* cells revealed a rapid turnover of the D-alanyl substituents of lipoteichoic acid (Fig. 2-45). At the onset of the chase, only a minor fraction was lost by spontaneous base-catalyzed hydrolysis; the major fraction seemed to turnover to wall teichoic acid. Such a transfer is of interest since so far the mode of alanine incorporation into teichoic acid has been unknown (Ward, 1981) and, topologically, the membrane-associated lipoteichoic acid would be a suitable donor for the esterification with alanine of the wall-linked polymer (see Fig. 2-1).

As can be seen from Fig. 2-45, the almost complete loss from lipoteichoic

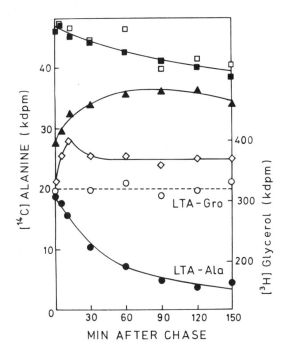

Figure 2-45. Pulse-chase kinetics of the D-alanine ester substituents of lipoteichoic acid on labeling of growing *Staphylococcus aureus* with D-[14C]alanine. The poly(glycerophosphate) chain of lipoteichoic acid was prelabeled with [3H]glycerol. [14C]Alanine of: (●) lipoteichoic acid; (▲) wall teichoic acid; (◇) peptidoglycan; (□, ■) total alanine ester (measured and calculated); (○) [3H]glycerol of lipoteichoic acid. (From Haas *et al.*, 1984.)

acid of [^{14}C]alanine after the chase in less than one bacterial doubling would rapidly produce alanine-free lipoteichoic acid unless the loss were compensated for by re-esterification. This possibility was substantiated using toluene-treated cells of *S. aureus* (Koch *et al.*, 1985). Unable to synthesize lipoteichoic acid, they possess the capacity of incorporating D-alanine into the preformed polymer. The rate of re-esterification correlated with the rate of alanine ester loss from lipoteichoic acid, which may explain how the high degree of substitution (Fischer and Rösel, 1980) is maintained in the living cell.

In the experiments with toluene-treated *S. aureus* transfer of alanine from lipoteichoic acid to completed wall teichoic acid was demonstrated (Koch *et al.*, 1985). Realanylation of completed wall teichoic acid was also seen in living *S. aureus* when the D-alanine ester loss by sublethal heating was compensated for during repair (Hurst *et al.*, 1975).

2.5.3.3. Lipoteichoic Acid Synthesis as Driving Force of Membrane Lipid Turnover

Although a rapid turnover of membrane lipids of gram-positive bacteria, particularly phosphatidylglycerol, has been known for 15 years (Short and White, 1970; 1971), it was not until recently that lipoteichoic acid synthesis has been recognized as the driving force for this turnover (Koch *et al.*, 1984; Koga *et al.*, 1984). Although in the membrane of *S. aureus* lipoteichoic acid constitutes only 6 mole % of the lipid amphiphiles, it contains approximately 50% of the membrane glycerol (Table 2-17). This, together with the other data in Table 2-17 indicates that in one bacterial doubling the total phosphatidylglycerol turns over three times for the lipoteichoic acid synthesized, and the diacylglycerol concomitantly formed is more than six times the amount present in the diacylglycerol pool. The excess diacylglycerol is not degraded (Koch *et al.*, 1984): approximately 15% is used for glycolipid synthesis (Table 2-17), the major part, however, recycles through phosphatidic acid to phosphatidylglycerol (Fig. 2-46). This recycling is reflected by the persistence of radioactivity after the chase in diacylglycerol, in the acylated glycerol moiety

Table 2-17. Composition of Lipid Amphiphiles of Logarithmically Growing Staphylococcus aureus[a]

Amphiphile	Mole %	Percent of glycerol
Phosphatidylglycerol	50.4	33.0
Lysylphosphatidylglycerol	9.9	6.5
Bisphophosphatidylglycerol	1.1	1.1
Phosphatidic acid	<0.1	—
Diacylglycerol	23.6	7.7
Glc(β1–3)acyl$_2$Gro	1.5	0.5
Glc(β1–6)Glc(β1–3)acyl$_2$Gro	7.6	2.5
GroP-6Glc(β1–6)Glc(β1–3)acyl$_2$Gro	0.1	—
(GroP)$_{25}$-6Glc(β1–6)Glc(β1–3)acyl$_2$Gro[b]	5.7	48.6

[a]From Koch *et al.* (1984).
[b]Lipoteichoic acid.

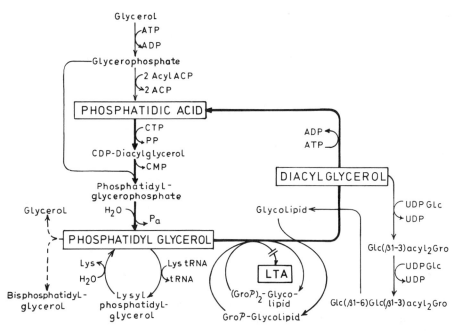

Figure 2-46. Interrelationship between lipoteichoic acid synthesis and membrane lipid metabolism in *Staphylococcus aureus*. (From Koch *et al.*, 1984.) Similar metabolic schemes have been designed for *Enterococcus faecalis* (Carson *et al.*, 1981) and *Lactobacillus casei* (Taron *et al.*, 1983; Ntamere *et al.*, 1987).

of phosphatidylglycerol, and particularly in phosphatidic acid (see Fig. 2-43). The recycling of diacylglycerol together with the utilization of diglucosyldiacylglycerol for lipoteichoic acid synthesis (Fig. 2-46) explains the earlier observation that in *S. aureus* the fatty acid residues of diglucosyldiacylglycerol seem to turnover more slowly than the glucosyl residues (Short and White, 1971).

Diacylglycerol kinase has been demonstrated in *L. casei* (Taron *et al.*, 1983), and recycling of diacylglycerol as a result of lipoteichoic acid synthesis has also been observed in this organism and other gram-positive bacteria (Lombardi *et al.*, 1980; Card and Finn, 1983; Koga *et al.*, 1984). In an alkalophilic *Bacillus* sp. strain that has no lipoteichoic acid, the radioactivity of phosphatidylglycerol turned over more slowly and was synchronously lost from the acylated and nonacylated glycerol moiety (Koga *et al.*, 1984).

2.5.3.4. Factors Affecting Lipoteichoic Acid Synthesis and Substitution

2.5.3.4a. Sporulation. As shown with *Bacillus megaterium*, the rate of lipoteichoic acid synthesis decreases at the end of logarithmic growth and is again accelerated when the spore membrane is synthesized (Johnstone *et al.*, 1982). It is of interest that spores contain lipoteichoic acids but, in contrast to vegetative cells, no wall teichoic acid.

2.5.3.4b. Energy Deprivation. When ATP synthesis in growing *S. aureus* cells is poisoned (NaF, carbonylcyanide *p*-trifluoromethoxyphenylhydrazone) lipoteichoic acid synthesis, although independent of energy supply, stops immediately (Koch *et al.*, 1984). Phosphatidylglycerol continues turning over and the label rapidly appears in bisphosphatidylglycerol, the synthesis of which is negligible under normal growth conditions (see Table 2-17). The physiological significance of this switch over in lipid amphiphile synthesis is not yet understood.

2.5.3.4c. Phosphate Limitation. Under conditions of phosphate limitation, various gram-positive bacteria cease wall teichoic acid synthesis and, in place of it, synthesize phosphate-free teichuronic acid (for review, see Ward, 1981). Reports on lipoteichoic acid synthesis under phosphate limitation have so far been conflicting. Unaffected synthesis was reported with *B. subtilis* (Ellwood and Tempest, 1968), whereas a decrease in the lipoteichoic acid content to 1/10 or less was seen with *B. licheniformis* (Button *et al.*, 1975) and group B streptococci (Nealon and Mattingly, 1984). Noteworthy in this context is the observation that phosphatidylglycerol constitutes 10% of the lipid in *B. subtilis* under the conditions of phosphate limitation as compared with 54% in magnesium limited culture (Minnikin *et al.*, 1972).

2.5.3.4d. Salt Concentration. With increasing salt concentration in the growth medium of *S. aureus* the D-alanyl content of wall teichoic acid and lipoteichoic acid gradually decreased (Table 2-18). Since toluene-treated cells, grown at high salt concentration, showed an accelerated incorporation of D-alanine into their low-substituted lipoteichoic acid, a direct inhibitory effect of salt concentration on the alanine-incorporating enzymes was suggested and experimentally established (Koch *et al.*, 1985). The sensitive component seems to be the D-alanine membrane acceptor ligase, which answered to changes of salt concentration with a decrease in activity (Reusch and Neuhaus, 1971). The varied alanine ester content of wall teichoic acid may be a direct result of the changes in lipoteichoic acid, which presumably functions as the donor in the alanylation of the wall polymer (see Section 2.5.2.2c).

Table 2-18. Effect of Salts in the Growth Medium on the Alanine Ester Content of Teichoic Acid and Lipoteichoic Acid of Staphylococcus aureus

Salt	Concentration (mole/liter)	Teichoic acid[a] $\dfrac{\text{Alanine}}{\text{Phosphate}}$	Lipoteichoic acid[b] $\dfrac{\text{Alanylglycerol}}{\text{Total glycerol}}$
NaCl	0–0.03	0.6	0.73
	0.9	—	0.55
	1.3	0.2	0.33
	1.7	—	0.29
KCl	1.3	—	0.46
	1.7	—	0.36

[a]Heptinstall *et al.* (1970).
[b]Fischer and Rösel (1980); Koch *et al.* (1985).

2.5.3.4e. Effect of pH. Glucose-limited chemostat cultures of *S. aureus* displayed a dramatic decrease in the D-alanine content of wall teichoic acid and lipoteichoic acid when the pH of the growth medium was raised from 5 to 8 (Archibald *et al.*, 1973; McArthur and Archibald, 1984). By contrast, ATP-supplied toluenized cells of *S. aureus* compensated the increased base-catalyzed loss by accelerated alanine incorporation (Koch *et al.*, 1985). The reason for this difference is unknown. Changes in pH may affect the lipoteichoic acid content of cells. In *Streptococcus mutans* Ingbritt, it increased fourfold when the pH was increased from 5 to 7.5 (Jacques *et al.*, 1979).

2.5.3.4f. Other Factors. Growth rate and carbohydrate source may also influence the lipoteichoic acid content of certain bacteria (Carson *et al.*, 1979; Hardy *et al.*, 1981; Knox and Wicken, 1981*b*). An increase in the glycosyl substitution of lipoteichoic acid was reported to occur during inhibition of protein synthesis in *E. faecalis* (Kessler *et al.*, 1983).

2.5.3.4g. L-Form Variants. Stabilized but osmotically fragile L-form cells of *Streptococcus pyogenes* retain the capacity of synthesizing lipoteichoic acid, but at a considerably diminished rate (Slabyj and Panos, 1976). Since they lost the ability to incorporate D-alanine (Chevion *et al.*, 1974) the lipoteichoic acid is unsubstituted in contrast to the parent strain (Slabyj and Panos, 1973).

2.6. Metabolic Fate of Lipoteichoic Acids

So far only two enzymes have been described that hydrolyze the hydrophilic chain of lipoteichoic acids: a phosphodiesterase from *Aspergillus niger* (Schneider and Kennedy, 1978), which, in conjunction with phosphomonoesterase, degrades nonsubstituted and substituted lipoteichoic acids (Childs and Neuhaus, 1980; Fischer *et al.*, 1980*b*), and a glycerophosphodiesterase from *Bacillus pumilis*, which requires nonsubstituted deacylated, rather than native lipoteichoic acids as substrate (Kusser and Fiedler, 1984). Since the latter enzyme is synthesized only under phosphate limitation, it is thought to supply the cell with phosphate from endogenous sources.

The fate of cellular lipoteichoic acid has been studied in a range of bacteria. In pulse-chase experiments using [2-^{3}H]glycerol as the label, after the chase there was no or negligible loss of lipoteichoic acid and lipids from growing cells of *Bacillus subtilis* (Koga *et al.*, 1984) and *Staphylococcus aureus* (Koch *et al.*, 1984; Raychaudhuri and Chatterjee, 1985). By contrast, in experiments with *E. faecalis* (Carson *et al.*, 1981) and *Bacillus stearothermophilus* (Card and Finn, 1983), approximately 75% of the radiolabel disappeared, within 60 min after the chase, from cellular lipoteichoic acid and lipids. Part of the label lost from *B. stearothermophilus* was recovered in the culture medium as apparently unchanged lipid amphiphiles (Card and Finn, 1983).

A number of gram-positive bacteria have been reported to lose lipoteichoic acid into the culture fluid (Joseph and Shockman, 1975; Markham *et al.*, 1975; Alkan and Beachy, 1978; Horne and Tomasz, 1979; Kessler *et al.*, 1981; Card and Finn, 1983). This so-called "excreted" or extracellular lipoteichoic acid may consist of acylated and deacylated forms in proportions ranging from almost entirely acylated form, as found with *Lactobacillus fermentum* (Markham *et al.*, 1975), to an entirely deacylated form, as found with *E. faecalis* (Joseph and Shockman, 1975). In certain bacteria, the amount of excreted lipoteichoic acid may far exceed the cell-associated fraction (Joseph and Shockman, 1975; Knox and Wicken, 1981*b*; Card and Finn, 1983). The excretion of lipoteichoic acid may be enhanced considerably under treatment with penicillin and other inhibitors of cell wall synthesis, without cells being lysed (Alkan and Beachy, 1978; Horne and Tomasz, 1979; Kessler and Van de Rijn, 1981; Brisette *et al.*, 1982; Utsui *et al.*, 1983).

The excreted deacylated lipoteichoic acid has been shown to be derived from the native acylated cellular form (Kessler and Shockman, 1979*a*), and enzymatic activities, catalyzing this conversion, have been detected in the membrane of *E. faecalis* (Kessler and Shockman, 1979*b*) and of group A streptococci (Kessler *et al.*, 1979). It is not yet known whether these enzymes have the capacity to discriminate between lipoteichoic acid and membrane lipids, which would be prerequisite for a potential role of these enzymes in regulating the cellular content of lipoteichoic acid.

Concerning the mechanism underlying the excretion of the native acylated form of lipoteichoic acid, observations are of interest which show that under normal growth conditions and on treatment with penicillin lipids (Horne *et al.*, 1977; Horne and Tomasz, 1977; Cabacungan and Pieringer, 1980; Brisette *et al.*, 1982; Brisette and Pieringer, 1985) and membrane proteins (Hakenbeck *et al.*, 1983) may be released from cells along with lipoteichoic acid. This suggests that shedding of membrane fragments may be involved in excretion, which hypothesis has been supported by the recent isolation from the growth medium of *Lactobacillus casei* of vesicles that contain lipoteichoic acid, lipids and protein. Interestingly, under treatment with penicillin, the excretion of these vesicles is greatly enhanced (Ntamere and Neuhaus, 1987; F. C. Neuhaus, personal communication). Whether acylated lipoteichoic acid and lipids may also be released from cells independently of each other, as suggested by other observations (Hurst *et al.*, 1975; Raychaudhuri and Chatterjee, 1985), requires further study.

2.7. Cellular Location

There are several lines of evidence that lipoteichoic acids are associated with the cytoplasmic membrane, most likely concentrated in the outer leaflet and, in some cases, extending through the wall to the surface of the cell (see Fig. 2-1; for reviews, see Wicken and Knox, 1975*a*; Shockman, 1981). Electron microscopic studies using ferritin-coupled antibody techniques with antibodies against poly(glycerophosphate) clearly demonstrated antibody adsorption on the surface of whole cells of *L. fermentum* and *L. casei*. This together with the heavy labeling by ferritin of the entire surface of intact protoplasts and, in disrupted cells, of membrane fragments, established that the main cellular location of lipoteichoic acid is the cell membrane (Van Driel *et al.*,

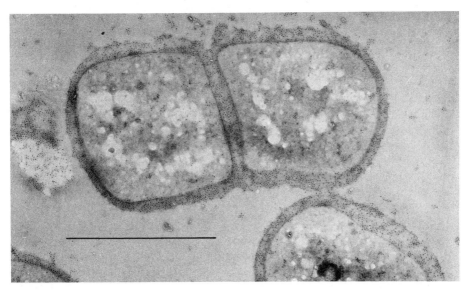

Figure 2-47. Localization of lipoteichoic acid in a thin section of *Lactobacillus plantarum* by immunoelectronmicroscopy. The section was treated with rabbit IgG specific for the poly(glycerophosphate) chain of lipoteichoic acid and then labeled with ferritin-conjugated goat antiserum to rabbit IgG. Bar = 1 μm. (Kindly provided by A. J. Wicken; from Wicken and Knox, 1975a.)

1973). Subsequent studies using thin sections of lactobacilli (Dickson and Wicken, 1974) confirmed the location of antibody binding material between the cytoplasmic membrane and the outside wall (Fig. 2-47). Where mesosomes appeared in sections, these too were labeled. Meantime similar results have been reported for *Staphylococcus aureus* (Aasjord and Grov, 1980a), groups A, B, C, and G streptococci (Miörner *et al.*, 1983; Orefici *et al.*, 1986), *Bifidobacterium bifidum* (Op den Camp *et al.*, 1985b), and *Streptococcus pneumoniae* (Sørensen *et al.*, 1988).

Difficult to reconcile with these observations was the finding that in *Staph. aureus*, *Strep. pneumoniae*, and other gram-positive bacteria, 80% of lipoteichoic acid was found associated with mesosomal vesicles (Huff *et al.*, 1974; Horne and Tomasz, 1985). These are usually formed by extrusion from the cell on plasmolysis in hypertonic solution (Salton, 1976) and were believed to be identical with intracellular membranes occasionally seen on electron microscopy (Reusch and Burger, 1973).

Although in *M. luteus* 80% of the lipomannan was also found in association with mesosomal vesicles (Owen and Freer, 1972), immunoadsorbent techniques showed it to be the major antigen at the surface of intact protoplasts (Owen and Salton, 1975c). Moreover, mesosomal vesicles of *M. luteus* were only capable of synthesizing lipomannan from D-mannosyl-1-phosphorylundecaprenol but, unlike cytoplasmic membranes, not from the precursor GDP-D-mannose (Owen and Salton, 1975b). In addition to these observations, there are a number of reasons for considering mesosomal vesicles to be artifacts, possibly formed by extrusion from the outer layer of the cytoplasmic membrane and therefore enriched with lipid macroamphiphiles but poor in enzymes of the inner side of the membrane (Salton and Owen, 1976).

The distribution of glycolipids and phosphoglycolipids between the two layers of the cytoplasmic membrane in gram-positive bacteria has not yet been studied, presumably mainly because of the lack of appropriate localization procedures (see Op den Kamp, 1979). The data on asymmetric distribution of phospholipids are limited and do not provide a coherent picture (Bishop *et al.*, 1977; Rothman and Kennedy, 1977*a*; Paton *et al.*, 1978; Demant *et al.*, 1979).

Using double-radiolabeling techniques together with chemical modification of the lipid specifically in the outer layer of the membrane, Rothman and Kennedy (1977*b*) demonstrated that in *Bacillus megaterium* newly synthesized phosphatidylethanolamine was found first on the cytoplasmic side of the membrane but a few minutes after synthesis appeared in the outer layer. The *trans*-bilayer movement was four orders of magnitude faster than in artificial membrane systems and did not require metabolic energy (Langley and Kennedy, 1979). Almost all enzymes involved in bacterial glycerolipid synthesis are membrane associated (see Pieringer, 1983) and may be located at the cytoplasmic layer, where they have access to water-soluble substrates, e.g., glycerophosphate, CTP, and UDP–glucose. For apparently the same reason, the active sites of glycerolipid synthesis in the endoplasmic reticulum of mammalian cells show an exclusively cytoplasmic surface location (Bell *et al.*, 1981).

Intact protoplasts and partly autolyzed, osmotically stabilized cells of *Bacillus subtilis* W23 proved excellent systems for studying the biosynthesis of extracellularly located polymers such as wall teichoic acid and its linkage unit (Bertram *et al.*, 1981) and peptidoglycan (Harrington and Baddiley, 1983). Both cell wall polymers were readily synthesized from externally supplied nucleotide precursors; enzymatic activity could be detected on the outer surface of the membrane. Since neither transfer of nucleotides across the membrane nor transphosphorylation from nucleotides to polyprenylphosphate carriers could be demonstrated, membrane-spanning enzyme complexes have been postulated that rotate or reorientate themselves such that the active sites are first exposed on the inner surface of the membrane where polymer synthesis, presumably in linkage to polyprenolphosphate, occurs from the nucleotide precursors. As loading of the lipid proceeds, the complex is thought to rearrange so that the growing polymer attached to its lipid emerges from the outer surface and the active sites become transiently accessible to externally added nucleotides (Baddiley, 1985).

The synthesis of lipoteichoic acid differs from the synthesis of wall teichoic acid and peptidoglycan, as nucleotide precursors and nucleotide by-products are not involved in chain synthesis. Lipoteichoic acid could therefore be directly synthesized on the outer layer of the membrane, its definitive location, if the lipid reactants can pass the membrane as does the newly synthesized phosphatidyl ethanolamine in *Bacillus megaterium*. In this model, diacylglycerol formed on lipoteichoic acid synthesis would move from the outer to the inner layer, be converted there via phosphatidic acid to phosphatidylglycerol, and this as the "glycerophosphate carrier" would move back to the site of synthesis. At the inner surface part of the diacylglycerol would be converted into glycolipid which after outward movement could serve as the starter in the synthesis of a novel lipoteichoic acid molecule. Not considered in this model is the addition to the poly(glycerophosphate) of glycosyl residues and D-alanine ester, both of which do require nucleotide substrates. If one takes into account that new alanine ester can be added to preexisting extracellularly located lipoteichoic acid (see Section 2.5.3.2c), transport of activated substrates for chain substitution through the membrane may be envisaged.

2.8. Biological Activities of Lipoteichoic Acids

2.8.1. Bacterial Systems

The widespread occurrence and structural homogeneity of lipoteichoic acid and its singular location on the membrane led, about ten years ago, to the suggestion that it serves important functions in cellular processes of gram-positive bacteria (for review, see Lambert *et al.*, 1977). These included carrier function in the biosynthesis of wall teichoic acid, regulation of autolytic enzymes, and control of the magnesium ion concentration for membrane associated enzymes. In the meantime, however, deviant lipoteichoic acid structures have been detected and, more important, glycosyl and alanyl substituents have been found to have a considerable effect on biological activities, so that the concept of structural and functional homogeneity in the previous sense is no longer tenable. Common functions of lipoteichoic acid would have to be compatible with the recently discovered structural diversities.

2.8.1.1. Carrier Activity in Wall Teichoic Acid Biosynthesis

When poly(ribitolphosphate) polymerase, the wall teichoic acid-synthesizing enzyme, was isolated from *S. aureus*, it required an amphiphilic carrier in linkage to which the polymer is synthesized (Fiedler and Glaser, 1974a,b). This carrier was thought to be similar to lipoteichoic acid (Fiedler and Glaser, 1974c), then to be a specialized fraction of it (Lambert *et al.*, 1977) and was later shown to be lipoteichoic acid itself (Fischer *et al.*, 1980a). The structural requirements for carrier function are a nonspecific lipid moiety and a tetra(glycerophosphate) sequence, at the glycerol terminus of which the growing poly(ribitolphosphate) is covalently attached (Fischer *et al.*, 1980a). Glycosyl and alanyl substituents interfere with carrier activity (Fischer *et al.*, 1980a,b). For binding to the polymerase four negatively charged phosphate groups are required on the tetra(glycerophosphate) terminus, so that a single alanyl residue anywhere on this sequence is sufficient to prevent binding and abolish carrier activity. Glycosyl residues are less inhibitory, possibly acting by steric hindrance (Koch *et al.*, 1982). Similar results were obtained with the poly(ribitolphosphate) polymerase from *Staphylococcus xylosus* (Fiedler, 1981).

At about the same time that the requirement of a carrier for wall teichoic acid synthesis was detected, it became evident that teichoic acid is attached to

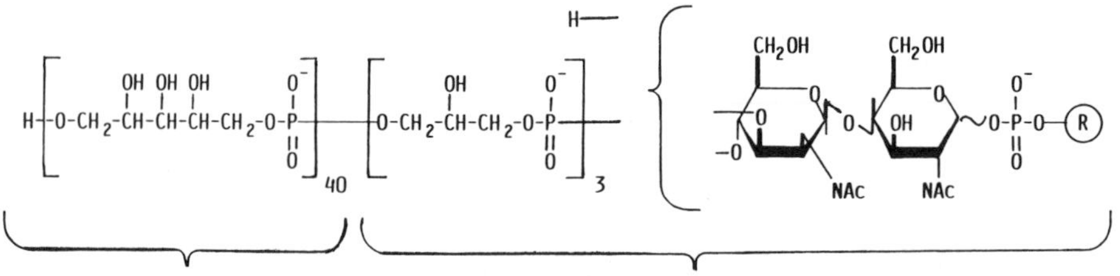

TEICHOIC ACID   LINKAGE UNIT

Figure 2-48. Wall teichoic acid–linkage unit complex. During biosynthesis of the complex poly-prenylphosphate, (R) is in the definite linkage to the cell wall C-6 of muramic acid in pep-tidoglycan. In the case of *Staphylococcus aureus* position C-2 and C-4 of the ribitol residues are substituted by D-alanine ester and *N*-acetyl-D-glucosaminyl residues. (From Coley *et al.*, 1978; Kojima *et al.*, 1983, 1985.)

the cell wall through a linkage unit as shown in Fig. 2-48. This linkage unit was then found to be synthesized on polyprenolphosphate (Fig. 2-49), and evidence was provided for a transfer of completed teichoic acid from lipotei-choic acid to the linkage unit lipid (McArthur *et al.*, 1981). Direct synthesis of poly(ribitolphosphate) on the linkage unit lipid could, however, also be dem-onstrated and as a consequence of this lipoteichoic acid was thought to be an *in vitro* analogue (Bracha *et al.*, 1978). This hypothesis is strongly supported by the fact that lipoteichoic acid from *S. aureus,* grown at low salt concentration, is *in vitro* completely inactive as a carrier, owing to its high alanyl content (Koch *et al.*, 1982). On the other hand the lipoteichoic acid of *S. aureus* displays some carrier activity when the alanyl content is reduced by growth on high salt concentration (Koch *et al.*, 1982). Another example for potential carrier ac-tivity in vivo is the lipoteichoic acid of *Micrococcus varians*. In the native state, it is unsubstituted (see Table 2-2), is active in vitro as a carrier for wall teichoic acid synthesis, and membrane preparations of this organism have the capacity to transfer teichoic acid, preformed on lipoteichoic acid, to the linkage unit lipid (McArthur *et al.*, 1981). Whether in *S. aureus* grown at high salt con-centration or in normally grown *M. varians*, pathway B (Fig. 2-49) is operative or lipoteichoic acid is in the native membrane inaccessible to the polymerase remains to be established. Attempts to isolate lipoteichoic acid loaded with teichoic acid from *M. varians* have so far failed (Fischer *et al.*, 1990).

2.8.1.2. Effect on Autolytic Enzymes

The first lipoteichoic acid shown to be inhibitory to autologous autolysin was the Forssman antigen of *Strep. pneumoniae* (Höltje and Tomasz, 1975*a,b*). The autolysin of this organism, a *N*-acetylmuramyl-L-alanine-amidase, re-quires pneumococcal teichoic acid for both conversion into a catalytically active form and for activity of the latter whereby the choline residues of the teichoic acid are essential (Höltje and Tomasz, 1975*c*; Tomasz *et al.*, 1975). The choline-containing pneumococcal lipoteichoic acid was found to prevent activation and to inhibit the activated enzyme. As a consequence of this antiautolytic effect,

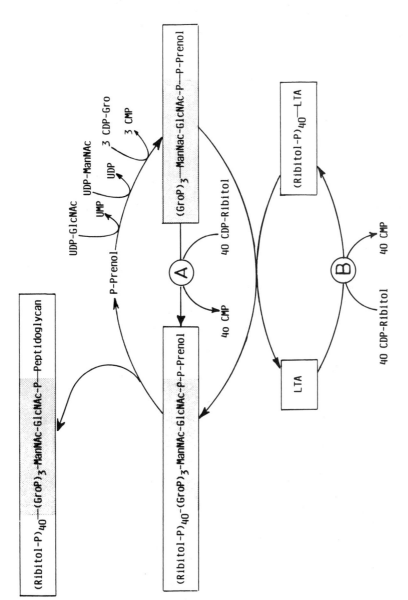

Figure 2-49. Biosynthesis of the linkage unit and the poly(ribitolphosphate)–linkage unit complex in *Staphylococcus aureus*. (For reviews, see Bracha *et al.*, 1978; Ward, 1981; Hancock and Baddiley, 1985.) Pathway A shows the direct assembly of poly(ribitolphosphate) on the linkage unit lipid (Bracha *et al.*, 1978), pathway B the assembly on lipoteichoic acid followed by the transfer of poly(ribitolphosphate) to the linkage unit lipid (McArthur *et al.*, 1981). The incorporation of N-acetyl-D-mannosamine into the linkage unit lipid has not been demonstrated until recently, because the requirement of this hexosamine was obscured by the presence of UDP–GlcNAc-2-epimerase in membrane preparations (Harrington and Baddiley, 1985).

lipoteichoic acid added to pneumococcal cultures caused chain formation, resistance to stationary phase lysis, and penicillin tolerance (Höltje and To-masz, 1975*a,b*). Analogous to their requirement for the activating effect of teichoic acid, choline residues are also essential for the inhibitory effect of lipoteichoic acid (Horne and Tomasz, 1985). This discrepancy may find an explanation in the recent observation that pneumococcal lipoteichoic acid covalently linked to Sepharose retained autolysin through specific binding to its choline residues but, in contrast to micellar lipoteichoic acid, had the capacity to convert the amidase to the active form and it did not suppress the wall-lytic activity (Briese and Hakenbeck, 1985). The antiautolytic effect was also abolished by deacylation of lipoteichoic acid (Briese and Hakenbeck, 1985). These observations led to the idea that micellar lipoteichoic acid may act as a topological barrier rather than an inhibitor of autolysin, whereby *in vivo* binding of the enzyme near the membrane may, as observed *in vitro,* prevent its access to the cell wall substrate.

Since the discovery of the antiautolytic effect of the Forssman antigen, lipoteichoic acids have been suggested generally to act on autolytic enzymes. Accordingly, it could be shown that poly(glycerophosphate) lipoteichoic acids have an inhibitory effect on autologous and heterologous autolysins, provided they were from bacteria that possess this type of lipoteichoic acid (Cleveland *et al.,* 1975, 1976*a,b;* Rogers *et al.,* 1984). Consequently, the structurally different lipoteichoic acid from *S. pneumoniae* (see Section 2.2.2) and the succinylated lipomannan from *M. luteus* (see Section 2.2.3) were inactive (Cleveland *et al.,* 1975). For autolysins of *S. aureus,* the range of specificity has been defined by the observations that a sequence of negatively charged glycerophosphate resi-dues is required and that structural differences in the glycolipid moiety do not alter the activity (Fischer *et al.,* 1981). Negatively charged phospholipids, such as cardiolipin and phosphatidylglycerol, are also inhibitory for a range of autolysins, whereas neutral glycolipids and zwitterionic phospholipids are not (Cleveland *et al.,* 1976*b;* Fischer *et al.,* 1981). Deacylation of lipoteichoic acids and phospholipids destroys their inhibitory activity and addition of Triton X-100 has a similar effect (Cleveland *et al.,* 1975, 1976*a,b*). This suggests that the amphiphiles interact with autolysins in the form of micelles and lipo-somes, respectively, and that the high density of negative charges on the surface of these particles may play a role.

Alanyl substituents interfere with the antiautolytic activity (Fischer *et al.,* 1981): In dose–response experiments, using autolysins of *S. aureus,* increas-ing alanine content of lipoteichoic acid systematically increased the concentra-tion required for 50% inhibition, and native lipoteichoic acid with an Ala-Gro/Gro molar ratio of 0.7 was virtually inactive (Table 2-19). In contrast to alanine ester, glycosyl substituents had little or no effect, even if 80% of the glycerophosphate residues are glycosylated (Table 2-19). These results sug-gest that alanine ester substituents act through charge compensation rather than steric hindrance.

The occurrence of lipoteichoic acid in acylated and deacylated form was thought to be appropriate for a role in *in vivo* regulation of autolytic activity (Cleveland *et al.,* 1975). Alanine-free lipoteichoic acid and lipoteichoic acid

Table 2-19. Relationship between Substitution of Lipoteichoic Acids and Their Inhibitory Activity on Extracellular Autolytic Enzymes (A) and Autolysis of S. aureus (B)[a]

Chain substititution	Concentration (µM) effecting 50% inhibition	
	A	B
None	1.6	0.8
AlaGro/Gro		
0.33	5.2	6.0
0.55	28.8	11.4
0.59	82.4	—
0.71	Inactive	32.5
GlycosylGro/Gro		
0.33	1.7	—
0.47	2.0	0.8
0.82	—	1.4

[a]From Fischer *et al.* (1981).

with a high alanine ester content represent another pair of inhibitory and noninhibitory forms and, by variation of the alanine ester content, the inhibitory effect would be variable in a stepless manner. There is, however, no evidence as to whether this potential is used in the living cell (Fischer *et al.*, 1981; Fischer and Koch, 1983). In low-salt culture of *S. aureus*, the bulk of lipoteichoic acid has no inhibitory capacity due to the high alanyl content (Tables 2-18 and 2-19).

Recently described mutant strains of *Lactobacillus casei* were defective in alanine ester content of lipoteichoic acid and showed defects in cell separation (Ntamere *et al.*, 1987). Although one might suggest that in these mutants the increased net negative charge of lipoteichoic acid could act as a trap of cell wall lytic enzymes resulting in the observed defective cell separation, there were no detectable differences in the rate of autolysis and cell wall turnover as compared with the parent strain. A number of attempts to demonstrate a regulation of autolysin by lipoteichoic acid in living cells of *E. faecalis* have so far also failed (Shungu *et al.*, 1980; Carson and Daneo-Moore, 1981; Wong *et al.*, 1981). It should be noted, however, that all these observations concern the bulk of cellular lipoteichoic acid, which does not preclude that a minor perhaps topologically defined fraction may be modified in terms of regulation.

2.8.1.3. Cation Binding and the Donnan Effect

Together with wall-linked teichoic acid, teichuronic acid, and carboxyl groups of peptidoglycan, membrane-associated lipoteichoic acid forms a negatively charged network between the membrane and the surface of the cell (see Fig. 2-1), which has been suggested to act as a reservoir of divalent cations and particularly to serve to supply various membrane associated enzymes with Mg^{2+} (Heptinstall *et al.*, 1970; Ellwood and Tempest, 1972; Hughes *et al.*,

1973; Archibald, 1974). It has further been proposed that alanine ester substituents of lipoteichoic acid and teichoic acid interfere with cation binding and variations in the alanine ester content might serve to regulate the cation binding capacity (Heptinstall *et al.*, 1970; Baddiley *et al.*, 1973; Lambert *et al.*, 1975*b*).

As recently pointed out (Fischer, 1988), differences in the association constant of Mg^{2+} binding, the binding capacity, and the effect of alanine ester on Mg^{2+} binding became apparent between experiments studying Mg^{2+} binding by poly(glycerophosphate) wall teichoic acid with equilibrium dialysis (Lambert *et al.*, 1975*a*) and ^{31}P-NMR studies on Mg^{2+} binding by poly(glycerophosphate) lipoteichoic acid (Batley *et al.*, 1987). Further experiments are required to harmonize these results and eventually reevaluate the role of lipoteichoic acid and teichoic acid in scavenging divalent cations from the surroundings.

Since the polyanionic chains of teichoic acid and lipoteichoic acid are anchored in the wall–membrane complex, Donnan equilibria will be established between the wall region and the bulk surroundings (Ou and Marquis, 1970). Little attention has been paid to Donnan effects, although at low external ionic strength, they may lead to considerable nonspecific accumulation of cations in the periplasmic and cell wall region and in addition elevate the osmotic pressure in this compartment over that of the surroundings. Whether the synthesis of lipoteichoic acid and teichoic acid is increased in response to low external ionic strength has not yet been studied. The increase of the alanine ester content of both polymers in *S. aureus* under these conditions (Table 2-18) does not, however, favor a role in osmoregulation.

2.8.1.4. Other Activities

Streptolysin S, produced by group A streptococci, requires for activation and stability a carrier which may be RNA, nonionic detergents, or lipoteichoic acid (Theodore and Calandra, 1981). *Enterococcus faecalis* spp. *zymogenes* produces a lysin with hemolytic and bacteriolytic properties and is protected from its action by an endogenous inhibitor (Davie and Brock, 1966; Basinger and Jackson, 1968). This was considered to be teichoic acid but, in view of its properties and cellular location, it might have been lipoteichoic acid. Interestingly, in this case alanyl substituents seem to be essential for the inhibitory activity (Davie and Brock, 1966).

2.8.1.5. Concluding Remarks

It remains to be established whether lipoteichoic acids play a vital role in bacterial physiology as previously proposed. Mutants that lost the capacity of lipoteichoic acid synthesis have not yet been described. However, mutants of *B. subtilis* and *S. aureus* that lack wall teichoic acid are known. They show pleiotropic alterations that concern wall morphology, growth characteristics, cell division and separation, as well as autolytic potentialities (Boylan *et al.*, 1972; Chatterjee *et al.*, 1969; Gilpin *et al.*, 1972).

2.8.2. Mammalian Organisms

2.8.2.1. Fate in the Mammalian Organism

As shown *in vitro*, lipoteichoic acid is bound by hydrophobic interaction with high affinity to serum albumin (Simpson *et al.*, 1980*a*) and plasma fibronectin (Stanislawski *et al.*, 1987; Courtney *et al.*, 1988). On fibronectin binding sites for lipoteichoic acid in its acylated form, other lipid amphiphiles and lipoproteins were localized on a 24-kDa domain from the amino terminus.

Although there are numerous experiments in which lipoteichoic acids were given parenterally to rabbits, rats, and mice, neither degradation nor excretion has so far been studied. When given parenterally to rabbits under conditions that do not elicit antibody formation, lipoteichoic acid was found deposited in the cortex-associated tubular regions of the kidneys (Waltersdorf *et al.*, 1977).

2.8.2.2. Immunological Properties

Isolated lipoteichoic acids have been considered nonimmunogenic (McCarty, 1959; Wicken and Knox, 1975*b*; Fiedel and Jackson, 1976) unless contaminated with protein (Wicken *et al.*, 1973) (for reviews, see Knox and Wicken, 1973; Wicken and Knox, 1975*a*, 1977). Antisera of high titer against purified lipoteichoic acids are obtained by the use of Freund's complete adjuvant (Knox *et al.*, 1970) or Freund's incomplete adjuvant in combination with methylated bovine serum albumin and a detergent such as Tween 80 or Triton X-100 (Fiedel and Jackson, 1976; Beachy and Ofek, 1976). Formation of antibody will also be effectively elicited by injection of whole bacteria if lipoteichoic acid is sufficiently exposed on the bacterial surface (Van Driel *et al.*, 1973) or, if not, by injection of disintegrated bacteria (Slade and Slamp, 1962; Wicken and Knox, 1975*b*).

Antilipoteichoic acid antibodies of the immunoglobulin IgM and IgG class have been induced and detected in rabbits, guinea pigs, and mice (Knox and Wicken, 1973, Frederick and Chorpenning, 1974; Jackson *et al.*, 1981*a,b*). In rabbits, IgM antibodies may predominate or be present solely (Aasjord and Grov, 1980*b*; Jackson *et al.*, 1981*a,b*). By application of bacterial cells, IgM memory against lipoteichoic acid is induced. In response to a second injection, there is a 10-fold increase in IgM-producing cells and a significantly higher increase of specific IgM antibodies as compared with the first injection. IgG antibodies are concomitantly formed but, in contrast to the classic picture, without causing cessation of IgM antibody synthesis (Jackson *et al.*, 1981*a,b*). Whether the immunogenicity of lipoteichoic acids is T-cell-dependent or independent remains to be clarified.

Immunodeterminants of lipoteichoic acids are the poly(glycerophosphate), glycosyl substituents (Wicken and Knox, 1975*a*) and alanine ester (McCarty, 1964). The specificity of antisera is variously directed against the glycosyl substituents, the poly(glycerophosphate) chain or both (Knox and Wicken 1972, 1973; Wicken and Knox, 1975*b*; Jackson *et al.*, 1981*b*). Mono-

Table 2-20. Glycosyl Lipoteichoic Acids Reacting as Group Antigen with Group-Specific Antisera[a]

Genus	Group	Determinant
Lactobacillus helveticus[b]	A	α-D-Glc1→
Lactobacillus fermenti[c]	F	α⇒D-Gal1→
Enterococcus faecalis[d]	D	α⇒D-Glc(1–2)α-D-Glc1→
Lactococcus lactis[e]	N	α⇒D-Gal1→
Streptococcus mutans[f]	Serotype a	β⇒D-Gal1→

[a]From Wicken and Knox (1975*a*), supplemented.
[b]Knox and Wicken (1971).
[c]Knox *et al.* (1970).
[d]Wicken *et al.* (1963), Toon *et al.* (1972), Schleifer and Kilpper-Bälz (1984).
[e]Wicken and Knox (1975*b*), Schleifer *et al.* (1985).
[f]Van de Rijn and Bleiweiss (1973).

clonal antibodies have been obtained which are monospecific for either the poly(glycerophosphate) or the glycosyl substituents (Jackson *et al.*, 1984). Alanyl lipoteichoic acids reacting with homologous antisera show little or no reaction with antipoly(glycerophosphate) antibodies (McCarty, 1964). Highly glycosylated lipoteichoic acids are likewise less reactive with antipoly(glycerophosphate) antisera (Kessler and Thivierge, 1983). Most of the examples of cross reaction are due to the presence of antibodies against poly(glycerophosphate), which is considered a common antigen of gram-positive bacteria. In a few bacteria, lipoteichoic acids fulfill the requirements of a group antigen (Wicken and Knox, 1975*a*). In the examples characterized chemically and serologically the group-specific determinant is a particular glycosyl substituent as shown in Table 2-20.

Frequently human sera contain natural antibodies to glycerol lipoteichoic acids, as detected by the hemagglutination reaction (Rantz *et al.*, 1952; Ginsburg *et al.*, 1972; Decker *et al.*, 1972; Markham *et al.*, 1973; Levine, 1982). These antibodies show varying specificities as outlined above. In healthy persons, they reach a maximum incidence in the late teens (Rantz *et al.*, 1952; Ginsburg *et al.*, 1972). Reports on the levels in patients with acute rheumatic diseases do not give a uniform picture (Rantz *et al.*, 1952; Harris and Harris, 1953; Ginsburg *et al.*, 1972).

2.8.2.3. Binding to Mammalian Cells

Like other amphiphiles lipoteichoic acids possess the ability to bind to mammalian cells (Moskowitz, 1966) as was shown with erythrocytes (Beachy *et al.*, 1979a; Weinreb *et al.*, 1986), platelets (Beachy *et al.*, 1977), lymphocytes (Beachy *et al.*, 1979*b*), polymorphnuclear leukocytes (Courtney *et al.*, 1981), and mucosal epithelial cells (Simpson *et al.*, 1980*b*). Binding requires the hydrophobic moiety of lipoteichoic acid (Hewett *et al.*, 1970; Ofek *et al.*, 1975; Beachy *et al.*, 1979a); is dependent on cell concentration, time, and temperature; is reversible; and is saturable with increasing concentrations of lipoteichoic acid (Beachy *et al.*, 1979*a*). Binding may be considered an equilibrium between the following three forms of lipoteichoic acid:

$$\text{Micellar LTA} \rightleftharpoons \text{monomolecular LTA} \rightleftharpoons \text{membrane-associated LTA} \quad (15)$$

Scatchard plots of binding data were interpreted to indicate a single population of binding sites with dissociation constants ranging from 4.5×10^{-6} M for erythrocytes to 8.9×10^{-5} M for lymphocytes (Beachy *et al.*, 1979*a*, 1983). Calculated numbers of binding sites per cell were 7.2×10^6 and 2.9×10^7 for sheep and human erythrocytes, respectively. That right-side-out vesicles of erythrocyte ghosts bound 10 times more lipoteichoic acids than the inside-out vesicles, was taken as evidence for the presence of specific binding sites (Chiang *et al.*, 1979). The nature of these binding sites remains, however, to be clarified. Possibly they simply reflect a limited number of foreign amphiphilic molecules that can be inserted between the native lipids of particular membranes or membrane layers.

2.8.2.4. *Bacterial Adhesion to Host Cells*

An early event in infection is the attachment of bacteria to the surface of host cells. Inhibition of attachment of streptococci and staphylococci to mammalian mucosal epithelial cells by pretreatment of the cells with lipoteichoic acid or of the bacteria with antibodies against lipoteichoic acid (Beachy and Ofek, 1976; Carruthers and Kabat, 1983; Teti *et al.*, 1987) led to the hypothesis that lipoteichoic acid, transiently located on the bacterial surface during excretion (see Section 2.6), might mediate the adherence by attachment of its lipid moiety to a receptor on the cytoplasmic membrane of host cells (Ofek *et al.*, 1982; Beachy *et al.*, 1983). In this model, single lipoteichoic acid molecules are thought to be held on the surface of the bacterial cell with energetically unfavorable exposure of their hydrophobic hydrocarbon chains through alignment of the anionic poly(glycerophosphate) chains with cationic surface-associated proteins, e.g., M protein in group A streptococci (Ofek *et al.*, 1982; Beachy *et al.*, 1983). On the surface of oropharyngeal epithelial cells, fibronectin has been suggested to play a role in the adhesion of group A streptococci through exposed lipoteichoic acid (Simpson *et al.*, 1987; Stanislawski *et al.*, 1987; Courtney *et al.*, 1988). Clearly, more direct evidence is required to substantiate this hypothetical role of lipoteichoic acid in pathogenicity.

2.8.2.5. *Sensitizing of Host Cells to Complement-Mediated and -Independent Destruction*

Lipoteichoic acids attached to mammalian cells are accessible to antibodies against the hydrophilic chain (Harris and Harris, 1953; Rantz *et al.*, 1956), and erythrocytes bearing lipoteichoic acid on their surface will lyse when incubated with antibodies and complement (Rantz *et al.*, 1952, 1956; Ne'eman and Ginsburg, 1972). Lysis of erythrocytes sensitized with poly(glycerophosphate) lipoteichoic acids does not occur in agammaglobulinemic and

C2-deficient sera which confirms the requirement of antibody and shows that activation of the classical complement pathway occurs (Weinreb *et al.*, 1986). In the presence of antilipoteichoic acid antibodies, lipoteichoic acid-bearing host cells are also sensitive to antibody-dependent cellular cytotoxicity (ADCC), which is mediated by effector cells possessing Fc receptors capable of recognizing and lysing antibody-coated target cells (Lopatin and Kessler, 1985).

Lipoteichoic acids themselves have the capacity of reacting with components of the complement system (Fiedel and Jackson, 1978; Silvestri *et al.*, 1979). Poly(glycerophosphate) lipoteichoic acids interact with C1, which may result in the activation of the classic complement pathway, as suggested by the concomitant consumption of C1, C2, and C4 in human and guinea pig serum (Loos *et al.*, 1986). The dose-dependent interaction of lipoteichoic acids with C1 occurs at the positively charged subunit C1q and appears to depend on the negative charge and accessibility of the phosphate groups of lipoteichoic acids. Accordingly, chain substitution with alanine ester and bulky glycosyl residues interferes with binding. The density of negative charges also seems to play a role because loss of micellar organization by deacylation of lipoteichoic acids drastically reduces the binding capacity.

The differently structured lipoteichoic acid of *S. pneumoniae* (see Section 2.2.2) is not capable of activating complement in micellar solution but acquires this capacity on the surface or erythrocytes (Hummell *et al.*, 1985; Hummell and Winkelstein, 1986). Significant lysis of sensitized erythrocytes in C2-deficient and C4-deficient sera indicates a role for the alternative complement pathway. Lysis is, however, facilitated by an intact classical pathway, as shown by enhancement of lysis when C4 is added to C4-deficient serum. Erythrocytes sensitized with pneumococcal lipoteichoic acid are also lysed in the blood circulation system of rats.

These results demonstrate that lipoteichoic acids are able to render the host cells susceptible to destruction by its own complement system and ADCC. Sensitizing of host cells by binding of lipoteichoic acid may contribute to pathogenicity of gram-positive bacteria if one considers that lipoteichoic acids are released from bacteria into surroundings and that this release is enhanced by bacterial exposure to certain antibiotics (see Section 2.6). In this context, it should be noted, however, that serum albumin and plasma fibronectin possess the capacity of binding lipoteichoic acid (see Section 2.8.2.1) and might thus protect host cells from lipoteichoic acid-mediated injury.

2.8.2.6. Interaction with Lipopolysaccharide-Recognizing Protein

Normal serum of mammalian species including man contains a low-molecular-weight protein that binds specifically to the inner core region of lipopolysaccharides of gram-negative bacteria presumably by recognizing 5-*O*-glycosylated 2-keto-3-deoxy-D-mannooctonic acid (Brade and Galanos, 1983*a,b;* Brade and Brade, 1985). Binding of the protein to lipopolysaccharides on the surface of erythrocytes results in complement-mediated lysis. Certain poly(glycerophosphate) lipoteichoic acids are also able to bind to this protein, whereas others are not. Binding occurs in a competitive manner to lipopolysaccharides, and erythrocytes coated with interacting lipoteichoic acids become, in the presence of lipopolysaccharide-binding protein, prone to lysis by complement. Up to now, no relationship between the structure of lipoteichoic acids and the interaction with lipopolysaccharide-binding protein has been recognized. It remains to be studied whether undiscovered minor structural elements of lipoteichoic acids might be responsible (H. Brade, L. Brade, and W. Fischer, unpublished data).

2.8.2.7. Effects on Leukocytes and Macrophages

The observation that lipoteichoic acid in acylated form can suppress the antibody response to sheep erythrocytes in mice and enhance antibody production against lipopolysaccharides (Miller and Jackson, 1973; Miller *et al.*, 1976) suggests modulating effects on the immune system. In these experiments, no mitogenic effect on murine spleen cells could be observed. Later on conflicting results were reported suggesting lipoteichoic acids to be selective T-cell (Beachy *et al.*, 1979*b*) or B-cell mitogens (Mishell *et al.*, 1981). Differences in the experimental conditions or contamination of lipoteichoic acids with protein (1–5%) may be responsible for these conflicting results (Mishell *et al.*, 1981). In support of the earlier observations, lipoteichoic acids from various sources purified by hydrophobic interaction chromatography (see Section 2.3.2) did not show mitogenic stimulation of mouse spleen cells (L. Brade, H. Brade, and W. Fischer, unpublished data). In contrast to lipoteichoic acids, lipopolysaccharides are potent B-cell mitogens.

Lipoteichoic acid will stimulate macrophages as measured by increases in the phagocytotic index in carbon clearance by the reticuloendothelial system (Miller *et al.*, 1976). Peritoneal macrophages from mice pretreated with lipoteichoic acid displayed enhanced bactericidal activity against *Listeria monocytogenes in vitro;* after *in vivo* infection of these mice, reduced colonization of lung and liver by *Listeria* was observed (Oshima *et al.*, 1988*a*). Moreover, to stimulation with zymosan peritoneal macrophages from lipoteichoic acid-treated mice reacted with an increase in chemiluminescence (Oshima *et al.*, 1988*c*), which is thought to be an indicator for the formation of bactericidal oxygen radicals. These mice also showed an enlargement of spleen, and after challenge with *L*-1 sarcoma cells a reduced pulmonary tumor colonisation as compared with untreated mice. Human monocytes directly responded to lipoteichoic acid with chemiluminescence, whereas human granulocytes did not (Oshima *et al.*, 1988*b*). In murine peritoneal macrophages, lipoteichoic acids were also reported to stimulate the secretion of hydrolytic enzymes (Harrop *et al.*, 1980), interleukin-1 (IL-1), and colony stimulating factor (CSF) (Mishell *et al.*, 1981).

2.8.2.8. Toxicity

Although lipoteichoic acids share the capacity of binding to mammalian cells with lipopolysaccharides of gram-negative bacteria, they do not possess endotoxic properties (Wicken and Knox, 1980). Since pyrogenicity and lethal toxicity of lipopolysaccharides reside in the lipid A moiety (Galanos *et al.*, 1985a), it is possibly the less complex lipid structure of lipoteichoic acids that renders them nonendotoxic. Since macrophages play a central role in the mediation of endotoxic effects (Peavy and Brandon, 1980; Maier and Ulevitch, 1981) there might be fundamental differences in the action of lipoteichoic acids and lipopolysaccharides on these cells.

Lipoteichoic acids have been reported to possess the capacity to produce the Shwartzman reaction in its localized and generalized form in rabbits, although higher doses of lipoteichoic acids are required than of lipopolysaccharides (Knox and Wicken, 1973; Wicken and Knox, 1975a). How this effect is brought about by lipoteichoic acids is unclear. For Shwartzman reactivity of lipopolysaccharides, a 3-acyloxyacyl residue on the lipid A moiety plays an important role (Galanos *et al.*, 1985b; Rietschel *et al.*, 1987), since a partial lipid A structure containing only the four sugar-linked 3-hydroxy fatty acids is comparatively ineffective (Galanos *et al.*, 1984).

Lipoteichoic acids, like lipopolysaccharides, stimulate bone resorption in tissue culture (Hausmann *et al.*, 1975). Severe inflammatory lesions are induced in the periodontal tissues of immunized and nonimmunized rats by intragingival injection of lipoteichoic acid (Bab *et al.*, 1979). Destructive capabilities of lipoteichoic acid in tissue culture have been reported for human kidney cells (De Vuono and Panos, 1978), heart cells (Simpson *et al.*, 1982; Courtney *et al.*, 1986), amnion and brain cells, HeLa cells (Goldschmidt and Panos, 1984), cultured mouse glomeruli (Tomlinson *et al.*, 1983), and mouse fibroblast monolayers (Leon and Panos, 1983, 1985). An increased toxicity of lipoteichoic acid was observed for established transformed versus primary normal human cell lines (Goldschmidt and Panos, 1984). In these experiments, the addition of serum to tissue cell culture negated the toxic effect of lipoteichoic acids in a dose-dependent manner, possibly by binding lipoteichoic acids to particular serum proteins (see Section 2.8.2.1). Whether cellular toxicity plays a role in the pathogenicity of infection with gram-positive bacteria is unclear, especially because, in the studies cited above, high concentrations of lipoteichoic acids were usually employed. Moreover, no signs of toxicity were seen in experiments in which rabbits and mice received injections of purified lipoteichoic acid for varying periods up to one year even at high single or total dosages (Miller and Jackson, 1973; Fiedel and Jackson, 1974).

2.8.2.9. Hypersensitivity

In hyperimmunized rabbits, arthritis can be produced by local injection of lipoteichoic acid (Ne'eman and Ginsburg, 1972). Severe hypersensitivity reactions have been observed in rabbits which were immunized and challenged with lipoteichoic acid–adjuvant complexes or bacterial cells with surface-associated lipoteichoic acid (Fiedel and Jackson, 1976, 1979; Aasjord *et al.*, 1980; Jackson *et al.*, 1981c). In the various studies, hypersensitivity manifested itself in different forms, i.e., symptoms suggesting anaphylactic shock (Fiedel and Jackson, 1976), nephropathy (Fiedel and Jackson, 1979), acute cor pulmonale with right-sided heart failure (Jackson *et al.*, 1981*c*), and encephalitis (Aasjord *et al.*, 1980). It is not yet clear whether these different manifestations were caused by differences in the lipoteichoic acid preparation or in the carrier complexes in which they were administered. In addition to type I hypersensitivity type III hypersensitivity seems to be involved, as indicated by the consumption of complement and a rapid decline in serum of antilipoteichoic acid antibodies after secondary challenge (Fiedel and Jackson, 1976, 1979).

ACKNOWLEDGMENTS. Over the years, the author has benefited from the collaboration of a number of co-workers, named in the reference list, to all of whom I am grateful. I thank my wife Elisabeth for doing the artwork and for help with the manuscript. The continuous financial support by the Deutsche Forschungsgemeinschaft of the work on phosphoglycolipids and lipoteichoic acids in the author's laboratory is gratefully acknowledged.

References

Aasjord, P., and Grov, A., 1980*a*, Immunoperoxidase and electron microscopy studies of staphylococcal lipoteichoic acid, *Acta Pathol. Microbiol. Scand.* [*B*] **88**:47.

Aasjord, P., and Grov, A., 1980*b*, Antibodies in rabbits immunized with staphylococcal lipoteichoic acid, *Acta Pathol. Microbiol. Scand.* [*B*] **88**:53.

Aasjord, P., Nyland, H., and Mørk, S., 1980, Encephalitis induced in rabbits by staphylococcal lipoteichoic acid, *Acta Pathol. Microbiol. Scand.* [*C*] **88**:287.

Adlam, C., Knights, J. M., Mugridge, A., Lindon, J. C., Williams, J. M., and Beesley, J. E., 1985, Purification, characterization and immunological properties of the capsular polysaccharide of *Pasteurella hemolytica* Serotype T15: Its identity with the K62 (K2ab) capsular polysaccharide of *Escherichia coli* and the capsular polysaccharide of *Neisseria meningitidis* serogroup H, *J. Gen. Microbiol.* **131**:1963.

Alkan, M. L., and Beachy, E. M., 1978, Excretion of lipoteichoic acid by group A streptococci: Influence of penicillin on excretion and loss of ability to adhere to human oral mucosal cells, *J. Clin. Invest.* **61**:671.

Ambron, R. T., and Pieringer, R. A., 1971, The metabolism of glyceride glycolipids. V. Identification of the membrane lipid formed from diglucosyl diglyceride in *Streptococcus faecalis* ATCC 9790 as an acylated derivative of glycerylphosphoryl diglucosylglycerol, *J. Biol. Chem.* **246**:4216.

Anderton, W. J., and Wilkinson, S. G., 1985, Structural studies of a mannitol teichoic acid from the cell wall of Bacterium N.C.T.C. 9742, *Biochem. J.* **226**:587.

Archibald, A. R., 1974, The structure, biosynthesis and function of teichoic acid, *Adv. Microbial Physiol.* **11**:53.

Archibald, A. R., and Baddiley, J., 1966, The teichoic acids, in *Advances in Carbohydrate Chemistry* (M. L. Wolfrom and R. S. Tipson, eds.), pp 323–375, Academic, New York.

Archibald, A. R., Baddiley, J., and Blumson, N. L., 1968, The teichoic acids, *Adv. Enzymol.* **30**:223.

Archibald, A. R., Baddiley, J., and Heptinstall, S., 1973, The alanine ester content and magnesium binding capacity of walls of *Staphylococcus aureus* H, grown at different pH values, *Biochim. Biophys. Acta* **291**:629.

Armstrong, J. J., Baddiley, J., Buchanan, J. G., Carrs, B., and Greenberg, G. R., 1958, Isolation and structure of ribitol phosphate derivatives (teichoic acids) from bacterial cell walls, *J. Chem. Soc.* **1958**:4344.

Armstrong, J. J., Baddiley, J., Buchanan, J. G., Davison, A. L., Kelemen, M. V., and Neuhaus, F. C., 1959, Teichoic acids from bacterial walls. Composition of teichoic acids from a number of bacterial walls, *Nature (Lond.)* **184**:247.

Arneth, D., 1984, Lipide und Fettsäuren aus motilen Streptococcen der Serogruppe N, Thesis, University Erlangen-Nürnberg.

Bab, I. A., Sela, M. N., Ginsburg, I., and Dishon, T., 1979, Inflammatory lesions and bone resorption induced in the rat periodontium by lipoteichoic acid of *Streptococcus mutans. Inflammation* **3**:345.

Baddiley, J., 1972, Teichoic acids in cell walls and membranes of bacteria, *Essays Biochem.* **8**:35.

Baddiley, J., 1985, Trans-membrane synthesis of cell wall polymers. *Biochem. Soc. Trans.* **13**:992.

Baddiley, J., Buchanan, J. G., Mathias, A. P., and Sanderson, A. R., 1956, Cytidine diphosphate glycerol, *J. Chem. Soc.* **1956**:4186.

Baddiley, J., Buchanan, J. G., and Sanderson, A. R., 1958, Synthesis of cytidine diphosphate glycerol, *J. Chem. Soc.* **1958**:3107.

Baddiley, J., Hancock, I. C., and Sherwood, P. M. A., 1973, X-ray photoelectron studies of magnesium ions bound to the cell walls of Gram-positive bacteria, *Nature (Lond.)* **243**:43.

Baer, E., and Kates, M., 1948, Migration during hydrolysis of esters of glycerophosphoric acid. The chemical hydrolysis of L-α-glyceryl-phosphoryl-choline, *J. Biol. Chem.* **175**:79.

Baer, E., and Kates, M., 1950, Migration of esters of glycerophosphoric acid. II. The acid and alkaline hydrolysis of L-α-lecithins, *J. Biol. Chem.* **185**:615.

Basinger, S. F. and Jackson, R. W., 1968, Bacteriocin (hemolysin) of *Streptococcus zymogenes*, *J. Bacteriol.* **96**:1895.

Batley, M., Packer, N., and Redmond, J., 1981, Determination of molecular structure of amphiphiles using N.M.R. spectroscopy, in *Chemistry and Biological Activities of Bacterial Surface Amphiphiles* (G. D. Shockman, and A. J. Wicken, eds.) pp. 125–136, Academic, New York.

Batley, M., Redmond, J. W., and Wicken, A. J., 1987, Nuclear magnetic resonance spectra of lipoteichoic acids, *Biochim. Biophys. Acta* **901**:127.

Beachy, E. H., and Ofek, I., 1976, Epithelial cell binding of group A streptococci by lipoteichoic acid on fimbriae denuded of M protein, *J. Exp. Med.* **143**:759.

Beachy, E. H., Chiang, T. M., Ofek, I., and Kang, A. H., 1977, Interaction of lipoteichoic acid of group A streptococci with human platelets, *Infect. Immun.* **16**:649.

Beachy, E. H., Dale, J. B., Simpson, W. A., Evans, J. D., Knox, K. W., Ofek, I., and Wicken, A. J., 1979a, Erythrocyte binding properties of streptococcal lipoteichoic acids, *Infect. Immun.* **23**:618.

Beachy, E. H., Dale, J. B., Grebe, S., Ahmed, A., Simpson, W. A., and Ofek, I., 1979b, Lymphocyte binding and T cell mitogenic properties of group A streptococcal lipoteichoic acid, *J. Immunol.* **122**:189.

Beachy, E. H., Simpson, W. A., Ofek, I., Hasty, D. L., Dale, J. B., and Whitnak, E., 1983, Attachment of *Streptococcus pyogenes* to mammalian cells, *Rev. Infect. Dis.* **5**:S670.

Bell, R. M., Ballas, L. M., and Coleman, R. A., 1981, Lipid topogenesis, *J. Lipid Res.* **22**:391.

Bennet, L. G., and Glaudemans, C. P. J., 1979, Binding studies with antibodies having phosphorylcholine specificity and fragments derived from their homologous *Streptococcus pneumoniae* Type 27 capsular polysaccharide, *J. Immunol.* **122**:2356.

Bertram, K. C., Hancock, I. C., and Baddiley, J., 1981, Synthesis of teichoic acid by *Bacillus subtilis* protoplasts, *J. Bacteriol.* **148**:406.

Bertsch, L., Bonsen, P. M., and Kornberg, A., 1969, Biochemical studies on bacterial sporulation and germination, *J. Bacteriol.* **98**:75.

Bishop, D. G., Op den Kamp, J. A. F., and Van Deenen, L. L. M., 1977, The distribution of lipids in the protoplast membranes of *Bacillus subtilis*, *Eur. J. Biochem.* **80**:381.

Bligh, E. G., and Dyer, W. J., 1959, A rapid method of total lipid extraction and purification, *Can. J. Biochem. Physiol.* **37**:911.

Bock, K., and Pedersen, C., 1983, Carbon-13 nuclear magnetic resonance spectroscopy of monosaccharides, *Adv. Carbohyd. Res.* **41**:27.

Boylan, R. J., Mendelson, N. H., Brooks, D., and Young, F. E., 1972, Regulation of the bacterial cell wall: analysis of a mutant of *Bacillus subtilis* defective in biosynthesis of teichoic acid, *J. Bacteriol.* **110**:281.

Bracha, R., Chang, M., Fiedler, F., and Glaser, L., 1978, Biosynthesis of teichoic acids, *Methods Enzymol.* **10**:387.

Brade, L., and Brade, H., 1985, A 28,000-Dalton protein of normal mouse serum binds specifically to the inner core region of bacterial lipopolysaccharide, *Infect. Immun.* **50**:687.

Brade, H., and Galanos, C., 1983a, A new lipopolysaccharide antigen identified in *Acinetobacter calcoaceticus:* Occurrence of widespread natural antibody, *J. Med. Microbiol.* **16**:203.

Brade, H., and Galanos, C., 1983b, Common lipopolysaccharide specificity: New type of antigen, residing in the inner core region of S- and R-form lipopolysaccharides from different families of gramnegative bacteria, *Infect. Immun.* **42**:250.

Brautigan, V. M., Childs W. C. III, and Neuhaus, F. C., 1981, Biosynthesis of D-alanyl-lipoteichoic acid in *Lactobacillus casei:* D-alanyl-lipophilic compounds as intermediates, *J. Bacteriol.* **146**:239.

Brennan, P. J., 1988, Mycobacterium and other actinomycetes, in: *Microbial Lipids* (C. Ratledge and S. G. Wilkinson, eds.), Vol 1, pp. 203–296, Academic Press, London.

Briese, T., and Hakenbeck, R., 1985, Interaction of the pneumococcal amidase with lipoteichoic acid and choline, *Eur. J. Biochem.* **146**:417.

Briles, E. B., and Tomasz, A., 1973, Pneumococcal Forssman antigen. A choline-containing lipoteichoic acid, *J. Biol. Chem.* **248**:6394.

Briles, E. B., and Tomasz, A., 1975, Membrane lipoteichoic acid is not a precursor to wall teichoic acid in pneumococci, *J. Bacteriol.* **122**:335.

Brisette, J. L., and Pieringer, R. A., 1985, The effect of penicillin on fatty acid synthesis and excretion in *Streptococcus mutans* BHT, *Lipids* **20**:173.

Brisette, J. L., Shockman, G. D., and Pieringer, R. A., 1982, Effects of penicillin on synthesis and excretion of lipid and lipoteichoic acid from *Streptococcus mutans* BHT, *J. Bacteriol.* **151**:838.

Brisette, J. L., Cabacungan, E. A., and Pieringer, R. A., 1986, Studies on the antibacterial activity of dodecylglycerol (Its limited metabolism and inhibition of glycerolipid and lipoteichoic acid biosynthesis in *Streptococcus mutans* BHT, *J. Biol. Chem.* **261**:6338.

Brotherus, J., Renkonen, O., Herrmann, J., and Fischer, W., 1974, Novel stereoconfiguration in lyso-bisphosphatidic acid of cultured BHK-cells, *Chem. Phys. Lipids* **13**:178.

Brown, D. M., and Todd, A. R., 1952, Nucleotides. Part X. Some observations on the structure and chemical behaviour of the nucleic acids, *J. Chem. Soc.* **1952**:52.

Brundish, D. E., and Baddiley, J., 1968, Pneumococcal C-substance, a ribitol teichoic acid containing choline phosphate, *Biochem. J.* **110**:573.

Bundle, D. R., Jennings. H. J., and Smith, I. C. P., 1973, The carbon-13 nuclear magnetic resonance spectroscopy of 2-acetamido-2-deoxy-D-hexoses and some specifically deuteriated, O-acetylated, and phosphorylated derivatives, *Can. J. Chem.* **51**:3812.

Bundle, D. R., Smith, I. C. P., and Jennings, H. J., 1974, Determination of the structure and conformation of bacterial polysaccharides by carbon-13 nuclear magnetic resonance, *J. Biol. Chem.* **249**:2275.

Burger, M. M., and Glaser, L., 1964, The synthesis of teichoic acids. I. Polyglycerophosphate, *J. Biol. Chem.* **239**:3168.

Button, D., and Hemmings, N. L., 1976a, Lipoteichoic acid from *Bacillus licheniformis* 6346 MH-1. Comparative studies on the lipid portion of the lipoteichoic acid and the membrane glycolipid, *Biochemistry* **15**:989.

Button, D., and Hemmings, N. L., 1976b, Teichoic acids and lipids associated with the membrane of a *Bacillus licheniformis* mutant and the membrane lipids of the parenteral strain, *J. Bacteriol.* **128**:149.

Button, D., Choudhry, M. K., and Hemmings, N. L., 1975, Lipoteichoic acid from *Bacillus licheniformis* and one of its mutants, *Proc. Soc. Gen. Microbiol.* **2**:45.

Cabacungan, E., and Pieringer, R. A., 1980, Excretion of extracellular lipids from *Streptococcus mutans* BHT and FA-1, *Infect. Immun.* **27**:556.

Cabacungan, E., and Pieringer, R. A., 1981, Mode of elongation of the glycerophosphate polymer of membrane lipoteichoic acid of *Streptococcus faecium* (*Streptococcus faecalis* ATCC 9790), *J. Bacteriol.* **147**:75.

Cabacungan, E., and Pieringer, R. A., 1985, Evidence for a tetraglucoside substituent on the lipoteichoic acid of *Streptococcus faecium* ATCC 9790, *FEMS Microbiol. Lett.* **26**:49.

Card, G. L., and Finn, D. J., 1983, Products of phospholipid metabolism in *Bacillus stearothermophilus*, *J. Bacteriol.* **154**:294.

Carruthers, M. M., and Kabat, W. J., 1983, Mediation of staphylococcal adherence to mucosal cells by lipoteichoic acid, *Infect. Immun.* **40**:444.

Carson, D. D., and Daneo-Moore, L., 1981, Cellular LTA content during growth and division in

Streptococcus faecium (ATCC 9790), in *Chemistry and Biological Activities of Bacterial Surface Amphiphiles* (G. D. Shockman, A. J. Wicken, eds.), pp. 259–262, Academic Press, New York.

Carson, D., Pieringer, R. A., and Daneo-Moore, L., 1979, Effect of growth rate on lipid and lipoteichoic acid composition in *Streptococcus faecium, Biochim. Biophys. Acta* **575**:225.

Carson, D. D., Pieringer, R. A., and Daneo-Moore, L., 1981, Effect of cerulenin on cellular autolytic activity and lipid metabolism during inhibition of protein synthesis in *Streptococcus faecalis, J. Bacteriol.* **146**:590.

Chatterjee, A. N., Mirelman, D., Singer, H. J., and Park, J. T., 1969, Properties of a novel pleiotropic bacteriophage-resistant mutant of *Staphylococcus aureus* H, *J. Bacteriol.* **100**:846.

Cheng, K.-J., and Costeron, J. W., 1977, Ultrastructure of *Butyrivibrio fibrisolvens* a gram-positive bacterium?, *J. Bacteriol.* **129**:1506.

Chevion, M., Panos, Ch., Linzer, R., and Neuhaus, F. C., 1974, Incorporation of D-alanine into the membrane of *Streptococcus pyogenes* and its stabilized L-form, *J. Bacteriol.* **120**:1026.

Chiang, T. M., Alkan, M. L., and Beachy, E. H., 1979, Binding of lipoteichoic acid of group A streptococci to isolated human erythrocyte membranes, *Infect. Immun.* **26**:316.

Childs, W. C. III, and Neuhaus, F. C., 1980, Biosynthesis of D-alanyl-lipoteichoic acid: Characterization of ester-linked D-alanine in the in vitro-synthesized product, *J. Bacteriol.* **143**:293.

Chiu, T., Arnold, B., Kim, S.-R., and Yeh, L. L., 1985, Phosphatidyl glycerolphosphate serves as glycerolphosphate donor in polymer synthesis, *Biochem. Biophys. Res. Commun.* **128**:906.

Clarke, N. G., Hazlewood, G. P., and Dawson, R. M. C., 1976, Novel lipids of *Butyrivibrio* spp., *Chem. Phys. Lipids* **17**:222.

Cleveland, R. F., Höltje, J. V., Wicken, A. J., Tomasz, A., Daneo-Moore, L., and Shockman, G. D., 1975, Inhibition of bacterial wall lysins by lipoteichoic acids and related compounds, *Biochim. Biophys. Res. Commun.* **67**:1128.

Cleveland, R. F., Wicken, A. J., Daneo-Moore, L., and Shockman, G. D., 1976a, Inhibition of wall autolysis in *Streptococcus faecalis* by lipoteichoic acid and lipids, *J. Bacteriol.* **126**:192.

Cleveland, R. F., Daneo-Moore, L., Wicken, A. J., and Shockman, G. D., 1976b, Effect of lipoteichoic acid and lipids on lysis of intact cells of *Streptococcus faecalis, J. Bacteriol.* **127**:1582.

Coley, J., Duckworth, M., and Baddiley, J., 1975, Extraction and purification of the lipoteichoic acids from gram-positive bacteria, *Carbohydr. Res.* **40**:41.

Coley, J., Tarelli, E., Archibald, A. R., and Baddiley, J., 1978, The linkage between teichoic acid and peptidoglycan in bacterial cell walls, *FEBS Lett.* **88**:1.

Collins, M. D., Farrow, J. A. E., Phillips, B. A., and Kandler, O., 1983, *Streptococcus garvieae* sp. nov, and *Streptococcus plantarum* sp. nov., *J. Gen. Microbiol.* **129**:3427.

Coulon-Morelec, M. J., and Douce, R., 1968, Le diphosphatidylglycerol des végétaux supérieurs. I. Structure du diphosphatidylglycérol végétal, *Bull. Soc. Chim. Biol.* **50**:1547.

Coulon-Morelec, M. J., Faure, M., and Maréchal, J., 1960, Etude du mécanisme de la libération des diglycérides des phosphatides sous l'action de l'acide acétique chaud, *Bull. Soc. Chim. Biol.* **42**:867.

Courtney, H., Ofek, I., Simpson, W. A., and Beachy, E. H., 1981, Characterization of lipoteichoic acid binding to polymorphonuclear leucocytes of human blood, *Infect. Immun.* **32**:625.

Courtney, H. S., Simpson, W. A., and Beachy, E. H., 1986, Relationship of critical micelle concentrations of bacterial lipoteichoic acids to biological activities, *Infect. Immun.* **51**:414.

Courtney, H., Stanislawski, L., Ofek, I., Simpson, W. A., Hasty, D. L., and Beachy, E. H., 1988, Localization of a lipoteichoic acid binding site to a 24-Kilodalton NH_2-terminal fragment of fibronectin, *Rev. Infect. Dis.* **10**:S360.

Crichtley, P., Archibald, A. R., and Baddiley, J., 1962, The intracellular teichoic acid from *Lactobacillus arabinosus* 17-5, *Biochem. J.* **85**:420.

Davie, J. M., and Brock, T. D., 1966, Effect of teichoic acid on resistance to the membrane-lytic agent of *Streptococcus zymogenes, J. Bacteriol.* **92**:1623.

Dawson, R. M. C., 1967, Analysis of phosphatides and glycolipids by chromatography of their partial hydrolysis products, in *Lipid Chromatographic Analysis* (G. V. Marinetti, ed.), Vol. 1, pp. 163–189, Marcel Dekker, New York.

De Boer, W. R., Kruyssen, F. J., Wouters, J. T. M., and Kruk, C., 1976, The structure of teichoic acid from *Bacillus subtilis* var. *niger* WM as determined by ^{13}C nuclear-magnetic-resonance spectroscopy, *Eur. J. Biochem.* **62**:1.

De Boer, W. R., Wouters, J. T. M., Anderson, A. J., and Archibald, A. R., 1978, Further evidence for the structure of the teichoic acids from *Bacillus stearothermophilus* B65 and *Bacillus subtilis* var. *niger* WM, *Eur. J. Biochem.* **85:**433.

Decker, G. P., Chorpenning, F. W., and Frederick, G. T., 1972, Naturally-occuring antibodies to bacillary teichoic acids, *J. Immunol.* **108:**214.

De Haas, G. H., Postema, N. M., Nieuwenhuizen, W., and Van Deenen, L. L. M., 1968, Purification and properties of phospholipase A from porcine pancreas, *Biochim. Biophys. Acta* **159:**103.

Demant, E. J. F., Op den Kamp, J. A. F., and Van Deenen, L. L. M., 1979, Localization of the phospholipids in the membrane of *Bacillus megaterium*, *Eur. J. Biochem.* **95:**613.

Deutsch, R., Engel, R., and Tropp, B. E., 1980, Effect of 3,4-dihydroxybutyl-1-phosphonate on phosphoglyceride and lipoteichoic acid synthesis in *Bacillus subtilis*, *J. Biol. Chem.* **255:**1521.

De Vuono, J., and Panos, C., 1978, Effect of L-form *Streptococcus pyogenes* and of lipoteichoic acid on human cells in tissue culture, *Infect. Immun.* **22:**255.

Dickson, M. R., and Wicken, A. J., 1974, Ferritin labelling of thin sections by a "sandwich" technique, in *Eighth International Congress Electron Microscopy*, Vol II, Canberra, pp. 114–115.

Dittmer, J. C., and Lester, R. L., 1964, A simple specific spray for the detection of phospholipids on thin-layer chromatograms, *J. Lipid Res.* **5:**126.

Duckworth, M., Archibald, A. R., and Baddiley, J., 1975, Lipoteichoic acid and lipoteichoic acid carrier in *Staphylococcus aureus* H, *FEBS Lett.* **53:**176.

Ellwood, D. C., and Tempest, D. W., 1968, The teichoic acids of *Bacillus subtilis* var. *niger* and *Bacillus subtilis* W23 grown in a chemostat, *Biochem. J.* **108:**40p.

Ellwood, D. C., and Tempest, D. W., 1972, Effects of environment on bacterial wall content and composition, *Adv. Microbiol. Physiol.* **7:**83.

Emdur, L. I., and Chiu, T. H., 1974, Turnover of phosphatidylglycerol in *Streptococcus sanguis*, *Biochem. Biophys. Res. Commun.* **59:**1137.

Emdur, L. I., and Chiu, T. H., 1975, The role of phosphatidylglycerol in the in vitro biosynthesis of teichoic and lipoteichoic acid, *FEBS Lett.* **55:**216.

Erbing, B., Kenne, L., and Lindberg, B., 1986, Structural studies of a teichoic acid from *Streptococcus agalactiae* Type III, *Carbohydr. Res.* **156:**147.

Estrada-Parra, S., Rebers, P. A., and Heidelberger, M., 1962, The specific polysaccharide of type XVIII *Pneumococcus* II, *Biochemistry* **1:**1175.

Fiedel, B. A., and Jackson, R. W., 1976, Immunogenicity of a purified and carrier-complexed streptococcal lipoteichoic acid, *Infect. Immun.* **13:**1585.

Fiedel, B. A., and Jackson, R. W., 1978, Activation of the alternative complement pathway by a streptococcal lipoteichoic acid, *Infect. Immun.* **22:**286.

Fiedel, B. A., and Jackson, R. W., 1979, Nephropathy in the rabbit associated with immunization to a group A streptococcal lipoteichoic acid, *Med. Microbiol. Immunol.* **167:**251.

Fiedler, F., 1981, On the participation of lipoteichoic acid in the biosynthesis of wall teichoic acids. in *Chemistry and Biological Activities of Bacterial Surface Amphiphiles* (G. D. Shockman and A. J. Wicken, eds.), pp. 195–208, Academic Press, New York.

Fiedler, F., and Glaser, L., 1974a, The synthesis of polyribitolphosphate. I. Purification of polyribitolphosphate polymerase and lipoteichoic acid carrier, *J. Biol. Chem.* **249:**2684.

Fiedler, F., and Glaser, L., 1974b, The synthesis of polyribitolphosphate. II. On the mechanism of polyribitol phosphate polymerase, *J. Biol. Chem.* **249:**2690.

Fiedler, F., and Glaser, L., 1974c, The synthesis of polyribitolphosphate. III. The attachment of polyribitolphosphate to lipoteichoic acid carrier, *Carbohydr. Res.* **57:**37.

Filgueiras, M. H., and Op den Kamp, J. A. F., 1980, Cardiolipin, a major phospholipid of grampositive bacteria that is not readily extractable, *Biochim. Biophys. Acta* **620:**332.

Fischer, W., 1970, A new phosphoglycolipid from *Streptococcus lactis*, *Biochem. Biophys. Res. Commun.* **41:**731.

Fischer, W., 1976, Newly discovered lipids from streptococci, in *Lipids*, Vol. 1: *Biochemistry* (R. Paoletti, G. Porcellati, and G. Jacini, eds.), pp. 255–266, Raven, New York.

Fischer, W., 1981, Glycerophosphoglycolipids presumptive biosynthetic precursors of lipoteichoic acids, in *Chemistry and Biological Activities of Bacterial Surface Amphiphiles* (G. D. Shockman, and A. J. Wicken, eds.), pp. 209–228, Academic, New York.

Fischer, W., 1982, D-Alanine ester-containing glycerophosphoglycolipids in the membrane of gram-positive bacteria, *Biochim. Biophys. Acta* **711**:372.

Fischer, W., 1984, Glycoglycerolipids, in *Handbook of Chromatography, Lipids* (H. K. Mangold, G. Zweig, and J. Sherma, eds.), Vol. I, pp. 555–587, CRC Press, Boca Raton, Florida.

Fischer, W., 1987, "Lipoteichoic acid" of *Bifidobacterium bifidum* subsp. *pennsylvanicum* DSM 20239: A lipoglycan with monoglycerophosphate side chains, *Eur. J. Biochem.* **165**:647.

Fischer, W., 1988, Physiology of lipoteichoic acids in bacteria, in *Advances in Microbial Physiology* (A. H. Rose and D. W. Tempest, eds.), Vol. 29, pp. 233–302, Academic, London.

Fischer, W., 1989, Structural analysis of lipoteichoic acids, in: *Proceedings of Session Lectures and Scientific Presentations on ISF-JOCS World Congress 1988*, Vol. II: *The 19th World Congress of the International Society for Fat Research (ISF) and the 27th Annual Meeting of the Japanese Oil Chemists' Society (JOCS)*, 7S01, pp. 851–857.

Fischer, W., and Koch, H. U., 1981, Alanine ester substitution and its effect on the biological properties of lipoteichoic acid, in *Chemistry and Biological Activities of Bacterial Surface Amphiphiles* (G. D. Shockman and A. J. Wicken, eds.), pp. 181–194, Academic Press, New York.

Fischer, W., and Koch, H. U., 1983, The influence of lipoteichoic acids on the autolytic activity of *Staphylococcus aureus*, in *The Target of Penicillin* (R. Hakenbeck, J.-V. Höltje, and H. Labischinski, eds.), pp. 179–184, Walter de Gruyter, Berlin.

Fischer, W., and Landgraf, H. R., 1975, Glycerophosphoryl phosphatidyl kojibiosyl diacylglycerol, a novel phosphoglucolipid from *Streptococcus faecalis*, *Biochim. Biophys. Acta* **380**:227.

Fischer, W., and Rösel, P., 1980, The alanine ester substitution of lipoteichoic acid (LTA) in *Staphylococcus aureus*, *FEBS Lett.* **119**:224.

Fischer, W., and Seyferth, W., 1968, 1-(0-α-D-glucopyranosyl-(1–2)-O-α-D-glucopyranosyl)-glycerin aus den Glykolipiden von *Streptococcus faecalis* und *Streptococcus lactis*, *Hoppe Seylers Z. Physiol. Chem.* **349**:1662.

Fischer, W., Ishizuka, I., Landgraf, H. R., and Herrmann, J., 1973*a*, Glycerophosphoryl diglucosyl diglyceride, a new phosphoglycolipid from streptococci, *Biochim. Biophys. Acta* **296**:527.

Fischer, W., Landgraf, H. R., and Herrmann, J., 1973*b*, Phosphatidyldiglucosyl diglyceride from streptococci and its relationship to other polar lipids, *Biochim. Biophys. Acta* **306**:353.

Fischer, W., Laine, R. A., Nakano, M., Schuster, D., and Egge, H., 1978*a*, The structure of acyl-α-kojibiosyldiacylglycerol from *Streptococcus lactis*, *Chem. Phys. Lipids* **21**:103.

Fischer, W., Nakano, M., Laine, A. R., and Bohrer, W., 1978*b*, On the relationship between glycerophosphoglycolipids and lipoteichoic acids in gram-positive bacteria. I. The occurrence of phosphoglycolipids, *Biochim. Biophys. Acta* **528**:288.

Fischer, W., Laine, A. R., and Nakano, M., 1978*c*, On the relationship between glycerophosphoglycolipids and lipoteichoic acids in gram-positive bacteria. II. Structures of glycerophosphoglycolipids, *Biochim. Biophys. Acta* **528**:298.

Fischer, W., Schuster, D., and Laine, A. R., 1979, Studies on the relationship between glycerophosphoglycolipids and lipoteichoic acids. IV. Trigalactosylglycerophospho-acylkojibiosyldiacylglycerol and related compounds from *Streptococcus lactis* Kiel 42172, *Biochim. Biophys. Acta* **575**:389.

Fischer, W., Koch, H. U., Rösel, P., Fiedler, F., and Schmuck, L., 1980*a*, Structural requirements of lipoteichoic acid carrier for recognition by the poly(ribitol phosphate) polymerase from *Staphylococcus aureus* H, *J. Biol. Chem.* **255**:4550.

Fischer, W., Koch, H. U., Rösel, P., and Fiedler, F., 1980*b*, Alanine ester-containing native lipoteichoic acids do not act as lipoteichoic acid carrier, *J. Biol. Chem* **255**:4557.

Fischer, W., Rösel, P., and Koch, H. U., 1981, Effect of alanine ester substitution and other structural features of lipoteichoic acids on their inhibitory activity against autolysins of *Staphylococcus aureus*, *J. Bacteriol.* **146**:467.

Fischer, W., Schmidt, M. A., Jann, B., and Jann, K., 1982, Structure of the *Escherichia coli* K2 capsular antigen. Stereochemical configuration of the glycerophosphate and distribution of galactopyranosyl and galactofuranosyl residues, *Biochemistry* **21**:1279.

Fischer, W., Koch, H. U., and Haas, R., 1983, Improved preparation of lipoteichoic acids, *Eur. J. Biochem.* **133**:523.

Fischer, W., Bauer, W., and Feigel, M., 1987, Analysis of the lipoteichoic acid-like macroamphiphile from *Bifidobacterium bifidum* subsp. *pennsylvanicum* by one- and two-dimensional

^{1}H- and ^{13}C-NMR spectroscopy, *Eur. J. Biochem.* **165**:639.

Fischer, W., Mannsfeld, T., and Hagen, G., 1990, On the basic structure of poly(glycerophosphate) lipoteichoic acids, *Biochem. Cell Biol.* **68** (in press).

Fleischmann-Sperber, T., 1982, Untersuchungen zur Struktur der Lipoteichonsäure aus *Lactobacillus plantarum* DSM 20314, Doctoral thesis, University of Erlangen-Nürnberg.

Folch, J., Lees, M., and Stanley, G. H. S., 1957, A simple method for the isolation and purification of total lipids from animal tissues, *J. Biol. Chem.* **226**:497.

Fox, G. E., Stackebrand, E., Hespell, R. B., Gibson, J., Maniloff, J., Dyer, T. A., Wolfe, R. S., Balch, W. E., Tanner, R. S., Magrum, L. J., Zablen, L. B., Blakemore, R., Gupta, R., Bonen, L., Lewis, B. J., Stahl, D. A., Luehrsen, K. R., Chen, K. N., and Woese, C. R., 1980, The phylogeny of prokaryotes, *Science* **209**:457.

Frederick, G. T., and Chorpenning, F. W., 1974, Characterisation of antibodies specific for polyglycerophosphate, *J. Immunol.* **113**:489.

Fujiwara, M., 1967, The Forssman antigen of *Pneumoccus, Jpn. J. Exp. Med.* **37**:581.

Fulco, A. J., 1983, Fatty acid metabolism in bacteria, *Prog. Lipid Res.* **22**:133.

Galanos, C., Lehmann, V., Lüderitz, O., Rietschel, E. Th., Westphal, O., Brade, H., Brade, L., Freudenberg, M. A., Hansen-Hagge, T., Lüderitz, Th., McKenzie, G., Schade, U., Strittmatter, W., Tanamoto, K., Zähringer, U., Imoto, M., Yoshimura, H., Yamamoto, M., Shimamoto, T., Kusomoto, S., and Shiba, T., 1984, Endotoxic properties of synthetic lipid A part structures. Comparison of bacterial free lipid A and the lipid A disaccharide precursor with chemically synthesized Lipid A precursor and analogues, *Eur. J. Biochem.* **140**:221.

Galanos, C., Lüderitz, O., Rietschel, E. Th., Westphal, O., Brade, H., Brade, L., Freudenberg, M., Schade, U., Imoto, M., Yoshimura, H., Kusomoto, S., and Shiba, T., 1985a, Synthetic and natural *Escherichia coli* free lipid A express identical endotoxic activities, *Eur. J. Biochem.* **148**:1.

Galanos, C., Hansen-Hagge, T., Lehmann, V., and Lüderitz, O., 1985b, Comparison of the capacity of two lipid A precursor molecules to express the local Shwartzman phenomenon, *Infect. Immun.* **48**:355.

Ganfield, M.-C. W., and Pieringer, R. A., 1975, Phosphatidylkojibiosyl diglyceride. The covalently linked lipid constituent of the membrane lipoteichoic acid from *Streptococcus faecalis (faecium)* ATCC 9790, *J. Biol. Chem.* **250**:702.

Ganfield, M.-C. W., and Pieringer, R. A., 1980, The biosynthesis of nascent membrane lipoteichoic acid of *Streptococcus faecium* (*S. faecalis* ATCC 9790) from phosphatidylkojibiosyl diacylglycerol and phosphatidyl glycerol, *J. Biol. Chem.* **255**:5164.

Ghuysen, J. M., Tipper, D. J., and Strominger, J. L., 1965, Structure of cell wall of *Staphylococcus aureus* strain Copenhagen. IV. The teichoic acid-glycopeptide complex, *Biochemistry* **4**:474.

Gilpin, R. W., Chatterjee, A. N., and Young, F. E., 1972, Autolysis of microbial cells: Salt activation of autolytic enzymes in a mutant of *Staphylococcus aureus, J. Bacteriol.* **111**:272.

Ginsburg, I., Rozengarten, O., Dishon, T., Szabo, M., Bergner-Rabinowitz, S., and Ofek, I., 1972, Antibodies to the cell-sensitizing antigen of group A streptococci in the sera of patients, in *Streptococcal Disease and the Community,* Proceedings of the Fifth International Symposium on *Streptococcus pyogenes* (International Congress Series 317) (M. J. Haverkorn, ed.), pp. 97–100, Excerpta Medica, Amsterdam.

Glaser, L., and Burger, M. M., 1964, The synthesis of teichoic acids. III. Glucosylation of polyglycerophosphate, *J. Biol. Chem.* **239**:3187.

Glaser, L., and Lindsay, B., 1974, The synthesis of lipoteichoic acid carrier, *Biochem. Biophys. Res. Commun.* **59**:1131.

Goebel, W. F., Shedlovsky, T., Lavin, G. I., and Adams, M. H., 1943, The heterophile antigen of *Pneumococcus, J. Biol. Chem.* **148**:1.

Goldschmidt, J. C. Jr., and Panos, C., 1984, Teichoic acids of *Streptococcus agalactiae:* Chemistry, cytotoxicity, and effect on bacterial adherence to human cells in tissue culture, *Infect. Immun.* **43**:670.

Gorin, P. A. J., 1981, Carbon-13 nuclear magnetic resonance spectroscopy of polysaccharides, *Adv. Carbohydr. Chem. Biochem.* **38**:13.

Gotschlich, E. C., Fraser, B. A., Nishimura, O., Robbins, J. B., and Liu, T.-Y., 1981, Lipid on capsular polysaccharides of gram-negative bacteria, *J. Biol. Chem.* **256**:8915.

Haas, R., Koch, H. U., and Fischer, W., 1984, Alanyl turnover from lipoteichoic acid to teichoic acid in *Staphylococcus aureus, FEMS Microbiol. Lett.* **21**:27.

Haberland, M. E., and Reynolds, J. A., 1975, Interaction of L-α-palmitoyl lysophosphatidylcholine with the AI polypeptide of high-density lipoprotein, *J. Biol. Chem.* **250**:6636.

Hakenbeck, R., Martin, C., and Morelli, G., 1983, *Streptococcus pneumoniae* proteins released into medium upon inhibition of cell wall biosynthesis, *J. Bacteriol.* **155**:1372.

Hamada, S., Tai, S., and Slade, H. D., 1976, Selective adsorption of heterophile polyglycerophosphate antigen from antigen extracts of *Streptococcus mutans* and other grampositive bacteria, *Infect. Immun.* **14**:903.

Hamada, S., Mizuno, J., Kotani, S., and Torii, M., 1980, Distribution of lipoteichoic acid and other amphipathic antigens in oral streptococci, *FEMS Microbiol. Lett.* **8**:93.

Hancock, I. C., and Baddiley, J., 1985, Biosynthesis of the bacterial envelope polymers teichoic acid and teichuronic acid, in *The Enzymes of Biological Membranes*, Vol. 2: *Biosynthesis and Metabolism* (A. N. Martonosi, ed.), pp. 279–307, Plenum, New York.

Hardy, L., Jacques, N. A., Forester, H., Campbell, L. K., Knox, K. W., and Wicken, A. J., 1981, Effect of fructose and other carbohydrates on the surface properties, lipoteichoic acid production, and extracellular proteins of *Streptococcus mutans* Ingbritt, grown in continuous culture, *Infect. Immun.* **31**:78.

Harrington, C. R., and Baddiley, J., 1983, Peptidoglycan synthesis by partly autolyzed cells of *Bacillus subtilis* W23, *J. Bacteriol.* **155**:776.

Harrington, C. R., and Baddiley, J., 1985, Biosynthesis of wall teichoic acid in *Staphylococcus aureus* H, *Micrococcus varians*, and *Bacillus subtilis* W23. Involvement of lipid intermediates containing the disaccharide *N*-acetylmannosaminyl *N*-acetylglucosamine, *Eur. J. Biochem.* **153**:639.

Harris, T. N., and Harris, S., 1953, Agglutination by human sera of erythrocytes incubated with streptococcal culture concentrates, *J. Bacteriol.* **66**:159.

Harrop, P. J., O'Grady, R. L., Knox, K. W., and Wicken, A. J., 1980, Stimulation of lysosomal enzyme release from macrophages by lipoteichoic acid, *J. Periodont. Res.* **15**:492.

Hausmann, E., Lüderitz, O., Knox, K. W., and Weinfeld, N., 1975, Structural requirements for bone resorption by endotoxin and lipoteichoic acid, *J. Dent. Res.* **54**:B94.

Heptinstall, S., Archibald, A. R., and Baddiley, J., 1970, Teichoic acids and membrane function in bacteria, *Nature (Lond.)* **225**:519.

Hether, N. W., and Jackson, L. L., 1983, Lipoteichoic acid from *Listeria monocytogenes, J. Bacteriol.* **156**:809.

Hewett, M. J., Knox, K. W., and Wicken, A. J., 1970, Studies on the group F antigen of lactobacilli: Detection of antibodies by haemagglutination, *J. Gen. Microbiol.* **60**:315.

Hewett, M. J., Wicken, A. J., Knox, K. W., and Sharpe, M. E., 1976, Isolation of lipoteichoic acids from *Butyrivibrio fibrisolvens, J. Gen Microbiol.* **94**:126.

Höltje, J. V., and Tomasz, A., 1975a, Lipoteichoic acid: A specific inhibitor of autolysin activity in *Pneumococcus, Proc. Nat. Acad. Sci. USA* **72**:1690.

Höltje, J. V., and Tomasz, A., 1975b, Biological effects of lipoteichoic acids, *J. Bacteriol.* **124**:1023.

Höltje, J. V., and Tomasz, A., 1975c, Specific recognition of choline residues in the cell wall teichoic acid by *N*-acetyl-muramyl-L-alanine amidase of *Pneumococcus, J. Biol. Chem.* **250**:6072.

Hohorst, H. J., 1970, L-(−)-Glycerin-3-Phosphat, in *Methoden der enzymatischen Analyse*, 2nd ed. (Bergmeyer, H. U., ed.), Vol. II, pp. 1379–1383, Verlag Chemie, Weinheim/Bergstrasse.

Horne, D., and Tomasz, A., 1977, Tolerant response of *Streptococcus sanguis* to beta-lactams and other cell wall inhibitors, *Antimicrob. Agents Chemother.* **11**:888.

Horne, D., and Tomasz, A., 1979, Release of lipoteichoic acid from *Streptococcus sanguis:* Stimulation of release during penicillin treatment, *J. Bacteriol.* **137**:1180.

Horne, D., and Tomasz, A., 1985, Pneumococcal Forssman antigen: Enrichment in mesosomal membranes and specific binding to the autolytic enzyme of *Streptococcus pneumoniae, J. Bacteriol.* **161**:18.

Horne, D., Hakenbeck, R., and Tomasz, A., 1977, Secretion of lipids induced by inhibition of peptidoglycan synthesis in streptococci, *J. Bacteriol.* **132**:704.

Houtsmuller, U. M. T., and Van Deenen, L. L. M., 1965, On the amino acid esters of phosphatidyl glycerol from bacteria, *Biochim. Biophys. Acta* **106**:564.

Huff, E., 1982, Lipoteichoic acid, a major amphiphile of gram-positive bacteria that is not readily extractable, *J. Bacteriol.* **149**:399.

Huff, E., Cole, R. M., and Theodore, T. S., 1974, Lipoteichoic acid localization in mesosomal vesicles of *Staphylococcus aureus, J. Bacteriol.* **120:**273.

Hughes, A. H., Hancock, I. C., and Baddiley, J., 1973, The function of teichoic acids in cation control in bacterial membranes, *Biochem. J.* **132:**83.

Hummell, D. S., and Winkelstein, J. A., 1986, Bacterial lipoteichoic acid sensitizes host cells for destruction by autologous complement, *J. Clin. Invest.* **77:**1533.

Hummell, D. S., Swift, A. J., Tomasz, A., and Winkelstein, J. A., 1985, Activation of the alternative complement pathway by pneumococcal lipoteichoic acid, *Infect. Immun.* **47:**384.

Hunter, S. W., Gaylord, H., and Brennan, P. J., 1986, Structure and antigenicity of the phosphorylated lipopolysaccharide antigens from leprosy and tubercle bacilli, *J. Biol. Chem.* **261:**12345.

Hurst, A., Hughes, A., Duckworth, M., and Baddiley, J., 1975, Loss of D-alanine during sublethal heating of *Staphylococcus aureus* s6 and magnesium binding during repair, *J. Gen. Microbiol.* **89:**277.

Ishizuka, I., and Yamakawa, T., 1968, Glycosyl glycerides from *Streptococcus hemolyticus* strain D-58, *J. Biochem.* **64:**13.

Ishizuka, I., and Yamakawa, T., 1985, Glycoglycerolipids, in *Glycolipids* (H. Wiegandt, ed.), pp. 101–196, Elsevier, New York.

Iwasaki, H., Shimada, A., and Ito, E., 1986, Comparative studies of lipoteichoic acids from several *Bacillus* strains, *J. Bacteriol.* **167:**508.

Jacin, H., and Mishkin, A. R., 1965, Separation of carbohydrates on borate-impregnated silica gel G plates, *J. Chromatogr.* **18:**170.

Jackson, D. E., Jackson, G. D. F., and Wicken, A. J., 1981a, Induction of IgM immunological memory to lipoteichoic acid in rabbits. Part I, *Int. Arch. Allergy Appl. Immunol.* **65:**198.

Jackson, D. E., Wicken, A. J., and Jackson, G. D. F., 1981b, Immune response to lipoteichoic acid: Comparison of antibody responses in rabbit and mice. Part II, *Int. Arch. Allergy Appl. Immunol.* **65:**203.

Jackson, D. E., Howlett, C. R., Wicken, A. J., and Jackson, G. D. F., 1981c, Induction of hypersensitivity reactions to *Lactobacillus fermentum* and lipoteichoic acid in rabbits (Part III), *Int. Arch. Allergy Appl. Immunol.* **65:**304.

Jackson, D. E., Wong, W., Largen, M. T., and Shockman, G. D., 1984, Monoclonal antibodies to immunodeterminants of lipoteichoic acids, *Infect. Immun.* **43:**800.

Jacques, N. A., Hardy, L., Campbell, L. K., Knox, K. W., Evans, J. D., and Wicken, A. J., 1979, Effect of carbohydrate source and growth condition on the production of lipoteichoic acid by *Streptococcus mutans* Ingbritt, *Infect. Immun.* **26:**1079.

Jann, K., and Schmidt, M. A., 1980, Comparative chemical analysis of two variants of the *Escherichia coli* K2 antigen, *FEMS Microbiol. Lett.* **7:**79.

Jann, B., Jann, K., Schmidt, G., Ørskov, I., and Ørskov, F., 1970, Immunochemical studies of polysaccharide surface antigens of *Escherichia coli* 0100:K?(B):H2, *Eur. J. Biochem.* **15:**29.

Jann, K., Jann, B., Schmidt, M. A., and Vann, W. F., 1980, Structure of *Escherichia coli* K2 capsular antigen, a teichoic acid-like polymer, *J. Bacteriol.* **143:**1108.

Jennings, H. J., 1983, Capsular polysaccharides as human vaccines, *Adv. Carbohydr. Res.* **41:**155.

Jennings, H. J., Rosell, K.-G., and Kenny, C. P., 1979, Structural elucidation of the capsular polysaccharide antigen of *Neisseria meningitidis* serogroup Z using ^{13}C nuclear magnetic resonance, *Can. J. Chem.* **57:**2902.

Jennings, H. J., Lugowski, C., and Young, N. M., 1980, Structure of the complex polysaccharide C-substance from *Streptococcus pneumoniae* type 1, *Biochemistry* **19:**4712.

Johnstone, K., Simion, F. A., and Ellar, D. J., 1982, Teichoic acid and lipid metabolism during sporulation of *Bacillus megaterium* KM, *Biochem. J.* **202:**459.

Joseph, R., and Shockman, G. D., 1975, Synthesis and excretion of glycerol teichoic acid during growth of two streptococcal species, *Infect. Immun.* **12:**333.

Kates, M., 1986, Techniques of lipidology, in *Laboratory Techniques in Biochemistry and Molecular Biology,* 2nd rev. ed. (R. H. Burdon and P. H. van Knippenberg, eds.), American Elsevier, New York.

Kelemen, M., and Baddiley, J., 1961, Structure of the intracellular glycerol teichoic acid from *Lactobacillus casei* ATCC 7469, *Biochem. J.* **80:**246.

Kennedy, D. A., 1964, Doctoral Thesis, University of Newcastle upon Tyne. Cited in Archibald, A. R., Baddiley, J., and Blumson, N. L., 1968, The teichoic acids, *Adv. Enzymol.* **30:**223.

Kennedy, L. T., and Shaw, D. R. D., 1968, Direction of poly(glycerol phosphate) chain growth in *Bacillus subtilis, Biochem. Biophys. Res. Commun.* **32:**861.

Kessler, R. E., and Shockman, G. D., 1979*a*, Precursor-product relationship of intracellular and extracellular lipoteichoic acid of *Streptococcus faecium, J. Bacteriol.* **137:**869.

Kessler, R. E., and Shockman, G. D., 1979*b*, Enzymatic deacylation of lipoteichoic acid by protoplasts of *Streptococcus faecium* (*Streptococcus faecalis* ATCC 9790), *J. Bacteriol.* **137:**1176.

Kessler, R. E., and Thivierge, B. H., 1983, Effects of substitution on polyglycerol phosphate-specific antibody binding to lipoteichoic acids, *Infect. Immun.* **41:**549.

Kessler, R. E., and Van de Rijn, I., 1979, Quantitative immunoelectrophoretic analysis of *Streptococcus pyogenes* membrane, *Infect. Immun.* **26:**892.

Kessler, R. E., and Van de Rijn, I., 1981, Effects of penicillin on group A streptococci: Loss of viability appears to precede stimulation of release of lipoteichoic acid, *Antimicrob. Agents Chemother.* **19:**39.

Kessler, R. E., Van de Rijn, I., and McCarty, M., 1979, Characterization and localization of the enzymatic deacylation of lipoteichoic acid in group A streptococci, *J. Exp. Med.* **150:**1498.

Kessler, R. E., Van de Rijn, I., and McCarty, M., 1981, Release of lipoteichoic acid by group A streptococci, in *Chemistry and Biological Activities of Bacterial Surface Amphiphiles,* (G. D. Shockman and A. J. Wicken, eds.), pp. 239–246, Academic Press, New York.

Kessler, R. E., Wicken, A. J., and Shockman, G. D., 1983, Increased carbohydrate substitution of lipoteichoic acid during inhibition of protein synthesis, *J. Bacteriol.* **155:**138.

Kilpper-Bälz, R., and Schleifer, K. H., 1984, Nucleic acid hybridization and cell wall composition studies of pyogenic streptococci, *FEMS Microbiol. Lett.* **24:**355.

Knox, K. W., and Wicken, A. J., 1971, Serological properties of the wall and membrane teichoic acids from *Lactobacillus helveticus* NCIB 8025, *J. Gen. Microbiol.* **63:**237.

Knox, K. W., and Wicken, A. J., 1972, Serological studies on the teichoic acids of *Lactobacillus plantarum, Infect. Immun.* **6:**43.

Knox, K. W., and Wicken, A. J., 1973, Immunological properties of teichoic acids, *Bacteriol. Rev.* **37:**215.

Knox, K. W., and Wicken, A. J., 1981*a*, Serological methods, in *Chemistry and Biological Activities of Bacterial Surface Amphiphiles* (G. D. Shockman and A. J. Wicken, eds.), pp. 95–100, Academic Press, New York.

Knox, K. W., and Wicken, A. J., 1981*b*, Effect of growth conditions on lipoteichoic acid production, in *Chemistry and Biological Activities of Bacterial Surface Amphiphiles* (G. D. Shockman and A. J. Wicken, eds.), pp. 229–237, Academic Press, New York.

Knox, K. W., Hewett, M. J., and Wicken, A. J., 1970, Studies on group F antigen of lactobacilli: antigenicity and serological specificity of teichoic acid preparations, *J. Gen. Microbiol.* **60:**303.

Koch, H. U., and Fischer, W., 1978, Acyldiglucosyldiacylglycerol-containing lipoteichoic acid with a poly(3-*O*-galabiosyl-2-*O*-galactosyl-*sn*-glycero-1-phosphate) chain from *Streptococcus lactis* Kiel 42172, *Biochemistry* **17:**5275.

Koch, H. U., Fischer, W., and Fiedler, F., 1982, Influence of alanine ester and glycosyl substitution on the lipoteichoic acid carrier activity of lipoteichoic acids, *J. Biol. Chem.* **257:**9473.

Koch, H. U., Haas, R., and Fischer, W., 1984, The role of lipoteichoic acid biosynthesis in membrane lipid metabolism of growing *Staphylococcus aureus, Eur. J. Biochem.* **138:**357.

Koch, H. U., Döker, R., and Fischer, W., 1985, Maintenance of D-alanine ester substitution of lipoteichoic acid by re-esterification in *Staphylococcus aureus, J. Bacteriol.* **164:**1211.

Koga, Y., Nishihara, M., and Morii, H., 1984, Products of phosphatidylglycerol turnover in two Bacillus strains with and without lipoteichoic acid in the cells, *Biochim. Biophys. Acta* **793:**86.

Kojima, N., Araki, Y., and Ito, E., 1983, Structure of linkage region between ribitol teichoic acid and peptidoglycan in cell walls of *Staphylococcus aureus* H, *J. Biol. Chem.* **258:**9043.

Kojima, N., Araki, Y., and Ito, E., 1985, Structure of the linkage units between ribitol teichoic acids and peptidoglycan, *J. Bacteriol.* **161:** 299.

Komaratat, P., and Kates, M., 1975, The lipids of a halotolerant species of *Staphylococcus epidermidis, Biochim. Biophys. Acta* **398:**464.

Kuo, J. S.-C., Doelling, V. W., Graveline, J. F., and McCoy, D. W., 1985, Evidence for covalent attachment of phospholipid to the capsular polysaccharide of *Haemophilus influenzae* type b, *J. Bacteriol.* **163**:769.

Kusser, W., and Fiedler, F., 1984, A novel glycerophosphodiesterase from *Bacillus pumilus*, *FEBS Lett.* **166**:301.

Labischinski, H., Barnickel, G., Bradaczek, H., Naumann, D., Rietschel, E. T., and Giesbrecht, P., 1985, High state of order of isolated bacterial lipopolysaccharide and its possible contribution to the permeation barrier property of the outer membrane, *J. Bacteriol.* **162**:9.

Laine, R. A., and Fischer, W., 1978, On the relationship between glycerophosphoglycolipids and lipoteichoic acids of gram-positive bacteria. III. Di(glycerophospho)-acylkojibiosyldiacylglycerol and related compounds from *Streptococcus lactis* NCDO 712, *Biochim. Biophys. Acta* **529**:250.

Lambert, P. A., Hancock, I. C., and Baddiley, J., 1975*a*, The interaction of magnesium ions with teichoic acid, *Biochem. J.* **149**:519.

Lambert, P. A., Hancock, I. C., and Baddiley, J., 1975*b*, Influence of alanyl ester residues on the binding of magnesium ions to teichoic acids, *Biochem. J.* **51**:671.

Lambert, P. A., Hancock, I. C., and Baddiley, J., 1977, Occurrence and function of membrane teichoic acids, *Biochem. Biophys. Acta* **472**:1.

Landgraf, H. R., 1976, Fettsäuremuster und positionsspezifische Verteilung der Fettsäuren in den polaren Lipiden von Streptococcen, Doctoral thesis, University Erlangen-Nürnberg.

Langley, K. E., and Kennedy, E. P., 1979, Energetics of rapid transmembrane movement and compositional asymmetry of phosphatidylethanolamine in membranes of *Bacillus megaterium*, *Proc. Natl. Acad. Sci. USA* **76**:6245.

Langworthy, T. A., 1983, Lipid tracers of *Mycoplasma* phylogeny, *Yale J. Biol. Med.* **56**:385.

Langworthy, T. A., Smith, P. F., and Mayberry, W. R., 1972, Lipids of *Thermoplasma acidophilum*, *J. Bacteriol.* **112**:1193.

Langworthy, T. A., Tornabene, T. G., and Holzer, G., 1982, Lipids of Archaebacteria, *Zbl. Bakt. Hyg.*, I. Abt. Orig. C **3**:228.

Lapper, R. D., Mantsch, H. H., and Smith, I. C. P., 1972, A carbon-13 and hydrogen-1 nuclear magnetic resonance study of the conformation of 3′,5′- and 2′,3′-cyclic nucleotides. A demonstration of the angular dependence of three bond spin–spin couplings between carbon and phosphorus, *J. Am. Chem. Soc.* **94**:6243.

Lapper, R. D., Mantsch, H. H., and Smith, I. C. P., 1973, A carbon-13 nuclear magnetic resonance study of the conformations of 3′,5′-cyclic nucleotides, *J. Am. Chem. Soc.* **95**:2878.

Lechevalier, M. P., 1977, Lipids in bacterial taxonomy—A taxonomist's view, *CRC Crit. Rev. Microbiol.* **5**:109.

Lennarz, W. J., and Talamo, B., 1966, The chemical characterization and enzymatic synthesis of mannolipids in *Micrococcus lysodeikticus*, *J. Biol. Chem.* **241**:2707.

Leon, O., and Panos, C., 1983, Cytotoxicity and inhibition of normal collagen synthesis in mouse fibroblasts by lipoteichoic acid from *Streptococcus pyogenes* type 12, *Infect. Immun.* **40**:785.

Leon, O., and Panos, C., 1985, Effect of streptococcal lipoteichoic acid on prolyl hydroxylase activity as related to collagen formation in mouse fibroblast monolayers, *Infect. Immun.* **50**:745.

Levine, M., 1982, Naturally occurring human serum precipitins specific for D-alanyl esters of glycerol teichoic acid, *Mol. Immunol.* **19**:133.

Lim, S., and Salton, R. J., 1985, Comparison of the chemical composition of lipomannan from *Micrococcus agilis* membranes with that of *Micrococcus luteus* strains, *FEMS Microbiol. Lett.* **27**:287.

Linzer, R., and Neuhaus, F. C., 1973, Biosynthesis of membrane teichoic acid. A role for the D-alanine-activating enzyme, *J. Biol. Chem.* **248**:3196.

Lipkin, D., Phillips, B. E., and Abrell, J. W., 1969, The action of hydrogen fluoride on nucleotides and other esters of phosphorus (V) acids, *J. Org. Chem.* **34**:1539.

Lombardi, F. J., Chen, S. L., and Fulco, A. J., 1980, A rapidly metabolizing pool of phosphatidylglycerol as a precursor for phosphatidylethanolamine and diglyceride in *Bacillus megaterium*, *J. Bacteriol.* **141**:626.

Loos, M., Clas, F., and Fischer, W., 1986, Interaction of purified lipoteichoic acid with the classical complement pathway, *Infect. Immun.* **53**:595.

Lopatin, D. E., and Kessler, R. E., 1985, Pretreatment with lipoteichoic acid sensitizes target cells to antibody-dependent cellular cytotoxicity in the presence of anti-lipoteichoic acid antibodies, *Infect. Immun.* **48:**638.

Lüderitz, O., Freudenberg, M. A., Galanos, C., Lehmann, V., Rietschel, E. T., and Shaw, D. H., 1982, Lipopolysaccharides of gram-negative bacteria, *Curr. Topics Membr. Transport* **17:**79.

Macfarlane, M. G., 1961, Isolation of a phosphatidylglycerol and a glycolipid from *Micrococcus lysodeikticus* cells, *Biochem. J.* **80:**45p.

Maier, R. V., and Ulevitch, R. J., 1981, The response of isolated rabbit hepatic macrophages to lipopolysaccharide, *Circ. Shock* **8:**165.

Mancuso, D. J., and Chiu, T.-H., 1982, Biosynthesis of glycosyl manophosphoryl undecaprenol and its role in lipoteichoic acid biosynthesis, *J. Bacteriol.* **152:**616.

Mancuso, D. J., Junker, D. D., Hsu, S. C., and Chiu, T.-H., 1979, Biosynthesis of glycosylated glycerolphosphate polymers in *Streptococcus sanguis*, *J. Bacteriol.* **140:**547.

Markham, J. L., Knox, K. W., Schamschula, R. G., and Wicken, A. J., 1973, Antibodies to teichoic acids in man, *Arch. Oral Biol.* **18:**313.

Markham, J. L., Knox, K. W., Wicken, A. J., and Hewett, M. J., 1975, Formation of extracellular lipoteichoic acid by oral streptococci and lactobacilli, *Infect. Immun.* **12:**378,

Mayer, H., and Schmidt, G., 1979, Chemistry and biology of enterobacterial common antigen (ECA), *Curr. Microbiol. Immunol.* **85:**99.

McArthur, A. E., and Archibald, A. R., 1984, Effect of the culture pH on the D-alanine ester content of lipoteichoic acid in *Staphylococcus aureus*, *J. Bacteriol.* **160:**792.

McArthur, H. A. I., Hancock, I. C., and Baddiley, J., 1981, Attachment of the main chain to the linkage unit in biosynthesis of teichoic acids, *J. Bacteriol.* **145:**1222.

McCarty, M., 1959, The occurrence of polyglycerophosphate as an antigenic component of various gram-positive bacterial species, *J. Exp. Med.* **109:**361.

McCarty, M., 1964, The role of D-alanine in the serological specificity of group A streptococcal glycerol teichoic acid, *Proc. Natl. Acad. Sci. USA* **52:**259.

Miller, G. A., and Jackson, R. W., 1973, The effect of a *Streptococcus pyogenes* lipoteichoic acid on the immune response of mice, *J. Immunol.* **110:**148.

Miller, G. A., Urban, J., and Jackson, R. W., 1976, Effects of a streptococcal lipoteichoic acid on host response in mice, *Infect. Immun.* **13:**1408.

Minnikin, D. E., Abdolrahimzadeh, H., and Baddiley, J., 1972, Variation of polar lipid composition of *Bacillus subtilis* (Marburg) with different growth conditions, *FEBS Lett.* **27:**16.

Miörner, H., Johanson, G., and Kronvall, G., 1983, Lipoteichoic acid is the major cell wall component responsible for surface hydrophobicity of group A streptococci, *Infect. Immun.* **39:**336.

Mishell, R. I., Chen, Y. U., Clark, G. C., Gold, M. R., Hill, J. L., Kwan, E., and Lee, D. A., 1981, Induction of murine lymphozytes and macrophages by components of bacterial cell walls and membranes, in *Chemistry and Biological Activities of Bacterial Surface Amphiphiles* (G. D. Shockman and A. J. Wicken, eds.), pp. 327–339, Academic Press, New York.

Moskowitz, M., 1966, Separation and properties of a red cell sensitizing substance from streptococci, *J. Bacteriol.* **91:**2200.

Nakano, M., and Fischer, W., 1977, The glycolipids of *Lactobacillus casei* DSM 20021, *Hoppe Seylers Z. Physiol. Chem.* **358:**1439.

Nakano, M., and Fischer, W., 1978, Trihexosyldiacylglycerol and acyltrihexosyldiacylglycerol as lipid anchors of the lipoteichoic acid of *Lactobacillus casei* DSM 20021, *Hoppe Seylers Z. Physiol. Chem.* **359:**1.

Naumann, D., Schultz, D., Born, J., Labischinski, H., Brandenburg, K., von Busse, G., and Seydel, U., 1987, Investigation into the polymorphism of lipid A from lipopolysaccharides of *Escherichia coli* and *Salmonella minnesota* by Fourier-transform infrared spectroscopy, *Eur. J. Biochem.* **164:**159.

Naumova, I. B., Kuznetsov, V. D., Kudrina, K. S., and Bezzubenkova, A. P., 1980, The occurrence of teichoic acids in streptomycetes, *Microbiology* **126:**71.

Nealon, T. J., and Mattingly, S. J., 1984, Role of cellular lipoteichoic acids in mediating adherence of serotype III strains of group B streptococci to human embryonic, fetal and adult epithelial cells, *Infect. Immun.* **43:**523.

Ne'eman, N., and Ginsburg, I., 1972, Red cell sensitizing antigen of group A streptococci. II. Immunological and immunopathological properties, *Israel J. Med. Sci.* **8**:1807.

Neuhaus, F. C., Linzer, R., and Reusch, V. M., Jr., 1974, Biosynthesis of membrane teichoic acid: Role of the D-alanine-activating enzyme and D-alanine : membrane acceptor ligase, *Ann. NY Acad. Sci.* **235**:501.

Ntamere, A. S., and Neuhaus, F. C., 1987, Lipoteichoic acid biosynthesis in vesicular structures released in the presence of penicillin from *Lactobacillus casei, Abstracts of the Annual Meeting of the American Society for Microbiology* **87**:226.

Ntamere, A. S., Taron, J. D., and Neuhaus, F. C., 1987, Assembly of D-alanyl-lipoteichoic acid in *Lactobacillus casei:* Mutants deficient in the D-alanyl ester content of this amphiphile, *J. Bacteriol.* **169**:1702.

Ofek, I., Beachy, E. H., Jefferson, W., and Campbell, G. L., 1975, Cell membrane-binding properties of group A streptococcal lipoteichoic acid, *J. Exp. Med.* **141**:990.

Ofek, I., Simpson, W. A., and Beachy, E. H., 1982, Formation of molecular complexes between a structurally defined M protein and acylated or deacylated lipoteichoic acid of *Streptococcus pyogenes, J. Bacteriol.* **149**:426.

Ogura, M., Kohama, T., Fujimoto, M., Kuninaka, A., Yoshino, H., and Sugiyama, H., 1974, Structure of malonogalactan: Carbon-13 nuclear magnetic resonance spectra of malonogalactan, *Agr. Biol. Chem. Tokyo* **38**:2563.

Op den Camp, H. J. M., Veerkamp, J. H., Oosterhof, A., and Van Halbeek, H., 1984, Structure of the lipoteichoic acids from *Bifidobacterium bifidum* spp. *pennsylvanicum, Biochim. Biophys. Acta* **795**:301.

Op den Camp, H. J. M., Oosterhof, A., and Veerkamp, J. H., 1985a, Phosphatidylglycerol as biosynthetic precursor for the poly(glycerol phosphate) backbone of bifidobacterial lipoteichoic acid, *Biochem. J.* **228**:683.

Op den Camp, H. J. M., Peeters, P. A. M., Oosterhof, A., and Veerkamp, J. H., 1985b, Immunochemical studies on the lipoteichoic acids of *Bifidobacterium bifidum* subsp. *pennsylvanicum, J. Gen. Microbiol.* **131**:661.

Op den Kamp, J. A. F., 1979, Lipid asymmetry in membranes, *Annu. Rev. Biochem.* **48**:47.

Orefici, G., Molinari, A., Donelli, G., Paradisi, S., Teti, G., and Arancia, G., 1986, Immunolocation of lipoteichoic acid on group B streptococcal surface, *FEMS Microbiol. Lett.* **34**:111.

Oshima, Y., Beuth, J., Ko, H. L., Roszkowski, K., Hauck, D., and Pulverer, G., 1988a, Immunomodulatory effects of staphylococcal lipoteichoic acid in early *Listeria monocytogenes* infection in Balb/c-mice, *Zbl. Bakt. Hyg. A* **269**:251.

Oshima, Y., Beuth, J., Yassin, A., Ko, H. L., and Pulverer, G., 1988b, Stimulation of human monocyte chemiluminescence by staphylococcal lipoteichoic acid, *Med. Microbiol. Immunol.* **177**:115.

Oshima, Y., Ko, H. L., Beuth, J., and Pulverer, G., 1988c, Immunostimulating staphylococcal lipoteichoic acid prevents pulmonary tumor colonization in Balb/c-mice, *Zbl. Bakt. Hyg. A* **270**:213.

Ou, L.-T., and Marquis, R. E., 1970, Electromechanical interactions in cell walls of gram-positive cocci, *J. Bacteriol.* **101**:92.

Owen, P., and Freer, J. H., 1972, Isolation and properties of mesosomal membrane fractions from *Micrococcus lysodeikticus, Biochem. J.* **129**:907.

Owen, P., and Salton, M. R. J., 1975a, A succinylated mannan in the membrane system of *Micrococcus lysodeikticus, Biochem. Biophys. Res. Commun.* **63**:875.

Owen, P., and Salton, M. R. J., 1975b, Distribution of enzymes involved in mannan synthesis in plasma membranes and mesosomal vesicles of *Micrococcus lysodeikticus, Biochim. Biophys. Acta* **406**:235.

Owen, P., and Salton, M. R. J., 1975c, Antigenic and enzymatic architecture of *Micrococcus lysodeikticus* membranes established by crossed immunoelektrophoresis, *Proc. Natl. Acad. Sci. USA* **72**:3711.

Paton, J. C., May, B. K., and Elliott, W. H., 1978, Membrane phospholipid asymmetry in *Bacillus amyloliquefaciens, J. Bacteriol.* **135**:393.

Peavy, D. L., and Brandon, C. L., 1980, Macrophages: Primary targets for LPS activity, in *Bacterial Endotoxins and Host Response* (M. K. Agarwal, ed.), pp. 299–309, Elsevier/North-Holland, Amsterdam.

Pieringer, R. A., 1968, The metabolism of glyceride glycolipids, *J. Biol. Chem.* **243**:4894.

Pieringer, R. A., 1972, Biosynthesis of the phosphatidyl diglucosyl diglyceride of *Streptococcus faecalis* from diglucosyl diglyceride and phosphatidyl glycerol or diphosphatidyl glycerol, *Biochem. Biophys. Res. Commun.* **49**:502.

Pieringer, R. A., 1983, Formation of bacterial glycerolipids, in *The Enzymes* (P. D. Boyer, ed.), Vol. XVI, pp. 255–306, Academic Press, New York.

Pieringer, R. A., Ganfield, M.-C. W., Gustow, E., and Cabacungan, E., 1981, Biosynthesis of the membrane lipoteichoic acid of *Streptococcus faecium* (*S. faecalis* ATCC 9790), in *Chemistry and Biological Activities of Bacterial Surface Amphiphiles* (G. D. Shockman and A. J. Wicken, eds.) pp. 167–179, Academic Press, New York.

Pless, D. D., Schmit, A. S., and Lennarz, W. J., 1975, The characterization of mannan of *Micrococcus lysodeikticus* as an acidic lipopolysaccharide, *J. Biol. Chem.* **250**:1319.

Potekhina, N. V., Streshinskaya, G. M., Novitskaya, G. V., and Naumova, I. B., 1983, Isolation of lipoteichoic acid from *Streptomyces levoris*, *Microbiology* **52**:340.

Powell, D. A., Duckworth, M., and Baddiley, J., 1974, An acylated mannan in the membrane of *Micrococcus lysodeikticus*, *FEBS Lett.* **41**:259.

Powell, D. A., Duckworth, M., and Baddiley, J., 1975, A membrane-associated lipomannan in Micrococci, *Biochem. J.* **151**:387.

Prottey, C., and Ballou, C. E., 1968, Diacylmyoinositol monomannoside from *Propionibacterium shermanii*, *J. Biol. Chem.* **243**:6196.

Raychaudhuri, D., and Chatterjee, A. N., 1985, Use of resistant mutants to study the interaction of Triton X-100 with *Staphylococcus aureus*, *J. Bacteriol.* **164**:1337.

Rajbhandary, U. L., and Baddiley, J., 1963, The intracellular teichoic acid from *Staphylococcus aureus* H, *Biochem. J.* **87**:429.

Rantz, L. A., Zuckerman, A., and Randall, E., 1952, Hemolysis of red blood cells treated by bacterial filtrates in the presence of serum and complement, *J. Lab. Clin. Med.* **39**:443.

Rantz, L. A., Randall, E., and Zuckerman, A., 1956, Hemolysis and hemagglutination by normal and immune serums of erythrocytes treated with a non-species-specific bacterial substance, *J. Infect. Dis.* **98**:211.

Reusch Jr., V. M., and Burger, M. M., 1973, The bacterial mesosome, *Biochim. Biophys. Acta* **300**:79.

Reusch, V. M., Jr., and Neuhaus, F. C., 1971, D-alanine: Membrane acceptor ligase from *Lactobacillus casei*, *J. Biol. Chem.* **246**:6136.

Rietschel, E. T., Brade, L., Schade, U., Galanos, C., Freudenberg, M., Lüderitz, O., Kusumoto, S., and Shiba, T., 1987, Endotoxic properties of synthetic pentaacyl lipid A precursor Ib and a structural isomer, *Eur. J. Biochem.* **169**:27.

Robbins, P. W., Bray, D., Dankert, M., and Wright, A., 1967, Direction of chain growth in polysaccharide synthesis, *Science* **158**:1536.

Roberts, M. F., Jacobson, G. R., Scott, P. J., Mimura, C. S., and Stinson, M. W., 1985, ^{31}P-NMR studies of the oral pathogen *Streptococcus mutans:* Observation of lipoteichoic acid, *Biochim. Biophys. Acta* **845**:242.

Rogers, H. J., Taylor, C., Rayter, S., and Ward, J. B., 1984, Purification and properties of autolytic endo-β-N-acetylglucosaminidase and the N-acetylmuramyl-L-alanine amidase from *Bacillus subtilis* strain 168, *J. Gen. Microbiol.* **130**:2395.

Rosan, B., 1978, Absence of glycerol teichoic acids in certain oral streptococci, *Science* **201**:918.

Rothman, J. E., and Kennedy, E. P., 1977a, Asymmetrical distribution of phospholipids in the membrane of *Bacillus megaterium*, *J. Mol. Biol.* **110**:603.

Rothman, J. E., and Kennedy, E. P., 1977b, Rapid transmembrane movement of newly synthesized phospholipids during membrane assembly, *Proc. Natl. Acad. Sci. USA* **74**:1821.

Rouser, G., Kritchevsky, G., and Yamamoto, A., 1967, Column chromatographic and associated procedures for separation and determination of phosphatides and glycolipids, in *Lipid Chromatographic Analysis* (G. V. Marinetti, ed.) pp. 99–162, Marcel Dekker, New York.

Ruhland, G. J., and Fiedler, F., 1987, Occurrence and biochemistry of lipoteichoic acids in the genus *Listeria*, *Syst. Appl. Microbiol.* **9**:40.

Saito, H., Ohki, T., Takasuka, N., and Sasaki, T., 1977, ^{13}C-Nuclear magnetic resonance spectral study of a gel-forming, branched (1-3)β-D-glucan (Lentinan) from *Lentinus edodes*, and its acid-degraded fractions, structure, and dependence of conformation on the molecular weight, *Carbohydr. Res.* **58**:293.

Salton, M. R. J., 1976, Methods of isolation and characterization of bacterial membranes, in *Methods in Membrane Biology* (E. D. Korn, ed.), Vol 6, pp. 101–150, Plenum, New York.

Salton, M. R. J., and Owen, P., 1976, Bacterial membrane structure, *Annu. Rev. Microbiol.* **30**:451.

Scher, M., and Lennarz, W. J., 1969, Studies on the biosynthesis of mannan in *Micrococcus lysodeikticus*, *J. Biol. Chem.* **244**:2777.

Scher, M., Lennarz, W. J., and Sweeley, C. C., 1968, The biosynthesis of mannosyl-1-phosphoryl-polyisoprenol in *Micrococcus lysodeikticus* and its role in mannan synthesis, *Proc. Natl. Acad. Sci. USA* **59**:1313.

Schleifer, K. H., and Kilpper-Bälz, R., 1984, Transfer of *Streptococcus faecalis* and *Streptococcus faecium* to the genus *Enterococcus* nom. rev. as *Enterococcus faecalis* comb. nov. and *Enterococcus faecium* comb. nov., *Int. J. Syst. Bacteriol.* **34**:31.

Schleifer, K. H., Kilpper-Bälz, R., Kraus, J., and Gehring, R., 1984, Relatedness and classification of *Streptococcus mutans* and "mutans-like" streptococci, *J. Dent. Res.* **63**:1046.

Schleifer, K. H., Kraus, J., Dvorak, C., Kilpper-Bälz, R., Collins, M. D., and Fischer, W., 1985, Transfer of *Streptococcus lactis* and related streptococci to the genus *Lactococcus* gen. nov., *Syst. Appl. Microbiol.* **6**:183.

Schmidt, M. A., and Jann, K., 1982, Phospholipid substitution of capsular (K) polysaccharide antigens from *Escherichia coli* causing extraintestinal infections, *FEMS Microbiol. Lett.* **14**:69.

Schneider, J. E., and Kennedy, E. P., 1978, A novel phosphodiesterase from *Aspergillus niger* and its application to the study of membrane derived oligosaccharides and other glycerol-containing biopolymers, *J. Biol. Chem.* **253**:7738.

Shabarova, Z. A., Hughes, N. A., and Baddiley, J., 1962, The influence of adjacent phosphate and hydroxyl groups on amino acid esters, *Biochem. J.* **83**:216.

Shaskov, A. S., Zaretskaya, M. Sh., Yarotsky, S. V., Naumova, I. B., Chizhov, O. S., and Shabarova, Z. A., 1979, On the structure of the teichoic acid from the cell wall of *Streptomyces antibioticus* 39. Localization of the phosphodiester linkages ad elucidation of the monomeric units structure by means of ^{13}C-nuclear-magnetic-resonance spectroscopy, *Eur. J. Biochem.* **102**:477.

Shaw, N., 1968, The detection of lipids on thin-layer chromatograms with the periodate-Schiff reagents, *Biochim. Biophys. Acta* **164**:435.

Shaw, N., 1974, Lipid composition as a guide to the classification of bacteria, *Adv. Appl. Microbiol.* **17**:63.

Shaw, J. M., and Pieringer, R. A., 1977, Phosphatidylmonoglucosyl diacylglycerol of *Pseudomonas diminuta* ATCC 11568, *J. Biol. Chem.* **252**:4395.

Shaw, N., and Stead, A., 1972, Bacterial glycophospholipids, *FEBS Lett.* **21**:249.

Shaw, N., and Stead, A., 1974, The reaction of phosphoglycolipids and other lipids with hydrofluoric acid, *Biochem. J.* **143**:461.

Shaw, N., Smith, P. F., and Verheij, H. M., 1972, The structure of a glycerylphosphoryldiglucosyl-diglyceride from the lipids of *Acholeplasma laidlawii* strain B, *Biochem. J.* **129**:167.

Shockman, G. D., 1981, Cellular localization, excretion, and physiological roles of lipoteichoic acid in gram-positive bacteria, in *Chemistry and Biological Activities of Bacterial Surface Amphiphiles* (G. D. Shockman and A. J. Wicken, eds.), pp. 21–40, Academic Press, New York.

Shockman, G. D., and Slade, H. D., 1964, The cellular location of the group D antigen, *J. Gen. Microbiol.* **37**:297.

Short, S. A., and White, D. C., 1970, Metabolism of the glucosyl diglycerides and phosphatidylglucose of *Staphylococcus aureus*, *J. Bacteriol.* **104**:126.

Short, S. A., and White, D. C., 1971, Metabolism of phosphatidylglycerol, lysylphosphatidylglycerol, and cardiolipin of *Staphylococcus aureus*, *J. Bacteriol.* **108**:219.

Shungu, D. L., Cornett, J. B., and Shockman, G. D., 1980, Lipids and lipoteichoic acid of autolysis-defective *Streptococcus faecium* strains. *J. Bacteriol.* **142**:741.

Silvestri, L. J., Craig, R. A., Ingram, L. O., Hoffmann, E. M., and Bleiweis, A. S., 1978, Purification of lipoteichoic acids by using phosphatidyl choline vesicles, *Infect. Immun.* **22**:107.

Silvestri, L. J., Knox, K. W., Wicken, A. J., and Hoffmann, E. M., 1979, Inhibition of complement-mediated lysis of sheep erythrozytes by cell-free preparations from *Streptococcus mutans* BHT, *J. Immunol.* **122**:54.

Simpson, W. A., Ofek, I., and Beachy, E. H., 1980a, Fatty acid binding sites of serum albumin as membrane receptor analogs for streptococcal lipoteichoic acid, *Infect. Immun.* **29**:119.

Simpson, W. A., Ofek, I., Sarasohn, C., Morrison, J. C., and Beachy, E. H., 1980b, Characteristics of the binding of streptococcal lipoteichoic acid to human oral epithelial cells, *J. Infect. Dis.* **141:**457.

Simpson, W. A., Dale, J. B., and Beachy, E. H., 1982, Cytotoxicity of the glycolipid region of lipoteichoic acid for cultures on human heart cells, *J. Lab. Clin. Med.* **99:**118.

Simpson, W. A., Courtney, H. S., and Ofek, I., 1987, Interaction of fibronectin with streptococci: The role of fibronectin as a receptor for *Streptococcus pyogenes, Rev. Infect. Dis.* **9:**S351.

Slabyj, B. M., and Panos, C., 1973, Teichoic acid of a stabilized L-form of *Streptococcus pyogenes, J. Bacteriol.* **114:**934.

Slabyj, B. M., and Panos, C., 1976, Membrane lipoteichoic acid of *Streptococcus pyogenes* and its stabilized L-form and the effect of two antibiotics upon its cellular content, *J. Bacteriol.* **127:**855.

Slade, H. D., and Slamp, W. C., 1962, Cell-wall composition and the grouping antigens of streptococci, *J. Bacteriol.* **84:**345.

Smith, P. F., 1972, A phosphatidyl diglucosyl diglyceride from *Acholeplasma laidlawii* B, *Biochim. Biophys. Acta* **280:**375.

Smith, P. F., Patel, K. R., and Al-Shammari, A. J. N., 1980, An aldehydophosphoglycolipid from *Acholeplasma granularum, Biochim. Biophys. Acta* **617:**419.

Smith, R., and Tanford, C., 1972, The critical micelle concentration of L-α-dipalmitoylphosphatidylcholine in water and water/methanol solutions, *J. Mol. Biol.* **67:**75.

Snyder, F., and Stephens, N., 1959, A simplified spectrophotometric determination of ester groups in lipids, *Biochim. Biophys. Acta* **34:**244.

Sørensen, U. B. S., Blom, J., Birch-Andersen, A., and Henrichsen, J., 1988, Ultrastructural localization of capsules, cell wall polysaccharide, cell wall proteins, and F antigen in pneumococci, *Infect. Immun.* **56:**1890.

Stanislawski, L., Courtney, H. S., Simpson, W. A., Hasty, D. L., Beachy, E. H., Robert, L., and Ofek, I., 1987, Hybridoma antibodies to the lipid-binding site(s) in the amino-terminal region of fibronectin inhibits binding of streptococcal lipoteichoic acid, *J. Infect. Dis.* **156:**344.

Sutcliffe, I. C., and Shaw, N., 1989, An inositol containing lipomannan from *Propionibacterium freudenreichii, FEMS Microbiol. Lett.* **59:**249.

Tanford, C., 1980, *The Hydrophobic Effect: Formation of Micelles and Biological Membranes,* Wiley, New York.

Taron, D. J., Childs, W. C. III, and Neuhaus, F. C., 1983, Biosynthesis of D-alanyl-lipoteichoic acid: Role of diglyceride kinase in the synthesis of phosphatidylglycerol for chain elongation, *J. Bacteriol.* **154:**1110.

Teti, G., Chiofalo, M. S., Tomasello, F., Fava, C., and Mastroeni, P. 1987, *Infect. Immun.* **55:**839.

Theodore, T. S., and Calandra, G. B., 1981, Streptolysin S activation by lipoteichoic acid, *Infect. Immun.* **33:**326.

Thomas, J., and Law, J. H., 1966, Biosynthesis of cyclopropane compounds. IX. Structural and stereochemical requirements for the cyclopropane synthetase substrate, *J. Biol. Chem.* **241:**5013.

Tomasz, A., Westphal, M., Briles, E. B., and Fletcher, P., 1975, On the physiological functions of teichoic acids, *J. Supramol. Struct.* **3:**1.

Tomlinson, K., Leon, O., and Panos, C., 1983, Morphological changes and pathology of mouse glomeruli infected with a streptococcal L-form or exposed to lipoteichoic acid, *Infect. Immun.* **42:**1144.

Toon, P., Brown, P. E., and Baddiley, J., 1972, The lipid-teichoic acid complex in the cytoplasmic membrane of *Streptococcus faecalis* NCIB 8191, *Biochem. J.* **127:**399.

Uchikawa, K., Sekikawa, I., and Azuma, I., 1986, Structural studies on lipoteichoic acids from four *Listeria* strains, *J. Bacteriol.* **168:**115.

Ukita, T., Bates, N. A., and Carter, H. E., 1955, Studies on the alkaline hydrolysis of lecithin: Synthesis of cyclic 1,2-glycerophosphate, *J. Biol. Chem.* **216:**867.

Ukita, T., Nagasawa, K., and Irie, M., 1957, Organic phosphates. II. Studies on hydrolysis of several cyclic phosphates, *Pharmacol. Bull. Tokyo* **5:**127.

Utsui, Y., Ohya, S., Takenouchi, Y., Tajima, M., and Sugawara, S., 1983, Release of lipoteichoic acid from *Staphylococcus aureus* by treatment with cefmetazole and other β-lactam antibiotics, *J. Antibiot.* **36:**1380.

Van de Rijn, I., and Bleiweis, A. S., 1973, Antigens of *Streptococcus mutans*. I. Characterization of a serotype-specific determinant from *Streptococcus mutans*, *Infect. Immun.* **7:**795.

Van der Kaaden, A., Van Doorn-Van Waakeren, J. I. M., Kamerling, J. P., Vliegenthart, J. F. G., and Tiesjema, R. H., 1984, Structure of the capsular antigen of *Neisseria meningitidis* serogroup H, *Eur. J. Biochem.* **141:**513.

Van Driel, D., Wicken, A. J., Dickson, M. R., and Knox, K. W., 1973, Cellular location of the lipoteichoic acids of *Lactobacillus fermenti* NCTC 6991 and *Lactobacillus casei* NCTC 6375, *J. Ultrastruct. Res.* **43:**483.

Van Schaik, F. W., and Veerkamp, J. H., 1976, Metabolic relationships between phospho(galacto)lipids in *Bifidobacterium bifidum* var. *pennsylvanicum*, *FEBS Lett.* **67:**13.

Veerkamp, J. H., 1972, Biochemical changes in *Bifidobacterium bifidum* var. *pennsylvanicum* after cell wall inhibition. V. Structure of the galactosyldiglycerides, *Biochim. Biophys. Acta* **273:**359.

Veerkamp, J. H., 1976, Biosynthesis of phosphogalactolipids and diphosphatidylglycerol in a membrane fraction of *Bifidobacterium bifidum* var. *pennsylvanicum*, *Biochim. Biophys. Acta* **441:**403.

Veerkamp, J. H., and Van Schaik, F. W., 1974, Biochemical changes in *Bifidobacterium bifidum* var. *pennsylvanicum* after cell wall inhibition. VII. Structure of the phosphogalactolipids, *Biochim. Biophys. Acta* **348:**370.

Waltersdorff, R. L., Fiedel, B. A., and Jackson, R. W., 1977, Induction of nephrocalcinosis in rabbit kidneys after long term exposure to a streptococcal teichoic acid, *Infect. Immun.* **17:**665.

Ward, J. B., 1981, Teichoic and teichuronic acids: Biosynthesis, assembly, and location, *Microbiol. Rev.* **45:**211.

Ward, J. B., and Perkins, H. R., 1973, The direction of glycan synthesis in a bacterial peptidoglycan, *Biochem. J.* **135:**721.

Weinreb, B. D., Shockman, G. D., Beachy, E. H., Swift, A. J., and Winkelstein, J. A., 1986, The ability to sensitize host's cells for destruction by autologous complement is a general property of lipoteichoic acid, *Infect. Immun.* **54:**494.

Westphal, O., Lüderitz, O., and Bister, F., 1952, Über die Extraktion von Bakterien mit Phenol/Wasser, *Z. Naturforsch.* **7b:**148.

Wicken, A. J., and Baddiley, J., 1963, Structure of intracellular teichoic acids from group D streptococci, *Biochem. J.* **87:**54.

Wicken, A. J., and Knox, K. W., 1970, Studies on the group F antigen of lactobacilli: Isolation of a teichoic acid–lipid complex from *Lactobacillus fermenti*, *J. Gen. Microbiol.* **60:**293.

Wicken, A. J., and Knox, K. W., 1975a, Lipoteichoic acids: A new class of bacterial antigen. Membrane lipoteichoic acids can function as surface antigens of gram-positive bacteria, *Science* **187:**1161.

Wicken, A. J., and Knox, K. W., 1975b, Characterization of group N Streptococcus lipoteichoic acid, *Infect. Immun.* **11:**973.

Wicken, A. J., and Knox, K. W., 1977, Immunological properties of lipoteichoic acids, *Prog. Immunol.* **3:**135.

Wicken, A. J., and Knox, K. W., 1980, Bacterial cell surface amphiphiles, *Biochim. Biophys. Acta* **604:**1.

Wicken, A. J., Elliott, S. D., and Baddiley, J., 1963, The identity of streptococcal group D antigen with teichoic acid, *J. Gen. Microbiol.* **31:**231.

Wicken, A. J., Gibbens, J. W., and Knox, K. W., 1973, Comparative studies on the isolation of membrane lipoteichoic acid from *Lactobacillus fermentum*, *J. Bacteriol.* **113:**365.

Wicken, A. J., Broady, K. W., Evans, J. D., and Knox, K. W., 1978, New cellular and extracellular amphipathic antigen from *Actinomyces viscosus* NY1, *Infect. Immun.* **22:**615.

Wicken, A. J., Evans, J. D., and Knox, K. W., 1986, Critical micelle concentrations of lipoteichoic acids. *J. Bacteriol.* **166:**72.

Wilkinson, S. G., and Bell, M. E., 1971, The phosphoglucolipid from *Pseudomonas diminuta*, *Biochim. Biophys. Acta* **248:**293.

Wilkinson, S. G., and Galbraith, L., 1979, Polar lipids of *Pseudomonas vesicularis*. Presence of a heptosyldiacylglycerol, *Biochim. Biophys. Acta* **575:**244.

Woese, C. R., 1987, Bacterial evolution, *Microbiol. Rev.* **51:**221.

Wong, W., Shockman, G. D., and Wicken, A. J., 1981, Depletion of lipoteichoic acid from intact cells of *Streptococcus faecium*, in *Chemistry and Biological Activities of Bacterial Surface Amphiphiles* (G. D. Shockman and A. J. Wicken, eds.), pp. 247–257, Academic Press, New York.

Yamamoto, T., Koga, T., Mizuno, J., and Hamada, S., 1985, Chemical and immunological characterization of a novel amphipathic antigen from biotype B *Streptococcus sanguis, J. Gen. Microbiol.* **131**:1981.

Yokoyama, K., Araki, Y., and Ito, E., 1987, Biosynthesis of poly(galactosylglycerol phosphate) in *Bacillus coagulans, Eur. J. Biochem.* **165**:47.

Yokoyama, K., Araki, Y., and Ito, E., 1988, The function of galactosylphosphorylpolyprenol in biosynthesis of lipoteichoic acid in *Bacillus coagulans, Eur. J. Biochem.* **173**:453.

Zalkin, H., Law, J. H., and Goldfine, H., 1963, Enzymatic synthesis of cyclopropane fatty acids catalyzed by bacterial extracts, *J. Biol. Chem.* **238**:1242.

Glycolipids of Higher Plants, Algae, Yeasts, and Fungi

Morris Kates

3.1. Introduction

In marked contrast to the extensive variety of glycoglycerolipids (GGroLs) present in bacteria (see Chapters 1 and 2, *this volume*), the plant glycolipids consist basically of three major components: monogalactosylDAG, digalactosylDAG, and sulfoquinovosylDAG (structures **1, 2,** and **4,** respectively) (Fig. 3-1). Small amounts of higher homologues of galactosylDAGs, i.e., trigalactosylDAG (structure **3**) (Fig. 3-1) and tetragalactosylDAG (structure **3′**) (Fig. 3-1) have also been detected in both photosynthetic and nonphotosynthetic tissues. Other minor components are the 6-acyl-monogalactosylDAG (structure **1′** (Fig. 3-1), and the 6-acyl-digalactosylDAG (structure **2′**) (Fig. 3-1), which may, however, be artifacts formed by reversible enzymatic acyl transfer from di- to monogalactosylDAG (see Section 3.8.1.4c).

Glycolipids not derived from diacylglycerol (DAG) are also present in plants: sterylglucoside (structure **5**) (Fig. 3-2), acylated sterylglucoside (structure **5′**) (Fig. 3-2), sterylcellobioside (**5″**), and sterylcellotrioside (**5‴**), and cerebrosides (structures **6–6‴** and **7–7‴**) (Fig. 3-3), as well as the more complex phytoglycolipid (structure **8**) (Fig. 3-4). Thus, plant glycolipids can be classified into three homologous series: glycosylDAGs, glycosylsterols, and glycosylceramides; each of these is covered in this chapter.

3.1.1. Nomenclature

In regard to the nomenclature of the glycolipids, the IUPAC–IUB (1978) system of nomenclature and abbreviations is followed in this chapter, as in Chapter 1 (*this volume*). For example, the plant monogalactosylDAG (**1**), 1,2-diacyl-3-β-D-galactopyranosyl-*sn*-glycerol, is abbreviated Gal*p*β1–3DAG; the digalactosylDAG (**2**), 1,2-diacyl-3(α-D-galactopyranosyl-1–6-β-D-galactopyranosyl)-*sn*-glycerol, is abbreviated Gal*p*α1–6Gal*p*β1–3DAG; and the plant sul-

Morris Kates • Department of Biochemistry, University of Ottawa, Ottawa, Ontario, Canada K1N 6N5.

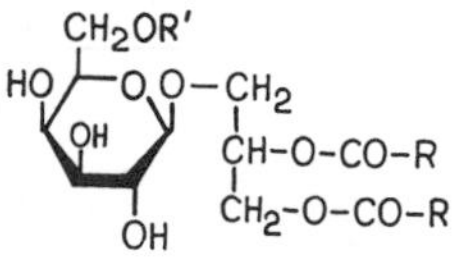

(1) R′ = H
(1′) R′ = acyl

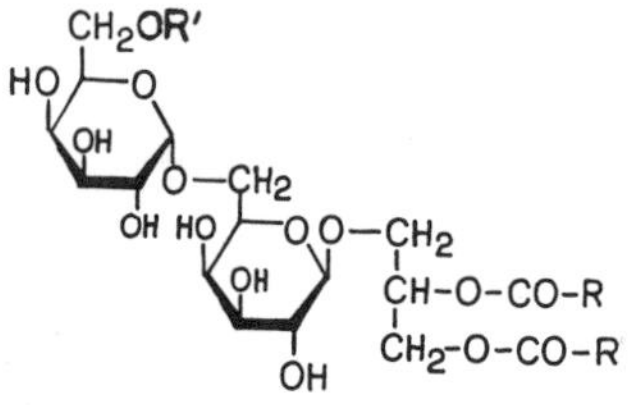

(2) R′ = H ; (2′) R′ = acyl

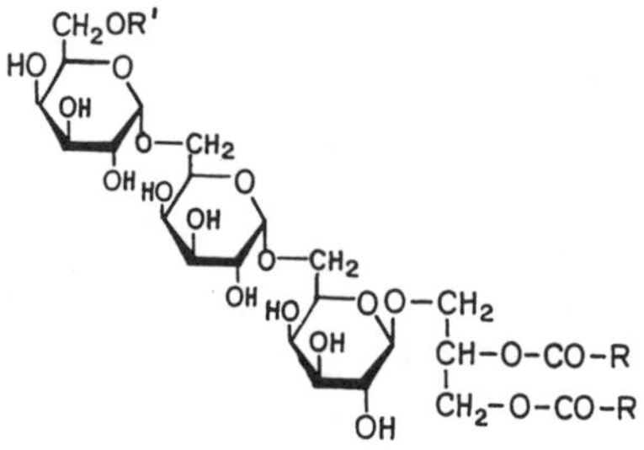

(3) R′ = H ; (3′)   R′ = Gal*p*α

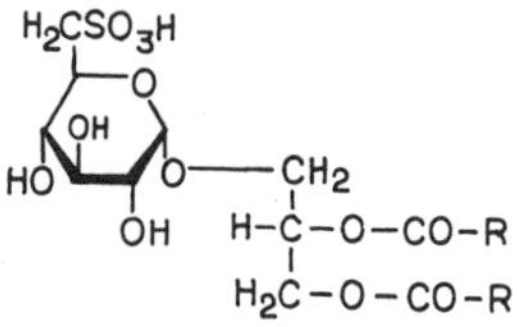

(4)

Figure 3-1. Structures of the major glycolipids of plants. (1) monogalactosyldiacylglycerol (1,2-diacyl-3-*O*-β-D-galactopyranosyl-*sn*-glycerol; Gal*p*β1–3DAG; GalDAG); (1′) acylmonogalactosyl-diacylglycerol (1,2-diacyl-3-*O*-[(6-0-acyl)-β-D-galactopyranosyl]-*sn*-glycerol; 6-acylGal*p*1–3βDAG; acylGalDAG); (2) digalactosyldiacylglycerol (1,2-diacyl-3-*O*-(α-D-galactopyranosyl-1-6-β-D-galac-topyranosyl)-*sn*-glycerol; Gal*p*α1–6Gal*p*β1–3DAG; Gal₂DAG; (2′) acyldigalactosyldiacylglycerol (1,2-diacyl-3-*O*-[(6-*O*-acyl)-α-D-galactopyranosyl-1–6-β-D-galactopyranosyl]-*sn*-glycerol; 6-acyl-Gal*p*α1–6Gal*p*β1–3DAG; acylGal₂DAG); (3) trigalactosyldiacylglycerol (1,2-diacyl-3-*O*-[(α-D-ga-lactopyranosyl-1-6)₂-β-D-galac topyranosyl]-*sn*-glycerol; (Gal*p*α1–6)₂Gal*p*β1–3DAG; Gal₃DAG; (3′) tetragalactosyldiacylglycerol (1,2-diacyl-3-*O*-[(α-D-galactopyranosyl-1–6)₃β-D-galactopyrano-syl]-*sn*-glycerol; (Gal*p*1–6)₃ Gal*p*β1–3DAG; Gal₄DAG). (4) sulfoquinovosyldiacylglycerol (1,2-di-acyl-3-0-α-D-6-sulfoquinovosyl-*sn*-glycerol; HSO₃Q*p*α1–3DAG; SQDAG).

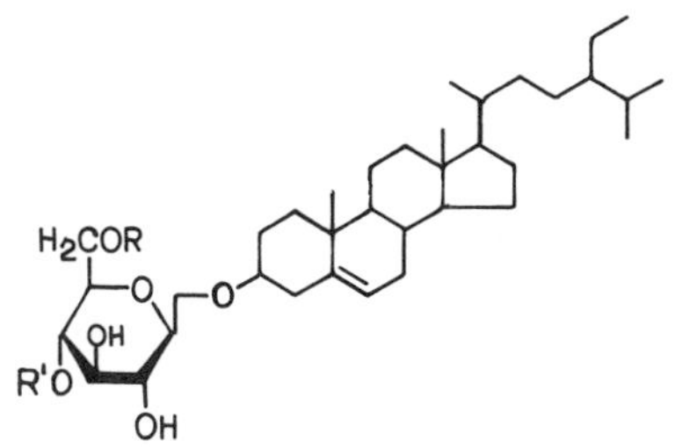

(5)   R = H ; R′= H
(5′) R = -COC$_{15}$H$_{31}$; R′= H
(5″) R = H ; R′= Glcpβ
(5‴) R = H ; R′= Glcpβ1–4Glcpβ

Figure 3-2. Structures of glycosyl and acylglycosyl sterols. (5) monoglucosyl sterol (β-D-glucopy-ranosyl-1–3-β-sitosterol); (5′) acylglucosyl sterol (6-*O*-acyl-β-D-glucopyranosyl-1–3-β-sitosterol); (5″) diglucosyl sterol (β-D-glucopyranosyl-1–4-β-D-glucopyranosyl-1–3-β-sitosterol; cellobiosyl-sitosterol). (5‴) triglucosylsterol (β-D-glucopyranosyl-1–4-β-D-glucopyranosyl-1–4-β-D-glucopy-ranosyl-1–3-β-sitosterol; cellotriosylsitosterol).

folipid (4), 1,2-diacyl-3-(6-sulfono-α-D-quinovosyl)-*sn*-glycerol, is abbreviated 6-HSO$_3$-Q*p*α1–3DAG.

3.1.2. Historical Background

The glycoglycerolipids were unknown until 1956, when Carter and col-leagues (1956) reported the isolation of a crude mixture of glycolipids from wheat flour later identified as mono- and digalactosylDAGs (structures 1 and 2) (Fig. 3-1) (Carter *et al.*, 1961*a,b*) (see Chapter 1, *this volume*). These "galac-tolipids" were subsequently shown to constitute a major class of membrane lipids in plants (see Table 3-6), being associated largely with photosynthetic tissues (Benson *et al.*, 1958, 1959; Wintermans, 1960). Higher homologues of galactolipids, e.g., the tri- and tetragalactosylDAG (structures 3 and 3′) (see Fig. 3-1), were later identified in both photosynthetic and nonphotosynthetic plant tissues (see review by Sastry, 1974).

About the same time, a sulfolipid, sulfoquinovosylDAG (structure 4) (see Fig. 3-1) was discovered in the alga *Chlorella* by Benson and colleagues (see Benson, 1963); this sulfoglycolipid proved to be a universal plant lipid compo-nent (Haines, 1973), which, together with the galactolipids, is involved in maintaining the structural integrity of the photosynthetic apparatus (Benson, 1963). The findings of Benson and his group greatly stimulated interest in the role of glycolipids in the structure and function of the thylakoid membrane, in their biosynthetic pathways, and in their distribution throughout the plant kingdom (see Table 3-6; see also reviews by Benson, 1963; Carter *et al.*, 1965; Kates, 1970; Hitchcock and Nichols, 1971; Haines, 1973, 1984; Mudd and Garcia, 1975; Sastry, 1974; Kenyon, 1978 and Ishizuka and Yamakawa, 1985), including yeasts and fungi (Brennan *et al.*, 1974; Wassef, 1977; Weete, 1974, 1980).

In addition to the galactolipids, Carter *et al.* (1960, 1961*a*) also reported

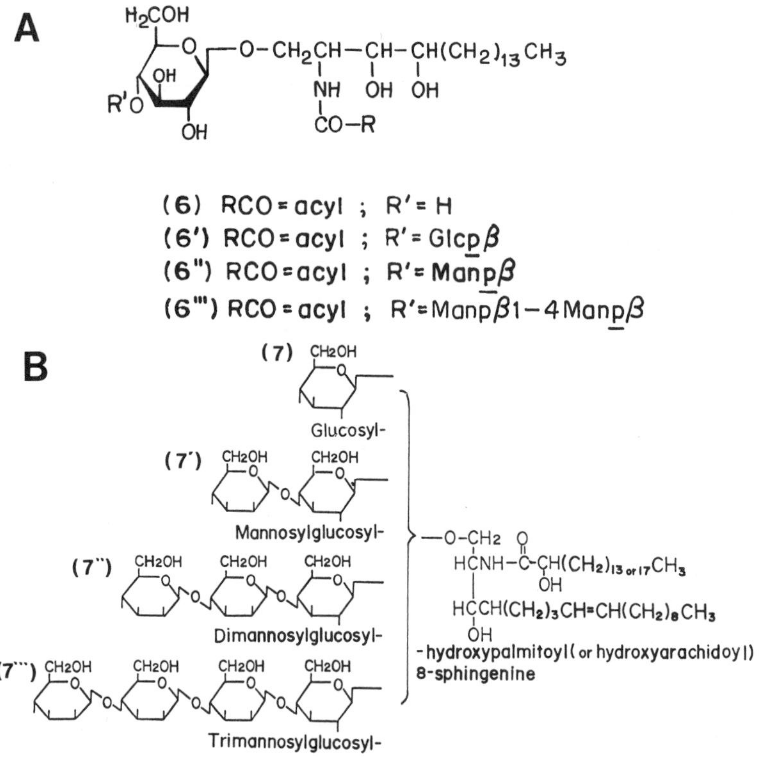

Figure 3-3. Structures of glycosylceramides. (A) Derived from phytosphingosine: (**6**) monoglu-cosylceramide (β-D-glucopyranosyl-1-1-*N*-acyl-phytosphingosine); (**6′**) diglucosylceramide (β-D-glucopyranosyl-1–4-β-D-glucopyranosyl-1-1-*N*-acylphytosphingosine; cellobiosylceramide); (**6″**) mannosylglucosylceramide (β-D-mannopyranosyl-1–4-β-D-glucopyranosyl-*N*-acylphytosphingo-sine); (**6‴**)′dimannosylglucosylceramide (β-D-mannopyranosyl-1–4-β-D-mannopyranosyl-β-D-glu-copyranosyl-*N*-acyl-phytosphingosine. (B) Derived from sphingosine: (**7**) monoglucosylceramide (β-D-glucopyranosyl-1-1-*N*-acyl-8-sphingenine); (**7′**) mannosylglucosylceramide (β-D-mannopy-ranosyl-1–4-β-D-glucopyranosyl-1-1-*N*-acyl-8-sphingenine); (**7″**) dimannosylglucosylceramide [(β-D-mannopyrannosyl-1–4)$_2$D-glucopyranosyl-1–1-*N*-acyl-8-sphingenine]; (**7‴**) trimannosyl-glucosylceramide [(β-D-mannopyranosyl-1–4)$_3$-β-D-glucopyranosyl-1–1-*N*-acyl-8-sphingenine].

the presence of cerebrosides (**6–6‴** and **7–7‴**) and sterylglycosides (**5–5‴**) in wheat flour. These components have also been found to be widely distributed among plants, including yeast and fungi (Kates, 1970; Weete, 1980; Harwood and Russell, 1984).

3.2. Lipid Extraction Procedures

3.2.1. Higher Plants

Lipid extraction of plant material, particularly photosynthetic tissue, re-quires special precautions to avoid the degradative action of phospholipases

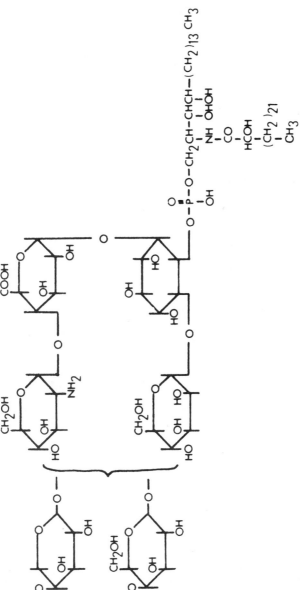

Figure 3-4. Partial structure of phytoglycolipid (**8**), from wheat flour and rice bran. (From Carter *et al.*, 1958*b*, 1969; Ito *et al.*, 1985*b*.)

and galactolipases, as well as to minimize peroxidation of the polyunsaturated fatty acid groups (Kates, 1970). A procedure that avoids these difficulties consists of fragmentation of fresh tissue under dry ice and homogenization of the powdered material in boiling isopropanol for 1–2 min in a Waring blender or Sorval Ominimixer; direct blending in isopropanol of leaves torn into small pieces is also effective (Kates 1970, 1986). The hot homogenate is filtered, and the isopropanol extract is concentrated *in vacuo* on a rotary evaporator; the residual crude lipids are taken up in chloroform–methanol (1 : 1, v/v) and partitioned into chloroform, following the modified procedure (Kates, 1986) of Bligh and Dyer (1959), by addition of 1% sodium chloride to give a mixture of chloroform–methanol–water (1 : 1 : 0.9, v/v), forming a two-phase system of chloroform and methanol–water (10 : 9, v/v). The lower chloroform phase is removed, then diluted with benzene and concentrated to dryness *in vacuo;* the residual total lipids (plus pigments) are immediately dissolved in chloroform–methanol (2 : 1, v/v) and stored at −10°C. Extraction of plant tissues may also be carried out with isopropanol at room temperature (Nichols, 1964; Galliard, 1968a) or with boiling 95% ethanol, followed by chloroform–methanol (Weenink, 1964).

Plant material that does not contain highly active lipolytic enzymes or in which these enzymes have been inactivated by dipping the tissue in boiling water for 1–2 min may be extracted by blending in chloroform–methanol (1 : 2, v/v), as follows (Kates, 1986): 100 g of fresh plant tissue (e.g., spinach leaves or carrot root) is torn or cut into small pieces and blended for 2 min with 300 ml of chloroform–methanol (1 : 2, v/v); the homogenate is filtered and the filter residue reblended with a mixture of 300 ml of chloroform–methanol (1 : 2) and 80 ml of water; the homogenate is filtered and the filter residue washed on the filter with 150 ml of chloroform–methanol (1 : 2). The combined filtrates, in a separatory funnel, are diluted with 250 ml of chloroform and 290 ml of water, mixed, and left overnight for the phases to separate. The lower chloroform layer is removed, diluted with benzene, and taken to dryness *in vacuo;* the residual lipids are dissolved in chloroform–methanol (2 : 1) and subjected to fractionation or stored at −10°C.

3.2.2. Algae, Diatoms, Yeasts, and Fungi

Algae, diatoms, yeast, or fungal cells are extracted by stirring a suspension of the cells (either fresh or freeze-dried) in water or 1% sodium chloride with chloroform–methanol in the ratio chloroform–methanol–water (1 : 2 : 0.8, v/v) at room temperature for 1 hr, followed by centrifugation or filtration to remove cell debris and partitioning of the lipids into chloroform as described for higher plants (Kates, 1986). Fungal mycelia and yeasts with rigid cell walls may also be extracted by the modified Bligh–Dyer procedure by homogenization in a Waring blendor (Jack, 1965) or sonicator (Wassef *et al.,* 1975) or Braun homogenizer (Itoh and Kaneko, 1974) with chloroform–methanol–water (1 : 2 : 0.9, v/v) as described for higher plants. To avoid the action of lipid-degrading enzymes, algae, diatoms, yeasts, and fungal cells may be extracted with hot isopropanol, followed by isopropanol–chloroform (1 : 1, v/v), as described elsewhere (Kates, 1986).

3.2.3. Seeds

Seeds (ground to a powder in a Wiley mill or Moulinex coffee grinder) are best extracted by vigorous shaking with hot isopropanol, followed by chloroform–methanol–water (1 : 2 : 0.8, v/v) (de la Roche *et al.*, 1973) or with hot 1-butanol saturated with water (Osagie *et al.*, 1984); 1-propanol is also an effective extracting solvent for certain seed material (Hoover *et al.*, 1988). After concentration of the extracts to dryness *in vacuo*, the lipid residues are freed from water-soluble impurities by extraction with chloroform–methanol–water (1 : 2 : 0.8, v/v), followed by formation of a biphasic system, i.e., chloroform–methanol–water (1 : 1 : 0.9, v/v) by addition of appropriate amounts of chloroform and water (Kates, 1986). The lower chloroform phase is diluted with benzene and brought to dryness *in vacuo*, the residual lipids being dissolved in chloroform–methanol (1 : 1, v/v), cleared by centrifugation if necessary, and stored at −10°C.

3.3. Glycolipid Isolation Procedures

Separation and isolation of glyco- and sulfoglycoDAG from total lipids of plants, yeasts, and fungi can be carried out essentially as described in Chapter 1 (Section 1.2.1.2) for bacteria. The presence of pigments (chlorophylls and carotenoids) may cause some difficulties in obtaining pure glycolipids, particularly with plant lipids, but these pigments can usually be removed either by a solvent partitioning step between petroleum ether and 95% methanol–water (the pigments partition largely into the petroleum ether phase), or by silicic acid column or thin-layer chromatography (TLC), or by a combination of these methods (Kates, 1986).

A convenient fractionation procedure is as follows. The total lipids (1 g/80 g silicic acid) in chloroform are applied to a column of Bio-Sil silicic acid (100 to 200-mesh) prepared in chloroform, and the column eluted with chloroform (10-column volumes, neutral lipid fraction), acetone (40-column volumes, glycolipid fraction), and methanol (10-column volumes, phospholipid fraction) to obtain the neutral lipids (e.g., triacylglycerols, pigments, sterols), glycolipids (galactolipids, sulfolipid, cerebrosides, sterolglycosides), and phospholipids (e.g., phosphatidylcholine, phosphatidylglycerol), respectively. The acetone eluates (glycolipid fraction) may then be further fractionated into individual components (mono- and digalactosylDAGs, sulfolipid, cerebrosides, sterylglycosides) by preparative TLC in one dimension, e.g., chloroform–methanol–28% ammonium hydroxide (65 : 35 : 5, v/v) or two dimensions, e.g., chloroform–methanol–28% ammonium hydroxide (65 : 35 : 5), then chloroform–acetone–methanol–acetic acid (10 : 4 : 2 : 2 : 1) (Kates, 1986) or by chromatography on DEAE cellulose (Rouser *et al.*, 1976). Preparative one-dimensional TLC on 1-mm-thick layers of silica gel G in acetone–acetic acid–water (100 : 2 : 1, v/v) is also effective in final purification of GalDAG and Gal₂DAG (Gardner, 1968).

Final purification of mono- and digalactosylDAG is best achieved by high-performance liquid chromatography (HPLC) on silica columns (Marion *et al.*, 1984; Demandre *et al.*, 1985). Each glycolipid can be further resolved into molecular species by reverse-phase HPLC (RP–HPLC) (Lynch *et al.*, 1983; Demandre *et al.*, 1985; Kesselmeier and Heinz, 1985); argentation TLC (Nichols and Moorehouse, 1969) has also been used for separation of molecular species of galactolipids (see Chapter 1, Section 1.2.1.2).

3.4. Characterization and Structure Analysis of Glycolipids

As was described for the bacterial glycosylDAGs (see Chapter 1, Section 1.2.2.1), the plant glycolipids may be characterized by TLC mobilities in several solvent systems (Table 3-1) and staining behavior with diagnostic reagents such as α-naphthol or periodic acid–Schiff, by spectral analysis, e.g., infrared (IR), nuclear magnetic resonance (NMR), and mass spectrometry (MS), and by analysis of molecular constituents.

3.4.1. Thin-Layer Chromatography

Thin-layer chromatography mobilities of plant galactolipids, sulfolipid, cerebrosides, and sterylglycosides in a few solvent systems are presented in Table 3-1. All the major plant glycolipids (**1, 2, 4**) and the minor galactolipids (**3, 3′**) are clearly well resolved by one-dimensional TLC and can be distinguished by their characteristic mobilities (Table 3-1). However, some of the minor glycolipids (**5, 5′, 6**) may not be well resolved from galactolipids by one-dimensional TLC in some solvent systems, such as solvent C (Table 3-1). Thus, the separation of complex natural mixtures of plant lipids is best carried out by two-dimensional TLC in suitable solvent pairs such as solvent systems A and B (Table 3-1). Examples of the separation of plant lipid mixtures are shown in Fig. 3-5A (soybean cell cultures), Fig. 3-5B (*Dunaliella* spp.), and Fig. 3-5C (*Nitzschia alba*).

3.4.2. Spectral Analysis

In regard to spectral analysis of the plant glycosylDAGs, IR spectra (Fig. 3-6), which are very similar to those of bacterial glycosylDAGs (see Chapter 1, Fig. 1-6), help confirm the basic structure of glycosylDAGs, such as the presence of ester linkages, sugar groups, sulfate, sulfonate, or phosphate groups. NMR spectra (Fig. 3-7A–E; see also Chapter 1, Fig. 1-6) provide more detailed structural information, particularly concerning the configuration of the anomeric carbon atoms (see also Chapter 1, Table 1-6, and Section 1.2.2.6).

Chemical ionization (CI) and fast-atom bombardment (FAB) mass spectrometry not only provide the molecular weight of the glycolipid but can also

Table 3-1. Thin-Layer Chromatography Mobilities of Plant Glycolipids

Glycolipid	Solvent[a] ($R_f \times 100$)								
	A	B	C	D[b]	E[c]	E'[f]	F[d]	G[j]	H[j]
GlycosylDAGs									
Galpβ monoacylGro	—	—	—	—	46	—	12	—	—
Galpβ DAG (**1**)	84[e,g]	84[e,g]	77[b] 82[c]	51	—	84	36	64	92
6-AcylGalpβ DAG (**1'**)	—	—	—	—	—	90	64	—	—
Galpα1–6GalpβDAG (**2**)	30[e,g]	38[e,g]	62[b] 54[c']	25	41	52	7	23	53
6-AcylGalpα1–6GalpDAG (**2'**)	—	—	—	—	—	83	24	—	—
[Galpα1–6]₂GalpβDAG (**3**)	—	—	32[c'] 38[c]	—	—	26	—	—	—
[Galpα1–6]₃GalpβDAG (**3'**)	—	—	15[c'] 21[c]	—	—	10	—	—	—
HSO₃-QpαDAG (**4**)	51[g] 46	37[g] 32	42[b,c]	22	—	38	—	41	43
Glycosylsterols									
Monoglycosylsterol (**5**)	64[e]	68[e]	70[c],73[b] 75[h]	48	—	—	—	—	—
Acyl–monoglycosylsterol (**5'**)	90[e]	80[e]	78[b]	65	—	—	44	—	—
Diglycosylsterol (**5''**)	—	—	53[h] 57	—	—	—	—	—	—
Triglycosylsterol (**5'''**)	—	—	29[h]	—	—	—	—	—	—
Glycosylceramides			62[c]						
Ceramidemonohexoside (**6**)	62[e]	58[e]	66[h]	—	43[i]	—	—	—	—
Ceramidedihexoside (**6'**)	—	—	42[h]	—	—	—	—	—	—
Ceramidetrihexoside (**6''**)	—	—	19[h]	—	—	—	—	—	—

[a]Solvents: A, chloroform–methanol–28% ammonium hydroxide (64 : 35 : 5, v/v) (Rouser *et al.*, 1976); B, chloroform–acetone–methanol–acetic acid–water (10 : 4 : 2 : 2 : 1, v/v) (Rouser *et al.*, 1976; Evans *et al.*, 1982); C, chloroform–methanol–water (65 : 25 : 4, v/v) (Lepage, 1964*a;* Rouser *et al.*, 1976); D, diisobutylketone–acetic acid–water (40 : 25 : 5, v/v) (Lepage, 1964*a;* Nichols, 1964); E, chloroform–methanol–water (65 : 16 : 2, v/v) (Fujino and Miyazawa, 1979); E', chloroform–methanol–water (70 : 30 : 4, v/v) (Siebertz *et al.*, 1979); F, chloroform–methanol (85 : 15, v/v) (Heinz *et al.*, 1974); G, chloroform–methanol–acetone–diethylamine–water (120 : 35 : 37.5 : 6 : 4.5, v/v) (Evans *et al.*, 1982); H, chloroform–90% acetic acid–methanol, 30 : 20 : 4, v/v) (Evans *et al.*, 1982).
[b]Lepage (1964*a*)
[c]Fujino and Miyazawa (1979); [c']Ito *et al.* (1985*a*).
[d]Heinz *et al.* (1974).
[e]Wilson *et al.* (1978).
[f]Siebertz *et al.* (1979).
[g]Anderson *et al.* (1978*a,b*).
[h]Fujino and Ohnishi (1979*a,b*).
[i]Ito and Fujino (1973).
[j]Evans *et al.* (1982).

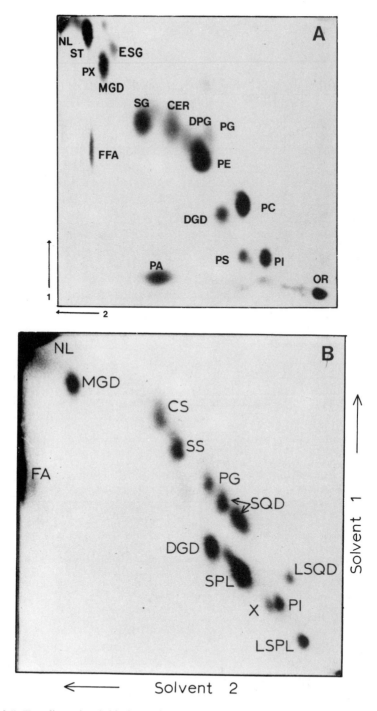

Figure 3-5. Two-dimensional thin-layer chromatograms of total lipids. (A) Soybean cell suspension culture in solvent systems (1) chloroform–methanol–conc. ammonia (65 : 35 : 5, v/v) and (2) chloroform–acetone–methanol–acetic acid–water (10 : 4 : 2 : 2 : 1, v/v). (From Wilson *et al.*, 1978.) (B) Nonphotosynthetic diatom *Nitzschia alba* in solvent systems as in A. (From Anderson *et al.*,

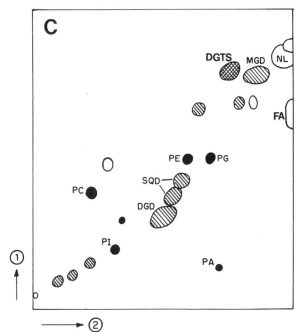

1978a.) (C) Halophilic *Dunaliella* species in solvent systems, (1) chloroform–methanol–conc. ammonia (65 : 25 : 5, v/v) followed by chloroform–methanol–conc. ammonia (65 : 35 : 5, v/v) and (2) chloroform–acetone–methanol–acetic acid–water (20 : 8 : 4 : 4 : 1, v/v); solid spots are phosphate-positive, hatched spots are sugar-positive, double-hatched spots are Dragendorff-positive, and open spots are nonstaining with these reagents. (From Evans and Kates, 1984.) PA, phosphatidic acid; PC, phosphatidylcholine; PE, phosphatidylethanolamine; PG, phosphatidylglycerol; PS, phosphatidylserine; DPG, diphosphatidylglycerol; PI, phosphatidylinositol; DGTS, diacylglyceryltrimethylhomoserine; MGD, monogalactosyldiacylglycerol; DGD, digalactosyldiacylglycerol; SQD, sulfoquinovosyldiacylglycerol; CS, ceramidesulfonic acid; SPL, phosphatidylsulfocholine; LSPL, lysophosphatidylsulfocholine; CER, cerebrosides; SG, sterylglycoside; ESG, esterified sterylglycoside; ST, sterols; SS, sterylsulfate; FA or FFA, fatty acids; NL, neutral lipids; O or OR, origin.

establish the sequence of sugars in the carbohydrate moiety as well as the molecular species composition (Budzikiewicz *et al.*, 1973; Rullkötter *et al.*, 1975) (see Chapter 1, Section 1.2.2.7). Mass spectra of mono-, di-, tri-, and tetragalactosylDAG derivatives from plants are shown in Fig. 3-8A–D.

3.4.3. Analysis of Molecular Constituents

Characterization of the plant glycosylDAGs by analysis of molecular constituents such as sugars, glycerol, ester groups, acyl groups, and the deacylated glycosylGros is carried out as described for the bacterial glycosylDAGs (see Chapter 1, Sections 1.2.2.1c, 1.2.2.2, and 1.2.2.3).

3.4.4. Molecular Species Analysis

Molecular species of glycosylDAGs can be separated by TLC on AgNO$_3$/SiO$_2$ plates into saturated, mono-, di-, tri-, tetra-, penta-, and hex-

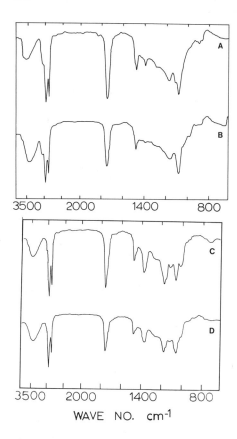

Figure 3-6. Infrared spectra (in KBr). (A) Monogalactosyldiacylglycerol (GalDAG). (B) Digalactosyldiacylglycerol (Gal$_2$DAG). (C) Sulfoquinovosyldiacylglycerol methyl ester (MeSQDAG). (D) Lysosulfoquinovosyl diacylglycerol methyl ester (Me-*lyso*SQDAG). (From R. Anderson and M. Kates, unpublished data.)

aenoic (or higher) classes (Nichols and Moorehouse, 1969). Similar separations into classes of molecular species can be more conveniently obtained by HPLC on RP-HPLC columns (Lynch *et al.*, 1983; Demandre *et al.*, 1985; Kesselmeier and Heinz, 1985). To determine the positional distribution of the fatty acids between the *sn*-1 and -2 positions, each separated class of molecular species is subjected to enzymatic hydrolysis by pancreatic lipase (Noda and Fujiwara, 1967; Safford and Nichols, 1970; Auling *et al.*, 1971) or by the lipase from *Rhizopus arrhizus delemar* (Fischer *et al.*, 1973; Tulloch *et al.*, 1973; Rullkötter *et al.*, 1975), which specifically hydrolyzes only the *sn*-1 fatty acid ester linkage (Fischer *et al.*, 1973). After separation of the liberated free acids and the resulting lyso compound by TLC, the fatty acids in each product are analyzed by gas–liquid chromatography (GLC) using an internal standard for quantitation. The molecular species composition is calculated from the total, *sn*-1 and *sn*-2 fatty acid compositional data (Kesselmeier and Heinz, 1985).

Molecular species analysis may also be carried out by mass spectrometry (electron impact or chemical ionization) (Rullkötter *et al.*, 1975), which indicates ion peaks corresponding to the individual diacylglycerol species present (Figs. 3-8 and 3-9). It should be noted that this method gives reliable qualitative analyses of molecular species but is not quantitatively accurate (Rullkötter *et al.*, 1975).

3.4.5. Identification of Deacylated Glycolipids

Mild alkaline deacylation of glycosylDAG and identification of the glycosylGro is an important procedure in the identification of glycosylDAG (see Chapter 1, Section 1.2.2.3). This procedure, which is based on the alkaline methanolysis procedure of Dawson (1954), was first introduced by Benson and Maruo (1958) for the identification of plant phospholipids and later used to identify the plant galactolipids and sulfolipid (Ferrari and Benson, 1961).

Total plant lipids are subjected to mild alkaline deacylation (as described in Chapter 1, Section 1.2.2.3) to yield chloroform-soluble products (fatty acid methyl esters and alkali-stable glycolipids) and the water-soluble deacylation products in the methanol–water phase of the reaction mixture. The latter are separated by two-dimensional paper chromatography and identified by their R_f values (Table 3-2) in comparison with authentic standards. Figure 3-10 shows a typical two-dimensional paper chromatograph (autoradiograph) indicating the clear separation of mono-GalGro, di-GalGro, and sulfoquinovosylGro from other deacylated lipids (e.g., phospholipids, acylglycerols). GlycosylGros can also be separated by GLC of their TMS ethers (Fujino and Miyazawa, 1979).

The minor plant glycolipids, sterolglycosides, and cerebrosides, are stable to alkaline methanolysis and are obtained together with fatty acid methyl esters as chloroform-soluble products of the deacylation reaction; the esterified sterolglycosides, however, are deacylated to the corresponding sterolglycosides. These chloroform-soluble alkali-stable glycolipids can then be separated and identified by TLC in a suitable solvent system (see Table 3-1). The fatty acid methyl esters are identified by GLC (see Kates, 1986).

3.4.6. Identification of Carbohydrate Structure and Configuration

Procedures for the structure determination of the carbohydrate moieties of the plant galactolipids and sulfolipid are essentially the same as described for the eubacterial glycolipids (see Chapter 1, Section 1.2.2), as follows: identification of sugar groups (Section 1.2.2.2); determination of sugar sequences (Section 1.2.2.4); determination of sugar linkage positions (Section 1.2.2.5), and anomeric configuration (Section 1.2.2.6). The molecular rotation of the intact glycosylDAG or glycosylGro (Table 3-3) can also be used to establish the anomeric configuration of the sugar groups (see Section 1.2.2.6).

For the plant sulfolipid, the position of the sulfonic acid group is determined by periodate/bromine oxidation of the free sulfosugar or the sulfosugar glycerolglycoside, which gives sulfoacetic acid and sulfolactic acid, respectively (Benson, 1963).

3.4.7. Identification of Nonpolar Moieties of Glycosylsterols and Glycosylceramides

The alkali-stable glycosylsterols and glycosyl ceramides in the chloroform-soluble products of mild alkaline degradation are isolated by a com-

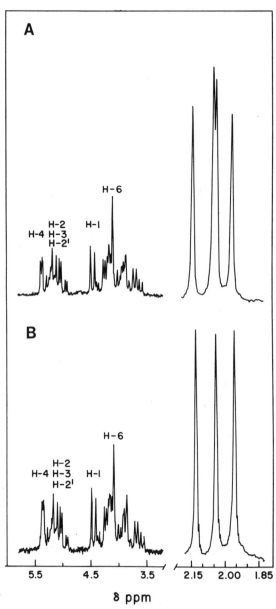

Figure 3-7. ^{1}H-NMR spectra (in deuterochloroform) of peracetylated derivatives of glycosyl-DAGs from spinach leaves. (A) GalDAG (**1**). (B) 6-Acyl-GalDAG (**1′**). (C) Gal$_2$DAG (**2**). (D) 6-Acyl-Gal$_2$DAG (**2′**). (E) SQDAG-Me-ester (**4**, methyl ester). (from Heinz and Tulloch, 1969; Heinz *et al.*, 1974; and Tulloch *et al.*, 1973, respectively.)

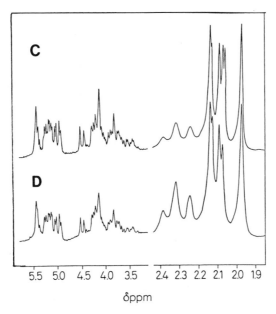

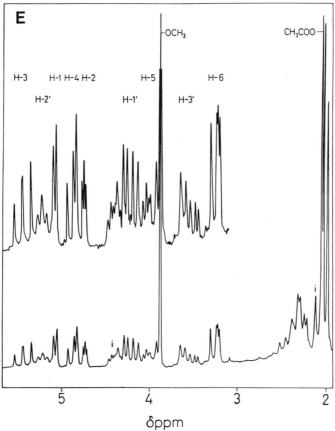

Figure 3-7. (*continued*)

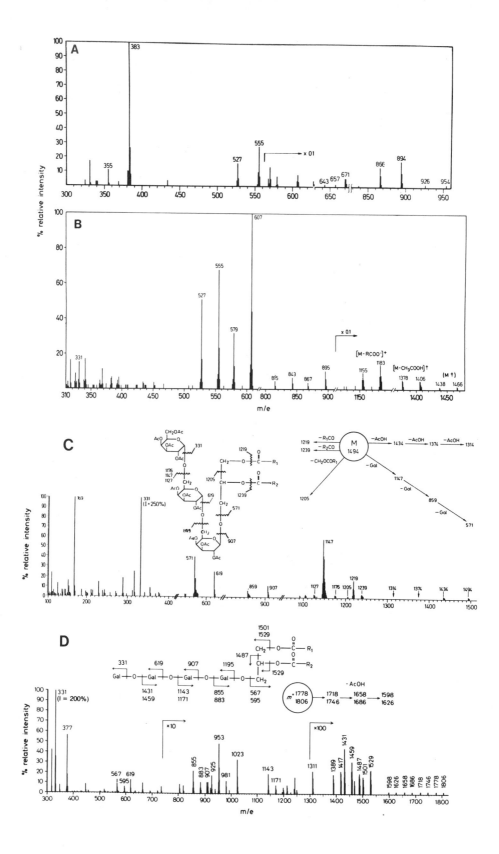

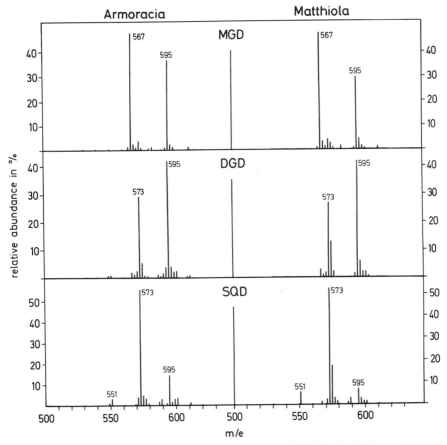

Figure 3-9. Molecular species analysis by mass spectrometry of GalDAG, Gal$_2$DAG, and SQDAG from leaves of *Armoracia rusticana* and *Matthiola incana* (angiosperms). Samples were peracetylated but not hydrogenated. Molecular species: 18:3/18:3, m/e 595; 18:3/16:3, m/e 567; 18:3/16:0, m/e 573; 16:0/16:0, m/e 551. (From Rullkötter *et al.*, 1975.)

bination of silicic acid column chromatography and TLC (Fujino and Ohnishi, 1979a,b; Fujino *et al.*, 1985a,b); their nonpolar moieties, free sterols, and long-chain bases + fatty acid methyl esters, respectively, are obtained after hydrolysis with methanolic 1 M HCl (Kates, 1986). The free sterols are analyzed and identified by AgNO$_3$–SiO$_2$ TLC and by GLC (Table 3-4).

Long-chain bases are separated by TLC in chloroform–methanol–2 M ammonium hydroxide (40:10:1, v/v) into sphinganine, sphingenine, and 4-hydroxysphinganine components (Ito and Fujino, 1973; Fujino and Ohnishi,

Figure 3-8. Mass spectra (electron impact) of peracetylated derivatives of glycosylDAGs. (A) 6-Acyl-Gal-monoacylglycerol (hydrogenated, from spinach leaves; spectrum from Critchley and Heinz, 1973). (B) 6-Acyl-Gal$_2$DAG (hydrogenated, from spinach leaves; spectrum from Heinz *et al.*, 1974.) (C) Gal$_3$DAG (from *Tolypothrix*; spectrum from Zepke *et al.*, 1978.) (D) Gal$_4$DAG (from spinach chloroplasts; spectrum from Siebertz *et al.*, 1979). Samples were introduced into the mass spectrometer (Varian-MAT 731) via direct inlet probe.

Table 3-2. Paper Chromatographic Mobilities of Deacylated Plant Glycolipids (Glycosylglycerols)

Glycosylglycerol	Parent glycosylDAG	Solvent[a] ($R_f \times 100$)				
		A	B	C	D	E
GalpβGro	(**1, 1'**)	62[b]	25[b]	30	26	46
		65[c]	24[c]	—	—	—
Galpα1–6Galpβ Gro	(**2, 2'**)	46[b]	13[b]	9	7	23
		50[c]	10[c]	—	—	—
(Galpα1–6)$_2$Galpβ Gro	(**3**)	38[b]	4[b]	—	—	—
		35[c]	4[c]	—	—	—
HSO$_3$·Qpα Gro	(**4**)	18[b]	8[b]	4	12	—

[a]Solvents: A, phenol–water (100 : 38 or 5 : 2, w/w) (Ferrari and Benson, 1961); B, butanol–propionic acid–water (15 : 7 : 10, v/v) (Ferrari and Benson, 1961); C, butanol–acetic acid–water (5 : 3 : 1, v/v) (Kates, 1960, 1986); D, pyridine–ethyl acetate–water (2 : 5 : 5, upper phase) (Sastry and Kates, 1964a; Kates, 1986); E, pyridine-1–butanol–water (3 : 2 : 1.5, v/v) (Carter *et al.*, 1956).
[b]Ferrari and Benson, 1961.
[c]Galliard, 1969.

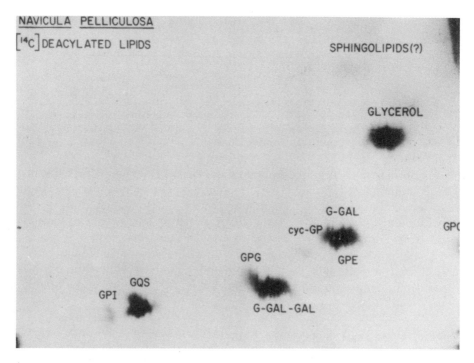

Figure 3-10. Autoradiograph of two-dimensional paper chromatogram of deacylated ^{14}C-total lipids of *Navicula pelliculosa*; solvent systems: in *x* direction, phenol–water (100 : 38, w/v); in *y* direction, butanol–propionic acid–water (142 : 71 : 100 v/v). (From Kates and Volcani, 1966.)

Table 3-3. *Molecular Rotations and Melting Points for Glycosylglycerols*

Glycosyl compound	Melting point[a] (°C)	$[\alpha]_D^b$ (degrees)	M_D^c (degrees)	References
GlycosylGros				
Galpβ1–3Gro	139–140	+3.77	+9.6	Carter et al.
Natural	139.5–142			(1956, 1961b)
	141.5	−7.7	−19.5	Heinz (1967a)
	137	−8.12	−20.6	Heinz (1971)
Synthetic	140.5–141.5 (corr)	−7	−17.8	Wickberg (1958a)
Galpα1–6Galpβ1–3Gro	182–184	+86.4	+359	Carter et al.
Natural				(1956)
	187–188	+88.8	+368	Heinz (1971)
	186–187	+86.9	+361	Heinz et al. (1974)
Synthetic				Wickberg (1958a)
HSO₃·Qpα1–3Gro[d]				
Natural	191–192	+74.5	+310	Lepage et al. (1961); Benson (1963)
Synthetic	191–193	—	—	Miyano and Benson (1962a)
GlycosylDAGs				
Galpβ1–3DAG (**1**)	151.5–153.5			Heinz (1971)
Hydrogenated	(91–92.5)	−2.28	—	
	162–163 (100–102)	−1.97	—	Heinz and Tulloch (1969)
6-Acyl–Galpα1–3DAG (**1′**)				
Hydrogenated	79 (67–69)	−1.30	—	Heinz and Tulloch (1969)
Galpα1–6Galpβ1–3DAG (**2**)			—	
Hydrogenated		+37.8		Heinz (1971)
6-Acyl-Galpα1–6Galpβ-1–3DAG (**2′**) Hydrogenated		+31.1	—	Heinz et al. (1974)
HSO₃·Qpα1–3DAG (**4**) (*Sinapis*) Free acid		+33.2		Tulloch et al. (1973)
Methyl ester		+45.2		Tulloch et al. (1973)

[a]Melting points are uncorrected unless otherwise indicated; values in parentheses are sintering points or transition temperatures.
[b]Specific rotation (degrees) in aqueous solution for glycosylGros; in pyridine solution for glycosylDAGs.
[c]$M_D = [\alpha]_D \times$ mol. wt/100.
[d]Cyclohexylamine salt.

Table 3-4. Thin-Layer Chromatographic and Gas–Liquid Chromatographic Properties of Plant and Fungal Sterols

Sterol	AgNO$_3$–SiO$_2$ TLC^a, R_f of acetate	1% SE-30^b acetate	GLC, Relative retention		1.5% OV-17^d acetate
			3% SE-30^c free sterol	acetate	
Cholesterol	0.52	1.00	1.00	1.00	1.00
Cholestanol	0.62	—	—	—	—
Brassicasterol	0.46	—	1.14	1.15	—
Tetrahydrobrassicasterol	0.61	—	—	1.39	—
Campesterol	0.58	—	1.36	1.38	1.33
Dihydrocampesterol	0.61	—	—	1.39	—
24-Methylenecholesterol	0.17	—	—	1.31	—
Stigmasterol	—	—	1.48	1.51	1.44
β-Sitosterol	—	—	1.79	1.81	1.64
Fucosterol	—	—	—	—	1.74
Zymosterol	—	1.13	—	—	—
Ergosterol	—	1.22	—	—	—
Episterol	—	1.43	—	—	—
Lanosterol	—	1.58	—	—	—
Δ^5-Avenasterol	—	—	—	—	1.84
Δ^7-Avenasterol	—	—	—	—	2.16
Δ^7-Stigmastenol	—	—	—	—	1.93

^{a}On 10% AgNO$_3$–silica gel H plates in solvent system CHCl$_3$–MeOH (99.6 : 0.4, v/v) (Kates *et al.*, 1978).
^{b}On a column of 1% SE-30 on Gas-Chrom Q (100–200 mesh) at 210°C (Työrinoja *et al.*, 1974).
^{c}On a column (46 × 0.6 cm) of 3% SE-30 on chromasorb W at 190°C and flow rate 110 ml/min; free sterol values are relative to cholesterol (ret. time 20.9 min); sterol acetate values are relative to cholesterol acetate (ret. time 29.2 min) (Kates *et al.*, 1978).
^{d}On 1.5% OV-17 on chromasorb W(AW–DMCS) or Gas Chrome-Z at 250°C (Kuroda *et al.*, 1977; Itoh *et al.*, 1973).

1983; Laine and Renkonen, 1973); each component (as the *N*-, *O*-acetyl derivatives) is further analyzed by silver nitrate–silica gel TLC according to the number and configuration of the double bonds (Laine and Renkonen, 1973) (Table 3-5). The chain type is then determined by GLC and GLC–MS analysis of the *N*-acetyl-TMS or TMS derivatives and by analysis of the aldehydes formed by periodate oxidation of the free base (Fujino and Ohnishi, 1976; Kates, 1986) (Table 3-5). Positioning of the double bonds is determined by GLC and GC–MS analysis of the fatty acids produced by periodate–permanganate oxidation of the free bases (Laine and Renkonen, 1976, 1974; Fujino and Ohnishi, 1976; Kates, 1986).

3.5. Structure Identification

3.5.1. Mono- and DigalactosylDAG

Mono- and digalactosylDAG were first isolated from the benzene extract of both bleached and unbleached wheat flour by Carter *et al.* (1956, 1961*a,b*) by a combination of solvent fractionation and silicic acid column chro-

Table 3-5. *Chromatographic Properties of Long-Chain Bases and Their Periodate Oxidation Products*

| Long-chain base | Chain type | TLC, R_f | | | | GLC, Rel. ret. | | | |
| | | Silica gel G[a] | | AgNO$_3$/SiO$_2$[b] | | TMS-N-acetyl bases[c] | TMS bases[d] | Aldehyde (free) | |
		(1)	(2)	(1)	(2)			Chain type	on DEGS[e]
Sphinganine (dihydrosphingosine)	18 : 0	0.54	0.66	1.00	1.00	1.00	1.00	16 : 0	1.00
trans-4-Sphingenine(sphingosine)	18 : 1	0.63	0.81	1.00	0.94	0.91	0.88	16 : 1(*trans*-2)	1.92
trans or *cis*-8-Sphingenine	18 : 1	0.54	0.66	0.66	0.76	0.93, 0.90	—	16 : 1(*trans* or *cis*-6)	1.15
trans-4,8-Sphingadienine	18 : 2	0.65	0.81	0.66	0.66	0.86	—	16 : 2(*trans*-2,6)	2.07
trans-4,*cis*-8-Sphingadienine	18 : 2	0.63	0.81	0.66	0.66	0.82	—	16 : 2(*trans*-2-*cis*-6)	2.14
4-Hydroxysphinganine (phytosphingosine)	18 : 0	0.37	0.44	—	—	—	1.64	15 : 0	0.66
4-Hydroxy-*trans*-8-sphingenine (dehydrophytosphingosine)	18 : 1	0.37	0.44	—	—	—	1.46	15 : 1(*trans*-5)	0.75
4-Hydroxy-*cis*-8-sphingenine	18 : 1	—	—	—	—	—	—	15 : 1(*cis*-5)	0.75

[a] Thin-layer chromatography (TLC) of the free bases on silica gel G analytical plates in solvent systesm: (1) chloroform–methanol–2N NH$_4$OH (40 : 10 : 1, v/v) (Ito and Fujino, 1973); (2) chloroform–methanol–7 N NH$_4$OH (60 : 20 : 3, v/v) (Laine and Renkonen, 1973).

[b] TLC of (1) the N-O-acetyl bases, and (2) the N-acetyl bases on 10% AgNO$_3$–SiO$_2$ plates in solvent system chloroform–methanol (98 : 2, v/v), relative to sphinganine (Laine and Renkonen, 1973; Renkonen and Hirvisalo, 1969).

[c] Gas–liquid chromatography (GLC) of N-acetyl–TMS bases on 4m column of 3% SE-30 at 220°C (Laine and Renkonen, 1973).

[d] GLC of TMS bases on a column (0.3 × 200 cm) of 5% SE-30 at 220°C (Ito and Fujino, 1973).

[e] GLC of free aldehydes, obtained by NaIO$_4$ oxidation, on 5% DEGS at 150°C (Ohnishi et al., 1983).

matography. It should be noted that the glycolipids from bleached flour contained substantial amounts of chlorine, and this source was not used in subsequent studies of wheat flour glycolipids. Structural studies of these galactolipids were begun with the deacylated glycolipids, mono- and digalactosylglycerols, respectively, obtained in pure crystalline form (Carter *et al.,* 1956). Both glycosides yielded only galactose and glycerol on acid hydrolysis, identified by paper chromatography; galactose was further identified as its α-methylphenylhydrazone, with melting point (189–191°C alone and mixed) and analytical data identical to those of the authentic galactose derivative.

Periodate oxidation of monogalactosylGro gave 1 mole each of formaldehyde and formic acid with an uptake of 3 moles of periodate, consistent with a galactopyranosyl-1-glycerol structure; with the digalactosylGro, 5 moles of periodate was consumed with formation of 2 moles of formic acid and one of formaldehyde, consistent with a galactopyranosyl-1,6-galactopyranosyl-1-glycerol structure. Hydrolysis of the monogalactosylGro with β-galactosidase, but not with α-galactosidase, established the presence of a β-galactoside linkage. Treatment with α-galactosidase cleaved the terminal galactose in digalactosylGro and formed monogalactosylGro, which was then cleaved by β-galactosidase. These results established the presence of an α-linked terminal galactose and a galactose β-linked to glycerol. These assignments were confirmed by the low specific rotation (+3.77°; reported for synthetic GalpβGro, −7°, Wickberg, 1958*b*) of the monogalactosylGro and the relatively high specific rotation (+86°; reported for synthetic Galpα1–6GalpβGro, +86°, Wickberg, 1958*a*) of the digalactosylGro (Carter *et al.,* 1956) (see Table 3-3; see also Chapter 1, Table 1-5). Finally, the mono- and digalactosylGros were found to be identical with respect to melting points, and IR spectra, as well as optical rotations, to the respective synthetic mono- and di-D-galactosyl-D-glycerols (Wickberg, 1958*a,b*).

It remained to determine the location of the acyl group(s); this was done by permethylation of the intact galactosylDAGs followed by alkaline hydrolysis yielding the corresponding methylated galactosylGros, which consumed 1 mole of periodate with formation of formaldehyde (Carter *et al.,* 1961*b*). The acyl groups were thus located on the 1- and 2-positions of the glycerol moiety, and not on the sugar group. Acid hydrolysis of the methylated monoGalGro gave 2,3,4,6-tetramethylGal plus glycerol, while the methylated diGalGro also gave 2,3,4-trimethylGal. These data thus establish the monoGalDAG as being 1,2-diacyl-3-β-D-galactopyranosyl-*sn*-glycerol (structure **1**, Fig. 3-1) and the diGalDAG as 1,2-diacyl-3-(α-D-galactopyranosyl-1-6-β-D-galactopyranosyl)-*sn*-glycerol (structure **2**, Fig. 3-1) (Carter *et al.,* 1961*b*). Structures (**1**) and (**2**) were confirmed in rice bran, but a GlcpβDAG analogue of (**1**) was also identified (Fujino and Miyazawa, 1979).

Acylmono- and diGalDAGs have been identified as artifacts in leaf homogenates (Heinz, 1967*a,b;* Heinz and Tulloch, 1969; Heinz *et al.,* 1974) (see Section 3.1.1). In both galactolipids, the acyl group has been shown by NMR (see Fig. 3-7B,C) and mass spectrometry (see Fig. 3-8B), and permethylation analysis to be located only on the 6-position of the terminal sugar (structures **1′** and **2′**, Fig. 3-1).

3.5.2. Tri- and TetragalactosylDAG

The presence of higher homologues of galactolipids in plants was first demonstrated in metabolic studies of *Chlorella* using [^{14}C]-CO$_2$ (Benson *et al.*, 1958), or spinach chloroplasts using UDP-[^{14}C]galactose (Neufeld and Hall, 1964; Ongun and Mudd, 1968), but their identification was based only on chromatographic properties of the deacylated radioactive lipids. Galliard (1969) isolated the trigalactosylDAG from potato tubers by a combination of DEAE-cellulose column chromatography, Florosil column chromatography, and preparative TLC. The pure triGalDAG had TLC mobilities lower than those of diGalDAG (Table 3-1) and had molar ratios of fatty acids–galactose–glycerol close to 2 : 3 : 1. Mild alkaline deacylation of triGalDAG gave the water-soluble triGalGro with mobilities on paper chromatography lower than those of diGalGro (Table 3-2). Periodate oxidation of triGalGro resulted in a consumption of close to 7 moles of periodate with liberation of 1 mole of formaldehyde, indicating the presence of two free hydroxyls in the glycerol moiety and three galactopyranosyl units linked 1–6 attached to C-1 of glycerol. Treatment with β-galactosidase did not release any galactose, suggesting that the terminal sugar unit was linked in the α-configuration. These findings suggested that the triGalDAG was a higher homologue of diGalDAG containing a third galactose unit linked α-(1–6) to the terminal galactose unit of diGalDAG (Galliard, 1969).

The triGalDAG was subsequently isolated from spinach chloroplasts (Webster and Chang, 1969; Joyard and Douce, 1976; Siebertz *et al.*, 1979), cassava tubers (Hudson and Ogunsua, 1974), pumpkin (Ito and Fujino, 1975), alfalfa leaves (Ito and Fujino, 1980), and rice bran (Fujino and Miyazawa, 1979) and its structure was determined unambiguously by Fujino and Miyazawa (1979) and Ito and Fujino (1980). Treatment of the deacylated triGalDAG with α-galactosidase liberated galactose and mono- and diGalGro, but β-galactosidase treatment gave no hydrolysis product; treatment of the diGalGro with α-galactosidase gave galactose and monoGalGro, but β-galactosidase gave no product; finally, the monoGalGro was hydrolyzed only by the β-galactosidase to give galactose and glycerol. These results established the anomeric sugar sequence as αGal-αGal-βGalGro and was confirmed by chromic acid oxidation (Fujino and Miyazawa, 1979). Permethylation of triGalDAG followed by methanolysis yielded 2,3,4,6-tetramethyl-Gal and 2,3,6-trimethyl-Gal, but some 2,3,4-trimethyl-Gal was also found. Thus, the structure of triGalDAG is established as 1,2-diacyl-3-(α-D-galactopyranosyl-1–6-α-D-galactopyranosyl-1–6-β-D-galactopyranosyl)-*sn*-glycerol (structure **3**, Fig. 3-1) and confirmed by mass spectrometry (Fig. 3-8C; Zepke *et al.*, 1978). However, the isomeric 1,2-diacyl-3-(α-D-galactopyranosyl-1–4-α-D-galactopyranosyl-1–4-β-D-galactopyranosyl)-*sn*-glycerol is also present in alfalfa leaves (Ito and Fujino, 1980).

The tetraGalDAG has been isolated from cassava tubers (Hudson and Ogunsua), spinach leaves (Siebertz *et al.*, 1979), and rice bran (Fujino and Miyazawa, 1979). Its structure was established as 1,2-diacyl-3-[(α-D-galactopyranosyl-1–6)$_3$-β-D-galactopyranosyl]-*sn*-glycerol (structure **3′**, Fig. 3-1; Fujino and Miyazawa, 1979) and confirmed by mass spectrometry (Fig. 3-8D; Siebertz *et al.*, 1979).

3.5.3. SulfoquinovosylDAG

The plant sulfolipid, sulfoquinovosylDAG was first detected in ^{35}S- and ^{14}C-labeling studies of a wide variety of photosynthetic tissues (Benson, 1963) and was assigned the general structure of a sulfoglycosylDAG on the basis of the ratio of ^{14}C in the glycerol–sulfosugar–fatty acid moieties being $1 : 2 : >11$; i.e., the mole ratios of glycerol–sulfohexose–fatty acids were $1 : 1 : 2$.

The early structural studies were carried out with the deacylated lipid, sulfoquinovosylGro (Lepage *et al.*, 1961), which was isolated from deacylated alfalfa leaf lipids by paper or ion-exchange chromatography as the crystalline cyclohexylamine salt. Periodate oxidation consumed about 3 moles of periodate, indicating a pyranoside structure for the sugar that must be attached to the 1-position of glycerol. After subsequent bromine oxidation, sulfolactic acid was identified, indicating that the sulfosugar was a sulfonic acid derivative of a deoxysugar. The complete resistance of the sulfosugar to acid hydrolysis and the fact that it was oxidized by periodate to sulfoacetic acid further indicated that it was a 6-sulfo-6-deoxyhexose (Benson, 1963). The 6-deoxy-D-glucose (D-quinovose) structure of the sulfosugar was suggested by the optical rotary shifts observed in cupra B solutions and the high dextrorotatory molecular rotation of the sulfoglycosylGro (Lepage *et al.*, 1961). The anomeric configuration of the latter was first erroneously assigned as a β-glycoside on the basis of hydrolysis with β-galactosidase, which cleaves α-glucosides anomalously (Shibuya and Benson, 1961).

The α-configuration for the glycoside linkage was consistent with the high dextrorotatory molecular rotation (Table 3-3) and rotary dispersion curve characteristic of α-glucosides. The structure of the sulfoquinovosylGro was established as 6-sulfo-6-deoxy-D-glucopyranosyl-α-1–3-*sn*-glycerol by elementary analysis, chemical synthesis (Miyano and Benson, 1962*b*), proton NMR, IR spectrometry, and X-ray crystallographic analysis of the rubidium salt (Benson, 1963). Confirmation of the *sn*-3-glycerol substitution was provided by X-ray analysis and also by NO_2 oxidation of the glycoside to give L(+)-glyceric acid (Miyano and Benson, 1962*a*).

The intact sulfoquinovosylDAG was isolated in pure form from alfalfa leaves and *Chlorella* and was shown to have molar ratios of fatty acids–sugar–S of $2 : 1 : 1$ (O'Brien and Benson, 1964). The sulfolipid structure is thus confirmed as 1,2-diacyl-6-sulfo-D-quinovopyranosyl-α-1–3-*sn*-glycerol (structure **4,** Fig. 3-1).

3.5.4. Mono- and Oligoglycosylsterols

A monoglycosylsterol, sitosterolglucoside, was first isolated by Carter *et al.* (1961*a*) from wheat flour glycolipids. The free sterylglycoside as well as its tetraacetyl derivative gave the expected elementary analytical data, and the latter compound had a melting point identical to that reported for authentic tetraacetyl-sitosterol-β-D-glucoside. Acid hydrolysis released glucose as the only sugar, and the sterol was identified by melting point as β-sitosterol. The structure of this glycosylsterol is thus β-D-glucopyranosyl-β-sitosterol (structure **5,** Fig. 3-2).

An acylated form of **5** was first isolated from potato tuber and appears to be widely distributed among plants (Lepage, 1964*a,b;* see Table 3-6). It was composed of fatty acid–sterol–and glucose in a molar ratio of 1 : 1 : 1, and yielded on mild alkaline deacylation fatty acid methyl esters and sterylglucoside, which was largely the β-sitosterolglucoside with some stigmasterolglucoside. Periodate oxidation of the intact acylated sterylglucoside released 1 mole of formic acid, indicating that the acyl group was attached at C-6 of the sugar moiety. The configuration of the glucosidic linkage could not be determined, since no hydrolysis was observed with β-galactosidase, maltase, or amylase, probably because of the insolubility of the glycolipid in water (Lepage, 1964*b*). Subsequently, the glucosylsterol from rice bran was shown to have the β-anomeric configuration by oxidation of the acetylated derivative with chromium trioxide (Fujino and Ohnishi, 1979*a,b;* Laine and Renkonen, 1975). The acylated monoglycosylsterol thus has the structure 6-acyl-β-D-glucopyranosyl-β-sitosterol (structure **5′**, Fig. 3-2) (Lepage, 1964*b;* Kuroda *et al.*, 1977; Fujino and Ohnishi, 1979*a,b*).

A series of oligoglycosylsterols has been detected in rice bran and leaves (Fujino, 1978; Ohnishi and Fujino, 1981), including di- and trihexosylsterols (Fujino and Ohnishi, 1979*a*) and tetra- and pentahexosylsterols (Fujino and Ohnishi, 1979*b*). These oligoglycosylsterols were isolated from the total glycolipid fraction by a combination of silicic acid column chromatography and preparative TLC, after mild alkaline deacylation to remove the glycosylDAGs, and finally were purified by preparative TLC of their peracetylated derivatives (Fujino and Ohnishi, 1979*a,b*). After deacetylation, the individual oligoglycosylsterols had TLC mobilities given in Table 3-1. D-Glucose, the only sugar present in each of these glycosylsterols, was linked only in the β-anomeric configuration. The major sterol present in all these glycosylsterols was sitosterol (65–73%), with smaller amounts of campesterol (11–19%) and stigmasterol (10–13%); very small amounts (1–2% or less) of cholesterol, Δ^5- and Δ^7-avenasterol, Δ^7-stigmasterol and other sterols were also detected by GLC and GC–MS (Fujino and Ohnishi, 1979*a,b*). The carbohydrate moieties of these oligoglycosylsterols represented a homologous series of β-1–4-linked glucans ranging from cellobiose to cellopentaose. The structure of the major species in this series are: Glcpβ1–4Glcpβ1–3-sitosterol (cellobiosylsitosterol, structure **5″**, Fig. 3-2); (Glcpβ1–4)$_2$Glcpβ1–3-sitosterol (cellotriosylsitosterol, structure **5‴**, Fig. 3-2); (Glcpβ1–4)$_3$Glcpβ1–3-sitosterol (cellotetraosylsitosterol); and (Glcpβ1–4)$_4$(Glcpβ1–3-sitosterol (cellopentaosylsitosterol) (Fujino and Ohnishi, 1979*a,b*).

3.5.5. Ceramide Mono- and Oligohexosides (Cerebrosides)

In their early studies on wheat flour glycolipids, Carter *et al.* (1956, 1961*a*) detected the presence of cerebrosides containing glucose and mannose. Subsequently, cerebroside fractions were obtained in pure form after alkaline hydrolysis to remove contaminating *O*-acyl lipids followed by silicic acid or ion-exchange column chromatography to remove fatty acids (Carter *et al.*, 1960, 1961*a,c*). Two main fractions were obtained: (1) a glucocerebroside (ceramidemonoglucoside) derived from the trihydroxy-base phytosphingosine (see Carter and Hendrickson, 1963), and the dihydroxy-bases di-hydrosphingosine and sphingosine isomer (later shown to be 8-sphingenine by MacMurray and Morrison, 1970), as major long-chain bases, with de-hydrophytosphingosine (4D-hydroxy-8-sphingenine) (see Carter and Hendrickson, 1963) as minor component, and 2-hydroxystearic acid as the major fatty acid (Carter *et al.*, 1960, 1961*c*); and (2) a mannocerebroside, most likely a trimannoside, the long-chain base and fatty acid composition of which was not determined (Carter *et al.*, 1961*a*). The low positive rotation of the mono-glucosylceramide suggested that the glucosyl group was β-linked (Carter *et al.*, 1961*c*); the phytosphingosine-derived cerebroside thus has structure **6** (Fig. 3-3A).

Later studies by Laine and Renkonen (1973) established the structures of the dihydroxy long-chain bases of the monoglucosylceramide in wheat flour as being *erythro-4-trans*-sphingenine, *erythro-4-trans-8-trans*-sphingadienine, *erythro-4-trans-8-cis*-sphingadienine, sphinganine, 8-*trans*-sphingenine and 8-*cis*-sphingenine. Laine and Renkonen (1974) also established the structure, stereochemistry, and anomeric configuration of the sugar moieties of the mono-, di-, and triglycosylceramides of wheat flour, by permethylation analysis, polarimetry, and CrO_3 oxidation. The ceramide monohexoside is Glc$p$β1–1-ceramide (structure **7**, Fig. 3-3B); the ceramide dihexoside is Man$p$β1–4Glc$p$β1–1-ceramide (structure **7′**, Fig. 3-3B); and the ceramide trihexoside is Man$p$β1–4Man$p$β1–4Glc$p$β1–1-ceramide (structure **7″**, Fig. 3-3B).

In a more detailed study of glycosylceramides in wheat grain Fujino and Ohnishi (1983) and Fujino *et al.* (1985*a*) established the structures of the major molecular species of the mono-, di-, tri-, and tetraglycosylceramide, respectively, as follows: Glc$p$β1–1-, Man$p$β1–4Glc$p$β1–1-, (Man$p$β1–4)$_2$Glc$p$β1–1-, and (Man$p$β1–4)$_3$Glc$p$β1–1-derivatives of *N*-2′-hydroxypalmitoyl (or hydroxyarachidoyl)-*cis*-8-sphingenine (structures **7**, **7′**, **7″**, and **7‴**, respectively, Fig. 3-3B). In addition, a minor diglycosylceramide, cellobiosylceramide (structure **6′**, Fig. 3-3A), the major species of which was Glc$p$β1–4Glc$p$β1–1-*N*-2′-hydroxypalmitoyl (or hydroxyarachidoyl)-*cis*-8-sphingenine was identified. It should be noted that neither Laine and Renkonen (1973, 1974) nor Fujino *et al.* (1985) found any trihydroxy long-chain bases in wheat grain or wheat flour glycosylceramides, in contrast to the finding of trihydroxybases by Carter *et al.* (1960, 1961*a,c*). This difference may perhaps be attributed to species or varietal differences in the wheat grains studied or to differences in the extraction and isolation procedures used. For example, it has been found that alkaline treatment of ceramides results in low recoveries of phytosphingosines (see Laine and Renkonen, 1973).

By contrast, the oligoglycosylceramides of rice grain were found to be derived from both trihydroxy and dihydroxy sphingoid bases (Fujino and Ohnishi, 1976; Fujino *et al.*, 1985*b*); the oligoglycosyl groups were also different in that they contained terminal glucose residues attached to mannose groups. The main molecular species found (Fujino *et al.*, 1985*b*) were as follows:

Glc$p$β1–1-*N*-2′-hydroxyarachidoyl-4,8-sphingadienine; Man$p$β1–4Glc$p$β1–1-, and Glc$p$β1–4Glc$p$β1–1-*N*-2′-hydroxylignoceryl-4-hydroxy-8-sphingenine; (Man$p$β1–4)$_2$Glc$p$β1–1-(**7″**), and Glc$p$β-1–4Man$p$β1–4Glc$p$β1–1-*N*-2′-

hydroxylignoceryl–4-hydroxy-8-sphingenine; and $(\text{Man}p1–4)_3\text{Glc}p\beta1–1-(\textbf{7}''')$ and $\text{Glc}p\beta1–4-(\text{Man}p\beta1–4)_2 \text{ Glc}p\beta1–1-N-2'-\text{hydroxylignoceryl-4-hydroxy-8-}$ sphingenine.

Early studies of glycosylceramides in photosynthetic tissue (runnerbean leaves) showed the presence of monoglucosylceramide containing 2-hydroxy fatty acids (chiefly 2-hydroxypalmitic, 2-hydroxybehenic, and 2-hydroxylig-noceric acids) and both dihydroxy(8-sphingenine, sphinganine) and tri-hydroxy(4-hydroxy-8-sphingenine and 4-hydroxysphinganine) long-chain bases (Sastry and Kates, 1964*a*). More detailed studies of the oligoglycosylcer-amides of spinach leaves (Ohnishi *et al.*, 1983) established the identity of two series of ceramide mono- and dihexosides derived from the ceramides N-2'-hydroxypalmitoyl-4,8-sphingadienine   and N-2'-hydroxylignoceryl-4-hy-droxy-8-sphingenine, the glycosyl groups being $\text{Glc}p\beta$ and $\text{Glc}p\beta1–4\text{Glc}p\beta$ (cellobiosyl) (structures **6** and **6'**, respectively).

3.5.6. Phytoglycolipid

A complex phosphoglycolipid, called phytoglycolipid (PGL), has been isolated from a variety of seeds (e.g., soybean, flax seed, corn, wheat) (Carter *et al.*, 1958*a*). It was composed of fatty acids, the long-chain base phy-tosphingosine, phosphate, inositol, glucuronic acid, glucosamine, galactose, arabinose, and mannose, in about equimolecular ratio. Barium hydroxide hydrolysis cleaved PGL into a chloroform-soluble product identified as ce-ramidephosphate and a water-soluble oligosaccharide containing inositol and other sugars mentioned above (Carter *et al.*, 1958*b*). Structural studies of the oligosaccharide established the sequence and configuration of sugars as shown in Fig. 3-4 (Carter *et al.*, 1969), in which the inositol is linked to an α-Man*p* residue and an α-D-glucuronyl group, linked to an α-D-glucosaminyl group; the positions of attachment of the terminal arabinose and galactose units have still not been established. The position of attachment to the phos-phate group is at the *sn*-1-position of inositol (structure **8**, Fig. 3-4). An identi-cal structure has recently been reported for the phytoglycolipid of rice bran (Ito *et al.*, 1985*b*).

3.6. Distribution of Glycolipids

The distribution of each of the three main classes of glycolipids, glycosylDAGs (structures **1, 2, 3,** and **3'**) and the sulfolipid (structure **4**), glycosylsterols (structures **5, 5', 5",** and **5"'**), and glycosylceramides (structures **6, 6', 6",** and **6"'**), among higher plants, algae, diatoms, cyanobacteria, and yeasts and fungi is presented in Table 3-6.

3.6.1. Higher Plants

3.6.1.1. Leaves

The leaves of all higher plants examined contain major proportions of GalDAG (**1**) and Gal$_2$DAG (**2**) and the sulfoquinovosylDAG (**4**) (Table 3-6) (see

Table 3-6. Distribution of GlycosylIDAGs, Glycosylsterols, and Glycosylceramides in Various Plants[a]

Plant or tissue	Total glyco-lipids	Composition, % of total lipids													References
		(1)	(2)	(3)	(3')	(4)	(5)	(5')	(5'')	(5''')	(6)	(6')	(6'')	(6''')	
Higher plants															
Leaves															
Spinach (*Spinacia oleracea*)	40	20	13	+	+	4	0.5	—	—	—	0.7	+	+	—	Allen et al. (1964, 1966a), Haverkate and van Deenen (1965a); Wintermans, 1963; Eichenberger and Menke (1966); Ohnishi et al. (1983)
Runner bean (*Phaseolus multiflorus*)	—	17	4	—	—	+	+	—			<1	—	—	—	Sastry and Kates (1964a); Kates (1960)
Red clover (*Trifolium pratense*)	(25)[b]	(16)	(8)			+									Weenink (1964)
Alfalfa (*Medicago sativa*)	10	4	9	0.2		2					0.1				O'Brien and Benson (1964), Ito and Fujino (1973, 1980)
Potato (*Solanum sativa*)		+	+			+	+	+							Lepage (1964a)
Lettuce and cabbage		+	+			+	+	+							Nichols (1963)
Fruits															
Cucumber		2	5												Kinsella (1971)
Green pepper		2	7												Kinsella (1971)
Pumpkin	(21)	(21)	(27)	(18)		(10)	(5)	(6)			(3)				Ito et al. (1974), Ito and Fujino (1975)
Apple															Galliard (1968b)
Preclimacteric	29.1	5	12			1	6	1			4				
Postclimacteric	19.8	1	5			1	5	2			5				

										Reference
Tubers, bulbs, roots										
Potato (*Solanum tuberosum*)	34	6	14	1	1	1	2	6	2	Galliard (1968a, 1969), Lepage (1964b)
Yam	35	+++	++	+	+	+	+		+	Osagie and Opute (1981)
Parsnip (*Pastinaca sativa*)		+	+	+	+	−	+			Roughan and Batt (1969)
Sweet potato	14	6	6			+				Walter *et al.* (1971)
Turnip		+	+	+		+				Lepage (1967)
Narcissus and daffodil bulbs		+	++	+	+	++				Nichols and James (1964)
Oat root	5	10	6	n.d.[c]	+	+	+	+	+	Liljenberg and Kates (1985) (data as mol %)
Seeds										
Briza spicata	2.6	29	49	n.d.	n.d.	n.d.	0.4	1.0	0.3	Smith and Wolff (1966)
Millet		0.3	0.4	n.d.	n.d.	0.4	1.0			Osagie and Kates (1984)
Azuka beans		+	+	+	+					Ito *et al.* (1985a)
Wheat		+	+	+	+	+	+	+	+	Carter *et al.* (1961a,b 1969), Ito *et al.* (1983), Fujino and Ohnishi (1983), Fujino *et al.* (1985a)
Rice bran (*Oryza sativa*)	3	+	+	+	−	+	+	+	+	Fujino and Miyazawa (1979), Fujino and Ohnishi (1979a,b), Fujino *et al.* (1985b)
Pea						+	++			Miyazawa *et al.* (1974)
Corn		+				+	++	+		Tanaka *et al.* (1984)
Algae										
Chlorella pyranoidosa	4		7	6						O'Brien and Benson (1964), Benson and Shibuya (1962)
Chlorella vulgaris		++	+		+					Sastry and Kates (1965)

(*continued*)

Table 3-6. (Continued)

Plant or tissue	Total glycolipids	(1)	(2)	(3)	(3')	(4)	(5)	(5')	(5")	(5"')	(6)	(6')	(6")	(6"')	References
Chlamydomonas reinhardtii		+	+			+									Eichenberger (1976), Janero and Barnett (1981)
Dunaliella tertiolecta or *parva*	41–52	21–22	11–20			7.9									Evans *et al.* (1982), Evans and Kates (1984), Kates (1987)
Halophilic *Dunaliella* spp.	54	23	17			14									Evans *et al.* (1982), Evans and Kates (1984), Kates (1987)
Fucus vesiculosus		7.3	6.4												Radunz (1969)
Batrachospermum monoliforme		12	12												Radunz (1969)
Euglena gracilis		+	+			+									Haverkate and van Deenen (1965b)
Ulva lactuca		+	+												Brush and Percival (1972)
Diatoms															
Nitzschia alba	3.7 (36)	0.9 (9)	0.9 (8)			1.6 (16)									Anderson *et al.* (1987a,b), Kates (1987, 1989)
Navicula pelliculosa		28	25			23									Anderson *et al.* (1987a,b), Kates (1987, 1989)
Cyanobacteria															
Anacystis nidulans[d]		30	12			10	0	0	0						Allen *et al.* (1966b), Nichols *et al.* (1965), Sato and Murata (1982a)
Anabena variabilis[d]		54	17			11	0	0	0						Nichols *et al.* (1965), Sato and Murata (1982a), Murata and Sato (1983)
Prochloron sp.[d]		55	10			26	0	0	0						Sato and Murata (1982a), Murata and Sato (1983)

					Reference
Ferns					
Adiantum capillus-veneris	++	+	+	+	Sato and Furuya (1984)
Mosses and liverworts					
Furoria and leptobryium pyriforme	+	+			Roughan and Batt (1969)
Marchantia bertoroana	+	+			Roughan and Batt (1969)
Yeast and fungi					
Saccharomyces cerevisiae	+	+	+		Weete (1980), Wassef (1977); Green *et al.* (1965), Barraud *et al.* (1970), Työrinoja *et al.* (1974)
Hansenula ciferri					Kaufman *et al.* (1971)
Rhodotorula glutinis	4–15	+	+	+	Yoon and Rhee (1983)
Phycomyces blakesleeanus	0.6			+	Weiss *et al.* (1973)
Aspergillus oryzea[e]				+	Fujino and Ohnishi (1977)
Fusarium lini	1			+	Weiss *et al.* (1973)
Agaricus bisporus	1–3			+	Weiss and Stiller (1972)
Amanita muscaria	1–3			+	Weiss and Stiller (1972)
A. rubescens	3			+	Weiss and Stiller (1972)
Macrophina phaseolina	48–61	+	+		Wassef *et al.* (1975)
Blastochladiella emersonii	11–16	+	+		Mills and Cantino (1974)

[a] Data taken in part from Kates (1970) and Sastry (1974); these references should be consulted for further details.
[b] Values in parentheses are % total polar lipids.
[c] n.d., not detected.
[d] Also present: 1% GlcDAG. Data for *A. nidulans* are given as molar ratios relative to sulfolipid (4).
[e] Cerebroside derived from sphingadienine.

Table 3-7. *Polar Lipid Composition of Cell Organelle Membranes*[a]

Organelle	Glycolipids					Phospholipids					
	(1)	(2)	(4)	(5+5′)	(6)	PC	PE	PI	PG	DPG	PS
Plastids											
Spinach chloroplasts[b]											
Thylakoids	52	26	7	0	0	5	0	2	10	0	—
Inner envelope	49	30	5	0	0	6	0	1	8	0	—
Outer envelope	17	29	6	0	0	32	0	5	10	0	—
Pea etioplasts[b]											
Prothylakoids	42	35	6	—	—	9	0	2	5	0	—
Envelope	34	31	6	—	—	17	0	4	5	0	—
Cauliflower bud[b]											
Plastids (nongreen)	30	29	6	—	—	19	2	5	8	0.6	—
Cyanobacteria											
Anabaena variabilis[c]	54	17	11	0	0	0	0	0	17	0	—
Anacystis nidulans[c]	32	12	10	0	0	0	0	0	9	—	—
Prochloron sp.[c]	55	10	26	0	0	0	0	0	5	—	—
Mitochondria											
Pea leaves[d]	0	0	0	0	0	40	46	6	2	14	—
Cauliflower buds[d]	tr	tr	0	0	0	37	38	8	2	13	—
Mungbean hypocotyls[d,e]											
Inner membrane	0	0	0	0	0	29	50	2	1	17	—
Outer membrane	0	0	0	0	0	68	24	5	2	0	—
Sycamore cells[d]											
Inner membrane	0	0	0	0	0	41	37	5	3	14.5	—
Outer membrane	0	0	0	0	0	54	30	11	5	0	—
Potato[e]											
Inner membrane	0	0	0	0	0	33	33	7	5	19	—
Outer membrane	0	0	0	0	0	36	64	n.d.	n.d.	n.d.	—
Yam tuber[f]											
Total mitochondria	0	0	0	0	0	35	27	7	3	9	—
Fungus (*Claviceps purpurea*)[g]											
Total mitochondria	0	0	0	0	0	27	18	28	—	7	13
Beef heart[e]	0	0	0	0	0	38	37	6	—	16	—
Peroxisomes											
Potatoe tuber[d]	0	0	0	0	0	52	48	0	0	0	—
Yam tuber[f,h]	7	8	n.d.	12	2	8	4	2	2	—	—
Plasma membrane											
Yam tuber[f]	7	17	n.d.	10	4	28	15	8	2	0.5	—
Yeast (*S. cerivisiae*)[i]	tr	—	—	+[j]	+	+	+	—	—	—	—
Oat root[k]	4	21	n.d.	+	+	18	36	—	—	—	—
Endoplasmic reticulum											
Yam tuber[f,l]	15	19	n.d.	10	8	18	8	3	4	—	—

[a]Data are given as wt. % of total lipids; n.d., not detected; tr, trace.
[b]Douce *et al.* (1973, 1987a); Allen *et al.* (1966b).
[c]Allen *et al.* (1966b); Sato and Murata (1982a); Murata and Sato (1983); 1% and 3% of GlcDAG is present in *A. variabilis* and *Prochloron*, respectively.
[d]Douce *et al.* (1987a).
[e]McCarty *et al.* (1973).
[f]Osagie and Kates (1986b).
[g]Anderson *et al.* (1964).
[h]Contains *ca.* 50% free sterols (mainly sitosterol, stigmasterol, and campesterol).
[i]Työrinoja *et al.* (1974).
[j]Includes cerebroside sulfate and phosphosphingolipids containing ceramide (P-inositol)$_2$ mannose.
[k]Liljenberg and Kates (1985).
[l]Contains 10% of neutral lipids.

Kates, 1970; Sastry, 1974, for further examples), as a result of the high concentration of these glycolipids in the thylakoid membranes of the chloroplasts (Table 3-7) (see Section 3.6.1.3). The concentration of monoGalDAG in photosynthetic tissues of most higher plants is generally much higher than that of diGalDAG or the sulfolipid, the molar ratio of GalDAG : Gal$_2$DAG, ranging from 1 : 1 to 3 : 1 or higher (Roughan and Batt, 1969; Sastry, 1974). As minor glycolipids, tri and tetraGalDAGs (structures **3** and **3′**, and even higher homologues have been detected in leaves of spinach (Webster and Chang, 1969; Siebertz *et al.*, 1979), and isolated from alfalfa leaves and identified structurally (Ito and Fujino, 1980); they are presumably fairly widespread in the leaves of other plants.

Sterolglucosides (structure **5**) are also minor glycolipids that have been detected in leaves of spinach, runner bean, alfalfa, and potato (Lepage, 1964*a*), and of other plants as well (Table 3-6); acylated sterolglucosides (**5′**) have also been detected in leaves of some plants (Lepage, 1964*a*).

Mono- and diglucosylceramides (structure **6**, **6′**) are also minor glycolipids in leaves and have been isolated from runner bean leaves (Sastry and Kates, 1964*a*), alfalfa leaves (Ito and Fujino, 1973), and spinach leaves (Ohnishi *et al.*, 1983). In spinach leaves, small proportions of mannosylglucosylceramides (structures **6″** and **7′**) were also identified (Ohnishi *et al.*, 1983).

3.6.1.2. Nonphotosynthetic Tissues

Only limited data are available on the distribution of glycolipids among nonphotosynthetic tissues. GalDAG and Gal$_2$DAG (**1** and **2**) have been identified in fruits such as apple, pumpkin, green pepper, and cucumber; in tubers such as potato, parsnip, sweet potato, and turnip; in oat root; and in narcissus and daffodil bulbs, which also contain the sulfolipid (Table 3-6). Gal$_3$DAG (**3**) has been isolated and characterized in potato tubers (Galliard, 1968*a*) and has also been detected in oat root (Table 3-6). In nonphotosynthetic tissues in general, the molar concentration of Gal$_2$DAG is much higher than that of GalDAG (Sastry, 1974), and the concentration of the plant sulfolipid (**4**) is generally low or not detectable because of the low content of chloroplasts or proplastids (see Section 3.6.1.3). Sterolglycosides (**5** and **5′**) have been isolated and identified in pumpkin fruit and in potato tuber and parsnip tuber and have also been detected in oat root. Ceramide hexosides (**6**) have been detected in oat root (Table 3-6).

3.6.1.3. Plant Cell Organelles

The glycolipid composition of photosynthetic and nonphotosynthetic tissues presented above (see Table 3-6) represents the summation of the compositions of individual plant organelle membranes, but in some tissues such as leaves, the composition may be dominated by that of the major organelle, the chloroplast. Isolation of intact, uncontaminated organelles has for many years posed seemingly insurmountable problems of isolation. These problems have now been essentially solved (Douce *et al.*, 1987*a*) and reliable analyses of most organelle membranes and their envelopes are now available (see Table 3-7).

Thylakoid membranes, which are the site of the photosynthetic apparatus, are characterized by high contents of GalDAG and Gal$_2$DAG in molar ratio close to 2 : 1 (Rawyler *et al.*, 1987), together with much lower contents of the sulfolipid and phosphatidylglycerol; phosphatidylcholine and phosphatidylinositol are minor phospholipids, but phosphatidylethanolamine is completely absent (Douce *et al.*, 1987*a*); sterols, sterolglycosides, and glycosylceramides are also totally absent. Total glycolipids account for 85% of total lipids and phospholipids for only 15% (Table 3-7). Except for the presence of phosphatidylcholine, the glycerolipid composition of chloroplasts is very similar to that for *Cyanobacteria* (Sato *et al.*, 1979; Sato and Murata, 1982*a*) and *Prochloron* species (Murata and Sato, 1983) (Table 3-7), which is consistent with a prokaryotic origin of the chloroplast. Virtually the same glycolipid and phospholipid composition as that of the thylakoids was found for the inner chloroplast envelope, but the outer envelope, although qualitatively similar in composition, has much higher contents of phosphatidylcholine and a much lower content of GalDAG, consistent with a eukaryotic origin for the outer envelope. Etioplast prothylakoids and envelope as well as nongreen plastids have qualitatively similar glycolipid and phospholipid compositions but differ from chloroplasts in having higher Gal$_2$DAG and lower GalDAG contents (in mole ratio ~ 1 : 1) (Table 3-7).

The transmembrane distribution of galactolipids and sulfolipid in chloroplast thylakoids was found to be asymmetric in a wide range of temperate climate plants (Rawyler *et al.*, 1985, 1987; Siegenthaler *et al.*, 1987*a*, 1988). In all plants examined, GalDAG was enriched in the outer leaflet (53–65%), while Gal$_2$DAG and SQDAG were enriched in the inner leaflet (78–90% and 60%, respectively) (Table 3-8).

In contrast to chloroplasts, mitochondria are characterized by a complete absence of glycolipids and by high phosphatidylcholine, phosphatidylethanolamine, and cardiolipin contents, with phosphatidylglycerol and phosphatidylinositol as minor components. The inner mitochondrial membrane is

Table 3-8. Galactolipid Mole Ratios in Inner and Outer Leaflets of Thylakoids from Various Plant Species[a]

Plant species	Total lipids (mol %)			Outer : inner (mol ratio)		
	GalDAG	Gal$_2$DAG	SQDAG	GalDAG	Gal$_2$DAG	SQDAG
Spinacia oleracea	55	27	10	65 : 35	15 : 85	40 : 60
Solanum nigrum (herbicide sensitive)	53	23	—	55 : 47	12 : 88	—
Lactuca sativa	43	35	—	57 : 43	18 : 82	—
Pisum sativum	42	29	—	59 : 41	18 : 82	—
Hordeum vulgare	52	26	—	56 : 44	22 : 78	—
Avena sativa	61	28	—	62 : 38	20 : 8	—

[a]Data from Rawyler *et al.* (1985, 1987); Siegenthaler *et al.* (1987*a*, 1988).

clearly distinguished from the outer mitochondrial membrane by its high content of cardiolipin, which is absent from the outer mitochondrial membrane (Table 3-7). The higher phosphatidylethanolamine and lower phosphatidylcholine contents are also characteristic of the inner membrane relative to the outer membrane. This would be consistent with a prokaryotic origin of mitochondria, and is supported by the striking similarity of mitochondrial lipid composition in both green and nongreen plant tissues as well as in human heart tissue (Table 3-7).

The question of the lipid composition of nonchloroplast membranes in photosynthetic tissues, such as plasmalemma and nuclear and endoplasmic reticulum, has not yet been resolved because of the difficulties in isolating pure uncontaminated membrane fractions. However, it is clear that sterol, sterol esters, sterolglycosides, and ceramideglycosides (cerebrosides) must be present in nonchloroplast membranes such as endoplasmic reticulum and perhaps plasmalemma and nuclear membranes, since these glycolipids are completely absent from thylakoid membranes. In regard to the galactolipids and the plant sulfolipid, the analyses of Wintermans (1960) on sugarbeet leaves suggest that the galactolipids may be absent from cytoplasmic membranes and may be specifically located in the chloroplast membranes, but the sulfolipid appears to be present both in chloroplast and cytoplasmic membranes. It should be noted, however, that these analyses were based on fresh (wet) weight of leaves and chloroplasts, introducing considerable ambiguity in the interpretation of the results. These analyses should be repeated on a dry weight basis and the results expressed as weight percent, but direct analysis of cytoplasmic membranes such as endoplasmic reticulum of leaf cells, should also be carried out.

In nonphotosynthetic tissues such as yam tuber, recent studies (Osagie and Kates, 1986a,b) with isolated cell fractions indicate the presence of high concentrations of glycolipids (including sterolglycosides, cerebrosides, and galactolipids) in peroxosomes, amyloplasts, plasmalemma, and endoplasmic reticulum. However, no glycolipids were found in mitochondria, in agreement with the compositions reported for mitochondria from various plants and other sources (Douce *et al.*, 1987a) (Table 3-7). The lipid composition of yam peroxisomes differs strikingly from that of potato peroxisomes, which are reported to have the unusual lipid composition of phosphatidylcholine and phospatidyl ethanolamine as the only lipid components (Douce *et al.*, 1987a) (Table 3-7).

The presence of galactolipids in all yam cell fractions (except mitochondria) might be interpreted as indicating the presence of plastid or proplastid contaminants in these fractions. However, such contamination would have to be very high (40–50%) to account for the high contents of galactolipids (20–35%), which would not be consistent with electron microscopic examination and marker enzyme activities of these fractions (Osagie and Kates, 1986a). Similar results were obtained for oat root plasmalemma and microsomal membrane fractions shown to be free from plastids by the absence of galactosyl transferase activity: galactolipids accounted for 30 and 36 mol %, respectively, of the total lipids, and cerebrosides were present in both membrane preparations (Liljenberg and Kates, 1982, 1985) (Table 3-7). Further studies are clearly needed to resolve these questions concerning the presence of galactolipids in nonplastid membranes. It would be surprising if galactolipids and sulfolipids were completely absent in such membranes as endoplasmic reticulum in view of the presence of these glycolipids in the outer (cytoplasmic) chloroplast envelope, which could exchange galactolipids with cytoplasmic organelles by membrane fusion or by a galactolipid transport mechanism.

3.6.1.4. Seeds

Plant seeds in general have relatively low contents of glycolipids compared to their contents of phospholipids and neutral lipids (largely triacylglycerols); the glycolipids account for only about 3% of the total lipids of all seed lipids examined (Table 3-6). The main glycolipids in seeds of wheat, rice, millet, pea, corn, and azuka bean are galactosylDAGs, sterol glycosides and to a lesser extent glycosyl ceramides, but sulfolipids have not been detected (Table 3-6). However, the composition of each of these glycolipid classes varies somewhat for different plants. Thus, all seeds examined contain GalDAG and Gal$_2$DAG but their mole ratios vary from about 1 (for millet and pea) to about 0.7 (for rice bran and *Briza spicata*); Gal$_3$DAG and higher homologs are present in rice bran and azuka beans but have not been detected in wheat, millet, or pea. All seeds examined also contain sterol glycosides and acylated sterolglycosides, but higher homologues (up to pentaglucosyl) have so far been identified only in rice bran and corn (Kuroda *et al.*, 1977; Fujino and Ohnishi, 1979a,b; Tanaka *et al.*, 1984). Ceramidemonohexosides also appear to be widespread among seeds of various plants, but so far only wheat and rice bran have been shown to contain higher homologues of ceramideglycosides (Fujino and Ohnishi, 1983; Fujino *et al.*, 1985a,b) (Table 3-6). It should also be mentioned that seeds of several plants examined (e.g., wheat, rice, soybean, linseed, flax) contain the complex glycolipid, phytoglycolipid (structure **8,** Fig. 3-4) (Carter *et al.*, 1969; Ito *et al.*, 1985b).

3.6.2. Algae, Diatoms, and Cyanobacteria

Unicellular algae in general have glycolipid compositions similar to those of leaves of higher plants, the major glycolipids being GalDAG, Gal$_2$DAG, and sulfoquinovosylDAG, reflecting the glycolipid composition of the chloroplast (Tables 3-6 and 3-7). However, the mole ratio of GalDAG to Gal$_2$DAG varies greatly for different algal species; for example, green algae (*Chlorophyceae*) such as *Chlorella*, *Chlamydomonas*, and *Dunaliella* species have ratios of 0.7, 4.2 and 1.6, respectively (see Sastry, 1974, and Brush and Percival, 1972, for more extensive data). Also the content of sulfolipid is generally much higher in green algae than in leaves of higher plants and is comparable to that of Gal$_2$DAG. Similar contents of galactolipids have been reported for brown algae, red algae, euglenids, acidophilic and thermophilic algae, and dinoflagellates (Table 3-6) (see Sastry, 1974). The presence of sterolglycosides or ceramidehexosides has not been reported so far in any alga, although the possible presence of sterolglycoside and acylated sterolglycoside may be inferred from the studies of Lepage (1964a). However, a more detailed and extensive survey of algae for the presence of these glycolipids is needed to settle this point.

Both photosynthetic and nonphotosynthetic diatoms contain galactolipids (**1** and **2**) in about equimolecular ratio and sulfoquinovosylDAG (**4**) in equal or even greater proportions (Kates and Volcani, 1966; Anderson *et al.*, 1978a) (Table 3-6). The low proportions of glycolipids in the total lipids are due to the high content of triacylglycerols (87% of total lipids) which are present as cytoplasmic oil droplets. Correcting for this, however, the total galactolipid content (17% of total polar lipids) is still very much lower than that of leaves, whereas the sulfoquinovosylDAG content (16% of polar lipids) is much higher than that of leaves (Table 3-6). The significance of these differences is unknown. No sterolglycosides or ceramidehoxosides have been reported so far in any diatom (Kates, 1987, 1989).

Cyanobacteria such as *Anacystis nidulans* and *Anaebena variabilis* have a glycolipid composition very similar to that of chloroplasts of leaves. In addition they contain small proportions of a precursor glycosylDAG, GlcDAG, and phosphatidylglycerol as the only phospholipid component (Sato and Murata, 1982a) (Tables 3-6 and 3-7) (see Section 3.6.1.3).

3.6.3. Yeasts

Relatively little information is available concerning the occurrence and composition of glycolipids in yeasts and fungi (see reviews by Weete, 1974, 1980; Brennan *et al.*, 1974; Rattray *et al.*, 1975; and Wassef, 1977). In general, glycolipids are relatively minor components of the total lipids in most yeasts and fungi and are rather restricted in composition, the major classes being glycosylceramides with smaller amounts of glycosylDAGs and glycosylsterols (Table 3-6).

The yeast *Saccharomyces cerevisiae* has been reported to contain only one glycerogalactolipid, GalDAG (**1**), together with sterol glycosides and sulfatides (presumably cerebroside sulfate) localized in the protoplast membrane (Baraud *et al.*, 1970). A cerebroside sulfate (sulfatide) was also detected in the plasma membrane fraction of baker's yeast (Nurminen and Suomalainen, 1971; Suomalainen and Nurminen, 1970; Työrinoja *et al.*, 1974) but no structural analysis was done. Only low amounts of unidentified cellular glycosylceramides have been reported present in *S. cerevisiae* (Green *et al.*, 1965), although it produces free ceramides, called cerebrins (*N*-acyl-phytosphingosines or *N*-acyl-sphingosines), which are widely distributed among yeasts and fungi (Carter *et al.*, 1954; Stanacev and Kates, 1963; Brennan *et al.*, 1974).

Subsequently, galactosylceramides were identified in *S. cerevisiae* and *Candida utilis* (Wagner and Zofcsik, 1966a), the major long-chain base and fatty acids being sphinganine and 2-hydroxy stearic acid (Table 3-9). These two yeasts have also been shown to contain a mannosyl-inositol phosphoryl-1-(*N*-acyl)phytosphingosine (Wagner and Zofcsik, 1966a,b; Smith and Lester, 1974), which may have been derived, during the alkaline treatment step in the

Table 3-9. Long-Chain Base and Fatty Acid Composition of Cerebrosides of Various Fungi

Species	Major long-chain base[a]		Major fatty acid	
	Chain type	%	Chain type	%
Sphingosines				
Candida utilis[b]	*n*-16 : 0	—	*n*-18 : 0–OH	100
Saccharomyces cerevisiae[b]	*n*-18 : 0	—		
Aspergillus niger[c] and	*br*-19 : 2 isomers	54	*n*-18 : 0–OH	52
A. *oryzae*[d]	(*n*-18 : 2)	2	*n*-18 : 1–OH	41
Hansenula ciferri[e]	*n*-17 : 1	—	*n*-16 : 0	17
	n-19 : 1	—	*n*-18 : 1	17
			n-18 : 2	31
			n-18 : 3	15
			n-18 : 0–OH	10
Phytosphingosines				
Phycomyes blakesleeamus[f]	*n*-18 : 0	32	*n*-16 : 0	18
	i-19 : 0	20	*n*-16 : 0–OH	42
	i-20 : 0	33	*n*-18 : 0	8
	i-21 : 0	4	*n*-18 : 1	15
	i-21 : X$_1$ and X$_2$	7	*n*-18 : 2	7
	n-22 : X$_1$ and X$_2$	6	*n*-20 : 0	1
Fusarium lini[f]	*i*-17 : 0	11	*n*-16 : 0–OH	2
	n-18 : 0	21	*n*-18 : 0–OH	94
	i-18 : 0	16	*n*-18 : 0	0.4
	i-20 : 0	38	*n*-18 : 1	0.4
	i-21 : X$_1$ and X$_2$ }	11	*n*-18 : 2	0.5
	i-21 : 0 }			
Amanita muscaria[g]	*n*-18 : 0	7	*n*-14 : 0–OH	8
	i-18 : 0	9	*n*-15 : 0–OH	16
	i-19 : 0	9	*n*-16 : 0	9
	i-20 : 0	15	*n*-16 : 0–OH	21
	i-21 : X$_1$ and X$_2$ }	36	*n*-18 : 0	9
	i-21 : 0 }		*n*-18 : 1	16
	n-22 : X$_1$ and X$_2$ }	22	*n*-18 : 2	12
	n-22 : 0 }		*n*-20 : 0	1
Amanita rubesceris[f]	*i*-17 : 0	6	*n*-16 : 0	11
	n-18 : 0	13	*n*-16 : 0–OH	33
	i-18 : 0	6	*n*-18 : 0	6
	i-19 : 0	22	*n*-18 : 1	12
	i-20 : 0	28	*n*-18 : 2	15
	i-21 : X$_1$ and X$_2$ }	15	*n*-18 : 0–OH	6
	i-21 : 0 }		*n*-19 : 0	2
	n-22 : X$_1$ and X$_2$ }	10		
	n-22 : 0 }			
Agaricus bisporus[f]	*n*-18 : 0	61	*n*-15 : 0–OH	8
	i-19 : 0	5	*n*-16 : 0	5
	i-21 : X$_1$ and X$_2$	34	*n*-16 : 0–OH	43
	n-22 : X$_1$	1	*n*-18 : 0	2
			n-18 : 1	6
			n-18 : 0–OH	13

[a]X$_1$ and X$_2$ indicate unidentified positions and geometry of chain unsaturation.
[b]Wagner and Zofcsik (1966*a,b*) galactocerebrosides.
[c]Wagner and Zofcsik, 1969.
[d]Fujino and Ohnishi (1977); only glucocerebrosides; traces of phytosphingosine cerebrosides also are present.
[e]Kaufman *et al.* (1971); glucocerebrosides.
[f]Weiss *et al.* (1973).
[g]Weiss and Stiller (1972).

isolation procedure, from a more complex sphingoglycolipid with tentative structure mannosyl(inositolphosphoryl)$_2$-1-(N-acyl)phytosphingosine (Steiner *et al.*, 1969). Other phytosphingoglycolipids containing inositol and P in 1 : 1 molar ratio and being stable to alkali have also been recognized in *S. cerevisiae* (Steiner and Lester, 1972).

Glucosylceramides have been identified in cells of the yeast *Hansenula ciferri*, the long-chain bases being normal C_{17} and C_{19}-sphingenines and the fatty acids being 16 : 0, 18 : 1, 18 : 2, 18 : 3, and HO-18 : 0 (Kaufman *et al.*, 1971) (Table 3-9). Glycolipids have also been reported present in the yeast *Rhodotorula glutinis*, in amounts varying from 6% under nitrogen-limited growth conditions to 35% under carbon-limited conditions (Yoon and Rhee, 1983). The main glycolipids were tentatively identified by TLC mobilities as GalDAG, Gal$_2$DAG, glycosylsterols, and cerebrosides, these four components accounting for 11–15% of total lipids.

In regard to the intracellular distribution of glycolipids in yeasts, (Table 3-7), it has been conclusively established that the glycosphingolipids of *S. cerevisiae*, including a mannosylinositolceramide, as well as other minor glycolipids, are localized in the plasma membrane (Nurminen and Suomalainen, 1971; Nurminen *et al.*, 1970; Työrinoja *et al.*, 1974), suggesting that these lipids may function in the process of cell–cell interaction or "recognition" in yeasts. Yeast mitochondria, however, have a similar lipid composition to that of typical mammalian or plant mitochondria and are essentially devoid of glycolipids (Anderson *et al.*, 1964) (see Table 3-7). No data are available on the lipid composition of endoplasmic reticulum or other cell organelles of yeasts. Many yeasts produce extracellular lipids, including carbohydrates esterified with fatty acids. These are covered in detail in Chapter 6 (*this* volume).

3.6.4. Fungi

As was mentioned for the yeasts, glycolipids are not major lipid components in most fungi. However, *Blastocladiella emersonii* has been found to contain a relatively high glycolipid content, 11–16% of total lipids depending on the stage of development, the major components being GalDAG (**1**) and Gal$_2$DAG (**2**) (Mills and Cantino, 1974). The Gal$_2$DAG may act as a glycosyl carrier intermediate in chitin biosynthesis rather than the more common phosphorylated polyisoprenoid carrier (Mills and Cantino, 1978).

High contents of glycolipids (61–48% of total lipids, depending on carbon–nitrogen ratio in growth medium) have been found in mycelia of *Macrophomina phaseolina*, and lower contents were found in the sclerotia (14–62%) (Wassef *et al.*, 1975). The major glycolipid components were tentatively identified as GalDAG and Gal$_2$DAG, but no other glycolipids were identified. However, the main glycolipids in most fungi are glycosphingolipids, largely D-glucosylceramides; these have been examined in detail in a few representative species of fungi (see reviews by Brennan *et al.*, 1974; Wassef, 1977; and Weete, 1974, 1980) (Table 3-6).

Aspergillus niger, a species of *Fungi imperfecti*, has been found to contain galactocerebrosides similar to those in *C. utilis* and *S. cerevisiae* (Wagner and Zofcsik, 1969) (Table 3-9).

Fusarium lini, another species of the *Fungi imperfecti* was found to contain about 0.1% of cell dry weight as glucose-containing cerebrosides (about 1% of total lipids), which are composed of a series of phytosphingosine homologs with 18:0 straight-chain and iso-17:0 to 21:0 branched saturated and unsaturated chain bases and hydroxystearate as the major fatty acid (Weiss *et al.,* 1973) (Table 3-9). Similar cerebrosides (0.6% of total lipid) were found in *Phycomycetes blakesleeanus,* a member of the most primitive class of true fungi (Eumycophyta), with the addition of the 22:0 normal chain base and a somewhat wider range of fatty acid constituents (*n*16:0, HO-16:0 and *n*18:1) (Weiss *et al.,* 1973) (Table 3-9).

Mushrooms, which are members of the most complex class (*Basidiomycetes*) of true fungi, also contain cerebrosides (about 1–3% of total lipids) with phytosphingosine base and fatty acid compositions similar to those of *P. blakesleeanus* and *F. lini* (Weiss and Stiller, 1972) (Table 3-9). The species of wild mushrooms (*Amanita muscaria* and *Amanita rubescens*) examined have overall similar patterns of long chain bases which resemble qualitatively that of the one cultivated mushroom (*Agricus bisporus*) studied. The major cerebroside bases in the wild species were saturated iso-C_{19}, and iso-C_{20}, and unsaturated iso-C_{21} and iso-C_{22} phytosphingosines, while in the cultivated species saturated n-C_{18} and unsaturated iso-C_{21} bases were predominant (Weiss and Stiller, 1972). The major cerebroside fatty acids in the wild species were hydroxy-15:0 and hydroxy-16:0 acids together with lesser amount of 16:0, 18:1, and 18:2 acids, while the major acid in the cultivated species was hydroxy-16:0 with lesser amounts of hydroxy-15:0, 16:0, 18:1, and 18:2 acids (Table 3-9). The cerebrosides in the wild species were exclusively glucocerebrosides, but the carbohydrate in the cultivated species was not identified.

3.7. Fatty Acid and Molecular Species Composition of Glycosyldiacylglycerols

3.7.1. Fatty Acid Composition

3.7.1.1. Higher Plants

The fatty acid composition of glycosylDAGs in photosynthetic tissues of higher plants and of green algae is characterized by high proportions of polyunsaturated fatty acids, mainly α-linolenic acid (9,12,15-octadecatrienoic acid, 18:3-ω-3) and its lower homolog 7,10,13-hexadecatrienoic acid (16:3-ω-3) (Table 3-10). However, there is a characteristic difference in the positional distribution of these two acids in galactolipids of various plants (Jamieson and Reid, 1971). In angiosperms such as members of the Fabaceae (e.g., runner bean, peas, red clover, alfalfa), Asteraceae (e.g., sunflower), and Poaceae (e.g., wheat, rice) and Euphorbaceae (castor) (Table 3-10), both the mono- and digalactosylDAGs have very high contents of the 18:3 acid, concentrated in both *sn*-1 and -2 positions of the galactolipids of chloroplast

Table 3-10. Fatty Acid Composition of GlycosylDAGs in Various Plants[a]

Plant, tissue, or organelle	Glycosyl DAG[b,c]	14:0	16:0	16:1 Δ9	16:2	16:3 Δ7,10,13	16:4	18:0	18:1	18:2 Δ9,12	18:3 Δ9,12,15	18:4	20:4	20:5	22:6	References
Photosynthetic tissue																
Spinach, leaves	(1)	—	tr	tr	—	30	—	1	1	1	67					Allen et al. (1964), Gardner (1968), Haverkate and van Deenen (1965a), Jamieson and Reid (1971)
	(2)	—	6	—	—	3	—	1	4	3	84					
	(4)	—	27	—	—	tr	—	—	6	39	28					
Lamellae	(1)	—	tr	tr	—	25	—	tr	tr	2	72					Allen et al. (1966a,b)
	(2)	—	3	—	—	5	—	—	2	2	87					
	(3)	—	9	—	—	15	—	—	1	1	70					
	(4)	—	39	tr	—	tr	—	tr	1	7	52					
Tobacco, leaves	(1)	—	11	5	—	11	—	4	8	6	57					Demandre et al. (1985)
	(2)	—	18	6	—	—	—	6	7	5	59					
Runner bean, leaves	(1)	—	2	tr	—	—	—	tr	tr	2	96					Sastry and Kates (1964a)
	(2)	—	5	tr	—	—	—	1	tr	1	93					
Red clover, leaves	(1)	—	17	tr	—	—	—	1	1	6	75					Weenink (1964)
	(2)	—	9	1	—	—	—	2	2	3	83					
Alfalfa	(1)	—	3	tr	—	—	—	tr	tr	2	95					O'Brien and Benson (1964)
	(2)	—	14	tr	—	—	—	3	tr	1	82					
	(4)	—	37	2	—	—	—	7	3	2	49					
Castor, leaves	(1)[d]	—	6	—	—	—	—	tr	1	1	91					James and Nichols (1966)
	(1)[e]	—	7	3	tr	—	—	4	7	14	67					
	(2)[d]	—	11	—	—	—	—	—	—	4	85					
	(2)[e]	—	14	2	—	—	—	7	8	20	51					
	(4)[d]	—	54	tr	—	—	—	5	4	5	33					
	(4)[e]	—	29	5	1	—	—	12	15	12	17					
Fern (Adiantum capillus-veneris)[f]	(1)	—	2	0	1	30	—	0	1	1	54	—	—	7	—	Sato and Furuya (1984)
	(2)	—	33	0	1	3	—	0	2	2	56	—	—	1	—	
	(4)	—	53	0	0	0	—	1	31	3	8	—	—	1	—	
Moss	(1)	—	2	1	2	11	—	tr	1	4	48	13	29	11	—	Nichols (1965b)
	(2)	—	6	2	1	2	—	2	3	5	62	6	14	4	—	
Conifers																
Pinus sylvestris	(1)	1	5	tr	0.6	13	—	0.5	2	4	55	13	tr	—	—	Jamieson and Reid (1972)
	(2)	2	10	tr	tr	1	—	1	1	5	63	6	0.6	—	—	

(continued)

Table 3-10. (Continued)

Plant, tissue, or organelle	Glycosyl DAG[b,c]	Major fatty acids (% of total)														References
		14:0[c]	16:0	16:1 Δ9	16:2	16:3 Δ7,10,13	16:4	18:0	18:1	18:2 Δ9,12	18:3 Δ9,12,15	18:4	20:4	20:5	22:6	
Algae																
Chlorella pyrenoidosa	(1)	—	3	10	—	—	12	tr	41	5	27	3	—	—	—	O'Brien and Benson (1969)
	(2)	—	12	10	—	—	3	tr	37	6	27	3	—	—	—	
	(4)	—	68	2	—	—	—	tr	18	2	10	1	—	—	—	
Chlorella vulgaris	(1)[g]	—	5	2	19	10	—	2	17	17	45	—	—	—	—	Nichols (1965a), Haverkate and van Deenen (1965b), Safford and Nichols (1970)
	(1)[h]	—	3	11	28	3	—	—	17	33	5	—	—	—	—	
	(2)[g]	—	8	3	6	1	—	3	3	35	37	—	—	—	—	
	(2)[h]	—	10	2	4	—	—	—	11	56	4	—	—	—	—	
	(4)[g]	—	32	5	3	3	—	4	10	25	15	—	—	—	—	
	(4)[h]	—	33	11	1	—	—	—	16	28	3	—	—	—	—	
Chlamydomonas reinhardii	(1)	<1	1	2	1	2	38	tr	tr	10	35	tr	—	—	—	Eichenberger (1976), Bloch et al. (1967), Haverkate and van Deenen (1965b)
	(2)	<1	28	8	1	0	tr	tr	20	14	21	—	—	—	—	
Duvaliella tertiolecta or parvai	(1)	<1	1	tr	tr	—	30	tr	10	4	51	tr	—	—	—	Evans et al. (1982), Evans and Kates (1984)
	(2)	<1	27	2	3	—	5	8	1	11	38	tr	—	—	—	
	(4)	<1	61	n.d.	n.d.	—	tr	1	3	6	28	—	—	—	—	
Halophilic Dunaliella spp.	(1)	<1	4	1	1	—	33	3	3	8	47	—	—	—	—	Evans et al. (1982), Evans and Kates (1984)
	(2)	<1	41	tr	3	—	2	1	9	18	23	—	—	—	—	
	(4)	<1	76	tr	n.d.	—	1	2	3	8	9	—	—	—	—	
Euglena gracilis	(1)[d]	2	5	8	7	20	11	1	3	8	24	2	3	1	—	Rosenberg et al. (1966); see Bloch et al. (1967), Haverkate and van Deenen (1965b)
	(1)[e]	2	7	14	8	10	6	3	13	4	18	—	—	—	—	
	(2)[d]	1	12	8	10	27	2	1	9	6	19	—	—	—	—	
	(2)[e]	2	11	14	7	12	—	4	28	5	12	—	—	—	—	
Red alga																
Porphyra yezoensis	(1)	—	14	—	—	—	—	—	5	2	tr	0.4	2	72	—	Araki et al. (1987)
	(2)	—	42	—	—	—	—	—	6	4	tr	0	0	45	—	
	(4)	—	48	—	—	—	—	—	1	0.5	0	0.8	1	46	—	
Diatoms																
Nitzschia alba[k]	(1)	36	15	1	—	—	—	—	45	2	0.5	—	—	0.6	—	Anderson et al. (1978a), Kates (1989)
	(2)	34	11	3	—	—	—	—	49	2	0.4	—	—	1	—	
	(4)	9	16	—	—	—	—	—	8	5	2	—	1	42	14	
Phaeodactylum tricornutum[l]	(1)	2	4	22	2	25	4	0	2	1	1	—	2	34	—	Arao et al. (1987)
	(2)	7	26	28	5	2	0	3	4	2	tr	—	1	19	—	
	(4)	34	23	31	3	1	0	0	3	1	tr	—	0	5	—	

													Reference
Cyanobacteria													
Anacystis nidulans	(1)	1	43	34	—	—	—	4	20	—	—	—	Nichols *et al.* (1965)
	(2)	2	52	28	—	—	—	4	16	—	—	—	
	(4)	5	43	28	—	—	—	4	16	—	—	—	
Anabaena variabilis	(1)	0	27	28	1	—	—	tr	12	19	15	—	Nichols *et al.* (1965)
	(2)	0	26	26	2	—	—	2	9	20	17	—	
	(4)	0	53	5	tr	—	—	2	19	14	7	—	
Prochloron sp.	(1)	1(17)	35	47	—	—	—	—	—	—	—	—	Murata and Sato (1983)
	GlcDAG	5 (7)	54	34	—	—	—	—	—	—	—	—	
	(2)	1(19)	26	54	—	1	—	—	—	—	—	—	
	(4)	17(2)	70	11	—	1	—	—	—	—	—	—	
Nonphotosynthetic tissues													
Potato tubers	(1)	—	1	—	—	—	—	—	1	58	40	—	Galliard (1968a), Demandre *et al.* (1985)
	(2)	—	14	—	—	—	—	6	8	47	25	—	
	(3)	—	25	—	—	—	—	11	10	43	12	—	
	(4)	—	28	—	—	—	—	8	8	47	9	—	
Turnip root	(1)	—	12	1	1	—	—	1	10	20	56	—	Lepage (1967)
	(2)	—	21	2	1	—	—	1	8	14	55	—	
Narcissus bulb	(1)	—	4	2	1	—	1	2	5	74	10	—	Nichols and James (1964)
	(2)	—	13	2	1	—	1	2	8	69	5	—	
Oat root													
Total	(1)	tr	19	—	—	—	—	—	—	41	40	—	Liljenberg and Kates (1985)
	(2)	1	32	—	—	—	—	tr	8	38	21	—	
	(3)	—	50	—	—	—	—	tr	2	32	15	—	
Plasmalemma	(1)	5	31	—	—	—	—	5	6	28	23	—	
	(2)	tr	41	—	—	—	—	1	5	36	17	—	
	(3)	1	49	4	—	—	—	6	6	24	10	—	
Apple pulp	(1)m	—	2	—	—	—	—	1	1	5	91	—	Galliard (1968b)
	(1)n	—	6	—	—	—	—	1	7	13	73	—	
	(2)m	—	17	—	—	—	—	4	2	17	60	—	
	(2)n	—	20	—	—	—	—	7	7	24	43	—	
	(4)m	—	32	—	—	—	—	7	8	31	23	—	
	(4)n	—	35	—	—	—	—	8	5	30	22	—	

(continued)

Table 3-10. (Continued)

Plant, tissue, or organelle	Glycosyl DAG[b,c]	Major fatty acids (% of total)														References
		14:0[c]	16:0	16:1 Δ9	16:2	16:3 Δ7,10,13	16:4	18:0	18:1	18:2 Δ9,12	18:3 Δ9,12,15	18:4	20:4	20:5	22:6	
Yam tuber																
Total	(1+2+3)	tr	23	—	—	—	—	1	4	52	18	—	—	—	—	Osagie and Opute (1981)
Plasmalemma	(1)	3	8	—	—	—	—	3	4	47	35	—	—	—	—	Osagie and Kates (1986b)
	(2)	2	39	—	—	—	—	3	4	41	10	—	—	—	—	
	(3)	44	22	—	—	—	—	24	10	—	—	—	—	—	—	
	(3')	56	23	—	—	—	—	11	9	—	—	—	—	—	—	
Microsomes	(1)	1	9	—	—	—	—	tr	9	67	15	—	—	—	—	Osagie and Kates (1986b)
	(2)	1	38	—	—	—	—	3	13	43	3	—	—	—	—	
	(3)	5	45	—	—	—	—	17	8	24	tr	—	—	—	—	
	(3')	5	49	—	—	—	—	9	7	28	3	—	—	—	—	
Pumpkin	(1)	tr	10	1	—	—	—	0.6	11	38	39	—	—	—	—	Ito et al. (1974)
	(2)	tr	12	1	—	—	—	1	6	30	50	—	—	—	—	
	(3)	0.2	8	1	—	—	—	0.2	4	27	60	—	—	—	—	
Wheat grain	(1)	0.2	8	0.6	—	—	—	0.6	9	78	4	—	—	—	—	Ito et al. (1983)
	(1')	—	29	0.2	—	—	—	1	7	59	3	—	—	—	—	
	(2)	0.3	26	0.6	—	—	—	2	8	61	2	—	—	—	—	
	(2')	1	17	2	—	—	—	2	13	64	2	—	—	—	—	

Source										Reference
Rice bran	(1)	0.3	7	—	—	—	3	45	25	Fujino and Miyazawa (1979),
	(2)	1	27	—	—	2	19	28	22	Fujino (1978)
	(3)	2	32	—	—	5	18	35	9	—
	(3')	2	33	—	—	3	9	47	6	—
Corn seed	(1)	—	10	3	—	2	38	30	18	Tanaka *et al.* (1984)
	(2)	—	15	3	—	4	22	32	24	—
Millet seed	(1)	—	20	—	—	6	26	34	14	Osagie and Kates (1984)
	(1')	—	28	—	—	8	22	36	6	—
	(2)	—	17	—	—	6	20	39	19	—
Field pea[o]	(1)	—	28	—	16	15	22	6	—	Hoover *et al.* (1988)
	(1')	—	27	—	17	20	45	—	—	—
	(2)	—	19	—	12	7	39	13	—	—

[a] See Kates, 1970 and Sastry (1974) for fatty acid analyses of a more extensive variety of plants.
[b] See Fig. 3-1.
[c] Values for (2), (3) and (3') are for total classes, including major galactosyl and minor glucosyl species.
[c'] Values in parentheses are for 14:1.
[d] Grown in the light.
[e] Grown in the dark.
[f] Values are for sporophytes.
[g] Grown in light on inorganic medium (photoautotrophic).
[h] Grown in dark on organic medium (heterotrophic).
[i] Mixotrophic cultures.
[j] Values are averages of two species.
[k] Nonphotosynthetic diatom.
[l] Photosynthetic diatom.
[m] Preclimacteric.
[n] Postclimacteric.
[o] 19:0 acid was present in (1), (1'), and (2) in amounts of 7%, 11% and 10%, respectively; (1) also contained 6% of 20:0.

Table 3-11. Positional Distribution (%) of Fatty Acids in Glycosyl DAGs of Various Plants[a]

Plant	Glycosyl DAG[b]	16:0	16:1	16:2	16:3	16:4	18:0	18:1	18:2	18:3	20:5
Spinach, leaves[c,d]	(1) sn-1	4			5			2	3	90	
	sn-2	4			38				2	56	
	(2) sn-1	7			1			4	2	88	
	sn-2				3				2	88	
	(4) sn-1	32						4	10	50	
	sn-2	56						0	2	40	
Vicia faba, leaves[e]	(2) sn-1	8							1	91	
	sn-2	0							3	97	
Chlorella vulgaris[f]	(1) sn-1	—	3	6	2				29	60	
	sn-2	5	2	30	19				13	30	
Chlamydomonas reinhardtii[g]	(1) sn-1	2	1	tr	tr	tr	1	27	15	52	
	sn-2	2	14	7	10	64	tr	—	tr	tr	
	(2) sn-1	4	tr	tr	—	tr	2	55	16	19	
	sn-2	76	14	4	2	2	1	tr	tr	tr	
	(4) sn-1	68	—	—	—	—	2	6	5	10	
	sn-2	99	tr	—	tr	—	—		tr	1	
Porphyra yezoensis[h]	(1) sn-1	4						1	0	0.3	43
	sn-2	10						4	3	0	29
	(2) sn-1	5						1	0	0	43
	sn-2	38						5	4	0.2	1
	(4) sn-1	3						0.2	0.4	0	46
	sn-2	46						1	0.1	0	0.4

Pheodactylum tricornutum[i]										
(1) sn-1	tr	17	2	4	1	0	1	0.4	0.5	66
sn-2	8	27	2	47	6	0	3	1	1	1
(2) sn-1	31	11	2	2	0	4	1	0	0	32
sn-2	17	49	8	1	0	1	7	5	6	3
(4) sn-1	2	19	4	1	0	0	1	1	0.4	10
sn-2	45	43	0	1	0	0	5	0.4	0	0
Cyanobacteria										
Anacystis nidulans[j]										
(1) sn-1	5	34								
sn-2	34	15								
(2) sn-1	4	39								
sn-2	36	12								
(4) sn-1	2	41								
sn-2	49	1								
Anabaena variabilis[j]										
(1) sn-1	0	4	1	0	0	7	15	23		
sn-2	21	24	3	0	0	0	0	1		
(2) sn-1	0	4	0	0	0	3	13	29		
sn-2	18	25	5	0	0	0	0	0		
(4) sn-1	3	6	1	1	13	13	13	14		
sn-2	49	1	0	0	0	0	0	0		

[a] See Sastry (1974), for more extensive coverage of fatty acid positional distribution.
[b] See Fig. 3-1.
[c] Safford and Nichols (1970); Tulloch et al. (1973); Auling et al. (1971).
[d] Data for (4) are for chloroplast envelopes (Siebertz et al., 1979).
[e] Tulloch et al. (1973).
[f] Safford and Nichols (1970).
[g] Giroud et al. (1988).
[h] Araki et al. (1987). Data are given as mole% of total fatty acids.
[i] Arao et al. (1987).
[j] Grown at 22°C; Sato et al. (1979).

lamellae, and lack the 16:3 acid; these plants are called 18:3 plants. By contrast, species of Bryophytes (mosses and liverworts), Gymnospermae (conifers), Pteridophytes (ferns), and some angiosperms such as Solanaceae (tobacco, tomato), Brassicaceae (broccoli, mustard), and Chenopodiaceae (spinach, sugar beet) have the 16:3 acid concentrated in the *sn*-2-position of GalDAG in chloroplast lamellae with the 18:3 in both positions; such plants are called 16:3 plants (see Tables 3-10 and 3-11).

Note that in 16:3 plants, the Gal$_2$DAG has only small proportions of 16:3, while the sulfoquinovosylDAG has only traces or none of this acid. In both 16:3 and 18:3 plants, the sulfolipid is much more saturated than the galactolipids, having much lower proportions of 18:3 acid and much higher proportions of 16:0 acid (Table 3-10).

3.7.1.2. Algae, Diatoms, and Cyanobacteria

In green algae such as *Chlorella* and *Chlamydomonas* and in *Euglenophyta* species, the GalDAG and to a lesser extent the Gal$_2$DAG contain appreciable amounts of the tetraenoic acid 16:4, as well as the 16:3, the 18:1, the 18:2 and the 18:3 acids. *Dunaliella* species, however, also have the 16:4 acid in their galactolipids but lack the 16:3 acid. As in higher plants, the sulfolipid of green algae is much more saturated than the galactolipids, with 16:0 as the major acid (Table 3-10). Green algae thus appear to have the characteristics of 16:3 plants. It should be noted that *Chlorella* and *Euglena*, when grown heterotrophically in the dark, contain more saturated species of galactolipids, having lower contents of 16:3, 16:4, 18:2, and 18:3 acids and correspondingly higher contents of 16:0, 16:1, 18:0, and 18:1 acids, than when grown in the light. Similar changes occur in etiolated leaves of higher plants (e.g., castor) (Table 3-10).

In contrast to the green algae, the GalDAG of red algae (e.g., *Porphyra yezoensis*) contains little or no C$_{16}$ or C$_{18}$ polyunsaturated acids, having instead high proportions of C$_{20}$ polyunsaturated acids, largely 20:5 acid. Both the Gal$_2$DAG and the sulfolipid are more saturated, having high proportions of 16:0 and correspondingly lower amounts of the 20:5 acid (Table 3-10).

Photosynthetic diatoms (e.g., *Phaeodactylum tricornutum*) also have fairly high amounts of the 20:5 acid and low contents of C$_{18}$ polyunsaturated acids in the galactolipids but have high contents of 16:1 and 16:3 acids (Arao *et al.*, 1987). However, the nonphotosynthetic diatom *Nitzschia alba*, lacks both the 16:3 and 18:3 acids in the galactolipids (having instead high contents of the 18:1 acid); surprisingly, the sulfolipid is highly unsaturated, having high contents of 20:5 and 22:6 acids (Anderson *et al.*, 1978*a*) (Table 3-10).

In cyanobacteria, the fatty acid composition of the galactolipids and the sulfolipid of *A. nidulans* resembles that of a prokaryote, with 16:0, 16:1, and 18:1 as major acids. Similar fatty acid compositions (except for the absence of C$_{18}$ acids) were found for the glycosylDAGs of the prokaryotic alga *Prochloron* (Murata and Sato, 1983), which has some properties similar to those of chloroplasts of green algae and higher plants (e.g., presence of both chlorophyll a and b) (Tables 3-10 and 3-11). However, the fatty acids of these same glycolipids in *A. variabilis* contain in addition the polyunsaturated acids 18:2 and 18:3, as found in green algae.

3.7.1.3. Nonphotosynthetic Tissue

The glycosylDAGs of nonphotosynthetic tissues of higher plants are characterized in general by their high contents of saturated fatty acids, mainly 16:0, and of 18:2 and to a lesser extent 18:1, in comparison with photosynthetic tissue (Table 3-10). However, fruits, tubers, and roots still contain appreciable amounts of the 18:3 acid, but seeds generally contain reduced amounts of this acid (with the obvious exception of linseed) but not of 18:2. Also, the Gal$_2$DAG and the sulfolipid, when present, are more saturated than the GalDAG, because of their higher contents of 16:0 acid (Table 3-10).

In regard to the fatty acid composition of glycosylDAGs in organelles of nonphotosynthetic tissues, the limited data available indicate that in the plasmalemma of oat root, the glycosylDAGs are somewhat more saturated than the corresponding glycolipids in the intact root, because of their higher contents of 16:0 and 18:0 acids and lower contents of 18:3 and 18:2 acids (Liljenberg and Kates, 1985). In yam tubers, the fatty acid composition of the galactolipids of plasmalemma and microsome fractions are quite similar, with the exception that the tri- and the tetragalactosylDAGs are more saturated in the plasmalemma (Table 3-10).

3.7.2. Positional Distribution of Fatty Acids and Molecular Species of GlycosylDAGs

3.7.2.1. Higher Plants

Positional distribution of fatty acids in glycosylDAGs is readily determined by enzymatic hydrolysis of the *sn*-1 fatty acid ester with a lipase from *Rhizopus arrhizus delemar* (Fischer *et al.*, 1973) and GLC analysis of the released fatty acids and those of the lyso compound formed (see Section 3.4.4). Molecular species are analyzed by fractionation of the mixture of species using AgNO$_3$/SiO$_2$ TLC or HPLC and GLC analysis of the fatty acids of the separated species or by chemical ionization mass spectrometry of the total mixture of species (see Sections 3.4.2 and 3.4.4). In leaves of 18:3 plants (e.g., *Vicia faba*), the major 18:3 acid in Gal$_2$DAG is equally distributed between the *sn*-1 and -2 positions, while the minor 16:0 acid occupies only the *sn*-1 position. The major molecular species as determined by mass spectrometry is clearly the dilinolenoyl (18:3/18:3) species with low amounts of 16:0/18:3 and 18:3/18:2 species (Tables 3-11 and 3-12). In 16:3 plants (e.g., spinach), the 16:3 acid is mostly located at the *sn*-2 position in the galactolipids, and the 18:3 acid is located at both positions. The main molecular species are thus 18:3/18:3 and 18:3/16:3 species, the latter being a minor species in the Gal$_2$DAG (Tables 3-11 and 3-12).

Table 3-12. Major Molecular Species Composition of GlycosylDAGs in Various Plants

Plant	Molecular species sn-1	sn-2	Mol% (1)	(2)	(3+3′)	(4)
Spinach, leaves[a]	18:3	18:3	56	88		
	18:3	16:3	38	3		
	16:0	18:3	4	7		
Tobacco, leaves[b]	18:3	18:3	72	74		
	18:3	16:3	22	—		
	18:2	18:3	4			
	16:0	18:3	3	26		
Vicia faba, leaves[c]	18:3	18:3	89	78		
	16:0	18:3	8	20		
	18:3	18:2	3	2		
Chlorella vulgaris[d]	18:3	18:3	30			
	18:3	16:3	19			
	18:2	18:2	13			
	18:2	16:2	30			
Chlamydomonas reinhardtii[e]	18:3	16:4	70	2		16
	18:3	16:3	16	11		15
	18:2	16:4	8			
	18:3	18:3		4		
	18:3	16:0		18		
	18:2	16:0		25		
	18:1	16:0		13		20
	16:0	16:0				49
Porphyra yezoensis[f]	20:5	20:5	++	—		—
	20:5	16:0	+	++		++
Phaeodactylum tricornutum[g]	20:5	16:3	+++	tr		tr
	20:5	16:1	++	+++		+
	20:5	16:0	+	++		+
	14:0	16:0 (16:1)	—	+		+++
Anabaena variabilis[h]	18:3	16:2	(i) 12 (ii) 0	(i) 4 (ii) 0		
	18:3	16:1	32 0	37 0		
	18:3	16:0	34 1	19 0		
	18:2	16:0	12 23	20 24		
	18:2	16:1	3 35	9 38		
	18:1	16:0	2 25	4 16		
Potato tubers[b]	18:3	18:3	17	5		
	18:3	18:2	36	21		
	18:2 16:0	18:2 18:3 }	47	51		
	16:0	18:2	tr	14		
	18:0	18:2	—	9		
Rice bran[i]	18:3	18:3	3	10	10	—
	18:2 18:3	18:3 18:2 }	15	15		2
	18:2	18:2	15	16		—
	18:1 18:3	18:3 18:1 }	16	15		—
	18:1	18:2	12	8		3

Table 3-12. (Continued)

Plant	Molecular species sn-1	sn-2	(1)	(2)	(3+3')	(4)
	16:0	18:3	12	10		2
	16:0	18:2	14	8		9
	18:1	18:1	6	6		2
	18:2	16:0	—	—		19
	16:0	18:1	5	8		7
	18:1	16:0	—	—		12
	16:0	16:0	—	—		38
Azuka bean[j]	18:3	18:3	+++	+++	+++	—
	18:3	18:2	+	+	+	—
	18:2	18:3	+	+	+	—
	18:2	18:2	+	+	+	—

[a]Values calculated from data of Safford and Nichols (1970), for (1) and from data of Tulloch *et al.* (1973).
[b]Demandre *et al.* (1985).
[c]Tulloch *et al.* (1973); Williams *et al.* (1983).
[d]Calculated from data of Safford and Nichols (1970).
[e]Giroud *et al.* (1988).
[f]Araki *et al.* (1987).
[g]Arao *et al.* (1987).
[h]Sato and Murata (1982b) (i) at 22°C, (ii) at 38°C.
[i]Fujino (1978).
[j]Ito *et al.* (1985a).

3.7.2.2. Algae, Diatoms, and Cyanobacteria

Much more complex molecular species compositions of glycosylDAGs are found in green algae (e.g., *Chlorella* and *Chlamydomonas*) because of the presence of considerable amounts of 18:2, 16:2, and 16:4 acids in addition to 18:3, 18:1, 16:0, and 16:3 acids found in higher plants. It is of interest that the C_{18} unsaturated acids are located at the *sn*-1 position, while the C_{16} acids, including 16:0, are located on the *sn*-2 position (Table 3-11). The major species of GalDAG include 18:3/16:4, 18:3/16:3, 18:2/16:4, as well as 18:3/18:3, but both the Gal_2DAG and the sulfolipid contain major amounts of the more saturated 18:3/16:0, 18:2/16:0, and 18:1/16:0, and the sulfolipid has in addition high amounts of the fully saturated 16:0/16:0 species (Table 3-12).

Red algae and photosynthetic diatoms have major molecular species of the glycosylDAGs containing eicosapentaenoic acid located mostly at the *sn*-1 position, and 16:0 located mostly at the *sn*-2 position (Table 3-11).

In red algae, the main molecular species of GalDAG is 20:5/20:5 while that of Gal_2DAG and the sulfolipid is 20:5/16:0. In the photosynthetic diatom, the major GalDAG species are 20:5/16:3 and 20:5/16:1 while those of Gal_2DAG and the sulfolipid are 20:5/16:1 and 20:5/16:0 (Table 3-12).

In cyanobacteria, the C_{16} fatty acids of glycosylDAGs are localized mostly at the *sn*-2 position and the C_{18} fatty acids at the *sn*-1 position as in 16 : 3 plants (Sato *et al.*, 1979, 1982*b*; Zepke *et al.*, 1978). In *A. variabilis* grown at 22°C, the major molecular species are 18 : 3/16 : 1, 18 : 3/16 : 0, 18 : 3/16 : 2, and 18 : 2/16 : 0 (Sato and Murata, 1982*b*) resembling the molecular species composition of spinach (Table 3-12).

3.7.2.3. Nonphotosynthetic Tissue

Few plant species have been examined for glycosylDAG molecular species in nonphotosynthetic tissues. Data available on rice bran glycosylDAGs (Fujino, 1978) show that the main species of mono- and diglycosylDAGs (both galactose and glucose varieties) are trienoic, mainly 16 : 0/18 : 3 and 18 : 1/18 : 2 (18 : 2/18 : 1); tetraenoic, mainly 18 : 2/18 : 2 and 18 : 1/18 : 3 (18 : 3/18 : 1); and pentaenoic, mainly 18 : 2/18 : 3 (18 : 3/18 : 2). The sulfolipid species, however, are mostly saturated (16 : 0/16 : 0), monoenoic (16 : 0/18 : 1, 18 : 1/16 : 0), and dienoic (16 : 0/18 : 2, 18 : 2/16 : 0, 18 : 1/18 : 1) (Table 3-12).

In azuka beans, the main molecular species of GalDAG, Gal_2DAG, Gal_3DAG, and Gal_4DAG is the 18 : 3/18 : 3 species, with lesser amounts of the 18 : 2/18 : 3 and 18 : 2/18 : 2 species (Ito *et al.*, 1985*a*).

3.8. Metabolism of Glycolipids

3.8.1. GlycosylDAGs

3.8.1.1. Biosynthesis of Galactosyldiacylglycerols

In higher plants and algae, the terminal step in the *de novo* biosynthesis of galactosylDAGs has been demonstrated to occur in intact chloroplasts by transfer of a galactosyl group from UDP–Gal to a suitable diacylglycerol acceptor or to a galactosylated diacylglycerol (Neufeld and Hall, 1964; Ongun and Mudd, 1968; Mudd *et al.*, 1969; Matson *et al.*, 1970; Eccleshall and Hawke, 1971), as follows:

$$1,2\text{-Diacyl-}sn\text{-glycerol} + \text{UDP-Gal} \rightarrow \text{GalDAG (1)} + \text{UDP} \qquad (1)$$

$$\text{GalDAG} + \text{UDP-Gal} \rightarrow \text{Gal-GalDAG (2)} + \text{UDP} \qquad (2)$$

$$\text{Gal-Gal-DAG} + \text{UDPGal} \rightarrow \text{Gal-Gal-Gal-DAG (3)} + \text{UDP} \qquad (3)$$

Evidence has been presented that separate enzymes are involved in galactosyl transfer reactions (1) and in (2) and (3) (Ongun and Mudd, 1968; Mudd *et al.*, 1969, 1971; Williams *et al.*, 1975, 1983), as would be expected for formation of a β-galactosidic bond in reaction (1) and an α-galactosidic bond in reactions (2) and (3). While the chloroplast has for some time been accepted as the chief if not the unique site of galactolipid synthesis in all photosynthetic cells (see Heemskerk and Wintermans, 1987), the location of the galactolipid synthesizing enzymes in the plastid envelopes has been established relatively recently (Douce and Joyard, 1980): the enzyme for reaction (1) on the inner envelope of spinach chloroplasts (Block *et al.*, 1983*b;* Heemskerk *et al.*, 1985) but in the outer envelope of pea chloroplasts (Cline and Keegstra, 1983), and that for galactosylation of GalDAG (reaction 2) on the outer envelope (Cline and Keegstra, 1983; Heemskerk *et al.*, 1986).

However, the formation of Gal_2DAG and Gal_3DAG by reactions (2) and (3), respectively, cannot account entirely for the biosynthesis of these galactolipids since spinach chloroplast envelope preparations can convert GalDAG into Gal_2DAG in the absence of UDPGal by the action of a galactolipid : galactolipid galactotransferase (van Besouw and Wintermans, 1978; Heemskerk *et al.*, 1983), involving reversible dismutation of two moles of GalDAG into Gal_2DAG and DAG:

$$\text{GalDAG} + \text{GalDAG} \leftrightarrow Gal_2DAG\ (\mathbf{2}) + \text{DAG} \tag{4}$$

Gal_3DAG and Gal_4DAG can be formed by similar reactions:

$$Gal_2DAG + \text{GalDAG} \leftrightarrow Gal_3DAG\ (\mathbf{3}) + \text{DAG} \tag{5}$$

$$Gal_3DAG + \text{GalDAG} \leftrightarrow Gal_4DAG\ (\mathbf{3'}) + \text{DAG} \tag{6}$$

Note that the DAG released in these reactions can be galactosylated by reaction (1) to form more GalDAG, which in turn can undergo dismutation by reaction (4), and so forth. Also, as was found in the early studies of Ongun and Mudd (1968) and Mudd *et al.* (1969), diolein or saturated/oligoene DAGs are efficient substrates for galactosylation by reaction (1), while polyene GalDAGs are more efficient for galactosylation by the dismutation reaction (4) (Heemskerk and Wintermans, 1987). The dismutase galactosyltransferase is localized in the outer chloroplast envelope (Dorne *et al.*, 1982; Heemskerk *et al.*, 1986, 1988), and is activated during isolation of chloroplast envelope membranes (Wintermans *et al.*, 1981; Dorne *et al.*, 1982), and by mono- and divalent cations, and is strongly inhibited by the proteinase thermolysin (Dorne *et al.*, 1982).

The question of the *in vivo* role of the galactolipid : galactolipid galactotransferase in galactolipid biosynthesis has not yet been settled, although the low pH optimum of this enzyme, its location in the outer chloroplast envelope and its production of unphysiologically high amounts of Gal_3DAG and Gal_4DAG would suggest its involvement in galactolipid catabolism rather than anabolism (Dorne *et al.*, 1982; Douce *et al.*, 1984, 1987*a*). By contrast, *in vivo* studies by Williams *et al.* (1975) showed that Gal_2DAG synthesis is a relatively slow process involving galactosylation of a separate pool of polyunsaturated GalDAG. This observation is not inconsistent with operation of the dismutase galactotransferase [reaction (4)] as well as the Gal_2DAG galactotransferase [reaction (2)]. As pointed out by Wintermans and Heemskerk (1987), the former enzyme might function in the eukaryotic pathway for GalDAG synthesis by catalyzing the reverse reaction of a Gal_2DAG in the outer chloroplast envelope with a DAG synthesized in the endoplasmic reticulum [reaction (4)].

In cyanobacteria, biosynthesis of galactolipids has been shown not to proceed as in higher plants and algae by galactosylation of diacylglycerol [reaction (1)] (Sato and Murata, 1982*a,c*). Instead, synthesis of the minor glycolipid, GlcDAG, occurs first by transfer of a glucosyl group from UDP–Glc to a diacylglycerol:

$$1,2\text{-Diacyl-}sn\text{-glycerol} + \text{UDP-Glc} \rightarrow \text{GlcDAG} + \text{UDP} \qquad (7)$$

This glucosyltransferase is associated with the membrane fraction of *Anabaena variabilis* and is strongly dependent on Mg^{2+} ions; it uses both saturated (e.g., $16:0/16:0$, $18:0/16:0$) and unsaturated (e.g., $18:1/16:0$) DAGs as glucosyl acceptor, the latter DAG being a better substrate (Sato and Murata, 1982*c*). No UDP–Gal-dependent galactotransferase activity was detected in the membrane fraction or in the soluble fraction.

Rapid conversion of GlcDAG to GalDAG was observed in whole cells of *A. variabilis* by epimerization at C-4 of the glycosyl group in GlcDAG and not by exchange of the glucosyl group with a galactosyl group (Sato and Murata, 1982*a*):

$$\text{GlcDAG} \xrightarrow{\text{epimerase}} \text{GalDAG} \qquad (8)$$

Labeling studies with intact cells of *A. variabilis* also showed that Gal_2DAG was probably synthesized by transfer of a galactose unit, presumably from UDP–Gal to GalDAG, as in reaction (2), rather than by the dismutation of GalDAG [reaction (4)] (Sato and Murata, 1982*a*). However, negligible incorporation of galactose from UDP–Gal into Gal_2DAG was observed with the membrane fraction, although a low rate of incorporation was detected with the soluble fraction of *A. variabilis* (Sato and Murata, 1982*c*). Further studies on the biosynthesis of Gal_2DAG are necessary to establish unambiguously the pathway for this galactolipid in cyanobacteria, which should have a bearing on the pathway for Gal_2DAG in higher plants and algae.

3.8.1.2. Biosynthesis of SulfoquinovosylDAG

An analogous reaction to reaction (1) has been suggested for biosynthesis of the plant sulfolipid, sulfoquinovosylDAG (Benson, 1963):

$$1,2\text{-Diacyl-}sn\text{-glycerol} + \text{NDP-sulfoquinovose} \rightarrow \text{SQDAG (4)} + \text{NDP} \qquad (9)$$

Indirect evidence for this terminal step in the synthesis of SQDAG in intact chloroplasts, such as competition of sulfolipid and galactolipid synthesis for the same diacylglycerol pool in the chloroplast envelope (Kleppinger-Sparace *et al.*, 1985) and incorporation into sulfolipid of DAGs formed by the envelope Kornberg–Pricer pathway (Joyard *et al.*, 1986, 1987) has been presented. However, more direct evidence for this reaction, by incubation of chloroplasts with labelled DAGs and UDP-sulfoquinovose has not yet been forthcoming.

The question of the origin of the sulfoquinovose moiety of the sulfolipid

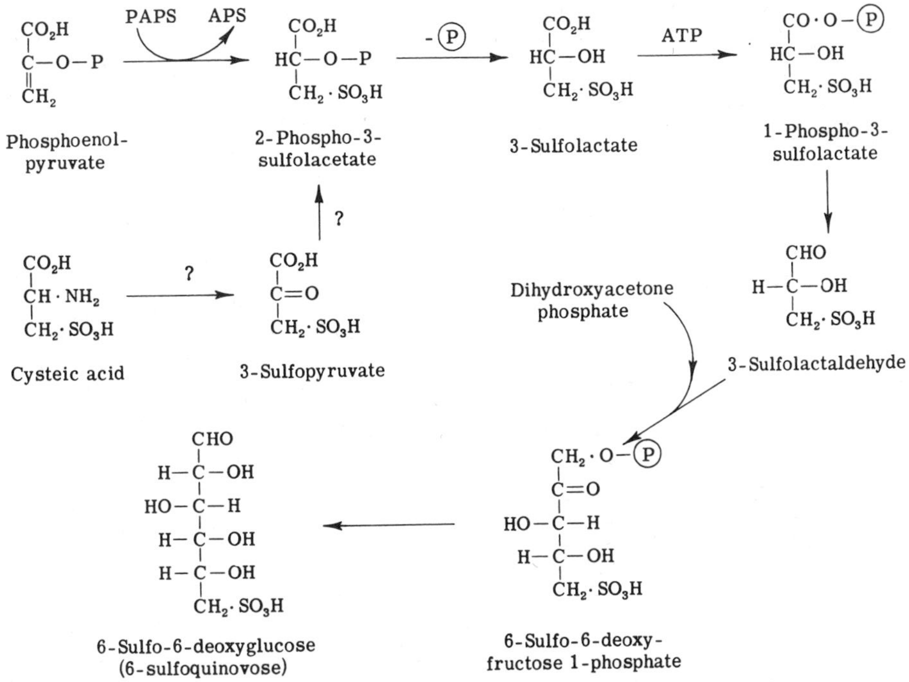

Figure 3-11. Proposed sulfoglycolytic pathway for biosynthesis of 6-sulfoquinovose. (After Davies *et al.*, 1966.)

has not yet been settled (see Kates, 1970; Haines, 1973; Mudd and Garcia, 1975; Kleppinger-Sparace *et al.*, 1985). In *Euglena*, biosynthesis of SQDAG appears to involve the incorporation of cysteic acid via a sulfoglycolytic pathway as shown in Fig. 3-11 (Davies *et al.*, 1966). However, studies with higher plants have shown that cysteic acid is not an effective precursor of SQDAG (Mudd *et al.*, 1980), thus casting doubt on the existence of a sulfoglycolytic pathway. Isolated chloroplasts have been found to incorporate ^{35}S from [^{35}S]sulfate into SQDAG, and this incorporation was light dependent (Haas *et al.*, 1980; Kleppinger-Sparace *et al.*, 1985). Further studies showed that this light dependence could be replaced by ATP or APS but not by PAPS (Kleppinger-Sparace and Mudd, 1987). A general scheme for biosynthesis of the sulfoquinovosylDAG which incorporates the information available up to the present is shown in Fig. 3-12. It should be emphasized that the steps leading to the synthesis of sulfoquinovose have still to be established with certainty. Studies with chloroplast membrane fractions must also be carried out to characterize reaction (9), particularly the diacylglycerol molecular species involved, and the identity of the nucleotide in NDP–sulfoquinovose.

3.8.1.3. Biosynthesis of Diacylglycerols

From the foregoing review, it is clear that diacylglycerols are the key substrates in the biosynthesis of the galactolipids and the sulfolipid, and that

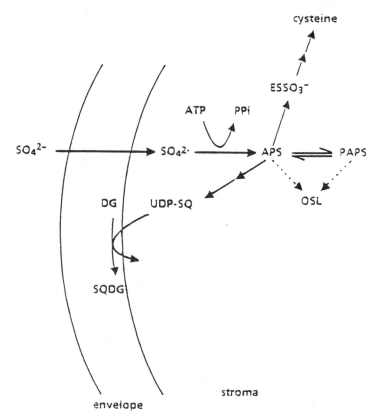

Figure 3-12. General scheme for biosynthesis of sulfoquinovosylDAG in the chloroplast. (After K. F. Kleppinger-Sparace, unpublished data.)

the fatty acid composition, the fatty acid distribution between *sn*-1 and -2 positions and ultimately the characteristic molecular species composition of each of these glycolipids is determined by the selectivity of the glycosyltransferases from among the available DAG molecular species. We must now consider the biosynthetic pathways for DAGs that are operative in plants (see reviews by Roughan, 1987; Heemskerk and Wintermans, 1987).

Two types of photosynthetic plants can be distinguished, based on the fatty acid composition of their glycosylDAGs: the 16 : 3 plants, with 16 : 3 concentrated in the *sn*-2 position of GalDAG, and the 18 : 3 plants, having 18 : 3 at both *sn*-1 and 2 positions (see Tables 3-10 and 3-11) (see Section 3.7.1). Since acyl lipids of cyanobacteria were found to have C_{16} acids predominantly in the *sn*-2 position, acyl-lpids with such a pattern are called prokaryotic, while those with C_{18} fatty acids at the *sn*-2 position are called eukaryotic (see Tables 3-10, 3-11, and 3-12). Prokaryotic-type DAGs are made in the chloroplast (see Fig. 3-13), while eukaryot-type DAGs are made in the cytosol of the plant cell, mainly in the endoplasmic reticulum (ER) (see Fig. 3-14).

Chloroplasts in higher plants and green algae can synthesize 16 : 0-ACP

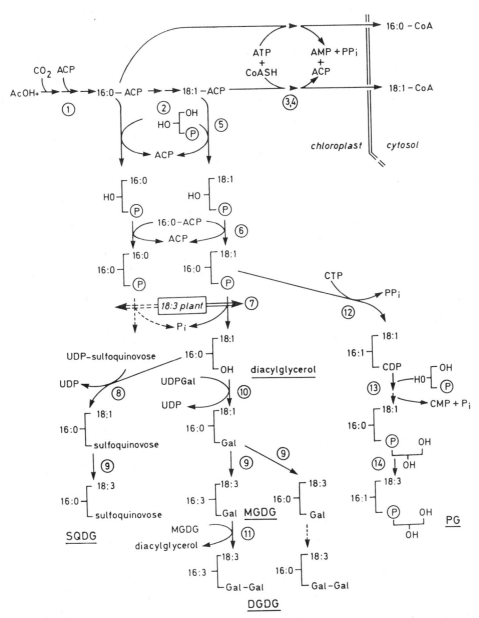

Figure 3-13. Prokaryotic pathway for acyl–lipid synthesis in 16 : 3 (and 18 : 3) plants. Enzymes involved: 1, fatty acid synthetase; 2, elongation and desaturation of 16 : 0-ACP; 3, acyl-ACP thioesterase; 4, acyl–CoA synthetase; 5, acyl-ACP: *sn*-glycerol 3-phosphate acyltransferase; 6, acyl-ACP: 1-acyl-*sn*-glycerol 3-phosphate acyltransferase; 7, phosphatidate phosphatase; 8, UDP–sulfoquinovose: *sn*-1,2-DAG sulfoquinovosyltransferase; 9, SQDAG and GalDAG desaturase; 10, UDP-Gal: *sn*-1,2-DAG galactosyl–transferase; 11, galactolipid:galactolipid galactosyltransferase; 12, CTP: phosphatidate cytidyltransferase; 13, CDP–diacylglycerol: *sn*-glycerol-3-phosphate phosphatidyltransferase; and PG-phosphate phosphohydrolase; 14, desaturation of PG (formation of 16:1-3*t*). (After Heemskerk and Wintermans, 1987.)

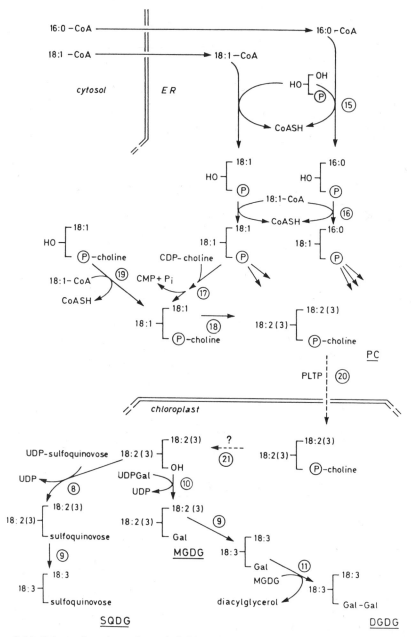

Figure 3-14. Eukaryotic pathway for acyl–lipid synthesis in 16:3 and 18:3 plants. Numbers refer to enzymes involved (see legend to Fig. 3-13): 15, acyl-CoA: *sn*-glycerol 3-phosphate acyltransferase; 16, acyl-CoA: 1-acyl-*sn*-glycerol 3-phosphate acyltransferase; 17, phosphatidate phosphatase and CDP–choline-*sn*-1,2-DAG choline phosphotransferase; 18, PC desaturase; 19, acyl-CoA: 1-acyl-*sn*-glycerol 3-phosphocholine acyltransferase; 20, phospholipid transfer protein (PTP); 21, conversion path from PC to eukaryotic diacylglycerol, as yet unknown. (After Heemskerk and Wintermans, 1987.)

in the chloroplast stroma, chain elongate it to 18 : 0-ACP and then desaturate it to 18 : 1-ACP. Some of these acyl-ACPs can then be converted to the corresponding acyl-CoAs on the outer envelope membrane and transported to the ER for use in eukaryotic synthesis of DAGs (Block *et al.,* 1983*a*) (see Fig. 3-14). The acyl-ACPs remaining in the stroma are used to acylate *sn*-glycero-3-phosphate (3-P-Gro) in two steps: first by the *sn*-1 specific stromal acyl-ACP: 3-P-Gro acyltransferase, using 18 : 1 or 16 : 0, to form 1-acyl-3-P-Gro (lysophosphatidic acid, lyso-PA), and then by acylation of the latter by the 16 : 0-ACP specific acyl-ACP: 1-acyl-3-P-Gro acyltransferase in the inner envelope to form phosphatidic acid (PA), as follows (see Fig. 3-13):

$$sn\text{-}3\text{-}P\text{-}Gro + \text{acyl-ACP} \rightarrow 1\text{-}acyl\text{-}sn\text{-}3\text{-}P\text{-}Gro + \text{ACP} \qquad (10)$$

$$1\text{-}acyl\text{-}sn\text{-}3\text{-}P\text{-}Gro + 16 : 0\text{-}ACP \rightarrow 1\text{-}acyl\text{-}2\text{-}16 : 0\text{-}3\text{-}P\text{-}Gro + \text{ACP} \qquad (11)$$

The phosphatidic acid formed is then hydrolyzed by a specific phosphatidate phosphatase present in the inner chloroplast envelope of 16 : 3 plants but not in that of 18 : 3 plants (Heinz and Roughan, 1983) to form diacylglycerols, as follows (see Fig. 3-13).

$$1\text{-}acyl\text{-}2\text{-}16 : 0\text{-}sn\text{-}3\text{-}P\text{-}Gro \rightarrow 1\text{-}acyl\text{-}2\text{-}16 : 0\text{-}sn\text{-}Gro + P_i \qquad (12)$$

These steps lead to the formation of 18 : 1/16 : 0-DAG as the major species, with only small amounts of 16 : 0/16 : 0-DAG, and these are used in the synthesis of GalDAG and SQDAG by reactions (1) and (9), respectively (see Fig. 3-13). Formation of the more unsaturated GalDAG species 18 : 3/16 : 3 and 18 : 3/16 : 0 and of the 18 : 3/16 : 0 species of SQDAG takes place through the action of specific desaturase enzyme systems for GalDAG and SQDAG, presumably located on the thylakoid membranes, which contain the required electron transport systems (Ohnishi and Yamada, 1982; Dubacq *et al.,* 1983, 1984; Cho and Thompson, Jr., 1987) (see Fig. 3-15). The polyunsaturated species of GalDAG (18 : 3/16 : 3 and 18 : 3/16 : 0) are then used in the synthesis of Gal$_2$DAG either by the dismutase galactotransferase reaction (4) or the UDP–Gal galactotransferase reaction (2) (see Fig. 3-13).

The prokaryotic pathway thus accounts for the predominance of 18 : 3/16 : 3 molecular species of galactolipids in 16 : 3 plants and in green algae (see Table 3-12). The absence of phosphatidate phosphatase in the inner chloroplast envelope of 18 : 3 plants (Heinz and Roughan, 1983) then accounts for the absence in these plants of molecular species of galactolipids and sulfolipid containing the 16 : 3 acid (Tables 3-10 and 3-12). To explain the presence of 18 : 3/18 : 3 and 16 : 0/18 : 3 species of galactolipids in both 16 : 3 and 18 : 3 plants (Table 3-12) we must now consider the eukaryotic pathway which is operative in both types of plants.

The eukaryotic pathway (see Fig. 3-14) is a more complicated pathway than the prokaryotic one, with which it is integrated at two points. First, acyl-CoAs (18 : 1- and 16 : 0-CoA) are exported by the chloroplast to the endo-

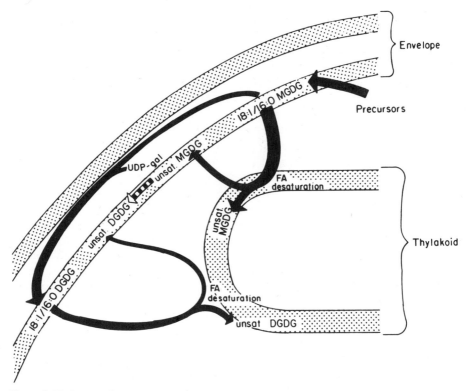

Figure 3-15. Proposed sites of galactolipid synthesis and desaturation in the chloroplast. (After Cho and Thompson, Jr., 1987.)

plasmic reticulum where they are used to acylate *sn*-3-P-Gro by the ER-bound acyl-CoA : 3-P-Gro acyltransferase to form 1-16 : 0(18 : 1)-*sn*-3-P-Gro (lysoPA):

$$sn\text{-}3\text{-}P\text{-}Gro + acyl\text{-}CoA \rightarrow 1\text{-}acyl\text{-}sn\text{-}3\text{-}P\text{-}Gro + CoA \tag{13}$$

This reaction is analogous to that occurring in the plastid stroma [reaction (10)] and gives rise to the same molecular species of lyso-PA (Fig. 3-13 and 3-14). The latter are then acylated by the 18 : 1-CoA specific ER-bound acyl-CoA : lyso-PA acyltransferase to form the PA species 18 : 1/18 : 1 and 16 : 0/18 : 1, thus:

$$1\text{-}Acyl\text{-}sn\text{-}3\text{-}P\text{-}Gro + 18:1\text{-}CoA \rightarrow 1\text{-}acyl\text{-}2\text{-}18:1\text{-}3\text{-}P\text{-}Gro + CoA \tag{14}$$

Acylation reactions (13) and (14) also occur in microsomes of developing oil seeds but acylation of the *sn*-2 position occurs better with the 18 : 2-CoA (Griffiths *et al.*, 1985).

The phosphatidic acid thus formed is then dephosphorylated by the ER-bound phosphatidate phosphatase to the 18 : 1/18 : 1 and 16 : 0/18 : 1 species of DAG, which are converted in the ER of leaves and oil seeds to the corresponding species of phosphatidylcholine (PC) (Kates and Marshall, 1975; Moore, 1982; Roughan and Slack, 1982; Griffiths *et al.*, 1985). Thus:

$$\text{diacylglycerol} + \text{CDP-choline} \rightarrow \text{phosphatidylcholine} + \text{CMP} \qquad (15)$$

The 18:1-PC species are then desaturated by a specific ER-bound Δ^{12}-desaturase system to the corresponding 18:2-PC (both at *sn*-1 and -2 positions) (Murphy *et al.*, 1985). Further desaturation of 18:2-PC to 18:3-PC may also take place by a specific Δ^{15}-desaturase in the ER (Appleby *et al.*, 1971; Norman and St. John, 1986). At this point, the desaturated PC, mainly di-18:2-PC is transported to the chloroplast envelope by means of a phospholipid transport protein (PLTP) (Kader, 1985) where it is converted sequentially to di-18:2-DAG, then to di-18:2-Gal-DAG and finally desaturated to di-18:3-GalDAG (Fig. 3-15) (Appleby *et al.*, 1971; Cho and Thompson, Jr., 1987; Ohnishi and Yamada, 1982; Dubacq *et al.*, 1983, 1984). The latter can be converted to 18:3-Gal$_2$DAG species by reactions (2) or (4), as discussed above (see Fig. 3-14). Conversion of the 18:2-DAG to the sulfolipid and its further desaturation to di-18:3-species also can take place in the chloroplast envelope (Joyard *et al.*, 1986, 1987; Ohnishi and Yamada, 1982).

The eukaryotic pathway thus can account readily for the presence of 18:3/18:3 molecular species of galactolipids and sulfolipid in both 18:3 plants and 16:3 plants. However, the mechanism by which PC is converted to diacylglycerol in the plastid envelope is still not known, since the phospholipase C necessary for this conversion has not been detected in envelope membranes (Douce *et al.*, 1987*b*). Further studies related to this question would be of great interest. Nevertheless, it is clear that the cooperative action of ER and chloroplast enzyme systems in the prokaryotic and eukaryotic pathways (two-pathway model) can completely explain the complicated molecular species compositions not only of glycosylDAGs but also of phospholipids in 16:3 and 18:3 plants (Roughan, 1987; Heemskerk and Wintermans, 1987). Furthermore, it is clear that the differentiation of the 18:3 plants from the 16:3 plants most probably is a consequence of the absence of the gene for phosphatidate phosphatase in the former plants (Heinz and Roughan, 1983; Roughan, 1987).

3.8.1.4. Catabolism of GlycosylDAGs

3.8.1.4a. Hydrolysis of Galactolipids. The enzymatic degradation of polar lipids is readily demonstrated in leaves of certain plants (e.g., *Phaseolus*) by comparing the lipids extracted from intact leaves with those from leaf homogenates. An almost complete disappearance of the phospholipids PC and PE, with concomitant formation of PA (due to the action of phospholipase D), and of the galactolipids is observed in the homogenates (Sastry and Kates, 1964*b*). The degradation of GalDAG and Gal$_2$DAG is due to the action of an acyl hydrolase, which appears to be associated with the chloroplast fraction but is also present in the cell-sap cytoplasm. This enzyme is distinct from any phospholipase, and has little or no activity toward diacyl- or triacylglycerols (Sastry and Kates, 1964*b*) and has therefore been termed a galactolipase (E.C. 3.1.1.26). However the term acyl hydrolase is more realistic in view of the action of this enzyme on lysophospholipids and monoglycerides (Galliard and Dennis, 1974). The enzyme is present in all *Phaseolus* species examined and in spinach leaves (Sastry and Kates, 1964*b;* Helmsing, 1967, 1969; Galliard, 1975). It has been purified 160-fold from *Phaseolus multiflorus* (Depery *et al.,* 1982) and ~50-fold from spinach (Helmsing, 1969). The reactions catalyzed by this acyl hydrolase are as follows (Sastry and Kates, 1964*b*; Depery *et al.,* 1982):

$$GalDAG \rightarrow Gal\text{-}monoacylGro + FA \qquad (16)$$

$$Gal\text{-}monoacylGro \rightarrow GalGro + FA \qquad (17)$$

$$Gal_2DAG \rightarrow Gal_2\text{-}monoacylGro + FA \qquad (18)$$

$$Gal_2\text{-}monoacylGro \rightarrow Gal_2Gro + FA \qquad (19)$$

The question whether one enzyme or two are responsible for these reactions has not yet been completely settled. Separate enzymes, specific for either the GalDAG or the Gal_2DAG have been postulated (Sastry and Kates, 1964*b*), on the basis of the following observations: (1) the optimum pH for the two substrates is different (7.0 and 5.6 for GalDAG and Gal_2DAG, respectively); (2) the enzyme preparation has a higher affinity for Gal_2DAG than for GalDAG; and (3) storage of the enzyme at 4°C results in considerable loss of activity toward the Gal_2DAG, but not toward the GalDAG. Attempts have been made to separate the two enzyme activities from spinach leaves, but only a single protein band with activity toward both GalDAG and Gal_2DAG and isoelectric point at pH 7.0 was obtained on 15% polyacrylamide gel electrophoresis (Helmsing, 1969). Depery *et al.* (1982) have purified the enzyme from *Phaseolus multiflorus* leaves and found it to have a molecular weight of about 60 kDa with a pI of 4.6. The enzyme preparation has activity toward both substrates, optimal conditions being pH 8 in the presence of 0.5 M deoxycholate and temperature in the range 35–50°C; it also shows activity toward phosphatidylcholine (PC) and lyso-PC. Further purification studies would be of interest with respect to the question of whether there are two enzymes or only one involved in galactolipid hydrolysis.

A lipolytic acyl hydrolase with similar but not identical properties to those of the *Phaseolus* enzyme has been isolated from potato tubers (Galliard, 1975). This enzyme is active toward both GalDAG and Gal_2DAG but is less specific than the *Phaseolus* enzyme in that it hydrolyzes monoolein and lyso-PC most readily and also acts on PC, methyl oleate, and diolein; it does not hydrolyze triolein, wax esters, or sterol esters. The pH optima for hydrolysis of GalDAG and Gal_2DAG are 6.0 and 4.6, respectively, differing distinctly from those of the *Phaseolus* enzyme. The partially purified enzyme has an apparent molecular weight of about 100 kDa but subunits of about 40 kDa can be detected on gels containing sodium dodecyl sulfate (SDS); at least three isozymes of this enzyme are present (Walcott *et al.*, 1982). This acyl hydrolase is extremely active in potato tubers, such that most of the endogenous membrane-bound polar lipids are degraded immediately the tissue is disrupted, even at 0°C. The enzyme also occurs in leaves and sprouts of potato and in a range of plant tissues (Galliard, 1975).

3.8.1.4b. Hydrolysis of SulfoquinovosylDAG. Enzymes that hydrolyze SQDAG are widespread, having been detected in green algae, alfalfa leaves and roots, and corn roots, and in emulsin prepared from almonds, as well as in nonplant sources such as beef pancreatin, but not in pig kidney, yeast, or snake venom (Yagi and Benson, 1962; Benson, 1963). Hydrolysis of SQDAG by cell-free extracts of *Scenedesmus* has been shown to occur by the action of two sulfolipases (A and B), as follows (Yagi and Benson, 1962):

$$\text{SQDAG} \xrightarrow{\text{A}} \text{SQ-monoacylGro} + \text{fatty acid} \tag{20}$$

$$\text{SQ-monoacylGro (lysosulfolipid)} \xrightarrow{\text{B}} \text{SQGro} + \text{fatty acid} \tag{21}$$

Further hydrolysis of SQGro was not observed, and the position of the acyl ester group attacked by sulfolipase A was not determined. Further purification studies on these enzymes would of interest.

3.8.1.4c. Action of Acyltransferases. When plant cells are broken, not only are galactolipids deacylated by the action of acyl hydrolases (see Section 3.8.1.4a), but they are subject also to the action of acyl transferases (Heinz, 1967*a,b*). Thus, in spinach leaf or bean leaf homogenates, rapid acyl transferase reactions occur involving the transfer of acyl groups from Gal_2DAG to the C-6 position of GalDAG, or from GalDAG to the C-6 position of GalDAG and Gal_2DAG, with the formation of 6-acyl-GalDag and 6-acyl-Gal_2DAG respectively (Heinz, 1967*a,b*, 1973; Heinz and Tulloch, 1969; Heinz *et al.*, 1974), as follows:

$$Gal_2DAG + GalDAG \rightarrow \text{6-acylGalDAG } (\mathbf{1'}) + \text{lyso-}Gal_2DAG \tag{22}$$

$$GalDAG + GalDAG \rightarrow \text{6-acylGalDAG} + \text{lyso-GalDAG} \tag{23}$$

$$GalDAG + Gal_2DAG \rightarrow \text{6-acyl-}Gal_2DAG \ (\mathbf{2'}) + \text{lyso-GalDAG} \tag{24}$$

Phospholipids such as phosphatidylethanolamine, phosphatidylcholine, and phosphatidylglycerol may also serve as acyl donors with partially purified spinach leaf preparations *in vitro* (Heinz, 1967*b*), but neither phospholipids nor SQDAG can serve as a donor with the purified enzyme from *Vicia faba* (Heinz, 1973) (see below).

The lysogalactolipids can also serve as acyl acceptors to form the corresponding 6-acyl-lysogalactolipids, as follows (Heinz, 1967*a*, 1973; Critchley and Heinz, 1973):

GalDAG + lyso-GalDAG → 6-acyl-lysoGalDAG + lyso-GalDAG   (25)

GalDAG + lyso-Gal$_2$DAG → 6-acyl-lyso-Gal$_2$DAG + lyso-Gal$_2$DAG   (26)

Lysogalactolipids may originate from reactions (22)–(24) and/or from partial deacylation of galactolipids by lipolytic acyl hydrolase activity (Critchley and Heinz, 1973).

Acylated galactolipids (6-acyl-GalDAG, 6-acyl-Gal$_2$DAG, structures **1′** and **2′**, respectively) (Heinz, 1967a; Heinz and Tulloch, 1969; Heinz *et al.*, 1974), as well as acylated Gal-monoacylGro (acyl-lysoGalDAG) (Critchley and Heinz, 1973; Heinz *et al.*, 1974), have been isolated and identified in spinach leaf homogenates. The 6-acyl-GalDAG (**1′**) has also been isolated from wheat flour (Myrhe, 1968; Clayton *et al.*, 1970), and acyl-Gal-monoacylGro, acyl-GalDAG (**1′**), and acyl-Gal$_2$DAG (**2′**) have been identified in wheat grain (Ito *et al.*, 1983).

Acyltransferase activity has been demonstrated in leaf homogenates of all higher plants so far studied, such as spinach, *Vicia faba*, and *Phaseolus vulgaris*, but not in homogenates of tissues of lower plants (e.g., *Phyllitis scopolopendrium, Spyrogyra*) (Heinz, 1967a).

Partial purification of the acyltransferase from *Vicia faba* leaves has been achieved by DEAE-cellulose column chromatography followed by isoelectric focusing in the pH range 5.1–5.2 (Heinz, 1973). This enzyme preparation catalyzed the transfer of acyl groups to GalDAG both by the dismutation reaction (23) and by the acyltransfer reaction (22) with an optimum pH of around 5.4 for both reactions, and with a stability optimum between pH 6–7. The enzyme appears to be associated only with photosynthetic tissue, being absent in etiolated seedlings, and is apparently a cytosol-soluble protein (Heinz, 1973). For optimal assay of *in vitro* acyltransferase activity, substrate dispersion in deoxycholate is essential, 3–6 mM giving maximal rates. The enzyme shows no positional specificity with respect to transferring the acyl group from the *sn*-1 or -2 positions of the donor molecules, in either the dismutase or acyltransfer reactions [(23) and (22), respectively], but specifically transfers the acyl group to the C-6 position of the acceptor molecule (Heinz, 1973; Heinz and Tulloch, 1969).

The question still remains whether the acyltransferase activity represents a separate protein distinct from the lipolytic acylhydrolase activity. Although these two activities have different isoelectric points (pH 5.1 and 7.0, respectively), further purification of the enzymes is necessary to settle this question.

3.8.2. Glycosylsterol and Acylglycosylsterol Biosynthesis

Glycosyl- and acylglycosylsterols have a widespread occurrence in higher plants, both in photosynthetic and in nonphotosynthetic tissues (see Table 3-6). Particulate enzyme preparations from a wide variety of plant tissues, such as spinach leaves or chloroplasts, lettuce leaves or chloroplasts, wheat root microsomes, avocado mitochondria, and immature soybean seeds (see reviews by Mudd and Garcia, 1975; Mudd, 1980; Wojciechowski, 1983;

Hartman, 1984; Axelos and Péaud-Lenoël, 1982), have been shown to incorporate [^{14}C]glucose from UDP-[^{14}C]glucose into steryl glucosides and acylated steryl glucosides. Neither ADP-glucose, CDP-glucose, GDP-glucose, IDP-glucose, GDP-mannose, nor UDP-glucuronic acid is effective as a glycosyl donor in sterylglycoside biosynthesis, but TDP-glucose can serve as a donor, although it is much less effective than UDP-glucose. UDP-galactose can also serve as a donor in the spinach leaf or pea root mitochondrial systems but the sugar incorporated into the glycosyl-sterols was always glucose, due to the presence of a UDP-Gal → UDP-Glc epimerase in these enzyme preparations (Eichenberger and Newman, 1968; Ongun and Mudd, 1970). The glycosyl transferase involved in glycosylsterol synthesis thus appears to have a high specificity for UDP-Glc as glycosyl donor, with an apparent K_m ranging from 0.22 mM for soybean seeds to 0.028 for wheat root microsomes (see Mudd and Garcia, 1975).

A requirement for free sterol in the biosynthesis of sterol glycoside has been demonstrated with acetone powder enzyme preparations or petroleum ether-extracted preparations from a variety of plant tissues, which were capable of sterol glycoside synthesis only in the presence of added free unlabelled sterol (Péaud-Lenoël and Axelos, 1972; Eichenberger and Grob, 1968; Ongun and Mudd, 1970) or [^{14}C]cholesterol (Bush and Grunwald, 1974). Addition to these systems of the acetone or petroleum ether lipid extract resulted in synthesis of the acylated sterol glycosides (see Mudd and Garcia, 1975). On the basis of these experiments, the reactions catalyzed by the enzyme preparations for synthesis of glycosylsterols and acylglycosyl-sterols are as follows:

$$\text{Sterol + UDP-Glc} \rightarrow \text{Glc-sterol (5) + UDP} \qquad (27)$$

$$\text{Glc-sterol + acyl donor} \rightarrow \text{acyl-Glc-sterol (5')} \qquad (28)$$

Reaction (27) is catalyzed by the membrane-bound enzyme UDP-Glc : sterol-β-D-glucosyl transferase, the preferred sterol substrate being sitosterol, although this may differ in various plants (Mudd and Garcia, 1975). Only sterols with the C-3 hydroxyl in the β-position and a planar ring system are glycosylated; the number and position of the double bonds in the ring system or the side chain, and the presence or absence of a C-24 alkyl group are of secondary importance (Hartman, 1984). The enzyme has been localized in the plasma membrane of maize coleoptiles and roots, *Phaseolus* cotyledons, and soybean protoplasts; mitochondria were inactive, but Golgi membranes could not be excluded (Harman, 1984; see also Mudd and Garcia, 1975). Attempts to solubilize and purity the glucosyl transferase from cotton seed by Triton X-100 solubilization followed by DEAE-cellulose chromatography gave a 23-fold purified enzyme preparation that was absolutely dependent on exogeneous sitosterol (Forsee *et al.*, 1972).

In reaction (28), the enzyme sterylglycoside acyltransferase catalyzes the transfer of an acyl group (palmitic or linoleic acid, depending on the plant system) (Mudd and Garcia, 1975) from an unidentified endogenous acyl donor to the C-6 position of the glycosyl group in sterolglycoside (Lepage, 1964*b*). Phosphatidylethanolamine and Gal$_2$DAG have been implicated as possible acyl donors, but further study is needed to identify the acyl donors with certainty (Mudd and Garcia, 1975). That reaction (27) precedes reaction (28) has been demonstrated in *in vitro* kinetic studies with wheat microsomes, showing that the time course for sterylglucoside formation has no lag period whereas the time course for acylated sterylglucoside formation has a definite lag period and shows a product–precursor type relation with that for steryl glucoside formation (Péaud-Lenoël and Axelos, 1971). The sterylglycoside acyltransferase has also been localized in the plasma membrane although some activity was also found in the ER fraction (Hartman, 1984).

3.8.3. Ceramidehexoside (Cerebroside) Biosynthesis in Fungi

Although the presence of cerebroside in higher plants has been known for many years (Carter *et al.,* 1956, 1961*a;* Sastry and Kates, 1964*a*), their biosynthesis in plants has so far not been studied. By contrast, the biosynthesis of long-chain bases, ceramides, and cerebrosides, has been completely elucidated in the yeast, *Hansenula ciferri* (see reviews by Snell *et al.,* 1970; Stoffel, 1970, 1971; Weete, 1980).

As had been shown earlier in animal systems (see Snell *et al.,* 1970; Stoffel, 1971), long-chain bases are formed by a condensation reaction between palmitoyl-CoA and L-serine to give 3-ketodihydrosphingosine, which is then reduced in the presence of NADPH to dihydrosphingosine:

$$\text{Palmitoyl-CoA} + \text{L-serine} \rightarrow \text{3-ketosphinganine} + CO_2 \qquad (29)$$

$$\text{3-Ketosphinganine} + \text{NADPH} + H^+ \rightarrow \text{sphinganine} + \text{NADP}^+ \quad (30)$$

These reactions have been demonstrated to occur with a microsomal preparation from *H. ciferri* (Snell *et al.,* 1970; Stoffel, 1970). The condensation reaction (29) is catalyzed by the enzyme 3-ketosphinganine synthetase, which contains firmly bound pyridoxal phosphate as coenzyme. The reaction gives rise to the formation of D- (or *S*)-3-ketosphinganine, involving inversion of the asymmetrical C-2 during the decarboxylation–condensation step (see Fig. 3-16). Reaction (30) is catalyzed by a stereospecific reductase requiring NADPH, which reduces the D-3-ketosphinganine to 2*S*, 3*R*-sphinganine (D-erythro-dihydrosphingosine) (see Fig. 3-16).

In animal systems, sphingosine is formed by desaturation of dihydrosphingosine by elimination of the 4*R* and 5*R* hydrogens to form the 4-*trans*-double bond (Stoffel, 1974). However, in *H. ciferrii* this desaturation reaction apparently does not occur. Instead, 4-*trans*-sphingenine is synthesized by an analogous pathway to reactions (29)–(30), but starting from *trans*-2-hexadecenoyl-CoA, as follows (DiMari *et al.,* 1971):

$$\text{R-CH=CH-CO-SCoA} + \text{L-serine} \rightarrow \text{R-CH=CH-CO-CH(NH}_2\text{)-CH}_2\text{OH}$$
$$+ CO_2 \qquad (31)$$

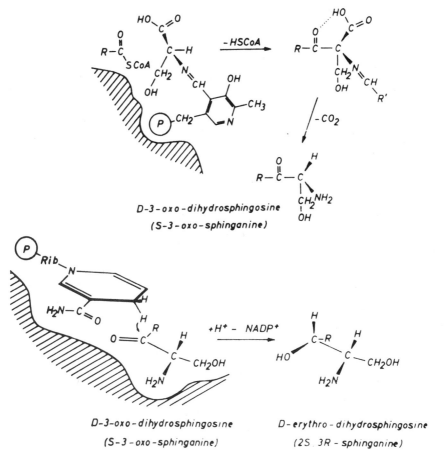

Figure 3-16. Stereospecificity of condensation and reduction reactions in the biosynthesis of sphinganine. (After Stoffel, 1970.)

$$R\text{-}CH{=}CH\text{-}CO\text{-}CH(NH_2)\text{-}CH_2OH + NADPH + H^+ \rightarrow$$
$$R\text{-}CH{=}CH\text{-}CH(OH)\text{-}CH(NH_2)CH_2OH + NADP^+ \qquad (32)$$

The *trans*-2-hexadecenoic acid is formed from palmitic acid by a specific desaturase present in yeast microsomes.

Formation of cerebrosides can occur by two pathways: *N*-acylation of a long-chain base followed by transfer of a glycosyl group to the resulting ceramide; or transfer of a glycosyl group to a long-chain base followed by *N*-acylation (Morell and Braun, 1972):

$$\text{Long-chain base} + \text{acyl-CoA} \rightarrow \textit{N}\text{-acyl-long-chain base (ceramide)}$$
$$+ \text{ CoA} \qquad (33)$$

$$\text{Ceramide} + \text{UDP-glycose} \rightarrow \text{glycosylceramide} + \text{UDP} \qquad (34)$$

$$\text{Long-chain base} + \text{UDP-glycose} \rightarrow \text{glycosyl long-chain base} \quad (35)$$

$$\text{Glycosyl long-chain base} + \text{acyl-CoA} \rightarrow \text{glycosylceramide} + \text{CoA} \quad (36)$$

The *N*-acyl transferases show some preference for certain acids, such as 2-hydroxy acids, depending on the organism, but there is no discrimination between dihydrosphingosine and sphingosine. Both glucose and galactose can be incorporated into cerebroside from their respective UDP derivatives by the glycosyltransferase-catalyzing reactions (34) or (35) (Basu *et al.*, 1968).

The pathway for biosynthesis of the trihydroxy base, phytosphingosine in yeasts and fungi has not yet been established unambiguously (Weete, 1980). Presumably the route for synthesis of dihydrosphingosine [reactions (29) and (30)] may be involved since [^{14}C]serine and [^{3}H]palmitate are incorporated into phytosphingosine in *H. ciferri* (Green *et al.*, 1965), but phytosphingosine cannot be made by hydroxylation of dihydrosphingosine or by hydration of sphingosine (Thorpe and Sweeley, 1967). However, the direct conversion of [1-^{14}C, 3-^{3}H]dihydrosphingosine to phytosphingosine has been shown to occur in *H. ciferri* with accompanying loss of a tritium atom at C-3 and the 4R hydrogen (Stoffel *et al.*, 1968; Weiss and Stiller, 1967).

3.9. Membrane Function of Plant Glycolipids

The high concentration of glycolipids, particularly galactolipids, in chloroplast membranes of higher plants, algae, and diatoms (Table 3-7) raises the question of their function in these membranes. As was discussed for bacterial glycolipids (see Chapter 1, Section 1.6), the plant glycolipids fulfill an important role in maintaining the structural integrity of chloroplast membranes (see reviews by Quinn and Williams, 1983; Murphy, 1986; Gounaris *et al.*, 1986; Curatolo, 1987*b*). In addition, the acyl lipids of thylakoids are being recognized as having a functional role in the activity of several integral membrane protein complexes (Gounaris *et al.*, 1986; Murphy, 1986; Bishop, 1983). Chloroplast lipids have also been implicated in physiological phenomena such as chilling sensitivity and resistance and salt stress (Gounaris *et al.*, 1986). These functions will be discussed in the following sections.

3.9.1. Structural Role

The lipid composition of thylakoid membranes of all chloroplasts examined is dominated by the presence of the galactolipids GalDAG and Gal$_2$DAG (Douce *et al.*, 1987*a*) (see Table 3-7), which together account for almost 80% of the total lipids. Furthermore, the mole ratio of GalDAG to Gal$_2$DAG is > 2 : 1 and the major molecular species of both lipids is 18 : 3/18 : 3 (Table 3-12). While aqueous dispersions of GalDAG form the hexagonal-II phase, and dispersions of Gal$_2$DAG form lamellar phases, mixed dispersions of these two lipids form bilayers with lipidic particles consisting of inverted micelles within the lipid bilayer (Quinn and Williams, 1983; Curatalo, 1987*a,b*). Unsaturated

(GalDAGs (with 5 or more double bonds per molecule) form hexagonal-II phases due to the conical shape of this molecule, with the glycosyl group at the small end of the cone; while the polyunsaturated Gal_2DAGs form lamellar phases due to the quasi cylindrical shape of this molecule (Quinn and Williams, 1983; Curatolo, 1987a) (see Chapter 1, Section 1.5.1.2). The sulfolipid also forms lamellar phases in water, since its ionized sulfoglycosyl head group imparts a cylindrical shape to its molecule (Quinn and Williams, 1983; Curatolo, 1987a). Total chloroplast lipid extracts (e.g., from broad bean), which contain the galactolipids, the sulfolipid, and phospholipids (see Table 3-7), show only lamellar phases when dispersed in water but exhibit inverse micelles on addition of cations or reduction of pH, resulting from the physical behavior of GalDAG (Quinn and Williams, 1983).

By contrast, no such inverted micelle structures can be detected in intact thylakoids (Quinn and Williams, 1983). However, cylindrical inverted micelles have been observed in thylakoid membranes of bean chloroplasts after heating above 45°C, a process that also resulted in destacking of grana, disruption of chlorophyll–protein complexes and perturbation of the rate of electron transport (Gounaris *et al.*, 1984). These observations are indicative of an interaction of the galactolipids with protein and protein–pigment complexes in the thylakoid membrane (Murphy, 1986).

It has been calculated (Murphy, 1986) that about 70% of the acyl lipids of appressed (i.e., stacked) regions of the thylakoid membranes and 30% of acyl lipids of nonappressed (i.e., nonstacked) regions are in direct contact with protein, forming a monomolecular lipid annulus surrounding each protein or protein–pigment complex. Furthermore, the appressed regions were enriched in GalDAG relative to Gal_2DAG (mole ratio, 2.4) while the non-appressed regions had equivalent amounts of these two lipids (mole ratio, 1.1) (Murphy, 1986; Gounaris *et al.*, 1986). Thus a large proportion of the GalDAG present in the thylakoid membrane is associated with protein and is not free to form hexagonal-II phase, and this would adequately explain the absence of nonlamellar structures in intact thylakoid membranes (Rawyler and Siegenthaler, 1985).

The question arises as to the purpose of the high content of galactolipids (GalDAG/Gal_2DAG) rather than phospholipids (PC/PE) in the thylakoid membrane. Although no definitive experimental data is available on this point, a theoretically plausible explanation would be the ability of glycolipids to form more stable membranes than phospholipids because of the capacity of glycolipids to form multiple intermolecular hydrogen bonding networks through their glycosyl polar head groups and also to hydrogen bond with water (Curatolo, 1987b). Furthermore, the presence of the anionic lipids, sulfoquinovosylDAG and phosphatidylglycerol, would provide anionic sites in the head group or hydrophilic region of the lipid bilayers. Such networks would provide an ionic and hydrophilic environment necessary for charge transfer, particularly proton conductance, at the thylakoid membrane surfaces (Murphy, 1986; Gounaris *et al.*, 1986; Curatolo, 1987b; Boggs, 1987).

Another structural function has been proposed for the galactolipids, i.e., that GalDAG, being cone-shaped, is involved in the inner concave monlayer of the curved membrane regions of the thylakoid discs and Gal_2DAG is involved in the convex outer monolayer (Murphy, 1986). This arrangement would appear to be inconsistent with the overall outside–inside ratios of GalDAG and Gal_2DAG reported for thylakoid membranes of various plants (see Table 3-8), but it has been pointed out that if most or all of the GalDAG in the inner monolayer is concentrated in the concave curved edges there would not be any inconsistencies with the observed ratios (Rawyler and Siegenthaler, 1985). However, further studies will be necessary to prove or disprove this proposal.

3.9.2. Functional Role in Activity of Protein Complexes

It has been proposed that the major galactolipid GalDAG is involved in the efficient packaging of the light harvesting protein–pigment complexes of photosystems I and II (PS I and PS II) in the thylakoid membrane bilayer (Quinn and Williams, 1983). Supporting this proposal is the finding that native GalDAG, but not saturated GalDAG, specifically restored energy transfer from the light harvesting chlorophyll a/b–protein complex to photosystems I and II in Triton X-100-solubilized thylakoid membranes (Seifermann-Harms *et al.*, 1982). Furthermore, it has been found that chlorophyll photobleaching of bean thylakoids beyond the 10% level is accompanied by a selective loss of GalDAG due to oxidation and/or peroxidation of the unsaturated acyl chains, suggesting that GalDAG is more closely associated with the P700–chlorophyll complex than Gal_2DAG (Dominy and Williams, 1987).

Evidence has also been presented showing that GalDAG localized in the outer monolayer of the thylakoid membrane is intimately involved in optimal electron-flow activities of both PS I and PS II (Siegenthaler *et al.*, 1987*b*). In this connection, the long-range transfer of electrons between the PS II in the appressed thylakoid membranes and PS I in the nonappressed thylakoid membranes is thought to occur via the lipophilic mobile electron carrier plastoquinone (rather than the bulkier hydrophilic protein plastocyanin) by diffusion through the hydrophobic bilayer core (Murphy, 1986). It is possible that the decreased microviscosity of the bilayer relative to the polar surface regions, a consequence of the highly unsaturated acyl chains of GalDAG and Gal_2DAG, may well lead to the rapid rate of diffusion of plastiquinone required for its role as the exclusive long-range electron carrier (Murphy, 1986). The importance of lipid unsaturation in this connection has been demonstrated by the finding that homogeneous catalytic hydrogenation of the photosynthetic membranes inhibits electron transport before inhibition of PS I and II activities (Horvath *et al.*, 1986).

GalDAD and SQDAG appear to be involved in the activity of the ATP–synthetase complex (see Gounaris *et al.*, 1986). Native GalDAG in its naturally occurring ratio to the other acyl lipids (but not saturated GalDAG) restored both the ATP hydrolysis and P_i–ATP exchange activities in reconstituted systems: small amounts (10%) of ionized lipids were required for optimal activity, but higher amounts were inhibitory. SQDAG is most likely the charged molecule required, since its complete removal from the ATP synthetase complex results in complete inactivity of the latter. SQDAG is thought to be an integral part of the complex because it copurifies with the enzyme and is not exchangeable (Gounaris *et al.*, 1986).

3.9.3. Role in Physiological Phenomena

3.9.3.1. Acclimation to Temperature Stress

3.9.3.1a. Chilling Sensitivity and Resistance. Higher plants on exposure to temperatures of 0–10°C for prolonged periods exhibit two kinds of behavior: chilling-sensitivity or chilling resistance. Chilling-sensitive plants undergo marked losses in water and electrolytes, sharp reduction in rates of respiration and photosynthesis, and changes in chloroplast structure, leading to necrosis, while chilling-resistant plants do not show these changes and can survive such low temperature exposures. Early studies of this phenomenon (Lyons *et al.*, 1964; Raison *et al.*, 1971) suggested that chilling sensitivity was correlated with the occurrence of lipid phase transitions around 10–12°C in the membranes of organelles from chilling-sensitive plants; such transitions were absent in chilling-resistant plants. It was suggested that these phase transitions were a reflection of more highly saturated lipids in the organelle membranes of chilling-sensitive plants relative to chilling-resistant plants. However, critical analysis of the results in the literature does not reveal convincing evidence for the occurrence of gel-to-liquid crystalline transitions in chilling-sensitive or resistant higher plants, or for any correlation between the overall degree of saturation of phospholipids or glycolipids in chloroplast or nonchloroplast membranes and chilling sensitivity or resistance (see reviews by Quinn and Williams, 1983, 1985; Bishop, 1983; Gounaris *et al.*, 1986).

However, a correlation has been found between chilling sensitivity or resistance and the molecular species composition of phosphatidylglycerol in chloroplasts of a variety of plants (Murata, 1983; Murata and Yamaya, 1984; Roughan, 1987; see Quinn and Williams, 1983, 1985). The phosphatidylglycerol component (but no other lipid class) in chilling-sensitive plants (e.g., sweet potato, castor bean, tobacco) contain high proportions of the saturated molecular species (16:0/16:0 and 16:0/16:1(t), in the range 26–65% compared with the low proportions of these species (3–19%) in chilling resistant plants (e.g., pea, spinach, cabbage). Such molecular species of PG could induce temperature-dependent phase separation below 20°C, since they undergo gel-to-liquid crystalline phase transitions in the range +20–40°C (Boggs, 1987; Bishop, 1983; Murata and Yamaya, 1984). It is this phase transition induced by the saturated PG molecular species that is believed to be the primary event leading to chilling damage in chilling-sensitive plants (Toriyama *et al.*, 1987). In rice, which is one of the most chilling-sensitive plants, it was found that the galactolipids and SQDAG as well as PG showed higher proportions of saturated fatty acids (16:0 + 18:0) in the anthers (particularly the anther walls) than in the leaves (Toriyama *et al.*, 1987). This finding is consistent with the fact that rice anthers are more sensitive to chilling than the leaves.

The effect of chilling temperatures in inducing phase separations in membrane lipids of photosynthetic cells above 0°C has been demonstrated conclusively in the *Cyanobacterium Anacystis nidulans,* but not in *Anabaena variabilis,* by freeze-fracture electron microscopy (Ono and Murata, 1982; Sato *et al.,* 1979). This finding can be attributed to the fact that *A. nidulans* contains high proportions of relatively saturated molecular species of galactolipids, sulfolipid and PG, as compared to *A. variabilis* which contains high proportions of polyunsaturated molecular species of these lipids (Sato *et al.,* 1979) (see Tables 3-11 and 3-12). Separation of domains of relatively high saturated lipids would thus be expected to occur in *A. nidulans* but not in *A. anabaena,* on exposure of cells to cold shock. Such changes in the thylakoid membranes of *A. nidulans* exposed to chilling temperatures did not correlate with losses in photosynthetic activity, including oxygen evolution and Hill activity to benzoquinone; however, with whole cells, the onset of chilling damage did coincide with the release of K^+ from chilled cells, suggesting that the lipid phase separation occurred in the plasma membrane, leading to increased permeability of this membrane (see Quinn and Williams, 1983, 1985; Bishop, 1983). Thus, the role of membrane lipids in inducing phase transitions appears to be well established in the response of *A. nudulans* to cold shock. Further study is however necessary to establish this phenomenon unambiguously as an explanation for chilling damage in higher plants.

3.9.3.1b. High- and Low-Temperature Acclimation. The phenomenon of changes in membrane lipid composition as a function of growth temperature has been extensively studied (see reviews by Bishop, 1983; Harwood and Russell, 1984; Quinn and Williams, 1983, 1985). As a general rule, growth at low temperature results in increases in lipid : protein ratios and phospholipid : galactolipid ratios, and to increases in overall lipid unsaturation, as has been shown, for example, in germinating spring and winter wheat (de la Roche *et al.,* 1972, 1973; de Silva *et al.,* 1975), in thylakoid membranes of frost-resistant and -sensitive wheat seedlings (Vigh, 1982), in wheat cell suspension cultures (Horvath and Vigh, 1984), and cucumber leaves (Kuiper, 1984). Cold-grown plants or seedlings that are able to withstand temperatures well below 0°C are termed cold-hardened and exhibit increased levels of polyunsaturated fatty acids. Increased synthesis of polyunsaturated fatty acids may be accounted for by low-temperature activation of acyl chain desaturase systems, as has been demonstrated in yeast microsomes (Ferrante *et al.,* 1983; Kates *et al.,* 1984) and in *Tetrahymena* (Thompson and Nozawa, 1984).

The apparent opposite correlation has been observed in conifer (*Pinus sylvestris*) chloroplasts during growth over a complete annual cycle (Oquist, 1982). Both the relative amount and degree of unsaturation of GalDAG are maximal over the summer and decrease rapidly in the fall to minimal values in the winter; whole chain and PS II-mediated electron transport activities show a similar pattern. These observations would seem to reflect the fact that thylakoid concentration in and photosynthetic activity of chloroplasts, and hence galactolipid concentration and its degree of unsaturation, are maximal during the summer months and minimal during the winter months; the decreases in the autumn are probably due to temperature-activated enzyme breakdown of the thylakoid membranes. However, it was observed that during short-term dehardening of cold-hardened pine plants, the pine needle chloroplasts showed a sharp increase in the relative amount of GalDAG and a corresponding decrease in Gal$_2$DAG (Bervaes *et al.*, 1972; DeYoe and Brown, 1979). The higher content of Gal$_2$DAG in cold hardened plants may be associated with the increased capacity of Gal$_2$DAG to bind intracellular water and thereby protect against frost injury by inhibiting ice crystal formation (Bervaes *et al.*, 1972).

As would be expected, growth of plants at high temperatures results in a decrease in unsaturation of the leaf lipids (see Quinn and Williams, 1985). Recent studies of the lipids of chloroplasts from *Nerium oleander* grown at a diurnal temperature regime of 20°C/15°C then shifted to 45°C/32°C for 14 days showed considerable decreases in proportions of 18:3 and increases in 16:0, 18:1, and 18:2 acids in GalDAG, Gal$_2$DAG and SQDAG but little change in 18:3 in the phospholipids (Raison *et al.*, 1982). The relatively small change in fatty acid composition of the phospholipids compared with those in the glycolipids suggests that the high temperature affects the site of synthesis of the glycolipids in the chloroplasts more than that of the phospholipids in the endoplasmic reticulum. The decrease in unsaturation accompanying acclimation to high temperature has been shown to result in a decreased fluidity of polar lipids extracted from the thylakoid membranes of *Nerium oleander* (Raison *et al.*, 1982).

Marked decreases in relative proportions of polyunsaturated molecular species of GalDAG and Gal$_2$DAG have been observed in the *Cyanobacteria Anabaena variabilis* and *Anacystis nidulans* when their growth temperature was shifted from 22°C to 38°C (Sato *et al.*, 1979; Sato and Murata, 1982*b*) (see Tables 3-11 and 3-12).

3.9.3.2. Salt Stress and Water Stress

Several studies have been made on the effect of growth of plants in saline environments on lipid composition (see Quinn and Williams, 1985; Harwood, 1984). In general, growth in the presence of NaCl (150–500 mM) results in a marked decrease in lipid content (particularly GalDAG) in chloroplast membranes of a variety of plants (e.g., barley seedlings, alfalfa, sunflower and olive; see Quinn and Williams, 1985; Harwood, 1984), together with decreases in the 18:3 content and corresponding increases in the 18:2 and 18:1 contents of the galactolipids. Such decreases in lipid unsaturation might help reduce the fluidity of chloroplast membranes with resulting decrease in salt permeability.

The limited number of studies of the effect of water-deficit stress or drought stress on the lipid composition of plants have indicated that a decrease in GalDAG together with a reduction in 18:3 and 16:1(t) occurred in drought-sensitive plants (see Quinn and Williams, 1985). In a more detailed study of the changes in lipid composition of oat root membranes as a function of water-deficit stress, it was found that the total acyl lipid content decreased with increasing degree of water stress, including decreases in sterylesters, acylated sterolglycoside, and, to a lesser extent, the galactolipids and phosphatidylethanolamine (Liljenberg and Kates, 1985). Decreases in the degree of unsaturation of the galactolipids in the plasmalema membrane fraction as well as in the total root cells were also observed, and electron microscopy showed that stressed root tip cells had much smoother plasma membranes than unstressed controls. These results suggest that oat root cells respond to water-deficit stress by reducing the total plasma membrane mass, hence its surface area, as well as the degree of lipid fluidity. These changes would help reduce the water permeability of the plasma membrane and help maintain cell turgidity (Liljenberg and Kates, 1985).

3.10. Summary and Conclusions

The foregoing discussion has shown that plants contain a unique set of membrane lipid components, particularly the glycolipid components that are not found in any other group of organisms. These unique lipids include the galactolipids (**1, 2, 3,** and **3′**), the sulfoquinovosyldiacylglycerol (**4**), and the phosphatidylglycerol molecular species $18:3/16:1(t)$. The uniqueness of these lipids arises from the fact that they are synthesized in the chloroplast membranes, with the involvement of the endoplasmic reticulum. Two types of pathways are used in the formation of the characteristic pattern of molecular species of galactolipids and sulfolipid, a "primitive" prokaryotic pathway located in the chloroplast and a more highly evolved eukariotic pathway associated with the endoplasmic reticulum. The resulting glycolipids have been shown to play a role in maintaining the structure of the thylakoid membranes and the optimal functioning of the photosynthetic apparatus. They also have been shown to play an important role in such physiological phenomena as high- and low-temperature acclimation and adaptation to salt and water stress. A more detailed understanding of the functions of plant glycolipids on the molecular level will require further experimentation on lipid–lipid and lipid–protein interaction. Such knowledge may one day make possible the reconstitution of the photosynthetic apparatus from its constituent protein, lipid, and pigment molecules.

ACKNOWLEDGMENTS. Support by the National Sciences and Engineering Research Council of Canada is acknowledged. The author is grateful to Mrs. E. Szabo for the drawings and to Mrs. Hélène Amyot for typing the manuscript.

References

Allen, C. F., Good, P., Davis, H. F., and Fowler, S. D., 1964, Plant and chloroplast lipids. I. Separation and composition of major spinach lipids, *Biochem. Biophys. Res. Commun.* **15:**424.

Allen, C. F., Good, P., Davis, H. F., Chisum, P., and Fowler, S. D., 1966a, Methodology for the separation of plant lipids and application to spinach leaf and chloroplast lamellae, *J. Am. Oil Chem. Soc.* **43:**223.

Allen, C. F., Hirayama, O., and Good, P., 1966*b*, Lipid composition of photosynthetic systems, in *Biochemistry of Chloroplasts*, Vol. I (T. W. Goodwin, ed.), pp. 195–200, Academic, New York.

Anderson, J. A., Sun, F. K., McDonald, J. K., and Cheldelin, V. H., 1964, Oxidase activity and lipid composition of respiratory particles from *Claviceps purpurea*, *Arch. Biochem. Biophys.* **107**:37.

Anderson, R., Livermore, B. P., Kates, M., and Volcani, B. E., 1978*a*, The lipid composition of the non-photosynthetic diatom *Nitzschia alba*, *Biochim. Biophys. Acta* **528**:77.

Anderson, R., Kates, M., and Volcani, B. E., 1978*b*, Identification of the sulfolipids in the non-photosynthetic diatom *Nitzschia alba*, *Biochim. Biophys. Acta* **528**:89.

Appleby, R. S., Safford, R., and Nichols, B. W., 1971, The involvement of lecithin and monogalac-tosyl diglyceride in linoleate synthesis by green and blue green algae, *Biochim. Biophys. Acta* **248**:205.

Araki, S., Sakurai, T., Kawaguchi, A., and Murata, N., 1987, Positional distribution of fatty acids in glycerolipids of the marine red alga, *Porphyra yezoensis*, *Plant Cell Physiol.* **28**:761.

Arao, T., Kawaguchi, A., and Yamada, M., 1987, Positional distribution of fatty acids in lipids of the marine diatoms, *Phaeodactylum tricornutum*, *Phytochemistry* **26**:2573.

Auling, G., Heinz, E., and Tulloch, A. P., 1971, Combination and positional distribution of fatty acids in plant galactolipids, *Hoppe-Seylers Z. Physiol. Chem.* **352**:905.

Axelos, M., and Péaud-Lenoël, C., 1982, Plant carbohydrates. 1. Intracellular carbohydrates, *Ency. Plant Physiol. New Ser.* **13A**:613.

Baraud, J., Maurice, A., and Napias, C., 1970, Composition et repartition des lipides sein des cellules de *Saccharomyces cerevisiae*, *Bull. Soc. Chim. Biol. (Paris)* **52**:421.

Basu, S., Kaufman, B., and Roseman, S., 1968, Enzymatic synthesis of ceramide–glucose and ceramide–lactose by glycosyltransferases from embryonic chicken brain, *J. Biol. Chem.* **243**:5802.

Benson, A. A., 1963, The plant sulfolipid, *Adv. Lipid Res.* **1**:387.

Benson, A. A., and Maruo, B., 1958, Plant phospholipids I. Identification of the phosphatidyl glycerols, *Biochim. Biophys. Acta* **27**:189.

Benson, A. A., and Shibuya, I., 1962, Surfactant lipids, in *Physiology and Biochemistry of Algae* (R. A. Lewin, ed.), pp. 371–383, Academic, New York.

Benson, A. A., Wiser, R., Ferrari, R. A., and Miller, J. A., 1958, Photosynthesis of galactolipids, *J. Am. Chem. Soc.* **80**:4740.

Benson, A. A., Wintermans, J. F. G. M., and Wiser, R., 1959, Chloroplast lipids as carbohydrate reservoirs, *Plant Physiol.* **34**:315.

Bervaes, J. C. A., Kuiper, P. L. C., and Kylin, A., 1972, Conversion of digalactosyl diglyceride (extra long carbon chain) into monogalactosyl diglyceride of pine needle chloroplasts upon dehardening, *Physiol. Plant* **27**:231.

Bishop, D. G., 1983, Functional role of plant membrane lipids, in *Biosynthesis and Function of Plant Lipids* (W. W. Thompson, J. B. Mudd, and M. Gibbs, eds.), pp. 81–103, American Society of Plant Physiologists, Rockville, Maryland.

Bligh, E. G., and Dyer, W. J., 1959, A rapid method of total lipid extraction and purification, *Can. J. Biochem. Physiol.* **37**:911.

Bloch, K., Constantopoulos, G., Kenyon, C., and Nagai, J., 1967, Lipid metabolism in algae in the light and in the dark, in *Biochemistry of Chloroplasts*, Vol. II (T. W. Goodwin, ed.), pp. 195–211, Academic, New York.

Block, M. A., Dorne, A.-J., Joyard, J., and Douce, R., 1983*a*, The acyl-CoA synthetase and the acyl-CoA thioesterase are located respectively on the outer and inner envelope membranes of the chloroplast envelope, *FEBS Lett.* **153**:377.

Block, M. A., Dorne, A.-J., Joyard, J., and Douce, R., 1983*b*, Preparation and characterization of membrane fractions enriched in outer and inner envelope membranes from spinach chloroplasts. II. Biochemical characterization, *J. Biol. Chem.* **258**:13281.

Boggs, J. M., 1987, Lipid intermolecular hydrogen bonding: Influence on structural organization and membrane function, *Biochim. Biophys. Acta* **906**:353.

Brennan, P. J., Griffin, P. F. S., Losei, D. M., and Tyrell, D., 1974, The lipids of fungi, *Prog. Chem. Fats Other Lipids* **14**:51.

Brush, P., and Percival, E., 1972, Glycolipids present in eight genera of *Chlorophyceae*, *Phytochemistry* **11**:1847.

Budzikiewicz, H., Rullkötter, J., and Heinz, E., 1973, Mass spectrometric investigation of glycosyl glycerides, *Z. Naturforsch.* **28c:**499.

Bush, P. B., and Grunwald, C., 1974, Steryl glycoside formation in seedlings of *Nicotinia tabacum, Plant Physiol.* **53:**131.

Carter, H. E., and Hendrickson, H. S., 1963, Biochemistry of the sphingolipids. XV. Structure of phytosphingosine and dehydro-phytosphingosine, *Biochemistry* **2:**389.

Carter, H. E., Celmer, W. D., Lands, W. E. M., Mueller, K. L., and Tomizawa, H. H., 1954, Biochemistry of sphingolipids. VIII. Occurrence of long chain-bases in plant phosphatides, *J. Biol. Chem.* **206:**613.

Carter, H. E., McCluer, R. H., and Slifer, E., 1956, Wheat flour lipids, *J. Am. Chem. Soc.* **78:**3735.

Carter, H. E., Celmer, W. D., Galanos, D. S., Gigg, R. H., Lands, W. E. M., Law, J., Mueller, K. L., Nakayama, T., Tomizawa, H. H., and Weber, E., 1958a, Biochemistry of the sphingolipids. X. Phytoglycolipid, a complex phytosphingosine-containing lipid from plant seeds, *J. Am. Oil Chem. Soc.* **35:**335.

Carter, H. E., Gigg, R. H., Nakayama, T., and Weber, E., 1958b, Biochemistry of sphingolipids. XI. Structure of phytoglycolipid, *J. Biol. Chem.* **233:**1309.

Carter, H. E., Hendry, R. A., Nojima, S., and Stanacev, N. Z., 1960, Isolation and structure of cerebrosides, *Biochem. Biophys. Acta* **45:**402.

Carter, H. E., Ohno, K., Nojima, S., Tipton, C. L., and Stanacev, N. Z., 1961a, Wheat flour lipids. II. Isolation and characterization of glycolipids of wheat flour and other plant sources, *J. Lipid Res.* **2:**215.

Carter, H. E., Hendry, R. A., and Stanacev, N. Z., 1961b, Wheat flour lipids. III. Structure determination of mono- and digalactosyl diglycerides, *J. Lipid Res.* **2:**223.

Carter, H. E., Hendry, R. A., Nojima, S., Stanacev, N. Z., and Ohno, K., 1961c, Biochemistry of the sphingolipids. XIII. Determination of the structure of cerebrosides from wheat flour, *J. Biol. Chem.* **236:**1912.

Carter, H. E., Johnson, P., and Weber, E. J., 1965, Glycolipids, *Annu. Rev. Biochem.* **34:**109.

Carter, H. E., Strobach, R., and Hawthorne, J. N., 1969, Biochemistry of the sphingolipids. XVIII. Complete structure of tetrasaccharide phytoglycolipid, *Biochemistry* **8:**383.

Cho, S. H., and Thompson, G. A., Jr., 1987, Metabolism of galactolipids in *Dunaliella salina,* in *The Metabolism, Structure and Function of Plant Lipids* (P. K. Stumpf, J. B. Mudd, and W. D. Nes, eds.), pp. 623–629, Plenum, New York.

Clayton, T. A., MacMurray, T. A., and Morrison, W. R., 1970, Identification of wheat flour lipids by thin-layer chromatography, *J. Chromatogr.* **47:**277.

Cline, K., and Keegstra, K., 1983, Galactosyltransferases involved in galactolipid biosynthesis are located on the outer membrane of pea chloroplast envelope, *Plant Physiol.* **71:**366.

Critchely, C., and Heinz, E., 1973, Characterization and enzymatic synthesis of acyl galactosyl monoglyceride, *Biochim. Biophys. Acta* **326:**184.

Curatolo, W., 1987a, The physical properties of glycolipids, *Biochim. Biophys. Acta* **906:**111.

Curatolo, W., 1987b, Glycolipid function, *Biochim. Biophys. Acta* **906:**137.

Davies, W. H., Mercer, E. I., and Goodwin, T. W., 1966, Some observations on the biosynthesis of the plant sulfolipid by *Euglena gracilis, Biochem. J.* **98:**369.

Dawson, R. M. C., 1954, The measurement of ^{32}P labelling of individual kephalins and lecithin in a small sample of tissue, *Biochim. Biophys. Acta* **14:**374.

de la Roche, I. A., Andrews, C. J., Pomeroy, M. K., Weinberger, P., and Kates, M., 1972, Lipid changes in winter wheat seedlings (*Triticum aestivum*) at temperatures inducing cold hardiness, *Can. J. Bot.* **50:**2401.

de la Roche, I. A., Andrews, C. J., and Kates, M., 1973, Changes in phospholipid composition of a winter wheat cultivar during germination at 2°C and 24°C, *Plant Physiol.* **51:**468.

Demandre, C., Tremolières, A., Justin, A.-M., and Mazliak, P., 1985, Analysis of molecular species of plant polar lipids by high performance and gas–liquid chromatography, *Phytochemistry* **24:**481.

Depery, F., Schurmann, P., and Siegenthaler, P.-A., 1982, Partial purification and properties of a lipolytic acyl hydrolase from *Phaseolus multiflorus* leaves, in *Biochemistry and Metabolism of Plant Lipids* (J. F. G. M. Wintermans and P. J. C. Kuiper, eds.), pp. 301–304, Elsevier, Amsterdam.

de Silva, N. S., Weinberger, P., Kates, M., and de la Roche, I. A., 1975, Comparative changes in hardiness and lipid composition in two near-isogenic lines of wheat (spring and winter) grown at 2°C and 24°C, *Can. J. Bot.* **53**:1899.

De Yoe, D. R., and Brown, G. N., 1979, Glycerolipid and fatty acid changes in eastern white pine chloroplast lamellae during onset of hardening, *Plant Physiol.* **64**:924.

DiMari, S. J., Brady, R. N., and Snell, E. E., 1971, Biosynthesis of sphingosine bases. IV. Biosynthetic origin of sphingosine in *Hansenula ciferri*, *Arch. Biochem. Biophys.* **143**:553.

Dominy, P. J., and Williams, J. P., 1987, Is monogalactosyl diacylglycerol involved in the packaging of light-harvesting chlorophyll proteins in the thylakoid membrane?, in *The Metabolism, Structure, and Function of Plant Lipids* (P. K. Stumpf, J. B. Mudd, and W. D. Nes, eds.), pp. 185–187, Plenum, New York.

Dorne, A.-J., Block, M. A., Joyard, J., and Douce, R., 1982, The galactolipid : galactolipid galactosyltransferase is located on the outer surface of the outer membrane of the chloroplast envelope, *FEBS Lett.* **145**:30.

Douce, R., and Joyard, J., 1980, Plant galactolipids, in *The Biochemistry of Plants. A Comprehensive Treatise*, Vol. 4 (E. E. Conn and P. K. Stumpf, eds.), pp. 321–362, Academic, Orlando, Florida.

Douce, R., Holtz, R. B., and Benson, A. A., 1973, Isolation and properties of the envelope of spinach chloroplasts, *J. Biol. Chem.* **248**:7215.

Douce, R., Block, M. A., Dorne, A.-J., and Joyard, J., 1984, The plastid envelope membranes: their structure, composition and role in chloroplast biogenesis, *Subcellular Biochemistry*, Vol. 10 (D. B. Roodyn, ed.), pp. 1–84, Plenum, New York.

Douce, R., Alban, C., Bligny, R., Block, M. A., Coves, J., Dorne, A.-J., Journet, E.-P., Joyard, J., Neuburger, M., and Rebeille, F., 1987a, Lipid distribution and synthesis within the plant cell, in *The Metabolism, Structure, and Function of Plant Lipids* (P. K. Stumpf, J. B. Mudd, and W. D. Nes, eds.), pp. 255–263, Plenum, New York.

Douce, R., Joyard, J., Dorne, A.-J., and Block, M. A., 1987b, Origin of the plastid envelope membranes, in *New Developments and Methods in Membrane Research and Biological Energy Transduction* (L. Packer, ed.), Plenum, New York.

Dubacq, J.-P., Drapier, D., and Tremolieres, A., 1983, Polyunsaturated fatty acid biosynthesis by a mixture of chloroplasts and microsomes from spinach leaves: Evidence for two distinct pathways for trienoic acids, *Plant Cell Physiol.* **24**:1.

Dubacq, J.-P., Miquel, M., Drapier, D., Tremolieres, A., and Kader, J.-C., 1984, Some aspects of the role of chloroplast envelope membranes in lipid metabolism, in *Structure, Function and Metabolism of Plant Lipids* (P. A. Siegenthaler and W. Eichenberger, eds.), pp. 311–314, Elsevier, Amsterdam.

Eccleshall, T. R., and Hawke, J. C., 1971, Biosynthesis of monogalactosyl diglyceride by chloroplasts from *Spinacia oleracea* and from some Graminae, *Phytochemistry* **10**:3035.

Eichenberger, W., 1976, Lipids of *Chlamydomonas reinhardii* under different growth conditions, *Phytochemistry* **15**:459.

Eichenberger, W., and Menke, W., 1966, Sterols in leaves and chloroplasts, *Z. Naturforsch.* **21b**:859.

Eichenberger, W., and Grob, E. C., 1968, Enzymic formation of sterol glycosides by particulate fractions from lettuce and spinach, *Chimia* **22**:46.

Eichenberger, W., and Newman, W. D., 1968, Hexose transfer from UDP-hexose in the formation of steryl glycoside and esterified steryl glucoside in leaves, *Biochem. Biophys. Res. Commun.* **32**:366.

Evans, R. W., Kates, M., and Ginzburg, B.-Z., 1982, Lipid composition of halotolerant algae, *Dunaliella parva* and *Dunaliella tertiolecta*, *Biochim. Biophys. Acta* **712**:186.

Evans, R. W., and Kates, M., 1984, Lipid composition of halophilic species of *Dunaliella* from the Sinai, *Arch. Microbiol.* **140**:50.

Ferrante, G., Ohno, Y., and Kates, M., 1983, Influence of temperature and growth phase on desaturase activity of the mesophilic yeast *Candida lipolytica*, *Can. J. Biochem. Cell Biol.* **61**:171.

Ferrari, R. A., and Benson, A. A., 1961, The path of carbon in photosynthesis of lipids, *Arch. Biochem. Biophys.* **93**:185.

Fischer, W., Heinz, E., and Zeus, M., 1973, The suitability of lipase from *Rhizopus arrhizus delemar* for analysis of fatty acid distribution in dihexosyl diglycerides, phospholipid and plant sulfolipids, *Z. Physiol. Chem.* **354**:1115.

Forsee, W. T., Laine, R. A., Chambers, J. P., and Elbein, A. D., 1972, *Methods Enzymol.* **28**:478.

Fujino, Y., 1978, Rice lipids, *Cereal Chem.* **55**:559.

Fujino, Y., and Miyazawa, T., 1979, Chemical structures of mono-, di-, tri- and tetraglycosyl glycerides in rice bran, *Biochim. Biophys. Acta* **572**:442.

Fujino, Y., and Ohnishi, M., 1976, Constituents of ceramide and ceramide monohexoside in rice bran, *Chem. Phys. Lipids* **17**:275.

Fujino, Y., and Ohnishi, M., 1977, Structure of cerebroside in *Aspergillus oryzae, Biochim. Biophys. Acta* **486**:161.

Fujino, Y., and Ohnishi, M., 1978, Characterization and composition of sterols in the free and esterified sterol fractions of *Aspergillus oryzae, Lipids* **14**:663.

Fujino, Y., and Ohnishi, M., 1979a, Isolation and structure of diglycosyl and triglycosylsterols in rice bran, *Biochim. Biophys. Acta* **574**:94.

Fujino, Y., and Ohnishi, M., 1979b, Novel sterylglycosides: Cellotetraosyl sitosterol and cellopentaosyl sitosterol in rice grain, *Proc. Jpn. Acad.* **55**:243.

Fujino, Y., and Ohnishi, M., 1983, Sphingolipids in wheat grain, *J. Cereal Sci.* **1**:159.

Fujino, Y., Ohnishi, M., and Ito, S., 1985a, Further studies on sphingolipids in wheat grain, *Lipids* **20**:337.

Fujino, Y., Ohnishi, M., and Ito, S., 1985b, Molecular species of ceramide and mono-, di-, tri- and tetraglycosylceramide in bran and endosperm of rice grains, *Agric. Bio. Chem.* **49**:2753.

Galliard, T., 1968a, Aspects of lipid metabolism in higher plants. I. Identification and quantitative determination of the lipids in potato tubers, *Phytochemistry* **7**:1907.

Galliard, T., 1968b, Aspects of lipid metabolism in higher plants. II. The identification and quantitative analysis of lipids from the pulp of pre- and post-climacteric apples, *Phytochemistry* **7**:1915.

Galliard, T., 1969, The isolation and characterization of trigalactosyl diglyceride from potato tubers, *Biochem. J.* **115**:335.

Galliard, T., 1975, Degradation of plant lipids by hydrolytic and oxidative enzymes, in *Recent Advances in the Chemistry and Biochemistry of Plant Lipids* (T. Galliard and E. I. Mercer, eds.), pp. 319–357, Academic, New York.

Galliard, T., and Dennis, S., 1974, Phospholipase, galactolipase and acyl gransferase activities of a lipolytic enzyme from potato, *Phytochemistry* **13**:1731.

Gardner, H. W., 1968, Preparative isolation of monogalactosyl and digalactosyl diglycerides by thin-layer chromatography, *J. Lipid Res.* **9**:139.

Giroud, C., Gerber, A., and Eichenberger, W., 1988, Lipids of *Chlamydomonas reinhardii*. Analysis of molecular species and intracellular site(s) of biosynthesis, *Plant Cell Physiol.* **29**:587.

Green, M. L., Kaneshiro, T., and Law, J. H., 1965, Studies on the production of sphingolipid bases by the yeast, *Hansenula ciferri, Biochim. Biophys. Acta* **98**:582.

Griffiths, G., Stobart, A. K., and Stymne, S., 1985, The acylation of *sn*-glycerol-3-phosphate and the metabolism of phosphatidate in microsomal preparations from the developing cotyledons of safflower seed, *Biochem. J.* **230**:379.

Gounaris, K., Brain, A. P., Quinn, P. J., and Williams, W. P., 1984, Structural reorganization of chloroplast thylakoid membranes in response to heat stress, *Biochim. Biophys. Acta* **766**:198.

Gounaris, K., Barber, J., and Harwood, J. L., 1986, The thylakoid membranes of higher plant chloroplasts, *Biochem. J.* **237**:313.

Haas, R., Siebertz, H. P., Wrage, K., and Heinz, E., 1980, Localization of sulfolipid labelling within cells and chloroplasts, *Planta* **148**:238.

Haines, T. H., 1973, Sulfolipids and halosulfolipids, in *Lipids and Biomembranes of Eukaryotic Microorganisms* (J. A. Erwin, ed.), pp. 197–232, Academic, New York.

Haines, T. H., 1984, Microbial sulfolipids, in *Handbook of Microbiology,* 2nd ed. (A. I. Laskin and H. A. Lechevalier, eds.), pp. 115–123, CRC Press, Boca Raton, Florida.

Hartman, M.-A., 1984, Sterol glucosylation in plasma membranes from maize coleoptiles, in *Structure, Function and Metabolism of Plant Lipids* (P. A. Siegenthaler and W. Eichenberger, eds.), pp. 315–318, Elsevier, Amsterdam.

Harwood, J. L., 1984, Effects of the environment on the acyl lipids of algae and higher plants, in *Structure, Function and Metabolism of Plant Lipids* (P. A. Siegenthaler and W. Eichenberger, eds.), pp. 543–550, Elsevier, Amsterdam.

Harwood, J., and Russell, N. J., 1984, *Lipids in Plants and Microbes*, Allen & Unwin, London.

Haverkate, F., and van Deenen, L. L. M., 1965a, Isolation and characterization of phosphatidylglycerol from spinach leaves, *Biochim. Biophys. Acta* **106**:78.

Haverkate, F., and van Deenen, 1965b, Phospholipids of photosynthetic microorganisms, *Proc. Kon. Ned. Akad. Wetensch. Ser. B* **68**:141.

Heemskerk, J. W. M., and Wintermans, J. F. G. M., 1987, Role of the chloroplast in the leaf acyllipid synthesis, *Physiol. Plantarum* **70**:558.

Heemskerk, J. W. M., Bogemann, G., and Wintermans, J. F. G. M., 1983, Turnover of galactolipids incorporated into chloroplast envelopes: An assay for galactolipid : galactolipid galactosyltransferase, *Biochim. Biophys. Acta* **754**:181.

Heemskerk, J. W. M., Bogemann, G., and Wintermans, J. F. G. M., 1985, Spinach chloroplasts: localization of enzymes involved in galactolipid metabolism, *Biochim. Biophys. Acta* **835**:212.

Heemskerk, J. W. M., Wintermans, J. F. G. M., Joyard, J., Block, M. A., Dorne, A. J., and Douce, R., 1986, Localization of galactolipid : galactolipid galactosyltransferase and acyltransferase in outer envelope membrane of spinach chloroplasts, *Biochim. Biophys. Acta* **877**:281.

Heemskerk, J. W. M., Bogemann, G., Helsper, J. P. F. G., and Wintermans, J. F. G. M., 1988, Synthesis of mono- and digalactosyldiacylglycerol in isolated spinach chloroplasts, *Plant Physiol.* **86**:971.

Heinz, E., 1967a, Acylgalactosyldiglycerid aus Blatthomogenaten, *Biochim. Biophys. Acta* **144**:321.

Heinz, E., 1967b, Über die Enzymatische Bildung von Acylgalactosyl diglyceride, *Biochim. Biophys. Acta* **144**:333.

Heinz, E., 1971, Semisynthetic galactolipids of plant origin, *Biochim. Biophys. Acta* **231**:537.

Heinz, E., 1973, Some properties of the acyl galactosyl diglyceride-forming enzyme from leaves, *Z. Pflanzen Physiol.* **69**:359.

Heinz, E., and Roughan, P. G., 1983, Similarities and differences in lipid metabolism of chloroplasts isolated from 18 : 3 and 16 : 3 plants, *Plant Physiol.* **72**:273.

Heinz, E., and Tulloch, A. P., 1969, Reinvestigation of the structure of acyl galactosyl diglyceride from spinach leaves, *Hoppe Seylers Z. Physiol. Chem.* **350**:493.

Heinz, E., and Rullkrötter, J., and Budzikiewicz, H., 1974, Acyldigalactosyldiglyceride from leaf homogenates, *Hoppe Seylers Z. Physiol. Chem.* **355**:612.

Helmsing, P. J., 1967, Hydrolysis of galactolipids by enzymes in spinach, *Biochim. Biophys. Acta* **144**:470.

Helmsing, P. J., 1969, Purification and properties of galactolipase, *Biochim. Biophys. Acta* **178**:519.

Hitchcock, C., and Nichols, B. W., 1971, *Plant Lipid Biochemistry*, Academic, New York.

Hoover, R., Cloutier, L., and Dalton, S., 1988, Lipid composition of field pea (*Pisum sativum*, cv Trapper) seed and starch, *Stärke* **40**:336.

Horvath, I., and Vigh, L., 1984, Self-adaptive modification of membrane lipids in cell culture of wheat (*Triticum monococcum* L.), in *Structure, Function and Metabolism of Plant Lipids* (P. A. Siegenthaler and W. Eichenberger, eds.), pp. 535–538, Elsevier, Amsterdam.

Horvath, G., Droppa, M., Szito, T., Mustardy, L. A., Horvath, L. I., and Vigh, L., 1986, Homogeneous catalytic hydrogenation of lipids in the photosynthetic membrane effects on membrane structure and photosynthetic activity, *Biochim. Biophys. Acta* **849**:325.

Hudson, J. F., and Ogunsua, A. O., 1974, Lipids of cassava tubers (*Manihot esculenta* Crantz), *J. Sci. Food Agric.* **25**:1503.

Ishizuka, I. and Yamakawa, T., 1985, Glycoglycerolipids, in *Glycolipids* (H. Wiegard, ed.), pp. 101–197, Elsevier, Amsterdam.

Ito, S., and Fujino, Y., 1973, Isolation of cerebrosides from alfalfa leaves, *Can. J. Biochem.* **51**:957.

Ito, S., and Fujino, Y., 1975, Trigalactosyl diglyceride of pumpkin, *Phytochemistry* **14**:1445.

Ito, S., and Fujino, Y., 1980, Structure of trigalactosyldiglyceride isolated from alfalfa leaves, *Agric. Biol. Chem.* **44**:1181.

Ito, S., Okada, S., and Fujino, Y., 1974, Glyceroglycolipids of Pumpkin, *Nippon Nogeikagaka Kaishi* **48**:431.

Ito, S., Ohba, K., and Fujino, Y., 1983, Glycosylglycerides in wheat grain, *Nippon Nogeikagaka Kaishi* **57**:1231.

Ito, S., Seki, K., Kojima, M., Ohnishi, M., and Y. Fujima, Y., 1985a, Novel glycerolglycolipids in Azuka bean seeds, *Proc. Jpn. Conf. Biochem.* **27**:49.

Ito, S., Kojima, M., and Fujino, Y., 1985*b*, Occurrence of phytoglycolipids in rice bran, *Agric. Biol. Chem.* **49**:1873.

Itoh, T., and Kaneko, H., 1974, Yeast lipids in species variation. I. A simple method for estimating the cellular lipids, *Yukagaku* **23**:350.

Itoh, T., Tamura, T., and Matsumoto, T., 1973, Sterol composition of 19 vegetable oils, *J. Am. Oil Chem. Soc.* **50**:122.

IUPAC-IUB Commission on Biochemical Nomenclature, 1978, The nomenclature of lipids, *Chem. Phys. Lipids* **21**:159.

Jack, R. C. M., 1965, Relation of triglycerides to phosphoglycerides in fungi, *J. Am. Oil. Chem. Soc.* **42**:1051.

James, A. T., and Nichols, B. W., 1966, Lipids of photosynthetic systems, *Nature (London)* **210**:372.

Jamieson, G. R., and Reid, E. H., 1971, The occurrence of hexadeca-7,10,13-trienoic acid in the leaf lipids of angiosperms, *Phytochemistry* **10**:1837.

Jamieson, G. R., and Reid, E. H., 1972, The leaf lipids of some conifer species, *Phytochemistry* **11**:269.

Janero, D. R., and Barnett, R., 1981, Cellular and thylakoid-membrane glycolipids of *Chlamydomonas reinhardii* 137+, *J. Lipid Res.* **22**:1119.

Joyard, J., and Douce, R., 1976, Mise en evidence et role des diacylglycerols de l'enveloppe des chloroplastes d'épinard, *Biochim. Biophys. Acta* **424**:125.

Joyard, J., Blée, E., and Douce, R., 1986, Sulfolipid synthesis from 35S sulfate and [1-^{14}C]acetate in isolated intact spinach chloroplasts, *Biochim. Biophys. Acta* **879**:78.

Joyard, J., Blée, E., and Douce, R., 1987, Sulfolipid synthesis by isolated intact spinach chloroplast, in *The Metabolism, Structure and Function of Plant Lipids* (P. K. Stumpf, J. B. Mudd, and W. D. Nes, eds.), pp. 313–315, Plenum, New York.

Kader, J.-C., 1985, Lipid binding proteins in plants, *Chem. Phys. Lipids* **38**:51.

Kates, M., 1960, Chromatographic and radioisotopic investigations of the lipid components of runner bean leaves, *Biochim. Biophys. Acta* **41**:315.

Kates, M., 1970, Plant phospholipids and glycolipids and glycolipids, *Adv. Lipid Res.* **8**:225.

Kates, M., 1986, *Techniques of Lipidology*, 2nd rev. ed., Elsevier, Amsterdam.

Kates, M., 1987, Lipids of diatoms and halophilic Dunaliella species, in *The Metabolism, Structure, and Function of Plant Lipids* (P. K. Stumpf, J. B. Mudd, and W. D. Nes, eds.), pp. 613–621, Plenum, New York.

Kates, M., 1989, Sulfolipids of diatoms, in *Marine Biogenic Lipids* (R. G. Ackman, ed.), pp. 389–427, CRC Press, Boca Raton, Florida.

Kates, M., and Marshall, M. D., 1975, Biosynthesis of phosphoglycerides in plants, in *Recent Advances in the Chemistry and Biochemistry of Plant Lipids* (T. Galliard and E. I. Mercer), pp. 115–159, Academic, New York.

Kates, M., and Volcani, B. E., 1966, Lipid components of diatoms, *Biochim. Biophys. Acta* **116**:264.

Kates, M., Tremblay, P., Anderson, R., and Volcani, B. E., 1978, Identification of the free and conjugated sterol in a non-photosynthetic diatom, *Nitzschia alba,* as 24-methylene cholesterol, *Lipids* **13**:34.

Kates, M., Pugh, E. L., and Ferrante, G., 1984, Regulation of membrane fluidity by lipid desaturases, in *Biomembranes,* Vol. 12: *Membrane Fluidity,* (M. Kates and L. A. Manson, eds.), pp. 379–395, Plenum, New York.

Kaufman, B., Basu, S., and Roseman, S., 1971, Isolation of glucosylceramides from yeast (*Hansenula ciferri*), *J. Biol. Chem.* **246**:4266.

Kenyon, C. N., 1978, Complex lipids and fatty acids of photosynthetic bacteria, in *The Photosynthetic Bacteria* (R. K. Clayton and W. R. Sistrom, eds.), pp. 281–313, Plenum, New York.

Kesselmeier, J., and Heinz, E., 1985, Separation and quantitation of molecular species from plant lipids by high-performance liquid chromatography, *Anal. Biochem.* **144**:319.

Kinsella, J. E., 1971, Composition of lipids of cucumber and peppers, *J. Food Sci.* **36**:865.

Kleppinger-Sparace, K. F., and Mudd, J. B., 1987, Biosynthesis of sulfoquinovosyldiacylglycerol in chloroplasts of higher plants, in *The Metabolism, Structure and Function of Plant Lipids* (P. K. Stumpf, J. B. Mudd, and W. D. Nes, eds.), pp. 309–311, Plenum, New York.

Kleppinger-Sparace, K. F., Mudd, J. B., and Bishop, D. G., 1985, Biosynthesis of sulfoquinovosyldiacylglycerol in higher plants: the incorporation of [^{35}S]sulfate by intact chloroplast, *Arch. Biochem. Biophys.* **240**:859.

Kuiper, P. J. C., 1984, Lipid metabolism of higher plants as a factor in environmental adaptation, in *Structure, Function and Metabolism of Plant Lipids* (P. A. Siegenthaler and W. Eichenberger, eds.), pp. 525–530, Elsevier, Amsterdam.

Kuroda, N., Ohnishi, M., and Fujino, Y., 1977, Sterol lipids in rice bran, *Cereal Chem.* **54**:997.

Laine, R., and Renkonen, O., 1973, 4-Sphingenine derivatives in wheat flour lipids, *Biochemistry* **12**:1106.

Laine, R., and Renkonen, O., 1974, Ceramide di- and trihexosides of wheat flour, *Biochemistry* **13**:2837.

Laine, R., and Renkonen, O., 1975, Analysis of anomeric configuration in glyceroglycolipids and sphingoglycolipids by chromium trioxide oxidation, *J. Lipid Res.* **16**:102.

Lepage, M., 1964a, The separation and identification of plant phospholipids and glycolipids by two-dimensional thin layer chromatography, *J. Chromatogr.* **13**:99.

Lepage, M., 1964b, Isolation and characterization of an esterified form of steryl glucoside, *J. Lipid Res.* **5**:587.

Lepage, M., 1967, Identification and composition of turnip root lipids, *Lipids* **2**:244.

Lepage, M., Daniel, H., and Benson, A. A., 1961, The plant sulfolipid. II. Isolation and properties of sulfoglycosyl glycerol, *J. Am. Chem. Soc.* **83**:157.

Liljenberg, C., and Kates, M., 1982, Effect of water stress on lipid composition of oat seedling root cell membranes, in *Biochemistry and Metabolism of Plant Lipids* (J. F. G. M. Wintermans and P. J. C. Kuiper, eds.), pp. 441–444, Elsevier, Amsterdam.

Liljenberg, C., and Kates, M., 1985, Changes in lipid composition of oat root membranes as a function of water-deficit stress, *Can. J. Biochem. Cell Biol.* **63**:77.

Lynch, D. V., Gundersen, R. E. and Thompson, G. A., Jr., 1983, Separation of galactolipid molecular species by high performance liquid chromatography, *Plant Physiol.* **72**:903.

Lyons, J. M., Wheaton, T. A., and Pratt, H. K., 1964, Relation between the physical nature of mitochondria membrane and chilling sensitivity in plants, *Plant Physiol.* **39**:264.

MacMurray, T. A., and Morrison, W. R., 1970, Composition of wheat flour, *J. Sci. Food Agric.* **21**:520.

Marion, D., Gandemer, G. and Douillard, R., 1984, Separation of plant phosphoglycerides and galactosylglycerides by high performance liquid chromatography, in *Structure, Function and Metabolism of Plant Lipids* (P. A. Siegenthaler and W. Eichenberger, eds.), pp. 139–143, Elsevier, Amsterdam.

Matson, R. S., Fei, M., and Chang, S. B., 1970, Comparative studies of biosynthesis of galactolipids in *Euglena gracilis* strain Z, *Plant Physiol.* **45**:531.

McCarty, R. E., Douce, R., and Benson, A. A., 1973, The acyl lipids of purified plant mitochondria, *Biochim. Biophys. Acta* **316**:266.

Mills, G. L., and Cantino, E. C., 1974, Lipid composition of the zoospores of *Blastocladiella emersonii*, *J. Bacteriol.* **118**:192.

Mills, G. L., and Cantino, E. C., 1978, The lipid composition of the *Blastocladiella emersonii* γ-particle and the function of the γ-particle lipid in chitin formation, *Exp. Mycol.* **2**:99.

Miyano, M., and Benson, A. A., 1962a, The plant sulfolipids. VI. Configuration of the glycerol moiety, *J. Am. Chem. Soc.* **84**:57.

Miyano, M., and Benson, A. A., 1962b, The plant sulfolipid. VII. Synthesis of 6-sulfo-alpha-D-quinovopyranosyl-(1-1′)-glycerol and radiochemical syntheses of sulfolipids, *J. Am. Chem. Soc.* **84**:59.

Miyazawa, T., Ito, S., and Fujino, Y., 1974, Sterol lipids isolated from pea seeds, *Cereal Chem.* **51**:623.

Moore, T. S., 1982, Phospholipid biosynthesis, *Annu. Rev. Plant Physiol.* **33**:235.

Morell, P., and Braun, P., 1972, Biosynthesis and metabolic degradation of sphingolipids not containing sialic acid, *J. Lipid Res.* **13**:293.

Mudd, J. B., 1980, in *The Biochemistry of Plants*, Vol. 4 (E. E. Conn and P. K. Stumpf, eds.), pp. 509–534, Academic, Orlando, Florida.

Mudd, J. B., and Garcia, R. E., 1975, Biosynthesis of Glycolipids, in *Recent Advances in Chemistry and Biochemistry of Plant Lipids* (T. Galliard and E. I. Mercer, eds.), pp. 161–201, Academic, New York.

Mudd, J. B., van Vliet, H. H. D. M., and van Deenen, L. L. M., 1969, Biosynthesis of galactolipids by enzyme preparations from spinach leaves, *J. Lipid Res.* **10**:623.

Mudd, J. B., McManus, T. T., Ongun, A., and McCulloch, T. E., 1971, Inhibition of glycolipid biosynthesis in chloroplasts by ozone and sulphydryl reagents, *Plant Physiol.* **48:**335.

Mudd, J. B., deZacks, R., and Smith, J., 1980, Studies on the biosynthesis of sulfo-quinovosyldiacylglycerol in higher plants, in *Biogenesis and Function of Plant Lipids* (P. Mazliak, P. Benveniste, C. Costes, and R. Douce, eds.), pp. 57–66, Elsevier, Amsterdam.

Murata, N., 1983, Molecular species composition of phosphatidylglycerols from chilling-sensitive and chilling-resistant plants, *Plant Cell Physiol.* **24:**81.

Murata, N., and Sato, N., 1983, Analysis of lipids in *Prochloron* sp.: occurrence of monoglucosyl diacylglycerol, *Plant Cell Physiol.* **24:**133.

Murata, N., and Yamaya, J., 1984, Temperature-dependent phase behavior of phosphatidylglycerols from chilling-sensitive and chilling-resistant plants, *Plant Physiol.* **74:**1016.

Murphy, D. J., 1986, The molecular organization of the photosynthetic membranes of higher plants, *Biochim. Biophys. Acta* **864:**33.

Murphy, D. J., Woodrow, I. E., and Mukherjee, K. D., 1985, Substrate specificities of the oleate desaturase system from photosynthetic tissue, *Biochem. J.* **225:**267.

Myrhe, D. V., 1968, Glycolipids of soft wheat flour. I. Isolation and characterization of 1-*O*-(6-*O*-acyl-β-D-galactopyranosyl)-2,3-di-*O*-acyl-D-glyceritol and phytosteryl 6-*O*-acyl-β-D-glucopyra-nosides, *Can. J. Chem.* **46:**3071.

Neufeld, E. H., and Hall, E. W., 1964, Formation of galactolipids by chloroplasts, *Biochem. Biophys. Res. Commun.* **14:**503.

Nichols, B. W., 1963, Separation of the lipids of photosynthetic tissues: Improvements in analysis by thin-layer chromatography, *Biochim. Biophys. Acta* **70:**417.

Nichols, B. W., 1964, in *New Biochemical Separations* (A. T. James and L. J. Morris, eds.), pp. 321–337, Van-Nostrand, New York.

Nichols, B. W., 1965a, Light induced changes in the lipids of *Chlorella vulgaris, Biochim. Biophys. Acta* **106:**274.

Nichols, B. W., 1965b, The lipids of a moss (*Hypnum cupressiforme*) and of the leaves of green holly (*Ilex aquifolium*), *Phytochemistry* **4:**769.

Nichols, B. W., and James, A. T., 1964, The lipids of plant storage tissue, *Fette Seife Anstrichmittel* **66:**1003.

Nichols, B. W., and Moorehouse, R., 1969, The separation, structure and metabolism of mono-galactosyl diglyceride species in *Chlorella vulgaris, Lipids* **4:**311.

Nichols, B. W., Harris, R. V., and James, A. T., 1965, The lipid metabolism of blue-green algae, *Biochem. Biophys. Res. Commun.* **20:**256.

Nichols, B. W., Stubbs, J. M., and James, A. T., 1967, The lipid composition and ultrastructure of normal developing and degenerating chloroplasts, in *Biochemistry of Chloroplasts*, Vol. II (T. W. Goodwin, ed.), pp. 677–690, Academic, New York.

Noda, M., and Fujiwara, N., 1967, Positional distribution of fatty acids in galactolipids of *Artemisia princeps* leaves, *Biochim. Biophys. Acta* **137:**199.

Norman, H. A., and St. John, J. B., 1986, Metabolism of unsaturated monogalac-tosyldiacylglycerol molecular species in *Arabidopsis thaliana* reveals different sites and sub-strates for linolenic acid synthesis, *Plant Physiol.* **81:**731.

Nurminen, T., and Suomalainen, H., 1971, Occurrence of long-chain fatty acids and glycolipids in the cell envelope fractions of baker's yeast, *Biochem. J.* **125:**963.

Nurminen, T., Oura, E., and Suomalainen, H., 1970, The enzymatic composition of the isolated cell wall and plasma membrane of baker's yeast, *Biochem. J.* **116:**61.

O'Brien, J. S., and Benson, A. A., 1964, Plant sulfolipid. VIII. Isolation and fatty acid composi-tion, *J. Lipid Res.* **5:**432.

Ohnishi, J., and Yamada, M., 1982, Glycerolipid synthesis in Avena leaves during greening of etiolated seedlings. III. Synthesis of linolenoylmonogalactosyl diacylglycerol from liposomal linoleoylphosphatidylcholine by *Avena* plastids in the presence of phosphatidyl exchange protein, *Plant Cell Physiol.* **23:**767.

Ohnishi, M., and Fujino, Y., 1981, Mono- and triglycosylsterols from leafy stem of rice, *Phy-tochemistry* **20:**1357.

Ohnishi, M., Ito, S., and Fujino, Y., 1983, Characterization of sphingolipids in spinach leaves, *Biochim. Biophys. Acta* **752:**416.

Ongun, A., and Mudd, J. B., 1968, Biosynthesis of galactolipids in plants, *J. Biol. Chem.* **239:**1558.

Ongun, A., and Mudd, J. B., 1970, Biosynthesis of steryl glucosides in plants, *Plant Physiol.* **45**:255.

Ono, T.-A., and Murata, N., 1982, Chilling-susceptibility of the blue-green alga *Anacystis nidulans.* III. Lipid phase of cytoplasmic membrane, *Plant Physiol.* **69**:125.

Öquist, G., 1982, Seasonally induced changes in acyl lipids and fatty acids of chloroplast thylakoids of *Pinus sylvestris.* A correlation between level of unsaturated monogalactosyl diacylglycerol and rate of electron transport, *Plant Physiol.* **69**:869.

Osagie, A. U., and Opute, F. I., 1981, Major lipid constituents of *Discorea rotundata* tubers during growth and maturation, *J. Exp. Bot.* **32**:737.

Osagie, A. U., and Kates, M., 1984, Lipid composition of millet (*Pennisetum americanum*) seeds, *Lipids* **19**:958.

Osagie, A. U., and Kates, M., 1986a, Isolation and properties of subcellular fractions from Yam (*Discorea rotundata* Poir) tubers, *Niger. J. Biol. Sci.* **1**:40.

Osagie, A. U., and Kates, M., 1986b, Lipid composition of subcellular fractions from yam (*Discorea rotundata* Poir) tubers, *Niger. J. Biol. Sci.* **1**:134.

Péaud-Lenoël, C., and Axelos, M., 1971, Uridine diphosphate glucose: sterol transglucosylase. Purification and activity of particulate and soluble preparations extracted from wheat, *C.R. Acad. Sci. Paris Ser. D.* **273**:1057.

Péaud-Lenoël, C., and Axelos, M., 1972, D-Glucosylation of phytosterols and acylation of steryl-D-glucosides in the presence of plant enzymes, *Carbohydr. Res.* **24**:247.

Quinn, P. J., and Williams, W. P., 1983, The structural role of lipids in photosynthetic membranes, *Biochim. Biophys. Acta* **737**:223.

Quinn, P. J., and Williams, W. P., 1985, Environmentally induced changes in chloroplast membranes and their effects on photosynthetic function, in *Photosynthetic Mechanisms and the Environment* (J. Barber and N. R. Baker, eds.), pp. 1–46, Elsevier, Amsterdam.

Radunz, A., 1969, Sulfoquionovosyl diacylglycerol from higher plants, algae and purple bacteria, *Hoppe Seylers Z. Physiol. Chem.* **350**:411.

Raison, J. K., Lyons, J. M., and Thomson, W. W., 1971, Influence of membranes on the temperature-induced changes in the kinetics of some respiratory enzymes of mitochondria, *Arch. Biochem. Biophys.* **142**:83.

Raison, J. K., Roberts, J. K. M., and Berry, J. A., 1982, Correlation between thermal stability of chloroplast (thylakoid) membranes and composition and fluidity of their polar lipids upon acclimation of the higher plant, *Nerium oleander*, to growth temperature, *Biochim. Biophys. Acta* **688**:218.

Rattray, J. B. M., Shibeci, A., and Kidby, D. K., 1975, Lipids of yeasts, *Bacteriol. Rev.* **39**:197.

Rawyler, A., and Siegenthaler, P.-A., 1985, Transversal localization of monogalactosylglycerol and digalactosylglycerol in spinach thylakoid membranes, *Biochim. Biophys. Acta* **815**:287.

Rawyler, A., Unitt, M. D., Giroud, C., Davies, H., Mayor, J.-P., Harwood, J. L., and Siegenthaler, P.-A., 1987, The transmembrane distribution of galactolipids in chloroplasts is universal in a wide variety of temperature climate plants, *Photosynth. Res.* **11**:3.

Renkonen, O., and Bloch, K., 1969, Biosynthesis of monogalactosyl diglycerides in photoauxotrophic *Euglena gracilis*, *J. Biol. Chem.* **244**:4899.

Renkonen, O., and Hirvisalo, E. I., 1969, Structure of plasma sphingadienine, *J. Lipid Res.* **10**:687.

Rosenberg, A., 1963, A comparison of lipid pattern in photosynthesizing and non-photosynthesizing cells of *Euglena gracilis*, *Biochemistry* **2**:1148.

Rosenberg, A., Gouaux, J., and Milch, P., 1966, Monogalactosyl and digalactosyl diglycerides from heterotrophic, heterotrophic-autotrophic, and photobiotic *Euglena gracilis*, *J. Lipid Res.* **7**:733.

Roughan, P. G., 1987, On the control of fatty acid compositions of plant glycerolipids, in *The Metabolism, Structure, and Function of Plant Lipids* (P. K. Stumpf, J. B. Mudd, and W. D. Nes, eds.), pp. 247–254, Plenum, New York.

Roughan, P. G., and Batt, R. D., 1969, Glycerolipids of various plants, *Phytochemistry* **8**:363.

Roughan, P. G., and Slack, C. R., 1982, Cellular organization of glycerolipid metabolism, *Annu. Rev. Plant Physiol.* **33**:97.

Rouser, G., Kritchevsky, G., and Yamamoto, A., 1976, Column chromatographic and associated procedures for separation and determination of phosphatides and glycolipids, in *Lipid Chromatographic Analysis*, Vol. 3, 2nd ed. (G. V. Marinetti, ed.), pp. 713–776, Marcel Dekker, New York.

Rullkötter, J., Heinz, E., and Tulloch, A. P., 1975, Combination and position distribution of fatty acids in plant digalactosyl diglycerides, *Z. Pflanzenphysiol.* **76:**163.

Safford, R., and Nichols, G. W., 1970, Positional distribution of fatty acids in mono-galactosyldiglyceride from leaves and algae, *Biochim. Biophys. Acta* **210:**57.

Sastry, P. S., 1974, Glycosyl glycerides, *Adv. Lipid Res.* **12:**251.

Sastry, P. S., and Kates, M., 1964a, Lipid components of leaves. V. Galactolipids, cerebrosides and lecithin of runner bean leaves, *Biochemistry* **3:**1271.

Sastry, P. S., and Kates, M., 1964b, Hydrolysis of monogalactosyl and digalactosyl diglycerides by specific enzymes in runner bean leaves, *Biochemistry* **3:**1280.

Sastry, P. S., and Kates, M., 1965, Biosynthesis of lipids in plants. I. Incorporation of orthophosphate-^{32}P and glycerophosphate-^{32}P into phosphatides of *Chlorella vulgaris, Can. J. Biochem.* **43:**1445.

Sato, N., and Murata, N., 1982a, Lipid biosynthesis in the blue-green alga, *Anabaena variabilis.* I. Lipid classes, *Biochim. Biophys. Acta* **710:**271.

Sato, N., and Murata, N., 1982b, Lipid biosynthesis in the blue-green alga, *Anabaena variabilis.* II. Fatty acids and lipid molecular species, *Biochim. Biophys. Acta* **710:**279.

Sato, N., and Murata, N., 1982c, Lipid biosynthesis in the blue-green alga, *Anabaena variabilis.* III. UDPglucose : diacylglycerol glucosyltransferase activity in vitro, *Plant Cell Physiol.* **23:**1115.

Sato, N., and Furuya, M., 1984, The composition of lipids and fatty acids determined at various stages of haploid and diploid generations in the fern *Adiantum capillus-veneris Physiol. Plant* **62:**139.

Sato, N., Murata, N., Miura, Y., and Ueta, N., 1979, Effect of growth temperature on lipid and fatty acid compositions in the blue-green algae, *Anabaena variabilis* and *Anacystis nidulans, Biochim. Biophys. Acta* **572:**19.

Shibuya, I., and Benson, A. A., 1961, Hydrolysis of α-sulfoquinovosides by β-galactosidase, *Nature* **192:**1186.

Siebertz, H. P., Heinz, E., Linscheid, M., Joyard, J., and R. Douce, 1979, Characterization of lipids from chloroplast envelopes, *Eur. J. Biochem.* **101:**429.

Siefermann-Harms, D., Ross, J. W., Kaneshiro, K. H., and Yamamoto, H. Y., 1982, Reconstitution by monogalactosyldiacylglycerol of energy transfer from light-harvesting chlorophyll a/b–protein complex to photosystems in Triton X-100-solubilized thylakoids, *FEBS Lett.* **149:**191.

Siegenthaler, P. A., Rawyler, A., and Giroud, C., 1987a, Spatial organization and functional roles of acyl lipids, in *The Metabolism, Structure and Function of Plant Lipids* (P. K. Stumpf, J. B. Mudd, and W. D. Nes, eds.), pp. 161–168, Plenum, New York.

Siegenthaler, P.-A., Giroud, C., and Smutny, J., 1987b, Evidence for different acyl lipid domains in spinach and oat thylakoid membranes supporting various photosynthetic functions, in *The Metabolism, Structure and Function of Plant Lipids* (P. K. Stumpf, J. B. Mudd, and W. D. Nes, eds.), pp. 177–179, Plenum, New York.

Siegenthaler, P.-A., Sutter, J., and Rawyler, A., 1988, The transmembrane distribution of galactolipids in spinach thylakoid inside-out vesicles is opposite to that found in intact thylakoids, *FEBS Lett.* **228:**94.

Smith, C. R., and Wolff, I. A., 1966, Glycolipids of *Briza spicata* seed, *Lipids* **1:**123.

Smith, S. W., and Lester, R. L., 1974, Inositol phosphorylceramide, a novel substance and the chief member of a group of yeast sphingolipids containing a single inositol phosphate, *J. Biol. Chem.* **249:**3395.

Snell, E. E., DiMari, S. J., and Brady, R. N., 1970, Biosynthesis of sphingosine and dihydrosphingosine by cell-free systems from *Hansenula ciferri, Chem. Phys. Lipids* **5:**116.

Stanacev, N. Z., and Kates, M., 1963, Constitution of Cerebrin from the yeast *Torulopsis utilis, Can. J. Biochem. Physiol.* **41:**1330.

Steiner, S., and Lester, R. L., 1972, Studies on the diversity of inositol-containing yeast phospholipids: Incorporation of 2-deoxyglucose into lipid, *J. Bacteriol.* **109:**81.

Steiner, S., Smith, S., Waechter, C. J., and Lester, R. L., 1969, Isolation and partial characterization of a major inositol-containing lipid in Baker's yeast, mannosyl diinositol diphosphorylceramide, *Proc. Natl. Acad. Sci. USA* **64:**1042.

Stoffel, W., 1970, Studies on the biosynthesis and degradation of sphingosine bases, *Chem. Phys. Lipids* **5:**139.

Stoffel, W., 1971, Sphingolipids, *Annu. Rev. Biochem.* **40:**57.

Stoffel, W., 1974, Sphingosine metabolism, in *Fundamentals of Lipid Chemistry* (R. M. Burton, ed.), pp. 339–346, BI-Science Publications Div., Webster Groves, Missouri.

Stoffel, W., Sticht, G., and Le Kim, D., 1968, Metabolism of sphingosine bases. VI. Synthesis and degradation of sphingosine bases in *Hansenula ciferri, Hoppe-Seylers Z. Physiol. Chem.* **349**:1149.

Suomalainen, H., and Nurminen, T., 1970, The lipid composition of cell wall and plasma membrane of baker's yeast, *Chem. Phys. Lipids* **4**:247.

Tanaka, H., Ohnishi, M., and Fujino, Y., 1984, On glycolipids in corn seeds, *Nippon Nogeikagaku Kaishi* **58**:17.

Thompson, G. A., Jr., and Nozawa, Y., 1984, The regulation of membrane fluidity in *Tetrahymena* in *Biomembranes*, Vol. 12: *Membrane Fluidity* (M. Kates and L. A. Manson, eds.), pp. 397–342, Plenum, New York.

Thorpe, S., and Sweeley, C. C., 1967, Chemistry and metabolism of sphingolipids on the biosynthesis of phytosphingosine by yeast, *Biochemistry* **6**:887.

Toriyama, S., Hinata, K., Nishida, I., and Murata, N., 1987, Lipids from rice anthers, in *The Metabolism, Structure and Function of Plant Lipids* (P. K. Stumpf, J. B. Mudd, and W. D. Nes, eds.), pp. 345–347, Plenum, New York.

Tulloch, A. P., Heinz, E., and Fischer, W., 1973, Combination and positional distribution of fatty acids in plant sulfolipids, *Hoppe-Seylers Z. Physiol. Chem.* **354**:879.

Työrinoja, K., Nurminen, T., and Suomalainen, H., 1974, The cell envelope glycolipids of Baker's Yeast, *Biochem. J.* **141**:133.

van Besouw, A., and Wintermans, J. F. G. M., 1978, Galactolipid formation in chloroplast envelopes. I. Evidence for two mechanism in galactosylation, *Biochim. Biophys. Acta* **529**:44.

Vigh, L., 1982, Adaptation of thylakoid membranes of wheat seedlings to low temperature, *Biochemistry and Metabolism of Plant Lipids* (J. F. G. M. Wintermans and P. J. C. Kuiper, eds.), pp. 401–415, Elsevier, Amsterdam.

Wagner, H., and Zofcsik, W., 1966a, Sphingolipide und Glykolipide von Pilzen und hoheren Pflanzen. I. Isolierung eines Cerebroside aus *Candida utilis, Biochem. Z.* **346**:333.

Wagner, H., and Zofcsik, W., 1966b, Sphingolipide und Glykolipide von Pilzen und hoheren Pflanzen. II. Isolierung eines Sphingoglycolipids aus *Candida utilis* und *Saccharomyces cerevisiae, Biochem. Z.* **346**:343.

Wagner, H., and Zofcsik, W., 1969, Sphingolipide und Glykolipide von Pilzen und hoheren Pflanzen. III. Isolierung eines Cerebrosids aus *Aspergillus niger, Z. Naturforsch.* **B 24**:359.

Walcott, P., Kenrick, J. R., and Bishop, D. G., 1982, Molecular properties of potato tuber lipolytic acyl hydrolase, in *Biochemistry and Metabolism of Plant Lipids* (J. F. G. M. Wintermans and P. J. C. Kuiper, eds.), pp. 297–300, Elsevier, Amsterdam.

Walter, M. W., Jr., Hansen, A. P., and Purcell, A. E., 1971, Lipids of cured centennial sweet potatoes, *J. Food Sci.* **36**:795.

Wassef, M. K., 1977, Fungal lipids, *Adv. Lipid Res.* **15**:159.

Wassef, M. K., Ammon, V., and Wyllie, T. D., 1975, Polar lipids of *Macrophomina phaseolina, Lipids* **10**:185.

Webster, D. E., and Chang, S. B., 1969, Polygalactolipids in spinach chloroplasts, *Plant Physiol.* **44**:1523.

Weenink, R. O., 1964, Lipids of the acetone-insoluble fraction from red clover (*Trifolium pratense*) leaves, *Biochem. J.* **93**:606.

Weete, J. D., 1974, *Fungal Lipid Biochemistry*, pp. 267–286, Plenum, New York.

Weete, J. D., 1980, *Lipid Biochemistry of Fungi and Other Organisms*, pp. 148–153, Plenum, New York.

Weiss, B., and Stiller, R. L., 1967, Biosynthesis of phytosphingosine. Hydroxylation of dihydrosphingosine, *J. Biol. Chem.* **242**:2903.

Weiss, B., and Stiller, R. L., 1972, Sphingolipids of mushrooms, *Biochemistry* **11**:4552.

Weiss, B., Stiller, R. L., and Jack, R. C. M., 1973, Sphingolipids of the fungi *Phycomycetes blakesleeanus* and *Fusarium lini, Lipids* **8**:25.

Wickberg, B., 1958a, Structure of a glyceritol glycoside from *Polysyphonia fastigiata* and *Corallina officinalis, Acta Chem. Scand.* **12**:1183.

Wickberg, B., 1985b, Synthesis of 1-glyceritol-D-galactopyranosides, *Acta Chem. Scand.* **12**:1187.

Wilson, A. C., Kates, M., and de la Roche, A. I., 1978, Incorporation of [1-^{14}C]acetate into lipids of soybean cell suspensions, *Lipids* **13**:504.

Williams, J. P., Kahn, M., and Leung, S., 1975, Biosynthesis of digalactosyl diglyceride in *Vicia faba* leaves, *J. Lipid Res.* **16:**61.

Williams, J. P., Khan, M. U., and Mitchell, K., 1983, Galactolipid biosynthesis in leaves of 16 : 3- and 18 : 3-plants, in *Biosynthesis and Function of Plant Lipids* (W. W. Thomson, J. B. Mudd, and M. Gibbs, eds.), pp. 28–39, American Society of Plant Physiologists, Rockville, Maryland.

Wintermans, J. F. G. M., 1960, Concentrations of phosphatides and glycolipids in leaves and chloroplasts, *Biochim. Biophys. Acta* **44:**49.

Wintermans, J. F. G. M., 1963, Phosphatides and glycolipids of the photosynthetic apparatus, *Coll. Int. Cent. Nat. Rech. Sci.* **119:**381.

Wintermans, J. F. G. M., and Heemskerk, J. W. M., 1987, On the synthesis of digalactosyldiacylglycerol in chloroplasts and its relation to monogalactolipid synthesis, in *The Metabolism, Structure and Function of Plant Lipids* (P. K. Stumpf, J. B. Mudd, and W. D. Nes, eds.), pp. 293–300, Plenum, New York.

Wintermans, J. F. G. M., van Besouw, A., and Bogemann, G., 1981, Galactolipid formation in chloroplast envelopes. II. Isolation induced changes in galactolipid composition of envelopes, *Biochim. Biophys. Acta* **663:**99.

Wojciechowski, Z. A., 1983, The biosynthesis of plant steryl glycosides and saponins, *Biochem. Soc. Trans.* **11:**565.

Yagi, T., and Benson, A. A., 1962, Plant sulfolipid. V. Lysosulfolipid formation, *Biochim. Biophys. Acta* **57:**601.

Yoon, S. H., and Rhee, J. S., 1983, Lipid from yeast fermentation: effect of cultural conditions on lipid production and its characteristics in *Rhodotorula glutinis, J. Am. Oil Chem. Soc.* **60:**1281.

Zepke, H. D., Heinz, E., Radunz, A., Linscheid, M., and Pesch, R., 1978, Combination and positional distribution of fatty acids in lipids of blue-green algae, *Arch. Microbiol.* **119:**157.

Chapter 4

Glycoglycerolipids of Animal Tissues

Robert K. Murray and Rajagopalan Narasimhan

4.1. Introduction

This chapter summarizes the principal features of the biochemistry of the glycoglycerolipids (GGroLs) present in animal tissues. The isolation from wheat flour and the elucidation of the structure of galactosyldiacylglycerol (GalDAG) and digalactosyldiacylglycerol (Gal$_2$DAG) by Carter and colleagues (1956, 1961*a,b*, 1965) is one of the landmarks of the history of the biochemistry of glycolipids, as it established the GGroLs as one of the major classes of glycolipids. Subsequently, it was shown relatively quickly that GalDAG and Gal$_2$DAG are present in many plants and algae (reviewed by Sastry, 1974; Ishizuka and Yamakawa, 1985; see also Chapter 3, *this volume*). Of particular relevance to the subject matter of this chapter was the report by Steim and Benson (1963) that GalDAG occurs in bovine brain, as this was the first demonstration that GGroLs are constituents of animal tissues. Over the succeeding years, it has become apparent that a minimum of some six galactoglycerolipids are found in animal tissues, principally, although perhaps not exclusively, located in either nervous tissue and/or testis and spermatozoa. These six lipids can be allotted to two subclasses of the galactoglycerolipids: the galactosyldiacylglycerols and the galactosylacylalkylglycerols. It appears that each of these two subclasses contains a monogalactosyl species, a sulfated monogalactosyl species, and a digalactosyl species. Formal and trivial nomenclature for these lipids, the abbreviations for them used in this chapter, and the references reporting their initial discoveries are presented in Table 4-1. Figure 4-1 illustrates the structures of these compounds; it should be noted that structural information available on the two digalactosyl-containing lipids is incomplete.

Information on a second class of GGroL present in animal tissues has been available since 1977, when the Slomianys and colleagues described the occurrence in human gastric content of a series of novel glyceryl ether-containing glucoglycerolipids (Slomiany and Slomiany, 1977; Slomiany *et al.*, 1977*a–c*). These compounds, containing anywhere from one to six glucosyl

Robert K. Murray and Rajagopalan Narasimhan • Departments of Biochemistry and Pathology, University of Toronto, Toronto, Ontario, Canada M5S 1A8

Table 4-1. Galactoglycerolipids of Animal Tissues[a]

Formal name	Trivial name	Abbreviation	References
Galactosyldiacylglycerols			
1,2-di-*O*-acyl[β-D-galactopyranosyl (1'→3)]-*sn*-glycerol (**1**)	Galactosyldiacylglycerol, monogalactosyldiglyceride	GalDAG (MGDG)	Steim and Benson (1963)
1,2-di-*O*-acyl[β-D-(3'-sulfatoxy)-galactopyranosyl-(1'→3)]-*sn*-glycerol (**2**)	Sulfatoxygalactosyl-diacylglycerol	Sulfo-GalDAG	Flynn *et al.* (1975)
1,2-di-*O*-acyl[α-D-galactopyranosyl (1→6)-β-D-galactopyranosyl (1'→3)]-*sn*-glycerol[b] (**3**)	Digalactosyldiacylglycerol	Gal$_2$DAG (DGDG)	Rouser *et al.* (1967); Wenger *et al.* (1970)
Galactosylacylalkylglycerols			
1-*O*-alkyl-2-*O*-acyl[β-D-galactopyranosyl (1'→3)]-*sn*-glycerol (**4**)	Galactosylacylalkyl-glycerol	GalAAG (GG)	Norton and Brotz (1963)
1-*O*-alkyl-2-O-acyl[-β-D-(3'-sulfatoxy)galactopyranosyl(1'→3)]-*sn*-glycerol (**5**)	Sulfatoxygalactosylacylalkylglycerol, seminolipid	Sulfo-GalAAG (SGG)	Kornblatt *et al.* (1972); Ishizuka *et al.* (1973)
1-*O*-alkyl-2-*O*-acyl-[α-D-galactopyranosyl (1→6)-β-D-galactopyranosyl (1'→3)]-*sn*-glycerol[c] (**6**)	Digalactosylacylalkyl-glycerol	Gal$_2$AAG	Levine *et al.* (1976)

[a]See Fig. 4-1 for structures (see also Murray *et al.*, 1980).
[b]The structure proposed for this compound is tentative, and is based mostly on analogy with the corresponding compound of plant origin.
[c]The structure proposed for this compound is also very tentative. It contains galactose, glyceryl ether and fatty acid in approximate molar ratios of 2 : 1 : 1; little else is known concerning its detailed structure.

residues, some of which are sulfated, have subsequently been reported to be quite widely distributed in various tissues and secretions (Slomiany and Slomiany, 1980). Representatives of the group are indicated in Table 4-2.

This chapter discusses the biochemical properties of the two types of GGroLs—galacto- and glucoglycerolipids—isolated from animal tissues and characterized at the time of writing. The emphasis is principally on metabolic and biological aspects of these lipids, as these topics are the authors' major interests; much less emphasis is placed on structural and physical studies. The principal methods used in the determination of their structures have been reviewed at a general level elsewhere (Murray *et al.*, 1980). Volume 3 of this series contains an extensive discussion by Hakomori (1983*a*) of the principal methods used to elucidate the structure of GSLs; most of these methods are

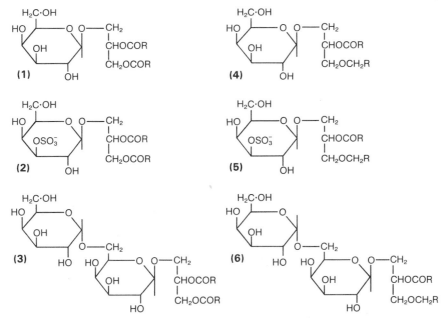

Figure 4-1. Structures of the six galactoglycerolipids isolated from animal tissues (see Table 4-1). **1**, Galactosyldiacylglycerol (GalDAG); **2**, Sulfatoxygalactosyldiacylglycerol (sulfo-GalDAG); **3**, Digalactosyldiacylglycerol (Gal₂DAG); **4**, Galactosylacylalkylglycerol (GalAAG); **5**, Sulfatoxygalactosylacylalkylglycerol (sulfo-GalAAG); **6**, Digalactosylacylalkylglycerol (Gal₂AAG). Compounds **3** and **6** have only been partially characterized, so that the structures shown for these lipids must be considered as quite tentative. The structure of compound **3** is based in part by analogy with the corresponding lipid found in plants. If the structures of compounds **3** and **6** prove to be as shown, there would be an exact analogy between the galactosyldiacylglycerols (compounds **1–3**) and the galactosylacylalkylglycerols (compounds **4–6**), each subclass consisting of a monogalactosyl species, a sulfated monogalactosyl species and a digalactosyl species.

also directly applicable to the GGroLs. By contrast, rather few physical studies have been performed on these lipids. The elegant application of methods of organic chemistry to synthesize GGroLs, such as sulfo-GalAAG (Gigg, 1978), is also not considered here, as these are covered in Chapter 7 (*this volume*). A comprehensive review of glycolipids, including material on both GSLs and GGroLs, was published by Sweeley and Siddiqui (1976). Various aspects of the biochemistry of the galactoglycerolipids of animals have been previously reviewed by Murray *et al.* (1976, 1980). Slomiany and Slomiany (1980) described progress made in the study of the glucoglycerolipids, and Egge (1983*a*) reviewed the subject of the glyceryl ether-containing glycolipids. Recently, the total field of GGroLs, including these compounds found in bacteria, plants, and animals, was reviewed by Ishizuka and Yamakawa (1985). As several of the GGroLs included in the present review contain sulfate, the review of sulfolipids by Farooqui (1981) and the reviews of sulfatide lipidosis by Moser and Dulaney (1978) and by Kolodny and Moser (1983) also provide relevant material on these particular compounds (see also Chapters 1 and 3, *this volume*).

Table 4-2. Structures of Some Glucoglycerolipids Isolated from Human Gastric Secretion[a]

Formal name	Abbreviation
1-*O*-alkyl-2-*O*-acyl[α-D-glucopyranosyl(1′→3)]-*sn*-glycerol (**7**)	Glc*p*AAG
1-*O*-alkyl-2-*O*-acyl[α-D-glucopyranosyl(1→6)-α-D-glucopyranosyl(1→6)-α-D-glucopyranosyl(1→6)-α-D-glucopyranosyl(1→6)-α-D-glucopyranosyl(1′→3)]-*sn*-glycerol (**8**)	(Glc*p*)$_6$AAG
1-*O*-alkyl-2-*O*-acyl-[α-D-glucopyranosyl(1→6)-α-D-glucopyranosyl(1→6)-α-D-glucopyranosyl(1→6)-α-D-glucopyranosyl(1→6)-α-D-glucopyranosyl(1→6)-α-D-glucopyranosyl(1′→3)]*sn*-gycerol (**9**)	(Glc*p*)$_8$AAG
1-*O*-alkyl-2-*O*-acyl-[α-D-(6′-sulfatoxy)-glucopyranosyl-α-D-glucopyranosyl(1→6)-α-D-glucopyranosyl(1′→3)]-*sn*-glycerol (**10**)	Sulfo-Glc*p*(Glc*p*)$_2$AAG
1-*O*-alkyl-2-*O*-acyl-[α-D-(6′-sulfatoxy)-glucopyranosyl-α-D-glucopyranosyl(1→6)-α-D-glucopyranosyl-(1→6)-α-D-glucopyranosyl(1′→3)]-*sn*-glycerol (**11**)	Sulfo-Glc*p*(Glc*p*)$_3$AAG

[a]From Slomiany and Slomiany (1980).

4.2. Isolation of Galactoglycerolipids

The isolation of galactoglycerolipids is considered only briefly; the reader should consult the original papers for details. However, a few points germane to this topic are emphasized. The extraction procedure described by Suzuki (1965) is suitable for the isolation of the known galactoglycerolipids of testis, spermatozoa, and brain. The lower-phase fraction of the Folch wash contains the neutral lipids, the phospholipids, the galactoglycerolipids, and the GSLs, excluding the gangliosides. The latter are found in the upper phase; appreciable amounts of certain of the more polar lipids (e.g., polyglycosylated GGroLs) may also appear in this phase. The lower-phase lipids are then separated into their major classes by silicic acid column chromatography, following the procedure of Vance and Sweeley (1967). Elution is performed with (1) chloroform, (2) acetone, (3) acetone–methanol (9 : 1 v/v), and (4) methanol. They yield fractions that contain mainly (1) cholesterol and triglycerides, (2) simple GSLs and the galactoglycerolipids, (3) more complex GSLs, and (4) phospholipids, respectively. The lipids present in each fraction are next resolved by thin-layer chromatography (TLC) and the glycolipids detected by spraying with aniline–diphenylamine or orcinol reagents. Figure 4-2 shows the major glycolipids present in the acetone fraction derived from the testes of rats of different ages. In all the channels shown (Fig. 4-2, channels 1–4), the slower migrating compound is sulfo-GalAAG (see Section 4.8.1), and the faster migrating compound is GalAAG (see Section 4.7.2). It is evident that no appreciable amounts of glycolipids other than these two galactoglycerolipids are observable in the acetone fraction obtained from rat testis. Sphingolipids react with the benzidine spray reagent of Bischel and Austin (1963), whereas GGroLs do not; use can be made of this property to distinguish among these two major lipid classes at this stage. Glycolipids of interest can then be isolated

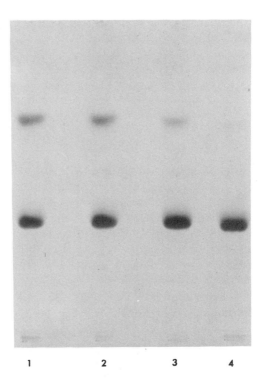

Figure 4-2. Analysis by thin-layer chromatography (TLC) of glycolipids from rat testes. Samples of acetone fractions from the testes of four groups: 1, 20-day-old rats; 2, 27-day-old rats; 3, 36-day-old rats; 4, a 350-g adult rat. Portions of the respective acetone fractions, each containing about 24 nmoles of lipid-bound sulfate, were spotted on a TLC plate, developed in chloroform–methanol–water–ammonia (65 : 25 : 3 : 1, v/v), and visualized with the aniline–diphenylamine reagent. The slower-moving compound is the major glycolipid, sulfo-GalAAG, and the faster-moving compound is GalAAG. (From Kornblatt *et al.*, 1974.)

1   2   3   4

by preparative TLC, spraying a guide strip with one of the above reagents to facilitate detection of the desired compounds.

Several points should be considered whenever one is attempting to isolate GGroLs from animal sources, particularly if no previous work has been reported on the tissue or cells under study:

1. It is not uncommon, when working with GSLs, to treat the lipid extract with mild alkali. This removes contamination by alkali-labile lipids (e.g., most phospholipids) and does not affect the great majority of GSLs. For reasons that will become apparent, such treatment with mild alkali must be avoided if GGroLs are to be isolated in their intact state.

2. Certain of the GGroLs are sulfated (see Tables 4-1 and 4-2). When possible, we have found it extremely useful for the assessment of the number of lipids sulfated in the tissue under analysis, to administer [^{35}S]sulfate some hours before collection of the tissue. A dose of approximately 1.0 mCi per 100 g body weight administered intraperitoneally (i.p.) ~3 hr before collecting the tissue has been found suitable in the rat for this purpose. Lipid fractions are then obtained as described earlier, subjected to TLC and the developed chromatograms covered with X-ray film for the purposes of autoradiography. Examination of the autoradiograms obtained should provide a fairly accurate indication of the number of sulfolipids present.

3. Because sulfation increases the polarity of lipids, it is important initially to analyze the upper-phase fraction obtained during extraction, as this may contain appreciable amounts of certain sulfolipids. As indicated in point (2) above, previous administration of [^{35}S]sulfate will greatly facilitate the detection by radiochromatography of sulfated lipids in both the upper and lower phases.

4.3. Methods Used for Analysis of Structure

Methods used for the study of the structures of GGroLs have been reviewed (Murray *et al.*, 1980). References to particular methods used in establishing the structures of the known GGroLs are provided later in this chapter when individual lipids are discussed. Hakomori (1983*a*) provides an excellent discussion on the principal methods used for elucidating the structures of GSLs. Most of these methods are directly applicable to the GGroLs, and the reader is referred to that source for relevant details. Sweeley and Nunez (1985) have recently reviewed the structural analysis of glycoconjugates by mass spectroscopy (MS) and nuclear magnetic resonance (NMR) spectroscopy. In addition, Egge (1983*b*) reviewed the application of MS to the analysis of ether-containing lipids, including the GGroLs. The major emphasis in this discussion is placed on certain relatively simple methods considered to be particularly useful when working with GGroLs.

Table 4-3 summarizes the principal methods used to analyze the structures of the GGroLs. The methods are listed under three categories: (1) simple but informative procedures involving analysis of intact lipids or their degradation products by TLC; (2) more complex procedures used on intact lipids, which, taken together, provide extensive structural information; and (3) application of a number of the techniques listed under (2) to analysis of the products of degradation of GGroLs. As the resolution afforded by methods such as high-resolution NMR spectroscopy is continually improving, it is possible that certain of the more classic and definitive, but time-consuming, methods of determining structure (e.g., permethylation) may become superfluous in the future.

Visual examination of a stained TLC can in itself provide useful general information that pertains to structure. Specifically, most simple GSLs (e.g., mono-, di-, tri-, and tetraglycosyl ceramides) resolve on TLC as double bands, probably because of the presence of hydroxy fatty acids in the slower migrating of the bands. By contrast, GGroLs have never been reported to resolve as double bands on TLC, presumably because they lack hydroxy fatty acids.

We believe that the most useful test for diagnosing the presence of a GGroL in a lipid fraction is to perform alkali-catalyzed methanolysis, using rather mild conditions (Levine *et al.*, 1975), on either the purified lipid or the lipid fraction. Generally speaking, most GSLs are stable under these conditions, as the amide bond linking the fatty acid to the long-chain base component of ceramide is not affected by mild alkali. However, the acyl bonds present in GGroLs are readily cleaved by mild alkali, resulting in the loss of

Table 4-3. Methods Used in Determining the Structure of Glycoglycerolipids[a]

Method	Conclusion from positive reaction	Reference
Methods based on TLC		
Staining with aniline-diphenylamine or orcinol	Contains sugar(s)	Kornblatt *et al.* (1972)
Spraying with benzidine or tetramethylbenzidine	Contains a long-chain base or an amino compound	Narasimhan, *et al.* (1982*b*)
Treatment of lipid with mild alkali followed by analysis by TLC	Can often deduce if lipid is a glycodiacylgycerol (completely deacylated), a glycoacylalkyl-glycerol (party deacylated) or a GSL (not affected)	Murray *et al.* (1980)
Label lipid *in vivo* by administration of [35S]sulfate	The lipid is a sulfolipid	Kornblatt *et al.* (1972)
Treatment of lipid with mild acid to desulfate it	The lipid is a sulfolipid	Levine *et al.* (1975)
Analysis of products of methanolysis	Can determine presence of sugars, fatty acids and glyceryl ethers by comparison with standards	Narasimhan *et al.* (1982*b*)
Analyses on intact lipids	*Information provided*	
IR spectrometry	A variety of structural information	Ishizuka *et al.* (1973)
HPLC–MS	A variety of structural information	
GLC–MS	A variety of structural information	
FAB–MS	A variety of structural information	
NMR spectroscopy	A variety of structural information	Ishizuka *et al.* (1973)
Enzyme degradation (e.g., various galactosidases)	Evidence regarding anomeric nature of glycosidic linkages	Ishizuka *et al.* (1973)
Permethylation studies	Detailed information about glycosidic linkages	Ishikuka *et al.* (1973)
Analyses on products of methanolysis	*Information provided*	
GLC	Identification of sugars, fatty acids and glyceryl ethers	Kornblatt *et al.* (1972)
GLC–MS	More specific identification of the above components	Ishizuka *et al.* (1978)
HPLC	Identification of sugars, fatty acids and glyceryl ethers	
HPLC–MS	More specific identification of the above components	

[a]A detailed evaluation of the structural information provided by each of the methods used is not provided here. The reader is directed to the references cited for additional information. Other methods used in analysing the structures of glycoglycerolipids have been described by Murray *et al.* (1980). Chapter 1 by Hakomori (1983a) in Volume 3 of this series presents a detailed analysis of the uses of many of the above methods, particularly those listed under the second and third categories. Where a reference is not cited, the method has not as yet been applied to the study of GGroLs to provide the information indicated.

either one or two fatty acids, depending on the type of GGroL under examination. For instance, in the case of GalAAG, only one fatty acid is present and, because the ether bond is stable to mild alkali, the resulting lyso-derivative still migrates appreciably in the usual chromatographic conditions employed. However, in the case of GalDAG, both fatty acids are removed, and the galactosylglycerol product, which is quite polar, barely leaves the origin of the chromatogram. The liberated galactosylglycerol can be readily recovered by washing the methanolysis mixture with one fifth volume of water. The galactosylglycerol partitions into the aqueous upper phase, which can be dried down, subjected to trimethylsilylation, and analyzed by gas–liquid chromatography (GLC) (Williams *et al.*, 1975). Similar considerations apply to the digalactosylglycerol product released from Gal_2DAG by treatment with mild alkali.

Reference has been made to the use of a benzidine reagent (Bischel and Austin, 1963) to distinguish between GSLs and GGroLs. We have found that the noncarcinogenic N,N,N',N'-tetramethylbenzidine may be substituted for the carcinogenic benzidine, retaining at least equal sensitivity.

If the tissue under study has been prelabeled by administration of [^{35}S]sulfate, a combination of treatment with mild alkali (the sulfate ester bond is stable to this treatment) and radiochromatographic analysis of the products can be very useful in sorting out the type of sulfolipids present in the tissues. Table 4-4 (modified after Farooqui, 1981) lists the principal sulfolipids known to be present in mammalian tissues; only the two galactoglycerolipids are affected by treatment with mild alkali. Under certain chromatographic conditions, cholesterol sulfate and sulfo-GalAAG exhibit rather similar R_f values. The use of a solvent system of chloroform–methanol–acetone–acetic acid–water (10 : 2 : 4 : 2 : 1 v/v) (Ishizuka and Tadano, 1982) affords an excellent separation of these two lipids. In confirmation of earlier work (Iwamori *et al.*, 1976), we have found cholesterol sulfate to be widely distributed in mammalian tissues. Recent studies in our laboratory have indicated that it is also present in the stomach, small and large intestines, pancreas, and salivary glands of rats, and it is our impression that it can be detected in most, if not all, tissues of that animal if labeling with [^{35}S]sulfate is employed. Thus, when studies of the sulfolipid profile of a tissue are contemplated, it would appear prudent to consider the possibility that cholesterol sulfate is present. This lipid elutes in the acetone fraction obtained by silicic acid column chromatography and is therefore present in the same fraction as sulfo-GalAAG, sulfo-GalDAG, and sulfo-GalCer. Like sulfo-GalCer, it is stable to mild alkali but does not yield a positive reaction with spray reagents, such as aniline–diphenylamine and orcinol, that detect sugars.

As shown in Table 4-4, recent studies (Slomiany and Slomiany, 1980; Tadano and Ishizuka, 1982*a,b*; Tadano *et al.*, 1982; Slomiany *et al.*, 1981) have enlarged the number of sulfated GSLs known to occur in mammalian tissues. These compounds should be relatively easy to distinguish from sulfated GGroLs, as they are stable to treatment with mild alkali and react positively with benzidine reagent.

A number of new techniques for analyses of glycolipids have been intro-

Table 4-4. Sulfolipids of Animal Tissues[a]

Sulfolipid	Site	References[b]
Sulfatoxygalactosylceramide (Sulfo-GalCer)	Nervous tissue Kidney Retina	
Sulfatoxylactosylceramide (Sulfo-LacCer)	Kidney	
Sulfatoxygalactosyllactosylceramide	Gastric mucosa	
Sulfated gangliotriaosylceramide	Rat kidney	Tadano and Ishizuka (1982a)
Bis-sulfated triglycosylceramide	Rat kidney	Tadano and Ishizuka (1982b)
Bis-sulfated gangliotetraosylceramide	Rat kidney	Tadano et al. (1982)
Sulfated ganglioside	Bovine gastric mucosa	Slomiany et al. (1981)
Sulfotoxygalactosylacylalkylglycerol (Sulfo-GalAAG)	Testis Spermatozoa Nervous tissue	
Sulfagalactosyldiacylglycerol (Sulfo-GalDAG)	Nervous tissue	
Sulfatoxytriglucosylacylalkylglycerol	Human gastric secretion	Slomiany and Slomiany (1980)
	Human saliva	Slomiany and Slomiany (1980)
Cholesterol sulfate	Kidney Spleen Adrenal Lung Brain	
Desmosterol sulfate	Testis Epididymis Spermatozoa	
Other steroid sulfates	Liver Kidney Brain Other tissues	

[a]Modified from Farooqui (1981).
[b]References to the original sources are given in Farooqui (1981), except where otherwise indicated. A number of other sulfated GSLs have been detected in hog gastric mucosa (Slomiany and Slomiany (1980)).

duced during recent years. McLuer and Evans (1973) and Ullman and McLuer (1977) pioneered the resolution of GSLs by high-performance liquid chromatography (HPLC), providing an ultrasensitive quantitative method for their analysis (McLuer and Ullman, 1980). Yahara and Kishimoto (1981) separated the perbenzoylated GalDAG and GalAAG species of calf brain by preparative TLC and subsequently by reverse-phase HPLC (RP-HPLC). Their results indicated the utility of this approach for separating and analyzing galactoglycerolipids; combination with MS could make this a very powerful tool for analysis of these compounds.

Recently, a number of investigators studied the suitability of fast atom bombardment mass spectrometry (FAB–MS) for the analysis of gangliosides and other GSLs. For example, Arita *et al.* (1984) and Egge *et al.* (1985) demonstrated the great power of negative-ion FAB–MS, used without any derivativization of the samples, for the analysis of gangliosides. The technique gave intense peaks of the molecular ion species of a number of gangliosides, and many fragment ions useful for the elucidation of their structure were detected. This appears likely to become a powerful tool for the analysis of various glycoconjugates, including GGroLs.

The specificity displayed by antibodies holds great potential both for analyzing the antigenic reactivity of glycolipids and for exploring their subcellular locations and possible functions. Just as specific proteins, subsequent to their resolution by electrophoresis, can be identified by antibodies, Magnani *et al.* (1980) demonstrated that the same principle can be applied to GSLs separated by TLC. This technique (immunoblotting) can demonstrate unsuspected cross-reactivity among various glycolipids (e.g., among glycolipids that contain a common nonreducing terminal sugar) and can also be used to study the interaction of various glycolipids with various proteins, such as cholera toxin. Application of this technique should prove useful for probing certain functional and immunological aspects of the GGroLs. Another very powerful immunological approach is that provided by the development of monoclonal antibodies (Kohler and Milstein, 1975, 1976). Hakomori and Young (1983) pointed out that, by use of this procedure, it should be possible to produce a battery of specific, high-titer antibodies to serve as probes of cell-surface carbohydrate structure and function. At least one monoclonal antibody to a galactoglycerolipid is now available (Eddy *et al.*, 1985) (see Section 4.8.1.7).

4.4. Galactosyldiacylglycerol (GalDAG)

Steim and Benson (1963) first described the presence of GalDAG (see Fig. 4-1) in brain, on the basis of paper chromatographic detection of its deacylation products. Their observation was confirmed by Norton and Brotz (1963), Rouser *et al.* (1963), and Rumsby and Gray (1965). Subsequently, Steim (1967) reported in detail the properties of GalDAG isolated from bovine spinal cord. It was shown to be identical with GalDAG (**1**) from spinach, except for its fatty acid analysis, with palmitic and oleic acids comprising 75% of the total. Unequivocal evidence was obtained for the presence of a β-glycosidic linkage, based on co-chromatography on paper of the deacylation products of the plant and animal lipids, co-chromatography in the vapor phase of the trimethylsilyl derivatives of monogalactosylglycerol of plant and animal origin and of the synthetic β-anomer, hydrolysis by β-galactosidase, and the splitting and chemical shift of the resonance signal of the anomeric proton. Infrared (IR) spectra of the monogalactosylglycerols of plant and animal origin and of their acetates indicated the D-configuration of the glyceryl moiety. Approximately 1.4 mg of GalDAG per g wet weight was found in bovine spinal cord, the lipid comprising about 0.6% of total lipids of the spinal cord. The lipid appeared to be located only in white matter, none being observed in gray matter, lactating bovine mammary gland, bovine liver, bovine spleen, or rat kidney. Steim (1967) concluded that GalDAG "may take its place with the better-known glycolipids as a structural component of myelin" and speculated about whether the mechanism of galactosylation proceeded through UDP-galactose, as had been shown for the GalDAG of green plants by Neufeld and Hall (1964) (see Chapter 3, Section 3.8.1.1).

In their comprehensive study of the postnatal changes in the concentration of the lipids of developing rat brain, Wells and Dittmer (1967) found that GalDAG belonged with a group of lipids comprising cerebroside, sphingomyelin, triphosphoinositide, phosphatidic acid, and inositol plasmalogen. All these lipids occurred at a concentration less than 10% of that found in the adult brain up to the onset of myelination and then rapidly increased in concentration throughout the period of active myelination.

Inoue *et al.* (1971) used a sensitive and specific assay by GLC involving the measurement of the trimethylsilyl derivative of galactosylglycerol, to determine the concentration of GalDAG in whole brain and in fractions of brains from rats of varying age. GalDAG was barely measurable before 10 days of age and then decreased rather quickly to adult values. In adult brain, most of the GalDAG was associated with myelin (64–68%), the next highest amount (15–18%) being recoverable from the microsomal fraction. In brains undergoing active myelination, the greater part of the GalDAG appeared in the microsomal fraction up to the age of 20 days. After 20 days, the amount of GalDAG in the microsomal fraction fell precipitously and increased sharply in a "small myelin" fraction. As the brains matured, progressively more GalDAG was located in a "large myelin" fraction. The results suggested that GalDAG was initially synthesized in the microsomal fraction and was then transferred to myelin.

Desmukh *et al.* (1971) showed that the brain of the Jimpy mouse, a mutant with a deficiency of myelin in the central nervous system (CNS), had little ability to synthesize GalDAG. The amount of GalDAG in the brain of the Jimpy (19 μmoles/g wet wt) was less than one twentieth that found in the brain of the normal mouse. Little or no ability to synthesize Gal_2DAG *in vitro* was observed in the brain of the Jimpy, whereas the activities of the enzymes involved in the degradation of the GalDAG and Gal_2DAG were not affected. These results, showing a relative inability to make and transfer galactosyldiglycerides to myelin, correlated with the relative absence of myelin in the brain of the Jimpy and support the contention that these lipids are associated with the process of myelination.

Wenger *et al.* (1968) investigated the pathway of biosynthesis of GalDAG. These investigators used microsomal enzyme preparations of brain from rats aged 13–20 days to effect the synthesis of GalDAG from UDP-galactose and 1,2-diglyceride. The enzyme was found to require the 1,2-isomer of the diglyceride substrate and preferred diglycerides with long chain-saturated fatty acid constituents. The activity of the enzyme was highest in preparations from brains of rats aged 14–18 days. The age of highest enzymatic activity corresponded to the age at which the concentration of GalDAG was found to increase most rapidly by Wells and Dittmer (1967) and to the age at which myelination in rat brain occurs at a maximal rate. Wenger *et al.* (1988) suggested that GalDAG and the galactosyltransferase responsible for its synthesis from diglyceride might play an important role in the process of myelination.

Subba Rao *et al.* (1970) examined the enzymatic hydrolysis by rat brain cell fractions of radiochemically pure GalDAG synthesized *in vitro* by enzymes derived from rat brain or spinach chloroplasts. Evidence was obtained for two distinct enzyme activities degrading GalDAG, as follows:

$$\text{GalDAG} \xrightarrow{\quad\beta\text{-galactosidase}\quad} 1,2\text{-diglyceride} + \text{galactose} \tag{1}$$

$$\text{GalDAG} \xrightarrow{\quad\text{galactolipase}\quad} \text{galactosylglycerol} + \text{fatty acids} \tag{2}$$

The activity (β-galactosidase) catalyzing the first reaction appeared to be located in the mitochondrial fraction and displayed a pH optimum of 4.4, whereas the activity (galactolipase) catalyzing the second reaction was located in the microsomal fraction and exhibited a pH optimum of 7.2. A study of the activity of galactolipase revealed that it increased with age (up to at least 40 days), except during the period of most active myelination (~ 10–20 days), at which time the activity was notably depressed; the activity of the galactosidase did not demonstrate this inverse correlation with the rate of myelination.

4.5. Sulfatoxygalactosyldiacylglycerol (Sulfo-GalDAG)

Flynn *et al.* (1975) reported the isolation of sulfo-GalDAG (see Fig. 4-1) following injection of [^{35}S]sulfate into the brains of rats of aged 22 days. The brain lipids were fractionated by chromatography on silicic acid and the fraction eluted with acetone was collected and resolved by TLC. Two radioactive glycolipids were detected, one of which appeared to be sulfo-GalCer. The other was studied further. It did not react with benzidine; on treatment with mild alkali, it was converted to a water-soluble product, characterized as sulfatoxygalactosylglycerol. Solvolysis with dioxane resulted in release of sulfate and formation of a product that co-chromatographed with a standard of GalDAG (**1**). Studies with periodate suggested that, like sulfo-GalCer, the sulfate was attached to position 3 of the galactose. Analysis by IR spectrometry indicated that the spectrum of this compound is very similar to that of sulfo-GalAAG (**5**) (Ishizuka *et al.*, 1973), except that the relative ester absorption (1720 cm^{-1}) is twice as strong. Levine *et al.* (1975), in their study demonstrating the presence of sulfo-GalAAG in rat brain, noted the presence of another minor [^{35}S]-containing glycolipid, the chromatographic migration of which was very similar to that of sulfo-GalAAG (**5**) but that was apparently completely deacylated by treatment with mild alkali. Levine and co-workers suggested that this compound was sulfo-GalDAG (**2**).

Flynn *et al.* (1975) also observed that incubation of GalDAG with 1% Triton X-100, [^{35}S]-labeled 3'-phosphoadenosine-5'-phosphosulfate (PAPS), and a microsomal fraction from the brains of 21-day-old rats yielded a radioactive lipid exhibiting properties, including identical migration in TLC, similar to those of the compound isolated from brain. In a subsequent study, Subba Rao *et al.* (1977) further investigated the biosynthesis of sulfo-GalDAG by extracts of the brains of 21-day-old rats *in vitro*. Triton X-100 extracts of the microsomal fraction of rat brain were found to catalyze the formation of sulfo-GalDAG (**2**) from GalDAG (**1**) and PAPS. The enzyme preparation did not catalyse sulfation of cholesterol or Gal$_2$DAG. Significant differences between the formation of sulfo-GalDAG and of sulfo-GalCer catalyzed by the same enzyme preparation were noted. For instance, ATP and Mg^{2+} were found to inhibit the formation of sulfo-GalDAG strongly but to stimulate the synthesis of sulfo-GalCer equally strongly. The apparent K_M for GalDAG was 200 µM, whereas that for GalCer was 45 µM. Also, GalDAG and GalCer were mutually inhibitory toward the synthesis of their respective sulfated derivatives. These results did not necessarily lead to the conclusion that two distinct sulfotransferases were present, but they did indicate a possible means of controlling the synthesis of these two sulfolipids.

Pieringer *et al.* (1977) demonstrated a close association of the sulfo-GalDAG of rat brain with the process of myelination, as shown by the following observations:

1. Sulfo-GalDAG is barely detectable in rat brain before 10 days of age, accumulates rapidly at 1–25 days of age and remains relatively constant in amount thereafter in adult life.
2. The activity of the sulfotransferase responsible for synthesis of sulfo-GalDAG (see above) is almost completely absent before 10 days of age, attains a maximum at age 20 days, and slowly decreases thereafter with increasing age. This particular pattern correlates well with that of other myelin-specific metabolites.
3. Both the concentration of sulfo-GalDAG and the activity of the sulfotransferase are greatly decreased in the brain of the nonmyelinating Jimpy mouse.
4. The myelin fraction of rat brain was found to contain most of the sulfo-GalDAG. Approximately equimolar amounts of sulfo-GalDAG and of sulfo-GalAAG were found in the brains of both 21-day-old and adult rats. Studies using [^{35}S]sulfate indicated that the turnover of sulfo-GalDAG was more rapid than that of sulfo-GalAAG. Sulfo-Gal-DAG contributed 16% of the total sulfolipid of the brains of 21-day-old rats and some 8.4% of that of adult rats.

Ishizuka *et al.* (1978) reported the results of a further study of the sulfated galactoglycerolipids of rat brain, one of the objectives of the study being to determine the relative ratios of sulfo-GalDAG and sulfo-GalAAG in rat brain at different stages of development. The sulfate moieties of the galactoglycerolipids were released by solvolysis in dioxane and, after deacylation, galactopyranosylglycerol and 1-*O*-alkyl-3-galactopyranosylglycerol were identified as their trimethylsilyl derivatives by GLC–MS. This finding substantiated the claim that sulfo-GalDAG and sulfo-GalAAG might coexist in rat brain (Levine *et al.*, 1975). The desulfated galactolipids were completely hydrolyzed by β-galactosidase, releasing D-galactose and 1,2-diacyl- and 1,2-al-kylacylglycerols. The synthesis of the sulfated galactoglycerolipids in rat brain was found to be most active at about 18 days of age. The amount of total sulfated galactoglycerolipids in rat brain increased with age and levelled off after 48 days of age. The molar ratio of the diacyl and the alkylacyl forms also changed with age; until 19 days of age, more of the diacyl type was present, but later the alkylacyl type was found to predominate (85% at 68 days).

Burkart *et al.* (1983*a*) studied the metabolism *in vivo* of sulfated galactoglycerolipids in the cerebrum and cerebellum of developing mice after intraperitoneal injection of [^{35}S]sulfate; because of the small amounts of material available, no effort was made to distinguish between sulfo-GalDAG and sulfo-GalAAG. Throughout development the rate of biosynthesis of these lipids was higher in the cerebellum than in the cerebrum. The developmental pattern of net synthesis of the sulfated galactoglycerolipids in both parts of the brain generally resembled that shown by sulfo-GalCer (Burkart *et al.*, 1981). The rate of synthesis of these lipids peaked somewhat earlier than that of sulfo-GalCer, however, at 14 days instead of 20 days. During postnatal development, 40–80% of the sulfated galactoglycerolipids synthesized daily disappeared within 24 hr, suggesting an important role for degradation in the regulation of the content of these lipids in myelinating brain. In a further study, Burkart *et al.* (1983*b*) analyzed the rate of synthesis of these lipids and the kinetics of their subcellular distribution subsequent to synthesis in the myelinating brains of 17-day-old mice. The results of this study indicated that the lipids are synthesized in the Golgi–endoplasmic reticulum complex and transferred in vesicles associated with lysosomes to the myelin membranes. During this transfer, an appreciable fraction of the lipids appears to be degraded, as was also found for sulfo-GalCer (Burkart *et al.*, 1982).

4.6. Digalactosyldiacylglycerol (Gal₂DAG)

Rouser *et al.* (1967) isolated a lipid exhibiting the chromatographic properties of Gal₂DAG from human brain (structure **3**, Fig. 4-1). Wenger *et al.* (1970) investigated the biosynthesis of this lipid by using particulate enzyme preparations from the brains of rats over 11 days of age. These preparations were found to catalyze the synthesis of Gal₂DAG from two precursors, UDP–galactose and diglyceride. GalDAG was implicated as an intermediate in the reaction. The use of specific galactosidases suggested that the terminal galactose was bonded to the inner galactose via an α-galactosidic linkage. The activity of the α-galactosyltransferase was dependent on the age of the rat from which the brain was taken; it was low or nonexistent before about 11 days but increased sharply after that time, reaching a peak at about 16–18 days postpartum. In like manner to the enzyme involved in the biosynthesis of GalDAG, a role in the process of myelination was also suggested for this enzyme.

In their study of the enzymatic pathways involved in the degradation of GalDAG in rat brain (see Section 4.4), Subba Rao *et al.* (1970) found that both galactosidase and galactolipase activities were present that degraded Gal_2DAG in a generally similar manner to that of GalDAG. Thus, the studies described in this and the two previous Sections revealed that brain possesses enzyme activities that can synthesize and degrade GalDAG, sulfo-GalDAG, and Gal_2DAG. All three of these galactoglycerolipids appear to be intimately associated with myelin.

4.7. Galactosylacylalkylglycerol (GalAAG)

4.7.1. GalAAG of Brain

Norton and Brotz (1963) isolated a glycolipid fraction from bovine brain of lower polarity than the cerebrosides and resolved it into four previously unreported crystalline galactolipids, one of which was shown to be GalAAG (structure **4**, Fig. 4-1). It was estimated that GalAAG was contaminated by GalDAG to an extent of approximately 20%; the three other compounds were possibly positional isomers of cerebroside fatty acid esters. Characterization of GalAAG was performed by isolating its products of deacylation, i.e., non-hydroxy fatty acids and a compound tentatively designated as lyso-GalAAG. The identity of the latter was established by treatment with mild acid, which resulted in liberation of galactose and a compound yielding an infrared spectrum identical with that of synthetic *rac*-1-octadecyl glyceryl ether. Analysis by IR spectroscopy indicated that the galactosidic linkage was of the β-configuration.

The presence of GalAAG in bovine brain was confirmed by Rumsby and Rossiter (1968). In addition, these workers showed that GalAAG is present in the brains of pigs and sheep. Singh (1973) demonstrated that both GalDAG and GalAAG are present in the peripheral nervous system, i.e., the sciatic nerve of rabbits.

4.7.2. GalAAG of Testis and Spermatozoa

The presence of GalAAG in rat testis and in porcine testis and sperm was shown by Kornblatt *et al.* (1972, 1974) and by Ishizuka *et al.* (1973), respectively. It has also been detected in mouse (Kornblatt *et al.*, 1972), rabbit (Kornblatt *et al.*, 1972), guinea pig (Suzuki *et al.*, 1973), and human (Levine *et al.*, 1976) testis. The mechanism of biosynthesis of GalAAG has not been established, although, by analogy with the pathway of biosynthesis of GalDAG (see Section 4.4), one would anticipate that it might be formed in the following manner:

$$\text{Alkylacylglycerol + UDP-galactose} \xrightarrow{\text{β-galactosyl transferase}} \text{GalAAG + UDP} \quad (3)$$

Our laboratory has attempted to demonstrate this reaction in extracts of rat testis (Levine, 1981), but to date all efforts have been unsuccessful. It is possible that the biosynthesis of GalAAG could involve some galactosyl donor other than UDP–galactose, but this is solely a matter for speculation. Perhaps pertinent to this line of conjecture is the fact that Suzuki *et al.* (1977) showed that a diet deficient in vitamin A markedly impairs the synthesis of sulfo-GalAAG in rat testis.

An important factor relating to the biosynthesis of GalAAG in testis and brain is obviously the availability of appropriate glyceryl ethers. The pathway of biosynthesis of the glyceryl ethers is not described here; rather, the reader is referred to recent reviews of this topic by Hajra (1983) and Snyder *et al.* (1985). Narasimhan *et al.* (1983) found that the amount of glyceryl ethers in rat testis increases approximately 40-fold at 10–20 days of age; the enzymatic basis of this phenomenon has not been elucidated.

Aspects of the metabolism of GalAAG in rat testis have been examined by Hsu *et al.* (1983), during their studies of the metabolism of sulfo-GalAAG in that tissue. The results of these studies (see Section 4.8.1.2) demonstrated that the kinetics of labeling of GalAAG *in vivo* were consistent with its being the precursor of sulfo-GalAAG. GalAAG did not show the remarkable metabolic stability of sulfo-GalAAG, but instead exhibited a turnover time ranging between 21–69 hr, depending on the precursor used. There are small amounts of GalAAG present in bovine spermatozoa (Narasimhan *et al.*, 1982*a*). When these sperm were incubated with radioactive galactose, palmitate, cetyl alcohol, or glycerol, no incorporation of radioactivity into GalAAG from any precursor was detected, as was also observed for sulfo-GalAAG (see Section 5.8.1.2). This result harmonizes with other data suggesting that the conversion of GalAAG to sulfo-GalAAG occurs at only one relatively early stage of spermatogenesis (the primary spermatocyte).

4.8. Sulfatoxygalactosylacylalkylglycerol (Sulfo-GalAAG)

4.8.1. Sulfo-GalAAG of Testis and Spermatozoa

4.8.1.1. Occurrence and Structure

The presence of sulfo-GalAAG in mammalian testis and spermatozoa was reported virtually simultaneously by Kornblatt *et al.* (1972) and Ishizuka *et al.* (1973) (structure **5**, Fig. 4-1). The former workers isolated the major (~0.5 mg/g wet wt. of testis) glycolipid of rat testis. Degradation studies revealed the presence of chimyl alcohol, palmitic acid, galactose, and sulfate in approximately equimolar amounts. The alkyl and acyl moieties were predominantly (>85%) C16 : 0. It was proposed that the compound was a monoalkylmonoacylglyceryl monogalactoside sulfate, termed sulfo-GalAAG.

A faster-migrating glycolipid, present in lower amounts, that was also detected in extracts of rat testis was proposed to be the nonsulfated analog, GalAAG; this designation was subsequently confirmed (Kornblatt *et al.*, 1974). A compound of similar migration to sulfo-galAAG was noted in the testes of other mammalian species (mouse, guinea pig, rabbit, and human) and also in ejaculated porcine spermatozoa; the occurrence of sulfo-GalAAG in the testis of various species has been reviewed by Murray *et al.* (1980).

Ishizuka *et al.* (1973) isolated the major glycolipid of boar testis and spermatozoa and showed that it contained equimolar amounts of fatty acid (>90% C16 : 0), galactose and sulfate. The presence of glyceryl ether (>80% C16 : 0) was revealed by analysis by GLC of the trimethylsilyl derivatives. IR spectroscopy suggested the presence of a sulfate ester at the equatorial hydroxyl group and also the presence of a β-anomeric configuration of the galactopyranoside. In addition, methylation studies revealed the presence of 2,4,6-tri-*O*-methylgalactosides, suggesting the presence of the sulfate on the 3'-position of the galactose moiety. Analyses by NMR spectroscopy indicated that the *O*-acyl group was attached to the C-2 of the glycerol. The structure proposed for the compound was 1-*O*-alkyl-2-*O*-acyl-3-[β-sulfogalactosyl]glycerol, with the trivial name of seminolipid also being conferred on it. Suzuki *et al.* (1973) isolated the major lipid of guinea pig testis and observed that it had the same structure as the seminolipid of boar testis. Ueno *et al.* (1977) demonstrated that the optical rotations of human and boar seminolipids were consistent with the presence of a β-galactosidic linkage. In addition, values obtained by measurement of the optical rotatory dispersion of the glyceryl ethers obtained from the sulfo-GalAAG species of human and boar testis suggested that they were D-isomers, corresponding to 1-*O*-alkyl-*sn*-glycerol. On the basis of these and earlier studies, the structure of sulfo-GalAAG was established as 1-*O*-alkyl-2-*O*-acyl-3-*O*-D-(3'-sulfo)galactopyranosyl-*sn*-glycerol (structure **5**, Fig. 4-1).

As described subsequently, sulfo-GalAAG has also been detected in small amounts in the brains of a number of species. Nervous tissue and testis are the only two sites known to contain sulfo-GalAAG. Lingwood *et al.* (1981) investigated the distribution of sulfo-GalAAG in a number of rat tissues by administering [^{35}S]sulfate and performing appropriate radiochromatographic analyses. Brain was not analyzed in their study. Whereas sulfo-GalAAG was found to be the major sulfolipid of the testis, it was observed that epididymis, heart, intestine, kidney, liver, lung, spleen, and stomach all lacked the capacity to incorporate [^{35}S]sulfate into sulfo-GalAAG. Moreover, immunofluorescence studies using an antiserum specific for sulfo-GalAAG also demonstrated that only the germinal cells of testis yielded a positive reaction.

An important point concerning the occurrence of sulfo-GalAAG is that it is not found in the testis of certain animals. Ueno *et al.* (1975) reported that the testis of the puffer fish contained sulfo-GalCer, but not sulfo-GalAAG. Levine *et al.* (1976) analyzed the glycolipids of the testes of a number of birds and fish. None of the testes from these species contained sulfo-GalAAG. Instead, sulfo-GalCer was found to be a major glycolipid of the testis of mature fowl, duck, and skate-fish and sulfatoxygalactosylglucosylceramide of the testis of mature salmon and trout. Immature duck testis contained only a trace of sulfo-GalCer. The biological meaning of these species differences in occurrence of sulfo-GalAAG, sulfo-GalCer, and sulfatoxygalactosylglucosylceramide is not clear. Perhaps it relates in some subtle manner to differences in testicular temperature or in ambient temperatures to which spermatozoa are exposed; however, this is quite a speculative consideration. What is clear is that the testes and spermatozoa of most chordates examined contain sulfated galactolipids as their principal glycolipids; this finding suggests that these compounds play an important role in the function of spermatozoa.

4.8.1.2. Biosynthesis

Knapp *et al.* (1973) described the existence in preparations of rat testis of an enzyme activity that catalyzed the transfer of sulfate from PAPS to GalAAG to form the sulfated lipid. The presence of a detergent (Triton X-100) was absolutely essential for demonstration of the enzyme activity. The enzyme preparation was found to sulfate GalCer and GalDAG, both of which contain a β-galactosyl residue, but not Gal_2DAG, which contains a terminal α-galactosyl residue. Substrate competition studies showed clear competition between GalCer and GalAAG, suggesting that the same PAPS–galactolipid sulfotransferase in rat testis acts on both galactosphingolipids and galactoglycerolipids. GalCer was the preferred substrate. It was suggested that the observed absence of sulfo-GalCer in rat testis may reflect the negligible amounts of GalCer in rat testis, rather than an inability to sulfate the latter lipid. The sulfotransferase was found to be markedly enriched in a Golgi apparatus fraction, as had been shown by Fleischer and Zambrano (1973) for the enzyme-catalyzing sulfation of GalCer in rat kidney.

Handa *et al.* (1974) described the results of studies investigating the biosynthesis of sulfo-GalAAG, both *in vivo* and *in vitro*. Mice were injected intraperitoneally with [^{35}S]sulfate and its incorporation into the sulfolipid was followed. Some incorporation was noted within 1 hr of injection, but the maximal amount was attained at approximately 24 hr, decreasing to one half maximum in 1 week. When mice of different age groups were used, the maximal incorporation was obtained ~10 days after birth.

A particulate fraction of boar testis was solubilized with Triton X-100 and was found to contain an enzyme catalyzing the transfer of sulfate from PAPS to GalAAG (**4**), yielding sulfo-GalAAG (**5**). Desulfated sulfo-GalAAG, GalCer and lactosylceramide (which also contains a terminal β-galactosyl residue) were all active as acceptors of sulfate; competition studies suggested that these glycolipids were sulfated by a single enzyme. Digalactosylceramide and trihexosylceramide (both of which contain a terminal α-galactosyl residue), glucosylceramide, cholesterol, and sulfo-GalAAG itself did not act as substrates.

Kornblatt (1979) injected [^{35}S]sulfate into the testes of adult rats. The loss of [^{35}S]-labeled sulfo-GalAAG from the testes and its appearance in the vas deferens plus epididymis were followed with time. DNA was also labeled by administration of [^{3}H]thymidine and the behavior of labeled DNA and sulfo-GalAAG was compared. The results were consistent with the concept that synthesis of the sulfolipid occurs only in very early spermatocytes. The results also suggested that once sulfo-GalAAG is made, it does not exhibit turnover, and that it is only lost from the testes when germinal cells die or mature into spermatozoa.

Hsu *et al.* (1983) studied several aspects of the synthesis of sulfo-GalAAG and of GalAAG in rat testis *in vivo*. They asked three questions:

1. Are the kinetics of incorporation of precursors *in vivo* into these two lipids consistent with the fact that sulfo-GalAAG can be formed from GalAAG *in vitro*, as described above?
2. Do the acyl, alkyl, and galactosyl moieties of sulfo-GalAAG exhibit metabolic stabilities similar to that demonstrated for its sulfate moiety by Kornblatt (1979)?
3. Does rat testis contain 1-*O*-alkyl-2-*O*-acylglycerol, a possible immediate precursor of GalAAG (see Section 4.7.2)?

To obtain at least partial answers to these questions, Hsu *et al.* (1983) injected radioactive palmitate, cetyl alcohol and galactose separately into the testes of rats, isolated GalAAG and sulfo-GalAAG at various times thereafter and determined their specific activities.

Concerning the first question, the results obtained were consistent with the hypothesis that GalAAG exhibits a precursor relationship to sulfo-GalAAG *in vivo*. It was found that for each of the three compounds used to label the galactoglycerolipids, the plot of the values of the specific activity of GalAAG peaked well before that of sulfo-GalAAG and its declining phase intersected the plot for the latter lipid at or very near to the commencement of its plateau value (see Fig. 4-3A–C). These results fit well with the criteria laid down by Zilversmit *et al.* (1943*a,b*) for demonstrating precursor–product relationships. The turnover time of GalAAG, as estimated from the use of each of the three precursors, was in the range of 21–69 hr. In marked contrast, a slow increase of radioactivity in sulfo-GalAAG was observed following injection of each of the three precursors, a plateau being reached 72–168 hr. After these times, the plots for sulfo-GalAAG remained at a relatively constant level until 28 days. Analysis by radiochromatography confirmed the relatively rapid labeling of GalAAG and the slow labeling of sulfo-GalAAG (Fig. 4-4).

In relation to the second question, it was found that, subsequent to an initial period of time during which equilibration of radioactivity in the acyl and alkyl moieties occurs, the acyl, alkyl, and galactosyl moieties of sulfo-GalAAG appear to display metabolic stabilities equivalent to that previously demonstrated for its sulfate moiety. The significance of this observation is discussed subsequently. The metabolic stability demonstrated by sulfo-GalAAG in rat testis compares with that of sulfo-GalCer in the brains of adult rats (Radin *et al.*, 1957; Green and Robinson, 1957), whereas the turnover of sulfatides in other tissues is apparently quite appreciable (Green and Robinson, 1957).

With regard to the third question, small amounts of a compound corresponding in its properties to monoalkylmonoacylglycerol were isolated from rat testis. The labeling properties of this compound were such that it could be a precursor of GalAAG. The demonstration of this compound in rat testis does not necessarily implicate it in the biosynthesis of GalAAG; however, it does add some support to the feasibility of the suggested pathway (see Section 4.7.2).

Figure 4-3. (A) Specific activity plotted versus time for the GalAAG (○) and sulfo-GalAAG (●) species of rat testis injected with [16-¹⁴C]palmitic acid. The results indicate the actual specific activities (horizontal bars) and the mean values (open or closed circles) obtained from analyses of the testes of three animals per time point. The values for specific activity were derived from the total radioactivity present in each lipid, independent of its distribution in the various moieties of the two lipids. (B) Specific activity plotted versus time for the GalAAG (○) and sulfo-GalAAG (●) species of rat testis injected with [1-¹⁴C]cetyl alcohol. The results indicate the actual specific activities (horizontal bars) and the mean values (open or closed circles) obtained from analyses of the testes of two animals per time point. The values for specific activity were derived from the total radioactivity present in each lipid, independent of its distribution in the various moieties of the two lipids. (C) Specific activities plotted versus time for the GalAAG (○) and sulfo-GalAAG (●) species of rat testis injected with D-[1-¹⁴C]galactose. The results indicate the actual specific activities (horizontal bars) and the mean values (open or closed circles) obtained from analyses of the testes of two animals per time point. The values for specific activity were derived from the radioactivity present in the galactosyl moiety of each lipid; more than 95% of the total radioactivity present in each lipid was present in galactose. (From Hsu *et al.*, 1983.)

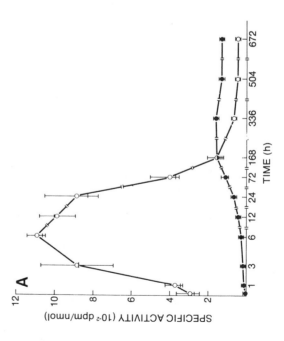

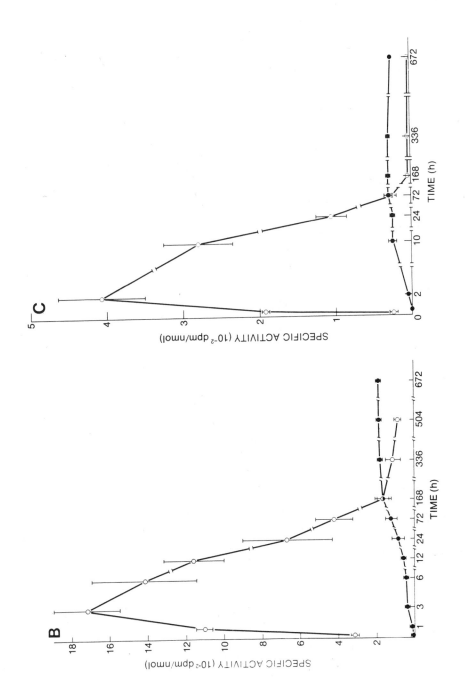

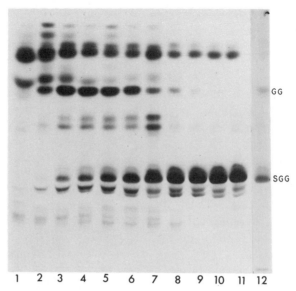

Figure 4-4. Analysis by thin-layer chromatography (TLC) and radioautography of the lipids of the acetone fraction of rat testes labeled by the intratesticular injection of [16-14]palmitic acid. 1, 10 min; 2, 60 min; 3, 3 hr; 4, 6 hr; 5, 12 hr; 6, 24 hr; 7, 72 hr; 8, 168 hr (7 days); 9, 336 hr (14 days); 10, 504 hr (21 days); 11, 672 hr (28 days); 12, nonradioactive standards of GalAAG(GG), and sulfo-GalAAG (SGG) detected by spraying with aniline–diphenylamine reagent. Both testes of each animal used were injected with 7.5 μCi of [16-14C]palmitic acid. One animal was then killed at each time interval indicated above, the two testes were pooled, and lipid extraction and fractionation were performed. A portion of the acetone fraction corresponding to 0.4 g wet weight of testis was applied to each channel and chromatography was performed in chloroform–methanol–water (64 : 25 : 4 v/v). The radioactivity applied to the various channels varied from 4 × 10^4 to 1 × 10^5 dpm. The radioautogram was developed for 3 weeks. (From Hsu *et al.*, 1983.)

4.8.1.3. Degradation

Yamato *et al.* (1974) purified arylsulfatase A from boar testes, monitoring its purification by using 4-nitrocatechol sulfate as substrate. Both [^{35}S]-labeled sulfo-GalCer and sulfo-GalAAG were found to act as substrates, with the former being desulfated somewhat more rapidly. The pH optimum in acetate buffer was 4.5 for both sulfolipids. The specific activities of the arylsulfatase A toward 4-nitrocatechol sulfate, sulfo-GalAAG and sulfo-GalCer increased proportionally throughout the various steps of the purification procedure used, suggesting that a single enzyme acted on the three different substrates. Yamato *et al.* suggested that seminolipid is a physiological substrate for arylsulfatase A in the testis.

Fluharty *et al.* (1974) isolated [^{35}S]-labeled rat testicular sulfo-GalAAG and studied its desulfation by arylsulfatase A obtained from human urine. By each parameter studied, including obligatory bile salt requirement, stimulation by $MnCl_2$, pH optimum and K_M, the conditions were identical to those for the hydrolysis of sulfo-GalCer. The latter lipid inhibited the hydrolysis of sulfo-GalAAG in a competitive manner. Neither sulfo-GalAAG nor sulfo-GalCer was hydrolyzed by arylsulfatase B. The authors, like Yamato *et al.* (1974), concluded that sulfo-GalAAG is a physiological substrate for arylsulfatase A.

Yamaguchi *et al.* (1975) studied the activity toward sulfo-GalAAG as substrate of brain tissue from two patients with a late infantile form of metachromatic leukodystrophy (MLD). This disease is known to be due to a deficiency of arylsulfatase A. The activity of this enzyme toward 4-nitrocatechol sulfate in brain tissue from the patients with MLD was only 1–2% of the control activities, and a similar deficiency in activity toward sulfo-GalAAG and sulfo-GalCer was also observed. This observation confirms in an interesting manner that arylsulfatase A acts on 4-nitrocatechol sulfate and both of the sulfatides studied. The authors reported that accumulation of sulfo-GalAAG in testis was not observed; however, as the lipid is not present in prepubertal human testis (Levine *et al.*, 1976), this result is not unexpected.

A lipase located in secondary lysosomes that acts on sulfo-GalAAG to deacylate it has been reported by Reiter *et al.* (1976). Iron-loaded secondary lysosomes were prepared from rat liver and incubated in acetate buffer (pH 4.5) with [^{35}S]-labeled sulfo-GalAAG or sulfo-GalCer. When the former was used as substrate, a product corresponding in migration to its lyso-derivative was formed. Evidence was also obtained that the lysosomes contained arylsulfatase A activity, acting to desulfate both sulfolipids. The observation that secondary lysosomes contain a lipase that deacylates sulfo-GalAAG could help explain why the lipid may not accumulate in tissues from adult patients with MLD.

4.8.1.4. Synthesis during Testicular Development

The testis of the newborn rat contains interstitial cells, Sertoli cells and gonocytes; the latter are the earliest cells in the spermatogenic series. During the first 40–50 days after birth, the gonocytes differentiate to produce the later stages of the spermatogenic series; the various cell types appear in the maturing testis in a sequential fashion (Clermont and Perey, 1957). Measurement of the level of sulfo-GalAAG in the testes of rats of varying ages should thus permit some insight into when it first appears in testis, and also possibly of the cell type responsible for its synthesis. Such a study was performed by Kornblatt *et al.* (1974) and some of the results are summarized in Fig. 4-5. It can be seen that the testis of the 7- to 10-day-old rat has a relatively low level of lipid-bound sulfate; at this stage the testis contains interstitial cells, Sertoli cells, gonocytes and spermatogonia. The level of sulfolipid (sulfo-GalAAG is by far the major sulfolipid present in rat testis) is seen to rise about ninefold 15–20 days after birth, corresponding to the appearance in rat testis of spermatocytes, which constitute the predominant cell type in the testis of the 20-day-old rat. The level of sulfolipid continues to increase at a lesser rate until day 30 and then declines to the adult level. The level of lipid-bound galactose (predominantly GalAAG plus sulfo-GalAAG) shows a similar pattern. During the period when the two glycolipids show dramatic variations in level, the total lipid content of the rat testis is seen to remain almost constant. These results

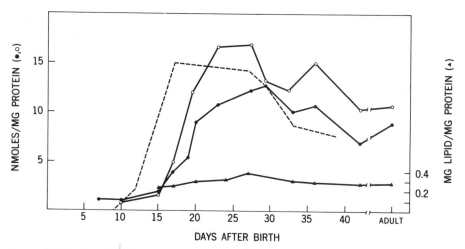

Figure 4-5. Levels of lipid-bound sulfate (●) and lipid-bound galactose (○) in the acetone fractions of lower-phase lipids from testes of rats at various times after birth. Duplicate assays were performed on lipids extracted from pooled testes. For each time point, data from two or three sets of animals were averaged. Total lipid content (▲) was determined only on a single set of animals. The discontinuous line represents the activity of the PAPS–galactolipid sulfotransferase of homogenates of rat testes prepared from animals of various ages. (Lipid extractions and assays were performed as described by Kornblatt *et al.*, 1974.)

are consistent with the hypothesis that sulfo-GalAAG first appears in early primary spermatocytes.

The results of assays of the activity of the PAPS–galactolipid sulfotransferase in testes obtained at various times after birth are also shown in Fig. 4-5. Enzymatic activities were found to be low in testes of 9 day old rats but rose dramatically 15 days after birth. The rise in enzymatic activity parallels the rise in level of sulfo-GalAAG but occurs several days earlier. The rise in enzymatic activity occurs when spermatocytes first begin to appear; the subsequent decrease occurs when spermatids appear in large numbers.

Recent studies by Lingwood (1985*a*) suggest that the decrease in activity of the PAPS–galactolipid sulfotransferase observed in maturing testis may have a very interesting basis. Lingwood obtained evidence that it may be mediated by an inhibitor of the activity of the testicular sulfotransferase. The inhibitor appears to be a soluble component that is first detected in rat testis about 25 days after birth. Rat testicular homogenates can sulfate GalAAG, GalCer, and LacCer *in vitro*. The inhibitor appears to be relatively substrate specific, in that it is most effective in preventing sulfation of GalAAG and inhibits sulfation of the two latter lipids to a lesser extent. The inhibitor also appeared to show tissue specificity, in that sulfation of GalCer by the testicular sulfotransferase was inhibited, but it did not appear to affect the sulfation of the same lipid by the transferases of kidney and brain. Elucidation of how the inhibitor functions should be facilitated by the finding that it can be purified approximately 9500-fold by HPLC.

That the above inhibitor may be implicated in the regulation of spermatogenesis is suggested by studies of Lingwood *et al.* (1985) on rats carrying the growth- and-reproduction-complex (*grc*) gene, which is linked to the major histocompatibility complex. The *grc* influences fertility and body size, with homozygous males being 30% smaller than the wild-type, their testes are 90% smaller, and spermatogenesis is blocked at the primary pachytene spermatocyte stage (see Gill *et al.*, 1983). Lingwood *et al.* (1985) found that the activity of the testicular galactolipid sulfotransferase was elevated relative to that in the wild-type animals and a concomitant deficiency of the sulfotransferase inhibitor was found. Spermatogenesis in the *grc* homozygotes was found to be blocked at a stage that correlates temporally with the earliest detection of the inhibitor in the wild-type animal. This biochemical abnormality is the first to be associated with a genetically regulated developmental defect linked to the major histocompatibility complex and appears to relate to the pathogenesis of the block in spermatogenesis exhibited by animals carrying the gene.

An interesting study of a different nature lends some support to the above conclusion that sulfo-GalAAG is not present in spermatogonia. Ishizuka and Yamakawa (1974) studied the glycolipid composition of three human seminomas, malignant tumors of the testis. Normal human testes were found to contain both sulfo-GalAAG and GalAAG, with only small amounts of ceramide mono- and dihexosides being detected. In contrast, no sulfo-GalAAG or GalAAG was detected in the seminomas, but relatively large amounts of the above GSLs and others were evident. Seminomas are thought to originate from testicular germ cells. Thus, the finding that the seminomas lack sulfo-GalAAG indirectly supports the concept that this lipid appears in testis after the germ cell stage.

4.8.1.5. Cellular Location of Sulfo-GalAAG

The results of the study by Kornblatt *et al.* (1974) suggested that sulfo-GalAAG was synthesized mainly, if not exclusively, in early spermatocytes. The problem of defining which testicular cells are responsible for the synthesis of sulfo-GalAAG was approached by Letts *et al.* (1978). These workers described a method for preparing late spermatocytes from immature rat testis. These cells were found to be highly active in glycoprotein biosynthesis. Studies on the incorporation of [^{35}S]sulfate into sulfo-GalAAG indicated that late spermatocytes were not the primary site of synthesis of this lipid, although these cells were enriched approximately fivefold in sulfo-GalAAG, as compared with the level in rat testis. It was found that [^{35}S]-labeled sulfo-GalAAG took 5 weeks to migrate from its site of synthesis to the epididymis. These results suggested that sulfation of sulfo-GalAAG occurs at a spermatocyte stage prior to the late (pachytene and diplotene) spermatocyte stage.

Lingwood (1985*b*) studied the timing of the biosynthesis of sulfo-GalAAG in rat testis by tissue radioautography. To do this, animals were fed [^{35}S]sulfate and frozen sections of testis were prepared. Sulfo-GalAAG was by far the major testicular component containing [^{35}S]sulfate under the conditions of radioautography employed. The results obtained were consistent with those of

Letts *et al.* (1978), indicating that the synthesis of sulfo-GalAAG occurred at the zygotene and early pachytene stages of spermatogenesis. Further synthesis of sulfo-GalAAG appeared to be prevented by the appearance at the mid-pachytene stage of an inhibitor of the activity of the sulfotransferase responsible for the sulfation of GalAAG.

4.8.1.6. Subcellular Location

Klugerman and Kornblatt (1980) studied the subcellular distribution of sulfo-GalAAG in rat testis. The sulfolipid was labeled by administration of $Na_2{}^{35}SO_4$; cell suspensions were homogenized and centrifuged on linear, continuous, sucrose gradients. The labeled lipid had an equilibrium density-gradient pattern identical with that of alkaline phosphatase, an enzyme of the plasma membrane. The pattern of distribution of the sulfolipid was different from the patterns of enzyme markers for the Golgi apparatus, lysosomes, mitochondria, and endoplasmic reticulum. Klugerman and Kornblatt concluded that sulfo-GalAAG is located on the plasma membrane of rat testicular germinal cells.

Shirley and Schachter (1980) prepared a plasma membrane fraction from adult rat testis that appeared to be free of contamination by Golgi membranes and other organelles. Sulfo-GalAAG was found to be enriched some 33-fold in this fraction. The plasma membrane fraction was probably contributed by spermatocytes, spermatids, and Sertoli cells. These results, combined with the results of analyses of the specific activity of sulfo-GalAAG at various time intervals after injection of [^{35}S]sulfate, were consistent with the hypothesis that sulfo-GalAAG is made in the Golgi apparatus of early primary spermatocytes and rapidly migrates to the Golgi apparatus, where it remains while the primary spermatocyte differentiates to the mature spermatozoon.

Recent studies (Metz and Radin, 1980; Brown *et al.*, 1985; Sasaki, 1985) indicate that specific proteins (glycolipid transfer proteins, GLTPs) may be involved in the transfer of glycolipids from one organelle to another in various organs. For instance, Sasaki (1985) showed that one particular GLTP can facilitate the transfer of various GSLs and of GalDAG, GlcDAG, and Gal$_2$DAG from liposomes to other liposomes or to mitochondria. It will be of great interest to determine whether such proteins are involved in the transfer of sulfo-GalAAG and of sulfo-GalDAG from one cellular compartment (e.g., the Golgi apparatus) to another (e.g., the plasma membrane and/or the myelin membrane).

Another approach that has been used quite extensively to study cell surface glycoconjugates employs galactose oxidase. Treatment with this enzyme, followed by reduction with sodium borotritide, will label specifically the terminal galactose and *N*-acetylgalactosamine residues of glycoproteins and glycolipids. Lingwood (1979) found this labeling procedure to be ineffective with GalAAG, under conditions that resulted in excellent labelling of GalCer. GalDAG was also poorly labeled by this procedure. Neither sulfo-GalAAG nor sulfo-GalCer was labeled by this procedure, indicating that sulfate substitution of the galactose moiety inhibits the action of galactose oxidase. Thus, apparently this method cannot be employed to attempt to define the presence of sulfo-GalAAG on the surface of spermatozoa and other spermatogenic cells.

4.8.1.7. Antibodies to Sulfo-GalAAG

Glycolipids are antigenic. Thus, antibodies to these molecules may prove invaluable in exploring certain aspects of these lipids, such as their subcellular locations and their functions. Lingwood *et al.* (1980) prepared a rabbit antiserum specific for sulfo-GalAAG. Female rabbits were immunized by repeated intravenous injections of liposomes containing sulfo-GalAAG. Anti-sulfo-Gal-AAG was detected by a complement fixation assay. All anti-sulfo-GalAAG activity was present in the IgG fraction. The anti-sulfo-GalAAG was purified by adsorption to, and subsequent elution from, sulfo-GalAAG-coated cholesterol particles. Affinity-purified anti-sulfo-GalAAG reacted with sulfo-GalAAG, and to a lesser extent with GalAAG, but not with sulfo-GalCer or GalCer. Lingwood and Schachter (1981) then used the anti-sulfo-GalAAG to locate sulfo-GalAAG in rat testicular germinal cells. An ionic interaction between sulfo-GalAAG and immunoglobulin was shown to occur at physiological pH, resulting in high fluorescence backgrounds for control cells treated with nonimmune sera. Immunofluorescence was therefore performed at alkaline pH such that this interaction was much reduced or eliminated. Sulfo-GalAAG was detected on the surfaces of spermatocytes and spermatids but was apparently absent from spermatogonia and spermatozoa. As spermatozoa are known to contain sulfo-GalAAG, the latter finding implies that a reorganization of the exposure of sulfo-GalAAG may occur on the cell surface during the late stages of spermatogenesis, such that it becomes nonreactive with antibody (i.e., "cryptic"). Evidence was also obtained indicating that sulfo-GalAAG was mobile in the plane of the membrane, undergoing ligand-induced "patching" and occasional "capping."

During the course of the above studies, Lingwood and Schachter (1981) observed that crosslinking of cell-surface proteins with glutaraldehyde affected the distribution of sulfo-GalAAG on the cell surface. This led Lingwood (1985c) to look for germ cell proteins that might interact with sulfo-GalAAG. By use of a new method for the production of affinity matrices using photoreactivatable heterofunctional crosslinking agents, sulfo-GalAAG was immobilized on agarose. Three proteins from rat spermatogenic cells with apparent molecular weights of 68,000, 34,000 and 24,000 were found to bind selectively and reversibly to the affinity matrix. An antiserum raised to the protein of apparent molecular weight 68,000 was found to be specific for this protein. The antiserum was then used to localize the antigen in frozen sections of testis. The protein was found to be present in the plasma membranes of all germ cells, but its expression was apparently elevated in testicular spermatozoa and in cells in or near the basal compartment of the seminiferous epithelium.

Eddy *et al.* (1985) prepared a monoclonal antibody to sulfo-GalAAG. Sulfo-GalAAG was immobilized on glass beads by means of a photoreactivatable, heterofunctional, crosslinking agent; the coated beads were used to stimulate lymphocytes antigenically *in vitro*. The resulting antibody was of the IgM class and appeared quite specific for sulfo-GalAAG. It is hoped that the availability of this and other specific monoclonal antibodies will facilitate efforts to determine a functional role for sulfo-GalAAG during male germ cell differentiation or during fertilization.

4.8.2. Sulfo-GalAAG of Brain

Norton and Brotz (1963) reported the presence of GalAAG in brain. Once the existence of both sulfo-GalAAG and GalAAG in testis had been shown (Kornblatt *et al.*, 1972; Ishizuka *et al.*, 1973), it appeared reasonable to investigate the possibility that brain also contained sulfo-GalAAG. Thus, Levine *et al.* (1975) injected [^{35}S]sulfate into the ventricles of young, adult rats. Radiochromatographic analyses revealed the presence of two minor ^{35}S-containing glycolipids, in addition to sulfo-GalCer. One of these was proposed to be a previously unrecognized minor species of sulfo-GalCer, but more recent studies in our laboratory indicate that it may have been cholesterol sulfate. The other compound was isolated, characterized, and shown to be sulfo-GalAAG; the only difference between the sulfo-GalAAG species isolated from brain and testis was that the alkyl and acyl compositions of the former were found to be more heterogeneous than those of the latter. The coexistence of both GalAAG and GalDAG in rat brain was also confirmed; these two lipids were found to be present in a mole ratio of approximately 3 : 2, whereas the mole ratio of GalAAG–sulfo-GalAAG was approximately 5 : 1. The amount of sulfo-GalAAG in rat brain was 0.19 μmoles/g wet wt., approximately one third of the amount in rat testis and approximately one fifteenth that of sulfo-GalCer in rat brain. Results obtained by treatment of the sulfatide fraction of rat brain with mild alkali, showing the presence of a highly polar product thought to be sulfatoxygalactosylglycerol, led Levine *et al.* (1975) to propose that rat brain might also contain a third sulfatide, sulfo-GalDAG. This hypothesis was verified independently by the work of Flynn *et al.* (1975) (see Section 4.5). Levine *et al.* (1975) also noted that rabbit brain contained sulfo-GalAAG at a concentration approximately 25% that of rat brain, whereas human brain apparently did not contain the lipid. These findings were interpreted as indicating that quite wide variations in the amount of sulfo-GalAAG could occur in the brains of different mammalian species. They also raised the possibility that sulfo-GalAAG might not be distributed uniformly throughout the brain, as analyses on human brain were performed solely on frontal lobes, whereas, in the case of the two other species they were performed on total brain.

Ishizuka *et al.* (1978) performed an extensive study of both the sulfo-GalAAG and sulfo-GalDAG species of rat brains of varying ages; their results have been discussed in Section 4.5. Burkart *et al.* (1983*a,b*) examined various aspects of the biosynthesis of these sulfated galactoglycerolipids in the myelinating brains of young rats; their findings and conclusions are reviewed in Section 4.5.

4.9. Digalactosylacylalkylglycerol (Gal₂AAG)

This lipid was detected in extracts of adult, human testis by Levine *et al.* (1976) (structure **6,** Fig. 4-1). It exhibited a chromatographic migration similar to that of Gal$_2$DAG but, unlike that lipid, its chromatographic migration was only partly retarded by treatment with mild alkali. Analysis by GLC of its constituents revealed that it contained glyceryl ethers, fatty acids and galactose in approximate molar ratios of 1 : 1 : 2 (R. Narasimhan and R. K. Murray, unpublished observations). The compound was also observed in ejaculated human sperm but not in immature human testis (Levine *et al.,* 1976). We have also detected a similar, if not identical compound, in lipid extracts of adult pig testis. On the basis of its composition and characteristics, we propose that this compound is Gal$_2$AAG (**6**). No information is available as to the stereochemical nature of the galactosidic linkages in this lipid. If the terminal galactose were to be joined to the inner galactose in an α-linkage, this would make the galactosylacylalkylglycerols exactly analogous to the galactosyldiacylglycerol series, each class containing a monogalactosyl, a sulfatoxygalactosyl, and a digalactosyl species (see Section 4.1).

4.10. Glucoglycerolipids

Reports from the laboratory of the Slomianys described the presence in human gastric juice of two novel neutral glucoglycerolipids (Slomiany and Slomiany, 1977; Slomiany *et al.,* 1977*b*) and of a sulfated glucoglycerolipid (Slomiany *et al.,* 1977*a,c*). A variety of procedures, including partial acid hydrolysis, oxidation with periodate, oxidation with chromium trioxide, and permethylation studies indicated that the structure of the sulfated glucoglycerolipid (**10**) was as indicated in Table 4-2. It was also proposed that the neutral glucoglycerolipids were monoalkylmonoacylglyceryl hexaglucoside (**8**) and monoalkylmonoacylglyceryl octaglucoside (**9**), respectively (see Table 4-2). The fatty acids of these three lipids were predominantly C16 : 0, C18 : 0, C20 : 0, and C22 : 0 and the glyceryl ethers were C16 : 0, C18 : 0, and C20 : 0. Evidence that the sulphate group was attached to the 6-position of glucose in (**10**) (Table 4-2) was obtained from the results of periodate oxidation of partially methylated and reduced hydrolysis products of the permethylated native glycolipid. Assignment of the anomeric nature of the glycosidic linkages was based on the results of analyses using chromium trioxide oxidation.

Subsequently, the lipid composition of the "mucous" barrier of rat stomach was investigated (Slomiany *et al.,* 1978). Cellular mucus of the mucous cells from gastric epithelium and surface mucus from gastric mucosa were obtained by perfusion *in vivo* of rat stomach with 2 M NaCl. The glycolipid fraction consisted of neutral and acidic glucoglycerolipids, similar to those just described, and was devoid of GSLs. The glucoglycerolipids were found to be absent from the gastric mucosa, and were therefore apparently absent from the plasma membrane and from the internal cellular membranes. These observations led to the proposal that glucoglycerolipids might be specific to certain secretions and not, unlike the case of most other lipids, predominantly membrane components.

A number of other interesting studies concerning these novel glucoglycerolipids have been reported. Glucoglycerolipids are major lipid components of both human submandibular (Slomiany *et al.*, 1980) and parotid (Slomiany *et al.*, 1981) saliva, and their amount is increased in the saliva of heavy calculus formers. The lipid composition of the submandibular saliva of normal individuals and of individuals with cystic fibrosis has been studied (Slomiany *et al.*, 1982); whereas the saliva of the former was found to contain only glucoglycerolipids, that of the latter contained both these compounds and small amounts of GSLs. The possibility was raised that the presence of GSLs in saliva might significantly affect its properties.

Studies on the biosynthesis *in vitro* of the sulfate moiety of the sulfated triglucosylmonoalklymonacylglycerol (**10**, Table 4-2) of rat gastric mucosa (Liau *et al.*, 1982) and of rat submandibular and parotid glands (Slomiany *et al.*, 1983) have also been described. The results of the two studies were very similar. Using triglucosylmonoalkylmonoacylglycerol as acceptor and PAPS as the sulfate donor, a sulfotransferase activity with a pH optimum of 7.8 that was capable of generating the sulfated lipid was detected in the cytosol. Triton X-100 was necessary for maximal activity. The enzyme would not use GalCer as substrate. Several lines of evidence indicated that the site of sulfation was C-6 of the terminal glucose. By contrast, both salivary glands (Zdebska *et al.*, 1983; Liau *et al.*, 1983) were also found to contain another sulfotransferase activity capable of sulfating GalCer of LacCer; this activity was concentrated mainly in the microsomal fraction, exhibited a lower pH optimum (6.8), and catalyzed sulfate attachment to C-3 of galactose.

Because of the interest of our laboratory in glyceryl ether-containing GGroLs, we have attempted to isolate the major members of the above class of compounds from human saliva and gastric juice (Narasimhan *et al.*, 1982). The glycolipids present in samples of saliva from 10 individuals and in samples of gastric juice from five individuals were analysed. In both fluids, compounds corresponding in the properties studied to standards of glucosyl- and lactosylceramides were found to be the major glycolipids. Other more complex GSLs were also present in smaller amounts. Human saliva was found to contain two glucoglycerolipids that were not detected in gastric juice. Analyses of these compounds indicated that they were mono- and diglucosyl diglycerides and were probably of bacterial origin. No compounds were detected that exhibited the chromatographic properties of glyceryl ether-containing glycolipids—*viz*. compounds whose migrations were retarded, due to formation of the lyso derivatives, following treatment with mild alkali. Methanolysis of the glycolipid fractions of saliva and gastric juice failed to reveal the presence of any more than traces of 1-*O*-alkyl glyceryl ethers. These results do not exclude the possibility that glyceryl ether-containing glucoglycerolipids may occur in human saliva and gastric juice; however, they do suggest that at most they would appear to be rather minor components of either fluid. In subsequent studies (R. Narasimhan and R. K. Murray, in preparation), we have not been able to detect any compounds corresponding in various chromatographic or other properties to glyceryl ether-containing glucoglycerolipids in the gastric mucosa, in the parotid or in the submandibular glands of rats. Interestingly, all of these tissues were found to contain small amounts of cholesterol sulfate, the detection of which was facilitated by prior injection of [^{35}S]sulfate (see Sections 4.2 and 4.3).

4.11. Possible Functions of GGroLs

The functions of the GGroLs discussed in this chapter are quite obscure. Slomiany and Slomiany (Slomiany and Slomiany, 1980) reviewed the possible functions of the glucoglycerolipids. They pointed out that these compounds, unlike most other lipids, do not occur in cell membranes, but rather in various secreted fluids. In the mucous secretions of the alimentary tract, they are postulated to be part of the protective lining of the surface epithelial cells. In saliva they may be involved in the process of tooth pellicle formation and in alveolar fluid they may participate in spreading of pulmonary surfactant.

Almost all of the postulated functions of the galactoglycerolipids are derived from three lines of speculation: (1) from analogy with knowledge of sulfo-GalCer; (2) from analogy with postulated functions of ether-containing lipids; (3) from known biological properties of individual members of the class. However, as documented below, little is known with certainty of the functions of sulfo-GalCer or of the functions of most ether-containing lipids. In addition, the major biological property of the galactoglycerolipids so far established is their occurrence in the nervous system and in testis and spermatozoa, so that postulates based on the third line of speculation represent little more than suggestions that specific galactoglycerolipids may play roles in myelination and in spermatogenesis and/or spermatozoal function. Despite the flimsiness of the available evidence, each of the three lines referred to above will now be discussed relatively briefly, more to stimulate future research than to pretend to offer any definitive insights as to the functions of the galactoglycerolipids.

Karlsson and colleagues made the intriguing suggestion that sulfo-GalCer may be involved in the active transport of sodium (Karlsson *et al.*, 1974). They based this suggestion primarily on findings made on the lipid composition and Na^+-K^+-ATPase activity of the salt (nasal) glands of the eider duck and herring gull. The salt gland, as the name suggests, is an organ with a high capacity for transporting sodium ions. Sulfo-GalCer was found to be the major GSL of the salt glands of both species. Most interestingly, the results of analyses of amounts of sulfo-GalCer and of the Na^+-K^+-ATPase revealed that their ratios were similar in salt glands of different origins and different functional states; other anionic lipids did not show this correlation. Also, a similar increase in the amount of sulfo-GalCer and of the ATPase activity was found for ducks given a load of salt water, but this increase was not found for other lipids. Bovine kidney medulla, brain gray matter and glial cells also showed similar ratios. However, little correlation of the amount of sulfo-GalCer with the ATPase activity was found in bovine kidney cortex and myelin; the authors offered reasonable explanations for the discrepancies found in these two tissues. Hansson *et al.* (1979) subsequently formulated a model for Na^+-K^+ translocation, in which sulfo-GalCer was postulated to be essential for K^+ influx. By contrast, as reviewed by Farooqui (1981), the results of several investigations, primarily using immunohistochemical methods, have found no apparent association of sulfo-GalCer with Na^+-K^+-ATPase. The involvement of sulfo-GalCer in the function of the Na^+-K^+-ATPase must still be considered as a provocative and challenging hypothesis. Corey–Pauling–Koltun molecular models of sulfo-GalAAG and sulfo-GalCer were found to be very similar (Levine *et al.*, 1975). It was thus suggested (Murray *et al.*, 1976), by analogy with sulfo-GalCer, that sulfo-GalAAG could be involved in Na^+-K^+ transport in testis and spermatozoa; this speculation has not been subjected to experimental analysis.

With regard to other possible functions of sulfo-GalCer, it has been noted that both Ca^{2+} and Mg^{2+} have a high affinity for sulfo-GalCer (Ishizuka and Tadano, 1982); presumably, this also applies to sulfo-GalAAG. Whether these particular sulfolipids could be involved in biologically significant processes involving binding or transport of Ca^{2+} and Mg^{2+} has not been investigated. Several studies have suggested that sulfo-GalCer may be a component of an opiate receptor in brain (Cho *et al.*, 1976; Loh *et al.*, 1978).

Roberts *et al.* (1985) studied the nature of the red blood component(s) involved in the agglutination of these cells by the basement membrane glycoprotein laminin. These investigators identified these components as monogalactosyl-containing sulfatides. Both the sulfatides from sheep red blood cells and from bovine brain were found to bind laminin with high affinity. Of various glycolipid and phospholipid standards tested, only sulfatides bound laminin with high affinity. The globular end regions of the 200-kDa subunits of laminin were probably involved in the agglutination of the red blood cells. The results suggested a physiological function of sulfatides in cell adhesion. It may be of interest to examine the appearance of laminin in testis during testicular differentiation and to determine whether there is any correlation with the appearance of sulfo-GalAAG.

As indicated above, the properties of the various ether-containing lipids of biological membranes are also generally poorly understood. The two principal properties (Paltauf, 1983) attributed to ether lipids are that they may be specific, structural components of membranes and that they may be more stable than their acyl-containing analogues to enzymatic and chemical hydrolysis. As reviewed by Paltauf (1983), some interesting information has become available from comparisons of the behavior in model membranes of certain ether-containing glycerophospholipids and their diacyl analogues (e.g., dialkylglycerophosphocholine and diacylglycerophosphocholine). In general, the molecular arrangements of the two classes of lipids resembled each other, as evidenced by similarities in the force–area curves of monolayers and in the dimensions of their bilayer structures. In addition, studies by NMR spectroscopy have shown that the conformations and segmental motions of their polar head groups are similar. However, one can anticipate finding differences in chain mobility, as ether lipids lack the bulky carbonyl oxygen and conformational freedom is greater around the ether bond. Also, effects such as tighter packing, lower surface potential, different permeability characteristics, and differences in interaction with cholesterol and membrane proteins have been noted.

Virtually nothing is known about the functions and physical behavior of the three glyceryl ether-containing galactoglycerolipids reviewed above. Certainly, the metabolic stability of sulfo-GalAAG is consistent with the presence of the ether bond conferring great metabolic stability (see Section 4.8.1.2). However, the turnover of GalAAG is quite rapid, so that it may be that the combination of both an ether bond and a sulfate residue are the determinants of the remarkable metabolic stability of sulfo-GalAAG. Moreover, the metabolic stability of sulfo-GalCer in mature myelin is also remarkable (Moser and Dulaney, 1978), and this compound does not contain an ether bond. Detailed studies of the transition temperatures of the various eukaryotic nonsulfated and sulfated galactoglycerolipids and of their effects on membrane microviscosity, on membrane permeability, and on the conformations of various membrane proteins can be anticipated. Some such information is already becoming available for sulfo-GalCer (Koshy and Boggs, 1983; Boggs *et al.,* 1984). As the galactoglycerolipids are usually relatively minor components of membranes, it will also be important to determine whether they occur randomly mixed throughout membranes or exhibit lateral or transverse segregation (Paltauf, 1983); monoclonal antibodies may prove invaluable in defining their localization in such studies.

A major biological finding with regard to the galactoglycerolipids is that they are almost solely confined to nervous tissue and to testis and spermatozoa. *Prima facie,* it would thus appear logical to assume that they may play relatively important roles in these tissues. Concerning nervous tissue, evidence has been presented above that the principal galactoglycerolipids are enriched in myelin and that they appear in nervous tissue as myelination commences (see Sections 4.5 and 4.8.2). However, there is no real understanding as to what these lipids are doing in myelin. This statement also applies to the much more studied lipid, sulfo-GalCer. Reference has been made above to the possible functions of sulfo-GalCer, such as involvement in cation transport, binding of cations, receptor function (opiates), and binding of laminin. None of these functions is well established. For instance, with regard to myelin, as discussed by Karlsson *et al.* (1974), there is no apparent correlation between the amount of sulfo-GalCer and the activity of the Na^+-K^+-ATPase.

It will be apparent to the reader that it will likely prove no easy matter to determine the possible functions of each of the galactoglycerolipids. Progress has been made in elucidating certain aspects of the functions of GSLs, such as their possible involvement in cell interaction and differentiation, in cell growth control and in oncogenic transformation (Hakomori, 1983*b*). There is also a considerable body of evidence implicating certain gangliosides as receptors (Kanfer, 1983). However, even in the latter case, "none of the evidence presented can be regarded as unequivocal but merely as suggestive" (Kanfer, 1983). The reasons for the difficulty in elucidating the functions of galactoglycerolipids include the following: (1) it is virtually impossible to extract a particular galactoglycerolipid from a membrane selectively and leave the functions of the other components (such as enzymes) intact; (2) in many cases, these lipids are rather minor components of the cells in which they occur; (3) considerable variations in the amount of certain of them may occur among species (e.g., the apparent absence of sulfo-GalAAG from human brain (Levine *et al.*, 1975); and (4) the general lack of mutant cells specifically deficient in individual members of this class. The nervous system is extremely complex, and it contains relatively small amounts of specific galactoglycerolipids; we believe that these considerations will make it difficult to unravel the functions of these lipids in neural tissues. The testis is also rather a complex organ, containing many cell types (Clermont and Perey, 1957). Interestingly, Teuscher *et al.* (1982) speculated that sulfo-GalAAG and/or GalAAG may be important autoantigens in the genesis of experimental immune orchitis. Spermatozoa themselves are comparatively simple cells, with only one unique biological function, and they contain high amounts of sulfo-GalAAG (Ishizuka *et al.*, 1973; Levine *et al.*, 1976; Selivonchick *et al.*, 1980). This may make them the ideal model system to study, using such new approaches as specific monoclonal antibodies and electron immunohistochemical techniques, in order to attempt to determine the function of at least this one particular galactoglycerolipid.

4.12. Summary

The biochemistry of the GGroLs present in mammalian tissues has been reviewed. Both galacto- and glucoglycerolipids occur. Six galactoglycerolipids, comprised of three galactosyldiacylglycerol species [GalDAG (**1**), sulfo-GalDAG (**2**), Gal_2DAG (**3**)] (Fig. 4-1) and three galactosylacylalkylglycerol species [GalAAG (**4**), sulfo-GalAAG (**5**) and Gal_2AAG (**6**)] (Fig. 4-1) have been detected; only partial characterizations of the digalactosyl-containing species are available. The galactoglycerolipids are present in nervous tissue and/or testis and spermatozoa. The galactoglycerolipids found in nervous tissue (all but Gal_2AAG, apparently) tend to be rather minor glycolipid components and are intimately associated with myelin. In testis and spermatozoa, the galactoglycerolipids (predominantly, if not exclusively, the galactosylacylalkylglycerol species) are prominent glycolipid components, with sulfo-GalAAG being the major glycolipid. The pathways of biosynthesis and degradation of the galactoglycerolipids have been elucidated at a general level, although a number of important points remain to be settled (e.g., the mechanism of addition of galactose to form GalAAG). Analyses of the appearance of certain of these lipids during development have shown dramatic increases of their amounts during myelination and at the time of appearance of primary spermatocytes in the testis. The metabolic stabilities of the sulfated galactoglycerolipids in mature myelin and in testis are quite remarkable.

A variety of nonsulfated and sulfated glyceryl ether-containing glucoglycerolipids, with glucose chain lengths of one to eight, have also been described. These lipids differ from most other glycolipids in being components of secretions, such as gastric juice, saliva, and alveolar fluid, rather than of membranes. The mechanism of sulfation of these lipids has been studied, but nothing is known concerning the mechanism of addition of their glucose residues.

Virtually nothing is known regarding the functions of the above GGroLs. Thus, the principal task for the future is to determine their biological significance. This will involve, among other approaches, detailed studies of their physical properties, the application of new techniques such as monoclonal antibodies, and possibly the development of mutant cells lacking one or other of these lipids. It is suggested that the occurrence of sulfo-GalAAG in relatively large amounts in spermatozoa may make it a particularly favorable candidate for studies aimed at elucidating function.

ACKNOWLEDGMENTS. Support at various times of the authors' research in the area of glycoglycerolipids by the Medical Research Council of Canada and by the National Cancer Institute of Canada is warmly acknowledged.

References

Arita, M., Iwamori, M., Higuchui, T., and Nagai, Y., 1984, Positive and negative fast atom bombardment mass spectrometry of glycosphingolipids. Discrimination of the positional isomers of gangliosides with sialic acids, *J. Biochem. (Tokyo)* **95**:971.

Bischel, M., and Austin, J., 1963, A modified benzidine method for chromatographic detection of sphingolipids and acid polysaccharides, *Biochim. Biophys. Acta* **70**:598.

Boggs, J. M., Koshy, K. M., and Rangaraj, G., 1984, Effect of fatty-acid chain-length, fatty-acid hydroxylation and various cations on phase-behavior of synthetic cerebroside sulfate, *Chem. Phys. Lipids* **36**:65.

Brown, R. E., Stephenson, F. A., Markello, T., Barenholz, Y., and Thompson, T. E., 1985, Properties of a specific glycolipid transfer protein from bovine brain, *Chem. Phys. Lipids* **38**:79.

Burkart, T., Caimi, L., and Wiesmann, U. N., 1983a, Synthesis and subcellular transport of sulfogalactosyl glycerolipids in the myelinating mouse brain, *Biochim. Biophys. Acta* **753**:294.

Burkart, T., Caimi, L., Herschkowitz, N. N., and Wiesmann, U. N., 1983b, Metabolism of sulfogalactosyl glycerolipids in the myelinating mouse brain, *Dev. Biol.* **98**:182.

Burkart, T., Caimi, L., Siegrist, H. P., Herschkowitz, N. N., and Wiesmann, U. N., 1982, Vesicular transport of sulfatide in the myelinating mouse brain. Functional association with lysosomes?, *J. Biol. Chem.* **257**, 3151.

Burkart, T., Hofmann, K., Siegrist, H. P., Herschkowitz, N. N., and Wiesmann, U. N., 1981, Quantitative measurement of *in vivo* sulfatide metabolism during development of the mouse brain: Evidence for a large rapidly degradable sulfatide pool, *Dev. Biol.* **83**:42.

Carter, H. E., McLuer, R. H., and Slifer, E. D., 1956, Lipids of wheat flour. I. Characterization of galactosyl glycerol components, *J. Am. Chem. Soc.* **78**:3735.

Carter, H. E., Ohno, K., Nojima, S., Tipton, C. L., and Stanacev, N. Z., 1961a, Wheat flour lipids. 2. Isolation and characterization of glycolipids of wheat flour and other plant sources, *J. Lipid Res.* **2**:215.

Carter, H. E., Hendry, R. A., and Stanacev, N. Z., 1961b, Wheat flour lipids. 3. Structure of mono- and digalactosylglycerol lipids, *J. Lipid Res.* **2**:223.

Carter, H. E., Johnson, P., and Weber, E. J., 1965, Glycolipids, *Annu. Rev. Biochem.* **34**:109.

Cho, T. M., Cho, J. S., and Loh, H. H., 1976, H^3-Cerebroside sulfate re-distribution induced by cation, opiate or phosphatidyl serine, *Life Sci.* **19**:117.

Clermont, Y., and Perey, B., 1957, Quantitative study of the cell population of the seminiferous tubules in immature rats, *Am. J. Anat.* **100**:241.

Desmukh, D. S., Inoue, T., and Pieringer, R. A., 1971, The association of the galactosyl diglycerides of brain with myelination. II. The inability of the myelin-deficient mutant, Jimpy mouse, to synthesize galactosyl diglycerides effectively, *J. Biol. Chem.* **246**:5695.

Eddy, E. M., Muller, C. H., and Lingwood, C. A., 1985, Preparation of monoclonal antibody to sulfatoxygalactosylglycerolipid by *in vitro* immunization with a glycolipid–glass conjugate, *J. Immunol. Methods* **81**:137.

Egge, H., 1983*a*, Ether glycolipids, in *Ether Lipids: Biochemical and Biomedical Aspects* (H. K. Mangold, and R. Paltauf, eds.), pp. 141–159, Academic, Orlando, Florida.

Egge, H., 1983*b*, Mass spectrometry of ether lipids, in *Ether Lipids: Biochemical and Biomedical Aspects* (H. K. Mangold and F. Paltauf, eds.), pp. 17–48, Academic, Orlando, Florida.

Egge, H., Peter-Katalinic, J., Reuter, G., Schauer, R., Ghidoni, R., Sonnino, S., and Tettamanti, G., 1985, Analysis of gangliosides using fast atom bombardment mass-spectrometry, *Chem. Phys. Lipids* **37**:127.

Farooqui, A. A., 1981, Metabolism of sulfolipids in mammalian tissues, *Adv. Lipid Res.* **18**:159.

Fleischer, B., and Zambrano, F., 1973, Localization of cerebroside-sulfotransferase activity in the Golgi apparatus of rat kidney, *Biochem. Biophys. Res. Commun.* **52**:951.

Fluharty, A. L., Stevens, R. L., Miller, R. T., and Kihara, H., 1974, Sulfoglycerogalactolipid from rat testis: A substrate for pure human arylsulfatase A, *Biochem. Biophys. Res. Commun.* **61**:348.

Flynn, T. J., Desmukh, G., Subba Rao, G., and Pieringer, R. A., 1975, Sulfogalactosyl diacylglycerol: Occurrence and biosynthesis of a novel lipid in rat brain, *Biochem. Biophys. Res. Commun.* **65**:122.

Gigg, R. H., 1978, Studies on the synthesis of sulfur-containing glycolipids ("sulfoglycolipids"), in *Carbohydrate Sulphates*, Vol. 77 (R. G. Schweiger, ed.), American Chemical Society Symposium Series, pp. 44–66, American Chemical Society, Washington, D. C.

Gill, T. J. III, Siew, S., and Kunz, H. W., 1983, Major histocompatibility complex (MHC)-linked genes affecting development, *J. Exp. Zool.* **228**:325.

Green, J. P., and Robinson, J. D., 1957, Cerebroside sulfate (sulfatide-A) in some organs of the rat and in a mast cell tumor, *J. Biol. Chem.* **235**:1621.

Hajra, A. K., 1983, Biosynthesis of O-alkylglycerol ether lipids, in *Ether Lipids* (H. K. Mangold and F. Paltauf, eds.), pp. 85–106, Academic, Orlando, Florida.

Hakomori, S.-I., 1983*a*, Chemistry of glycosphingolipids, in *Handbook of Lipid Research* (D. J. Hanahan, ed.), Vol. 3: *Sphingolipid Biochemistry* (J. N. Kanfer and S.-I. Hakomori, eds.), pp. 1–166, Plenum, New York.

Hakomori, S.-I., 1983*b*, Glycosphingolipids in cellular interaction, differentiation and oncogenesis, in *Handbook of Lipid Research* (D. J. Hanahan, ed.), Vol. 3: *Sphingolipid Biochemistry* (J. N. Kanfer and S.-I. Hakomori, eds.), pp. 327–380, Plenum, New York.

Hakomori, S.-I., and Young, W. W., 1983, Glycolipid antigens and genetic markers, in *Handbook of Lipid Research* (D. L. Hanahan, ed.), Vol. 3: *Sphingolipid Biochemistry* (J. N. Kanfer and S.-I. Hakomori, eds.), pp. 381–436, Plenum, New York.

Handa, S., Yamato, K., Ishizuka, I., Suzuki, A., and Yamakawa, T., 1974, Biosynthesis of seminolipid: Sulfation *in vivo* and *in vitro*, *J. Biochem. (Tokyo)* **75**:77.

Hansson, G. C., Heilbronn, E., Karlsson, K. A., and Samuelsson, B. E., 1979, The lipid composition of the electric organ from the ray, *Torpedo marmorata*, with special reference to sulfatides and Na⁺–K⁺–ATPase, *J. Lipid Res.* **20**:509.

Hsu, L.-H., Narasimhan, R., Levine, M., Norwich, K. H., and Murray, R. K., 1983, Studies of the biosynthesis and metabolism of rat testicular galactoglycerolipids, *Can. J. Biochem. Cell Biol.* **61**:1272.

Inoue, T., Desmukh, D. S., and Pieringer, R. A., 1971, The association of the galactosyl diglycerides of brain with myelination. I. Changes in the concentration of monogalactosyl diglyceride in the microsomal and myelin fractions of brain of rats during development, *J. Biol. Chem.* **246**:5688.

Ishizuka, I., and Tadano, K., 1982, The sulfoglycolipid, highly acidic amphiphiles of mammalian renal tubules, *Adv. Exp. Biol. Med.* **152**:195.

Ishizuka, I., and Yamakawa, T., 1974, Absence of seminolipid in seminoma tissue with concomitant increase of sphingoglycolipids, *J. Biochem. (Tokyo)* **76**:221.

Ishizuka, I., and Yamakawa, T., 1985, Glycoglycerolipids, in *New Comprehensive Biochemistry*, Vol. 10: *Glycolipids* (A. Neuberger, and L. L. M., van Deenen, eds.), pp. 101–197, Elsevier, Amsterdam.

Ishizuka, I., Suzuki, M., and Yamakawa, T., 1973, Isolation and characterization of a novel sulfoglycolipid, "seminolipid," from boar testis and spermatozoa, *J. Biochem. (Tokyo)* **73**:77.

Ishizuka, I., Inomata, M., Ueno, K., and Yamakawa, T., 1978, Sulfated glyceroglycolipids in rat brain: Structure, sulfation *in vitro*, and accumulation in whole brain during development, *J. Biol. Chem.* **253**:898.

Iwamori, M., Moser, H. W., and Kishimoto, Y., 1976, Cholesterol sulfate in rat tissues. Tissue distribution, developmental change and brain subcellular localization, *Biochim. Biophys. Acta* **441**:268.

Kanfer, J. N., 1983, Glycosphingolipids as receptors, in *Handbook of Lipid Research* (D. J. Hanahan, ed.), Vol. 3: *Sphingolipid Biochemistry* (J. N. Kanfer and S.-I. Hakomori, eds.), pp. 437–471, Plenum, New York.

Karlsson, K.-A., Samuelsson, B. E., and Steen, G. D., 1974, The lipid composition and Na$^+$K$^+$-dependent adenosine triphosphatase activity of the salt (nasal) gland of eider duck and herring gull: A role for sulphatides in sodium-ion tranport, *Eur. J. Biochem.* **46**:243.

Klugerman, A., and Kornblatt, M. J., 1980, The subcellular localization of testicular sulfogalactoglycerolipid, *Can. J. Biochem.* **58**:225.

Knapp, A., Kornblatt, M. J., Schachter, H., and Murray, R. K., 1973, Studies on the biosynthesis of testicular sulfoglycerogalactolipid: Demonstration of a Golgi-associated sulfotransferase activity, *Biochem. Biophys. Res. Commun.* **55**:179.

Kohler, G., and Milstein, C., 1975, Continuous cultures of fused cells secreting antibody of predefined specificity, *Nature (Lond.)* **256**:495.

Kohler, G., and Milstein, C., 1976, Derivation of specific antibody-producing tissue culture and tumor lines by cell fusion, *Eur. J. Immunol.* **6**:511.

Kolodny, E. H., and Moser, H. W., 1983, Sulfatide lipidosis, in *The Metabolic Basis of Inherited Disease*, 5th ed. (J. B. Stanbury, J. B. Wyngaarden, D. S. Fredrickson, J. L. Goldstein, and M. S. Brown, eds.), pp. 881–905, McGraw-Hill, New York.

Kornblatt, M. J., 1979, Synthesis and turnover of sulphogalactoglycerolipid, a membrane lipid, during spermatogenesis, *Can. J. Biochem.* **57**:255.

Kornblatt, M. J., Schachter, H., and Murray, R. K., 1972, Partial characterization of a novel glycerogalactolipid from rat testis, *Biochem. Biophys. Res. Commun.* **48**:1489.

Kornblatt, M. J., Knapp, A., Levine, M., Schachter, H., and Murray, R. K., 1974, Studies on the structure and formation during spermatogenesis of the sulfoglycerogalactolipid of rat testis, *Can. J. Biochem.* **52**:689.

Koshy, K. M., and Boggs, J. M., 1983, Partial synthesis and physical properties of cerebroside sulfate containing palmitic acid or alpha-hydroxy palmitic acid, *Chem. Phys. Lipids* **34**:41.

Letts, P. J., Hunt, R. C., Shirley, M. A., Pinteric, L., and Schachter, H., 1978, Late spermatocytes from immature rat testis. Isolation, electron microscopy, lectin agglutinability and capacity for glycoprotein and sulfogalactoglycerolipid biosynthesis, *Biochim. Biophys. Acta* **541**:59.

Levine, M., 1981, Studies on Glycolipids of Testis, Spermatozoa and Brain, Ph.D thesis, University of Toronto.

Levine, M., Kornblatt, M. J., and Murray, R. K., 1975, Isolation and partial characterization of a sulfogalactoglycerolipid from rat brain, *Can. J. Biochem.* **53**:679.

Levine, M., Bain, J., Narasimhan, R., Palmer, B., Yates, A. J., and Murray, R. K., 1976, A comparative study of the glycolipids of human, bird and fish testes and of human sperm, *Biochim. Biophys. Acta* **441**:134.

Liau, Y. H., Zdebska, E., Slomiany, A., and Slomiany, B. L., 1982, Biosynthesis *in vitro* of a sulfated triglucosyl monoalkylmonoacylglycerol by rat gastric mucosa, *J. Biol. Chem.* **257**:12019.

Liau, Y. H., Zdebska, E., Aono, M., Slomiany, A., and Slomiany, B. L., 1983, *In vitro* biosynthesis of sulphatoglycosphingolipids by rat submandibular salivary glands, *Arch. Oral Biol.* **28**:1001.

Lingwood, C. A., 1979, Action of galactose oxidase on galactolipids, *Can. J. Biochem.* **57**:1138.

Lingwood, C. A., 1985*a*, Developmental regulation of galactoglycerolipid sulphation during mammalian spermatogenesis: Evidence for a substrate-selective inhibitor of testicular sulphotransferase activity in the rat, *Biochem. J.* **231**:393.

Lingwood, C. A., 1985*b*, Timing of sulphogalactolipid biosynthesis in the rat testis studied by tissue autoradiography, *J. Cell Sci.* **75**:329.

Lingwood, C. A., 1985*c*, Protein–glycolipid interactions during spermatogenesis. Binding of specific germ cell proteins to sulfatoxygalactosylacylalkylglycerol, the major glycolipid of mammalian male germ cells, *Canad. J. Biochem. Cell Biol.* **63**:1077.

Lingwood, C., and Schachter, H., 1981, Localization of sulfatoxygalactosylacylalkylglycerol at the surface of rat testicular germinal cells by immunocytochemical techniques: pH dependence of a nonimmunological reaction between immunoglobulin and germinal cells, *J. Cell Biol.* **89**:621.

Lingwood, C. A., Murray, R. K., and Schachter, H., 1980, The preparation of rabbit antiserum specific for mammalian testicular sulfogalactoglycerolipid, *J. Immunol.* **124**:769.

Lingwood, C., Hay, G., and Schachter, H., 1981, Tissue distribution of sulfolipids in the rat. Restricted location of sulfatoxygalactosylacylalkylglycerol, *Can. J. Biochem.* **59**:556.

Lingwood, C. A., Kunz, H. W., and Gill, T. J. III, 1985, Deficiency in the regulation of testicular galactolipid sulphotransferase in rats carrying the growth- and reproduction-complex (grc) gene, *Biochem. J.* **231**:401.

Loh, H. H., Law, P. Y., Ostwald, T., Cho, T. M., and Way, E. L., 1978, Possible involvement of cerebroside sulfate in opiate receptor binding, *Fed. Proc. Fed. Am. Soc. Exp. Biol.* **37**:147.

Magnani, J. F., Smith, D. F., and Ginsburg, V., 1980, Detection of gangliosides that bind cholera toxin: Direct binding of ^{125}I-labelled toxin to thin-layer chromatograms, *Anal. Biochem.* **109**:399.

McLuer, R. H., and Evans, J. E., 1973, Preparation and analysis of benzoylated cerebrosides, *J. Lipid Res.* **14**:611.

McLuer, R. H., and Ullman, M. D., 1980, Preparation and analytical high performance liquid chromatography of glycolipids, in *Cell Surface Glycolipids*, Vol. 128 (C. C. Sweeley, ed.), pp. 1–14, Amererican Chemical Society Symposium Series, Washington, D.C.

Metz, R. J., and Radin, N. S., 1980, Glucosylceramide uptake protein from spleen cytosol, *J. Biol. Chem.* **255**:4463.

Moser, H. W., and Dulaney, J. T., 1978, Sulfatide lipidosis: Metachromatic leukodystrophy, in *The Metabolic Basis of Inherited Disease*, 4th ed. (J. B. Stanbury, J. B. Wyngaarden, and D. S. Fredrickson, eds.), pp. 770–809, McGraw-Hill, New York.

Murray, R. K., Levine, M., and Kornblatt, M. J., 1976, Sulfatides: Principal glycolipids of the testes and spermatozoa of chordates, in *Glycolipid Methodology* (L. A. Witting, ed.), pp. 305–327, American Oil Chemical Society, Champaign, Illinois.

Murray, R. K., Narasimhan, R., Levine, M., Pinteric, L., Shirley, M., Lingwood, C., and Schachter, H., 1980, Galactoglycerolipids of mammalian testis, spermatozoa and nervous tissue, in *Cell Surface Glycolipids*, Vol. 128 (C. C. Sweeley, ed.), pp. 105–125, American Chemical Society Symposium Series, Amer. Chem. Soc. Washington, D.C.

Narasimhan, R., Hsu, L.-H., and Murray, R. K., 1982*a*, Inability of bovine spermatozoa to synthesize galactolipids, *Fed. Proc. Fed. Am. Soc. Exp. Biol.* **41**:1170 (abst.).

Narashimhan, R., Bennick, A., Palmer, B., and Murray, R. K., 1982*b*, Studies on the glycolipids of human saliva and gastric juice, *J. Biol. Chem.* **257**:15122.

Narasimhan, R., Hsu, L.-H., and Murray, R. K., 1983, An analysis of the relationship between the synthesis of glyceryl ethers and the synthesis of galactoglycerolipid in rat testis, *Fed. Proc. Fed. Am. Soc. Exp. Biol.* **42**:2020 (abst.).

Neufeld, E. F., and Hall, C. W., 1964, Formation of galactolipids by chloroplasts, *Biochem. Biophys. Res. Commun.* **14**:503.

Norton, W. T., and Brotz, M., 1963, New galactolipids of brain: A monoalkyl-monoacyl-glyceryl galactoside and cerebroside fatty acid esters, *Biochem. Biophys. Res. Commun.* **12**:198.

Paltauf, R., 1983, Ether lipids in biological and model membranes, in *Ether Lipids: Biochemical and Biomedical Aspects* (H. K. Mangold, and R. Paltauf, eds.), pp. 309–355, Academic, Orlando, Florida.

Pieringer, J., Subba Rao, G., Mandel, P., and Pieringer, R. A., 1977, The association of the sulphogalactosylglycerolipid of rat brain with myelination, *Biochem. J.* **166**:421.

Radin, N. S., Martin, F. B., and Brown, J. R., 1957, Galactolipide metabolism, *J. Biol. Chem.* **224**:499.

Reiter, S., Fischer, G., and Jatzkewitz, H., 1976, Degradation of seminolipid by a lipase in secondary lysosomes, *FEBS Lett.* **68**:250.

Roberts, D. D., Rao, N. C., Magnani, J. L., Spitalnik, S. L., Liotta, L. A., and Ginsburg, V., 1985, Laminin binds specifically to sulfated glycolipids, *Proc. Natl. Acad. Sci. USA* **82**:1306.

Rouser, G., Kritchevsky, G., Heller, D., and Lieber, E., 1963, Lipid composition of beef brain, beef liver, and sea anemone—two approaches to quantitative fractionation of complex lipid mixtures, *J. Am. Oil Chem. Soc.* **40**:425 (abst.).

Rouser, G., Kritchevsky, G., Simon, G., and Nelson, G. J., 1967, Quantitative analysis of brain and spinach leaf glycolipids employing silicic acid column chromatography and acetone for elution of glycolipids, *Lipids* **2**:37.

Rumsby, M. G., and Gray, I. K., 1965, A monogalactolipid component in extracts of sheep brain, *J. Neurochem.* **12**:1005.

Rumsby, M. G., and Rossiter, R. J., 1968, Alkyl ethers from the glycerogalactolipid fraction of nerve tissue, *J. Neurochem.* **15**:1473.

Sasaki, T., 1985, Glycolipid-binding proteins, *Chem. Phys. Lipids* **38**:63.

Sastry, P. S., 1974, Glycosyl glycerides, *Adv. Lipid Res.* **12**:251.

Selivonchick, D. P., Schmid, P. C., Natarajan, V., and Schmid, H. H. O. 1980, Structure and metabolism of phospholipids in bovine epididymal spermatozoa, *Biochim. Biophys. Acta* **618**:242.

Shirley, M. A., and Schachter, H., 1980, Enrichment of sulfogalactosylalkylacylglycerol in a plasma membrane fraction from adult rat testis. *Can. J. Biochem.* **58**:1230.

Singh, H., 1973, Glycolipids of peripheral nerve: Isolation and characterization of glycolipids from rabbit sciatic nerve, *J. Lipid Res.* **14**:41.

Slomiany, A., and Slomiany, B. L., 1977, Neutral glyceroglucolipids of the human gastric content, *Biochem. Biophys. Res. Commun.* **76**:115.

Slomiany, B. L., and Slomiany, A., 1980, Glycosphingolipids and glyceroglucolipids of glandular epithelial tissue, in *Cell Surface Glycolipids*, Vol. 128 (C. C. Sweeley, ed.), pp. 149–176, American Chemical Society Sympoisum Series, American Chemical Society, Washington, D.C.

Slomiany, B. L., Slomiany, A., and Glass, G. B. J., 1977*a*, Partial characterization of a novel sulfated glyceroglucolipid of the human gastric content, *FEBS Lett.* **77**:47.

Slomiany, B. L., Slomiany, A., and Glass, G. B. J., 1977*b*, Characterization of two major neutral glyceroglucolipids of the human gastric content, *Biochemistry* **16**:3954.

Slomiany, B. L., Slomiany, A., and Glass, G. B. J., 1977*c*, Glycolipids of the human gastric content: Structure of the sulfated glyceroglucolipid, *Eur. J. Biochem.* **78**:33.

Slomiany, B. L., Slomiany, A., and Mandel, I. D., 1980, Lipid composition of human submandibular gland secretion from light and heavy calculus formers, *Arch. Oral Biol.* **25**:749.

Slomiany, A., Yano, S., Slomiany, B. L., and Glass, G. B. J., 1978, Lipid composition of the gastric mucous barrier in the rat, *J. Biol. Chem.* **253**:3785.

Slomiany, A., Slomiany, B. L., and Mandel, I. D., 1981, Lipid composition of human parotid saliva from light and heavy dental calculus-formers, *Arch. Oral Biol.* **26**:151.

Slomiany, B. L., Liau, Y. H., Zdebska, E., Murty, V. L. N., and Slomiany, A., 1983, Enzymatic sulfation of triglucosyl monoalkylmonoacylglycerol in rat salivary glands, *Biochem. Biophys. Res. Commun.* **113**:817.

Slomiany, B. L., Aono, M., Murty, V. L. N., Slomiany, A., Levine, M. J., and Tabak, L. A., 1982, Lipid composition of submandibular saliva from normal and cystic fibrosis individuals, *J. Dent. Res.* **61**:1163.

Slomiany, B. L., Kojima, K., Banas-Gruszka, Z., Murty, V. L. N., Galicki, N. I., and Slomiany, A., 1981, Characterization of the sulfated monosialyltriglycosylceramide from bovine gastric mucosa, *Eur. J. Biochem.* **119**:647.

Snyder, F., Lee, T-c., and Wykle, R. L., 1985, Ether-linked glycerolipids and their bioactive species: Enzymes and metabolic regulation, in *The Enzymes of Biological Membranes*, Vol. 2, 2nd ed. (A. N. Martonosi, ed.), pp. 1–58, Plenum, New York.

Steim, J. M., 1967, Monogalactosyl diglyceride, *Biochim. Biophys. Acta* **144**:119.

Steim, J. M., and Benson, A. A., 1963, Galactosyl diglyceride in brain, *Fed. Proc. Fed. Am. Soc. Exp. Biol.* **22**:299 (abst.).

Subba Rao, K., Wenger, D. A., and Pieringer, R. A., 1970, The metabolism of glyceride glycolipids. IV. Enzymatic hydrolysis of monogalactosyl and digalactosyl diglyceride in rat brain, *J. Biol. Chem.* **245:**2520.

Subba, Rao, G., Norcia, L. N., Pieringer, J., and Pieringer, R. A., 1977, The biosynthesis of sulphogalactosyldiacylglycerol of rat brain *in vitro, Biochem. J.* **166:**429.

Suzuki, A., Ishizuka, I., Ueta, N., and Yamakawa, T., 1973, Isolation and characterization of seminolipid (1-*O*-alkyl-2-*O*-acyl-3-[β3′-sulfogalactosyl]glycerol) from guinea pig testis and incorporation of ³⁵S-sulfate into seminolipid in sliced testis, *Jpn. J. Exp. Med.* **43:**435.

Suzuki, A., Sato, M., Handa, S., Muto, Y., and Yamakawa, T., 1977, Decrease of seminolipid content in the testes of rats with vitamin A deficiency determined by high performance liquid chromatography, *J. Biochem. (Tokyo)* **82:**461.

Suzuki, K., 1965, The pattern of mammalian brain gangliosides. II. Evaluation of the extraction procedures, post-mortem changes and the effect of formalin preservation, *J. Neurochem.* **12:**629.

Sweeley, C. C., and Nunez, H. A., 1985, Structural analysis of glycoconjugates by mass spectrometry and nuclear magnetic resonance spectroscopy, *Annu. Rev. Biochem.* **54:**765.

Sweeley, C. C., and Siddiqui, B., 1976, Chemistry of mammalian glycolipids, in *The Glycoconjugates,* Vol. 1, pp. 459–540. (M. Horowitz and W. Pigman, eds.), Academic, New York.

Tadano, K., and Ishizuka, I., 1982*a,* Isolation and characterization of the sulfated gangliotriaosylceramide from rat kidney, *J. Biol. Chem.* **257:**1482.

Tadano, K., and Ishizuka, I., 1982*b,* Bis-sulfoglycosphingolipid containing a unique 3-*O*-sulfated *N*-acetylgalactosamine from rat kidney, *J. Biol. Chem.* **257:**9294.

Tadano, K., Ishizuka, I., Matsuo, M., and Matsumoto, S., 1982, Bis-sulfated gangliotetraosylceramide from rat kidney, *J. Biol. Chem.* **257:**13413.

Teuscher, C., Wild, G. C., and Tung, K. S. K., 1982, Immunochemical analysis of guinea pig sperm autoantigens, *Biol. Reprod.* **26:**218.

Ueno, K., Ishizuka, I., and Yamakawa, T., 1975, Glycolipids of the fish testis, *J. Biochem. (Tokyo)* **77:**1223.

Ueno, K., Ishizuka, I., and Yamakawa, T., 1977, Glycolipid composition of human testis at different ages and the stereochemical configuration of seminolipid, *Biochim. Biophys. Acta* **487:**61.

Ullman, M. D., and McLuer, R. H., 1977, Quantitative analysis of plasma neutral glycosphingolipids by high performance liquid chromatography of their perbenzoyl derivatives, *J. Lipid Res.* **18:**371.

Vance, D. E., and Sweeley, C. C., 1967, Quantitative determination of the neutral glycosyl ceramides in human blood, *J. Lipid Res.* **8:**621.

Wells, M. A., and Dittmer, J. C., 1967, A comprehensive study of the postnatal changes in the concentration of the lipids of developing rat brain, *Biochemistry* **6:**3169.

Wenger, D. A., Petitpas, J. W., and Pieringer, R. A., 1968, The metabolism of glyceride glycolipids. II. Biosynthesis of monogalactosyl diglyceride from uridine diphosphate galactose and diglyceride in brain, *Biochemistry* **7:**3700.

Wenger, D. A., Subba Rao, K., and Pieringer, R. A., 1970, The metabolism of glyceride glycolipids. III. Biosynthesis of digalactosyl diglyceride by galactosyl transferase pathways in brain, *J. Biol. Chem.* **245:**2513.

Williams, J. P., Watson, G. R., Khan, M., Leung, S., Kuksis, A., Stachnyk, O., and Myher, J. J., 1975, Gas–liquid chromatography of plant galactolipids and their deacylation and methanolysis products, *Anal. Biochem.* **66:**110.

Wong, M., Brown, R. E., Barenholz, Y., and Thompson, T. E., 1984, Glycolipid transfer protein from bovine brain, *Biochemistry* **23:**6498.

Yahara, S., and Kishimoto, Y., 1981, Characterization of alkylgalactolipids from calf brain by high performance liquid chromatography, *J. Neurochem.* **36:**190.

Yamaguchi, S., Aoki, K., Handa, S., and Yamakawa, T., 1975, Deficiency of seminolipid sulfatase activity in brain tissue of metachromatic leukodystrophy, *J. Neurochem.* **24:**1087.

Yamato, K., Handa, S., and Yamakawa, T., 1974, Purification of arylsulfatase A from boar testis and its activities toward seminolipid and sulfatide, *J. Biochem. (Tokyo)* **75:**1241.

Zdebska, E., Liau, Y. H., Slomiany, A., and Slomiany, B. L., 1983, Enzymatic sulfation of glycosphingolipids in rat parotid salivary glands, *J. Dent. Res.* **62:**1026.

Zilversmit, D. B., Entenman, C., and Fishler, M. C., 1943*a*, On the calculation of "turnover time" and "turnover rate" from experiments involving the use of labeling agents, *J. Gen. Physiol.* **26**:325.

Zilversmit, D. B., Entenman, C., Fishler, M. C., and Chaikoff, I. L., 1943*b*, The turnover rate of phospholipids in the plasma of the dog as measured with radioactive phosphorus, *J. Gen. Physiol.* **26**:333.

Chapter 5

Mycobacterial Fatty Acid Esters of Sugars and Sulfosugars

Mayer B. Goren

5.1. Introduction

The chemistry, immunochemistry, and immunology of the mycobacterial fatty acid esters of sugars and of sulfated sugars have found their principal focus in the simple nonreducing disaccharide α,α-trehalose, i.e., α-D-glucopyrano-syl-α-D-glucopyranoside (**1**). During the past three decades, among the multi-

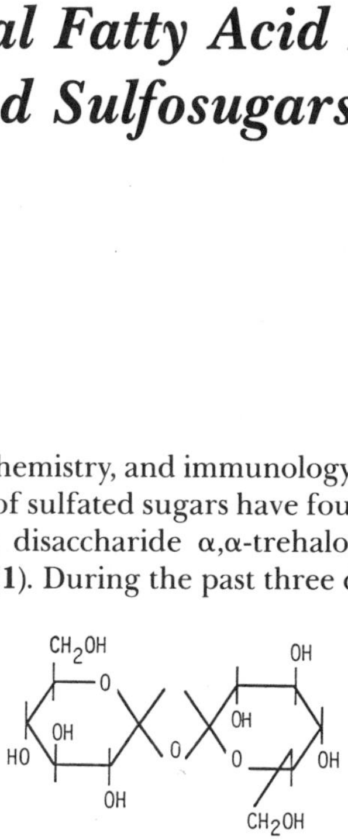

(1)

tude of mycobacterial components, it was the trehalose glycolipids that were implicated in the most engrossing chemical challenges, in diverse biological activities, and probably in the pathogenesis of mycobacterial disease. Some trehalose glycolipids exhibit high toxicity for mice, granulomagenic activity, immunostimulant behavior, antitumor activity, chemotactic, and/or chemotaxigenic activity for macrophages. Some are putatively implicated in inhibiting phagosome–lysosome fusion in macrophages, in mycobacterial escape from phagosomes, in the phenomenon of cord formation (i.e., the serpentine-like morphology assumed by virulent tubercle bacilli in culture or even in tuberculosis lesions), and they behave as antigens to elicit a humoral immune

Mayer B. Goren • Division of Molecular and Cellular Biology, Department of Pediatrics, National Jewish Center for Immunology and Respiratory Medicine, Denver, Colorado 80206, and Department of Microbiology and Immunology, University of Colorado Health Sciences Center, Denver, Colorado 80262.
This review is dedicated to the memories of treasured friends and colleagues, who figured so prominently in many of the scientific developments described here: Masahiko Kato, Edgar Lederer, Gardner Middlebrook, Herbert Rapp, Edgar Ribi, Werner Schaefer, and Raoul Toubiana.

response recognizable in a broad spectrum of mycobacterioses. This litany may well be expected to grow still further.

This chapter deals with mycobacterial fatty acid esters of sugars and sulfosugars. As was evident very early, in mycobacteria these substances embrace principally derivatives of trehalose, since this disaccharide is ubiquitous in the genus and in related taxa (*Corynebacterium, Nocardiae*); it plays a central role in the biology and biochemistry of these cells. Thus, this chapter is devoted largely to a description of fatty acid esters of trehalose.

A number of excellent reviews and chapters in monographs have been written that deal at least in part with the chemistry of these glycolipids. H. Noll (1956) reviewed the early history of the discovery of cord factor, trehalose-6,6'-dimycolate, its purification, and the determination of its structure. Lederer (1967, 1977, 1979, 1984) updated both the chemical and the emerging biochemical and immunological properties, with special emphasis recently on the latter (Lemaire *et al.*, 1986). Comprehensive reviews concerning mycobacteria that contain extensive sections on the trehalose glycolipids have also been published by Goren and Brennan (1979), Goren (1982), Barksdale and Kim (1977), and Minnikin (1982). A definitive exposition by Cécile and Jean Asselineau (1978) reviews in great detail the chemistry, biochemistry, biological, and immunological properties of the various trehalose glycolipids as they were known at the time of publication.

5.2. Trehalose Mycolate Esters

5.2.1. History and Nomenclature

Trehalose (**1**) was recognized even during the earliest investigations by R. J. Anderson and his group (see review by Anderson, 1940) in studies on the lipids of *Mycobacterium tuberculosis,* the etiologic agent of the disease from which the bacterium derives its name. These investigators expressed their astonishment at the tubercle bacillus's facility for synthesis of this sugar even when glycerol was the sole carbon source in the growth medium. Several decades later, a careful study by Winder and colleagues (1972) established that during the steady-state growth of *M. smegmatis,* free trehalose amounted to somewhat more than 3% of the bacterial mass. It was found to have a fairly low turnover. By contrast, the acylated forms of trehalose (not characterized) constituted about a tenth as much of bacterial mass (0.3%) and had a much higher turnover number (~13). It was estimated that about 10% of the bacterial weight could pass through the trehalose fraction, and about 3.8% through the acylated trehalose components (some of which are mycolate esters), suggesting that neither one represented major metabolic intermediates nor had major roles in the bacterial economy. Thus, in 1972 "the metabolic role of these acylated sugars, and with it an explanation of their high degree of turnover, remain[ed] unexplained" (Winder *et al.*, 1972).

Only a year later, at the Joint Tuberculosis Panel meeting of the U.S.–Japan Cooperative Medical Science Program, R. W. Walker described his studies on the biosynthesis of corynomycolic acid by cell free extracts of *Corynebacterium diphtheriae*. This low-molecular-weight (32-carbon) α-branched β-hydroxy acid has been a useful model for studying late steps in the biosynthesis of the much more complex mycolic acids of mycobacteria (60–90 carbons)—also α-branched, β-hydroxy acids (Gastambide-Odier and Lederer, 1964). For reviews, see Goren and Brennan (1979); Minnikin (1982); Takayama and Qureshi (1984).

According to Walker *et al.* (1973), a cell-free extract from *C. diphtheriae* carried out the condensation of two molecules of [1-^{14}C]palmitate via a Claisen-type reaction to produce 2-tetradecyl-3-keto-octadecanoic acid, which was reduced chemically to a mixture of *erythro-* and *threo-*corynomycolic acids:

$$2C_{15}H_{31}-\overset{*}{C}OOH \rightarrow C_{15}H_{31}\text{-}\underset{\overset{|}{C_{14}H_{29}}}{CO}\overset{*}{CH}-\overset{*}{C}OCOH \xrightarrow{\text{NaBH}_4} C_{15}H_{31}\text{-}\overset{*}{C}H(OH)\text{-}\underset{\overset{|}{C_{14}H_{29}}}{CH}\text{-}\overset{*}{C}OOH \quad (1)$$

erythro/*threo*-corynomycolic acid

But when *C. diphtheriae* in culture condensed the labeled palmitate, the intermediate keto acid was immediately incorporated into a 6-monoester of trehalose, which was subsequently reduced biologically to trehalose 6-corynomycolate (Promé *et al.*, 1974; Ahibo-Coffy *et al.*, 1978).

It seems likely that this sequence must also be valid for mycobacteria. Thus, the completely assembled carbon skeleton of mycobacterial mycolic acid might also be expected to appear immediately as a trehalose derivative, very likely the precursor of trehalose-6-mycolate, i.e., trehalose monomycolate (TMM) (**2**) (Fig. 5-1) (Kato and Maeda, 1974) and therefore of cord factor, trehalose-6,6'-dimycolate (TDM) (**3**) (Kilburn *et al.*, 1982; for reviews, see Asselineau and Asselineau, 1978; Lederer, 1977, 1984).

If all the mycolic acid synthesized by mycobacteria accumulated as trehalose derivatives, then the steady-state level of these glycolipids and their constituent mycolic acids would be very high. In fact, the level of these glycolipids is very low. The bulk of the mycolic acids is instead found mainly esterified to the mycobacterial cell wall arabinogalactan (Azuma and Yamamura, 1963; Kanetsuna *et al.*, 1969; Bruneteau and Michel, 1968; Vilkas and Markovits, 1968). The implicit speculation that could therefore be advanced on the basis of the studies just reviewed was that the principal function of trehalose mycolates is to serve, either as such or as intermediate carriers of newly synthesized mycolic acid for transfer to the growing cell wall. Important evidence in support of a trehalose mycolate carrier, potentially implicating trehalose-6-monoacetate-6'-monomycolate (MAT) (**4**) (Fig. 5-1) in this function was reported by Takayama and Armstrong (1976). *M. tuberculosis* H37Ra incorporated [^{14}C]acetate to produce MAT labeled in the mycolate moiety. Following a chase with cold acetate, the radiolabel of this glycolipid declined with time and appeared instead in cell wall-associated mycolate. It seems reasonable to anticipate that all mycobacteria synthesize mycolic acid and incorporate it into cell wall via trehalose mycolate intermediates. Thus, in-

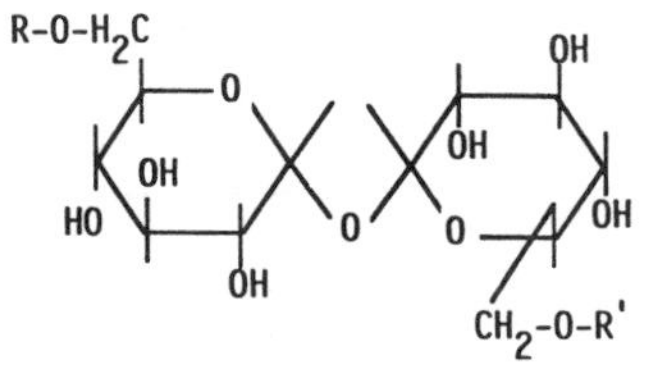

(2) R = mycolate, R' = H

(3) R = R' = mycolate

(4) R = mycolate; R' = acetate

Representative examples of mycolates

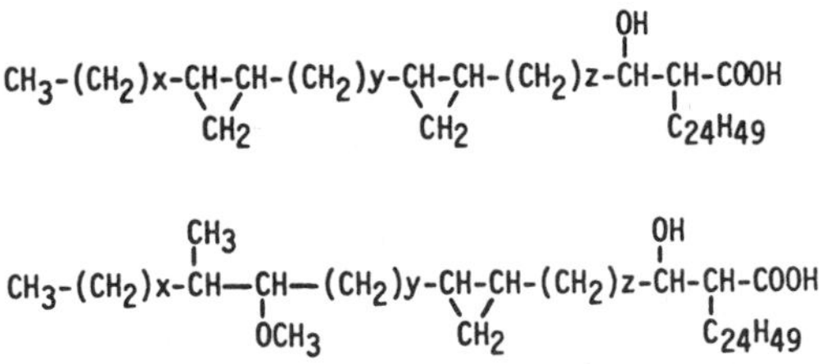

$$CH_3-(CH_2)x-\underset{\underset{CH_2}{\diagup\diagdown}}{CH-CH}-(CH_2)y-\underset{\underset{CH_2}{\diagup\diagdown}}{CH-CH}-(CH_2)z-\underset{OH}{CH}-CH-COOH \quad C_{24}H_{49}$$

$$CH_3-(CH_2)x-\underset{\underset{OCH_3}{|}}{CH}-CH-(CH_2)y-\underset{\underset{CH_2}{\diagup\diagdown}}{CH-CH}-(CH_2)z-\underset{OH}{CH}-CH-COOH \quad C_{24}H_{49}$$

In general x, y, z are odd, even, odd; e.g., 17, 14, 17

Figure 5-1. Structures of important trehalose mycolates: trehalose-6-mycolate (TMM-**2**); trehalose-6,6′-dimycolate (TDM-**3**); 6-mycoloyl-6′-acetyl trehalose (MAT-**4**).

volvement of some form of trehalose mycolates in mycolic acid biosynthesis may be required absolutely and would explain their evident ubiquity, albeit as minor components, in the genus *Mycobacterium* (Dhariwal *et al.,* 1987) (cf. Section 5.3, on biosynthesis).

If it is assumed, probably correctly, that most of Winder's acylated trehalose (0.3% of cell mass) is in fact some form of trehalose mycolate, with a turnover number of 13 involving perhaps 3.5% of the cell mass in a genera-tion, then the calculated amount of mycolic acid that is conveyed to cell wall actually agrees reasonably closely with that found, i.e., about 3% of the dried cell mass. E. Ribi and J. Cantrell (personal communication) found that dried cells of *M. bovis* bacillus Calmette–Guérin (BCG) contain about 15% "cell wall skeleton," which in turn contains about 20% mycolic acid esterified to the arabinogalactan. A composite of these data with those from Kotani *et al.* (1959) and Davidson *et al.* (1982) leads to a figure somewhat higher than 3%.

5.2.2. History of Trehalose-6,6′-Dimycolate

Although the Anderson group (Anderson, 1940) described trehalose as a prominent carbohydrate among the hydrolysis products of extracts of *M. tuberculosis,* the original intact glycolipids containing trehalose were not char-acterized. The implication of trehalose esters in the pathogenicity of *M. tuber-culosis* stemmed from Robert Koch's recognition of a curious morphological property of the tubercle bacillus when grown in culture: its propensity for

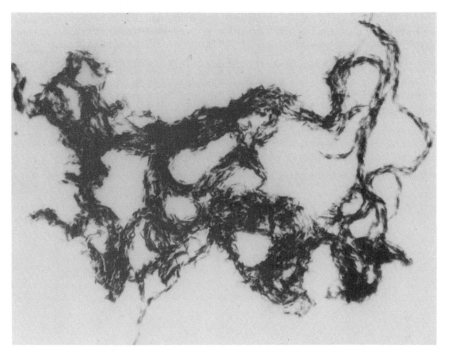

Figure 5-2. Cord formation in surface culture of *Mycobacterium tuberculosis.*

growing in the form of ropy serpentine cords (Fig. 5-2). This type of morphology was found to be relatively restricted to virulent strains of *M. tuberculosis* by Middlebrook *et al.* (1947). Dubos and Middlebrook (1948) also observed that cord-forming strains invariably adsorb and are considerably colored by the cationic phenazine dye neutral red. Most attenuated strains of *M. tuberculosis* and saprophytic mycobacteria as well are ordinarily indifferent to this dye. Adsorption of this dye from neutral or even from slightly alkaline solutions appears to implicate an acidic surface component of the reactive species. The observations led Dubos (1948) to the hypothesis that some peripherally located substance(s) of virulent tubercle might be involved in cording, in neutral red fixation (or in both) and perhaps be implicated in virulence as well. The hypothesis stimulated a search for the relevant substance(s) that led Hubert Bloch to the isolation, characterization, and structural analysis of the glycolipid cord factor (trehalose-6,6'-dimycolate); it also led Middlebrook to the sulfolipid, or more correctly "sulfatides" of *M. tuberculosis.* Both are trehalose glycolipids. The neutral red reactivity of many virulent strains is quite appropriately attributable to their sulfatide content. But an exclusive and convincing role for either cord factor or for the sulfatides in cord formation has not yet been demonstrated (see, however, Section 5.2.9.3b).

Bloch found that exposure of "corded" cultures of *M. tuberculosis* to solvents such as petroleum ether disrupted the serpentine cords without killing the bacteria, while at the same time extracting some component that even in crude form exhibited a peculiar toxicity for mice. In a classic series of studies, collaborative at first with H. Noll and later with J. Asselineau and E. Lederer as well, this toxic substance (provisionally named cord factor) was isolated and purified by column chromatography (Noll *et al.*, 1956; review: Noll, 1956). The purification was followed by infrared spectrophotometry, by chemical analysis, and by assessment of mouse toxicity. Recently, I examined an early sample of cord factor (BCG) from H. Bloch, by infrared (IR) spectrophotometry and by thin-layer chromatography (TLC) and found it to be as free of contaminants as most examples prepared today by much more sophisticated techniques. In studies by Noll *et al.* (1956), the structure of cord factor was established as trehalose-6,6'-dimycolate (TDM). The most probable conformation of this molecule as rendered by J. Asselineau is shown in (**5**).

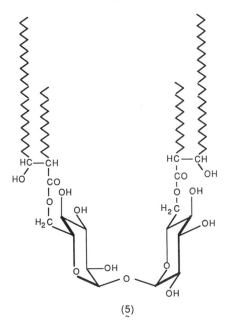

(5)

5.2.3. Summary of Early Structural Studies

The structure of the purified cord factor was determined by the following procedures: (1) IR spectrophotometry, which showed it to be an esterified carbohydrate; (2) alkaline hydrolysis, which showed that the carboxylate components were typical mycolic acids, esterified to a nonreducing glucoside (the crucial piece of evidence that an elusive aglycone was also a glucose residue led to the identification of α,α-trehalose as the core of the glycolipid, and that it was esterified with 2 moles of mycolate); and (3) permethylation and alkaline saponification of the permethylated cord factor afforded a hexa-*O*-methyl trehalose, giving an acidic hydrolysis only 2,3,4-tri-*O*-methyl glucose (Fig. 5.3).

On the basis of these results, it can be concluded that in cord factor trehalose is symmetrically mycolated at the 6- and 6'-positions. Permethylation, albeit by improved methods, still remains a method of choice for the

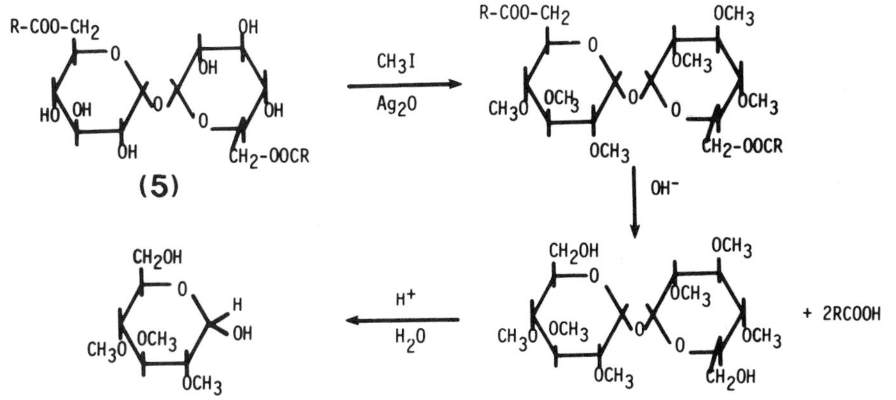

(5)

2,3,4-tri-0-methylglucose

Figure 5-3. Determination of the structure of cord factor by permethylation analysis. The symmetrical structure was revealed by the final product, 2,3,4-tri-*O*-methyl glucose.

structural analysis of new cord factors and analogues. The methods employed by Noll *et al.* (1956) of permethylation with methyl iodide–silver oxide (see also Takayama and Armstrong, 1976) or the variation using dimethyl formamide (DMF) as solvent (Kuhn *et al.*, 1955) are tedious and rarely give complete methylation of the trehalose glycolipids. Goren and Toubiana (1979) described a facilitated methylation of cord factor in a solvent mixture of DMF and anhydrous ether carried out in the presence of molecular sieve, with CH_3I and the commercial oil-coated NaH as catalyst. Still, with these strongly basic substances, the possibility of partial hydrolysis or of acyl migration (as from the 4- to the 6-position) is always a hazard. Indeed, base-catalyzed acetate migration has even been used as a means of synthesizing 4-*O*-methyl glucose derivatives (Bouveng *et al.*, 1957). We recently examined an elegant procedure described by Arnarp and Lönngren (1978). It uses di-tert-butyl pyridine as the basic catalyst, with CH_3I and CH_2Cl_2 as solvent and can be carried out efficiently simply under reflux to achieve essentially complete methylation without either hydrolysis or acyl migration (Liav and Goren, 1986; Dhariwal *et al.*, 1987). An attractive alternative to permethylation, methyl replacement, may be used for locating the acylated positions in, e.g., acylated trehalose. This procedure very adroitly avoids hydrolysis/acyl migration as summarized in Fig. 5-4. The unsubstituted hydroxyls in the glycolipid (5) are blocked with dihydropyran under slightly acidic conditions. In the method of choice, the product (6) is then deacylated and the freed hydroxyls simultaneously methylated by treatment first with dimsyl sodium (CH_3SO $CH_2^- Na^+$) and then with methyl iodide (7). The protective tetrahydropyranyl (oxan-2-yl) groups are readily removed with dilute acid to release a methylated trehalose (8), in which the methyl groups are in the positions that were originally esterified in the parent glycolipid. The method was elegantly exploited by Promé *et al.* (1976) to establish the structures of three types of cord factors isolated from *M. phlei.*

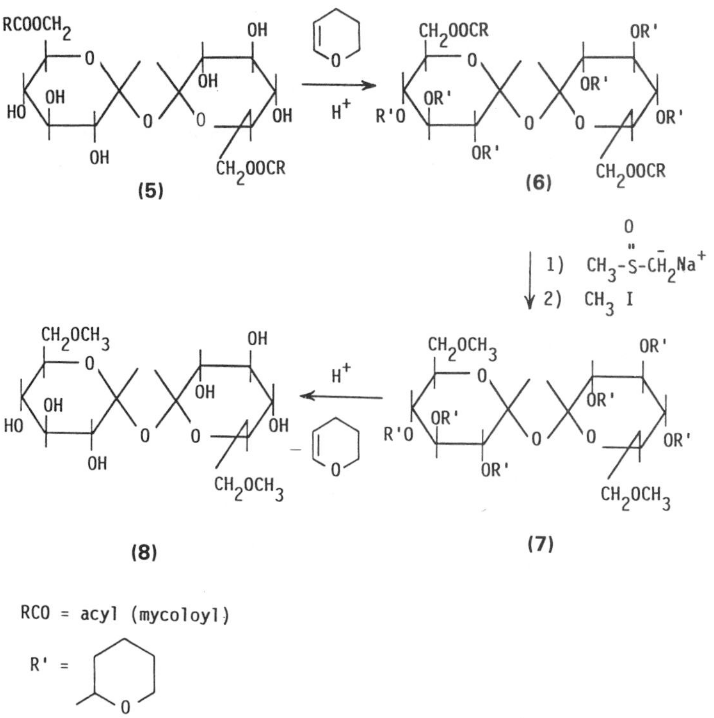

Figure 5-4. Determination of cord factor structure by methyl replacement: the methyl groups in product (8) are in the positions that were originally acylated in the native glycolipid (5).

5.2.4. Heterogeneity of Trehalose Dimycolates and of Mycolic Acids

I have intentionally avoided the term "homogeneous" in describing cord factor samples; and "pure" is meant to convey only that the materials under discussion are almost exclusively trehalose-6,6'-dimycolates. The awkwardness stems from the inhomogeneity of the mycolate substituents. More than a half-dozen "classes" of mycolic acids have been described among mycobacteria. Many of these have also been found in samples of TDM—the kinds reflecting the mycobacterial species from which the TDM was recovered. A uniform nomenclature for the mycolates has not yet, to my knowledge, been adopted by agreement. But a practical step toward meeting this need was recently suggested by Daffé *et al.* (1983) (see their Table II). The usage adopted by D. E. Minnikin and colleagues also recommends itself (see, e.g., Davidson *et al.*, 1982). Perhaps a synthesis of these two approaches could lead to an agreeable concensus.

Some of the mycolate varieties are depicted by the structures given in Table 5-1. Many of these have been seen in examples of cord factor (TDM) recovered from various mycobacteria. Accordingly, TDM homogeneity is compromised by both the classes of mycolate substituents they contain and by homology within them. My anticipation still seems valid that "because of the

Table 5-1. Some Varieties of Mycolic Acids

Name/structure	Homology	Mycobacterium sp. source	Reference
Smegmamycolic acids $$CH_3(CH_2)_A CH=CH(CH_2)_5 CH-CH-COOH$$ with $\overset{OH}{\mid}$ and $\underset{C_{22}H_{45}}{\mid}$	$C_{60}-C_{66}$	*M. smegmatis*	Wong and Gray (1979)
$$CH_3(CH_2)_A CH=CH(CH_2)_5-CH=CH-CH(CH_2)_2 CH-CH-COOH$$ with $\underset{CH_3}{\mid}$ $\overset{OH}{\mid}$ and $\underset{C_{22}H_{45}}{\mid}$	$C_{75}-C_{83}$	*M. smegmatis*	Davidson and Gray (1982)
α-Kansamycolic acid $$CH_3(CH_2)_A-CH-CH-(CH_2)_5 CH-CH-(CH_2)_2-CHCH-COOH$$ with $\underset{CH_2}{\overset{\diagdown\diagup}{}}$, $\underset{CH_2}{\overset{\diagdown\diagup}{}}$ $\overset{OH}{\mid}$ and $\underset{C_{22}H_{45}}{\mid}$	$C_{76}-C_{84}$	*M. kansasii*	Étemadi *et al.* (1964)
Methoxymycolic acids $$CH_3(CH_2)_{15,17} CH(CH_3) CH(CH_2)_5-CH-CH(CH_2)_2 CH-CH-COOH$$ with $\underset{CH_2}{\overset{\diagdown\diagup}{}}$ $\overset{OCH_3}{\mid}$ $\overset{OH}{\mid}$ and $\underset{C_{24}H_{49}}{\mid}$	$C_{77}-C_{89}$	*M. microti*	Davidson *et al.* (1982)
Wax ester mycolic acids $$CH_3(CH_2)_{17}-CH-OOC(C_nH_{2n-2}) CH-CH-COOH$$ with $\underset{CH_3}{\mid}$ $\overset{OH}{\mid}$ and $\underset{C_{22}H_{45}}{\mid}$	$C_{78}-C_{85}$	*M. paratuberculosis*	Lanéelle and Lanéelle (1970)
Ketomycolic acids $$CH_3(CH_2)_A CH-CO(CH_2)_5 CH-CH(CH_2)_2 CH CH-COOH$$ with $\underset{CH_2}{\overset{\diagdown\diagup}{}}$ $\overset{CH_3}{\mid}$ $\overset{OH}{\mid}$ and $\underset{C_{24}H_{49}}{\mid}$	$C_{80}-C_{90}$	*M. bovis* (BCG)	Adam *et al.* (1967)

Table 5-2. Some Examples of Mycobacterium *Species from Which Trehalose Dimycolates Were Isolated*

Mycobacterium sp.	Identity	M.p.(°C)	[α]ᴅ	Reference
M. tuberculosis PN, DT, C	Mixture of human strains	36–37	+31°	Noll and Bloch (1955)
M. tuberculosis H37Rv	Human: laboratory strain	41–42	+30°	Noll and Bloch (1955)
H37Rv (Ciba)	Isoniazid-resistant H37Rv	42	—	Noll and Bloch (1955)
M. tuberculosis Bré-vannes	Human strain	40–41	+31°	Noll et al. (1956)
M. tuberculosis Aoyama B	Human strain			Azuma et al. (1974)
M. bovis BCG	Attenuated bovine strain			Asselineau and Lederer cited in Noll et al. (1956)
M. bovis Vallée	Virulent bovine strain	38–39	+29°`	Noll and Bloch (1955)
M. tuberculosis Peurois	Human strain			Adam et al. (1967)
M. microti	Vole bacillus			K. Dhariwal and M. Goren (unpublished data)
M. phlei	Saprophyte		+36.5°ᵃ +41°ᵃ +38°ᵃ	Promé et al. (1976)
M. smegmatis	Saprophyte	45–46	+33°	Mompon et al. (1978)
M. fortuitum	Nontuberculosis strain	43–45		Azuma and Yamamura (1962)
M. tuberculosis H37Ra	Human, attenuated laboratory strain			Bekierkunst et al. (1979)
M. lepraemurium	Mouse pathogen in vivo harvest			Goren et al. (1979)
M. lepraemurium	In vitro grown harvest			Nakamura et al. (1984)

ᵃIndividual symmetrical and dissymmetrical TDMs separated after petrimethylsilylation and regenerated.

structural heterogeneity of the [mycolates], amplified by the multitude of homologs . . . it is unlikely that an entirely homogeneous example of natural cord factor will be isolated in the near future" (Goren, 1975). It should also be noted, respecting Bloch's dedication to the thesis that TDM was restricted in distribution to only cord-forming virulent tubercle bacilli, that this anticipation has not withstood the test of even casual scrutiny. Cord factor is essentially ubiquitous among mycobacteria (Table 5-2), the sole exception to date probably being *M. leprae*; and even this exception may be relieved, if sufficient gross lipids of *M. leprae* are examined (Dhariwal *et al.,* 1985, 1987).

Although homogeneity may not be improved, the TDM preparations designated as P₃ by E. Ribi and colleagues are probably among the cleanest so far obtained. By accepted usage, P₃ is understood to describe highly purified cord factor products, ordinarily obtained from somewhat impure TDM concentrates by high-pressure liquid chromatography (HPLC) on microparticulate silica gel (Ribi *et al.*, 1974) followed by elimination of residual traces of mycolic acids on DEAE cellulose (Goren and Brokl, 1974). P_3 is a proprietary product produced weekly in multigram quantities by the recently formed Ribi ImmunoChem Research, Inc. and used in the preparation of antitumor vaccines, as will be described in a later section. The current commercial methods for recovering and purifying P_3 have not been disclosed.

R. Toubiana (unpublished data) found that multiple TLC development of quite "pure" cord factor samples achieved a separation of otherwise unresolved TDM components into a form of nested "spear-pointed hats" (Toubiana's description, personal communication), considered different species of TDM. In studies by Strain *et al.* (1977) and independently by Promé *et al.* (1976), a facilitated resolution of individual cord factor components was achieved via pertrimethylsilylation of TDM samples. The polar contributions due to the free trehalose hydroxyls were therefore masked, and chromatographic mobilities of the per-TMS products were conferred by the subtle differences in structures of the mycolate substituents.

In this fashion, pertrimethylsilylated cord factor of the noncording saprophytic *M. phlei* was separated by preparative TLC into three principal fractions from which the mycolate components were recovered by alkaline methanolysis (Promé *et al.*, 1976). Trehalose-6,6'-dimycolate was characterized (Table 5-2) as consisting of a (symmetrical) dicyclopropane dimycolate, a (symmetrical) wax ester dimycolate, and a mixed wax ester (dicyclopropanoid dimycolate). Further heterogeneity conferred by homology within each mycolate class should also be borne in mind. Strain *et al.* (1977) described the mycolate components of TDM from virulent human and bovine tubercle bacilli. Six separable mycolate derivatives of trehalose were recognized: symmetric methoxy-, symmetric dicyclopropane-, symmetric keto-, and the three unsymmetric permutations that these can generate. Cord factor of *M. bovis* BCG lacked the methoxy-substituted mycolic acids. Whether this is due to a complete absence of methoxy mycolate in BCG or whether it is present but is not used for TDM synthesis has evidently not been studied.

Although the trehalose mycolates of lower molecular weight, e.g., from *Corynebacterium* and *Nocardiae* species, are not extensively described in this chapter, a paper from the prominent group at Toulouse (Puzo *et al.*, 1978) is particularly relevant to the topic of TDM heterogeneity. The natural cord factor originally (and quite appropriately) characterized as a trehalose dicorynomycolate (Ioneda *et al.*, 1963), in the study by Puzo was purified to homogeneity via pertrimethylsilylation and preparative TLC. The regenerated TDM was examined by cesium cationization field desorption mass spectrometry (FDMS), in which each molecular species gives rise to a unique molecular ion $[M + Cs]^+$. Thus, the homogeneous trehalose dicorynomycolate was found to contain at least 22 molecular species differing in carbon number from C_{66} to C_{76}. These were attributable to differences in the sizes of the β-hydroxylated acyl groups, which ranged from C_{24} to C_{32} species. Additional heterogeneity was conferred by degrees of acyl group unsaturation: 0, 1, and 2 double bonds in most examples, and very possibly in the location of the double bonds as well. Table 5-2 presents a compilation of trehalose dimycolates recovered from a variety of mycobacteria. It is evident that the glycolipid is almost ubiquitous in this genus.

5.2.5 Cord Factor Purification: Current Practices, Infrared Spectrometry

The waxes C, and more recently D, of mycobacterial lipids, prepared according to the scheme of Anderson as modified by Aebi *et al.* (1953), have been rich sources for recovery of TDM. In the earlier studies (Noll *et al.,* 1956) cord factor was purified by a series of tedious column chromatographies on silicic acid and magnesium silicate. Mycolic acid comprised the most tenacious contaminant. Goren and Brokl (1974) demonstrated that this was easily and quantitatively removable by selective adsorption of the contaminant on DEAE cellulose (free base form). The purification was readily followed by TLC and by IR spectrophotometry. Figure 5-5 shows the expected changes in the IR spectrum that accompany the facile recovery of a purified cord factor from a cord factor concentrate contaminated with mycolic acid.

The arrows in the spectrum (Fig. 5-5) define the important peaks and bands characteristic of cord factor. Several of these were originally defined by Noll and Bloch (1955) in their classic study. They comprise a peak at about

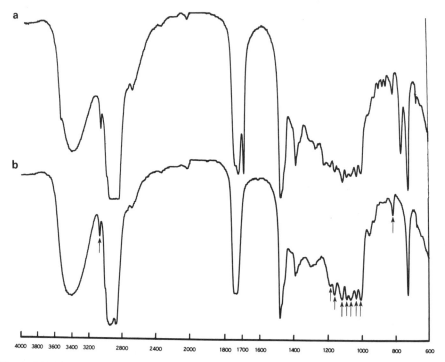

Figure 5-5. Changes in the infrared spectrum of a cord factor contaminated with mycolic acid (a) when it is purified by adsorption of the contaminant on DEAE cellulose (free base). Absorption bands attributable to the mycolic acid disappear. The spectrum of the purified TDM (b) exhibits specific identifying peaks indicated by arrows (see text).

807 cm^{-1} that we have identified in every trehalose derivative examined so far (cf. comments regarding the spectrum of 6-corynomycoloyl-3',6'-anhydrotrehalose, discussed in Section 5.2.7.2c). The peak at about 3070 cm^{-1} is associated with the cyclopropane methylene groups in mycolic acid substituents and, save for the usual OH and carbonyl absorption bands, the important fingerprint region shows seven closely spaced and unique peaks at 997, 1022, 1057, 1079, 1104, 1152, and 1175 cm^{-1}. These are in close agreement with bands originally described by Noll and Bloch (1955). We found that the sharpness and prominence of these peaks reflect the relative purity of the TDM sample.

In our laboratory, TDM has been recovered in various ways over the years and is currently prepared as follows: mycobacterial cells are extracted according to sequences described by Khuller and colleagues (see Prabhudesai et al., 1981) for securing and purifying the sulfatides of *M. tuberculosis*. As we described, however, we found this approach unsuitable for this purpose (see Dhariwal *et al.*, 1984; see also Section 5.4.6) but a useful one instead for cord factor recovery. In brief, the bacteria are extracted many times with cold methanol, then boiling methanol and finally several times with chloroform–methanol. A moiety of cord factor is "lost" in the hot methanol extract, (along with much of the sulfatides, if present); the bulk of the TDM is found in the chloroform–methanol. This is treated with methanol to extract most of the contaminating phospholipids and the residue is chromatographed on silica gel. Chloroform–methanol (95 : 5) elutes almost all of a slightly contaminated cord factor. Acidic lipids are then removed on DEAE cellulose (free base form), and the TDM product is rechromatographed a final time on silica gel.

Promé *et al.* (1976) described an effective purification of TDM (and trehalose monomycolate, TMM) of *M. phlei*. In this example, these two lipids were recovered from crude wax A (J. Asselineau, 1966) by chromatography on neutral silicic acid: CHCl$_3$ with 2–5% CH$_3$OH eluted TDM, and CHCl$_3$ with 10 to 15% CH$_3$OH eluted TMM. TDM was purified further on florisil and TMM on DEAE-cellulose (acetate).

Trehalose dimycolate has become an article of commerce because of its important immunostimulant properties (see the review by Lemaire *et al.*, 1986; see also Section 5.2.9). Therefore, more practical methods for TDM recovery have been developed, but because of proprietary concerns these have been shrouded in secrecy. Ribi and colleagues originally prepared highly purified TDM, designated P$_3$, by high-pressure liquid chromatography (HPLC) on a microparticulate silica gel of concentrates obtained from wax D (Ribi *et al.*, 1974) with a final purification on DEAE cellulose (free base) to remove residual mycolic acids (E. Ribi, personal communication). We quote further from an article by Lemaire *et al.* (1986):

> The TDM preparation available in France (Institut Choay, 92120 Montrouge) is extracted from autoclaved residues of the production of tuberculin at the Pasteur Institute. After extraction with dichloromethane, it is purified by two successive column chromatographies on silica gel, with a gradient of methanol in dichloromethane. It still contains a few percent of . . . inactive contaminants [which include TMM and mycolic acid].

5.2.6. Determination of Structure: Mass and NMR Spectrometry

Trehalose esters of lower moleular weight can be studied effectively by mass spectrometry to gain (sometimes complete) structural information. Lederer (1971, 1976) and Asselineau and Asselineau (1978) have reviewed in detail the kinds of data and their interpretation—derived from electron impact mass spectrometry on simple and also on more complex trehalose esters and their peracetylated derivatives: viz., a curious unsymmetrical 4,6-diester from *M. fortuitum* (Vilkas *et al.*, 1968), trehalose dicorynomycolates from *Corynebacterium diphtheriae* (Senn *et al.*, 1967); and cord factors from *M. tuberculosis* Peurois, from *M. bovis* BCG and from *M. butyricum* (now classified as *M. smegmatis*) (Adam *et al.*, 1967). Except for trehalose esters of quite modest molecular weight, such as the peracetylated trehalose dicorynomycolates (see Ledrerer, 1971), much of the information obtained from mass spectra of e.g., the "true" cord factors has been derived from first analyzing the mycolic acids that they yield on hydrolysis (Senn *et al.*, 1967; Adam *et al.*, 1967). Afterward, electron impact mass spectrometry (EIMS) of, e.g., peracetylated derivatives of the native glycolipid yields important fragments, especially the primary oxonium ions that arise from cleavage at the anomeric oxygen, viz.

These subsequently undergo fragmentations that can be quite characteristic, hence informative, of more detailed structures. In the study by Senn *et al.* (1967) of the dicorynomycolate from *Corynebacterium diphtheriae*, the recovered corynomycolic acids were found to consist of a mixture of the saturated corynomycolic acid ($C_{32}H_{64}O_3$) (see Section 5.2.1) and of an unsaturated corynomycolenic acid ($C_{32}H_{62}O_3$). Mass spectrometry of the peracetylated glycolipid showed this product to be at about the upper limit in molecular weight for the trehalose glycolipids to give recognizable molecular ions by electron impact mass spectrometry. Three very weak molecular ions at m/z 1634, 1632, and 1630 showed that the original glycolipid is a mixture of diesters: with two saturated acyl substituents; a species with one saturated and one unsaturated function; and a species with two unsaturated substituents (Senn *et al.*, 1967). More sophisticated methodology (described later) resolved this mixture of glycolipids into at least 22 molecular species (Puzo *et al.*, 1978).

When the approach used by Senn *et al.* was applied instead to the much larger peracetylated cord factors from, e.g., *M. bovis*, BCG, or *M. tuberculosis*, with molecular weight above 3000, no molecular ions could be detected. Instead, fragment oxonium ions (see above) arising from cleavage at the glycosidic linkage accompanied by loss of two acetic acid molecules were found in the higher mass region (m/z above 1450) (Adam *et al.*, 1967).

Goren and Brennan (1979) briefly reviewed and analyzed some of the mass spectrometric data obtained during the structural studies on the principal sulfatide (SL-I) of *M. tuberculosis* strain H37RV (Goren *et al.*, 1976a). A portion of this study, useful in illustrating the mass spectrometric details, is summarized in Section 5.4.4.4.

More sophisticated methodologies (field desorption or plasma desorption mass spectrometry) have not been particularly useful for studying TDM from *M. tuberculosis* and related species. They have been fruitfully applied, however, to studies of a variety of trehalose monomycolates, including those of the true mycobacteria. These are discussed in Section 5.2.10.4.

Although 'H-NMR has been useful for studying the individual acyl substituents of the trehalose glycolipids, it has not provided important data concerning their positions on the trehalose core. However, ^{13}C-NMR has been exploited in one study by E. Wenkert (see Polonsky *et al.*, 1978a) to establish essentially complete structures for samples of cord factor from *M. smegmatis, M. tuberculosis*, Brévannes, *M. tuberculosis* Peurois, and other species. From the data it was possible to define the structural patterns of the mycolic acid substituents as well as the (6,6') sites of attachment of the acids to the carbohydrate core. This elegant technique is limited, however, because of its requirement for large samples of the lipid—several hundred milligrams—which renders the methodology unavailable for rare examples of the glycolipid.

5.2.7. Synthesis of Cord Factors and Cord Factor Analogues

5.2.7.1. Early Synthesis Schemes

Interest in the synthesis of cord factors and analogues was intense, even before the exact structure of TDM was firmly established. The enthusiasm stemmed from both a desire to supplement the structural studies by anticipating their outcome, and also from the challenge of synthesizing substances that might be endowed with the toxic properties of cord factor (see Section 5.2.9.1).

These were evidently the first of such dramatic activity documented for an identifiable lipid component of mycobacteria. The biological activities of the phosphatide fractions and of the branched unsaturated phthienoic acids were originally also of great interest (see review by Goren, 1984), but they were subsequently (and probably justifiably) eclipsed by cord factor.

The current interest has been stimulated by the practical exploitation of trehalose dimycolate in commercial products for treatment of certain malignant tumors in domestic animals, in experimental (stage 1) studies in man and also because of immunostimulant properties that have been documented for these glycolipids (see Sections 5.2.9.2, 5.2.9.4, 5.2.9.5).

An early scheme for the synthesis of trehalose-6-mycolates involves the substitution of the 6,6'-hydroxyls with toluenesulfonyl (tosyl) groups followed by nucleophilic displacement with, e.g., K-mycolate in a suitable solvent such as DMF at 125°C for 80 hr. The reactions have been carried out with the remainder of the sugar hydroxyls either protected, as by acetylation (Polonsky *et al.*, 1956), or free (Brocheré-Ferréol and Polonsky, 1958; Polonsky *et al.*, 1978). The former route (OH groups acetylated) led to difficulties in subsequent deacetylation owing to simultaneous loss of mycolate groups. The alternative route posed difficulties in the preparation of a homogeneous 6,6′-ditosylate (Brocheré-Ferréol and Polonsky, 1958; Toubiana *et al.*, 1973) and in the displacement reaction with unprotected hydroxyls because of the formation of putative 6-mycoloyl-3′,6′-anhydrotrehalose, a troublesome side reaction (Polonsky *et al.*, 1978). According to these investigators, this problem was largely alleviated when the displacement was carried out in toluene at lower temperature and in the presence of a suitable crown ether.

Tocanne (1975) and Toubiana *et al.* (1975) independently developed almost identical syntheses for cord factor and analogues that circumvent several of the aforementioned problems. They selectively prepared 6,6′-di-deoxy-6,6′-dihalo derivatives of trehalose according to the methods of Hanessian *et al.* (1972), blocked the unsubstituted hydroxyls by trimethylsilylation and condensed this blocked 6,6′-dihalide (**9**) with K-mycolates in hexamethylphosphoric triamide (HMPT). In still another variant, the primary hydroxyls of trehalose were converted to the alkoxy tris(dimethylamino)phosphonium hexafluorophosphates (Castro *et al.*, 1973), the free hydroxyls then blocked by trimethylsilylation and the product (**10**) condensed in HMPT with K-corynomycolate. According to this procedure, the phosphonium salts are obtained in quantitative yield and are evidently readily displaced by the carboxylate anion (Toubiana *et al.*, 1978). These schemes have the advantage of selectivity for the primary hydroxyls, in contrast to the less selective tosylation, as well as of a facile blocking by the readily removed trimethylsilyl groups, thus avoiding formation of the troublesome presumed 3,6-anhydro derivative (see Section 5.2.7.2*c*).

(9) R = halide (bromide, iodide)

(10) R = $-O\overset{+}{P}(NMe_2)_3(\overset{-}{PF_6})$

5.2.7.2. Recent Synthesis Schemes

5.2.7.2a. Hexabenzyl Trehalose as Intermediate. The search for suitably blocked trehalose derivatives led to the successful exploitation of 2,3,4,2′,3′,4′-hexa-*O*-benzyl trehalose, as shown in Fig. 5-6: (**11**), for the synthesis of simple 6,6′-diacyl trehaloses, to cord factor itself, and ultimately to a variety of still more challenging analogues and pseudo-cord factors. In the first of these, as summarized in Fig. 5-6, the well-described 6,6′-ditrityl trehalose (Bredereck, 1930) was perbenzylated with benzyl chloride and alkali and detritylated to

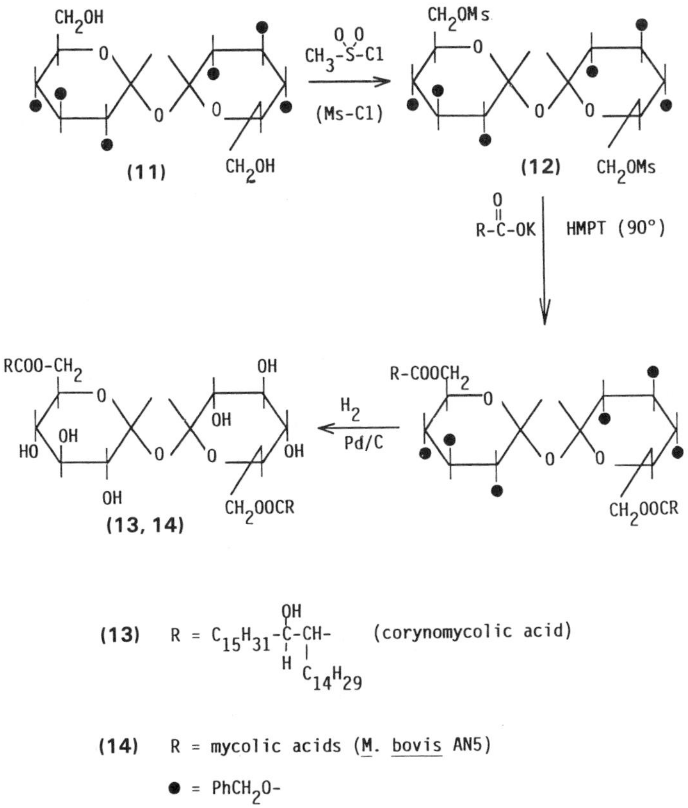

Figure 5-6. Synthesis of cord factors and analogues via hexa-*O*-benzyl trehalose (**11**). The benzyl blocking groups prevent formation of 3,6-anhydro products.

(**11**); the freed primary hydroxyls were quantitatively mesylated with methanesulfonyl chloride in pyridine to the reactive 6,6′-dimesylate (**12**). Reaction with synthetic K-corynomycolate (Polonsky and Lederer, 1954) followed by debenzylation yielded trehalose-6,6′-dicorynomycolate (**13**). With the K salt of mycolic acid from *M. bovis* AN5, the hexabenzyl dimycolate was obtained, which on hydrogenolysis catalyzed by Pd/charcoal yielded a highly pure cord factor product (**14**), (Liav and Goren, 1980*a,b*).

5.2.7.2b. Tetrabenzyltrehalose as Intermediate. Utilization of 2,3,2′,3′-tetra-*O*-benzyl trehalose also provides a useful alternative route to a variety of trehalose esters. The initial steps are based on an earlier-described scheme of Hough *et al.* (1972) that gave first 4,6,4′,6′-dibenzylidene trehalose, and on perbenzylation and debenzylidenation yielded the key substance 2,3,2′,3′tetra-*O*-benzyl trehalose (see also Yoshimoto *et al.*, 1982; Numata *et al.*, 1985). Owing to the considerably greater reactivity of the hydroxyls at the 6-positions as compared with those at C-4, this tetrabenzyl trehalose was directly convertible into 6-stearoyl, 6,6′-distearoyl, and 4,6,4′,6′-tetrastearoyl

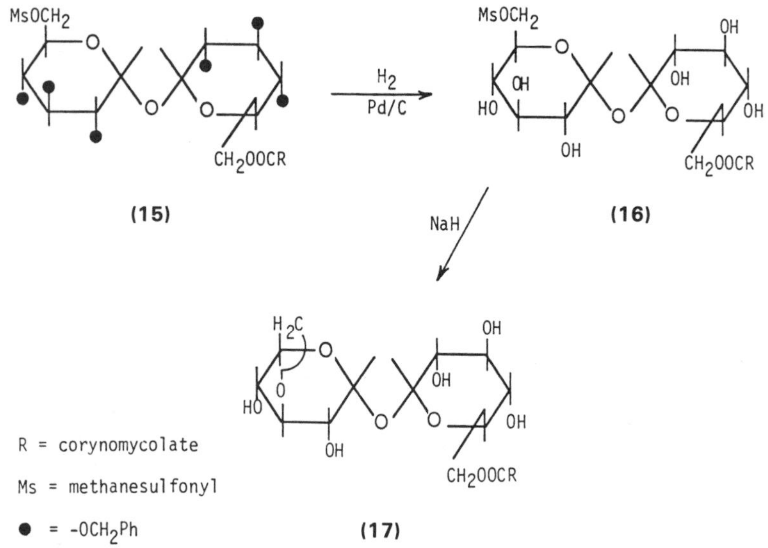

Figure 5-7. Proof of structure of 6-*O*-corynomycoloyl-3',6'-anhydrotrehalose (**17**) by synthesis from the 6-ester-6'-mesylate (**15**).

trehalose—by acylation with appropriate amounts of stearoyl chloride followed by hydrogenolytic debenzylation (Yoshimoto *et al.*, 1982).

The differential reactivity alluded to above also permitted the preparation of the tetrabenzyl 6,6'-ditosylate (Hough *et al.*, 1972). This intermediate was recently used to prepared trehalose-6,6'-dimycolates as well as simpler diesters—by the usual metathesis reaction with potassium salts of carboxylic acids followed by hydrogenolysis (Numata *et al.*, 1985). This scheme also clearly prevents formation of the 3,6-anhydro 6'-mycoloyl by-product that was believed to form when the metathesis reaction is conducted with unblocked 6,6'-ditosyltrehalose (Polonsky *et al.*, 1978*b*).

5.2.7.2c. Derivatives of 3,6-anhydrotrehalose. In examining a semicommercial trehalose dicorynomycolate product that allegedly had been prepared according to the "modified" Polonsky (1978*b*) method, but not by Polonsky and colleagues, Liav and Goren (unpublished; Goren, 1982) recognized that it was grossly contaminated with a more polar product, detected by TLC. It was separated only with difficulty from the cord factor analogue. The infrared spectrum of this product resembled that of most 6,6'-diacyltrehaloses, except that the weak absorption band ordinarily seen at about 807 cm^{-1} was shifted to about 790 cm^{-1}. We judged this to be the 6-acyl-3',6'-anhydrotrehalose (**17**) whose generation in the reaction of the K-mycolate with ditosyl trehalose had long been suspected. Referring to Figs. 5-6 and 5-7, when the metathesis of the hexabenzyl-6,6'-di-*O*-methanesulfonyl trehalose (**12**) was conducted with less than an equivalent of K-corynomycolate, the major product obtained was 2,3,4,2',3',4'-hexa-*O*-benzyl 6-*O*-corynomycoloyl-6'-*O*-methanesulfonyl trehalose (**15**). On Pd/C reduction of this monocorynomycolate, it was debenzylated to the 6-mesyl-6'-corynomycoloyl trehalose (**16**). This was transformed to the monoanhydro analogue (**17**) upon treatment with NaH in

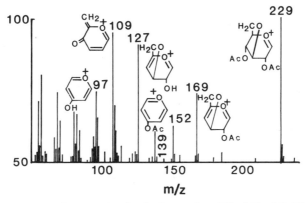

Figure 5-8. Mass spectrum of peracetylated anhydro product (**17**) of Fig. 5-7. The peaks at m/z 229 (base peak), 169 and 127 are definitive for fragmentations of the anhydro ring.

toluene for several hours (Liav and Goren, 1982). This product had the aberrant infrared spectrum (790 cm^{-1} band) of the earlier suspected anhydro byproduct. Its structure was confirmed by mass spectrometry of the peracetylated derivative (Fig. 5-8). The base peak (m/z 229) is definitive for the *oxonium fragment* derived from the ring bearing the 3,6-anhydro group.

5.2.7.3. Syntheses via the Mitsunobu Reaction

Perhaps the most fruitful efforts at synthesis of cord factors have evolved from investigations leading to application of the Mitsunobu reaction (Mitsunobu, 1981) to this system. In the simplest example, Bottle and Jenkins (1984) synthesized 6,6′-dipalmitoyl trehalose and 6,6′-dipalmitoyl sucrose in good yields directly from the free disaccharides, palmitic acid and the Mitsunobu reagents triphenylphosphine (TPP) and diisopropylazodicarboxylate (DIAD). However, when the reaction was attempted with corynomycolic acid, under the influence of the reagents, this hydroxy acid underwent decarboxylative β-elimination, without giving any cord factor product. This was avoided when α,β-mycolenoic acid (lacking the β-hydroxyl group) (**18**) was substituted for natural mycolate, or when suitably β-blocked mycolates were used: the β-*O*-methyl ether (**19**) on one hand or the β-*O*-tetrahydropyranyl-(oxan-2-yl) derivative (**20**) on the other (Jenkins and Goren, 1986). Al-

$$CH_3(CH_2)_x-CH-CH-(CH_2)_y-CH-CH-(CH_2)_z-CH = C \begin{smallmatrix} COOH \\ \\ C_{24}H_{49} \end{smallmatrix}$$
$$\overset{\diagdown/}{CH_2} \qquad \overset{\diagdown/}{CH_2}$$

(**18**)

$$CH_3(CH_2)_x-CH-CH-(CH_2)_y-CH-CH-(CH_2)z-\overset{\overset{\displaystyle OR}{|}}{CH}-CHCOOH$$
$$\overset{\diagdown/}{CH_2} \qquad \overset{\diagdown/}{CH_2} \qquad \underset{C_{24}H_{49}}{|}$$

(**19**)   R = CH$_3$

(**20**)   R =

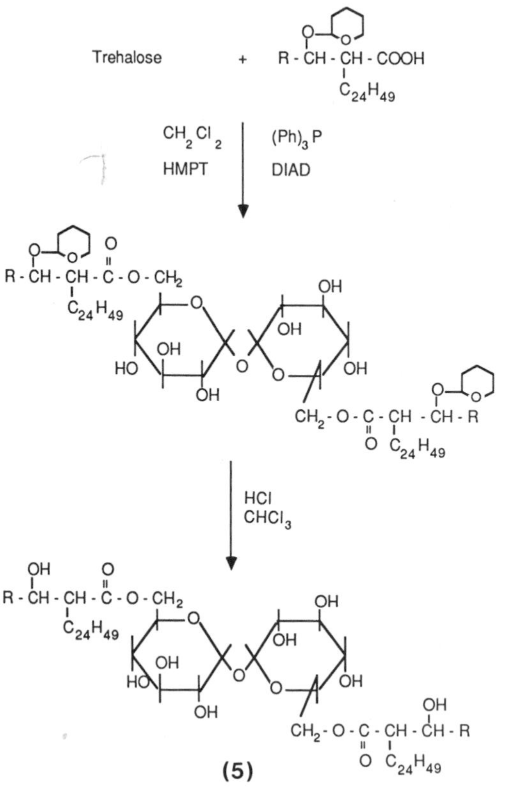

Figure 5-9. Synthesis of cord factor via the Mitsunobu reaction. The β-*O*-oxan-2-yl mycolic acid is stable during the esterification and the intermediate diester is readily deblocked.

though the β-*O*-methyl ether is a permanently blocked derivative, the β-*O*-oxanyl blocking group was easily removed after the esterification with trehalose by mild treatment with HCl in chloroform. The synthesis of cord factor (**5**) is given in Fig. 5-9.

The reaction sequences are remarkably simple: by use of 1 : 1 (v/v) mixture of methylene chloride and hexamethylphosphoric triamide (HMPT) as solvent, a homogeneous reaction mixture is obtained with such diverse reagents as anhydrous trehalose and free mycolic acid. After esterification, the reaction mixture is poured into a large excess of methanol with stirring to dissolve all the reagents except for residual mycolic acid and the glycolipid product, which is separated and purified by chromatography.

The mechanism of the Mitsunobu reaction has recently been shown (Von Itzstein and Jenkins, 1983; Grochowski *et al.*, 1982) to involve the formation of dialkoxytriphenylphosphorane intermediates. The selectivity (e.g., with trehalose) is presumably a result of the preferential formation of such phosphorans on the primary hydroxyl groups. This interpretation is depicted in the reaction scheme of Fig. 5-10.

5.2.7.4. Synthesis of Diastereoisomeric Cord Factor Analogues

In order to probe the effects of subtle stereochemical variations on the biological activities of cord factor, dimycolate analogues based on diaster–

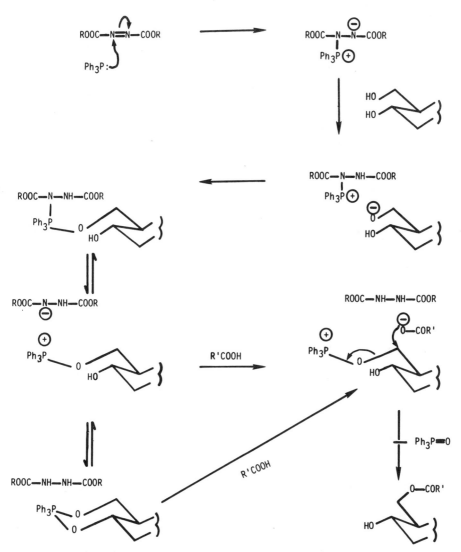

Figure 5-10. Interpretation of mechanisms in cord factor synthesis by the Mitsunobu reaction. Participation of both an initial betaine and a penultimate dialkoxyphosphorane have been documented.

eoisomers of trehalose have been synthesized. Early on, in collaborative studies with M. Kato, J. Asselineau prepared dimycoloyl sucrose (Kato and Asselineau, 1971) as well as various mycolates of simpler monosaccharides such as 6-mycolates of methyl glucoside, alloside, mannoside, and galactoside (Asselineau and Kato, 1973, 1975) (see also Table 6 from Asselineau and Asselineau, 1978). The mycolates were prepared according to the displacement reactions of Brocheré-Ferréol and Polonsky (1958) from K-mycolate and the appropriate 6-tosyl monosaccharides or the 6,6'-ditosyl disaccharide.

In the more recent studies concerning diastereoisomers of trehalose, Liav and Goren (1984a) prepared the cord factor analogue corresponding to α,α-

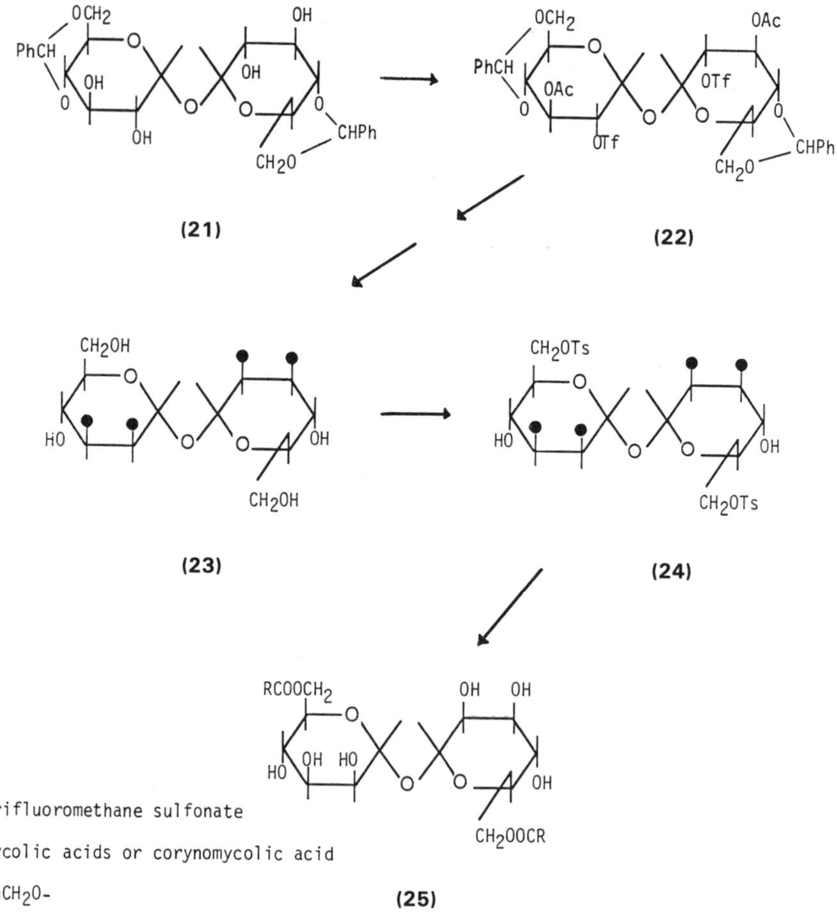

Tf = trifluoromethane sulfonate

R = mycolic acids or corynomycolic acid

● = $PhCH_2O$-

Figure 5-11. Synthesis of the mannosyl–mannoside diastereoisomer of TDM (**25**). See text for details.

mannosyl-mannoside, as well as of the α,α′-galactosyl galactoside analogue (Liav *et al.*, 1984*a*). In preparing the mannose analogue (Fig. 5-11) 4,6:4′,6′-dibenzylidene trehalose (**21**) (Hough *et al.*, 1972) was converted with trifluoromethanesulfonyl chloride into the 2,2′-ditrifluoromesylate (ditriflate) derivative. However, before solvolysis of the ditriflate to achieve inversion, it was necessary to block the free 3-hydroxyls by acetylation to give (**22**), in order to avoid complicating side reactions. Treatment of the diacetate-ditrifluoromesylate with $NaNO_2$ in hexamethylphosphorictriamide (HMPT) achieved the desired inversion at the 2,2′-positions. The product was then simultaneously deacetylated and *O*-benzylated with benzyl chloride and KOH, and then freed of the benzylidene groups with 80% aqueous acetic acid to give (**23**). Selective tosylation gave the 6,6′-ditosylate (**24**), which was transformed into the diastereoisomeric (mannosyl) cord factor (**25**) in the usual manner. The 6,6′-dicorynomycolate was also prepared (Liav and Goren, 1984*a*).

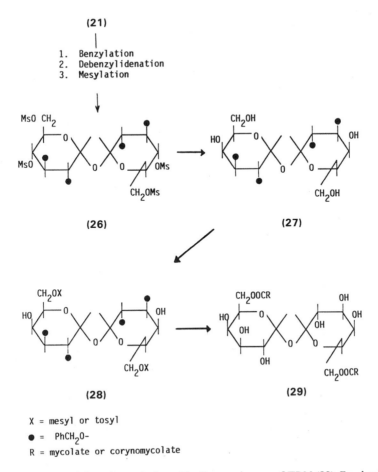

Figure 5-12. Synthesis of the galactosylgalactoside diastereoisomer of TDM (**29**). For details, see text. Because of the unblocked (axial) hydroxyl groups at the 4,4'-positions, formation of 4,6-anhydro by-products from (**28**) could not be avoided.

The galactosyl–galactoside diastereoisomer was prepared by a somewhat similar scheme (Liav *et al.*, 1984). As shown in Fig. 5-12, dibenzylidene trehalose (**21**, Fig. 5-11) was converted into 2,3,2',3'-tetra-*O*-benzyl-4,6,4',6'-tetra-*O*-mesyl trehalose (**26**) as described earlier by Hough *et al.* (1972). Solvolysis with Na benzoate, accompanied by inversion at the 4,4'-positions, and catalytic deacylation yielded 2,3,2',3'-tetra-*O*-benzyl (α-D-galactopyranosyl α-D-galactopyranoside (**27**). Selective tosylation or mesylation gave the 6,6'-ditosylate (or mesylate) (**28**), which was then converted with K-mycolate in HMPT, followed by Pd/C-catalyzed hydrogenolysis, into the galactosylgalactoside diastereoisomer of cord factor (**29**). An undesired side reaction involving the 6-*O*-tosyl or mesyl group was found attributable to facile formation of a 6-acyl-4',6'–anhydro byproduct (**30**, not shown) (Liav and Goren, 1985; Liav *et al.*, 1986).

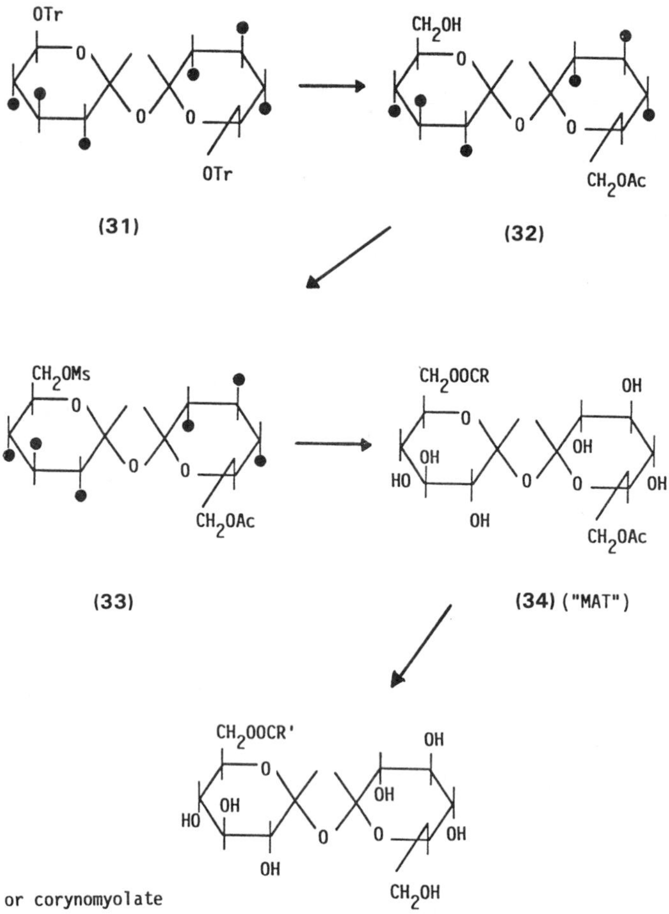

R = mycolate
R' = mycolate or corynomyolate
● = $PhCH_2O$-

Figure 5-13. Synthesis of trehalose monomycolate (TMM) (**35**) and of 6-mycoloyl-6′-acetyl trehalose (MAT) (**34**). See text for details.

5.2.7.5. Synthesis of Trehalose-6-Mycolates

Some of the preceding studies led to simple routes (Fig. 5-13) for the synthesis of the increasingly important trehalose-6-monomycolate (TMM) (more correctly, trehalose-6-mycolate). It was recognized that the mild detritylation of 6,6′-ditrityl hexa-*O*-benzyltrehalose (**31**) with 80% acetic acid in H_2O resulted in a modest conversion of the product into a 6-monoacetate (**32**). A similar curious acetylation of methylglucoside had been described earlier by Duff (1957). Treatment of the hexabenzyl-6-monoacetate with methanesulfonyl chloride yielded the 6′-mesylate (**33**). By metathesis with K-mycolate followed by hydrogenolysis, trehalose-6-acetate-6′-mycolate (**34**) (MAT) (Takayama and Armstrong, 1974) was obtained. If the 6-acetate-6′-mesylate was first deacetylated with sodium methoxide, the hexabenzyl-6-

mesylate was readily convertible into trehalose-6-mycolate (35) (Liav and Goren, 1984*b*, 1986). This glycolipid was first described by Kato and Maeda (1974). Because of its growing importance, trehalose monomycolate is considered in much greater detail in Section 5.2.10.

5.2.8. Pseudo-Cord Factors

Several other structures have been proposed as suitable variants for synthesizing cord-factor-like lipids termed pseudo-cord factors. Simplified structures for these are depicted in Fig. 5-14. True cord factors and analogues are 6,6′-diesters of trehalose with various carboxylic acids. As an alternative, mirror pseudo-cord factors may be constructed if the functionality of the components at the ester linkages is reversed, i.e., a trehalose analogue to function as the carboxyic acid with the lipid portion derived from a fatty alcohol. These may be prepared from (α-D-glucopyranosyluronic acid)—(α-D-glucopyranosiduronic acid)—obtainable as described below. If this acid, referred to as trehalose dicarboxylic acid (TDA) is instead converted to a *bis*(alkyl) amide, the product is termed a mirror amide pseudo-cord factor, whereas the corresponding analogue (functionality reversed at the amide linkages) is designated an amide pseudo-cord factor. It is derived from *N*-acylation of 6,6′-diamino-6,6′-dideoxy trehalose, also synthesized as described below. It may be mentioned at this stage that many of these pseudo-cord factors possess biological activities that closely resemble those of cord factors of similar gross structures and molecular weights.

5.2.8.1. Mirror Pseudo-Cord Factors: Derivatives of Trehalose Dicarboxylic Acid

The parent carbohydrate for this series (α-D-glucopyranosyluronic acid)–(α-D-glucopyranosiduronic acid) was first synthesized by platinum-catalyzed oxidation of trehalose in aqueous sodium bicarbonate with gaseous oxygen (Goren and Jiang, 1980). Over a period of about 6 hr, several oxidation products formed including the monocarboxylic acid (α-D-glucopyranosyluronic acid)-(α-D-glucopyranoside) comprising about 35% of the mixture and "trehalose dicarboxylic acid" (about 40%). They were recovered, and purified by chromatography on a column of powdered cellulose.

Trehalose dicarboxylic acid [Fig. 5-15 (36)] was characterized as a monohydrate in the form of its dipotassium salt $[\alpha]_D^{25}$ + 123° (H$_2$O). The free acid obtained by percolation of the salt through AG 50 WX-8 resin (H$^+$ form) appeared to be readily converted into presumed lactonized products when the ion exchange effluent was taken to dryness. The dimethylester (37), prepared by mild "Fisher esterification" (Fieser and Fieser, 1944) or with diazomethane, was recovered as a hemihydrate that melted partially at 120° with foaming, and resolidified to melt with decomposition at 180–182°; $[\alpha]_D^{23}$ + 127° (methanol). It was obtained anhydrous on drying for 4 hr at 100°.

Conversion of trehalose dicarboxylic acid into the 2,3,4,2′,3′,4′-hexa-*O*-acetate (38) was difficult owing to formation of a variety of by-products. The transformation was most successful when the diacid was suspended

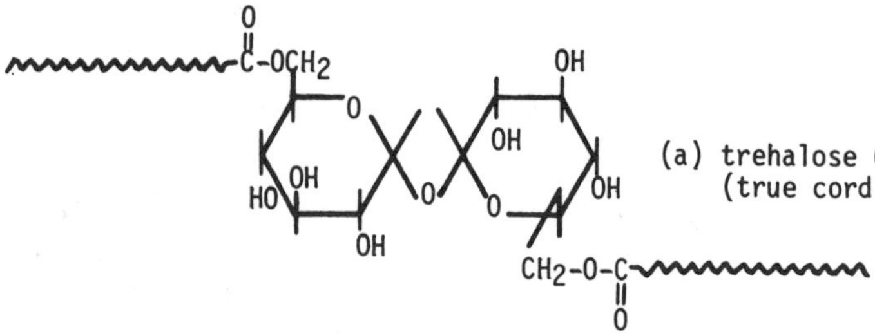

(a) trehalose 6,6' diesters
(true cord factors)

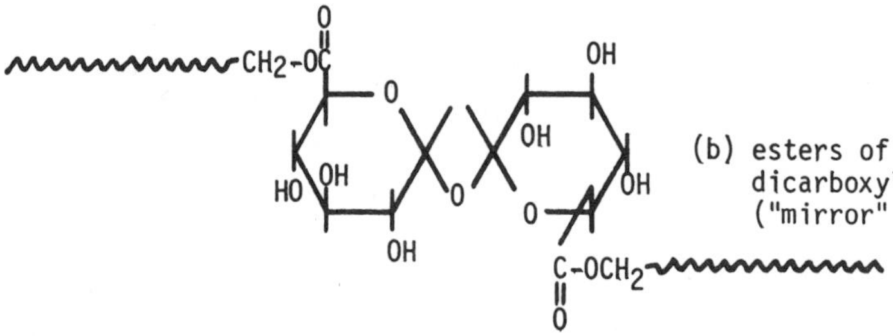

(b) esters of trehalose-6,6'-dicarboxylic acid
("mirror" pseudo CF)

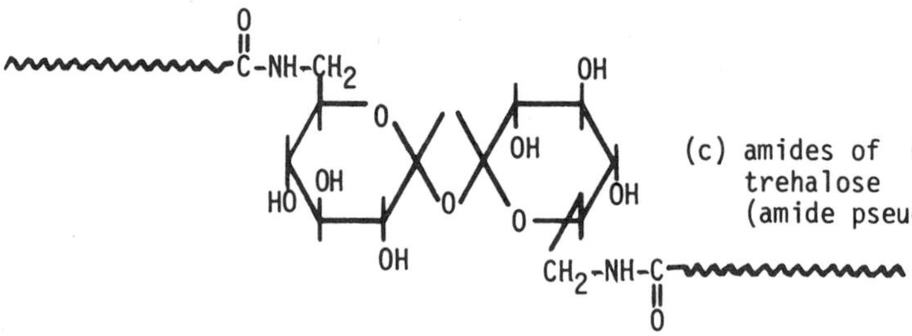

(c) amides of 6,6'-diamino trehalose
(amide pseudo CF)

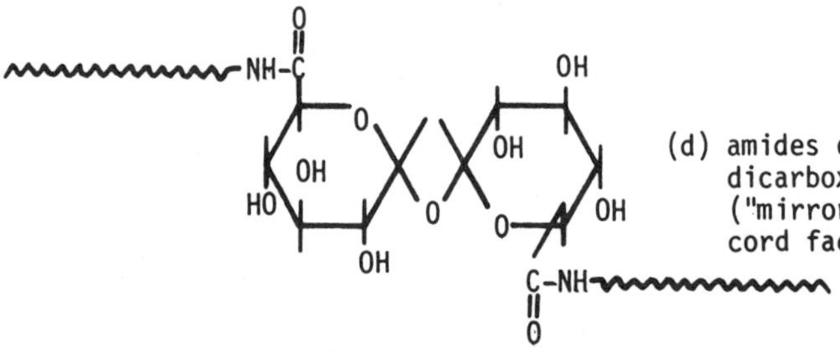

(d) amides of trehalose-6,6'-dicarboxylic acid
("mirror" amide pseudo cord factors)

Figure 5-14. Abbreviated structures of cord factor and three pseudo-cord factors. The mirror classes derive from trehalose dicarboxylic acid.

Trehalose

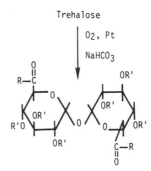

(36) R = OH, R' = H ("TREHALOSE DICARBOXYLIC ACID")

(37) R = OCH₃, R' = H

Figure 5-15. Trehalose dicarboxylic acid
(**36**) and simple derivatives. The hexaacetate diacid chloride (**39**) is a useful
intermediate for synthesis of various
pseudo-cord factors (ψCF).

(38) R = OH, R' = CH₃CO—

(39) R = Cl, R' = CH₃CO—

in acetic anhydride and the mixture was treated with a trace of concentrated
H_2SO_4. [In later studies, the same product was readily obtained by Jones
oxidation (Bowden *et al.*, 1946) of trehalose hexaacetate with CrO_3; m.p. 143–
145°C, $[\alpha]_D^{24}$ + 157°C (chloroform) (Liav & Goren, 1980c)]. On treatment with
thionyl chloride, (**38**) was converted into a reactive diacid chloride (**39**) useful
for preparing specific derivatives of TDA. The IR spectra of most of these
products showed considerable departures from the absorption bands that
were expected for such compounds, as was discussed in considerable detail
(Goren and Jiang, 1980).

Figure 5-16 summarizes some simple conversions of the hexaacetate diacid chloride (**39**) into a "mirror amide" pseudo-cord factor (ψCF) (**41**) and a
"mirror" ψCF (**42**) (Goren and Jiang, 1980). For the former, the diacid chloride was condensed with di-*n*-octadecylamine to the symmetrical diamide hexaacetate (**40**), which was gently deacetylated by solvolysis in CH_3OH on a
column of DEAE cellulose (free base form) to give the bis-*N*-di-*n*-oc-
tadecylamide of "trehalose dicarboxylic acid" (BDA-TDA) (**41**). The 36-carbon
secondary amine was chosen for the acylation in order to mimic more closely
the α-branched structures of the mycolates in natural cord factors. For preparing esters of TDA of intermediate molecular weight, and in other applications as well, *p*-hexadecyloxyphenylbutanol (see Fig. 5-14) has been used frequently. Not unexpectedly, conversion of the diacid chloride hexa-*O*-acetate
into "mirror" pseudo-cord factors was beset with difficulties. With the aromat-
ic alcohol, formation of the 6,6'-diester hexaacetate (**42**) proceeded readily,
but removal of the blocking acetate groups was accompanied by loss of the
alcohol esterifying the glucuronic acid functions. For the bis-hexadecyloxy-
phenylbutanol derivative, the deacetylation was achieved by an unusual ploy:

Figure 5-16. Conversion of the hexaacetate diacid chloride (**39**) into the *bis*-di-*n*-octadecylamide (**41**) and into a diester mirror ψCF (**43**).

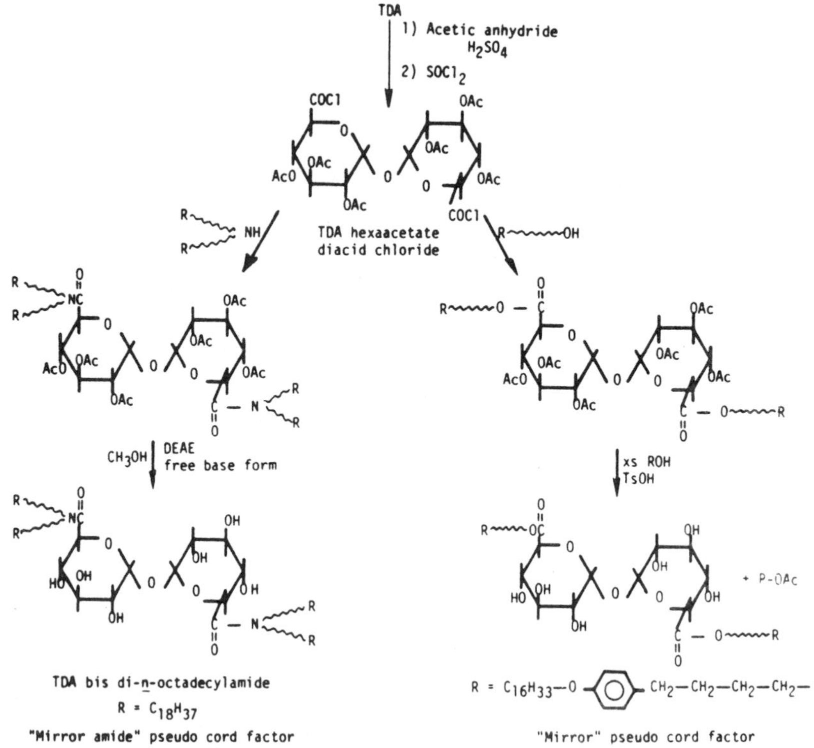

the acetates were removed by slow acid-catalyzed transesterification with this same alcohol to yield **43** (see Fig. 5-14). However, this ploy failed for the diester formed from hexa-*O*-acetyl TDA diacid chloride and mycolyl alcohol—derived by lithium aluminum hydride reduction of methyl mycolate.

It was the need for an *O*-blocked trehalose dicarboxylic acid that could be esterified to alcohols of higher molecular weight and subsequently deblocked without hydrolyzing the ester that led to the development of hexa-*O*-benzyl trehalose as the intermediate for such syntheses and then to the synthesis of true cord factors and analogues, as described in the preceding section. Thus Jones oxidation of hexa-*O*-benzyl trehalose provides 2,3,4-tri-*O*-benzyl(α-D-glucopyranosyluronic acid) 2,3,4-tri-*O*-benzyl(α-D-glucopyranosiduronic acid) hexabenzyl TDA in good yield. This serves for preparation of the corresponding diacid chloride and thence for preparing intermediate esters leading to the mirror pseudo-cord factors of Figs. 5-14 and 5-16 (A. Liav and M. Goren, unpublished data).

5.2.8.2. Amide Pseudo-Cord Factors

These products are derivatives of 6,6′-diamino-6,6′-dideoxy trehalose (**44**) (see Section 5.2.8). This substance was synthesized by Goren and Jiang

6,6'-diamino-6,6'-dideoxy trehalose **(44)**

(unpublished) from 6,6'-di-*p*-toluenesulfonyl trehalose by conversion to the diazide followed by catalytic hydrogenation. The resulting diamino trehalose was acylated by vigorous stirring of an aqueous bicarbonate solution of the amine with *p*-hexadecyloxyphenylbutanoylchloride according to the method of Fieser *et al.* (1956). In the alternative scheme of Liav *et al.* (1981), as shown in Fig. 5-17, hexa-*O*-benzyl trehalose was converted to the 6,6'-diamino-6,6'-dideoxy derivative by the reaction sequence:

$$6,6'\text{-di-methanesulfonyl} \rightarrow 6,6'\text{-diazido} \rightarrow 6,6'\text{-diamino} \qquad (2)$$

and the product was acylated in the presence of *N,N'*-dicyclohexylcarbodiimide (DCCD) with *p*-hexadecyloxyphenylbutanoic acid, with corynomycolic acid, or with mycolic acid to yield the corresponding hexabenzyl "amide" pseudo-cord factors. These were deblocked by hydrogenolysis (Pd/C), as described earlier (see Section 5.2.7.2*a*) to yield the amide ψCFs (**45, 46** and **47**) (Fig. 5-17). As an alternative route to the diamide ψCF constructed with mycolic acids, Numata *et al.* (1985) transformed 6,6'-ditosyl-2,3,2',3'-tetra-*O*-benzyltrehalose (described in Section 5.2.7.2*b*) successively into the 6,6'-diazido-6,6'-dideoxytetrabenzyl intermediate and thence to the 6,6'-diamino product. After acylation with mycolic acid (via the activated *N*-succinimide derivative) the product was hydrogenolyzed to give 6,6'-dideoxy-6,6'-bismycoloylaminotrehalose in good yield.

In summary, true cord factors, diastereoisomers of TDM, and all types of the pseudo cord factors depicted in Fig. 5-14 have been synthesized. From the various structural permutations, considerable, although incomplete, evidence has been obtained concerning the structural requirements for expression of various of the biological activities that were first recognized for TDM itself (see Section 5.2.9.1).

5.2.9. Biological Activities

5.2.9.1. Toxicity of Cord Factor

The toxicity of cord factor for mice played a central role in the discovery and characterization of the glycolipid. Nevertheless, after some 30 years, the mechanism of its action still remains to be convincingly elucidated. Bloch recognized a peculiar toxicity in even the crude substances extracted from virulent tubercle bacilli (and from BCG) by petroleum ether: a few repeated intraperitoneal injections into mice of small amounts dissolved in about 0.1 ml paraffin oil ultimately killed most of the animals—particularly C57 Black and

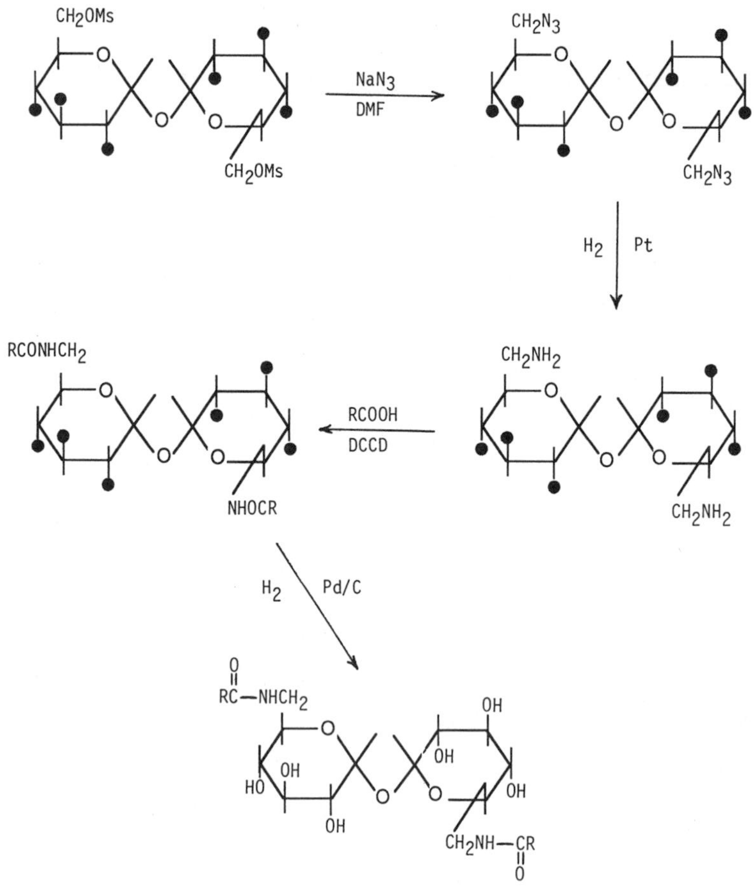

Amide Ψ Cord Factor

RC(=O) = Corynomycoloyl ("C76-amide") **(45)**

Mycoloyl **(46)**

p $C_{16}H_{33}O$-C_6H_4-$(CH_2)_3$C(=O)- **(47)**

● = $-OCH_2Ph$

Figure 5-17. Synthesis of pseudo-cord factors (**45–47**) based on 6,6′-diamino-6,6′-dideoxy trehalose (**44**), i.e., amide ψCF. Some of these exhibit very high toxicity in mice.

Dba mice. With purified material, 5–10 μg injected about 3–4 times was usually lethal. Although single injections of quite large amounts (50–100 μg) were much less effective, the smaller doses given at 2- to 3-day intervals began to kill the mice after about a week, following a precipitous weight loss that is evident even after the first injection. In Bloch's interpretation, the initial exposure(s) sensitized the mice in some manner to the subsequent injections.

The most striking feature seen at necropsy (aside from an intense peritonitis accompanied by multiple adhesions) is what appeared to be acute pulmonary hemorrhage—reminiscent of anaphylaxis on one hand or of an unusual pulmonary "Schwartzman reaction" on the other. Bloch and Noll (1955) sought for evidence in support of a Schwartzman contribution to the pulmonary lesions, but without success. C. Tihon and the present author (unpublished results) were also unable to find such evidence.

We have found recently that pulmonary hemorrhage, albeit less extensive, is elicited in most susceptible mice by even the first intraperitoneal injection of cord factor (Seggev *et al.*, 1982). This is seen principally as isolated hemorrhagic foci but they may be almost confluent in some mice. Without further injections these lesions ultimately resolve with essentially complete healing. In our experience four injections each of 10 µg of highly purified TDM in 0.1 ml mineral oil killed all of the mice (C57BL/6, ~10–12 g wt.) in a group within ten to fifteen days.

We have found further, however (Goren, 1982), that sensitization of the mice for a lethal response to a single small injection (10 µg) of cord factor can be established by several spaced prior injections of 0.1 ml mineral oil alone or, as seen more recently (Liav *et al.*, 1984*b*) and without exhausting the variety of possible protocols, quite similar results were achieved with a single injection of 0.6 ml mineral oil alone followed in 6 days by a single dose of 0.1 ml oil containing the cord factor. Under these conditions, 10 µg TDM killed all the mice within about 15 days; the LD_{50} was found to be 7.5 µg. [In studies by Kato and Asselineau (1971), the LD_{50} for natural cord factor (Aoyama B) injected as a single dose in 0.1 ml Bayol F was 50 µg.] Thus, prior oil injection reduces the glycolipid requirement about six- to sevenfold. This protocol allows a facile determination of the LD_{50} toxicity of a given cord factor or analog, and it is preferred over subjecting the animals to 4 or 5 intraperitoneal injections. Figure 5-18 compares the results obtained with multiple

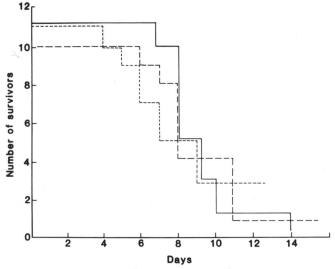

Figure 5-18. Testing cord factor toxicity in mice. Solutions of TDM in oil (100 µg/ml) were injected intraperitoneally according to three protocols. (---) 10 µg TDM on days 0,2,4,6. (————) 0.1 ml oil on days -12,-10,-8,-6; 10 µg TDM on day 0. (—) 0.6 ml oil on day -6, 10 µg TDM on day 0. The "presensitization" with oil reduces LD_{100} from about 40 µg TDM to 10 µg.

Table 5-3. Toxicity of Cord Factors and Pseudo-Cord Factors

6,6'-acyl substituents	$LD_{50}{}^a$ (μg)	$LD_{100}{}^a$ (μg)
Trehalose core (true cord factors and analogues)		
Mycolic acids	7.5	10
Corynomycolic acid (C_{32})	50	75–100
Hydroxyphthioceranic acid[b] (C_{40})	50	>100
Palmitic acid (C_{16})	>100	N.D.
β-O-Methylmycolic acid	10	35
β-O-Tetrahydropyran-2-yl mycolic acid	15	N.D.
α,β-Mycolenoic acid	75	N.D.
Diastereoisomers of cord factor		
Dimycoloylmannosyl mannoside	30	50
Dimycoloylgalactosyl galactoside	>>100	N.D.
Diaminotrehalose core		
Dicorynomycoloyl	<10	<10
Dimycoloyl	N.D.	10–15
Derivatives of "trehalose dicarboxylic acid"		
bis-n-Octadecylamide	>100	N.D.
bis-di-n-Octadecylamide	≈40	≈100
Mycoloyl tetraamide[c]	>>100	N.D.
Aromatic tetraamide[c]	>>100	N.D.
Corynotetraamide[c]	>100	N.D.
Miscellaneous cord factor derivatives		
Peracetylated cord factor	Not toxic	
Permethylated cord factor	Not toxic	
Sulfatide 1 or 3 of *M. tuberculosis*	Not toxic	

[a]Approximate values.
[b]See section on sulfatides.
[c]See Fig. 5-19.

injections of 10 μg TDM; with multiple injections of oil followed by a single cord factor injection (10 μg); and by presensitization with 0.6 ml oil followed by a single TDM injection. The results are substantially the same.

We examined many of our synthetic cord factor analogues and ψ cord factors by the three protocols, but eventually selected the simple presensitization with 0.6 ml oil followed by one-dose elicitation. Some of the results are reviewed in Table 5-3. From these data, certain conclusions may be drawn:

1. For the 6,6'-diesters of trehalose (true cord factors and analogues), increasing molecular weight of the acyl function is reflected in increased toxicity: mycolate > > corynomycolate = hydroxyphthioceranate > > palmitate.

2. Blocking of the free hydroxyls in the trehalose core abolishes activity (peracetylation, permethylation, or cluttering of the core as in the sulfatides).

3. For acyl groups of comparable molecular weight (corynomycolate − C_{32}, and hydroxyphthioceranate, principally C_{40} (see Section 5.4.4.3a),

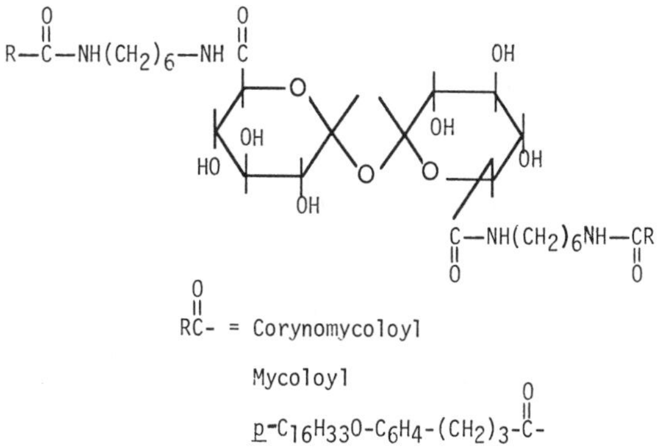

Figure 5-19. Structures of tetraamide pseudo-cord factors developed as a means of increasing the length of lipid substituent attached to "trehalose dicarboxylic acid" (TDA). All were of surprisingly low toxicity in mice. They are prepared by reaction of TDA dimethyl ester with excess hexamethylenediamine and acylating the product.

the *large α-branch* of the former does not enhance toxicity above that of the multiple methyl branches in the latter. It cannot be predicted however that a waxy straight-chain dotriacontanoic acid (C_{32}) diester of trehalose would exhibit the same toxicity as shown by corynomycolate. According to Numata *et al.* (1985), the triacontanoic diester was not toxic, probably because of poor solubility in oil.

4. The toxicity of the mirror amide pseudo-cord factors such as BOA·TDA (*bis-n*-octadecylamide) and BDA·TDA (*bis*-di-*n*-octadecylamide) of trehalose dicarboxylic acid is similar to that of comparably structured trehalose diesters. [But tetraamide pseudo-cord factors, as shown in Fig. 5-19, are remarkably devoid of toxicity. The rationale for their synthesis was discussed by Goren and Jiang (1979). We have speculated (unpublished) that the hexamethylene diamine "spacer" between the TDA and the acyl groups may stabilize a tertiary structure of the glycolipid in which the acyl functions loop over the carbohydrate core and mask it from the environment. We expect (and have some evidence) that other biological activities of these interesting pseudo cord factors may also be abrogated (in progress).

5. The diasteroisomeric cord factors are notably attenuated in toxicity, the galactose analogue much more so than the mannose analogue. These results confirm in the disaccharide analogues the incisive conclusions of Asselineau and Kato (1973) derived from studies with 6-mycoloylmethyl mannoside, alloside, and galactoside.

6. In the esters of trehalose, the β-hydroxyl group in the mycolates seems to have some relevance to toxicity: blocking of the OH by methylation or dihydropyranylation reduces the toxicity modestly. Elimination of the OH to yield the α,β-unsaturated *mycolenoic* acid reduces the toxicity much more significantly. Retention of antitumor activity (see Section 5.2.9.5) by this analogue could have important practical significance. *N*-acyl derivatives of "trehalose amine" (Lemaire *et al.*, 1986), i.e., the amide pseudo-cord factors (Fig. 5-17) seem to have very high toxicity. The dicorynomycoloylamide (**41**) ("C-76" amide, for its content of 76 carbons) is probably more toxic than TDM of *M. tuber-*

tuberculosis. [As a useful shorthand notation, some of the trehalose 6,6'-diesters are referred to by their carbon numbers: viz. the dicorynomycolate is "C-76" (Yarkoni *et al.,* 1978); a dibehenate is C-56; and a dibehenylbehenate is C-100 (Lemaire *et al.,* 1986)]. Whereas "C-76" amide was very toxic in our system, it is noteworthy that Numata *et al.* (1985) found it to be of quite low toxicity when tested as an oil solution emulsified in saline and injected intravenously. The mechanics of presenting the lipids are therefore sometimes of unrecognized importance.

5.2.9.2. Biological Properties: Mechanics

The preceding section describes testing of trehalose esters for toxicity by intraperitoneal injections of the glycolipids dissolved in mineral oil. Vegetable oils are usually less effective, and mice have so far been found to be the only susceptible animals. Rats (Saito *et al.,* 1975) and guinea pigs (J. S. Seggev and M. B. Goren, unpublished observations) seem to be essentially insensitive to the toxic effects, even after high doses of the lipid. Although oil solutions and even thick water-in-oil emulsions may be cautiously injected intravenously into animals (Moore *et al.,* 1972), so far as I am aware this has not been done for cord factor solutions in oil, but only for such solutions dispersed (emulsified) in aqueous saline (Bekierkunst, 1968; Bekierkunst *et al.,* 1969; Moore *et al.,* 1972; Yarkoni *et al.,* 1978; Numata *et al.,* 1985).

The mechanics of presenting TDM to animals, or to cells of the immune system, has a profound influence on the response that is achieved. We adopt and extend the notations used by Lemaire *et al.* (1986) to define the vehicles used for preparing cord factor solutions, dispersions or emulsions as follows:

Cord factor dissolved in mineral oil (TDM/m.o.), in vegetable oil (TDM/v.o.), in squalane (TDM/sqa) or in squalene (TDM/sqe). These are suitable for intraperitoneal injections in mice and usually exhibit the kind of toxicity originally described by Bloch. They are not suitable for intravenous injection.

Important variations derive from emulsifying an oil solution of TDM in saline containing a small amount of Tween 80. These may be prepared by grinding the ingredients in a homogenizer or by sonication, to yield the very fluid oil-in-water type of emulsions rather than the very viscous water-in-oil types exemplified by Freund adjuvant preparations. These may be designated TDM/oil/aq.emul. We know of no important evidence that such emulsions are toxic (e.g., for mice) when injected *intraperitoneally.* Instead, presented in this manner they may exhibit important immunostimulant and antitumor properties (Sukumar *et al.,* 1981; Numata *et al.,* 1985). By contrast, depending on how the emulsion is prepared, granuloma formation, immunostimulation, or overt toxicity can be elicited by intravenous injection of these aqueous emulsions of TDM dissolved in oils. Although the effects are undoubtedly exacerbated by the TDM contents, still formation of oil emboli in lung capillaries probably contributes to toxicity and granulomagenicity, which depend in part on the size of the oil droplets and on the quantity of oil in the emulsion

Table 5-4. Influence of Preparative Methods on the Effects of Glycolipids/Oil/Aqueous Emulsions in Mice

Composition[a]	Method of preparation	Results	Reference
2 μg TDM/1% m.o.	Grind	Maximum lung granulomas	Yarkoni and Rapp (1977)
	Ultrasonicate	No gransulomas	Yarkoni and Rapp (1977)
10 μg TDM/1% m.o.	Grind	No granulomas	Yarkoni and Rapp (1977)
3% m.o.	Grind	Few granulomas	Yarkoni and Rapp (1977)
10% m.o.	Grind	Maximum granulomas	Yarkoni and Rapp (1977)
1 mg C-39[b]/10% m.o.	Grind	10/10 mice died	Yarkoni et al. (1979)
	Ultrasonicate	0/10 mice died	Yarkoni et al. (1979)
100 μg TDM/10% m.o.	Grind	10/10 mice died	Yarkoni et al. (1978)
150 μg TDM/9% m.o.	Grind	9/10 mice died	Numata et al. (1986)
150 μg TDM/10% m.o.	Ultrasonicate	Protects mice against *K. pneumoniae* or *L. monocytogenes*	Parant et al. (1978)

[a]Usually 0.2 ml emulsion was injected i.v.; m.o. = mineral oil.
[b]C39: methyl-6-*O*-(2-tetradecyl-3-hydroxyoctadecanoyl)-α-D-glucopyranoside.

(Yarkoni and Rapp, 1977, 1978). Certain antitumor effects are also dependent on these variables (Yarkoni *et al.*, 1977, 1978; Yarkoni, 1982).

It is regrettable that the importance of these variables was not recognized earlier (and may still be ignored in practice). Undoubtedly many complexities, indeed major discrepancies, in results obtained with glycolipids/oil/ aqueous emulsions may be largely attributable to physical qualities of the emulsions, which we suspect may be very difficult to control. Reliable information that is available stems almost solely from the studies by E. Yarkoni (Yarkoni and Rapp, 1977, 1978, 1979; Yarkoni *et al.*, 1976, 1977a,b, 1978, 1979; Yarkoni, 1982). A few of the sometimes disparate results are reviewed in Table 5-4.

Briefly summarized, for emulsions injected intravenously in mice, minute amounts (2–10 μg) of TDM or other active glycolipids in small (1%) or large (9%) amounts of oil emulsified by grinding to produce large oil droplets were potently granulomagenic in the lungs of the recipient mice. Similar preparations emulsified by ultrasonication and giving very small oil droplets are neither granulomagenic or toxic. These are probably more efficiently trapped in liver and spleen. Large amounts of TDM (100–200 μg) in 10% oil and emulsified by ultransonication are immunostimulatory, protecting against i.v. challenge with various pathogens, but when prepared by grinding are lethally toxic. It is noteworthy that our C-76 amide (**41**) (Fig. 5-17), which was found extremely toxic on i.p. injection of oil solutions (see Table 5-3), exhibited very low toxicity when injected i.v. as an oil-in-saline emulsion (150 μg in 9% oil) (Numata *et al.*, 1985).

Cord factor can also be satisfactorily dispersed in aqueous media directly, without first dissolving in oil, and even in the complete absence of stabilizing agents as shown by Kato (1967b), by Kierszenbaum and Walz (1981), and by Retzinger *et al.* (1981). When injected intraperitoneally, as in mice, these dispersions of neat cord factor do not exhibit the toxic qualities of the oil solutions presented in the same manner. They are essentially nontoxic even at very high doses, and instead can activate effector cells of the immune system (Lepoivre *et al.*, 1982; Orbach-Arbouys *et al.*, 1983; Numata *et al.*, 1985; Sharma *et al.*, 1985). If given intravenously they can exhibit impressive adjuvant and immunostimulant properties (reviewed by Lemaire *et al.*, 1986).

Finally, TDM may be adsorbed on polystyrene-divinylbenzene beads, by evaporating solutions of the glycolipid in organic solvents onto the beads. Dispersed in aqueous media, they can be injected into mice either i.v. or i.p., sometimes with toxic and sometimes with immunostimulant consequences (Retzinger *et al.*, 1981, 1982). Accordingly, the manner in which the TDM is presented is invariably crucial to the response that is achieved. Indeed, these variables contribute enormously to the seemingly bewildering spectrum of biological responses that has been reported for TDM. And it is equally crucial to dissection of the activities that the manner of presentation of the glycolipid be clearly, accurately and unequivocally documented.

The assembly of a comprehensive summary of the various activities that have been ascribed to cord factor and analogues is a painstaking task that has recently been completed by Lemaire *et al.* (1986). Their review and the various tables therein are recommended to the interested reader. Owing to the enormity of this task, it is not surprising that some sections regrettably contain a number of errors. These are mentioned, with sincere apologies to the authors, only to alert the reader who may need a rigorous treatment of the topic. Not being without similar sins (see especially Section 5.4.4.4, which deals with structures of the mycobacterial sulfatides), it is hardly my intention to cast stones!

The discussion by Lemaire *et al.* (p. 250) "IV. How to Use Trehalose Diesters" is introduced by a section titled "Water-in-oil Emulsions." This is very likely a "Freundian slip." Indeed, almost none of the examples in which TDM or the like are utilized employ these kinds of heavy viscous emulsions. Instead they involve solutions of TDM in oil that have been emulsified into aqueous saline by grinding or sonication to give mobile aqueous dispersions.

The statement (p. 251) "TDM/m.o. injected i.v. into mice," etc. is misleading in conveying the notion that neat mineral oil solutions of the glycolipid were injected to reveal that TDM was "an efficient immunomodulator." In the studies to which these statements allude, aqueous emulsions of TDM dissolved in mineral oil were used—for i.v. and i.p. injections (Bekierkunst *et al.*, 1969). In their Table III (p. 252), the vehicles for presenting TDM to mice are similarly inadequately described: it is certain that all of the oily TDM solutions (mineral oil, vegetable oil, or squalane) that are described as being injected either intraperitoneally or intravenously are given as aqueous emulsions of the oily solution. In short, a footnote added to this table and stating that the injections were given as oil-in-water emulsions would unambiguously define the vehicle for the entries on their lines 1, 2, 3, 5, 7, 8, and 21. Their Table III is important for illustrating induction of nonspecific resistance in mice by TDM to a host of pathogenic bacteria, parasites and viruses: *Salmonella typhi, S. typhimurium, Klebsiella pneumoniae, Listeria monocytogenes, Mycobacterium tuberculosis, Schistosoma mansoni, Babesia microti, Trypanosoma cruzi, Plasmodium berghei, Mesocestoides corti,* and an influenza virus. The list is not exhaustive.

5.2.9.3. Mechanisms in Cord Factor Toxicity

This topic has been reviewed at length in a number of publications (Goren, 1975; Asselineau and Asselineau, 1978; Goren, 1982) and so the earlier studies are summarized only briefly. The reviews by Lederer (1976, 1979; Lemaire *et al.,* 1986) have usually been more concerned with the chemistry and the immunostimulant properties of the glycolipid. Mechanisms that have been invoked to account for cord factor toxicity include a stimulation of mammalian NADase activity to deplete NAD levels in lung, liver, and spleen (Artman *et al.,* 1964; Bekierkunst and Artman, 1966); depression of muscle and liver glycogen synthesis (Shankaran and Venkitasubramanian, 1970); and in a prodigious series of biochemical studies with mitochondrial preparations *in vitro,* Kato (1967a, 1968, 1970) recognized severe derangement in the function of these organelles under the influence of cord factor. These involved loss of respiratory control, deranged electron transport, and inhibition of oxidative phosphorylation. The results appeared to be attributable to a direct attack by cord factor on mitochondrial membranes with resulting swelling and distortion and fragmentation of cristae as seen by electron microscopy (Kato and Fukushi, 1969). Quite similar results were obtained with either liver mitochondria preparations from mice injected with cord factor-in-oil or with healthy liver mitochondria exposed *in vitro* to aqueous finely dispersed cord factor. It is certain, however, that the attack by cord factor is not merely the random assault of a detergent on membranes. The consequences are much more specific. The techniques developed by Kato have currently been adopted by the French investigators at Toulouse, to complement synthetic and physicochemical studies on cord factors and analogues in order to gain additional insight on possible mechanisms of action and on the influence of structure on activity.

5.2.9.3a. Structural Requirements.

To quote from C. and J. Asselineau (1978), "As the structure of cord factor is relatively simple, it was possible to synthesize cord factor analogues and thus to study the relationships between *in vivo* toxicity, *in vitro* inhibition of mitochondrial oxidative phosphorylation, and chemical structure." Indeed, these considerations inspired a fruitful collaboration between J. Asselineau and M. Kato that lasted until Kato's tragic and untimely death in 1976. They found *in vivo* toxicity, in parallel with *in vitro* activity of cord factor analogues on mitochondria to be very sensitive to structural changes in the glycolipid molecule: substitution of sucrose or glucose for trehalose resulted in considerable attenuation of both *in vivo* and *in vitro* activities. Acylation at the 6-position of the carbohydrate is stringently required; and the stereochemical disposition of the hydroxyl groups at positions 2, 3, and 4 (as studied with half-molecules, e.g., 6-mycoloylmethylglycosides) critically affected both kinds of activities (Asselineau and Kato 1973, 1975). Whereas the 6-mycoloylglucoside has about one-third the toxicity of cord factor, the mannoside analogue is about one tenth as toxic and the galactoside is essentially inert. Our comparable diastereoisomeric analogs of *trehalose* (described in Section 5.2.7.4) permitted a more precise comparison of the toxic effects, since the severe attenuation already characteristic of the half-molecules was avoided. The results (Table 5-3) confirmed the earlier conclusions of Asselineau and Kato: toxicity of the mannose–mannose analogue was considerably reduced, and the galactose–galactose isomer was almost inert. Attempts to correlate this *in vivo* behavior of these diastereoisomers with their effects on isolated rat liver mitochondria have not yet given unequivocal results (G. Lanéelle and M. Goren, unpublished data). However, the consequences of an even more subtle structural alteration of the glycolipid molecule were dramatic: 6,6'-dicorynomycololyl-β,β'-trehalose lost most of the activity shown by the α,α analogue for uncoupling of mitochondrial oxidative phosphorylation (Gillois *et al.*, 1984).

A cautious interpretation suggests then that the association between these spatial orientations and the expression of toxicity on one hand and attack on mitochondria on the other bespeaks more than the random attack of a surface active molecule on a membrane. Rather, it may reflect much more specific interactions between the glycolipid hydroxyls and specific enzyme sites targeted for attack. In support of this interpretation, it is noteworthy that the *in vitro* inhibition of oxidative phosphorylation by cord factor and by closely related analogues was found to affect only so-called coupling site II–the span of the mitochondrial respiratory chain from the cytochrome *b*–coenzyme Q complex to cytochrome *c*. Site I was only slightly affected. But site III phosphorylation was left entirely intact. According to Asselineau *et al.* (1981), whereas a moderate effect at site I has been "noted with hydrophobic substrates," the magnitude of site II uncoupling appears to be a distinct feature of cord factor activity, hence the quality of specificity. [For detailed analyses of these interpretations, see Kato *et al.* (1971). For the influences of glycolipid structure on mitochondrial respiration, see Asselineau *et al.* (1981).] It is also noteworthy that a number of glycolipid substances, simple as well as more complex, also physically attack mitochondria, with drastic consequences on oxidative phosphorylation, but at both coupling sites II and III. Moreover, these agents rarely exhibit toxicity *in vivo*. As a paradigm, a principal sulfatide of *Mycobacterium tuberculosis* (SL-I, Section 5.4) was known to be completely innocuous *in vivo*—even in doses exceeding several hundred micrograms given to mice intraperitoneally in mineral oil. And yet this sulfated trehalose glycolipid disrupted oxidative phosphorylation *in vitro* even more vigorously than cord factor—and at both coupling sites (Kato and Goren, 1974). The paradox of *in vitro* inertness in the face of a vigorous mitochondrial attack was resolved when it was recognized that the strongly acidic sulfatide is almost completely neutralized *in vitro* (probably complexed) by bovine serum albumin (BSA) so that it no longer attacked mitochondrial membranes. And by extension, it is reasonable that the neutralization would occur with various serum proteins *in vivo*. By contrast, *in vitro* activity of cord factor is not altered by BSA, and it is, of course, toxic *in vivo*.

A number of simple straight chain fatty acid (C_{14}–C_{22}) esters of α-methylglucoside also mount a vigorous random attack on mitochondria *in vitro* that also affects both coupling sites II and III (Kato *et al.*, 1971). They are also essentially nontoxic *in vivo*. BSA antagonizes their *in vitro* activity, albeit to a lesser extent than in the case of the strongly acidic sulfatide. But it seems likely that their *in vivo* inactivation may be promoted through hydrolysis of these low-molecular-weight detergent-like molecules by tissue enzymes (Kato *et al.*, 1971).

Studies on liquid–gel phase transitions and on the influence of the organizational state of the glycolipids in antagonizing oxidative phosphorylation showed that the fluid state of the dispersion was the most active (Durand *et al.*, 1979*a,b*). High-molecular-weight condensed (not fluid) cord factors, however, "melt" on interaction with phospholipid membranes (Lanéelle and Tocanne, 1980) and thus acquire the more active state.

The French investigators therefore postulate that TDM binds nonspecifically to the mitochondrial membrane lipid matrix and then diffuses within the lipid phase, disturbing its organization. Cord factor could then interact with components of the phosphorylation system, the interactions depending upon the (correct stereochemical) nature of the carbohydrate moiety. (Contrast the behavior of the β,β′-trehalose analogue.) They also suggest that "effects on respiration could result from nonspecific dislocation of the lipid matrix of the membrane (coupled with) specific interactions with proteins" (Asselineau *et al.*, 1981; Gillois *et al.*, 1984).

Although Numata *et al.* (1985) tacitly accept this sequence of events as a valid interpretation of how cord factor may interact with mitochondrial membranes, they nevertheless argue that this mechanism "seems not to be sufficient to account for the death of the mice"—the ultimate expression of cord factor intoxication; it should be noted that the death of the mice is invariably associated with confluent pulmonary hemorrhages. It is curious that Kato never commented on this pathology.

5.2.9.3b. Behavior of TDM Adsorbed on Hydrophobic Beads. A provocative hypothesis concerning mechanisms in cord factor toxicity, in expression of several other biological activities, and that may account for the pulmonary hemorrhages, was recently proposed by Retzinger and colleagues. In their pioneering studies, Retzinger *et al.* (1981) recognized first that at low surface concentration of TDM at an air–water interface, "much of the lipid–water interface consists of hydrophobic domains to which proteins readily adsorb." A fruitful model system that was developed consisted of hydrophobic polystyrene-divinylbenzene beads (5-μm diameter) onto the surface of which various amounts of TDM were coated by evaporating appropriate solutions. Beads coated with small amounts of TDM were readily dispersible into uniparticulate suspensions in phosphate-buffered saline (PBS) or in dilute protein solutions of bovine serum albumin (BSA). Beads coated with sufficient

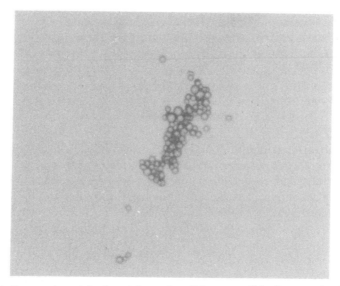

Figure 5-20. Presumed model of cord formation. Polystyrene-divinylbenzene beads, 5 μm in diameter, were coated with SL-I, the principal sulfatide of *Mycobacterium tuberculosis* and dispersed in saline containing low levels of Ca^{2+}. Agglomeration was spontaneous. A similar phenomenon was first reported for beads coated with TDM by Retzinger *et al.* (1981).

TDM to form a collapsed monolayer on the bead surface are dispersible in PBS but quickly coalesce into small clumps. Beads coated with a large excess of TDM immediately formed large aggregates—even in the presence of considerable protein—suggestive to the authors of cord formation in tubercle bacilli. And indeed the coated beads avidly adsorb proteins, like BSA. In their interpretation, at equilibrium spreading, most of the cross-sectional area of TDM consists of a hydrophobic domain in direct contact with water and the adsorption of protein is attributable to this feature. The amount of protein adsorbed is somewhat diminished when the beads are coated with an excess of TDM over that required for forming a collapsed monolayer on the bead surface. This excess is held in the form of large cylindrical micelles; the micellar form adsorbs protein poorly. Moreover, since such beads form large aggregates, it was inferred that the micelles *link* the beads and, by analogy, probably form the links between adjacent *M. tuberculosis* cells during cording. [It is noteworthy, however, that in our experience, beads similarly coated with sulfatide from *M. tuberculosis* also aggregate to form "pseudo cords" in PBS containing a small amount of calcium ion (see Fig. 5-20).]

Many of the biologic effects of TDM can be reproduced by monomolecular films of the glycolipid on the surface of the hydrophobic beads; e.g., adjuvant, inflammatory and granulomagenic activities; and induction of pulmonary hemorrhages (Retzinger *et al.*, 1982). But the cylindrical micelles that TDM forms in aqueous media are biologically inert. The latter claim should be accepted with caution: aqueous dispersions of neat TDM show important immunostimulant properties. It may be argued, however, that the activity is due to nonmicellar TDM in these dispersions!

These studies indicated that several activities of the beads were greatest when they were coated with TDM near the equilibrium spreading density, which is 2.5×10^{-3} molecule/Å^2. According to Retzinger, the biological effects of the TDM-coated beads are intimately associated with the ability of the monolayer to bind *fibrinogen* (my italics) almost to the exclusion of other plasma proteins. And fibrinogen emerges as a necessary cofactor of the TDM monolayer for expression of the biological activities. The scenario envisaged is as follows (Retzinger, 1984):

The TDM monolayer orients the adsorbed fibrinogen in a fashion that facilitates the thrombin-catalyzed formation of fibrin but with preferential release of fibrinopeptide B, a potent macrophage and neutrophil chemoattractant, with resulting recruitment of these inflammatory cells. Release of the B peptide enhances the stability of fibrin strands to attack by proteases. Retzinger suggests that this greater stability may sustain the local inflammation and thus promote granuloma formation, with all the associated consequences. The pathologies that Bloch reported following repeated i.p. injections of cord factor in oil are evidently reproducible on multiple injections of TDM-coated beads. The pulmonary hemorrhage may be explained by an elevated level of thrombin activity that is generated from inflammatory cells contacting the TDM-beads and that elevated thrombin can account for reduced coagulability of the blood and therefore of hemorrhaging (presumably in the vicinity of the TDM-coated beads or near TDM–oil emboli). This behavior may be exacerbated by the documented anticoagulant effects of fibrinogen–fibrin degradation products (Retzinger, 1984).

Although these interpretations are unquestionably persuasive, it is nevertheless paradoxical that not only are animal species other than mice entirely resistant to cord factor intoxication but that various strains of mice too are resistant, including nude mice (Seggev *et al.*, 1984), and of the susceptible mouse strains, even somewhat older mice, e.g. 10 weeks of age (see Retzinger *et al.*, 1982), are already resistant.

Most recently, a fascinating and provocative alternative interpretation was suggested to account for the toxic effects of cord factor. Silva and Faccioli (1988) proposed that cord factor stimulates macrophages to secrete cachectin—tumor necrosis factor (TNF)—and that it is the "complex repertoire of metabolic reactions" that this cytokine elicits that accounts for the various expressions of cord factor intoxication. In accord with their interpretation, TNF was found in the plasma of mice injected i.p. with TDM-in-oil solution. This was assayed by measuring cytotoxicity of the plasma to L929 cells in culture and its antagonism by specific antibody to TNF. Moreover, transfer of the plasma to recipient control mice elicited the usual CF-intoxication symptoms in the recipient: weight loss, diarrhea, and interstitial pneumonitis. These symptoms were also abolished if the plasma was pretreated with antibody to cachectin.

As a personal comment, I must note that my good friend, the late Masahiko Kato, told me that his own studies on cord factor toxicity were largely prompted by his desire to understand why so many of his tuberculosis patients showed wasting and emaciation, thus leading to the biochemical studies on the influence of cord factor on mitochondrial function. Silva and Faccioli were similarly stimulated to understand the cachexia in tuberculosis patients, and thus now speculate that the liberation of large amounts of cachectin by CF stimulation may indeed be sufficient to explain these catastrophic consequences (of the disease).

5.2.9.4. Transition: The Primed Macrophage

Even from the earliest studies, evidence has been accumulating that supports a role for macrophages in the expression of the immunostimulating properties of TDM. Lemaire *et al.* (1986) extensively review the involvement of macrophages and possibly of other cells as the potential targets of TDM. The chemotaxigenic or chemotactic properties of TDM for macrophages have been well documented (Kelly, 1977; Ofek and Bekierkunst, 1976) and very probably contribute to the granulomagenic properties described earlier. According to Retzinger *et al.* (1981, 1982), the special reactivity of TDM for fibrinogen may also contribute to a sustained generation of substances that are chemoattractive for macrophages. And activation of the alternative complement pathway, noted by Ramanathan *et al.* (1980), would also contribute to influx of monocytes/macrophages. It is especially interesting that i.p. injection of TDM/w dispersions (see Section 5.2.9.2) in mice results in an unusual *renewal* of the peritoneal population (Lepoivre *et al.*, 1982) in which peroxidase-positive (therefore newly recruited) macrophages become predominant (45%) in the cell population but without simultaneously eliciting a strong inflammation with its marked increase in cell numbers. Although these elicited cells were described in early reports as tumoricidal, various studies by the French groups (Lepoivre *et al.*, 1982; Orbach-Arbouys *et al.*, 1983; Reisser *et al.*, 1984) demonstrated that these recruited macrophages, when assayed *in vitro* in the *absence of endotoxin* (lipopolysaccharide, LPS) were unable to limit growth of, e.g., P815 mastocytoma cells. The effect of TDM *in vivo*, therefore, is not to produce "fully activated" macrophages, but to prime them to respond *in vitro* to a second agent (low dose of LPS), which provides the final stimulus to "antitumor competence" (Grand-Perret *et al.*, 1986). This resembles the earlier findings by Ruco and Meltzer (1978) in which peroxidase-positive mononuclear cells elicited in a sterile inflammation (by starch, latex beads, fetal calf serum) were in a primed state in which they responded to lymphokines to develop tumoricidal capacity. As impressive as these findings are, still, only rarely is cord factor alone effective in eradicating an established tumor in an experimental animal.

5.2.9.5. Antitumor Activities of Cord Factor: History and Current Status

In a series of heuristic investigations the group of Herbert J. Rapp developed a guinea pig model for tumor immunotherapy (Rapp, 1973) that is still being exploited at this writing. In the culminating successful effort, a hepatoma induced in a strain-2 guinea pig with diethyl nitrosamine proved to be highly transplantable; of stable, low antigenicity; and highly metastatic and virulent. Converted into an ascites form it is easily harvested and standardized as a single cell suspension for immunotherapy studies. When 10^6 of these line 10 cells are injected intradermally into syngeneic recipients, a tumor nodule is formed which metastasizes within about 6 days, grows progressively, and kills the animal in about 8–9 weeks. However, as found by the Rapp group if viable *M. bovis* BCG was admixed with the tumor inoculum, or if BCG was injected into the established tumor at a sufficiently early stage, the tumor and metastases were eradicated in most of the animals and they were rendered immune to a second challenge with line 10 cells (Zbar *et al.*, 1972; Bartlett and Zbar, 1972).

The events that are triggered in the tumor environment were analyzed by Bast *et al.* (1976). An intense chronic granulomatous inflammation is produced by the infection with BCG with subsequent development of activated macrophages. The abundant and persistent mycobacterial antigens stimulate lymphocytes to inhibit growth of tumor cells by direct contact or by release of lymphokines. The latter embrace factors that activate macrophages nonspecifically to a cytotoxic or cytostatic level for neoplastic but not for normal cells. Thus, at first, tumor cells are killed as "innocent bystanders." But macrophage processing of tumor antigens, amplified by the adjuvant qualities of the mycobacterial components (e.g., cord factor, muramyl dipeptide, Wax D) adds a specific dimension to the immunologic events. This is believed to provide the tumor-specific systemic immunity that ultimately destroys the metastatic foci.

The recognition that cord factor might be endowed with antitumor properties is appropriately attributable to pioneering studies by A. Bekierkunst (1968) on one hand and to E. Ribi* on the other. Bekierkunst elicited pulmonary granulomas in mice by intravenous injections of TDM-in-oil solutions emulsified into saline. He drew the analogy that these resembled the lung lesions induced by viable intravenous BCG (Bekierkunst *et al.*, 1969). Despite alternative interpretations by Barksdale and Kim (1977), by Moore *et al.* (1972) and by Merrill Chase (personal communication) that the lesions induced by TDM/oil were more like foreign body granulomas, still the cellular reaction induced some protection against an intravenous challenge with *M. tuberculosis*. Moreover, elicitation of the granulomas in mice also induced a level of protection against induction of certain low-grade pulmonary malignancies (adenomas) following challenge of the animals with urethane (Bekierkunst *et al.*, 1971); whereas intraperitoneal injection of the cord factor-in-oil aqueous emulsion protected mice to some extent against proliferation of subsequently injected Erlich ascites cells. This antitumor behavior seems to have been sought on the basis of the inflammatory granulomagenic reactions that Bekierkunst recognized—probably even in his earliest studies on cord factor toxicity—alluded to before (Artman *et al.*, 1964; Bekierkunst and Artman, 1966). Indeed, intratumor injection of similar aqueous emulsions of cord factor-in-oil are entirely effective in completely eradicating intradermal syngeneic transplants of a murine fibrosarcoma (1023) originally established in C3H/HeN mice with methyl cholanthrene (see especially Yarkoni *et al.*, 1977*b*,

---

*Deceased, 1986.

1979*b*). However, these preparations are ineffective in treatment of a variety of other tumors, e.g., the line 10 guinea pig tumors.

Elicitation of the kind of granulomas more akin to those induced by viable BCG was achieved by E. Ribi and colleagues with a mineral oil–mycobacterial cell wall "vaccine": cell walls (as of BCG) were wet with a little mineral oil and emulsified by grinding into aqueous saline containing a small amount of Tween 80. This vaccine, when injected intravenously into mice or even subhuman primates elicits extensive splenic and pulmonary granulomas. And the vaccine confers a high degree of protection against an aerosol challenge with virulent *M. tuberculosis,* much more than achieved with TDM/oil alone (Barclay *et al.,* 1967; Ribi *et al.,* 1968, 1971). Delipidation and deproteinization of the cell wall to yield "cell wall skeleton" leaves a product from which most of the granulomagenic activity (undesirable in a tuberculosis vaccine) has been abolished, but with a concomitant major loss in immunogenic activity. Both were restored in parallel if the cell wall skeleton was combined with some tuberculoprotein and especially with some cord factor before preparing the oil–wet mixture and emulsifying (Anacker *et al.,* 1973). Either the intact cell wall–oil vaccine or that prepared with cord factor and CW skeleton was a more than adequate substitute for viable BCG in the line 10-strain-2 guinea pig model, and in other systems (Meyer *et al.,* 1974; Gray *et al.,* 1975; Ribi *et al.,* 1976*a,b*; Schwartzman *et al.,* 1984). Other developments followed apace: all of the following, wet with mineral oil and emulsified in Tween–saline were effective: nonliving BCG with TDM (Bekierkunst *et al.,* 1974); deep rough (Re) mutant endotoxins (for preparation, see Ribi *et al.,* 1979) of various *Salmonellae* species coated with TDM (McLaughlin *et al.,* 1978*a,b*; Ribi *et al.,* 1975; 1976*a–c*); endotoxins from other gram-negative organisms (Ribi *et al.,* 1976*b*); and from *Serratia* (Ribi *et al.,* 1980). Currently combinations of TDM, CW skeleton plus mutant endotoxin are described as the most effective (McLaughlin *et al.,* 1978; Ribi *et al.,* 1981). Wild-type endotoxins, with their full complement of hydrophilic O-antigens have not been (and evidently cannot be) used successfully in these vaccines.

The evidence is persuasive that a principal function of cord factor in these active preparations is in promoting the association of the second agent (CW skeleton, endotoxin) with the oil phase (McLaughlin *et al.,* 1978; Ribi *et al.,* 1976*d,* 1981). Thus, polar–polar interactions of TDM with either substance promotes adsorption of the glycolipid with a concomitant outward orientation of the hydrophobic acyl chains. It is this hydrophobic "paint" that probably facilitates association of the cell wall skeleton or the endotoxin with the oil phase—even after emulsification. The association with oil phase is evidently essential so as to localize the inflammatory substances in paracortical areas of regional lymph nodes where a brisk granulomatous reaction is elicited; e.g., "cell walls that are not attached to oil droplets pass through the same nodes, producing little change in their architecture" (Bast *et al.,* 1976).

5.2.9.5a. Therapy for Autochthonous Tumors. Various of these combinations have already found practical exploitation in treatment of autochthonous tumors in domestic animals: "cancer eye" in cattle (bovine squamous cell carcinoma), and equine sarcoid, described as the most common tumor found in horses. Schwartzman *et al.* (1984) described a proprietary combination of mycobacterial cell wall skeleton (CWS) with TDM (Ribigen: Ribi ImmunoChem) which has proven very effective in treatment of these spontaneous tumors. In cancer eye of cattle, cure rates of 73% have been achieved in the treatment of tumors <7 cm in diameter; equine sarcoids were treated with an overall regression rate of 90%, with minimal incidence of recurrence (5%). Studies are underway on equine melanomas and canine mammary tumors and adenomas.

Phase I–II studies have also been conducted in humans with some levels of success. Richman *et al.* (1978) injected a saline emulsion of CWS/TDM/m.o. into a total of 99 tumor nodules in 23 patients with metastatic melanoma and carcinoma of the breast; 34% of all injected nodules resolved, most particularly in nodules ≤1 cm in size. Response correlated with immunocompetence of the patient and with minimal tumor burden. Toxicity (not serious) was manifested in ulceration, fever, and pain. In other phase I studies, Vosika and Gray (1983) reported that i.v. injection of saline suspensions of CWS combined with TDM were of only moderate toxicity (chills, fever at doses of 1 mg CWS/m^2) but without antitumor effects. On the other hand, weekly intravenous injections of aqueous suspensions of oil-coated CWS/TDM were of like toxicity as the oil-free dispersions, but they achieved complete regression of a bronchial squamous cell carcinoma in one of three patients receiving the maximum 2 mg CWS/m^2 (Vosika *et al.*, 1984).

5.2.9.5b. Potentials for Improvements. The immunotherapeutic approaches described immediately preceding made no mention of the promising Re mutant endotoxin preparations that were discussed earlier. They were not used because of potentially severe synergistic toxic effects with other vaccine components. Cord factor and mycobacterial cell wall constituents (e.g., muramyl dipeptide) are well known to enhance susceptibility to endotoxin (Suter and Kirsanow, 1961; Parant *et al.*, 1977; Yarkoni and Rapp, 1979; Ribi *et al.*, 1979). Humans, horses, dogs, and rabbits are highly susceptible to the toxic effects of endotoxin; therefore, its potential for clinical application (especially in humans) has not been realized (Ribi *et al.*, 1984). A "refined" (but still highly toxic) endotoxin can be prepared by extracting purified cell walls of heptose-deficient (Re) mutants of *Salmonella typhimurium* rather than whole cells. This reduces protein and nucleic acid contamination, and facilitates subsequent chromatographic purification. On weak acid hydrolysis the product undergoes a well-known transformation—loss of its complement of ketodeoxyoctulosonic acid (KDO) to yield, in this instance, a more homogeneous "lipid A" than is ordinarily obtained. It is a diphosphorylated multiacylated diglucosamine, still of high toxicity. However, careful acid hydrolysis converts this to monophosphoryl lipid A (MPL) by loss of the more labile phosphate at the C-1 position (Ribi *et al.*, 1984). This product is remarkably free of toxicity—at least 1000 times less lethal and pyrogenic for rabbits than diphosphorylated lipid A or the parent glycolipid. A single injection of 150 µg endotoxin can be lethal to a horse, whereas 20,000 µg of MPL have no detectable adverse effects (Ribi *et al.*, 1985). Phase I drug trials in humans have shown that MPL can be safely administered in doses of 100 μg/m² (Vosika et al., 1984). Like the parent glycolipid, MPL retains the ability to synergistically enhance the antitumor activity of mycobacterial cell wall skeleton. The triple mixture of CWS, TDM and MPL is very effective in the treatment of the bovine ocular squamous cell carcinoma and equine sarcoid (E. Ribi, personal communication). Cyclophosphamide, to inhibit T-cell-induced suppression is a beneficial adjunct. It is expected that the inherently weak responses against autochthonous tumors may be enhanced significantly by the ability of the MPL to activate macrophages, and to stimulate production of various lymphokines, e.g., interleukin-1 (IL-1), colony stimulating factor (CSF), interferon (IFN), hence that the MPL will find important applications for immunotherapy of certain human tumors (Ribi *et al.*, 1985).

5.2.10. Trehalose Monomycolates

In Section 5.2.1, two additional mycolate esters of trehalose were mentioned briefly and their gross structures given: "trehalose monomycolate" (TMM), i.e., 6-mycoloyltrehalose (**2**) (Fig. 5-1), discovered by Kato and Maeda (1974) and 6-mycoloyl-6'-acetyl trehalose (MAT) (**4**) (Fig. 5-1) discovered by Takayama and Armstrong (1976). Both glycolipids have been putatively implicated in mycobacterial cell wall biosynthesis (Takayama and Armstrong, 1976; Takayama and Qureshi, 1984). Although MAT has not been recognized in other studies, TMM has been more frequently encountered in an evergrowing spectrum of mycobacteria (see, e.g., Dhariwal *et al.*, 1984) and including even *M. leprae* (Dhariwal *et al.*, 1987). Accordingly, this trehalose glycolipid is evidently truly ubiquitous in this genus.

Trehalose-6-mycolate was purified by Kato and Maeda (1974) from Wax D of *M. tuberculosis* by chromatography on magnesium silicate-Celite (recovered in the more polar fractions), followed by chromatography on silica gel and again on magnesium silicate-Celite. Careful elution of "tail fractions" with ether–methanol afforded purified TDM, mixtures of TDM with TMM, and finally "homogeneous" (see below) TMM. Chemical analyses including permethylation under nonisomerizing conditions, revealed its structure as 6-mycoloyltrehalose. Kilburn *et al.* (1982) exhaustively extracted *M. smegmatis* with chloroform–methanol (C/M) (2 : 1), obtained a partially purified preparation by silica gel preparative TLC, and further purified the TMM by column chromatography on Sephadex LH 20.

In an improved method a concentrate of the lipids extracted with C/M was added to cold acetone to precipitate a product containing both TDM and TMM. The trehalose glycolipids were enriched when this precipitated material was dissolved in C/M (4 : 1), and passed through a column of DEAE cellulose (acetate form) with chloroform/methanol/water (2 : 3 : 1) as the eluting solvent: the TMM and TDM fractions eluted from the column in the wash. This product was then fractionated on a Bio-Sil HA column (325 mesh) using a gradient of 0–35% methanol in $CHCl_3$ to separate and recover both TDM and TMM. The monomycolate was purified further by reverse-phase

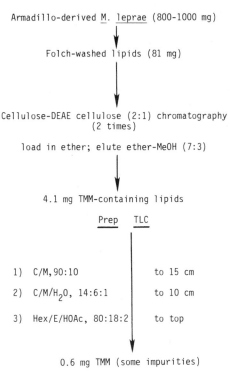

Figure 5-21. Isolation of trehalose monomycolate from *Mycobacterium leprae* recovered from infected armadillo tissue.

high-performance liquid chromatography (RP–HPLC) (Takayama and Sathyamoorthy, 1986; Kilburn *et al.*, 1982 and below).

Promé *et al.* (1976) isolated both TMM and TDM from crude Wax A (see Asselineau, 1966) of *Mycobacterium phlei* (see Section 5.2.1), which was chromatographed on a column of Silicar CC7 (Mallinckrodt). TMM was eluted with $CHCl_3$ containing 10–15% CH_3OH, and it was rechromatographed on DEAE cellulose (acetate form). Elution with C/M (1 : 1) afforded the homogeneous monomycolate (the "homogeneity" is of course compromised by the varieties of mycolate substituents and by homology).

Dhariwal *et al.* (1987) isolated and identified trehalose-6-mycolate from harvests of armadillo-derived *Mycobacterium leprae,* as well as from *M. lepraemurium* harvested from livers of infected mice. Purified lyophilized *M. leprae* recovered from irradiated infected armadillo livers (Draper, 1979; Hunter and Brennan, 1981) were extracted with 2 : 1 (v/v) chloroform/methanol at 50°C for 15–18 hr, partitioned according to Folch *et al.* (1957) and "crude *M. leprae* lipids" recovered from the $CHCl_3$ phase. TMM was recovered and purified as summarized in Fig. 5-21. To remove stubbornly persistent contaminants from the isolated product it was purified further by HPLC (in the laboratory of K. Takayama, and as described above). About 400 µg of product was obtained from 1 g lyophilized bacilli. The structure was established by infared spectrophotometry, by Cf-252 plasma desorption mass spectrometry and by permethylation analysis. Identification of TMM in *M. lepraemurium* (Dhariwal *et al.*, 1987) as well as in *M. leprae* would seem therefore to establish the ubiquity of this trehalose glycolipid across the spectrum of the genus *Mycobacterium*: it has been found in every mycobacterial species in which it has been sought.

Although trehalose dimycolate was earlier recognized among the lipids of *in vivo*-derived *M. lepraemurium* (Goren *et al.*, 1979) and of *in vitro*-grown organisms (Nakamura *et al.*, 1984), it has not yet been found in what is surely the paradigm of mycobacterial obligate parasitism: *M. leprae*. Dhariwal *et al.* (1987) speculated that the miniscule amount of TMM recovered from *M. leprae* suggests that fulfilling its probable anabolic role in cell wall biosynthesis almost exhausts the amount of the glycolipid available and that the cell's economy may not permit any diversion of the monomycolate for TDM biosynthesis.

5.2.10.1. Biosynthesis of Trehalose Mycolates: Biological Activities of TMM

Investigations by K. Takayama and colleagues dedicated to studies on biosynthesis of mycolic acids, on the mode of action of isoniazid (INH) and of ethambutol, and the involvement of cell wall biosynthesis (reviewed by Takayama and Qureshi, 1984) inevitably led to the recognition of roles for trehalose mycolates in essential anabolic processes for mycobacteria. INH has been convincingly implicated in inhibiting mycolic acid biosynthesis (reviewed by Winder 1982; Takayama and Qureshi, 1984), while the antitubercular effect of ethambutol seems to be to antagonize the transfer of preformed mycolate to growing cell wall (Takayama *et al.*, 1979; Kilburn and Takayama, 1981).

Although the evidence is insufficient to justify an unequivocal interpretation, still the data might implicate a putative enzyme that should be involved at least in transferring mycolate to cell wall, as well as a hypothetical donor lipid as a source of newly formed mycolic acid. Indeed, an enzyme with this and still other functions was recognized—a "mycolyl transferase" as well as a partially characterized "activated mycolic acid"—the donor lipid. With a cell-free enzyme system from *Mycobacterium smegmatis* three partial reactions were identified that are directly involved in the synthesis of trehalose dimycolate (Kilburn and Takayama, 1983):

$$\text{Nucleotide diphosphate-glucose} + \text{glucose 6-phosphate} \rightarrow \text{trehalose}$$
$$\text{6-phosphate} \rightarrow \text{trehalose} \tag{3}$$

$$\text{Trehalose} + \text{activated mycolic acid} \rightarrow \text{trehalose monomycolate} \tag{4}$$

$$\text{Trehalose monomycolate} + \text{activated mycolic acid} \rightarrow \text{trehalose}$$
$$\text{dimycolate} \tag{5}$$

When a 17,000*g* supernatant fraction from sonically disrupted *M. smegmatis* was incubated with UDP-[^{14}C]glucose and glucose 6-phosphate,

[^{14}C]trehalose was produced. The reaction was previously studied in *M. tuberculosis* H37Ra by Goldman and Lornitzo (1962) and Lornitzo and Goldman (1965). Incubation of [^{14}C]trehalose with the crude cell-free preparation resulted in the rapid formation of [^{14}C]trehalose monomycolate along with some dimycolate.

In a separate study, a 105,000g membrane fraction also prepared from sonically disrupted *M. smegmatis* converted [^{14}C]trehalose monomycolate into the labeled dimycolate—shown by TLC and by HPLC. Quantitative data on the amount of labeled TDM produced from labeled TMM showed that the enzymatically active pellet contained at least "166 p moles of activated mycolic acid (possibly a protein-bound thioester) per μg of protein and an acyl transferase enzyme" (Kilburn *et al.*, 1982). Thus, the activities of a mycolate donor and of acyl *transferases* were recognized. The biosynthesis of TDM was envisaged (Kilburn and Takayama, 1983) as proceeding by

> two successive additions of mycolic acid to the 6 and 6' positions of trehalose via two acyltransferases. One molecule . . . is added to trehalose by a trehalose acyltransferase, forming . . . [TMM]. The second molecule of mycolic acid is added to (the monomycolate) by a trehalose monomycolate acyltransferase to form [TDM] or the final product. . . . We postulate that an endogenous mycolic acid exists in an activated form, perhaps as a thioester of coenzyme A or acyl carrier protein. It appears that the transfer enzymes and the activated mycolic acid are closely associated . . . [they] are found in the membrane and cytoplasmic (105,000g supernatant) fractions.

5.2.10.2. Donor Lipid and Trehalose Mycolyltransferase

A partially purified trehalose mycolyltransferase was prepared from *M. smegmatis* cells that were lysed by sonication in the presence of 1,2,3-heptanetriol (Takayama, 1985). The enzyme was present in a 105,000g supernatant in which the donor lipid was depleted. Thus, the enzyme required the addition of the mycolyl donor lipid in order to reconstitute mycolate transfer activity. It could therefore be used to monitor various products, as of chromatographic fractions, for the presence of the donor lipid. This elusive material was isolated in minute amounts (<1 mg from about 500 mg of a $CHCl_3/CH_3OH$ lipid extract of sonicated *M. smegmatis*). When partially purified, it was shown to contain mycolic acid; and because the donor activity eluted in a chromatographic fraction that coincided with a 280-nm absorbance peak, it was considered likely to be a mycolyl–acyl carrier protein complex.

These studies culminated (for the present) in the isolation of a considerably purified trehalose mycolyltransferase that "catalyzes the exchange of a mycolyl group between trehalose, trehalose (monomycolate) . . . and trehalose 6,6'-dimycolate" (Takayama and Sathyamoorthy, 1986). When unlabeled semisynthetic TMM (Liav and Goren, 1984*b*) was added to this enzyme along with [^{14}C]trehalose, [^{14}C]-TMM was formed. Thus,

$$\text{TMM} + [^{14}\text{C}]\text{trehalose} \xrightarrow{\substack{\text{trehalose mycolyl} \\ \text{transferase}}} [^{14}\text{C}]\text{-TMM} \qquad (6)$$

In addition, TDM was able to serve (but less effectively) as a mycolate donor in this exchange reaction to produce [^{14}C]-TMM from [^{14}C]trehalose.

It was postulated that the role of this new mycolate-exchange enzyme in the physiology of the mycobacteria might be as follows:

1. The mycoloyl enzyme may be the *actual initial product* (italics added) in the biosynthesis of mycolic acid rather than TMM as inferred from the studies by Walker *et al.* (1973) (reviewed in Section 5.2.1).
2. Such an intermediate is utilized in the transfer of the newly synthesized mycolic acid to form TDM and the *cell wall murein-arabinogalactan mycolate*. This exchange reaction may be a two-step process:

$$\text{Protein} + \text{TMM} \rightarrow \text{mycolyl protein} + \text{trehalose} \qquad (7)$$

$$\text{Mycolyl protein} + [^{14}\text{C}]\text{trehalose} \rightarrow [^{14}\text{C}]\text{-TMM} + \text{protein} \qquad (8)$$

"The postulated intermediate in this equilibrium could be the key to both mycolic acid synthesis and utilization functions" (Takayama and Sathyamoorthy, 1986).

It should be noted that any putative role for 6,6'-mycoloyl acetyl trehalose (MAT) in these "utilization functions" remains unclear. Perhaps this glycolipid has relevance to only *M. tuberculosis* H37Ra (Takayama and Armstrong, 1976, 1977).

5.2.10.3. Central Role of Trehalose Monomycolate

In the interpretation of Takayama (see Takayama and Sathyamoorthy, 1986), the complex of trehalose mycolyltransferase with mycolic acid can transfer mycolate to either the arabinogalactan of cell wall, or to trehalose or to trehalose monomycolate—to produce TMM and TDM, respectively. Although Kato and Maeda (1974) described toxic properties for TMM resembling those of TDM, a more intriguing biological activity that secures a particularly significant role for TMM is implicit in the remarkable finding that an infection with a variety of pathogenic mycobacteria elicits a powerful humoral immune response for TMM. This was recognized in studies with a serologically active glycolipid (SAG) antigen designated A$_1$ isolated from *M. bovis* BCG by the late wife and husband team of Reggiardo and Middlebrook (1975*a*,*b*; see also Reggiardo and Vazquez, 1981; Reggiardo *et al.*, 1981). The various mycobacterioses included infections with *M. tuberculosis*, with *M. leprae*, and with various "atypical," i.e., nontuberculosis, mycobacteria. The *lowest* incidence of false negative serologic reactions was given with antigen A$_1$ among the several SAG that were studied. Two of these were identified as belonging to the group of phosphatidylinositol mannosides, that have long been known for their serological activity (see Pangborn and McKinney, 1966; Takahashi and Ono, 1961; Sasaki and Takahashi, 1974; Goren, 1984). In collaborative unpublished studies by Reggiardo, Polonsky, Lederer, and Goren, SAG antigen A$_1$ proved to be *6-mycoloyltrehalose* (TMM), by TLC and field-desorption mass spectrometry (see Section 5.2.10.4). Samples of naturally derived and synthetic TMM were equally active in completely substituting for SAG A_1 in sensitizing sheep red blood cells (SRBC) for measuring indirect passive hemagglutination reactions with a variety of antisera from patients with mycobacterial infections. The known natural and synthetic samples were also equally as active as authentic SAG A_1 in adsorption of the reactive antibodies from human sera and thus in inhibiting the PHA reactions for the sensitized RBC. It is of special interest that TDM was not useful as either the coating antigen or in inhibiting the PHA reaction. But trehalose 6-corynomycolate served in the inhibition assay somewhat more effectively than either natural or synthetic trehalose 6-mycolate. Therefore, the specific antigen that is being recognized is most likely a trehalose-6-ester rather than 6-mycoloyl trehalose exclusively. Testing of this inference is planned. [Copies of the laboratory notebooks of Dr. Zulema Reggiardo as well as of relevant correspondence with her collaborators in this program were kindly made available to this author by Dr. Blaine Beaman from the University of California School of Medicine at Davis and the late Dr. Gardner Middlebrook. It is likely that the results will be compiled for publication.]

5.2.10.3a. Antigenicity of Glycolipids: Speculation. An abundance of evidence exists that purified simple glycolipids are notoriously poor antigens unless they are complexed with a native protein, or more effectively, with methylated BSA (MBSA). Thus, trehalose dimycolate is essentially inert as an immunogen and elicited humoral antibodies only when presented in swine serum as a "schlepper" or carrier protein (Ohara *et al.*, 1957) or when complexed with MBSA (Kato, 1972). The antigenic determinant appears to be the trehalose core. The serologically specific surface antigens (Schaefer, 1965, 1980) from various *Mycobacterium avium–Mycobacterium intracellulare–Mycobacterium scrofulaceum* serovars have been characterized as polar variants of the well known glycopeptidolipid C-mycosides and the complete structures of many of them determined (see Goren and Brennan, 1979; Brennan and Goren, 1979; Brennan *et al.*, 1981; Brennan, 1984). The antigenic specificity resides in small oligosaccharides that are attached to an *allo* threonine in the lipopeptide core (Fig. 5-22). Although whole killed cells injected into rabbits elicit good humoral responses of specific antibodies, the isolated purified antigens are essentially immunogenically inert unless presented, for example, as components of liposomal membranes (Kinsky and Nicolotti, 1972; see Brennan, 1984), or when complexed with MBSA (Barrow and Brennan, 1982; Tereletsky and Barrow, 1983). Similarly, infection (in man or armadillos) with the pathogen, or injection of whole cells of *Mycobacterium leprae* into rabbits elicits an important humoral response to the major specific phenolic glycolipid (PGL-I) antigen of *M. leprae* (Fig. 5-23) (Hunter *et al.*, 1982; Tarelli *et al.*, 1984; Chatterjee *et al.*, 1985). The specificity resides in the glycosidically linked trisaccharide. However, the purified PGL-I is itself immunogenically essentially inert. But a good response is achieved with an MBSA complex (described in Hunter *et al.*, 1982).

We are stimulated to ask, therefore, "Whence the antigenicity of these

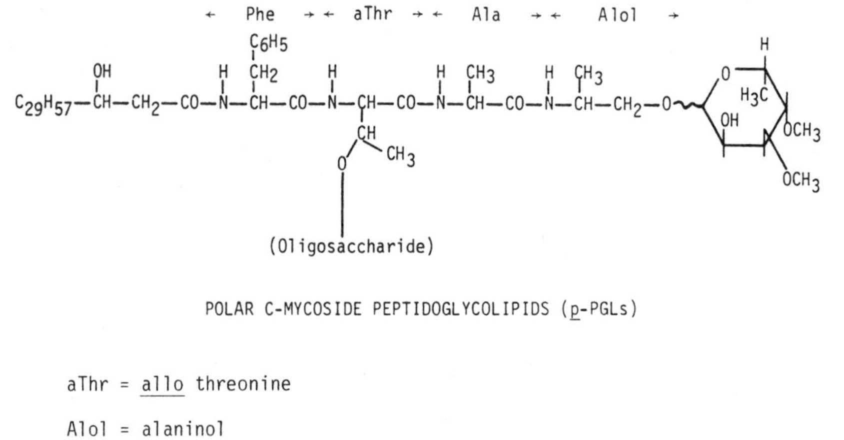

$\leftarrow$ Phe $\rightarrow$ $\leftarrow$ aThr $\rightarrow$ $\leftarrow$ Ala $\rightarrow$ $\leftarrow$ Alol $\rightarrow$

$$C_{29}H_{57}-\overset{\overset{OH}{|}}{CH}-CH_2-CO-\overset{\overset{H}{|}}{N}-\overset{\overset{CH_2}{\overset{|}{C_6H_5}}}{CH}-CO-\overset{\overset{H}{|}}{N}-\overset{\overset{CH}{\overset{|}{\underset{O}{\diagup}\diagdown CH_3}}}{CH}-CO-\overset{\overset{H}{|}}{N}-\overset{\overset{CH_3}{|}}{CH}-CO-\overset{\overset{H}{|}}{N}-\overset{\overset{CH_3}{|}}{CH}-CH_2-O$$

(Oligosaccharide)

POLAR C-MYCOSIDE PEPTIDOGLYCOLIPIDS (p-PGLs)

aThr = allo threonine

Alol = alaninol

Figure 5-22. Gross structure of surface Schaefer antigens recovered from various serovars of the *Mycobacterium avium–Mycobacterium intracellulare* complex. Serological specificity is conferred by small oligosaccharides attached to the *allo*-threonine.

native lipids during infection or in response to injections of cells"? Is it due to a fortuitous formation of loose complexes between the glycolipids and protein components of the bacteria? Or to some antigenically powerful configuration in which the glycolipid is maintained by the bacteria and presented to the host? It is possible that such factors indeed contribute. But we would like to suggest, simply as speculation, that the haptenic carbohydrate determinants in the penultimate stages of assembly, whether linked to the lipid or not, are probably present in some form covalently linked to an anabolic enzyme that is involved in the construction of the glycolipid, and that such complexes might contribute significantly to immunogenicity. Thus, trehalose or TMM linked to trehalose mycolyl transferase (see Section 5.2.10.2) might serve as an immunogenic complex that elicits a humoral response with 6-acyltrehalose specificity.

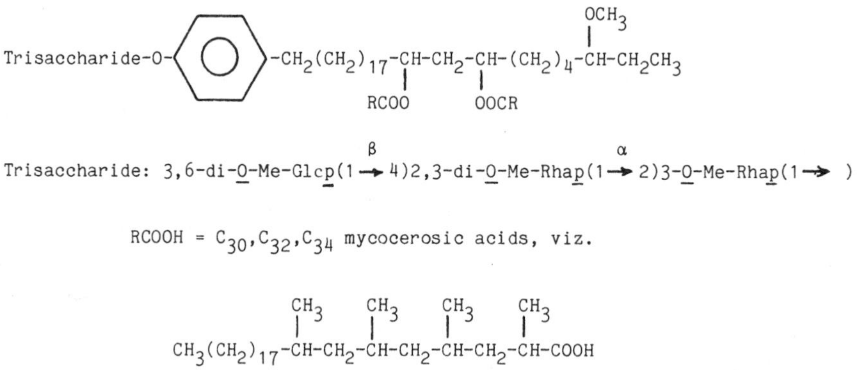

Trisaccharide-O-⟨○⟩-$CH_2(CH_2)_{17}$-$\overset{\overset{}{|}}{\underset{RCOO}{CH}}$-$CH_2$-$\overset{\overset{}{|}}{\underset{OOCR}{CH}}$-$(CH_2)_4$-$\overset{\overset{OCH_3}{|}}{CH}$-$CH_2CH_3$

Trisaccharide: 3,6-di-O-Me-Glcp(1 $\xrightarrow{\beta}$ 4)2,3-di-O-Me-Rhap(1 $\xrightarrow{}$ 2)3-O-Me-Rhap(1 $\xrightarrow{\alpha}$)

RCOOH = C_{30}, C_{32}, C_{34} mycocerosic acids, viz.

$$CH_3(CH_2)_{17}-\overset{\overset{CH_3}{|}}{CH}-CH_2-\overset{\overset{CH_3}{|}}{CH}-CH_2-\overset{\overset{CH_3}{|}}{CH}-CH_2-\overset{\overset{CH_3}{|}}{CH}-COOH$$

Figure 5-23. Structure of serologically active phenolic glycolipid (PGL-I) of *Mycobacterium leprae*. The principal specificity resides in the terminal 3,6-di-O-methyl glucose moiety.

5.2.10.4. Field Desorption and Plasma Desorption Mass Spectrometry

More sophisticated methodology especially relevant to trehalose mono-mycolates now permits retrieval of much more information from mass spectra of these glycolipids. Puzo *et al.* (1978) exploited a combination of mass spectral ionization methods to establish the composition and structure(s) of a preparation of trehalose dicorynomycolate from *C. diphtheriae* (see Section 5.2.4). Field desorption mass spectrometry and cationization with cesium iodide allows emission of cationized molecules of the individual species with minimal fragmentation. The procedure revealed the molecular weights of at least 22 components that were present in the natural glycolipid isolate. These differed in chain length of the acyl substituents and in their degrees of unsaturation (from 0 to 4 in examples with the same carbon number). From electron impact mass spectrometry (EIMS) of *pertrimethylsilylated* model compounds and of the *C. diphtheriae* cord factor, fragments were identified that revealed the 6-positions of the sugar to be the only ones bearing the acyl substituents. Thus the fragment structured as in (**48**),

$$
\begin{array}{l}
6 \\
CH_2 - OOC - R \\
| \qquad\qquad\qquad (48) \\
CH = 0 - TMS \\
5 \qquad +
\end{array}
$$

where R is defined by the acyl substituent, still contained the whole chain and the C_6–C_5 of the sugar rings. These powerful tools were subsequently used by Shimakata *et al.* (1984, 1985) to identify a spectrum of trehalose 6-mycolates synthesized by a cell-free system of *Bacterionema matruchotii*. (Their claim to priority for the first documentation of *in vitro* synthesis of TMM from exogenous fatty acids and endogenous trehalose seems to be valid.) Harvested *B. matruchotii* was disrupted in a cold French pressure cell and the product centrifuged to provide a pellet, fluffy layer, and supernatant fractions. Mycolate biosynthesis was initiated by adding palmitate to a portion of the fluffy layer along with buffer and 10 mM $MgCl_2$. On one hand (Shimakata *et al.*, 1984) the mycolic acids that were synthesized were recovered as methyl mycolates and were examined by gas chromatography coupled with EIMS. About a dozen species were recognized of saturated and monounsaturated mycolates similar to those synthesized by *C. diphtheriae* (from C_{24} to C_{36}). By contrast (Shimakata *et al.*, 1985), when the fluffy layer–palmitate reaction mixture was directly extracted with chloroform–methanol, it yielded a principal glycolipid, which was examined by a combination of FDMS, SIMS and EIMS. In the first mode, with sodium cationization, the results indicated the product to be a trehalose monocorynomycolate of mass 820: $[M + Na]^+ = 843$. This composition was supported by SIMS, in which the underivatized glycolipid yielded a protonated quasimolecular ion $(M + H^+)$ at m/z 821 and ions derived from this by cleavage at the glycoside bond and at the ester linkage. Fragments m/z 641 and m/z 479 were observed, in accordance with the inferred structures (**49**) and (**49a**), respectively:

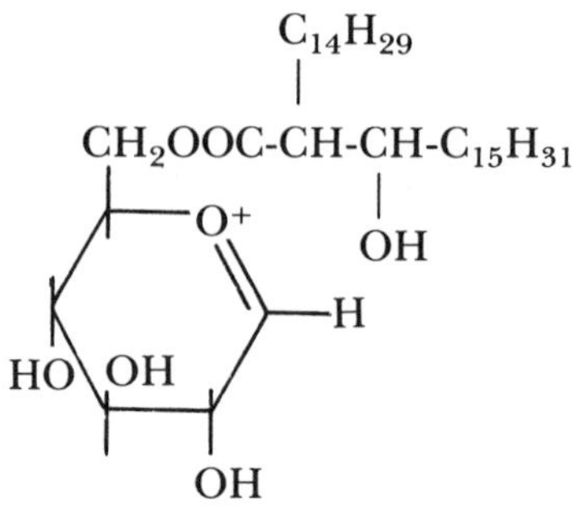

$$C_{14}H_{29}$$
$$|$$
$$CH_2OOC-CH-CH-C_{15}H_{31}$$

(**49**; m/z = 641)

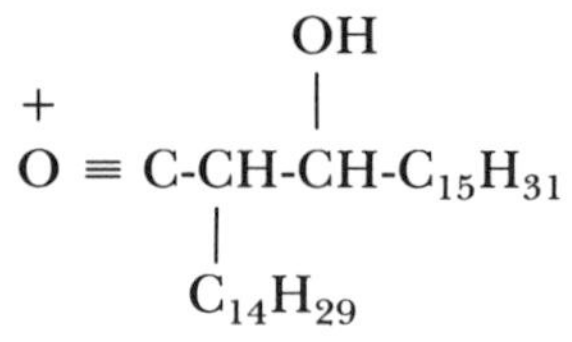

OH
|
O $\equiv$ C-CH-CH-C$_{15}$H$_{31}$
|
C$_{14}$H$_{29}$

(**49a**; m/z = 479)

After the isolated product was saponified, trehalose was recovered and identified. And from adopting the methods of Puzo *et al.* (1978) (above), electron impact mass spectrometry of the *pertrimethylsilylated* glycolipid gave the fragment ion (**48**) m/z = 583, confirming that the corynomycolyl group is linked to the 6-position of the trehalose molecule.

5.2.10.5. Trehalose Monomycolate of Mycobacterium Leprae

In the studies by Dhariwal *et al.* (1987) the presumed TMM from *M. leprae* was analyzed by permethylation, alkaline solvolysis, and acid hydrolysis of the sugar to establish the acylation at the 6-position. The permethylated glycolipid on solvolysis yields a mixture of the β-*O*-methyl mycolate and a demethanolated $\Delta^{2,3}$-mycolenoate, both of which are of limited use in defining the structure of the mycolate component by mass spectrometry. Instead, underivatized TMM was hydrolyzed to provide the free mycolic acid, which was examined as the methyl ester by EIMS. At the high temperature of the probe, mycolates undergo a reverse Claisen reaction (Fig. 5-24) to generate a meroaldehyde and an ester. The latter provides the base peak in the spectrum and so defines the size of the alpha branch (Etemadi, 1964). For the *M. leprae* methyl mycolate this fragmentation gave the base peak m/z = 354 (methyl docosanoate) with a smaller complement of tetracosanoate. This established that the alpha branch in the mycolate of *M. leprae* is principally a C$_{20}$ constituent, in agreement with earlier reports (Kusaka *et al.*, 1981; C. Asselineau *et al.*, 1981; Draper *et al.*, 1982).

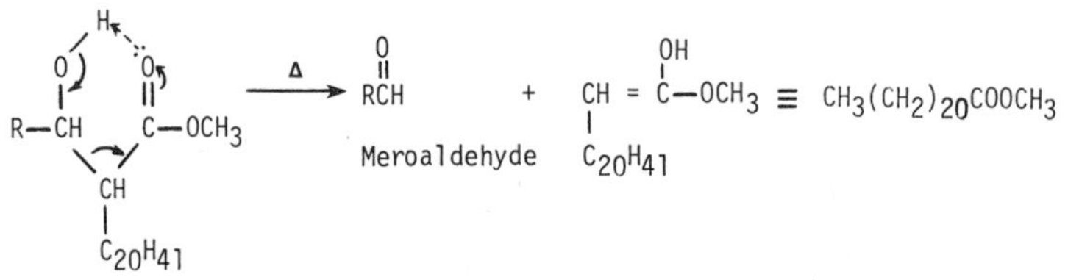

Figure 5-24. Pyrolytic decomposition of mycolic (α-branched, β-hydroxy) acid esters (a reverse Claisen condensation). A meroaldehyde and an ester are formed. The size of the α-branch is revealed by the ester moiety.

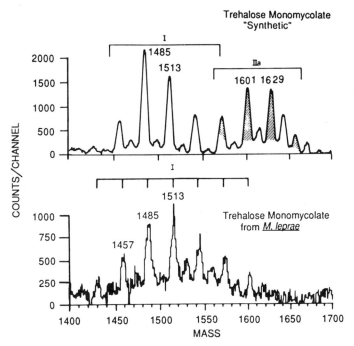

Figure 5-25. Plasma desorption mass spectrum of a semisynthetic trehalose-6-mycolate (upper panel) and TMM of *Mycobacterium leprae* (lower panel). The semisynthetic glycolipid comprises at least seven species, containing dicyclopropane and also methoxysubstituted mycolate. The *M. leprae* lipid has only the dicyclopropane variety (see text).

5.2.10.5a. Plasma Desorption Mass Spectrometry (PDMS). For the *M. leprae* TMM, PDMS pinpointed the spectrum of component substances contained by the purified glycolipid. The results, and those with a semisynthetic TMM (Liav and Goren, 1984*b*, 1986) served as a model that dramatically illustrates the utility of the technique in revealing the heterogeneity of trehalose mycolates, as alluded to several times in preceding sections.

The instrument and techniques for its use were described by Macfarlane (1983). Samples dissolved in polar solvents were applied to the conductive surface of an aluminized Mylar film in a sample holder either by evaporation or electrospraying. The sample holder was then placed through a vacuum lock in front of a ^{252}Cf source [~10 μCi (370 kBq)]. The chamber is maintained at about 10^{-6} mmHg. Very high energy particles from the Cf source (200 MeV) impinge on the sample and thus generate a plasma containing desorbed molecules with attached sodium cations, and these are accelerated down a 42-cm flight tube by a 10-kV grid. Flight times for the individual cationized molecules, which are proportional to the square root of the mass, are digitized, converted into masses and stored in a computer. Each such ion is collected and added to others in the same channel to provide the mass spectrum. As indicated in Fig. 5.25, this mode of operation is useful principally to determine the molecular masses of the compounds under investigation. Figure 5-25 (upper panel) is the mass spectrum of a semisynthetic TMM prepared from trehalose and a mycolic acid mixture derived from *M. tuberculosis* H37Rv (see Liav and Goren, 1984*b*, 1986 for details).

The spectrum shows that the semisynthetic TMM is characterized by two prominent series of mycolate substituents that are identified as types I and IIa according to the scheme of D. E. Minnikin (Davidson *et al.*, 1982). The series of type I are of the (*cis,cis*)-dicyclopropane types, also referred to as "alpha" mycolates (see below and reviews mentioned earlier). The IIa series (shaded peaks) contain a methoxyl group with an adjacent methyl branch as depicted by the structure

$$
\begin{array}{cccc}
CH_3 & OCH_3 & CH_2 & OH\quad O \\
| & | & |\,\backslash & |\qquad \| \\
\end{array}
$$
$$
CH_3(CH_2)_xCH{-}CH{-}(CH_2)_y{-}CH{-}CH{-}(CH_2)_z{-}CH{-}CH{-}C{-}
$$
$$
\begin{array}{c}
| \\
C_{24}H_{49}
\end{array}
$$

For the peaks at m/z 1601 and 1629, $x + y + z = 50, 52$ and the molecular weights of the glycolipids are 1578 and 1606, respectively.

The spectrum of the *M. leprae* lipid (Fig. 5-25, lower panel) shows essentially only one principal series of peaks (derived from the sodium cationized molecules) at $M + 23 = 1457, 1485, 1513, 1541$ that correspond to trehalose esterified with a type I (α) mycolate of the structure

$$
\begin{array}{cccc}
CH_2 & & CH_2 & OH\quad O \\
/\;\backslash & & /\;\backslash & |\qquad \| \\
\end{array}
$$
$$
CH_3{-}(CH_2)_x{-}CH{-}CH{-}(CH_2)_y{-}CH{-}CH{-}(CH_2)_z{-}CH{-}CH{-}C{-}
$$
$$
\begin{array}{c}
| \\
C_{20}H_{41}
\end{array}
$$

in which $x + y + z = 46, 48, 50, 52$ (principal components **48, 50**), in accordance with a $C_{20}H_{41}$ α-branch as described above. These results agree with those obtained by Draper *et al.* (1982) for the type I mycolic acids recovered from defatted cells of (armadillo-grown) *M. leprae*, with principal $x + y + z = 48$ (molecular weight 1462). Although *M. leprae* elaborates ketomycolates as well (Draper *et al.*, 1982), these were not discernible from the spectrum of the *M. leprae* TMM. The peaks that were analyzed therefore define almost exactly the identities of the individual mycolate substituents that are esterified to the trehalose. For the semisynthetic glycolipid, at least seven components are easily identifiable and it is likely that an additional equal number are represented by some of the minor peaks. Very possibly, these may even represent species derived from still other types of mycolates. Accordingly, if a 6,6'-dimycolate were being analyzed, with all the possible permutations of mycolate types and homologues, the spectrum would be inordinately complex. We have not yet succeeded in obtaining a spectrum from a natural mycobacterial trehalose-6,6'-dimycolate. But a spectrum of the dicorynomycolate, con-

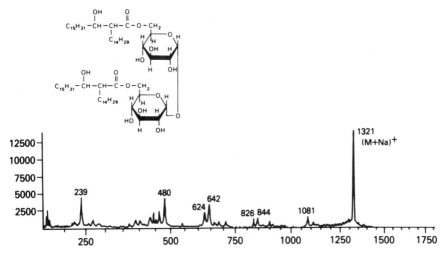

Figure 5-26. Californium-252 plasma desorption mass spectrum of semisynthetic trehalose-6,6'-dicorynomycolate. Only a single cationized molecular ion peak (M + 23) is seen at m/z 1321.

structed from trehalose and synthetic erythro/threo corynomycolic acid, was readily obtained and, as depicted in Fig. 5-26, it is remarkably "clean." The principal peak m/z 1321 represents the cationized molecular ion that corresponds exactly to the mass of the dicorynomycolate (1298) plus Na⁺. As expected, no other (M + Na)⁺ ions are evident because no homologues of the acyl substituent are present. The other peaks represent fragments derived from specific cleavages of the cationized molecules.

These data define the complete structure(s) of representative trehalose-6-mycolates from *Mycobacterium leprae,* and would appear to establish the ubiquity of this glycolipid across the spectrum of the genus *Mycobacterium.* The finding is in accord with the anticipation expressed at the beginning of this chapter, and in effect closes the circle.

5.3. Miscellaneous Acylated Trehalose Derivatives of Mycobacteria

5.3.1. Trehalose Phleates

Extraction of cells of *Mycobacterium phlei* with hexane leaves the bacteria alive and extracts, among other lipids, multiacylated trehalose derivatives in which so-called trehalose phleates (from *M. phlei*) are prominent (C. Asselineau *et al.,* 1969, 1972; Asselineau and Asselineau, 1978). The phleic acids, derived from hydrolysis of the trehalose derivatives are *multiunsaturated* straight-chain carboxylic acid, and their discovery invalidated the longstanding notion that mycobacteria do not produce polyenoic acids. They comprise two principal series having the structures represented as shown in **50:**

$$CH_3\text{-}(CH_2)_m\text{-}(CH = CH\text{-}CH_2\text{-}CH_2)_n\text{-}COOH$$

(50)

for $m = 14$ $n = 4, 5,$ or 6
for $m = 12$ $n = 5$ or 6

The main component of the mixture is hexatriaconta-4,8,12,16,20-pentaenoic acid (**50**: m = 14, n = 5).

Separation of the mixture of trehalose phleates gave a principal trehalose octa-phleate product, of ~4400 kDa, and having about 40 double bonds. It was obtained in chromatographically homogeneous form but nevertheless was still heterogeneous in its phleate components. This is yet another example of the recurrent theme first expressed in Section 5.2.4 concerning the inhomogeneity of "purified" mycobacterial lipids. Trehalose polyphleates have not been found in other *Mycobacterium* species. The acids are evidently synthesized from myristate and palmitate, which are the precursors of the two series; the unsaturated portions of the molecules derive from acetate, probably via preformed crotonate units (Asselineau and Montrozier, 1976; see also Goren and Brennan, 1979).

5.3.2. Pyruvylated Glycolipids from Mycobacterium Smegmatis

An unusual class of acylated oligosaccharides composed of only glucose residues and containing a terminal trehalose unit occurs in *M. smegmatis*. The oligosaccharide moieties are acidic because of the presence of pyruvic acid in the form of carboxyethylidene function(s). They therefore bear a resemblance to one of the type-specific serologically active glycopeptidolipids (polar C-mycosides) of the *Mycobacterium avium–intracellulare* groups alluded to in Section 5.2.10.3a. This polar C-mycoside was found in *M. intracellulare* serovar 8 (Brennan *et al.*, 1981).

According to Saadat and Ballou (1983) a crude glycolipid extract from *M. smegmatis* was first deacylated in alkali to yield (water-soluble) oligosaccharides, an important first step long practiced by the Ballou school for studying various mycobacterial lipopolysaccharides and oligosaccharides (see Goren, 1984). After conversion to the free acids, the pyruvylated components were isolated by adsorption on Dowex-1 (acetate) and elution with a linear gradient of ammonium acetate in water. Additional purifications of the acidic oligosaccharides yielded three pyruvylated products: A (principal), B_1, and B_2. The products were studied by complete acid hydrolysis, which showed them to contain only glucose and 3-*O*-methyl glucose. Further analysis involved permethylation of the intact oligosaccharides, as well as of the depyruvylated oligosaccharides, acid hydrolysis, $NaBH_4$ reduction and acetylation to the alditol acetates, which were identified by a combination of gas chromatography and mass spectrometry. Coupled with the results of ^{1}H-NMR analysis, fast atom bombardment mass spectrometry and periodate oxidation, the structures of the three acidic oligosaccharides were established. It could be deduced that acidic oligosaccharide B_1, which contains no methyl glucose, is a

(51)

monopyruvylated tetraglucose that is very likely a precursor of acidic oligosaccharide B_2, a pentasaccharide containing one pyruvyl residue and one 3-*O*-methyl glucose. Some structural heterogeneity in the position of the single pyruvate group was suspected. B_2 is in turn the precursor for acidic oligosaccharide A, which is doubly pyruvylated B_2. Structure (**51**), depicting acidic oligosaccharide A, indicates in dashed lines the portions of the molecule corresponding to B_1 and the transformations that lead to B_2 (arrow) and thence to A (broken arrow), respectively.

The nonreducing natures of the three oligosaccharides is consistent with their structural assignment as glycosylated trehalose derivatives; and the double pyruvylation of A as distinguished from the singly pyruvylated B_2 accounts for the separation of the compounds by anion-exchange chromatography.

In a succeeding study, Kamisango *et al.* (1985) achieved an apparently facile purification of the native pyruvylated glycolipids by silicic acid-celite chromatography. The purified glycolipid contained 2 moles of ester and 2 moles pyruvic acid for 5 moles of hexose. The fatty acid methyl esters obtained by alkaline solvolysis were identified as methyl myristate/methyl palmitate and methyl-2,4-dimethyl-2-eicosenoic acid. The latter was identified by a combination of EI mass spectrometry on the native and reduced esters; by oxidation to locate the double bond, and by NMR spectrometry. Fast atom bombardment (FAB) spectrometry in the positive mode revealed essentially two glycolipids based on the dipyruvylated oligosaccharide A: they contained a single molecule of the docosenoic acid and differed in that one also contained palmitate while the other had a myristate substituent. An important fragment was attributed to loss of a single hexose unit along with either myristate or palmitate and was judged to be possible only as a fragmentation from the trehalose end of the oligosaccharide, specifically of the terminal glucose unit. The C_{22} fatty acid is associated with the remaining tetrasaccharide fragment. FAB in the negative ion mode revealed considerably more data which were analyzed (see Kamisango *et al.*, 1985) to show that the C_{22} enoic acid was located on the penultimate glucose of the trehalose moiety.

The native glycolipids could not be completely permethylated without some deacylation: Hakomori permethylation (used either to study the deacylated oligosaccharide) was not attempted because of expected losses of acyl groups. The more gentle method of Kuhn *et al.* (1955) discussed in Section 4.2.3, using Ag_2O and CH_3I in dimethylformamide gave incomplete methylation during 3 days at ambient temperature. Under more drastic conditions, 10 days under reflux followed by 15 days at 23°C, considerable deacylation occurred. Thus, permethylation of glycolipids for structural analysis remains beset with pitfalls. Methyl replacements or di-tert-butyl pyridine catalysis with CH_3I were not examined (cf. Section 5.2.3).

Instead, a carefully crafted and excellently detailed Smith degradation (periodate oxidation, $NaBH_4$ reduction, saponification and gentle acid hydrolysis) indicated that the terminal trehalose very probably contained palmitate/myristate at the 6-position of the ultimate glucose moiety and eicosenoate at the 4-position of the penultimate glucose. The native glycolipid(s) therefore have the structure

4,6-(1-carboxyethylidene))-3-*O*-Me-β-D-Glc*p*(1 → 3)-4,6-(1-carboxyethylidene)-β-D-Glc*p*(1 → 4)-β-D-Glc*p*-

$$(1 → 6)\text{-}\alpha\text{-}D\text{-}Glpc\text{-}(1 → 1)\alpha\text{-}D\text{-}Glpc$$
$$|4 \qquad\qquad |6$$
$$FA_2 \qquad\qquad FA_1$$

where FA_1 is myristic or palmitic acid, and FA_2 is 2,4-dimethyl-2-eicosenoic acid.

5.3.3. Serologically Active Trehalose-Containing Lipooligosaccharides

Mycobacterium kansasii, which produces pulmonary disease in man that resembles infection with *M. tuberculosis*, elaborates a series of about seven species-specific neutral lipooligosaccharide (LOS) antigens. All have a tetra-glucose "core" terminating in trehalose, thus again glycosylated trehaloses as developed in the preceding section. Hunter *et al.* (1983) isolated crude lipids extracts from lyophilized *M. kansasii* by several extractions with mixed chloroform and methanol solvents and separated some seven fractions of LOS by multiple column chromatographies and preparative TLC. Designated LOS I–VII, of increasing polarity, all were reactive by enzyme-linked immunosorbent assay (ELISA) with specific antiserum from immunized rabbits. The most polar LOS (VI, VII) also showed bands of precipitation by agar gel immunodiffusion.

In later studies (Hunter *et al.*, 1985) trehalose dimycolate, trehalose monomycolate, and eight LOS (I–VIII) were isolated from the "crude" lipids by long column chromatography and many LOS were characterized. Upon acetolysis, all produce the tetra glucose core:

β-D-Glc*p*(1 → 3)-β-D-Glc*p*(1 → 4)-α-D-Glc*p*
(1 → 1)-α-D-Glc*p*

The simplest LOS, designated I′ contains an additional α (1 → 3) linked 3-*O*-Me-L-Rha*p*. LOS I–III, of intermediate complexity contain additionally a β(1

→ 4) linked *D*-Xyl*p* or Xylobiose; and those of ultimate complexity (LOS IV–VIII) contain still more *D*-Xyl*p* residues with a distal disaccharide—*N*-acyl" kansosamine" (1 → 3)Fuc*p*. Thus the oligosaccharide obtained from alkaline deacylation of LOS VI has the structure

Kan *N* acyl(1 → 3)Fuc*p*(1 → 4)[− β -*D*-Xyl*p*(1 → 4)]$_6$ α-*L*-3-*O*-Me-Rha*p*(1 →3)Glc$_4$

In a separate study, Hunter *et al.* (1984) proposed for the novel *N*-acyl kansosamine the structure (**52**): 4,6-dideoxy-2-*O*-methyl 3-C-methyl-4-(2'-methoxypropionamido)- α and β − *L-manno*-hexopyranose.

$$ CH_3O-CH-C-HN \quad CH_3 \quad CH_3 \quad H, OH $$

(**52**)

In the native LOS, the acyl functions, whose positions are not yet established, comprise principally acetates and 2,4-dimethyl tetradecanoate. The most potent serologically reactive LOS, of higher complexity, all contain the *N*-acyl-kansosamine residue, which is therefore judged to be the principal antigenic determinant, not the trehalose terminus.

5.3.4. Trehalose Esters of Intermediate Molecular Weight

The complex array of trehalose esters was rendered even more so by recent findings of Minnikin *et al.* (1985) of two novel *families* of trehalose glycolipids differing widely in polarity. They were isolated from a variety of *M. tuberculosis* strains, but studies were facilitated when the least polar and the more polar families were isolated in considerable amounts from *M. tuberculosis*, strain C and H37Rv, respectively. The intimate structures of the various glycolipids were not sought, but all were based on trehalose acylated with a great variety of simple straight chain and more complex methyl-branched acids—further evidence of the kinds of heterogeneity I have alluded to as a recurring theme.

Fifteen straight-chain, unsaturated acids of carbon number up to C24 were recognized, and 10-methyl branched as well as about nine multimethyl-branched acids ($>C_{24}$). The former group contained, among other components, tetradecanoate, hexa- and hepta-decenoates, tuberculopalmitate, and tuberculostearate (17Me and 19Me); 2,4-dimethyldocosanoate, and tetracosanoate. Among the higher, multimethyl branched acids, C_{27} mycolipenate (C_{27} phthienoic acid) (**53**), and 3-hydroxy-2,4,6-trimethyltertracosanoic acid (C_{27} mycolipanolic acid (**54**) were found. Both are dextrorotatory and of the *L* configuration and are therefore related to the phthioceranic acids of considerably higher molecular weight, which are found in the sulfatides of *Mycobacterium tuberculosis* (succeeding sections):

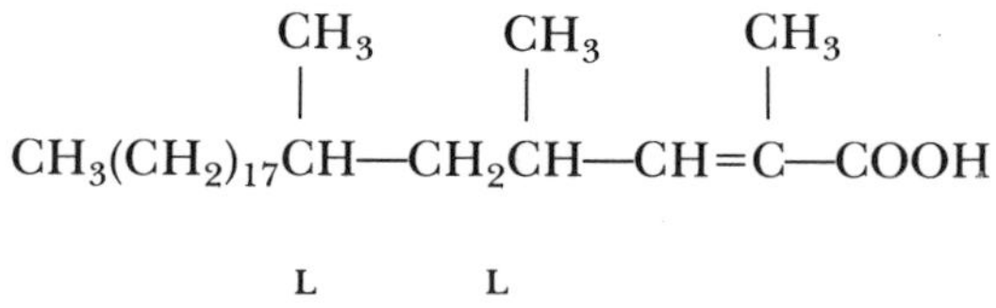

$$\begin{array}{ccccc} & CH_3 & CH_3 & CH_3 \\ & | & | & | \\ CH_3(CH_2)_{17}CH & -CH_2CH & -CH=C & -COOH \end{array}$$

L   L

(**53**) (C$_{27}$ mycolipenic or phthienoic acid)

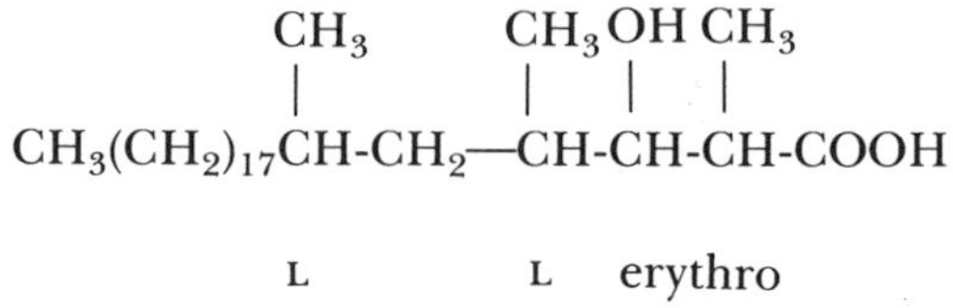

$$\begin{array}{ccc} & CH_3 & CH_3\ OH\ CH_3 \\ & | & |\quad |\quad | \\ CH_3(CH_2)_{17}CH\text{-}CH_2 & -CH\text{-}CH\text{-}CH\text{-}COOH \end{array}$$

L   L   erythro

(**54**) (C$_{27}$ mycolipanolic acid).

On the basis of TLC mobilities, Minnikin *et al.* (1985) speculate that certain of their polar glycolipids, designated C and D, "correspond to an antigenic phosphorus-free glycolipid fraction studied by Reggiardo *et al.* (1981). . . ." However, since mycolates were not recognized in the Minnikin study, it is more likely that C and D are simply trehalose multiesters of lower-molecular-weight fatty acids, which, from their data, embrace palmitate/stearate, and 2,4-dimethyl docosanoate as principal components. These could mimic the TLC mobility of Reggiardo's phosphorus-free glycolipid (TMM). However according to our own speculations (see Section 5.2.10.3*a*), during steps in their biosynthesis these newly recognized trehalose esters could contribute to the antigenicity that elicits antibodies that recognize TMM.

5.4. Sulfatides of Mycobacterium tuberculosis

5.4.1. History

The sulfatides of *M. tuberculosis* were discovered in the search for neutral red-reactive substances located at or near the surface of virulent human and bovine strains of the pathogen. As briefly described in Section 5.2.2., Dubos and Gardner Middlebrook* observed that virulent strains of tubercle bacilli adsorb the weakly basic phenazine dye from neutral or even slightly alkaline medium, and are colored a pink-red hue. In contrast mutant strains of the organisms, which do not manifest this behavior are invariably found to be highly attenuated or avirulent. Most saprophytic mycobacteria are also indifferent to the dye. Treatment of harvests of *M. tuberculosis* H37Rv with hexane containing 0.05% decylamine extracted lipid substances that, in solution in hexane, retained neutral red reactivity—complexing and solubilizing the dye from aqueous solutions into the organic phase (Middlebrook *et al.*, 1959). This

---

*Deceased, 1986.

property seems consistent with the behavior of a strongly acidic lipid of fairly high molecular weight. In the early studies, the most abundant neutral red-fixing component(s), separated by solvent fractionation and silicic acid or silica gel chromatography were found to be depleted in phosphorus content and instead enriched in sulfur, believed present as a sulfonic acid. Other deductions from the earliest studies suggested the presence of multimethyl-branched fatty acids, probably as esters, little or no unsaturation, and that the sulfur-containing material probably consisted of a group of closely related sulfolipids with slight differences in polarity, an average molecular weight of about 3000 and S content of about 1.1%. And in closely following unpublished studies, it was recognized that the sulfolipid was probably a multiacylated trehalose containing a sulfate ester substituent. When compared with the now known structures of the principal, and even several minor, sulfatides of *M. tuberculosis* and the varieties of closely related analogues, these were remarkably prescient findings.

5.4.2. Separation and Purification of Sulfatides

In our succeeding investigations, originally begun in collaboration with Middlebrook, the "sulfolipid" of H37Rv (referred to herein as various SL) was found to consist of several groupings or families of trehalose-containing sulfatides, closely related in composition and structure, and separable to various extents by column chromatography on diethylaminoethyl cellulose (DEAE)—acetate form. In brief, after the lipid-containing hexane solutions were freed of excess decylamine by extraction with concentrated aqueous citric acid, the crude lipids in chloroform–methanol (C/M) were loaded on DEAE (acetate). Nonpolar, neutral, weakly acidic and many polar contaminants were eluted by a succession of neutral and then acidic solvents (formic acid in C/M, glacial acetic acid), and finally by methanol before the sulfatides were eluted (either singly or as mixtures) usually with C/M containing increasing amounts of NH_4OH, and ultimately ammonium acetate as well. These ammoniacal solvents eluted tetraacylated trehalose-2-sulfates SL-II', SL-II, SL-I (principal), and SL-I'—all as ammonium salts. These were followed by more polar products in small amounts, only one of which, a triacylated trehalose-2-sulfate designated SL-III was characterized completely (Goren *et al.*, 1978). Additional rechromatography was required in all instances to achieve the separations desired. The methodology for achieving both most complete separations of the tetraacylated glycolipids on one hand and for a considerably abbreviated separation on the other, were detailed by Goren (1970a) and Goren *et al.* (1978). The latter procedure was adopted for facilitating recovery of the minor polar SL-III that at one time appeared to exhibit unusual antitumor properties.

The kinds of separations that have been achieved are illustrated in the TL chromatograms of Fig. 5-27A,B. The former shows the behavior of a mixture of SL-II, SL-I, and SL-III previously separated by the most elaborate procedures and then reconstituted (Goren, 1970a). By contrast, Fig. 5-27B was obtained from successive DEAE column effluent fractions obtained in the

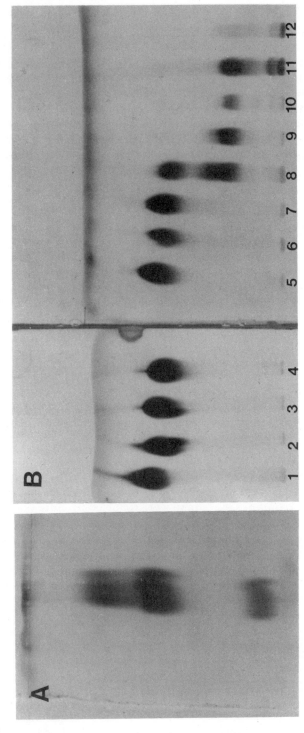

Figure 5-27. Thin layer chromatograms of sulfatide products separated from extracts of *Mycobacterium tuberculosis.* (A) Sulfatides II, I, and III (in descending order) purified by extensive chromatography on DEAE (acetate) and reconstituted for TLC (Goren, 1970*a*). (B) Products obtained in the DEAE cellulose abbreviated chromatography. The succession of samples 1–7 contains unseparated mixtures of increasingly polar tetraacylated SL-II′, II, I, I′ and leads to the triacylated SL-III seen in rows 8–10 (Goren *et al.,* 1978). Silica gel plates were developed 10 cm in chloroform–acetone–methanol–water–acetic acid (158 : 83 : 1 : 6 : 32 V/V), sprayed with 60% W/W sulfuric acid (dichromate) and charred at 180°C.

"abbreviated chromatography" during elution with ammoniacal solvents (Goren *et al.,* 1978). All the least polar SL-II'-II-I-I' products are present as unseparated mixtures in the first several effluent samples obtained. [It should be noted that because of the very complex heterogeneity exhibited by the fatty acids within each SL family and their permutations (detailed below), that if all of the components could be obtained truly *homogeneous* from fractions 1–7 and from the top spot of fraction 8, they would blend into a single overlapping continuum of spots if examined by TLC. A similar second continuum is implicit in the behavior of the lower spot of row 8, rows 9 and 10 and the most mobile components of row 11.]

5.4.3. SL Characterization

The principal NH_4 sulfatides, appearing homogeneous in TLC and entirely free of phosphorus are colorless, very viscous oils that are unstable under most conditions of storage because of spontaneous desulfation (see below). Elemental analysis and optical rotations have been obtained only on the most abundant sulfatide SL-I (about 0.7% based on dry bacillary harvest) and SL-III obtained after a prodigious effort dedicated to its recovery (about 0.02%).

$$NH_4SL\text{-}I:\ C_{145}H_{283}O_{20}NS:\ [\alpha]_D + 40.5°$$

$$NH_4SL\text{-}III:\ C_{108}H_{221}O_{20}NS:\ [\alpha]_D + 57.7°$$

5.4.3.1. Infrared Spectrometry

The infrared spectrum (Fig. 5-28) of a neat sample of NH_4SL-I on a KBr plate is best interpreted on the basis of its now established structure as a

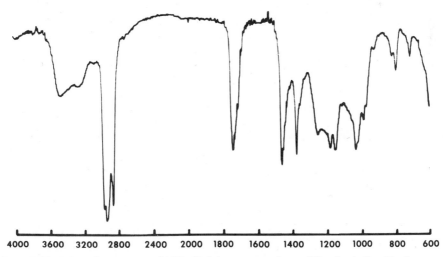

Figure 5-28. Infrared spectrum of NH_4 SL-I (a neat sample on KBr plate). Specific features involve the regions 3200–3500 cm^{-1} (OH and NH); 1660–1745 cm^{-1} (carbonyl); 824 cm^{-1} (equatorial secondary sulfate ester) and 808 cm^{-1} (a weak band specifically associated with trehalose).

tetraacyl trehalose-2-sulfate, as developed below. Peaks attributable to NH and H-bonded hydroxyl functions are evident from about 3200 to 3500 cm^{-1}. The carbonyl region exhibits shouldering and band broadening in the region 1660–1745 cm^{-1} and is associated with the multiplicity of ester functions and inferred intermolecular interactions between carbonyl and OH functions (Goren *et al.*, 1971). The sulfate ester occupying a secondary equatorial position is characterized by a specific peak at 824 cm^{-1} as well as by the (S = O) absorption at 1250 cm^{-1}. Desulfation of the SL is accompanied by loss of these bands as well as those associated with the NH absorptions. The weak band at about 808 cm^{-1} specifically associated with α,α-trehalose and discussed earlier in Section 5.2.5. is also prominent in all SL spectra. The deep absorption band at 1370 cm^{-1} is due to multiply methyl-branched acyl substituents derived from phthioceranic (**55**) and hydroxyphthioceranic (**56**) acids esterified to the trehalose core (Goren *et al.*, 1971). These are discussed in considerable detail in Section 5.4.4.3.

5.4.3.2. Desulfation of the Sulfatides

All of the sulfatides from *M. tuberculosis* are subject to facile desulfation when dissolved in essentially anhydrous aprotic solvents (ether or dioxane) or when stored as neat samples after evaporation of solvents. An earlier finding that SL-III was resistant to desulfation was ultimately proved misleading, when a pure neat sample stored for several years in the cold was found to have undergone complete desulfation. In anhydrous ether (which contains the trace of moisture necessary for hydrolysis), spontaneous desulfation of pure NH$_4$SL-I usually occurs within hours. It is remarkably catalyzed by even traces of mineral acid. Thus, the spotaneous desulfation is autocatalytic owing to the bisulfate ion produced; and it is essentially inhibited by water. This behavior was studied in considerable detail and led to the elaboration of exceptionally gentle methods for desulfation of appropriately structured sulfate esters (Goren, 1970a, 1971; Goren and Kochansky, 1973). In current practice, the sulfatides are stored as solutions in hexane that are kept over a small amount of *aqueous* NaHCO$_3$ to suppress the hydrolysis and to continuously neutralize any traces of acid that are formed from slow degradation of the lipid.

Practically, the spontaneous desulfation was employed in a precise method for determining the molecular weight of the principal sulfatide. NH$_4$SL-I was allowed to degrade completely in ether. The water-soluble NH$_4$HSO$_4$ was recovered by complete extraction with water, percolated through Dowex 50 (H$^+$ form) and the effluent titrated with standard base. The calculated molecular weight of NH$_4$SL-I is 2398 (Goren, 1970a).

5.4.4. Structural Studies on SL-I

The structural studies on the principal sulfatide can be broken down into four parts:

1. Location of the position of the sulfate substituent and identification of trehalose

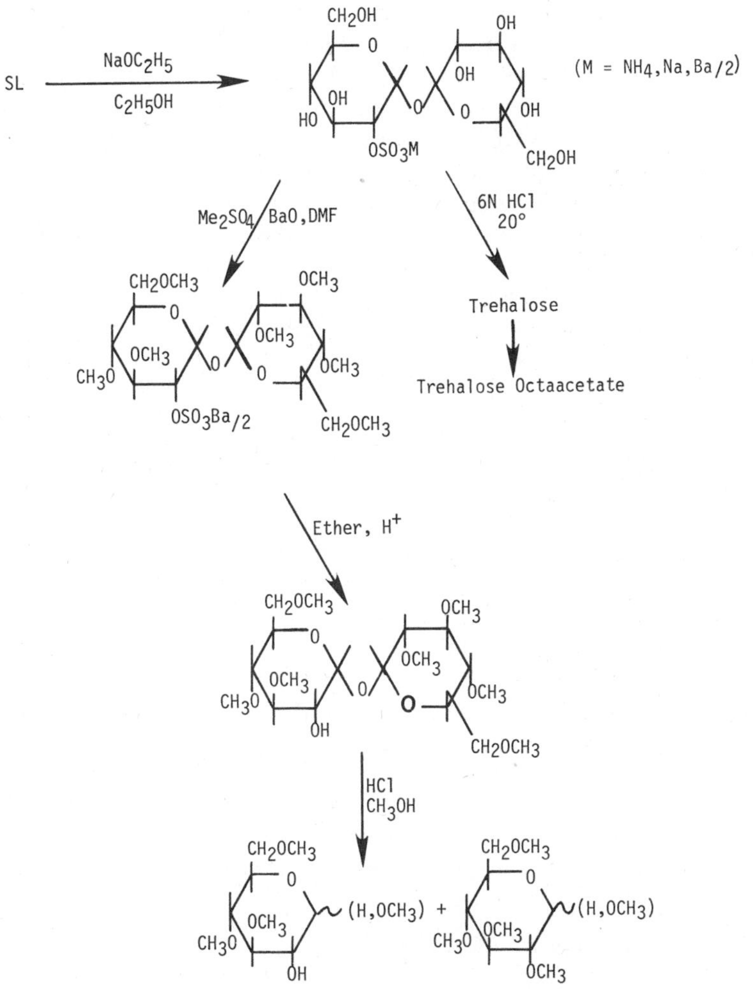

Figure 5-29. Location of the sulfate ester function in SL-I. Alkaline solvolysis released a carbohydrate sulfate salt purified as the Ba derivative. It gave trehalose on prolonged acid-catalyzed hydrolysis. On permethylation and desulfation of the product, a hepta-*O*-methyl trehalose was obtained, which yielded 2,3,4,6-tetra-*O*-methyl glucose and 3,4,6-tri-*O*-methylglucose. Thus, the lone secondary OH in the latter depicts the location of the sulfate.

2. Identification of the acylated positions on the trehalose core
3. Identification of the acyl substituents
4. Correlation of acyl functions with specific positions on the trehalose core

5.4.4.1. Location of the Sulfate Ester: Identification of Trehalose-2-sulfate

As summarized in Fig. 5-29, solvolysis of SL-I with $NaOC_2H_5$ in ethanol resulted in the rapid precipitation of a semicrystalline nonreducing carbohydrate sulfate. It was purified and characterized through the crystalline NH_4^+

and Ba^{2+} salts as a monosulfate ester of trehalose: Ba salt, $[\alpha]_D$ + 154°; m.p. 230–232°(d). The salt was further characterized by conversion to the free acid [percolation through Dowex 50 (H$^+$)] and titration of the effluent with standard base. The molecular weight so determined was within 1.5% of the theoretical. At 20°C in 6 NHCl the NH$_4$ salt was slowly converted to α,α-D-trehalose, which was identified by comparison with an authentic sample: melting point, mixture melting point, infra red spectrum, gas chromatography of the trimethylsilyl derivative; by elemental analysis and by conversion to the crystalline octaacetate.

As shown further in Fig. 5-29, a pure salt of the trehalose sulfate was permethylated to define the position of the sulfate ester group. In dimethylformamide, complete methylation was achieved at low temperature with methylsulfate in the presence of BaO–BaCO$_3$. The sulfate ester is notably resistant to cleavage by alkali: radiolabel (^{35}S) was quantitatively retained during the permethylation. Indeed the Ba heptamethyl trehalose sulfate was recovered from the reaction mixture (after drying) by extraction into ether, in which the permethylated Ba salt is surprisingly soluble. In some studies the carbohydrate sulfate salt was then desulfated in ether with a trace of HCl, as shown in Fig. 5-29. The product was homogeneous by both TLC and by GLC and appeared to be a single substance. Acid methanolysis of the permethylated sugar yielded methyl 3,4,6-tri-O-methyl glucoside and methyl 2,3,4,6-tetra-O-methyl glucoside as the only products and proved, therefore, that in SL-I the sulfate is located solely in a 2-position. The initial product from alkaline solvolysis is, therefore, Na trehalose-2-sulfate. The K-salt was also prepared and was similarly recovered from KOH-catalyzed solvolysis of several of the other sulfatides (Goren *et al.*, 1976, 1978). Therefore, all of the sulfatides are structured on a trehalose-2-sulfate core. This is a secondary equatorial sulfate and confirms the deduction from infrared spectrometry.

5.4.4.2. *Determination of Positions of Acylation*

The acylated positions in all of the sulfatides examined were identified as detailed for SL-I. In Fig. 5-30, SL-I is shown in an abbreviated formulation as 2,3,6,6'-tetraacyl-trehalose-2'-sulfate, a gross structure that was actually determined. In each structure the solid arrow identifies the 2-position that is originally sulfated in the intact lipid in order to allow the fate of this carbon to be followed in the reaction scheme. The native glycolipid was desulfated in ether to yield (**57**) (Fig. 5-30), a product termed SL-I CF (for SL-I cord factor). Desulfated products from the other known SL are designated according to the same scheme. Asselineau and Asselineau (1978) recommended desulfolipid-I, II, and so on, as an alternative and it is attractive. But since all our structural studies have used the SL-CF designation, it will be retained here.

SLI-CF was permethylated with CH$_2$N$_2$ and BF$_3$, under nonisomerizing conditions; the product (**58**) (Fig. 5-30) was quantitatively deacylated by solvolysis. This gave the tetra-O-methyl trehalose **59**, whose structure was determined from the methyl glucose fragments **60** and **61** (Fig. 5-30) that were generated from **59** on acid hydrolysis. The results were fortuitous. In frag-

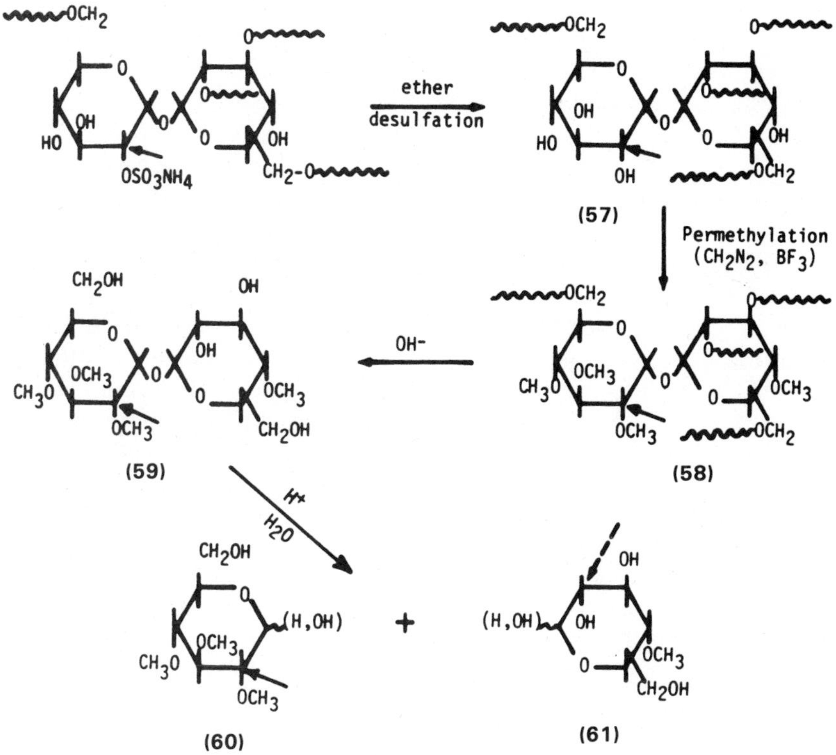

Figure 5-30. Location of the acylated positions in SL-I. NH$_4$ SL-I was desulfated in ether for a few hours at 23°C to free the OH at the 2'-position (solid arrow). Permethylation under nonisomerizing conditions was followed by alkaline solvolysis to yield a tetra-O-methyl trehalose (**59**). Acid hydrolysis gave the fragments (**60**) and (**61**). In (**60**), the solid arrow defines the 2-position originally bearing the sulfate, but now methylated. The 2-position in Fragment (**61**) (broken arrow) is free along with three other unmethylated positions. Therefore, these were blocked by four acyl groups in the native sulfatide.

ment **60,** the 2-position (solid arrow) is methylated indicating this position to have been unsubstituted during the permethylation of SLI-CF. By contrast, in fragment **61** the 2-position (broken arrow) is open and shows that it was blocked (by an acyl group) during the permethylation, since one 2-position originally carried the sulfate group. Accordingly, fragment **60** derives from this glucose moiety that was originally sulfated. Only the 6-position is unmethylated and it was therefore blocked by an acyl function. Fragment **61** derives from the second glucose and is methylated only in the 4-position. C-2, C-3, and C-6 were therefore blocked by acyl functions during the permethylation. Therefore, SL-I as shown in Fig. 5-30 is a 2,3,6,6'-tetra-O-acyl-trehalose-2'-sulfate. If the triacylated glucose had been substituted 3,4,6, this procedure would not have revealed the glucose bearing the sulfate group, since both 2-positions would have been open during premethylation. The locations of the acyl functions in the remaining SL families were derived in a similar manner.

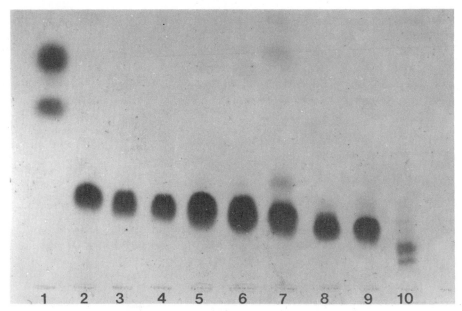

Figure 5-31. Silica gel thin-layer chromatogram of methyl ester fractions obtained from total methanolysis of desulfated SL-I followed by column chromatography on silicic acid–Celite 2 : 1 (Goren *et al.*, 1971). Components **A** and **B** eluted together (row 1). Rows 2–9 form a continuum of homologues of the **C** series. The minor **D** components (row 10) are the lowest members of the **C** series. Chromatogram developed 10 cm in hexane–ether–acetic acid (90 : 10 : 1, v/v), sprayed with 60% H_2SO_4 (dichromate) and charred at 125°C.

5.4.4.3. Nature of the Acyl Substituents

Ethyl or methyl esters of the carboxylic acid substituents were prepared from solvolysis of the various sulfatides. The principal studies were dedicated to SL-I. Early on (Goren, 1970*b*), ethyl esters were quite cleanly separated in TLC examination and appeared to contain four components (designated A, B, C, D) with increasing polarity.

A similar behavior was seen with methyl esters, and these were separated relatively easily by preparative column chromatography on silicic acid (Goren *et al.*, 1971). Figure 5-31 is a TL chromatogram of the products obtained. Esters A and B (row 1) eluted together and were subsequently separated by re-chromatography and also by preparative TLC. Figure 5-31 suggests that the components belonging to the "C" series (rows 2–9) apparently form a continuum, very probably of homologs, an inference ultimately found to be correct. Moreover, a careful rechromatography of the A component showed that it too separated into another continuous series of increasing polarity. The carboxylic acids of component B consist principally of palmitic and stearic acids, about equally, with minor amounts of both lower and higher homologs as recognized in GLC.

Infrared spectra of the methyl esters A, C, and D (row 10) indicated that these were probably closely related fatty acids with multiple methyl branches (intense absorption at 1370 cm^{-1}) and that C and D were apparently hydroxylated variants of the A series. To anticipate, mass spectrometry revealed that the D components were the lowest molecular weight examples of the C continuum, but separated from the remainder by an abrupt discontinuity in the homology.

D represents about 3% of the total hydroxylated branched acids; the principal components of c (Fig. 5-31, fractions 6 and 7) amount to about 80%. Thus, in effect SL-I contains three kinds of acyl substituents: palmitate/ stearate, A and C. Components A and C have been given the trivial names phthioceranic (**55**) and hydroxyphthioceranic (**56**) acids.

$$\begin{array}{ccc} R & CH_3 & CH_3 \\ | & | & | \\ \end{array}$$
$$C_{15}H_{31}\text{-}CH\text{-}[CH\text{-}CH_2]_n\text{-}CH\text{-}COOH$$

(**55**) R = H; n = 4–9; 5–6 principal

(**56**) R = OH; n = 4–9; 7 principal

5.4.4.3a. Hydroxyphthioceranic Acids: Mass Spectrometry. Mass spectrometric examination of the fractions of the C series presented a cleavage pattern (Fig. 5-32) that is consistent with that expected for a series of methyl esters having essentially a constant (C_{16}) portion bearing the hydroxyl group and a variable multibranched fragment, with a pattern of homology that progressively increases by 42 mass units (i.e., 3-carbon units) terminating in the carbomethoxyl function. This spectrum was obtained from sample #7 of Fig. 5-31—a major fraction. The principal component in this fraction gave a molecular ion peak at m/z 622 ($C_{41}H_{82}O_3$), therefore a C_{40} carboxylic acid. The very prominent peaks at m/z 88 and 101 are in accord with the presence of an alpha methyl branch. The multiplicity of methyl branches inferred from typical IR spectra (compare Fig. 5-28) and depicted in Fig. 5-32 was confirmed with ^{1}H-NMR spectrometry. All NMR spectra of C (and of A components as well) were characterized by prominent doublets at 9.1 and 9.2τ attributable to secondary methyl groups. And the mass spectra of various samples of A and C were also in accord with this multimethyl branched structure with homology by 42 mass units.

With reference to Fig. 5-32 the most prominent characteristic of the C series spectra was due to cleavage of the molecule at x with H transfer to the charged right-hand fragment. The residue at the left was, in almost all examples, of mass 240, i.e., palmitaldehyde. Thus, the principal charged fragments are always of mass M-240. For Fig. 5-32, this peak is at m/z 382, corresponding to a fragment with a total of 8 methyl branches and the terminal carbomethoxyl group. The small peak at m/z 424 is due to the M-240 fragment from the next higher homologue, a C_{43} hydroxyphthioceranic acid methyl ester. High resolution examination confirmed the identities of the fragments indicated in Fig. 5-32.

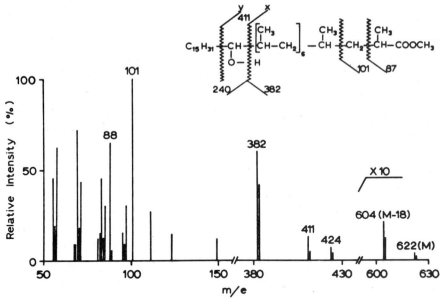

Figure 5-32. Electron impact mass spectrum of a C ester (hydroxyphthioceranate) fraction. The most prominent characteristic is due to cleavage as indicated at x to yield the principal charged fragments of mass M-240, e.g., m/z 382, 424. With similar fragments from other C esters, homology by 3-carbon (42-mass) units is discerned.

Still higher homologues (from C_{43} and C_{46} acids) were recognized in fraction 6 (Fig. 5-31); whereas spectra from more polar samples were attributed to C_{37} and C_{34} acids. The most polar C fraction (sample 12 of Fig. 5-31) was almost homogeneous C_{31} hydroxyphthioceranic methyl ester (5-methyl branches). There was no evidence of a C_{28} component. The most prominent component of D esters (sample 14) discerned by MS was the C_{25} hydroxyphthioceranate, a structure with 3 methyl branches—thus the abrupt discontinuity from the C to the D series.

Mass spectrometry of the O-methyl ether derived from C_{40} hydroxyphthioceranate not only served to confirm the structures of the C series, but was also useful in defining the complete structures of SL I and SL III (Goren *et al.*, 1976, 1978). In the spectra of the native hydroxylated members of the C series, the important peaks derive from the portions of the molecules bearing the branched methyl and terminal carbomethoxyl groups. But, as shown for C_{40} methoxy-phthioceranic methyl ester (Fig. 5-33), the methoxylated C_{16} moiety is dominant in the cleavage of the methyl ether and generates the base peak at m/z 255. Indeed, this pattern of scission is so strong that a low level of CH_2 homology in this fragment, not discernible before, could be inferred from the spectrum of the methyl ether. The prominence of the peak m/z 255 was ultimately found to be unequivocal evidence of methoxyphthioceranates in various samples derived from the sulfatides after a permethylation step (see Section 5.4.4.4).

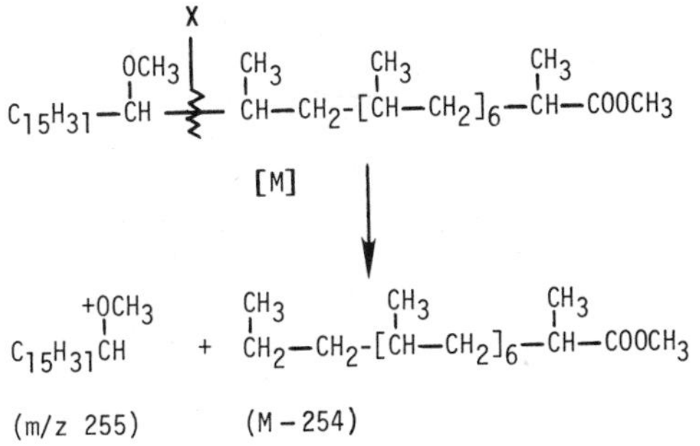

$$\overset{X}{\underset{}{\text{OCH}_3}}$$

$$C_{15}H_{31}-\overset{\text{OCH}_3}{\underset{|}{\text{CH}}} \overset{\xleftarrow{\hspace{0.5cm}}}{\underset{\xleftarrow{\hspace{0.5cm}}}{}} \overset{\text{CH}_3}{\underset{|}{\text{CH}}}-\text{CH}_2-[\overset{\text{CH}_3}{\underset{|}{\text{CH}}}-\text{CH}_2]_6-\overset{\text{CH}_3}{\underset{|}{\text{CH}}}-\text{COOCH}_3$$

Figure 5-33. The *O*-methyl ether of methyl C_{40} hydroxyphthioceranate cleaves at x during electron impact mass spectrometry to generate the charged fragment m/z 255, the base peak. This peak is definitive for methoxyphthioceranate substituents in SL products.

[M]

$$C_{15}H_{31}\overset{+\text{OCH}_3}{\underset{|}{\text{CH}}} \quad + \quad \overset{\text{CH}_3}{\underset{|}{\text{CH}_2}}-\text{CH}_2-[\overset{\text{CH}_3}{\underset{|}{\text{CH}}}-\text{CH}_2]_6-\overset{\text{CH}_3}{\underset{|}{\text{CH}}}-\text{COOCH}_3$$

(m/z 255)　　　　(M − 254)

5.4.4.3b. The Phthioceranic Acids—A Components. The composition of these products was suggested by the i.r. spectrum discussed earlier, i.e., the close relationship to the members of the C series, but lacking the hydroxylic function. This expectation was confirmed from ^{1}H-NMR spectroscopy that supported the important contributions of secondary methyl functions and essentially duplicated the spectrum gotten for members of the C series, save for the OH contribution in the latter. The complex mixture of homologues found in the whole sample of A esters (after separation from the B components) (Fig. 5-31) was examined by mass spectrometry. The results (see Goren *et al.*, 1971) supported the structure in (55)—containing an α-methyl branch, multiple methyl branches associated with mass 42 homology, and molecular ions of mass 480, (C_{31}-phthioceranic ester, principal), 522 (C_{34}), 564 (C_{36}), and 606 (C_{40}) and ultimately 648 and 690 (C_{43}, C_{46}). Separation of the mixture of A esters by careful chromatography afforded almost homogeneous examples of the members of highest molecular weight. All the fractions also provided evidence of CH_2 homology, which we interpret as being due to the participation of both palmitate and stearate in the biosynthesis of these branched acids.

Further examination of the A and C series involved dehydration of hydroxyphthioceranates followed by catalytic hydrogenation, which generated the corresponding phthioceranate (55); oxidation of the hydroxylated form to the ketone; and deuterium exchange to confirm the presence of a tertiary hydrogen on the alpha carbon. All these studies buttressed by mass spectrometric evidence (Goren *et al.*, 1971) supported the structures proposed for the phthioceranic (55) and hydroxyphthioceranic (56) acids and confirmed the relationship between them. As judged from the mass spectrometric evidence the major component of the A series is 2,4,6,8,10,12,14-heptamethyltriacontanoic acid and of the C series, 17-hydroxy-2,4,6,8,10,12,14,16-octamethyldotriacontanoic acid (C_{37} phthioceranic and C_{40} hydroxyphthioceranic acids).

5.4.4.3c. Compositions of the Various Sulfatides. From the estimated mean molecular weights of the components derived from solvolysis of SL-I, and the quantities of each recovered in preparative column chromatography, the components of which are depicted in the TL chromatogram of Fig. 5.31, a close approximation indicates SL-I to contain 1 mole each of palmitate/ stearate and phthioceranate and 2 moles of hydroxyphthioceranate. Even a TLC com-

Table 5-5. Gross Structures of Five SL Families

Sulfatide	Trehalose positions substituted	Moles acyl substituent		
		Palmitate/ stearate	Phthioceranate	Hydroxyphthioceranate
SL-II'	2,4,6,6'	1	0	3
SL-II	2,3,6,6'	1	0	3
SL-1 (principal)	2,3,6,6'	1	1	2
SL-I'	2,3,6,6'	1	2	1
SL-III	2,3,6	1	0	2

parison of the esters obtained as solvolysis products from other sulfatides with those from SL-I suggests that in the SL variants the acyl substituents may be the same as in SL-I, but present in different proportions. Permethylation analysis also showed that only in SL-II' are they in a different positional distribution on the trehalose 2-sulfate core. The additional information on the gross structural features of five SL "families" is given in Table 5-5. Four of the sulfatides are tetraacylated; SL-III alone is a triacyltrehalose-2-sulfate. Complete structural expressions were developed only for SL-I and SL-III (Goren *et al.*, 1976, 1978).

It is curious, and still unresolved, why the tetraacyl sulfatides in which the unhydroxylated phthioceranates are more prominent than the hydroxyphthioceranates have the more polar qualities of the pairs: SL-I more polar than SL-II or II'; SL-I' more polar than SL-I.

5.4.4.3d. Biogenesis of the Phthioceranates: Stereochemistry. All the fractions of branched acids represented in Fig. 5-31 were dextrorotatory: the sample of mixed phthioceranates had $[\alpha]_D^{21} + 7.9°$ and individual fractions from almost complete separations were all dextrorotatory. The relatively homogeneous portion of C esters (fraction 7) had $[\alpha]_D^{21} + 21.8°$. Biosynthesis of these branched acids very likely involves condensation of successive units of methylmalonyl CoA upon a palmitate or stearate acceptor root. This, therefore, is equivalent to incorporations of propionate in accord with the hypothesis of Polgar and Robinson (1951) respecting biogenesis of the (dextrorotatory) phthienoic (mycolipenic) acids (vide **53**). The scheme can account for all the variants in homology that are implicit in the various mass spectra: by 14, 28, and 42 mass units. We documented (Goren *et al.*, 1976, 1978) and regularly have exploited incorporation of [^{14}C]Na propionate to label the sulfatides elaborated by *M. tuberculosis* H37Rv in culture. This is very useful for monitoring the isolation and chromatographic purification of the various sulfatides. Hydrolysis of the individual lipids leads to isolation of heavily labeled phthioceranates and hydroxyphthioceranates, whereas the palmitate/stearate fraction is almost free of marker. In accord with this biogenesis and the dextrorotatory properties of the acids, we judge that they have the same configuration as the phthienoic acids, i.e., L, as established by Ställberg-Sten-

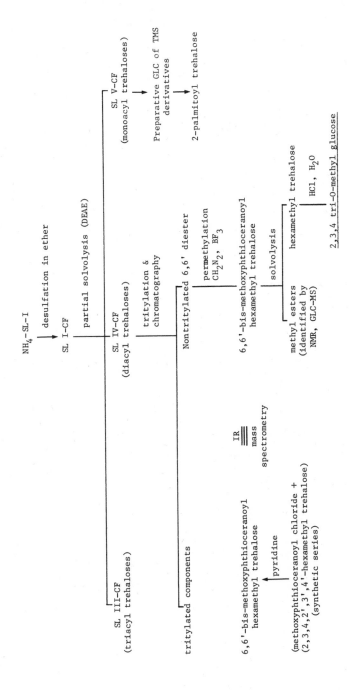

Figure 5-34. Summary of studies leading to assignment of individual carboxylic acids to the acylated positions in SL-I.

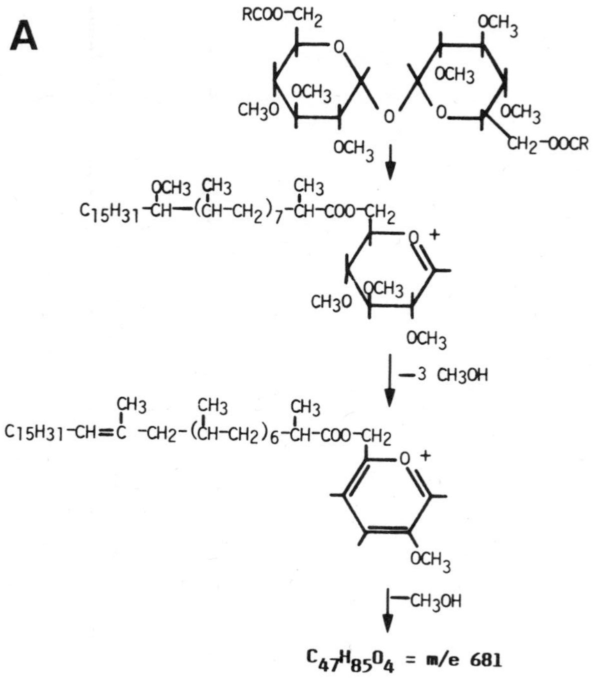
A
—3 CH3OH
—CH3OH
C47H85O4 = m/e 681

(62)

(63) m/e 809

(64) m/e 713

Figure 5-35. (A) Cleavage of (**62**) generates the primary oxonium ion m/z 809, which loses successive CH_3OH (32 mass units) to produce the series 809,777,713,681 (composition confirmed by high-resolution analysis). (B) The simple mass spectrum of permethylated symmetrical SL-IV CF (**62**).

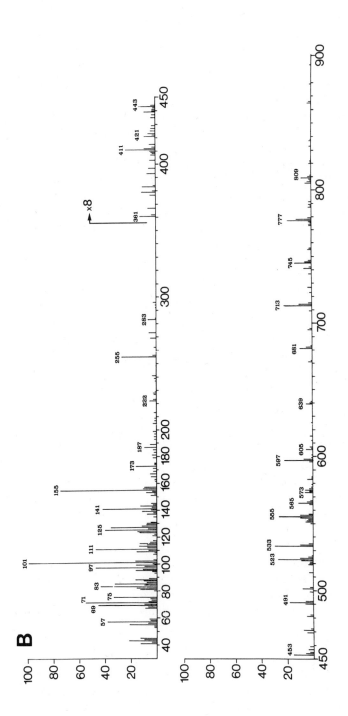

hagen (1954) and Fray and Polger (1956). The phthioceranates are, therefore, of opposite configuration to the well-known seemingly closely related but levorotatory (D) mycocerosic (also called mycoceranic) acids (reviewed by Asselineau, 1966; by Minnikin, 1982 and by Goren and Brennan, 1979). Incorporation of labeled propionate into these branched acids was demonstrated by Gastambide-Odier *et al.* (1963). These considerations led to the choice of "phthioceranic" as the trivial name for these branched acids, a term originally suggested by C. Coleman (see Goren *et al.*, 1971). Although the configuration of the hydroxyl function in the hydroxyphthioceranates has not been resolved, Asselineau and Asselineau (1978) suggest that the D configuration is most probable, based on the positive shift in rotation from phthioceranic to hydroxyphthioceranic acids and provided that the asymmetric center bearing the hydroxyl has one methylene group on one side and a L-methyl branched group on the other. Although the hydroxyphthioceranates have evidently not been recognized in other species of *Mycobacterium,* the unhydroxylated variety were described in lipids of *M. ulcerans* (Daffé *et al.,* 1984).

5.4.4.4. Location of the Individual Acyl Substituents: Complete Structures of SL-I and SL-III

The specific locations of the acyl substituents in SL-I were identified (Goren *et al.,* 1976) as summarized in Fig. 5-34 from Goren and Brennan (1979). Desulfated SL-I was gently and partially solvolyzed with methanol on a column of DEAE (free base form) and the spectrum of products generated was separated as mixtures of triacyl (SL-III CF), diacyl (SL-IV CF), and monoacyl (SL-V CF) trehalose derivatives, respectively. The major group (mixed SL-IV CF) afforded a principal fraction that was unreactive for trityl chloride, therefore consisting of product(s) blocked at the 6- and 6'-positions. It was almost uniform 6,6'-bis-hydroxyphthioceranoyl trehalose (Goren *et al.,* 1976). On permethylation and alkaline solvolysis SL-IV CF yielded 2,3,4,2'3'4-hexa-O-methyl trehalose, identical with a synthesized sample. Examination of the products cleaved from the trehalose core by TLC, NMR and combined GLC-MS showed them to be almost entirely the methyl esters of methoxyphthioceranic acids. Thus, permethylated SL-IV CF is principally 6,6'-bis(methoxyphthioceranoyl)2,3,4,2',3',4'-hexa-O-methyltrehalose (**62**) (Fig. 5-35A). This was synthesized from hexa-O-methyltrehalose and methoxyphthioceranoyl chloride and mass spectra of the synthetic and naturally derived products compared. The spectra were simple (Fig. 5-35B) and essentially identical. As interpreted in Fig. 5-35A, the initial fragmentation of the symmetrical diester **62** yields the primary oxonium ion **63** (m/z 809). By successive losses of CH_3OH (32 mass units) from the ring positions and from the methoxyphthioceranoyl substituent, the ion **63** gives rise to a principal series of peaks m/z 777, 745, 713, and 681. The fragment m/z 713 has the probable structure **64,** a composition confirmed by high-resolution analysis. Peaks attributable to lower (and higher) homologues of methoxyphthioceranates were discernible, but no peaks derivable from unmethoxylated phthioceranates were evident. The m/z = 255 fragment peak described earlier, which is highly specific for methoxyphthioceranate was prominent. And both this peak and the series of peaks m/z 777, 745, 713 were important in the mass spectrum of permethylated SL-I CF, impressive evidence that one glucose moiety in desulfated SL has only a single (hydroxyphthioceranoyl) group, specifically at the 6-position. All the evidence confirms that **62** is symmetrical as deduced before. For SL-I, this accounted for two of the four acyl substituents and their positions. Since only two more substituents are present (palmitate/stearate and phthioceranate) positional assignment to either one would substantially define the complete SL-I structure.

Examination of mixed SL-V CF (Fig. 5-34) the penultimate monoacyl trehalose products from SL-I CF solvolysis showed it to be almost entirely 6-hydroxyphthioceranoyl trehalose. But preparative gas chromatography of the pertrimethylsilyated products allowed recovery and subsequent identification of 2-palmitoyl (stearoyl) trehalose from this sample, and thus to fix the position of this component. A unique confirmation of this assignment was obtained from examining the products derived from gentle *acid-catalyzed* solvolysis of SL-I CF. This too afforded identifiable trehalose 2-palmitate in the monoacyl trehalose fraction. But the diacyl trehalose obtained during the solvolysis under acid conditions (SL-IV′ CF) was different from the SL-IV CF gotten before. The acyl components in this product consisted of palmitate/stearate and *phthioceranate* (**55**). As described below, it has the structure (**65**): 2-palmitoyl/stearoyl-3-phthioceranoyltrehalose. The mass spectrum of

$$CH_2OH \qquad OOCCH[CH_2-CH]_nCH_2C_{15}H_{31}$$

(65)

RCOOH = palmitic or stearic acid

n = 4 to 7

permethylated SL-IV′ CF (Fig. 5-36) confirmed this acyl group distribution, and it provided unequivocal evidence that the 3-position is occupied almost exclusively by phthioceranates, albeit with a small contribution by hydroxyphthioceranates (**56**). This *small* contribution is deduced in part from the minor aspect of the peak m/z = 255, and it was in accord with the recognition of a minor amount of hydroxyphthioceranate in the solvolysis product of SL-IV′ CF. A complete analysis of the mass spectrum, detailed by Goren *et al.* (1976) implicated the presence of at least four homologs of pthioceranate along with palmitate and stearate.

In a brief recapitulation, the oxonium ion generated from permethylated SL-IV CF has the structure (**66**) (Fig. 5-37) deduced from the mass spectrum. Scissions that this ion undergoes at *x* generate an abundance of informative fragments. As summarized in Table 5-6, the peaks m/z 1031, 1003, 989, 961, 947, 919, 905, and 877 are attributable to structure **66** with 42 mass unit

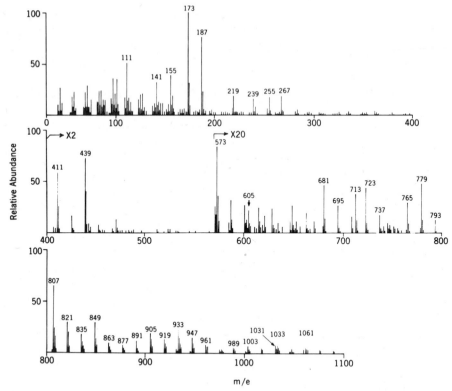

Figure 5-36. The mass spectrum of permethylated SL-IV' CF, much more complex than that of permethylated SL-IV CF, Fig. 5-35B, with which it should be compared. Many of the peaks in the higher mass region can be accounted for by fragmentation of the primary oxonium ion (**66**) (see text, Table 5-6 and Fig. 5-37).

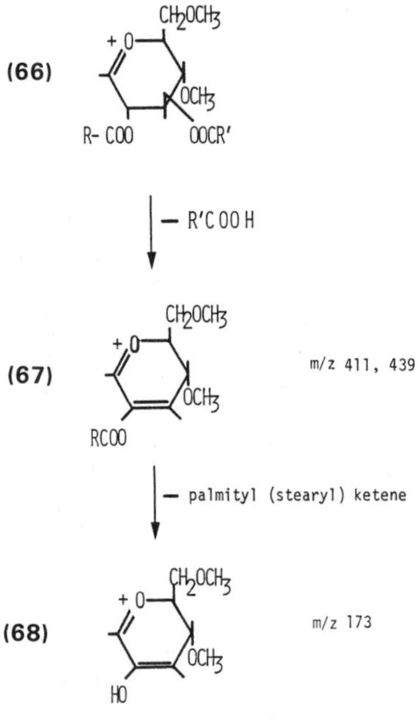

(66)

(67)   m/z 411, 439

(68)   m/z 173

Figure 5-37. Fragmentation of a 2,3-diacyl oxonium ion structured like (**66**). The 3-acyl function is eliminated first as a carboxylic acid giving the important peaks m/z 411 and 439 (Fig. 5-36), followed by expulsion of palmityl or stearyl ketene to generate the base peak m/z 173. R, palmitate or stearate; R', principally phthioceranate homologues.

Table 5-6. Analysis of Fragments 66 and Their Scission Products

2-Acyl functions	Fragment 66 n(phthioceranate)	m/z	m/z-224	Components[a]
Stearate	7	1031	807	$C_{50}H_{95}O_7$
Palmitate	7	1003	779	$C_{48}H_{91}O_7$
Stearate	6	989	765	$C_{47}H_{89}O_7$
Palmitate	6	961	737	$C_{45}H_{85}O_7$
Stearate	5	947	723	$C_{44}H_{83}O_7$
Palmitate	5	919	695	$C_{42}H_{79}O_7$
Stearate	4	905	681	$C_{41}H_{77}O_7$
Palmitate	4	877	653	$C_{39}H_{73}O_7$

[a]Composition confirmed by high-resolution analysis.

homology in the phthioceranate moiety decreasing from $n = 7$ to $n = 4$, and with contributions from stearate on one hand and palmitate on the other. From scissions at x that result in the loss of 224 mass units (x-H), structure **66** gives rise to the peaks m/z 807, 779, 765, 737, 723, 695, 681, and 653, whose compositions were confirmed by high resolution analysis. In accord with these assignments, the fragments do not accommodate contributions from methoxyphthioceranate. Therefore, the principal contributions to SL-IV' CF are from phthioceranate, with palmitate and stearate.

(66)

The crucial assignment of the 2,3-diacyl structure for SL-IV CF and identification of the specific substituents located there also derive from the mass spectrum. As analyzed in detail earlier (Goren *et al.*, 1976, 1978) and summarized in Fig. 5-37, in the further fragmentations of oxonium ion **66,** an acyl group at the 3-position is eliminated as the carboxylic acid to leave (in this instance) fragment **67** that retains the substituent(s) at the 3-position. From the important peaks m/z 411, 439 these are recognized as palmitate and stearate, respectively. They are then expelled as the respective ketenes to generate the same fragment m/z 173 (**68**). This is the base peak in the spectrum of permethylated SL-IV' CF.

The second glucose moiety, not acylated, of permethylated SL-IV' CF is recognized in the mass spectrum from the important fragments to which it gives rise (Fig. 5-38). Therefore, the different diacyl trehalose products derived in the gentle alkaline and in the weakly acidic solvolysis of SL-I CF provide the composite evidence from which the principal structure for SL-I is derived: 2-palmitoyl(stearoyl)-3-phthioceranoyl-6,6'-bis(hydroxyphthioceranoyl)trehalose-2'-sulfate (Fig. 5-39).

The structure of the minor sulfatide SL-III, 2-palmitoyl(stearoyl)-3,6'-

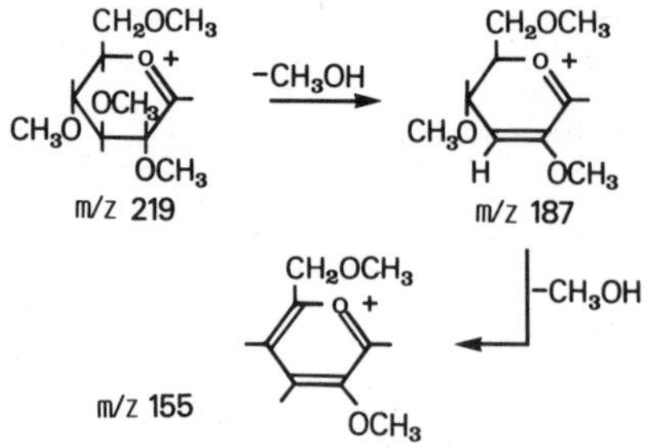

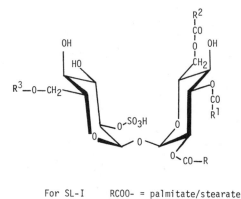

Figure 5-38. Analysis of peaks in Fig. 5-36 attributed to the nonacylated glucose moiety of SL-IV′ CF.

bis(hydroxyphthioceranoyl)trehalose-2′-sulfate was elucidated in a somewhat simpler fashion (Goren *et al.*, 1978). The mass spectrum of permethylated SL-III CF confirmed that the glucose moiety originally bearing the sulfate was otherwise unsubstituted. For the second glucose moiety (acylated with one palmitate/stearate and two hydroxyphthioceranates) a sequence of eliminations as described for the primary oxonium ion from SL-IV′ CF (**66**) (Fig. 5-37) reveals a methoxyphthioceranate and palmitate/stearate at the 3- and 2-positions, respectively. The second hydroxyphthioceranate is at the 6-position. The specific cleavage of methoxyphthioceranates that generates the characteristic peak m/z = 255 was very prominent in this spectrum and, occurring in higher mass fragments, generated informative peaks in the inter-

```
For SL-I      RCOO-  = palmitate/stearate

              R¹COO- = phthioceranate

    R²COO-, R³COO- = hydroxyphthioceranate

For SL-III    RCOO-  = palmitate/stearate

       R¹COO- = R²COO- = hydroxyphthioceranate

              R³ = H
```

Figure 5-39. Structures of SL-I and SL-III according to the conformation of Asselineau and Asselineau (1978).

mediate mass range. This spectrum was thoroughly analyzed in the earlier study (Goren *et al.*, 1978). In sum, these investigations led to structural expressions for SL-I and SL-III, as given in Fig. 5-39, according to the conformation deduced by J. and C. Asselineau.

[I have always expressed my reticence at pointing to and correcting errors in the writings of others. In any complex undertaking, these inevitably occur. And I have stated (in Section 5.2.9.2) that, not being without similar sins, I am reluctant to cast stones. Therefore, it is appropriate to mention here that Professor Jean Asselineau once informed me, as kindly as possible, that the structures for the sulfatides that I had published for years (Goren, 1970*a*; Goren *et al.*, 1970, 1976, 1978) were unfortunately incorrect in that they depicted one of the glucose moieties as D and the other as L! Truly, "Writing maketh an exact man!"]

5.4.5. Association of Sulfatides with Virulence

Shortly after discovery of the sulfatides, when Middlebrook held the opinion that "this material plays a prominent role in the biology of tuberculosis" (Middlebrook *et al.*, 1959), examination of a limited number of British and Indian strains of *M. tuberculosis* by Gangadharam *et al.* (1963) suggested a significant correlation between sulfolipid content and their order of infectivity for guinea pigs (see Mitchison, 1963, 1964). We subsequently tested this correlation in a much larger group of strains (forty) for which "root indices of virulence" had earlier been established by Mitchison and colleagues (Goren *et al.*, 1974*a*). Most of the strains examined were of phage types A and I (see Grange *et al.*, 1978; Grange, 1982). Type A strains were usually rich in strongly acidic lipids (SAL), most often due to SL content, and were virulent. Phage-type I strains, principally from India and surrounding countries usually were of low SAL and SL contents and carried a phenotypic attenuation indicator lipid (AIL) marker—a *p*-methoxyphenylphthiocerol diester. These I-type strains were invariably of low virulence and thus reinforced the correlation between attenuation and low levels of SAL established earlier. But a small group of phage-type B, also low SAL strains, but from Britain, were aberrant in being of high virulence. Accordingly, if this were usual for type B strains, then it seemed unlikely that the similarly low content of SAL (and SL) in type I strains was the direct cause of attenuation.

In an attempt to settle this question, a much larger group of type "B" strains and still another group of type I strains were examined. Among Indian-type strains (usually I) low SAL and SL values were associated with attenuation in the guinea pig. But in the Western-type strains—surprisingly both B and I—low SAL concentration and high virulence occurred together (Goren *et al.*, 1982). Thus, the possibility that SL is a major determinant of virulence seems unlikely.

5.4.6. Distribution of Mycobacterial Sulfatides

The previous section indicated that the sulfatides are not necessarily abundant in, and may be almost absent from, even virulent strains of *M.*

tuberculosis. Reports of the occurrence of SL in considerable amounts in such strains as *M. smegmatis* and *M. phlei* were therefore indeed provocative (Prahbudesai *et al.,* 1981; Malik *et al.,* 1982). Elaboration of the sulfatides by saprophytic mycobacteria would simplify by orders of magnitude studies of SL biosynthesis by essentially eliminating the operational hazard of undertaking such studies with fully virulent strains. However, Dhariwal *et al.* (1984) were unsuccessful in recognizing any sulfatides of *M. tuberculosis* in the lipid extracts obtained from several different strains of *M. smegmatis* or *M. phlei.* We were unable to account for the difference in results but postulated that in the studies from the group of Khuller, ubiquitous trehalose dimycolate may have been mistaken for the principal sulfatide of *M. tuberculosis,* possibly because of inadvertent desulfation of the reference SL-I sample during storage. Desulfated SL-I has TLC mobility similar to that of TDM in some solvent systems. Although sulfolipids have been found in other species of mycobacteria (McCarthy, 1976; Tsukamura *et al.,* 1984; Tsukamura and Mizuno, 1986) these sulfur-containing products, recognized only through incorporated ^{35}S, have been only minimally characterized, and in my judgment do not resemble the sulfatides of *M. tuberculosis.* All evidence currently available suggests that sulfatides such as represented in Fig. 5-39 are restricted to *M. tuberculosis* and indeed are not ubiquitous in even this species.

5.4.7. Synthesis, Biosyntheis

Elaboration of the phthioceranates and hydroxyphthioceranates very likely proceeds from condensation of successive molecules of propionate, via methyl malonyl CoA, onto a palmitate or stearate receptor. The incorporation of propionate is in accord with the interpretation of Polgar and Robinson concerning biogenesis of the (L) phthienoic acids and Gastambide-Odier's documentation of labeled propionate incorporation into the branched (D) mycocerosic acids (see Section 5.4.4.3d). The phthioceranates are evidently higher-molecular-weight enantiomers of these mycocerosic (mycoceranic) acids. Their biosyntheses are judged to occur in the same fashion but probably involve different enzymes:

$$
\begin{array}{c}
\mathrm{COOH} \\
| \\
\mathrm{C_{15}H_{31}\text{-}CO\text{-}SACP} + n\ \mathrm{CH_3\text{-}CH\text{-}COSCoA} \\[4pt]
+\ 2n\ \mathrm{NADPH} + 2n\ \mathrm{H^+} \qquad\qquad \longrightarrow \\[8pt]
\begin{array}{cc}
\mathrm{CH_3} & \mathrm{CH_3} \\
| & | \\
\end{array} \\
\mathrm{C_{15}H_{31}\text{-}CH_2\text{-}[CH\text{-}CH_2]_{n-1}\text{-}CH\text{-}COOH} \\[4pt]
+\ 2n\ \mathrm{NADP^+} + (n-1)\ \mathrm{H_2O} + n\ \mathrm{CoA\text{-}SH} + \mathrm{HS\text{-}ACP}
\end{array}
\tag{9}
$$

where $n = 4.9$. In the instance of the hydroxyphthioceranates, it is not known at what stage of synthesis the hydroxyl group is generated; but it probably is a product of the initial Claisen condensation of palmitate with methyl malonyl CoA. In that event, it seems likely that two parallel biosynthetic pathways are required for the two series (Asselineau and Asselineau, 1978).

5.4.7.1. Synthesis of Model Compounds

Earlier we described synthesis of 6,6'-bis(methoxyphthioceranoyl)-hexa-O-methyltrehalose as a reference compound for mass spectrometric analysis of SL-IV CF. Liav and Goren (unpublished) also synthesized trehalose-6,6'-bishydroxyphthioceranate via 6,6'-di-O-methanesulfonyl-hexa-O-benzyl trehalose (see Section 5.2.9.1) in order to compare its toxicity in mice with that of trehalose dicorynomycolate, which bears the relatively bulky α-branch. As noted in Table 5-3, the two were of comparable toxicity, suggesting that the long α-branch was not required, except possibly in influencing the temperature of phase transition from the gel to the liquid state (Durand *et al.*, 1979).

Synthesis of the core trehalose 2-sulfate was readily achieved via 4,6,4',6'-dibenzylidene trehalose as summarized in Fig. 5-40 (Liav and Goren, 1984c). The methodology also proved useful for the synthesis of 3,4,6-tri-O-methyl glucose (Liav and Goren, 1984d), exploiting a 2-sulfate ester as a blocking group stable to alkali during permethylation, but easily removed subsequently by gentle acid-catalyzed desulfation in slightly moist dioxane.

5.4.8. Biological Activities

Owing to their amphipathic qualities, the sulfatides, like cord factor, derange mitochondrial function *in vitro*; but probably because of antagonism by serum albumin, this activity was not expressed *in vivo*: injected intraperitoneally in oil solution into mice, the sulfatides are entirely innocuous (Kato and Goren, 1974). Furthermore, also because of their amphipathic qualities, it is not surprising that the sulfatides can effectively replace cord factor as components of Ribi's antitumor vaccines: in the line 10 tumor system, SLI or SL-III in combination with Re mutant endotoxin in oil (aqueous emulsion) gives very high cure rates after intralesional injection. Neither SLI nor SL-III alone is effective (Yarkoni *et al.*, 1979).

Early studies correlating root indices of virulence with sulfatide content (see Section 5.4.5) suggested that these strongly acidic lipids might have a recognizable role in the pathogenesis of tuberculosis—perhaps supporting the intracellular survival of the pathogen. We speculated (Goren *et al.*, 1974) that the demonstrable surface active properties of SL might be directed against lysosomal membranes, just as mitochondrial membranes are attacked—with possible disruption and intracellular release of damaging lysosomal enzymes. It also seemed possible that within the phagosomal environment the strongly acidic lipid at the surface of the tubercle bacilli might ionically adsorb and possibly distort, and so antagonize, lysomal enzymes delivered to the phagosome or, by orientation within the phagosomal membrane, interfere with its functional properties.

A collaborative study with P. D. Hart, M. R. Young, and J. A. Armstrong

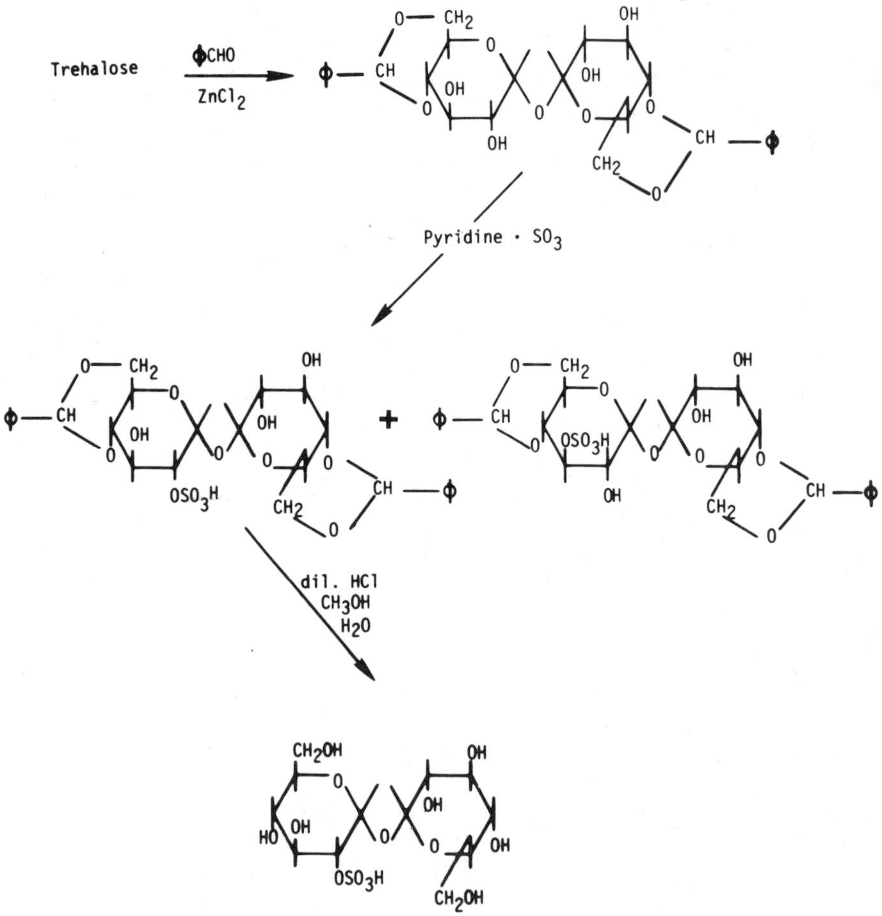

Figure 5-40. Synthesis of trehalose-2-sulfate. Sulfation of 4,6,4',6'-di-*O*-benzylidene trehalose with pyridine•SO₃ complex leads to both 2- and 3-sulfates. The blocking groups are removed by dilute aqueous/methanolic HCl, toward which the sulfate ester is stable.

using cultured mouse peritoneal macrophages showed that the sulfatides were notably lysosomotropic and appeared to inhibit fusion of these organelles with newly formed phagosomes (Goren *et al.*, 1976). Indeed viable virulent tubercle bacilli from within the confines of the phagosome also prevent fusion of this organelle with normal lysosomes, a facility that may promote survival of this intracellular pathogen (Armstrong and Hart, 1971; Goren, 1977). However, a long series of studies that followed suggested that the intervention of sulfatides in antagonizing the fusion behavior is uncertain (Goren and Brennan, 1979; Goren *et al.*, 1987a,b).

Studies by Myrvik *et al.* (1984) show that virulent tubercle bacilli (H37Rv) phagocytosed by rabbit alveolar macrophages can "escape" from the potentially hostile environment of the phagosome to survive in the cytoplasmic compartment, thus shielded from the lysosomal hydrolases. The avirulent

H37Ra does not exhibit like behavior, but H37Rv in macrophages from rabbits immunized with bacillus Calmette–Guérin (BCG) show a much lower incidence of the escape phenomenon (Leake *et al.*, 1984). As noted above, we speculated earlier that the sulfatides might contribute to membrane disruption; and R. L. Hunter (unpublished results) earlier invoked the participation of trehalose dimycolate coated onto polystyrene beads as a mediator of phagosomal escape (see below). In recent unpublished studies, Q. Myrvik, E. Leake, and M. Goren have examined the intraphagosomal behavior of 5 μm polystyrene divinyl benzene beads coated with various quantities of TDM, or SL-I, or of combinations of both glycolipids, but found no evidence that these can attack and disrupt phagosomal membranes. However, Hunter *et al.* recently reported that coating of 1-μm polystyrene beads (not cross-linked) with TDM produces a target for phagocytosis that disrupts the phagosomal membranes and ultimately irreversibly damages the cell (Hunter *et al.*, 1987).

An entirely new role for the sulfatides in the pathogenicity of tubercle bacilli is implicit in current studies by M. J. Pabst *et al.* (1987) showing that SL-I blocks macrophage activation in cultured human monocytes. Monocytes activated *in vitro* with either lipopolysaccharide or interferon γ release increased amounts of superoxide anion when stimulated with *N*-formyl-methionyl-leucyl-phenylalanine (F-met-leu-phe) or phorbol myristate acetate (PMA). But if the activated monocytes were treated with 10 μg SLI/ml, the enhanced superoxide response was entirely abrogated, along with other expressions of the activated state. The sulfatide had no effect if it was added after the cells had been stimulated by F-met-leu-phe or PMA, suggesting that it does not inhibit enzymes involved in superoxide release, but rather that it inhibited the process of activation. These provocative findings may lead to important new areas of research. If the antagonism to expression of the activated state in monocytes/macrophages is a necessary component of pathogenicity in *M. tuberculosis*, then it is certain that additional mycobacterial substances having this activity will surely be found.

5.5. Epilogue

The pioneering investigations by R. J. Anderson and colleagues on mycobacterial lipids (Review: Anderson, 1940) were originally sponsored by the Medical Research Committee of the National Tuberculosis Association:

> The objects of this . . . investigation were two-fold: first, to secure compounds of reasonable purity for biological studies, and secondly to determine the chemical composition of the various substances as nearly quantitatively as possible" (Anderson, 1940). The results of earlier investigations to which Anderson alluded "all agreed on one point at least, namely that the bacillus contained an unusually high percentage of lipids ranging from 20 to 40 percent."

And surely it was this unusual property of the tubercle bacillus that, through studies of the biological activities, sought to link some of these components with the ability of the microorganism to produce disease.

Although this specific goal has not been achieved in the 50 years since

Anderson began his studies, it seems increasingly likely that promoting the intracellular survival of the tubercle bacillus will prove an important component of the virulence armamentarium of the pathogen and that important roles for the trehalose glycolipids in contributing to this survival will ultimately be firmly established.

ACKNOWLEDGMENTS. I thank Shirley Downs and Marjorie McCormick for their care and diligence in preparing the manuscript, and Nadia de Stackelburg and Ethel Goren for the many figures of complex chemical structures. Judith Warren Fiscus, Olga Brokl, Drs. Avraham Liav, K.-S. Jiang, K. R. Dhariwal and Jiro Maeda have been valuable colleagues in our research efforts described herein. I thank W. B. Saunders Co. for permission to reproduce Figs. 5-2, 5-28, 5-29, 5-30, 5-34, 5-38; Springer-Verlag for Fig. 5-5; *Carbohydrate Research* for Fig. 5-8; *Chemistry and Physics of Lipids* for Figs. 5-10, 5-14, 5-16; *American Review of Respiratory Diseases* for Fig. 5-17; Marcel Dekker for Fig. 5-29; the *Journal of General Microbiology* for Figs. 5-21, 5-25, 5-26. Figures 5-31, 5-32, 5-35, 5-36, and 5-37 were reproduced with permission from *Biochemistry 10*:72–79 and *Biochemistry 15*:2728–2735, copyright 1971 and 1976, respectively, by the American Chemical Society. Work of the author cited in this chapter was supported in part by U.S. Public Health Service grant AI-08401 from the U.S.–Japan Cooperative Medical Science Program, administered by the National Institute of Allergy and Infectious Diseases. The author is the Margaret Regan Investigator in Chemical Pathology, National Jewish Center for Immunology and Respiratory Medicine.

References

Adam, A., Senn, M., Vilkas, E., and Lederer, E., 1967, Spectrométrie de masse de glycolipides. 2. Diesters de tréhalose naturels et synthétiques, *Eur. J. Biochem.* **2**:460.

Aebi, A., Asselineau, J., and Lederer, E., 1953, Sur les lipides de la souche humaine "Brévannes" de *Mycobacterium tuberculosis, Bull. Soc. Chim. Biol.* **35**:661.

Anacker, R. L., Matsumoto, J., Ribi, E., Smith, R. F., and Yamamoto, K., 1973, Enhancement of resistance to tuberculosis by purified components of mycobacterial lipid fractions, *J. Infect. Dis.* **127**:357.

Anderson, R. J., 1940, The chemistry of the lipids of tubercle bacilli, *The Harvey Lecturers* **35**:271.

Arnarp, J., and Lönngren, J., 1978, Alkylation of carbohydrates with alkyl trifluoromethanesulfonates, *Acta. Chem. Scand.* **B32**:465.

Artman, M., Bekierkunst, A., and Goldenberg, I., 1964, Tissue metabolism in infection: Biochemical changes in mice treated with cord factor, *Arch. Biochem. Biophys.* **105**:80.

Asselineau, C., and Asselineau, J., 1978, Trehalose-containing glycolipids, *Prog. Chem. Fats Other Lipids* **16**:59.

Asselineau, C. P., and Montrozier, H. L., 1976, Étude du processus de biosynthèses des acides phléiques, acides polyinsaturés synthetisés par *Mycobacterium phlei, Eur. J. Biochem.* **63**:509.

Asselineau, C., Montrozier, H., and Promé, J. C., 1969, Présence d'acides polyinsaturés dans une bactèrie: Isolement, à partir des lipides de *Mycobacterium phlei*, d'acide hexatriacontapentaène-4,8,12,16,20, öique et d'acides analogues, *Eur. J. Biochem.* **10**:580.

Asselineau, C. P., Montrozier, H. L., Promé, J.-C., Savagnac, A. M., and Welby, M., 1972, Étude d'un glycolipide polyinsaturé synthétisé par *Mycobacterium phlei, Eur. J. Biochem.* **28**:102.

Asselineau, C., Clavel, S., Clérment, F., Daffé, M., David, H., Lanéelle, M. A., and Promé, J. C., 1981, Constituants lipidiques de *Mycobacterium leprae* isolé de tatou infecté expérimentalment, *Ann. Microbiol. Inst. Pasteur* **132(A)**:19.

Asselineau, J., 1966, *Bacterial Lipids*, Holden-Day, San Francisco.

Asselineau, J., and Kato, M., 1973, Chemical structure and biochemical activity of cord factor analogs. II. Relationships between activity and stereochemistry of the sugar moiety, *Biochimie* **55**:559.

Asselineau, J., and Kato, M., 1975, Relationships between chemical structure and biochemical activity of trehalose-6,6'-dimycolate (cord factor) of *Mycobacterium tuberculosis, Jpn. J. Med. Sci. Biol.* **28**:94.

Asselineau, J., Durand, E., Lanéèlle, G., Rouanet, J-M., and Tocanne, J-F., 1981, Structure–activity relationships of wall glycolipids of mycobacteria and corynebacteria, in *Actinomycetes Zbl. Bakt. Suppl.* **11** (K. P. Schaal and G. Pulverer, eds.), pp. 407–414, Fischer Verlag, Stuttgart.

Azuma, I., and Yamamura, Y., 1962, Studies on the toxic substances isolated from mycobacteria. I. Toxic glycolipids of *Mycobacterium fortuitum* and atypical mycobacteria strain P_{16}, *J. Biochem.* **52**:82.

Azuma, I., and Yamamura, Y., 1963, Studies on the firmly bound lipids of human *Tubercle bacillus.* II. Isolation of arabinose mycolate and identification of its chemical structure, *J. Biochem.* **53**:275.

Azuma, I., Ribi, E., Meyer, T. J., and Zbar, B., 1974, Biologically active components from mycobacterial cell walls. I. Isolation and composition of cell wall skeleton and component P_3, *J. Natl. Cancer Inst.* **52**:95.

Barclay, W. R., Anacker, R., Brehmer, W., and Ribi, E., 1967, Effects of oil-treated mycobacterial cell walls on the organs of mice, *J. Bacteriol.* **94**:1736.

Barksdale, L., and Kim, K-S., 1977, *Mycobacterium, Bacteriol. Rev.* **41**:217.

Barrow, W. W., and Brennan, P. J., 1982, Immunogenicity of the type-specific C-mycoside glycopeptidolipids of mycobacteria, *Infect. Immun.* **36**:676.

Bartlett, G. L., and Zbar, B., 1972, Tumor-specific vaccine containing *Mycobacterium bovis* and tumor cells: Safety and efficacy, *J. Natl. Cancer Inst.* **48**:1709.

Bast, R. C., Jr., Bast, B. S., and Rapp, H. J., 1976, Critical review of previously reported animal studies of tumor immunotherapy with nonspecific immunostimulants. International Conference on Immunotherapy of Cancer, *Ann. NY Acad. Sci.* **277**:60.

Bekierkunst, A., 1968, Acute granulomatous response produced in mice by trehalose-6,6'-dimycolate, *J. Bacteriol.* **96**:958.

Bekierkunst, A., and Artman, M., 1966, Tissue metabolism in infection: DPNase activity, DPN levels, and DPN-linked dehydrogenases in tissues from normal and tuberculosis mice, *Am. Rev. Respir. Dis.* **86**:832.

Bekierkunst, A., Levij, I. S., Yarkoni, E., Vilkas, E., Adam, A., and Lederer, E., 1969, Granuloma formation induced in mice by chemically defined mycobacterial fractions, *J. Bacteriol.* **100**:95.

Bekierkunst, A., Levij, I. S., Yarkoni, E., Vilkas, E., and Lederer, E., 1971, Suppression of urethane-induced lung adenomas in mice treated with trehalose-6,6'-dimycolate (cord factor) and living bacillus Calmette–Guérin, *Science* **174**:1240.

Bekierkunst, A., Wang, L., Toubiana, R., and Lederer, E., 1974, Immunotherapy of cancer with nonliving BCG and fractions derived from mycobacteria: Role of cord factor (trehalose-6,6'-dimycolate) in tumor regression, *Infect. Immun.* **10**:1044.

Birch, G., and Richardson, A. C., 1970, Chemical modification of trehalose. Part II. Synthesis of the *galacto* analogue of trehalose, *J. Chem. Soc.* **1970**:749.

Bloch, H., 1950, Studies on the virulence of tubercle bacilli. Isolation and biological properties of a constituent of virulent organisms, *J. Exp. Med.* **91**:197.

Bloch, H., and Noll, H., 1955, Studies on the virulence of tubercle bacilli. The effect of cord factor on murine tuberculosis, *Br. J. Exp. Pathol.* **36**:8.

Bottle, S., and Jenkins, I. A., 1984, Improved synthesis of "cord factor" analogues, *J. Chem. Soc. Chem. Commun.* **1984**:385.

Bouveng, H., Lindberg, B., and Theander, O., 1957, New synthesis of 4-*O*-methyl-D-glucose, *Acta Chem. Scand.* **11**:1788.

Bowden, K., Heilbron, I. M., Jones, E. R. H., and Weedon, B. C. L., 1946, Researches on acetylenic compounds. Part I. The preparation of acetylenic ketones by oxidation of acetylenic carbinols and glycols, *J. Chem. Soc.* **1946**:39.

Bredereck, H., 1930, Zur konstitution des trehalose, *Ber* **63(B)**:959.

Brennan, P. J., 1984, Antigenic peptidoglycolipids, phospholipids and glycolipids, in *The Mycobacteria. A Source Book* (G. P. Kubica and L. G. Wayne, eds.), pp. 467–489, Marcel Dekker, New York.

Brennan, P. J., and Goren, M. B., 1979, Structural studies on the type-specific antigens and lipids of the *Mycobacterium avium–Mycobacterium intracellulare–Mycobacterium scrofulaceum* serocomplex, *Mycobacterium intracellulare* serotype 9, *J. Biol. Chem.* **254**:4205.

Brennan, P. J., Aspinall, G. O., and Nam-Shin, J. E., 1981, Structures of the specific oligosaccharides from the glycopeptidolipid antigens of serovars in the *Mycobacterium avium–Myobbacterium intracellulare–Mycobacterium scrofulaceum* complex, *J. Biol. Chem.* **256**:6817.

Brocheré-Ferréol, G., and Polonsky, J., 1958, Sur la synthèse de substances à activité de "cord factor." Synthèses de diesters de tréhalose en position 6,6′, *Bull. Soc. Chim. Fr.* **1958**:714.

Bruneteau, M., and Michel, G., 1968, Structure d'un dimycolate d'arabinose isolé de *Mycobacterium marianum*, *Chem. Phys. Lipids* **2**:229.

Castro, B., Chapleur, Y., and Gross, B., 1973, Sels d'alkyloxytris(dimethyl-amino)phosphonium. IV. Activation selective de l'hydroxyle primaire de quelques hexosides, *Bull. Soc. Chim. Fr.* **1973**:3034.

Chatterjee, D., Cho, S-N., and Brennan, P. J., 1986, Chemical synthesis and seroreactivity of *O*-(3,6-di-*O*-methyl-β-D-glucopyranosyl)-(1 → 4)-*O*-(2,3-di-*O*-methyl-α-L-rhamnopyranosyl)-(1 → 9)-oxynonanoyl–bovine serum albumin– the leprosy-specific, natural disaccharide octylneoglycoprotein, *Carbohydr. Res.* **156**:39.

Davidson, L. A., Draper, P., Minnikin, D. E., 1982, Studies on the mycolic acids from the walls of *Mycobacterium microti*, *J. Gen. Microbiol.* **128**:823.

Daffé, M., Lanéélle, M. A., Asselineau, C., Lévy-Frébault, and David, H., 1983, Intéret taxonomique des acides gras des mycobactéries: Proposition d'une méthode d'analyse, *Ann. Microbiol. Inst. Pasteur* **134(B)**:241.

Daffé, M., Lanéélle, M. A., Roussel, J., and Asselineau, C., 1984, Lipides spécifiques de *Mycobacterium ulcerans*, *Ann. Microbiol. Inst. Pasteur* **135(A)**:191.

Dhariwal, K. R., Dhariwal, G., and Goren, M. B., 1984, Observations on the ubiquity of the *Mycobacterium tuberculosis* sulfatides in mycobacteria, *Am. Rev. Respir. Dis.* **130**:641.

Dhariwal, K. R., Yang, Y.-M., Fales, H. M., and Goren, M. B., 1987, Detection of trehalose monomycolate in *Mycobacterium leprae* grown in armadillo tissues, *J. Gen. Microbiol.* **133**:201.

Draper, P., 1979, Annex 1, protocol 1/79. Report of the enlarged supervisory council meeting for research on the immunology of leprosy, *Geneva World Health Organization.*

Draper, P., Dobson, G., Minnikin, D. E., and Minnikin, S. M., 1982, The mycolic acids of *Mycobacterium leprae* harvested from experimentally infected nine-banded armadillos, *Ann. Microbiol. (Paris) (Inst. Pasteur)* **133B**:39.

Dubos, R. J., 1948, Cellular structures and functions concerned in parasitism, *Bacteriol. Rev.* **12**:173.

Dubos, R. J. and Middlebrook, G., 1948, Cytochemical reaction of virulent tubercle bacilli, *Am. Rev. Tuberc.* **58**:698.

Duff, R. B., 1957, Esterification of primary alcoholic groups of carbohydrates with acetic acid: A general reaction, *J. Chem. Soc.* **1957**:4730.

Durand, E., Welby, M., Lanéélle, G., and Tocanne, J.-F., 1979a, Phase behaviour of cord factor and related bacterial glycolipid toxins, *Eur. J. Biochem.* **93**:103.

Durand, E., Gillois, M., Tocanne, J.-F., and Lanéelle, G., 1979b, Property and activity of mycoloyl esters of methyl glucoside and trehalose, *Eur. J. Biochem.* **94**:109.

Étemadi, A. H., 1964, Techniques microanalytiques d'étude de structure d'esters α-ramifiés β-hydroxylés. Chromatographie en phase vapeur et spectrométrie de masse, *Bull. Soc. Chim. Fr.* **1964**:1537.

Étemadi, A. H., Miquel, A. M., Lederer, E., and Barbier, M., 1964, Sur la structure des acides α-mycoliques de *Mycobacterium kansasii*. Spectromètrie de masse à haute résolution pour des masses de 750 à 1200, *Bull. Soc. Chim. Fr.* **1964**:3274.

Fieser, L. F., and Fieser, M., 1944, *Organic Chemistry*, pp. 133–134. D.C. Heath, Boston.

Fieser, M., Fieser, L. F., Toromanoff, E., Hirata, Y., Heymann, H., Tefft, M., and Bhattacharya, S., 1956, Synthetic emulsifying agents, *J. Am. Chem. Soc.* **78**:2825.

Folch, J., Lees, M., and Sloane-Stanley, G. H., 1957, A simple method for isolation and purification of total lipids from animal tissues, *J. Biol. Chem.* **226**:497.

Fray, G. I., and Polgar, N., 1956, Synthesis of (+)-2(L):4(L)-dimethyldocosanoic acid, an oxidation product of mycolipenic acid, *Chem. Ind.* **1956**:23.

Gangadharam, P. R. J., Cohn, M. L., and Middlebrook, G., 1963, Infectivity, pathogenicity and sulpholipid fraction of some Indian and British strains of tubercle bacilli, *Tubercle* **44**:452.

Gillois, M., Silve, G., Asselineau, J., and Lanéelle, G., 1984, Dicorynomycoloyl trehalose activity: Comparison of the activity of α,α'- and β,β'-trehalose derivatives on mitochondrial oxidative phosphorylation, *Ann. Microbiol. (Paris)* **135(B)**:13.

Goldman, D. S., and Lornitzo, F. A., 1962, Enzyme systems of mycobacteria. The inhibition of the transglycosidase-catalyzed formation of trehalose-6-phosphate, *J. Biol. Chem.* **237**:3332.

Goren, M. B., 1970a, Sulfolipid I of *Mycobacterium tuberculosis*, strain H37Rv. Purification and properties, *Biochim. Biophys. Acta* **210**:116.

Goren, M. B., 1970b, Sulfolipid I of *Mycobacterium tuberculosis*, strain H37Rv. II. Structural studies, *Biochim. Biophys. Acta* **210**:127.

Goren, M. B., 1971, Mycobacterial sulfolipids: Spontaneous desulfation, *Lipids* **6**:40.

Goren, M. B., 1972, Mycobacterial lipids: Selected topics, *Bacteriol. Rev.* **36**:33.

Goren, M. B., 1975, Cord factor revisited: A tribute to the late Dr. Hubert Bloch, *Tubercle* **56**:65.

Goren, M. B., 1977, Phagocyte lysosomes: Interaction with infectious agents, phagosomes, and experimental perturbations in function, *Annu. Rev. Microbiol.* **31**:507.

Goren, M. B., 1982, Immunoreactive substances of mycobacteria, *Am. Rev. Respir. Dis.* **125**(pt. 2):50.

Goren, M. B., 1984, Biosynthesis and structures of phospholipids and sulfatides, in *The Mycobacteria, A Source Book* (G. P. Kubica and L. G. Wayne, eds.), pp. 379–415. Marcel Dekker, New York.

Goren, M. B., and Brennan, P. J., 1979, Mycobacterial lipids: Chemistry and biologic activities, in *Tuberculosis* (G. P. Youmans, ed.), pp. 63–193, W. B. Saunders, Philadelphia.

Goren, M. B., and Brokl, O., 1974, Separation and purification of cord factor (6,6'-dimycoloyl trehalose) from wax C or from mycolic acids, *Recent Results Cancer Res.* **47**:251.

Goren, M. B., and Jiang, K.-S., 1979, Pseudo cord factors: Derivatives of α-D-glucopyranuronosyl (1-1) α-D-glucopyranuronoside, *Chem. Phys. Lipids* **25**:209.

Goren, M. B., and Jiang, K.-S., 1980, (α-D-Glucopyranosyluronic acid) (α-D-glucopyranosiduronic acid) and simple derivatives, *Carbohydr. Res.* **79**:225.

Goren, M. B., and Kochansky, M., 1973, The stringent requirement for electrophiles in the facile solvolytic hydrolysis of neutral sulfate ester salts, *J. Org. Chem.* **38**:3510.

Goren, M. B., and Toubiana, R., 1979, A facile permethylation of cord factor, *Biochim. Biophys. Acta* **574**:64.

Goren, M. B., Brokl, O., Das, B. C., and Lederer, E., 1971, Sulfolipid I of *Mycobacterium tuberculosis* strain H37Rv. Nature of the acyl substituents, *Biochemistry* **10**:72.

Goren, M. B., Brokl, O., and Schaefer, W. B., 1974a, Lipids of putative relevance to virulence in *Mycobacterium tuberculosis:* Correlation of virulence with elaboration of sulfatides and strongly acidic lipids, *Infect. Immun.* **9**:142.

Goren, M. B., Brokl, O., and Schaefer, W. B., 1974b, Lipids of putative relevance to virulence in *Mycobacterium tuberculosis:* Phthiocerol dimycocerosate and the attenuation indicator lipid, *Infect. Immun.* **9**:150.

Goren, M. B., Brokl, O., Roller, P., Fales, H. M., and Das, B. C., 1976a, Sulfatides of *Mycobacterium tuberculosis:* The structure of the principal sulfatide (SL-I), *Biochemistry* **15**:2728.

Goren, M. B., Hart, P. D., Young, M. R., and Armstrong, J. A., 1976b, Prevention of phagosome-lysosome fusion in cultured macrophages by sulfatides of *Mycobacterium tuberculosis*, *Proc. Natl. Acad. Sci. USA* **73**:2510.

Goren, M. B., Das, B. C., and Brokl, O., 1978, Sulfatide III of *Mycobacterium tuberculosis* strain H37Rv, *Nouv. J. Chim.* **2**:379.

Goren, M. B., Brokl, O., and Roller, P., 1979, Cord factor (trehalose-6,6'-dimycolate) of *in vivo*-derived *Mycobacterium lepraemurium*, *Biochim. Biophys. Acta* **574**:70.

Goren, M. B., Grange, J. M., Aber, V. R., Allen, B. S., and Mitchison, D. A., 1982, Role of lipid content and hydrogen peroxide susceptibility in determining the guinea-pig virulence of *Mycobacterium tuberculosis*, *Br. J. Exp. Pathol.* **63**:693.

Goren, M. B., Vatter, A. E., and Fiscus, J., 1987, Polyanionic agents as inhibitors of phagosome-lysosome fusion in cultured macrophages: Evolution of an alternative interpretation, *J. Leukocyte Biol.* **41**:111.

Grand-Perret, T., Lepoivre, M., Petit, J.-F., and Lemaire, G., 1986, Macrophage activation by trehalose dimycolate. Requirement for an expression signal *in vitro* for antitumoral activity; biochemical markers distinguishing primed and fully activated macrophages, *Eur. J. Immunol.* **16**:332.

Grange, J. M., 1982, The genetics of mycobacteria and mycobacteriophages, in *The Biology of Mycobacteria* (C. Ratledge and J. Stanford, eds.), pp. 309–351, Academic, London.

Grange, J. M., Aber, V. R., Allen, B. W., Mitchison, D. A., and Goren, M. B., 1978, The correlation of bacteriophage types of *Mycobacterium tuberculosis* with guinea-pig virulence and *in vitro* indicators of virulence, *J. Gen. Microbiol.* **108**:1.

Gray, G. R., Ribi, E., Granger, D., Parker, R., Azuma, I., and Yamamoto, K., 1975, Immunotherapy of cancer: Tumor suppression and regression by cell walls of *Mycobacterium phlei* attached to oil droplets, *J. Natl. Cancer Inst.* **55**:727.

Grochowski, E., Hilton, B. D., Kupper, R. J., and Michejda, C. J., 1982, Mechanism of the triphenylphosphine and diethyl azodicarboxylate induced dehydration reactions (Mitsunobu Reaction). The central role of pentavalent phosphorus intermediates, *J. Am. Chem. Soc.* **106**:6876.

Hanessian, S., Ponpipom, M. M., and Lavallée, P., 1972, Procedures for the direct replacement of primary hydroxyl groups in carbohydrates by halogen, *Carbohydr. Res.* **24**:45.

Hough, L., Palmer, A. K., and Richardson, A. C., 1972, Chemical modification of trehalose. Part XI. 6,6'-dideoxy-6,6'-difluoro-α,α-trehalose and its *galacto* analog, *J. Chem. Soc. Perkin I* **1972**:2513.

Hunter, R. L., Hunter, R. L., and Bennett, B., 1987, Mycobacterial cord factor kills macrophages by a membrane adhesive mechanism, Abstracts, *Annual Meeting Am. Soc. Microbiol.* U-22.

Hunter, S. W., and Brennan, P. J., 1981, A novel phenolic glycolipid from *Mycobacterium leprae* possibly involved in immunogenicity and pathogenicity, *J. Bacteriol.* **147**:728.

Hunter, S. W., Fujiwara, T., and Brennan, P. J., 1982, Structure and antigenicity of the major specific glycolipid antigen of *Mycobacterium leprae*, *J. Biol. Chem.* **257**:15072.

Hunter, S. W., Murphy, R. C., Clay, K., Goren, M. B., and Brennan, P. J., 1983, Trehalose-containing lipooligosaccharides, *J. Biol. Chem.* **258**:10481.

Hunter, S. W., Fujiwara, T., Murphy, R. C., and Brennan, P. J., 1984, N-acylkansosamine—A novel N-acylamino sugar from the trehalose-containing lipooligosaccharide antigens of *Mycobacterium kansasii*, *J. Biol. Chem.* **259**:9729.

Hunter, S. W., Jardine, I., Yanagihara, D., and Brennan, P. J., 1985, Trehalose-containing lipooligosaccharides from mycobacteria: Structures of the oligosaccharide segments and recognition of a unique N-acylkansosamine-containing epitope, *Biochemistry* **24**:2798.

Ioneda, T., Lenz, M., and Pudles, J., 1963, Chemical constitution of a glycolipid from *C. diphtheria* PW-8, *Biochem. Biophys. Res. Commun.* **13**:110.

Jenkins, I. D., and Goren, M. B., 1986, Improved synthesis of cord factor analogues via the Mitsunobu reaction, *Chem. Phys. Lipids* **41**:225.

Kamisango, I.-I., Saadat, S., Dell, A., and Ballou, C. E., 1985, Pyruvylated glycolipids from *Mycobacterium smegmatis*. Nature and location of the lipid components, *J. Biol. Chem.* **260**:4117.

Kanetsuna, F., Imaeda, T., and Cunto, G., 1969, On the linkage between mycolic acid and arabinogalactan in phenol treated mycobacterial cell walls, *Biochim. Biophys. Acta* **173**:341.

Kato, M., 1967a, Studies of a biochemical lesion in experimental tuberculosis in mice. VI. Effects of toxic bacterial constituents of tubercle bacilli on oxidative phosphorylation in host cells, *Am. Rev. Respir. Dis.* **96**:998.

Kato, M., 1967b, Procedure for the preparation of aqueous suspension of cord factor, *Am. Rev. Respir. Dis.* **96**:553.

Kato, M., 1968, Studies of a biochemical lesion in experimental tuberculosis in mice. VII. Structural and functional damage in mouse liver mitochondria under the toxic action of cord factor, *Am. Rev. Respir. Dis.* **98**:260.

Kato, M., 1970, Site II-specific inhibition of mitochondrial oxidative phosphorylation by trehalose-6,6'-dimycolate (cord factor) of *Mycobacterium tuberculosis*, *Arch. Biochem. Biophys.* **140**:379.

Kato, M., 1972, Antibody formation to trehalose-6,6'-dimycolate (cord factor) of *Mycobacterium tuberculosis*, *Infect. Immun.* **5**:203.

Kato, M., and Asselineau, J., 1971, Chemical structure and biochemical activity of cord fator analogs 6,6'-dimycoloyl sucrose and methyl 6-mycoloyl-α-D-glucoside, *Eur. J. Biochem.* **22**:364.

Kato, M., and Fukushi, K., 1969, Studies of a biochemical lesion in experimental tuberculosis in mice. X. Mitochondrial swelling induced by cord factor *in vivo* and accompanying biochemical change, *Am. Rev. Respir. Dis.* **100**:42.

Kato, M., and Goren, M. B., 1974, Synergistic action of cord factor and mycobacterial sulfatides on mitochondria, *Infect. Immun.* **10**:733.

Kato, M., and Maeda, J., 1974, Isolation and biochemical activities of trehalose-6-monomycolate of *Mycobacterium tuberculosis*, *Infect. Immun.* **9**:8.

Kato, M., Tamura, T., Silve, G., and Asselineau, J., 1978, Chemical structure and biochemical activity of cord factor analogs. A comparative study of esters of methyl glucoside and nonhydroxylated fatty acids, *Eur. J. Biochem.* **87**:497.

Kelly, M. T., 1977, Plasma-dependent chemotaxis of macrophages toward BCG cell walls and the mycobacterial glycolipid P_3, *Infect. Immun.* **15**:180.

Kierszenbaum, F., and Walz, D. R., 1981, Proliferative responses of central and peripheral rat lymphocytes elicited by cord factor (trehalose-6,6'-dimycolate), *Infect. Immun.* **33**:115.

Kilburn, J. O., Takayama, K., 1981, Effects of ethambutol on accumulation and secretion of trehalose mycolates and free mycolic acid in *Mycobacterium smegmatis*, *Antimicrob. Agents Chemother.* **20**:401.

Kilburn, J. O., Takayama, K., 1983, Cell-free synthesis of trehalose monomycolate and dimycolate by *Mycobacterium smegmatis*, in *Eighteenth Joint Conference on Tuberculosis*, pp. 305–319. U.S.–Japan Cooperative Medical Science Program, Bethesda.

Kilburn, J. O., Takayama, K., and Armstrong, E. L., 1982, Synthesis of trehalose dimycolate (cord factor) by a cell-free system of *Mycobacterium smegmatis*, *Biochem. Biophys. Res. Commun.* **108**:132.

Kinsky, S. C., and Nicolotti, F. A., 1972, Immunological properties of model membranes, *Annu. Rev. Biochem.* **46**:49.

Kotani, S., Kitaura, Hirano, T., and Tanaka, A., 1959, Isolation and chemical composition of the cell walls of BCG, *Biken J.* **2**:129.

Kuhn, R., Trischmann, H., and Low, J., 1955, Zur permethylierung von Zuckern und Glykosiden, *Angew. Chem.* **67**:32.

Kusaka, T., Kohsaka, K., Fukunishi, Y., and Akimori, H., 1981, Isolation and identification of mycolic acids in *Mycobacterium leprae* and *Mycobacterium lepraemurium*, *Int. J. Leprosy* **49**:406.

Lanéelle, M. A., and Lanéelle, G., 1970, Structure d'acides mycoliques et d'un intermediaire dans la biosynthèse d'acides mycoliques dicarboxyliques, *Eur. J. Biochem.* **12**:296.

Lanéelle, G., and Tocanne, J. F., 1980, Evidence for penetration in liposomes and in mitochondrial membranes of a fluorescent analogue of cord factor, *Eur. J. Biochem.* **109**:177.

Leake, E. S., Myrvik, Q. N., and Wright, M. J., 1984, Phagosomal membranes of *Mycobacterium bovis* BCG-immune alveolar macrophages are resistant to disruption by *Mycobacterium tuberculosis* H37Rv, *Infect. Immun.* **45**:443.

Lederer, E., 1976, Cord factor and related trehalose esters, *Chem. Phys. Lipids* **16**:91.

Lederer, E., 1977, Natural and synthetic immunostimulants related to the mycobacterial cell wall, *Med. Chem.* **V**:257.

Lederer, E., 1979, Cord factor and related synthetic trehalose diesters, *Springer Semin. Immunopathol.* **2**:133.

Lederer, E., 1984, Chemistry of mycobacterial cord factor and related natural and synthetic trehalose esters, in *The Mycobacteria. A Source Book* (G. Kubica and L. G. Wayne, eds.), pp. 361–378, Marcel Dekker, New York.

Lemaire, G., Tenu, J.-P., Petit, J.-F., and Lederer, E., 1986, Natural and synthetic trehalose diesters as immunomodulators, *Med. Res. Rev.* **6**:243.

Lepoivre, M., Tenu, J-P., Lemaire, G., and Petit, J.-F., 1982, Antitumor activity and hydrogen peroxide release by macrophages elicited by trehalose diesters, *J. Immunol.* **129**:860.

Liav, A., and Goren, M. B., 1980a, Synthesis of 6,6'-di-O-acylated α,α-trehaloses via 2,3,4,2',3',4'-hexa-O-benzyl-α,α-trehalose, *Carbohydr. Res.* **81**:C1.

Liav, A., and Goren, M. B., 1980b, A new synthesis of cord factors and analogs, *Chem. Phys. Lipids* **27**:345.

Liav, A., and Goren, M. B., 1980c, An improved synthesis of (2,3,4-tri-*O*-acetyl-α-D-glucopyranosyl)uronic acid (2,3,4-tri-*O*-acetyl-α-D-glucopyranosid)uronic acid, *Carbohydr. Res.* **84**:171.

Liav, A., and Goren, M. B., 1982, 3′,6′-anhydro-6-*O*-corynomycoloyl trehalose: Synthesis and identification as the impurity in synthetic cord factor preparations, *Chem. Phys. Lipids* **30**:27.

Liav, A., and Goren, M. B., 1983, Synthesis of α-D-mannopyranosyl α-D-mannopyranoside from α,α-trehalose: A route to cord factor analogs, *Carbohyrd. Res.* **123**:C22.

Liav, A., and Goren, M. B., 1984a, Cord-factor analogs: Synthesis of 6,6′-di-*O*-mycoloyl- and -corynomycoloyl-(α-D-mannopyranosyl α-D-mannopyranoside), *Carbohydr. Res.* **129**:121.

Liav, A., and Goren, M. B., 1984b, Synthesis of 6-*O*-mycoloyl and 6-*O*-corynomycoloyl α,α-trehalose, *Carbohydr. Res.* **125**:323.

Liav, A., and Goren, M. B., 1984c, Sulfatides of *Mycobacterium tuberculosis.* Synthesis of the core α,α-trehalose 2-sulfate, *Carbohydr. Res.* **127**:211.

Liav, A., and Goren, M. B., 1984d, Sulfate as a blocking group in alkali-catalyzed permethylation: an alternative synthesis of 3,4,6-tri-*O*-methyl-D-glucose, *Carbohydr. Res.* **131**:C8.

Liav, A., and Goren, M. B., 1985, Synthesis of 4,6,4′,6′- and 3,6,3′,6′-dianhydro-(α-D-galactopyranosyl α-D-galactopyranoside), *Carbohydr. Res.* **141**:323.

Liav, A., and Goren, M. B., 1986, An improved synthesis of 6-*O*-mycoloyl- and 6-*O*-corynomycoloyl-α,α-trehalose with observations on the permethylation analysis of trehalose glycolipids, *Carbohydr. Res.* **155**:229.

Liav, A., Das, B. C., and Goren, M. B., 1981, Diamide pseudo cord-factors: bis-N-acyl derivatives of 6,6′-diamino-6,6′-dideoxy-α,α-trehalose, *Carbohydr. Res.* **94**:230.

Liav, A., Flowers, H. M., and Goren, M. B., 1984a, Cord-factor analogs: Synthesis of 6,6′-di-*O*-mycoloyl- and -corynomycoloyl-(α-D-galactopyranosyl α-D-galactopyranoside), *Carbohydr. Res.* **133**:53.

Liav, A., Flowers, H., Blancquaert, A.-M., and Goren, M. B., 1984b, Relation of carbohydrate stereochemistry to toxicity of cord factor: Mycoloylated trehalose diastereoisomers, in *Nineteenth Joint Conference on Tuberculosis,* pp. 253–262, U.S.–Japan Cooperative Medical Science Program, Tokyo.

Liav, A., Goren, M. B., Yang, Y-M., and Fales, H. M., 1986, Synthesis of 4,6-anhydro-6′-*O*-mycoloyl and 4,6-anhydro-6′-*O*-corynomycoloyl-(α-D-galactopyranosyl α-D-galactopyranoside), *Carbohydr. Res.* **155**:223.

Lornitzo, F., and Goldman, D. S., 1965, Reversible effect of bicarbonate on the inhibition of mycobacterial and yeast transglucosylase by mycoribnin, *J. Bacteriol.* **89**:1086.

Macfarlane, R. D., 1983, Californium-252 plasma desorption mass spectrometry, *Anal. Chem.* **55**:1247A.

Malik, U., Prabhudesai, A. V., Khuller, G. K., and Subrahmanyam, D., 1982, Sulfolipids of mycobacteria, *IRCS Med. Sci. Biochem.* **10**:910.

McCarthy, C., 1976, Synthesis and release of sulfolipid by *Mycobacterium avium* during growth and cell division, *Infect. Immun.* **14**:1241.

McLaughlin, C. A., Bickel, W. D., Kyle, J. S., and Ribi, E., 1978a, Synergistic tumor regressive activity observed following treatment of Line-10 hepatocellular carcinomas with deproteinized BCG cell walls and mutant *Salmonella typhimurium* glycolipid, *Cancer Immunol. Immunother.* **5**:45.

McLaughlin, C. A., Ribi, E. E., Goren, M. B., and Toubiana, R., 1978b, Tumor regression induced by defined microbial components in an oil-in-water emulsion is mediated through their binding to oil droplets, *Cancer Immunol. Immunother.* **4**:109.

Meyer, T. J., Ribi, E., Azuma, I., and Zbar, B., 1974, Biologically active components from mycobacterial cell walls. II. Suppression and regression of strain-2 guinea pig hepatoma, *J. Natl. Cancer Inst.* **52**:103.

Middlebrook, G., Coleman, C. M., and Schaefer, W. B., 1959, Sulfolipid from virulent tubercle bacilli, *Proc. Natl. Acad. Sci. USA* **45**:1801.

Middlebrook, G., Dubos, R. J., and Pierce, C., 1947, Virulence and morphological characteristics of mammalian tubercle bacilli, *J. Exp. Med.* **86**:175.

Minnikin, D. E., 1982, Lipids: Complex lipids, their chemistry, biosynthesis and roles, in *The*

Biology of the Mycobacteria, Vol. 1 (C. Ratledge and J. Stanford, eds.), pp. 95–184, Academic Press, London.

Minnikin, D. E., Dobson, G., Sesardic, D., and Ridell, M., 1985, Mycolipenates and mycolipanolates of trehalose from *Mycobacterium tuberculosis, J. Gen. Microbiol.* **131:**1369.

Mitchison, D. A., 1963, The infectivity and pathogenicity of tubercle bacilli, *Am. Rev. Respir. Dis.* **88:**267.

Mitchison, D. A., 1964, The virulence of tubercle bacilli from patients with pulmonary tuberculosis in India and other countries, *Bull. Int. Union Tuberc.* **35:**287.

Mitsunobu, O., 1981, The use of diethyl azodicarboxylate and triphenylphosphine in synthesis and transformation of natural products, *Synthesis* **1981:**1.

Mompon, B., Federici, C., Toubiana, R., and Lederer, E., 1978, Isolation and structural determination of a "cord-factor" (trehalose-6,6'-dimycolate) from *Mycobacterium smegmatis, Chem. Phys. Lipids* **21:**97.

Moore, V. L., Myrvik, Q. N., and Kato, M., 1972, Role of cord factor (trehalose-6,6'-dimycolate) in allergic granuloma formation in rabbits, *Infect. Immun.* **6:**5.

Myrvik, Q. N., Leake, E. S., and Wright, M. J., 1984, Disruption of phagosomal membranes of normal alveolar macrophages by the H37Rv strain of *Mycobacterium tuberculosis.* A correlate of virulence, *Am. Rev. Respir. Dis.* **129:**322.

Nakamura, M., Itoh, T., Yoshitake, Y., Sengupta, U., and Goren, M. B., 1984, Biochemical characteristics of *M. lepraemurium* propagated in cell-free liquid medium, *Int. J. Lepr. Other Mycobact. Dis.* **52**(4 suppl.):720.

Noll, H., 1956, The chemistry of cord factor, a toxic glycolipid of *M. tuberculosis, Adv. Tuberc. Res.* **7:**149.

Noll, H., and Bloch, H., 1955, Studies on the chemistry of the cord factor of *Mycobacterium tuberculosis, J. Biol. Chem.* **214:**251.

Noll, H., Bloch, H., Asselineau, J., and Lederer, E., 1956, The chemical structure of the cord factor of *Mycobacterium tuberculosis, Biochim. Biophys. Acta* **20:**299.

Numata, F., Nishimura, K., Ishida, H., Ukei, S., Tone, Y., Ishihara, C., Saiki, I., Sekikawa, I., and Azuma, I., 1985, Lethal and adjuvant activities of cord factor (trehalose-6,6'-dimycolate) and synthetic analogs in mice, *Chem. Pharm. Bull.* **33:**4544.

Ofek, S., and Bekierkunst, A., 1976, Chemotactic responses of leukocytes to cord factor (trehalose-6,6'-dimycolate), *J. Natl. Cancer Inst.* **57:**1379.

Ohara, T., Shimmyo, Y., Sekikawa, I., Morikawa, K., and Sumikawa, E., 1957, Studies on the cord factor, with special reference to its immunological properties, *Jpn. J. Tuberc.* **5:**128.

Orbach-Arbouys, S., Tenu, J. P., and Petit, J.-F., 1983, Enhancement of *in vitro* and *in vivo* antitumor activity by cord factor (6,6'-dimycolate of trehalose) administered suspended in saline, *Int. Arch. Allergy Appl. Immunol.* **71:**67.

Pabst, M. J., Gross, J. M., Brozna, J. P., and Goren, M. B., 1987, Inhibition of macrophage activation by sulfatide from *Mycobacterium tuberculosis J. Immunol.* **140:**643.

Pangborn, M. C., and McKinney, J. A., 1966, Purification of serologically active phosphoinositides of *Mycobacterium tuberculosis, J. Lipid Res.* **7:**627.

Parant, M., Parant, F., Chedid, L., Drapier, J. C., Petit, J.-F., Wietzerbin, J., and Lederer, E., 1977, Enhancement of nonspecific immunity to bacterial infection by cord factor (6,6'-trehalose dimycolate), *J. Infect. Dis.* **135:**771.

Parant, M., Audibert, F., Parant, F., Chedid, L., Soler, E., Polonsky, J., and Lederer, E., 1978, Nonspecific immunostimulant activities of synthetic trehalose-6,6'-diesters (lower homologs of cord factor), *Infect. Immun.* **20:**12.

Polonsky, J., and Lederer, E., 1954, Synthèses de quelques acides mycoliques, *Bull. Soc. Chim. Fr.* **1954:**504.

Polonsky, J., Ferréol, G., Toubiana, R., and Lederer, E., 1956, Sur le "cord factor" lipide toxique de *Mycobacterium tuberculosis.* Synthèses de substances à activité de "cord factor" (esters de tréhalose et d'acides ramifiés synthetique), *Bull. Soc. Chim. Fr.* **1956:**1471.

Polonsky, J., Soler, E., Toubiana, R., Takayama, K., Raju, M. S., and Wenkert, E., 1978a, A carbon-13 nuclear magnetic resonance spectral analysis of cord factors and related substances, *Nouv. J. Chimie.* **2:**317.

Polonsky, J., Soler, E., and Varenne, J., 1978b, Sur la synthèse du cord-factor et de ses analogues, *Carbohydr. Res.* **65:**295.

Prabhudesai, A. V., Malik, U., and Subrahmanyam, D., 1981, Isolation and purification of sulpholipids of *Mycobacterium tuberculosis*, H37Rv, *Ind. J. Biochem. Biophys.* **18:**71.

Promé, J.-C., Lacave, C., Ahibo-Coffy, A., and Savagnac, A., 1976, Separation et étude structurale des espèces moleculaires de monomycolates et de dimycolates de α-D-tréhalose présents chez *Mycobacterium phlei, Eur. J. Biochem.* **63:**543.

Puzo, G., Tissie, G., Lacave, C., Aurelle, H., and Promé, J.-C., 1978, Structural determination of "cord factor" from a *Corynebacterium diphtheriae* strain by a combination of mass spectral ionization methods: Field desorption cesium cationization and electron impact mass spectrometry studies, *Biomed. Mass Spectrometry* **5:**699.

Qureshi, N., Takayama, K., and Ribi, E., 1982, Purification and structural determination of nontoxic lipid A obtained from lipopolysaccharide of *Salmonella typhimurium, J. Biol. Chem.* **251:**11808.

Ramathan, V. D., Curtis, J., and Turk, J. L., 1980, Activation of the alternative pathway of complement by mycobacteria and cord factor, *Infect. Immun.* **29:**30.

Rapp, H. J., 1973, A guinea pig model for tumor immunology. A summary, *Israel J. Med. Sci.* **9:**366.

Reggiardo, Z., and Middlebrook, G., 1975a, Serologically active glycolipid families from *Mycobacterium bovis* BCG. I. Extraction, purification and immunologic studies, *Am. J. Epidemiol.* **100:**469.

Reggiardo, Z., and Middlebrook, G., 1975b, Serologically active glycolipid families from *Mycobacterium bovis* BCG. II. Serological studies on human sera, *Am. J. Epidemiol.* **100:**477.

Reggiardo, Z., and Vazquez, E., 1981, Comparison of enzyme-linked immunosorbent assay and hemagglutination test using mycobacterial glycolipids, *J. Clin. Microbiol.* **13:**1007.

Reggiardo, Z., Aber, V. R., Mitchison, D. A. and Devi, S., 1981, Hemagglutination tests for tuberculosis with mycobacterial glycolipid antigens, *Am. Rev. Respir. Dis.* **124:**21.

Reisser, D., Jeannin, J.-F., Martin, F., 1984, Effet *in vivo* et *in vitro* du dimycolate de trehalose (TDM) sur l'activité tumoricide des macrophages péritonéaux de rats, *C.R. Acad. Sc. Paris, t. 298, Ser. III* **1984:**181.

Retzinger, G. S., 1984, An experimental model of miliary tuberculosis and diffuse intravascular coagulation, in *Proceedings of the Nineteenth Joint Research Conference on Tuberculosis*, pp. 295–313, U.S.–Japan Cooperative Medical Science Program, Tokyo.

Retzinger, G. S., Meredith, S. C., Takayama, K., Hunter, R. L., and Kezdy, F. J., 1981, The role of surface in the biological activities of trehalose 6,6'-dimycolate, *J. Biol. Chem.* **256:**8208.

Retzinger, G. S., Meredith, S. C., Hunter, R. L., Takayama, K., and Kezdy, F. J., 1982, Identification of the physiologically active state of the mycobacterial glycolipid trehalose-6,6'-dimycolate and the role of fibrinogen in the biologic activities of trehalose-6,6'-dimycolate monolayers, *J. Immunol.* **129:**735.

Ribi, E., Anacker, R. L., Barclay, W. R., Brehmer, W., Middlebrook, G., Milner, K. C., and Tarmina, D. F., 1968, Structure and biological functions of mycobacteria, *Ann. NY Acad. Sci.* **154:**41.

Ribi, E., Anacker, R. L., Barclay, W. R., Brehmer, W., Harris, S. C., Leif, W. R., and Simmons, J., 1971, Efficacy of mycobacterial cell walls as a vaccine against airborne tuberculosis in the Rhesus monkey, *J. Infect. Dis.* **123:**527.

Ribi, E., Parker, R., and Milner, K., 1974, Microparticulate gel chromatography accelerated by centrifugal force and pressure, *Methods Biochem. Anal.* **22:**355.

Ribi, E., Granger, D. L., Milner, K. C., and Strain, S. M., 1975, Tumor regression caused by endotoxins and mycobacterial fractions, *J. Natl. Cancer Inst.* **55:**1253.

Ribi, E., Milner, K. C., Granger, D. L., Kelly, M. T., Yamamoto, K., Brehmer, W., Parker, R., Smith, R. F., and Strain, S. M., 1976a, Immunotherapy with nonviable microbial components, *Ann. NY Acad. Sci.* **277:**228.

Ribi, E., Milner, K., Kelly, M. T., Granger, D., Yamamoto, K., McLaughlin, C. A., Brehmer, W., Strain, S. M., Smith, R. F., and Parker, R., 1976b, Structural requirements of microbial agents for immunotherapy of the guinea pig line-10 tumor, in *BCG in Cancer Immunotherapy* (G. S. Lamoureux, R. Turcotte, and V. Portelance, eds.), pp. 51–61, Grune & Stratton, New York.

Ribi, E., Takayama, K., Milner, K., Gray, G. R., Goren, M., Parker, R., McLaughlin, C. A., and Kelly, M., 1976c, Regression of tumors by an endotoxin combined with trehalose mycolates of differing structure, *Cancer Immunol. Immunother.* **1:**265.

Ribi, E., Parker, R., Strain, S. M., Mizuno, Y., Nowotny, A., Von Eschen, K. B., Cantrell, J. L., McLaughlin, C. A., Hwang, K. M., and Goren, M. B., 1979, Peptides as requirement for immunotherapy of the guinea-pig line-10 tumor with endotoxins, *Cancer Immunol. Immunother.* **7:**43.

Ribi, E., Cantrell, J. L., Nowotny, A., Parker, R., Schwartzman, S. M., Von Eschen, K. B., Wheat, R. W., and McLaughlin, C. A., 1980, Tumor regression caused by endotoxins combined with trehalose dimycolate, in *Natural Toxins* (D. Eaker and T. Wadstrom, eds.), pp. 327–335, Pergamon, Oxford.

Ribi, E., Cantrell, J., Schwartzman, S., and Parker, R., 1981, BCG cell wall skeleton, P3, MDP and other microbial components—structure activity studies in animal models, in *Augmenting Agents in Cancer Therapy* (E. M. Hersh, ed.), pp. 15–31, Raven, New York.

Ribi, E., Cantrell, J. L., Takayama, K., Qureshi, N., Peterson, J., and Ribi, H. O., 1984, Lipid A and immunotherapy, *Rev. Infect. Dis.* **6:**567.

Ribi, E., Cantrell, J., and Takayama, K., 1985, A new immunomodulator with potential clinical applications: Monophosphoryl lipid A, a detoxified endotoxin, *Clin. Immunol. Newsl.* **6:**33.

Richman, S. P., Gutterman, J. U., Hersh, E. M., and Ribi, E. E., 1978, Phase I–II study of intratumor immunotherapy with BCG cell wall skeleton plus P3, *Cancer Immunol. Immunother.* **5:**41.

Ruco, L. P., and Meltzer, M., 1978, Macrophage activation for tumor cytotoxicity: Increased lymphokine responsiveness of peritoneal macrophages during acute inflammation, *J. Immunol.* **120:**1054.

Saito, R., Tanaka, A., Sugiyama, K., and Kato, M., 1975, Cord factor not toxic in rats, *Am. Rev. Respir. Dis.* **112:**578.

Saito, R., Tanaka, A., Sugiyama, K., Azuma, I., Yamamura, Y., Kato, M., and Goren, M. B., 1976, Adjuvant effect of cord factor, a mycobacterial lipid, *Infect. Immun.* **13:**776.

Saito, R., Nagao, S., Sugiyama, K., and Tanaka, A., 1977, Adjuvanticity (immunity-inducing property) of cord factor in mice and rats, *Infect. Immun.* **16:**725.

Sasaki, A., and Takahashi, Y., 1974, Activité serologiqne des phospholipides purifiés de *Mycobacterium tuberculosis*, *C. R. Soc. Biol. (Paris)* **168:**626.

Sathyamoorthy, N., Qureshi, N., and Takayama, K., 1986, Enzymatic exchange of mycolic acid in *Mycobacterium smegmatis* involving free trehalose and trehalose-6-monomycolate, *Fed. Proc.* **45:**1823.

Schaefer, W. B., 1965, Serologic identification and classification of the atypical mycobacteria by their agglutination, *Am. Rev. Respir. Dis.* **92**(suppl.):85.

Schaefer, W. B., 1980, Serological identification of atypical mycobacteria, in *Methods in Microbiology* (T. Bergan and J. R. Norris, eds.), Vol. 13, pp. 324–344, Academic, London.

Schwartzman, S. M., Cantrell, J. L., Ribi, E., and Ward, J., 1984, Immunotherapy of equine sarcoid with cell wall skeleton (CWS)-trehalose dimycolate (TDM) biologic, *Equine Pract.* **6:**13.

Seggev, J., Goren, M. B., Carr, R. I., and Kirkpatrick, C. H., 1982, The pathogenesis of interstitial and hemorrhagic pneumonitis induced by mycobacterial trehalose dimycolate, *Am. J. Pathol.* **106:**348.

Seggev, J. S., Goren, M. B., and Kirkpatrick, C. H., 1984a, The pathogenesis of trehalose dimycolate-induced interstitial pneumonitis, *Cell. Immunol.* **85:**428.

Seggev, J. S., Kirkpatrick, C. H., Goren, M. B., 1984b, Interstitial pneumonitis induced by trehalose dimycolate. II. Reserpine prevents formation of lesions, *Am. Rev. Respir. Dis.* **129:**840.

Senn, M., Ioneda, T., Pudles, J., and Lederer, E., 1967, Spectrométrie de masse de glycolipids. I. Structure du "cord factor" de *Corynebacterium diphtheriae*, *Eur. J. Biochem.* **1:**353.

Sharma, A., Haq, A., Ahmad, S., and Lederer, E., 1985, Vaccination of rabbits against *Entamoeba histolytica* with aqueous suspensions of trehalose-dimycolate as the adjuvant, *Infect, Immun.* **48:**634.

Shimakata, T., Iwaki, M., and Kusaka, T., 1984, *In vitro* synthesis of mycolic acids by the fluffy layer fraction of *Bacterionema matruchotii*, *Arch. Biochem. Biophys.* **229:**329.

Shimakata, T., Tsubokura, K., Kusaka, T., and Shizukuishi, K.-I., 1985, Mass spectrometric identification of the trehalose 6-monomycolate synthesized by the cell-free system of *Bacterionema matruchotii*, *Arch. Biochem. Biophys.* **238:**497.

Silva, C., and Faccioli, L. H., 1988, *Infect. Immunol.* **56:**3067.

Ställberg-Stenhagen, S., 1954, Optically active higher aliphatic compounds. XII *trans* -2,5-D-dimethyl-$\Delta^{2:3}$-heneicosenoic acid and *trans*-2,4 D-dimethyl-$\Delta^{2:3}$-heneicosenoic acid, *Ark. Kemi* **6**:537.

Strain, S. M., Toubiana, R., Ribi, E., and Parker, R., 1977, Separation of the mixture of trehalose-,6,6′-dimycolates comprising the mycobacterial glycolipid fraction, "P3," *Biochem. Biophys. Res. Commun.* **77**:449.

Sukumar, S., Hunter, J. T., Yarkoni, E., Rapp, H. J., Zbar, B., and Lederer, E., 1981, Efficacy of mycobacterial components in the immunotherapy of mice with pulmonary tumor deposits, *Cancer Immunol. Immunother.* **11**:125.

Suter, E., and Kirsanow, E. M., 1961, Hyperreactivity to endotoxin in mice infected with mycobacteria. Induction and elicitation of the reactions, *Immunol.* **4**:354.

Takahashi, Y., and Ono, K., 1961, Study of the passive hemagglutination reaction by the phosphatides of *M. tuberculosis*. I. The reaction and its specificity, *Am. Rev. Respir. Dis.* **83**:381.

Takayama, K., 1985, Enzymatic transfer of mycolic acid to trehalose from a new mycobacterial lipid to form the trehalose mycolates, in *Proceedings of the Twentieth Joint Conference on Tuberculosis*, pp. 9–20, U.S.–Japan Cooperative Medical Science Program, Bethesda.

Takayama, K., and Armstrong, E. L., 1976, Isolation, characterization, and function of 6-mycolyl-6′-acetyltrehalose in the H37Ra strain of *Mycobacterium tuberculosis*, *Biochemistry* **15**:441.

Takayama, K., and Qureshi, N., 1984, Structure and synthesis of lipids, in *The Mycobacteria, A Source Book*, Part A (G. P. Kubica and L. G. Wayne, eds.), pp. 315–344, Marcel Dekker, New York.

Takayama, K., and Sathyamoorthy, N., 1986, Isolation, purification and characterization of trehalose mycolyltransferase from *Mycobacterium smegmatis*, in *Proceedings of the Twenty-first Joint Research Conference on Tuberculosis*, pp. 69–79, U.S.–Japan Cooperative Medical Science Program, Osaka.

Takayama, K., Wang, L., and David, H. L., 1972, Effect of isoniazid on the *in vivo* mycolic acid synthesis, cell growth, and viability of *Mycobacterium tuberculosis*, *Antimicrob. Agents Chemother.* **2**:29.

Takayama, K., Wang, L., and Merkal, R. S., 1973, Scanning electron microscopy of the H37Ra strain of *Mycobacterium tuberculosis* exposed to isoniazid, *Antimicrob. Agents Chemother.* **4**:62.

Takayama, K., Qureshi, N., Raetz, C. R. H., Ribi, E., Peterson, J., Cantrell, J. L., Pearson, F. C., Wiggins, J., and Johnson, A. G., 1984, Influence of fine structure of lipid A on limulus amebocyte lysate clotting and toxic activities, *Infect. Immun.* **45**:350.

Tarelli, E., Draper, P., and Payne, S., 1984, Structure of the oligosaccharide component of a serologically active phenolic glycolipid isolated from *Mycobacterium leprae*, *Carbohydr. Res.* **131**:346.

Tereletsky, M. J., and Barrow, W. W., 1983, Postphagocytic detection of glycopeptidolipids associated with superficial L layer of *Mycobacterium intracellulare*, *Infect. Immun.* **41**:1312.

Tocanne, J.-F., 1975, Sur une nouvelle voie de synthèse du cord-factor glycolipide toxique de *Mycobacterium tuberculosis* (esters du trehalose et d'acides gras α-ramifiés, β-hydroxylés), *Carbohydr. Res.* **44**:301.

Toubiana, R., Toubiana, M.-J., Das, B. C., and Richardson, A. C., 1973, Synthèse d'analogues du cord-factor. I. Étude de l'esterification selective du tréhalose par le chlorure de para-toluene sulfonylé. Mise en evidence d'un ester dissymetrique le 2,6-ditosylate de tréhalose, *Biochimie* **55**:569.

Toubiana, R., Das, B. C., Defaye, J., Mompon, B., and Toubiana, M.-J., 1975, Étude du cord-factor et de ses analogues. III. Synthèse du cord-factor (6,6′-di-*O*-mycoloyl-α,α-tréhalose) et du 6,6′-di-*O*-palmitoyl-α,α-tréhalose, *Carbohydr. Res.* **44**:308.

Toubiana, R., Pizza, C., Chapleur, Y., and Castro, B., 1978, Étude du cord-factor et de ses analogues. IV. Preparation du di-*O*-corynomycoloyl-6,6′-α,α-tréhalose par l'intermediaire du bisel d'alkyloxytris (dimethyl-amino) phosphonium (ATDP), *J. Carbohydr. Nucleosides Nucleotides* **5**:127.

Tsukamura, M., and Mizuno, S., 1986, Test for sulfolipid formation from [^{35}S]sulfate as an aid to differentiate mycobacteria from Rhodococci and Nocardiae, *Microbiol. Immunol.* **30**:589.

Tsukamura, M., Mizuno, S., and Toyama, H., 1984, Differentiation of mycobacterial species by investigation of petroleum ether-soluble sulfolipids using thin layer chromatography after incubation with [^{35}S]sulfate, *Microbiol. Immunol.* **28**:965.

Vilkas, E., and Markovits, J., 1968, Isolément d'un digalactoside et d'un mycolate de diarabino-side à partir de cires D d'une souche humaine virulente de *Mycobacterium tuberculosis, FEBS Lett.* **2**:20.

Vilkas, E., Adam, A., and Senn, M., 1968, Isolément d'un nouveau type de diester de tréhalose à partir de *Mycobacterium fortuitum, Chem. Phys. Lipids* **2**:11.

Von Itzstein, M., and Jenkins, I. D., 1983, The mechanism of the Mitsunobu reaction. II. Dialkoxytriphenylphosphoranes, *Aust. J. Chem.* **36**:557.

Vosika, G. J., and Gray, G. R., 1983, Phase I study of i.v. mycobacterial cell wall skeleton and cell wall skeleton combined with trehalose dimycolate, *Cancer Treatm. Rep.* **67**:785.

Vosika, G., Giddings, C., and Gray, G. R., 1984a, Phase I study of intravenous mycobacterial cell wall skeleton and trehalose dimycolate attached to oil droplets, *J. Biol. Resp. Mod.* **3**:620.

Vosika, G., Barr, J. C., and Gilbertson, D., 1984b, Phase-I study of intravenous modified lipid A, *Cancer Immunol. Immunother.* **18**:107.

Walker, R. W., Promé, J.-C., and Lacave, C. S., 1973, Biosynthesis of mycolic acids. Formation of a C_{32} β-keto ester from palmitic acid in a cell-free system of *Corynebacterium diphtheriae, Biochim. Biophys. Acta* **326**:52.

Walker, R. W., Promé, J. C., and Lacave, C., 1973, Mycolic acid biosynthesis: Formation of the β-keto precursor of corynomycolic acid in a cell free system by extracts of *Corynebacterium diphtheriae*, in *Proceedings, Eighth Joint Conference on Tuberculosis*, pp. 261–278, U.S.–Japan Cooperative Medical Science Program, New Orleans.

Wang, L., and Takayama, K., 1972, Relationship between the uptake of isoniazid and its action on *in vivo* mycolic acid synthesis in *Mycobacterium tuberculosis, Antimicrob. Agents Chemother.* **2**:438.

Winder, F. G., 1982, Mode of action of the antimycobacterial agents and associated aspects of the molecular biology of the mycobacteria, in *The Biology of the Mycobacteria*, Vol. 1 (C. Ratledge and J. Stanford, eds.), pp. 354–438, Academic, London.

Yarkoni, E., 1982, Factors affecting the antitumor activity of mycobacterial components emulsi-fied in hexadecane, *J. Clin. Hematol. Oncol.* **12**:105.

Yarkoni, E., and Rapp, H. J., 1977, Granuloma formation in lungs of mice after intravenous administration of emulsified trehalose-6,6'-dimycolate (cord factor): Reaction intensity de-pends on size distribution of the oil droplets, *Infect. Immun.* **18**:552.

Yarkoni, E., and Rapp, H. J., 1978, Toxicity of emulsified trehalose-6,6'-dimycolate (cord factor) in mice depends on size distribution of mineral oil droplets, *Infect. Immun.* **20**:856.

Yarkoni, E., and Rapp, H. J., 1979, Influence of oil and Tween concentrations on enhanced endotoxin lethality in mice pretreated with emulsified trehalose-6,6'-dimycolate (cord fac-tor), *Infect. Immun.* **24**:571.

Yarkoni, E., Meltzer, M. S., and Rapp, H. J., 1976, Failure of trehalose-6,6'-dimycolate (P_3 or cord factor) to enhance endotoxin lethality in mice, *Infect. Immun.* **14**:1375.

Yarkoni, E., Rapp, H. J., and Zbar, B., 1977a, Immunotherapy of a guinea pig hepatoma with ultrasonically prepared mycobacterial vaccines, *Cancer Immunol. Immunother.* **2**:143.

Yarkoni, E., Meltzer, M. S., and Rapp, H. J., 1977b, Tumor regression after intralesional injection of emulsified trehalose-6,6'-dimycolate (cord factor): efficacy increases with oil concentra-tion, *Int. J. Cancer* **19**:818.

Yarkoni, E., Rapp, H. J., Polonsky, J., and Lederer, E., 1978, Immunotherapy with an intrale-sionally administered synthetic cord factor analogue, *Int. J. Cancer* **22**:564.

Yarkoni, E., Goren, M. B., and Rapp, H. J., 1979a, Regression of a transplanted guinea pig hepatoma after intralesional injection of an emulsified mixture of endotoxin and mycobac-terial sulfolipid, *Infect. Immun.* **24**:357.

Yarkoni, E., Rapp, H. J., Polonsky, J., Varenne, J., and Lederer, E., 1979b, Regression of a murine fibrosarcoma after intralesional injection of a synthetic C39 glycolipid related to cord factor, *Infect. Immun.* **26**:462.

Yoshimoto, K., Wakamiya, T., and Nishikawa, Y., 1982, Chemical and biochemical studies on carbohydrate esters. XIII. Synthesis of 6-*O*-, 6,6'-di-*O*-, and 4,6,4',6'-tetra-*O*-stearoyl-α,α-trehaloses, *Chem. Pharm. Bull.* **30**:1169.

Zbar, B., Bernstein, I. D., Bartlett, G. L., Hanna, M. G., and Rapp, H. J., 1972, Immunotherapy of cancer: Regression of intradermal tumors and prevention of growth of lymph node metasta-ses after intralesional injection of living *Mycobacterium bovis, J. Natl. Cancer Inst.* **49**:119.

Chapter 6

Glycosides of Hydroxy Fatty Acids

Alexander P. Tulloch†

6.1. Introduction

Hydroxy fatty acid glycosides are probably the least well known of all the glycolipids, certainly less so than any of the other glycolipids discussed in this volume. They never seem to have been mentioned in any textbooks on lipid biochemistry. The other, better known glycolipids, also composed of sugars and hydroxy fatty acids, are those in which the sugar is acylated by the acid such as the trehalose esters (Asselineau and Asselineau, 1978) (see Chapter 5, *this volume*). Mannitol esters of 3-hydroxy acids have also been reported (Tulloch and Spencer, 1964). The explanation for this neglect appears to be that hydroxy fatty acid glycosides have been reported from only a few species of microorganisms and plants. They are, however, present in these species in relatively large amounts, sometimes very large, this being the reason for their discovery in the first place. Hitherto, there has been no indication that they may be more widespread.

6.2. Hydroxy Fatty Acid Glycosides from Bacteria

Cell walls of a number of species of bacteria have been known for a long time to contain 3-hydroxy fatty acid esters of polysaccharides and a lipid portion of one was recently synthesized (Imoto *et al.*, 1985). Glycosides of hydroxy fatty acids, however, have only been obtained from a single bacterial species, *Pseudomonas aeruginosa*.

6.2.1. Characterization of Rhamnolipids from P. aeruginosa

Bacteria of the genus *Pseudomonas* are gram-negative, asporogenous, and rod-shaped. The species *P. aeruginosa* is generally identified by formation of a

---

†Dr. Tulloch died on February 14, 1987. This chapter is NRCC contribution No. 24522.

Alexander P. Tulloch • Plant Biotechnology Institute, National Research Council of Canada, Saskatoon, Saskatchewan, Canada S7N 0W9.

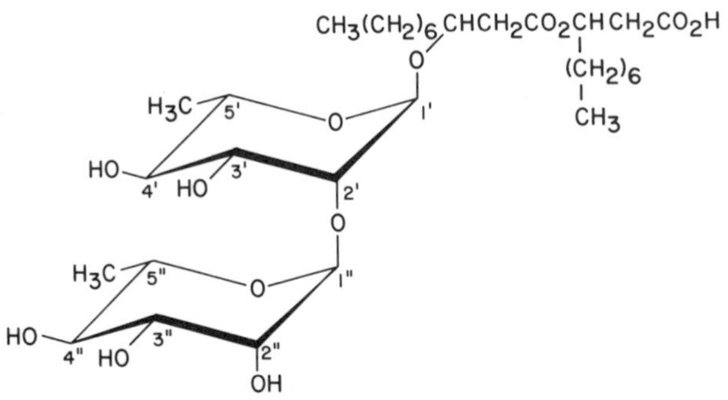

Figure 6-1. Rhamnolipids produced by *Pseudomonas aeruginosa*.

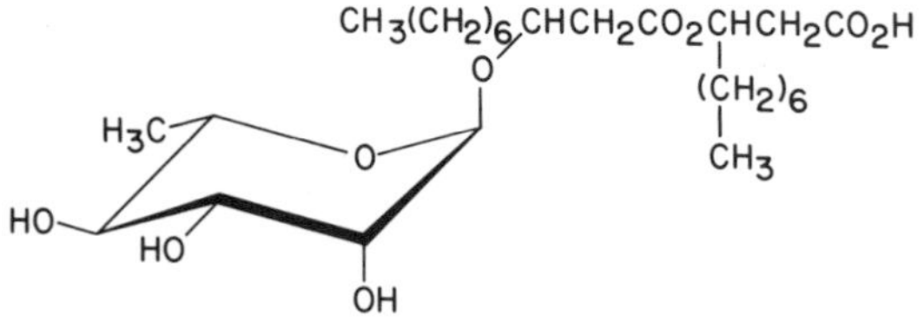

blue-green pigment (Bergan, 1981). The crystalline glycolipid separated from stationary cultures of the bacteria grown on 3% glycerol in a yield of 2.5 g/liter (Jarvis and Johnson, 1949). Acid hydrolysis gave two molecules of L-rhamnose and two molecules of a hydroxy acid. Chromic acid oxidation of the latter gave octanoic acid indicating a 3-hydroxydecanoic acid and the large negative rotation in chloroform showed that the configuration was $3R$ (formerly D). Another product was an acid ester formed from two molecules of the hydroxy acid. This finding and an apparent consumption of 2 moles of periodate per mole of glycolipid suggested that the structure was 1,3-rhamnosylrhamnose glycosidically linked to the hydroxyl of the acid ester. 3-Hydroxydecanoic acid was also among the acids obtained by hydrolysis of the lipopolysaccharide from *P. aeruginosa* (Key et al., 1970).

Later it was shown that the rhamnolipid has structure (1) (Fig. 6.1), 3-{3-[(2'-O-α-L-rhamnopyranosyl-α-L-rhamnopyrano syl)oxy]decanoyloxy}-decanoic acid, with an α-1,2 linkage (Edwards and Hayashi, 1965). Prolonged periodate oxidation led to an uptake of 3 moles of oxidant and 1,2-propanediol was the only product isolated after borohydride reduction and hydrolysis. Methylation, followed by acid hydrolysis gave a 1 : 1 mixture of 2,3,4-tri-O-methyl rhamnose and 3,4-di-O-methylrhamnose. The α-linkages were indicated by the high negative optical rotation and the infrared (IR) spectrum.

Hydrocarbon-using strains of *P. aeruginosa* have also been isolated, that will grow, in a stirred fermentor, on a mixture of C_{12}–C_{14} alkanes as the sole carbon source (Itoh et al., 1971). Yields of rhamnolipid as high as 8.5 g/liter were obtained, about 50% was lipid (1) and the rest was a second lipid (2),

3-{3-[(α-L-rhamnopyranosyl)oxy]decanoyloxy}decanoic acid (Fig. 6-1). The two products were separated by column chromatography. It is important to note, in view of the results on the utilization of hydrocarbons by the yeast *Torulopsis bombicola* (see Section 6.4.1.1) that although the hydrocarbon substrate contained C_{12}–C_{14} components, the hydroxy acid still has a 10-carbon chain (C_{10}).

More recently, production of two rhamnolipids related to (1) and (2) by other strains of *P. aeruginosa* has been reported. The first of these, obtained in a yield of 14 g/liter by shake culture on a medium containing 5% alkanes, was stated to be a mixture of lipids (1) and (2) but modified in each case by acylation of the free C-2 hydroxyl of the rhamnose portion with (*E*)-2-decenoic acid (Yamaguchi *et al.*, 1976). This structure was based on the results of periodate oxidation, which is known to be difficult to interpret (Jarvis and Johnson, 1949; Edwards and Hayashi, 1965), and the isolation of (*E*)-2-decenoic acid after alkaline hydrolysis of the glycolipids.

Without further confirmation, the structures of these acylated rhamnolipids cannot be considered established because in all previous studies, acid hydrolysis of rhamnolipid gave 3-hydroxydecanoic acid, alkaline hydrolysis was never used. Since it is well known that alkaline hydrolysis of acylated 3-hydroxy acids leads to considerable elimination of the acyl group with formation of (*E*)-2-acids (Linstead *et al.*, 1953; Tulloch and Spencer, 1964; Tulloch, 1971), the (*E*)-2-decenoic acid may be simply an artifact of alkaline hydrolysis.

The other modified rhamnolipids are the methyl esters of (1) and (2), which were obtained in relatively low yields of less than 0.1 g/liter (Hirayama and Kato, 1982). Structures were based on the isolation of methyl 3-(3-hydroxydecanoyloxy)decanoate and methyl 3-hydroxydecanoate from the products of acid hydrolysis.

6.2.2. Biosynthesis of Rhamnolipids

The biosynthetic pathway to (1) has been examined using enzymes isolated from extracts of *P. aeruginosa* (Burger *et al.*, 1963). Two moles of 3-hydroxydecanoyl-CoA were first converted to 3-(3-hydroxydecanoyloxy)decanoic acid and thymidine diphosphate L-rhamnose transferred first one rhamnose to the hydroxy acid ester and then a second rhamnose to the rhamnosyl lipid to produce glycolipid (1). Thus the monorhamnosyl lipid (2) later found as a product in some fermentations is, as might be expected, an intermediate in the formation of (1).

6.2.3. Function and Application of Rhamnolipids

It has been proposed that one function of the rhamnolipids, at least in hydrocarbon-using strains, is to act as an emulsifier and thereby promote consumption of alkane (Hisatsuka *et al.*, 1971; Itoh and Suzuki, 1972). The rhamnolipids are active against bacteria, *Mycoplasma*, and viruses (Itoh *et al.*, 1971) and recently they have been considered as possible biosurfactants. They reduced the interfacial tension between hexadecane and salt solutions at concentrations of 30 mg/liter (Lang *et al.,* 1984). Conditions for continuous culture of *P. aeruginosa* in a medium containing glucose were established and gave a yield of 1.5 g/liter of glycolipid, which could be used as biosurfactant (Guerra-Santos *et al.,* 1984).

6.3. Hydroxy Fatty Acid Glycosides from Fungi

During an investigation of fungi that would use glucose well, it was found that *Ustilago zeae,* the corn smut fungus, could be grown in shake culture in a yeastlike form (Haskins, 1950). In mineral solution, containing 5–10% glucose, the sugar was rapidly used and a crystalline glycolipid accumulated in the medium; yields of 15 g/liter were obtained in stirred fermentors corresponding to conversion of 23% of available glucose (Thorn and Haskins, 1951).

6.3.1. Characterization of Cellobiosides from Ustilago zeae

Alkaline hydrolysis of the product (known as ustilagic acid) liberated acetic, 3-hydroxyhexanoic, and 3-hydroxyoctanoic acids, which were attached as esters, in the molar ratio of 1.0 : 0.7 : 0.3 per mole of glycolipid (Lemieux and Charanduk, 1951). The two short-chain hydroxy acids had strong positive rotations so that, in contrast with 3-hydroxydecanoic acid produced by *P. aeruginosa* (Jarvis and Johnson, 1949), they had the S (formerly L) configuration (Lemieux and Giguere, 1951). Methylation of the glycolipid, remaining after alkaline hydrolysis, followed by acid hydrolysis, gave 2,3,6-tri-*O*-methyl-D-glucose, 2,3,4,6-tetra-*O*-methyl-D-glucose and partially methylated hydroxy fatty acids in the approximate ratio 1 : 1 : 1 (Lemieux *et al.,* 1953). Acetolysis of the acetylated glycolipid gave octa-*O*-acetyl-α-D-cellobioside. This, together with the negative optical rotation, showed that the glycolipid was a β-cellobioside (**3**) (Fig. 6-2).

Acid methanolysis of the deacylated glucolipid, or of ustilagic acid, yielded the methyl esters of the aglycone hydroxy acids (Lemieux, 1953*a*).

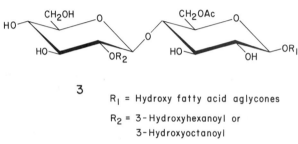

3

R₁ = Hydroxy fatty acid aglycones

R₂ = 3-Hydroxyhexanoyl or 3-Hydroxyoctanoyl

Figure 6-2. Cellobiosides produced by *Ustilago zeae.*

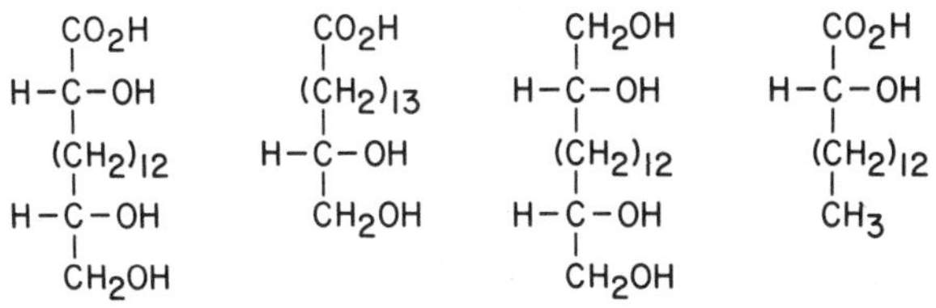

```
   CO2H            CO2H            CH2OH           CO2H
    |               |               |               |
 H-C-OH          (CH2)13         H-C-OH          H-C-OH
    |               |               |               |
 (CH2)12         H-C-OH          (CH2)12         (CH2)12
    |               |               |               |
 H-C-OH          CH2OH           H-C-OH           CH3
    |                               |
 CH2OH                            CH2OH

    4               5               6               7
```

Figure 6-3. Hydroxy fatty acids from *Ustilago zeae* and related compounds.

These were 2,15,16-trihydroxyhexadecanoic acid (**4**) and 15,16-dihydroxy-hexadecanoic acid (**5**) (Fig. 6-3). Dihydroxy acid **5** formed 70% of the aglycone mixture. The two chiral centers in (**4**) were shown to have the same configuration by reduction to the mesotetrol (**6**). Conversion of (**4**) to (2R) 2-hydroxypentadecanoic acid (**7**) established the configuration. This was formerly described as 2D,15D, but precedence rules require 2R,15S. The particular hydroxyl group to which the β-cellobioside portion of the molecule is attached has apparently not been determined.

The positions of the acyl groups on cellobiose were shown to be as in (**3**) (Fig. 6-2) (Bhattacharjee *et al.*, 1970). The free hydroxyls were protected as the methoxy acetals by reaction with methyl vinyl ether, deacylated and methylated so that the originally acylated hydroxyls became methoxyl groups. Identification of the methyl glucoses obtained on hydrolysis showed that the 6'- and 2''-hydroxyls had been acylated. The ^{1}H-NMR spectrum of the product of oxidation with lead tetraacetate in pyridine showed that the 3-hydroxy acids were lost as a result of oxidation presumably because they were attached to the hydroxyl adjacent to the cleaved 3'',4''-diol. The acetate group at C-6' was unaffected.

6.3.2. Biosynthesis of Cellobiosides from Ustilago zeae

Labeled ustilagic acid was produced by growing *U. zeae* on glucose containing [1-^{14}C]glucose (Boothroyd *et al.*, 1955). Degradation of the product showed that most of the glucose in the glycolipid was directly incorporated. The aglycones were labeled mainly at odd carbons, showing that they were formed from [1-^{14}C]acetate derived from labeled glucose. There seems to have been no investigation of the use of long-chain substrates in the production of ustilagic acids.

6.3.3. Function and Application of Cellobiosides

Ustilagic acid had a broad antibiotic spectrum and was active against both bacteria and fungi (Haskins and Thorn, 1951), which simplified its production by fermentation (Haskins, 1950). The acid was not very toxic to mice, but activity was lost on deacylation. The low solubility of ustilagic acid in water appears to have prevented its further development as an antibiotic.

Conversion of dihydroxy acid (**5**) to 15-hydroxypentadecanoic acid, which could be converted to the macrocyclic lactone, having considerable value as a perfume, has been proposed (Lemieux, 1953*b*). Glycol cleavage of (**5**) with sodium bismuthate or lead tetraacetate followed by catalytic hydrogenation of the aldehyde group was used. Improved strains of *U. maydis*, giving yields of ustilagic acid of 30 g/liter, have been developed for this purpose (Kamibayashi and Matsui, 1963). To avoid using a mixture of acids (**4**) and (**5**) in the reaction, which yields 14-hydroxytetradecanoic acid as well as the C_{15} acid (Lemieux, 1953*b*), fermentation at a higher pH, which leads almost entirely to the formation of glycolipid from acid (**5**) only, has been reported (Tokyo Yakuruto Co. Ltd., 1970).

6.4. Hydroxy Fatty Acid Glycosides from Yeasts

Discovery of formation of hydroxy fatty acid glycosides by yeast species was the result of observations that many yeasts produced droplets of extracellular, heavier-than-water "oils" (Stodola *et al.*, 1967). Thus, an osmophilic *Torulopsis* species from sow thistle nectar was observed in 1954 to produce unusually large amounts of extracellular product (Gorin *et al.*, 1961; Spencer *et al.*, 1979). Also, a species of *Candida*, isolated from the phyllosphere in Indonesia, produced a mixture of crystalline and "oily" extracellular material (Stodola *et al.*, 1967). Formation of extracellular glycolipids by yeasts has been reviewed (Tulloch, 1976; Spencer *et al.*, 1979).

6.4.1. Characterization of Sophorolipids from Torulopsis bombicola

Further investigation of the *Torulopsis* species mentioned above showed it to be a new species and it was named *T. bombicola* (Spencer *et al.*, 1970). It was earlier thought to be a strain of *T. magnoliae* (Gorin *et al.*, 1961) or of *T. apicola* (Tulloch and Spencer, 1968). The usual habitat of *T. bombicola* is probably in nests of various bumblebee species in Western Canada and a number of strains have been isolated from the honey pots (Spencer *et al.*, 1979).

The yeast was grown in stirred fermentors on 20% glucose solution, and the glycolipid separated as a viscous liquid containing its own weight of water and formed 5% of the volume of the medium. Removal of water yielded a gummy glycoside in yields of 25 g/liter (Gorin *et al.*, 1961). During a number of investigations, particularly after it was found, as discussed later, that the lipid portion could be modified by the addition of long-chain substrates to the medium, the product was shown to be a mixture of partially acetylated lactonic and acidic sophorosides (**8**), (**9**), and (**10**) (Fig. 6-4) (Gorin *et al.*, 1961; Tulloch *et al.*, 1967, 1968*a*). Originally a mixture of glycosides of several hydroxy acids was obtained (Gorin *et al.*, 1961; Tulloch *et al.*, 1962), but only those of 17-hydroxyoctadecanoic acid are shown in Fig. 6-4. The two lactones are 17-[(2'-*O*-β-D-glucopyranosyl-β-D-glucopyranosyl)oxy]octadecanoic acid 1,4″-lactone 6',6″-diacetate (**8**) and 6″-acetate (**9**), and the acid is 17-[(2'-*O*-β-D-glucopyranosyl-β-D-glucopyranosyl)oxy]octadecanoic acid 6',6″-diacetate (**10**).

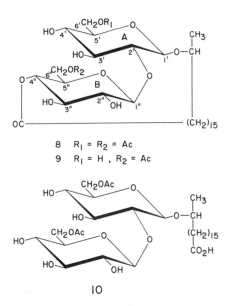

$$8 \quad R_1 = R_2 = Ac$$
$$9 \quad R_1 = H, \ R_2 = Ac$$

10

Figure 6-4. Lactonic and acidic sophorosides produced by *Torulopsis bombicola*.

Alkaline methanolysis removed the acetate groups and opened the lactone ring. Methylation and acid hydrolysis gave equimolar amounts of 2,3,4,6-tetra-*O*-methyl-D-glucose and 3,4,6-tri-*O*-methyl-D-glucose. Susceptibility to β-glucosidase and the large negative molecular rotation indicated β-glycoside linkages showing that the glycolipid was a sophoroside (Gorin *et al.*, 1961). The principal hydroxy fatty acids were 17-hydroxy-(Z)-9-octadecenoic and 17-hydroxyoctadecanoic acids (Gorin *et al.*, 1961) together with minor amounts of 15- and 16-hydroxy hexadecanoic acids and 18-hydroxy C_{18} acids (Tulloch *et al.*, 1962). The 17S configuration (formerly 17L) for the hydroxy acids was established by conversion to (2S)-2-octadecanol (Tulloch, 1968).

Later the positions of the acyl groups were established. The two lactones (8) and (9), which have two and one acetate groups, respectively, were highly crystalline and could be readily separated from amorphous acid (10). The points of attachment of the three acyl groups in (8) were determined by oxidation experiments. One mole of periodate was consumed, giving, after reduction and hydrolysis, 2-*O*-β-D-glucopyranosylglycerol, showing that the 3′,4′-diol of ring A had been cleaved. Oxidation with lead tetraacetate in pyridine, which cleaves resistant diols, and reduction and hydrolysis gave tetritols formed by cleavage of the relatively resistant 2″,3″-diol of ring B. Rates of oxidation of the two lactones and of model compounds indicated that lactone (9) lacked an acyl group at C-6′ and, since both lactones gave the same hexaacetate, the missing group in (9) is an acetate and the fatty acyl linkage is to either C-6″ or C-4″.

Acetobrominolysis of lactone hexaacetate gave principally the acetobromosophorose derivative (11, Fig. 6-5), but treatment of mother liquors from (11) with silver acetate led to isolation of the glucose tetraacetate with acetylated hydroxy fatty acid attached at C-4′ (12, Fig. 6-5). These reac-

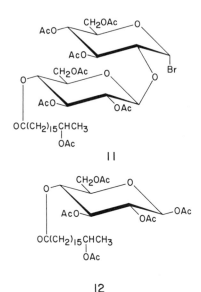

Figure 6-5. Acetobrominolysis product (**11**) and derived acetate (**12**) from lactonic sophoroside of *Torulopsis bombicola*.

tions showed that the fatty acid was esterified to the C-4″ hydroxyl in lactone (**8**) (Tulloch *et al.*, 1967, 1968*a*). The structures were completely confirmed by ¹H-NMR spectroscopy (see Section 6.6.1).

6.4.1.1. Formation of Sophorolipids by Torulopsis bombicola

Production of hyroxy fatty acid glycosides by *T. bombicola* has one unique feature that distinguishes it from all the other microbial reactions. When long straight-chain fatty acids or alkanes are added to the fermentation, they are hydroxylated and, in the case of alkanes, oxidized to acids, and directly incorporated into glycolipid resulting in an up to fivefold increase in yield and some modification of the hydroxy acid portion depending on the chain length of the added component (Tulloch *et al.*, 1962; Spencer *et al.*, 1962). Tables 6-1 and 6-2 summarize the results obtained when various methyl esters were added. Esters of C_{14} and C_{15} acids were poorly incorporated, and the chain length was largely increased by two or four carbons; C_{16}, C_{17}, and C_{18} esters were 60–80% incorporated without a change in chain length. Ester of C_{19} acid was directly hydroxylated, but there was also some loss of two carbons to form 17-hydroxheptadecanoic acid. Esters of the C_{20} and C_{22} acids were shortened by two or four carbons, respectively, from the carboxyl end, giving the glycoside of 17-hydroxyoleic acid (Tulloch *et al.*, 1962; Tulloch and Spencer, 1968).

The point of hydroxylation, either on the penultimate or terminal carbon, was very dependent on chain length. Substrates with 17 or more carbons (except linoleate) were hydroxylated mainly at the penultimate carbon, but

Table 6-1. Composition of Hydroxy Fatty Acids Obtained from Fermentation by Torulopsis bombicola of Fatty Acid Methyl Esters with an Even Number of Carbon Atoms[a]

Ester added	Products (%)												Conversion to OH acid (%)
	13-OH 14:0	14-OH 14:0	15-OH 16:0	15-OH 16:1	16-OH 16:0	16-OH 16:1	17-OH 18:0	17-OH 18:1	17-OH 18:2	18-OH 18:1	18-OH 18:2	Others	
None	—	—	8	—	9	4	19	48	—	4	—	8[b]	—
Tetradecanoate	13	14	10	5	11	11	6	25	—	3	—	2	10
Hexadecanoate	—	—	36	1	42	4	5	10	—	2	—	—	70
(Z)-9-Hexadecenoate	—	—	4	13	4	55	9	10	—	2	—	—	70
Octadecanoate	—	—	2	—	2	—	80	9	—	—	—	—	65
(Z)-9-Octadecenoate	—	—	2	—	2	—	6	77	—	13	—	—	80
(Z,Z)-9,12-Octadecadienoate	—	—	1	—	2	—	4	?	37[c]	?	56[c]	—	60
(Z)-11-Icosenoate	—	—	2	—	3	—	9	36	—	11	—	19[d]	55
(Z)-13-Docosenoate	—	—	2	—	4	—	5	65	—	12	—	12	50

[a] See Tulloch et al. (1962) and Spencer et al. (1962).
[b] Contains 4% 18-OH 18:0.
[c] Contains some monoenoic acid.
[d] Contains some 19-OH 20:1.

Table 6-2. Composition of Hydroxy Fatty Acids Obtained from Fermentation by Torulopsis bombicola of Fatty Acid Methyl Esters with an Odd Number of Carbon Atoms[a]

Ester added	14-OH 15:0	15-OH 15:0	15-OH 16:0	16-OH 16:0	16-OH 17:0	17-OH 17:0	17-OH 18:0	17-OH 18:1	18-OH 18:1	18-OH 19:0	Others	Conversion to OH acid (%)
					Products (%)							
None	—	—	8	9	—	—	19	48	4	—	12	—
Pentadecanoate	13	14	1	2	42[b]	3	6	12	2	—	5	10
Heptadecanoate	—	—	—	—	87[b]	13	—	—	—	—	—	65
Nonadecanoate	—	—	3	5	24[b]	9	6	—	—	53	—	60
Henicosanoate	—	—	3	4	54[b]	17	5	—	—	8	3[c]	55

[a]See Tulloch et al. (1962) and Spencer et al. (1962).
[b]Contains some monoenoic acid.
[c]20-OH 21:0.

linoleate and C_{14}–C_{16} esters were hydroxylated to a slightly greater extent at the terminal carbon than at the penultimate carbon. The two *cis* double bonds effectively shorten the length of the linoleate molecule, so that it is about the same length as hexadecanoate, and this may be why they are both about 50% hydroxylated at the terminal carbon (Tulloch *et al.*, 1962).

Straight-chain hydrocarbons were hydroxylated (after conversion to acids) and incorporated into glycolipid to nearly the same extent as methyl esters of the same chain length (Tulloch *et al.*, 1962; Tulloch and Spencer, 1968). Thus, in a typical fermentation of octadecane in stirred fermentors, the yeast was grown on 10% glucose supplemented with yeast extract and urea; 630 g of octadecane was added over 4 days; the yield of water-free glycolipid was 712 g (Tulloch *et al.*, 1968a). The effects of fermentation conditions on yields were also studied (Spencer *et al.*, 1962).

The ability of the yeast to hydroxylate fatty acids made it possibly to examine the hydroxylation mechanism. Using [$^{18}O_2$]oxygen and [^{18}O]water, it was shown that molecular oxygen, not water or glucose, was the source of the hydroxyl oxygen (Heinz *et al.*, 1969). By means of octadecanoates specifically deuterated at C-16, C-17, or C-18, it was found that no deuterium was lost from C-16 or C-18 and that only one deuterium was lost from C-17, thus eliminating unsaturated or ketonic intermediates. When (17S)-[17-^{2}H]octadecanoate was used the deuterium was lost and (17S)-17-hydroxyoctadecanoate was formed. There was no loss of deuterium, however, when the 17R isomer was fermented, demonstrating that the hydroxyl is introduced with retention of configuration (loss of a 17S hydrogen) involving no inversion or possibly an even number of inversions (Heinz *et al.*, 1969).

Oleic acid was also converted to 17-hydroxyoleic acid by a cell-free system from *T. bombicola* (Heinz *et al.*, 1970). The activity was in the particulate fraction sedimenting at 48,000*g*. Since NADPH was required for reaction, the enzyme system is probably a mixed-function oxidase. The glycoside was not formed, perhaps because of the absence of the appropriate glucose nucleotide (Heinz *et al.*, 1970).

6.4.1.2. *Function and Application of Sophorolipids from Torulopsis bombicola*

The advantages to *T. bombicola* of producing sophorolipids have not been clearly shown, but it appears that they facilitate hydrocarbon utilization by this species (Ito and Inoue, 1982), yet inhibit the growth of other species on alkanes (Ito *et al.*, 1980). Possibly it is a means of reducing competition by other microorganisms as in the example of *U. zeae* discussed earlier. The sophorolipid has also been proposed as a possible biosurfactant mainly because of the high yield and a new strain, isolated from a cabbage leaf, has given yields of up to 130 g/liter when grown on safflower oil (Inoue and Ito, 1982). Other studies have investigated the factors influencing economic production of the glycolipid as a biosurfactant (Cooper and Paddock, 1984; Kosaric *et al.*, 1984).

Other investigations concentrated on chemical production by fermentation. Dicarboxylic acids with relatively limited chain-length ranges were prepared by nitric acid oxidation or potassium hydroxide fusion of hydroxy acids from glycolipid produced by fermentation of crude hydrocarbons or oleic acid (Tulloch and Spencer, 1966). Because C_{15} compounds give a complex mixture of products in low yield (Tulloch and Spencer, 1968), 15-hydroxypentadecanoic acid cannot be easily prepared for possible production of perfume components (as described in Section 6.3.3 on hydroxy acids from *U. zeae*), but 16-hydroxyhexadecanoic acid can be isolated from fermentation of palmitate (Tulloch *et al.*, 1962). The formation of macrolides with a musk aroma by distillation of polyol esters of hydroxy acids prepared in such a fermentation has been reported (Kao Soap Co., 1979).

Fermentation of a long-chain alcohol was also examined in the hope that oxidation of the terminal methyl group to carboxyl would lead to a higher proportion of ω-hydroxy acids. However, when oleyl alcohol was fermented, it appeared that the alcohol was first oxidized to oleic acid and then hydroxylated, mainly at the penultimate carbon, as usual; there was no increase in ω-hydroxy acid (Tulloch and Spencer, 1972). A major product of the fermentation was, unexpectedly, the oleyl alcohol ester of an acid similar to acid (**10**) (Fig. 6-4) but with a 9,10-double bond in the hydroxy acid chain (Tulloch and Spencer, 1972). Use of oleyl and other long-chain alcohol esters of the glycolipid in cosmetic preparations has been described (Kao Soap Co., 1980*a*). Hydroxyalkyl ethers of the glycolipid have been used as humectants in cosmetics (Kao Soap Co., 1980*b*).

6.4.2. Characterization of Sophorolipids from Candida bogoriensis

The yeast *C. bogoriensis*, isolated from the leaf surface of *Randia malleifera* grown in Indonesia (Deinema, 1961), also produced a partly crystalline extracellular glycolipid (Stodola *et al.*, 1967), but the yields were only 1–2 g/liter, much lower than those from *T. bombicola*. The product was shown to be another sophoroside 13-[(2-O-β-D-glucopyranosyl-β-D-glucopyranosyl)oxy]-docosanoic acid 6′,6″-diacetate (**13**) (Fig. 6-6), similar to acid (**10**) (Fig. 6-4), but with 13-hydroxydocosanoic acid as aglycone in place of 17-hydroxyoctadecanoic acid (Tulloch *et al.*, 1968*b*). The positions of the acetate groups

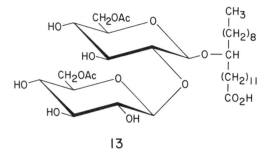

13

Figure 6-6. Acidic sophoroside produced by *Candida bogoriensis*.

followed from the rate of periodate oxidation, and the isolation of α-acetobromosophorose by acetobrominolysis of the fully acetylated glycolipid methyl ester showed the structure of the disaccharide portion. The structure was confirmed by ^{1}H-NMR and methylation–hydrolysis studies. The structure of the hydroxy fatty acid was shown by conversion to, and synthesis of, methyl 13-oxodocosanoate (Tulloch *et al.*, 1968*b*). Synthesis of methyl esters of (13*R*) and (13*S*) 13-hydroxydocosanoic acid showed that the acid from the glycolipid has the 13*S* form (Tulloch, 1968).

6.4.2.1. Formation of Sophorolipids by Candida bogoriensis

Unlike *T. bombicola, C. bogoriensis* was apparently incapable of hydroxylating long-chain acids and incorporating them into glycolipids. There have been a number of investigations of the biosynthesis of this sophorolipid. A glucosyl transferase that transferred glucose from uridine diphosphate glucose to hydroxy acid and then to glucosyl hydroxy acid was isolated from *C. bogoriensis* cells (Esders and Light, 1972*a*). All attempts to separate the enzyme into two, one for each stage, proved unsuccessful and indicated that both transferases were the function of the same polypeptide chain (Breithaupt and Light, 1982).

Diacetate (**13**) was usually accompanied by lesser amounts of monoacetate and glycolipid with no acetyl groups (Tulloch *et al.*, 1968*b*); the mixture was analyzed by gas–liquid chromatography (GLC) after formation of the trimethylsilyl glycolipid methyl ester (Esders and Light, 1972*b*). The acetylated glycolipid is formed by the action of an acetyl transferase (Bucholtz and Light, 1976*a*); the monoacetate is then produced later by an acetyl esterase (Bucholtz and Light, 1976*b;* Esders and Light, 1972*a*). The effect of glucose level in the medium on glycolipid formation was studied. At 0.5% glucose, very little glycolipid was produced, but at 3% glucose, the yield was greater than 2 g/liter (Cutler and Light, 1982).

6.5. Hydroxy Fatty Acid Glycosides from Plants

While formation of this type of glycolipid by the four species of microorganism discussed above has been investigated only during the past 40 years, study of what were later found to be hydroxy fatty acid glycosides began early in the present century (see Shellard, 1961; Kawasaki *et al.*, 1971). Most investigators of the microbiologically produced glycosides were apparently unaware that similar, but much more complex, compounds were also produced by plants.

6.5.1. Hydroxy Fatty Acid Glycosides from Convolvulus, Ipomoea, and Pharbitis Species

These glycosides have only been reported from plants in the family Convolvulaceae and only from species of three genera, *Ipomoea, Convolvulus,* and

Pharbitis. Formerly extracts of certain species in these genera were used in medicine as strong purgatives (Osol and Farrar, 1947) under such names as *Ipomoea* resin and Jalap resin. Early investigations were greatly hampered not only by the complexity of the product, but by confused botanical nomenclature and by the need to use commercial samples of obscure origin as well (Shellard, 1961; Osol and Farrar, 1947). For these reasons, the present review of these glycolipids is limited to studies reported during the past 21 years.

6.5.2. Characterization of Hydroxy Fatty Acid Glycosides from Plants

Table 6-3 lists most of the glycosides from convolvulaceous species that have recently been examined. Glycosides have been obtained from all parts of the plants, more often from the roots, as these may be large and tuberous, but also from leaves and seeds. Amounts isolated were 5% from roots, 3–5% from seeds, and 2% from leaves. In many cases, the glycosides were complex mixtures; purification of the crude extract presented considerable difficulties. In earlier work, much use was made of solubility in various solvents, the product being soluble in ethanol but insoluble in water or ether (Shellard, 1961). These properties would be much affected by the degree of acylation, which might vary with the age of the plant, and would not necessarily indicate variation in hydroxy fatty acid or sugar portions. More than one type of oligosaccharide portion, containing up to seven sugar units, might be present.

Alkaline hydrolysis removed the acyl groups, giving a large number of short-chain acids, including acetic, propionic, butyric, valeric, isovaleric, 2-methylbutyric, tiglic, and 3-hydroxy-2-methylbutyric acids (Smith *et al.*, 1964; Legler, 1965; Kawasaki *et al.*, 1971). The total amount of these acids corresponded approximately to one molecule of acid per sugar unit. The fatty acid was also sometimes esterified to a sugar unit, resulting in a polyester with an average molecular weight of 25,000 (Legler, 1965).

The hydroxy fatty acid portions of the glycolipid were obtained as usual by acid hydrolysis. The most common monohydroxy acid was 11-hydroxyhexadecanoic acid (Table 6-3), but 11-hydroxytetradecanoic and 7-hydroxydecanoic acids have also been identified (Legler, 1965; Singh and Stacey, 1973). A number of dihydroxy acids, most containing a 3-hydroxyl group, have been isolated. These include 3,11-dihydroxytetradecanoic acid (Legler, 1965; Kawasaki *et al.*, 1971), 3,11-dihydroxyhexadecanoic acid (Sarin *et al.*, 1973), 3,12-dihydroxyhexadecanoic acid (Wagner and Kazmaier, 1971, 1977), and an assortment of 3,12- and 4,12-dihydroxy C_{15} and C_{16} acids (Wagner *et al.*, 1978). Although specific rotations have been reported for some of these hydroxy acids, the results cannot be employed to decide on the absolute configuration (Smith, 1970). It is not clear how the known effects of solvent and concentration on the rotation of 3-hydroxy acids would affect rotations of the dihydroxy acids.

Besides the above-described variation in structure, there may be more than one arrangement of sugar units in the product from one plant species. Chromatography of the glycosides, obtained after hydrolysis of the acyl groups, as a suitable derivative, has given fractions containing different

Table 6-3. Hydroxy Fatty Acid Glycosides from Plants[a]

Plant (tissue)	Hydroxy acid	Sugar (no. of units)	Reference
I. parasitica (seeds)	11-OH C_{16}	Fucose (1), quinovose (2), 6-deoxygulose (1)	Smith *et al.* (1964)
I. fistulosa (leaves)	7-OH C_{10}; 11-OH C_{14}; 11-OH C_{16}; 3,11-diOH C_{14}	Glucose (2), fucose (2), rhamnose (1), quinovose (1)	Legler (1965)
P. nil (seeds)	3,11-diOH C_{14}	Glucose (2), quinovose (1), rhamnose (2 or 3)	Okabe and Kawasaki (1970, 1972); Kawasaki *et al.* (1971); Okabe *et al.* (1971)
I. leari (whole plant)	3,11-diOH C_{16}	Glucose (1), rhamnose (2), fucose (1)	Sarin *et al.* (1973)
I. purga (root)	11-OH C_{14}	Quinovose (1)	Singh and Stacey (1973)
I. operculata (root)	3,12-diOH C_{16}	Glucose (4), rhamnose (2)	Wagner and Kazmaier (1971, 1977)
C. microphyllus (whole plant)	11-OH C_{16}	Glucose (1), rhamnose (3), fucose (1)	Wagner and Schwarting (1977)
I. muriatica (seeds)	11-OH C_{16}	Rhamnose (2)	Khanna and Gupta (1967); Liptak *et al.* (1978)
I. turpethum (roots)	11-OH, 3,12-diOH; 4,12-diOH C_{16}; 3,12-diOH, 4,12-diOH C_{15}	Glucose (3), rhamnose (1)	Wagner *et al.* (1978)
I. quamoclit (roots)	11-OH C_{16}	Rhamnose (4), glucose (1)	Wagner *et al.* (1983)
I. lacunosa (roots)	11-OH C_{16}	Rhamnose (5), glucose (2)	Wagner *et al.* (1983)
I. pandurata (roots)	11-OH C_{16}	Rhamnose (3), glucose (1)	Wagner *et al.* (1983)
C. al-sirensis (roots)	11-OH C_{16}	Glucose (3), rhamnose (1)	Wagner *et al.* (1983)

[a]Glycosides are also partially acylated with short-chain acids.

oligosaccharides (Okabe *et al.*, 1971; Wagner and Kazmaier, 1977; Wagner and Schwarting, 1977; Wagner *et al.*, 1978). The sugars most commonly found were D-glucose and L-rhamnose and less often other 6-deoxysugars such as D-fucose and D-quinovose. Two of the glycosides in Table 6-3 had simpler structures. The most simple consisted, after deacylation, of D-quinovose (6-deoxyglucose) linked to 11-hydroxytetradecanoic acid as a β-glycoside (Singh and Stacey, 1973). The other consisted of a 4-*O*-L-rhamnopyranosyl-L-rhamnopyranose linked to 11-hydroxyhexadecanoic acid (Khanna and Gupta, 1967). Both sugar linkages were shown to have the α-configuration by synthesis of the glycosidic link (Liptak *et al.*, 1978).

Complete structures, sometimes only those of the principal components, have been derived from some of the glycosides presented in Table 6-3 and are

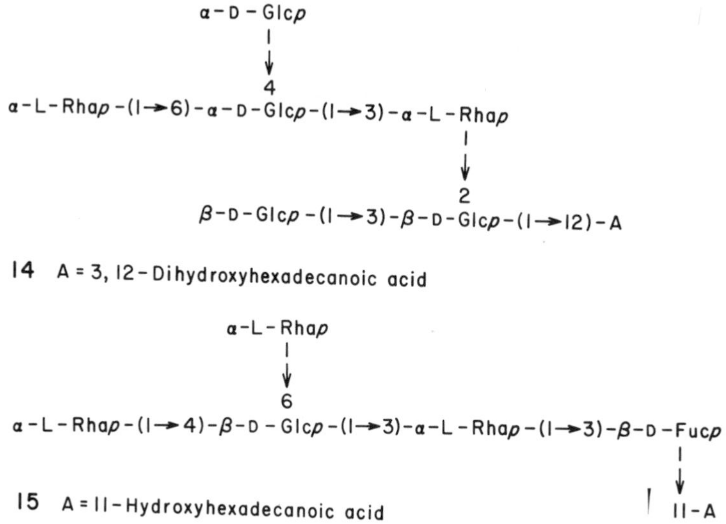

α – D – Glc*p*

4

α–L–Rha*p* –(I→6)–α–D–Glc*p*–(I→3)–α–L–Rha*p*

2

β–D–Glc*p*–(I→3)–β–D–Glc*p*–(I→I2)–A

14   A = 3, I2–Dihydroxyhexadecanoic acid

α–L–Rha*p*

6

α–L–Rha*p*–(I→4)–β–D–Glc*p*–(I→3)–α–L–Rha*p*–(I→3)–β–D–Fuc*p*

15   A = II–Hydroxyhexadecanoic acid   II–A

Figure 6-7. Hydroxy fatty acid glycosides (14) from *Ipomoea operculata* and (15) from *Convolvulus microphyllus*.

shown in Figs. 6-7 and 6-8. The deacylated glycoside (14) (Fig. 6-7) was purified by column chromatography as the octadecaacetate methyl ester (Wagner and Kazmaier, 1971, 1977). The deacetylated product, on methanolysis, gave methyl 3,12-dihydroxyhexadecanoate; methanolysis after chromic acid oxidation gave methyl 12-hydroxy-3-oxohexadecanoate, showing that the C-12 hydroxyl was linked to the sugar portion. The glycoside was fully methylated, hydrolyzed, reduced, and acetylated, giving a mixture of partially methylated alditol acetates, which was analyzed by GLC, thus indicating the substituent sugars and the points of attachment of the glycosidic linkages. To determine the order of the sugars, the glycoside was subjected to partial acetolysis and the acetylated fragments separated by thin-layer chromatography (TLC). These were identified by comparison with known di- and trisaccharides by methylation as described above.

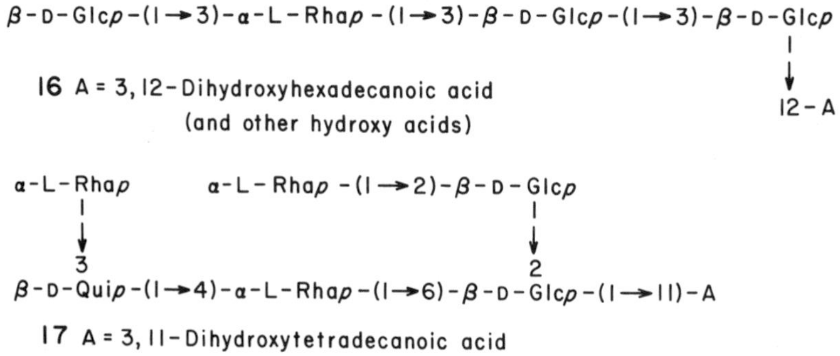

β–D–Glc*p*–(I→3)–α–L–Rha*p* –(I→3)–β–D–Glc*p*–(I→3)–β–D–Glc*p*

16   A = 3, I2–Dihydroxyhexadecanoic acid   I2–A
        (and other hydroxy acids)

α–L–Rha*p*   α–L–Rha*p* –(I→2)–β–D–Glc*p*

3   2

β–D–Qui*p*–(I→4)–α–L–Rha*p*–(I→6)–β–D–Glc*p*–(I→II)–A

17   A = 3, II–Dihydroxytetradecanoic acid

Figure 6-8. Hydroxy fatty glycosides (16) from *Ipomoea turpethum* and (17) from *Pharbitis nil.*

The structure of glycoside (**15**) (Fig. 6-7) was derived in a similar way, but in addition rhamnose was isolated from the product of periodate oxidation, confirming the 1,3-linkage (Wagner and Schwarting, 1977). Also, partial hydrolysis with 70% formic acid (Okabe and Kawasaki, 1970) gave a mixture of glycosides of fucose and rhamnosylfucose and some disaccharides. The glycoside from *I. turpethum* (Table 6-2) and (**16**) (Fig. 6-8) was unusual in having four dihydroxy acids, two of them C_{15} acids, among the aglycones (Wagner *et al.*, 1978). Structure (**16**) was again determined by methylation and conversion to methylated alditol acetates and by partial hydrolysis.

Glycoside (**17**) (Fig. 6-8) is one of several deacylated glycosides isolated from *Pharbitis nil* (Okabe *et al.*, 1971; Okabe and Kawasaki, 1972). Oxidation of (**17**) and hydrolysis yielded breakdown products of 11-hydroxy-3-oxotetradecanoic acid and showed that the sugar portion was linked to the hydroxyl at C-11. The glycosidic mixture, of which (**17**) was one of the major components, was separated by column chromatography of the *p*-phenylphenacyl esters. Methylation followed by acid methanolysis gave partially methylated methyl pyranosides, which were identified by GLC, and methyl 11-hydroxy-3-methoxytetradecanoate (confirming the point of attachment of the glycoside). Partial hydrolysis with 80% formic acid gave a number of products separated by chromatography and identified by methylation hydrolysis. Further information concerning the sugar sequence was obtained from the products of partial acetolysis and from the mass spectrum of the fully methylated glycosides. The ^{1}H-NMR spectra of the methylated glycosides and some of the breakdown products indicated the configuration of some of the linkages, since the splitting of the anomeric protons could be clearly seen. Besides the hexasaccharide (**17**), a pentasaccharide with a related structure, but lacking the terminal rhamnose unit linked to quinovose, was also identified.

The other glycosides presented in Table 6-3 have not been fully characterized, but some unusual results were obtained during examination of the glycoside from seeds of *I. parasitica* (Smith *et al.*, 1964). One of the sugar components was 6-deoxygulose, which has not been reported elsewhere as a glycoside component. This sugar and fucose were liberated by very mild acid treatment; for this reason, they were thought to be attached in the furanose form. The quinovose unit had the pyranose form and was strongly bound directly to the aglycone (11-hydroxyhexadecanoic acid). This particular behavior was not observed with any of the other glycosides.

6.5.3. Function and Application of Hydroxy Fatty Acid Glycosides from Plants

Because of the strong purgative action of these plant glycosides on mammals (Shellard, 1961) and their presence in relatively large amounts in the seeds, leaves, and roots of plants, it seems reasonable to suppose that one function is to repel herbivores. The glycoside is presumably absent from the sweet potato (*I. batatas*), so that not all (or even perhaps not many) species of *Ipomoea* contain the glycoside. The deacylated glycoside does not act as a purgative (Power and Rogerson, 1910).

The complexity and branched nature of the oligosaccharide portion of the glycoside has a general resemblance to the chains of sugars in structures involved in cell-recognition processes. Lipopolysaccharides of bacteria are examples and frequently contain rhamnose units (Elbein, 1981). Molecules of this type are not usually present in such large amounts, but the possibility of plant hydroxy fatty glycosides being involved in cell recognition should, perhaps, not be overlooked.

6.6. Determination of Structure of Hydroxy Fatty Acid Glycosides

Now that a number of different structures have been reviewed, it is useful to discuss and compare the methods employed to elucidate structure. As with other natural products, it is important to examine the product as far as possible before hydrolyzing or otherwise breaking down the molecule. Examination of the oligosaccharide portion has employed the usual methods of carbohydrate chemistry that have been reviewed (Bouveng and Lindberg, 1960). The hydroxy fatty acid has also been investigated by fairly standard procedures. The processes used are discussed in the following sections.

6.6.1. Spectroscopy

Probably the most valuable spectroscopic technique that can be applied to glycosides is nuclear magnetic resonance (NMR) spectroscopy (Kotowycz and Lemieux, 1973). Proton NMR is particularly useful in locating acyl groups and in determining the nature of the glycosidic linkages. As an example, Fig. 6-9 shows the ^{1}H-NMR spectrum of lactonic sophoroside (8) from *T. bombicola* (Tulloch *et al.*, 1968a). Much of the structure can be deduced from Fig. 6-9, although other glycosides may not give as useful spectra. The H-1 signals were assigned from changes occurring on acetylation. The H-6 signals, which have markedly different splittings, were assigned from the spectra of model compounds (Tulloch and Hill, 1968). Spin decoupling experiments related the low-field triplet to H-6″ through H-5″, demonstrating that the triplet is due to H-4″. The splittings of the resolved triplets show that only β-glucopyranose units are present; the splittings of the anomeric protons confirm that both linkages have the β-configuration. The two acetate methyl groups are at 2 ppm and the terminal methyl doublet, showing that the fatty acid is hydroxylated at the penultimate carbon, is at 1.2 ppm.

In spectra of other glycosides, such as (10) (Tulloch *et al.*, 1968a) and (13) (Tulloch *et al.*, 1968b), the anomeric protons are obscured by the H-6 signals but can be seen in the spectrum of the deacylated product. Splitting of signals due to H-1 protons can also be measured in the spectra of fully methylated glycosides where all other signals are at higher field (Tulloch *et al.*, 1968b; Okabe and Kawasaki, 1972). The small splittings ($J = 2$ Hz) of the H-1 signals in the spectrum of a synthetic dirhamnoside confirmed that the sugars were α-linked (Liptak *et al.*, 1978). NMR spectroscopy is most useful if model com-

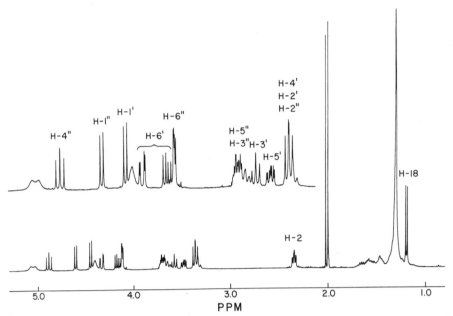

Figure 6-9. ^{1}H-NMR spectrum (360 MHz, [^{2}H$_6$]acetone) of lactone (**8**). (Inset) The 3.2- to 5.2-ppm region expanded and amplified. (From Tulloch *et al.*, 1968*a*.)

pounds are available for comparison (Tulloch *et al.*, 1968*a*; Tulloch and Hill, 1968).

Proton NMR spectra of all the monohydroxy octadecanoic acids have been examined, and hydroxylation near the ends of the chain was clearly indicated (Tulloch, 1966). Structures of the 3,11-dihydroxyhexadecanoic acid from *I. leari* (Sarin *et al.*, 1973) and the 3,12-dihydroxyhexadecanoic acid from *I. operculata* (Wagner and Kazmaier, 1977), which are less conveniently examined by oxidative degradation, have been investigated by ^{1}H-NMR. The hydroxyl group at C-3 was recognized by the splitting and relatively low field position of the protons α to the carboxyl group. The 3-oxo acids obtained by methanolysis after oxidation of the glycoside were also examined (Sarin *et al.*, 1973; Wagner and Kazmaier, 1977). NMR was also used to identify the (*E*)-2-unsaturated acid formed by elimination of water from a 3-hydroxy acid during degradation (Okabe *et al.*, 1971).

Mass spectroscopy (Lonngren and Svensson, 1974) has not been applied to the larger intact hydroxy fatty acid glycosides, but some methylated acetolysis products of the glycolipid from *P. nil* were identified from their spectra (Okabe and Kawasaki, 1972). The 3,11-dihyroxyhexadecanoic acid from *I. leari*, and the derived dioxo acid, were examined by mass spectroscopy (Sarin *et al.*, 1973). Structures of the four dihydroxy acids from *I. turpethum* were established by the mass spectra of the trimethylsilyl methyl esters (Wagner *et al.*, 1978). Application of mass spectroscopy to methylation products is discussed in Section 6.6.3.

6.6.2. Oxidation

As described earlier (see Section 6.4.1) in connection with the determination of the structure of lactone (**8**), unhindered glycols are oxidized with periodate and with lead tetraacetate in acetic acid, but "resistant diols" are only oxidized with the latter reagent in pyridine (Tulloch *et al.*, 1968*a*). Difficulties may arise with underoxidation (Jarvis and Johnson, 1949) and sometimes with overoxidation. Isolation of products after reduction and hydrolysis indicates structure (Edwards and Hayashi, 1965). Isolation of rhamnose after periodate oxidation, borohydride reduction, and glycoside hydrolysis of (**15**) indicated a 1,3-linkage to that sugar (Wagner and Schwarting, 1977). Similarly a 1,3-linkage to quinovose was shown in glycoside (**17**) by the isolation of this sugar after oxidation with excess periodate (Okabe *et al.*, 1971). A second glycoside, which lacked this linkage, did not yield quinovose on oxidation.

Glycol cleavage was also employed in establishing the structure of trihydroxy acid (**4**) from the glycolipid of *U. zeae* (Lemieux, 1953*a*). Chromic acid oxidation of a monohydroxy acid to the corresponding oxo acid was an essential part of the identification of the aglycone in earlier work; the oxo acid was identified by comparison with a synthetic sample. Thus 17-hydroxyoctadecanoic acid was converted to 17-oxooctadecanoic acid (Gorin *et al.*, 1961), 11-hydroxytetradecanoic acid to 11-oxotetradecanoic acid (Legler, 1965), and 13-hydroxydocosanoic acid to 13-oxodocosanoic acid (Tulloch *et al.*, 1968*b*). Oxidative cleavage at the hydroxylated carbon to a mixture of two mono- and two dicarboxylic acids was also employed (Smith *et al.*, 1964).

6.6.3. Methylation

Methylation of a polysaccharide followed by hydrolysis, separation, identification, and estimation of the partially methylated sugars has long been a standard procedure in carbohydrate chemistry (Bouveng and Lindberg, 1960). In earlier work, the products were isolated as the free sugars and separated by column or paper chromatography (Lemieux *et al.*, 1953; Gorin *et al.*, 1961), but later the methylated methyl glycosides were separated by GLC (Edwards and Hayashi, 1965). More recently, these degradation products have been examined by gas chromatography–mass spectrometry (GC–MS) (Dutton, 1974). The partially methylated sugars were reduced to alditols, with borohydride, acetylated, and analyzed by GLC (Wagner and Kazmaier, 1977) or by GC–MS (Wagner and Schwarting, 1977). Mass spectroscopy was employed to identify the methylated acetolysis products from the glycoside of *P. nil* (Okabe and Kawasaki, 1972).

The signals of the anomeric protons are readily seen in ^{1}H-NMR spectra of methylated oligosaccharides. Also, the 8 methoxy signals seen in the spectrum of the heptamethylsophoroside methyl ester from (**13**) (Tulloch *et al.*, 1968*b*) suggest the possibility of identifying isomeric structures, particularly if model compounds are available.

6.6.4. Hydrolysis and Related Reactions

Alkaline hydrolysis to remove acyl groups and acid hydrolysis to liberate the constituent sugars and aglycones have been applied to all the glycosides reviewed (see Sections 6.2.1 and 6.3.1) Partial hydrolysis, with dilute mineral acid, of the sophorolipid from *T. bombicola* gave a rather low yield of sophorose (Gorin *et al.*, 1961) but, except for the glycolipid from *I. parasitica* (Smith *et al.*, 1964), was not useful with the larger glycosides from plants (Okabe *et al.*, 1971). Partial hydrolysis with 80% formic acid (Okabe and Kawasaki, 1970; Okabe *et al.*, 1971) gave useful results (see Section 6.5.2).

Partial acetolysis with acetic anhydride containing sulfuric acid was also valuable; octaacetyl-α-D-cellobiose was obtained from the glycoside from *U. zeae* (Lemieux *et al.*, 1953a) and parts of the oligosaccharide were obtained from the glycosides of *C. microphyllus* (**15**) (Wagner and Schwarting, 1977) and *I. opeculata* (**14**) (Wagner and Kazmaier, 1977) (Figure 6-7). The success of the reaction seems to depend on the sugar linkages present. The fully acetylated lactone from (**8**) gave no useful products on acetolysis (A. P. Tulloch, unpublished work).

Acetobrominolysis, in acetic acid containing 3% hydrogen bromide, which has so far been applied to acetylated sophorosides only, was quite successful (Tulloch *et al.*, 1968a). The presence of the lactone ring accelerated the reaction considerably, the acetobromosophorose derivative (**11**) (Fig. 6-5) was obtained in about 50% yield in 4 days at 25°C, but acetobromosophorose was produced in only 10–15% yield in 7 days from acetylated methyl ester from (**10**) (Tulloch *et al.*, 1968a) and (**13**) (Tulloch *et al.*, 1968b). This reaction also inverts the glucose to hydroxy fatty acid linkage to the α-form within a few hours (Tulloch *et al.*, 1968a), which in conjunction with ¹H-NMR further confirms the structure. Acetobrominolysis, like acetolysis, has no effect on the configuration of the sugar to sugar linkage.

6.7. Summary

When the structures of all the hydroxy fatty acid glycosides are compared, a number of resemblances can be seen both in the hydroxy fatty acid portion and in the oligosaccharide portion. Thus, 3-hydroxyl groups are common in the hydroxy acids; for example, there is 3-hydroxydecanoic acid in the rhamnolipid from *P. aeruginosa*, 3-hydroxyhexanoic and octanoic acids esterified to the cellobiosides from *U. zeae* and most of the dihydroxy acids from the plant glycolipids contain a 3-hydroxyl group. Acids with penultimate and/or terminal hydroxylation were present in the glycosides from *U. zeae* and *T. bombicola*, although not in the plant glycosides.

Glucose, as a disaccharide, formed the sugar portion of three of the four microbial lipids and was present in the oligosaccharide portion of most of the plant glycosides. The glycolipid from *P. aeruginosa* was a dirhamnoside and rhamnose was also a major sugar constituent of the plant glycolipids. Almost all the glycosides also contained acyl groups, about one per sugar residue. These acids were generally C_2 to C_5 acids except for the glycoside from *U. zeae*, where 3-hydroxy C_6 and C_8 acids were attached as esters.

No clear function has been established for the glycosides except that of bioemulsifier or surfactant for the microbial products. The resemblance of the oligosaccharide portion of the plant glycosides to the branched chains of sugars present in compounds involved in recognition processes in plants seems worth noting. In fact, synthetic antigens have linked oligosaccharide to protein as a glycoside of 9-hydroxynonanoic acid (Lemieux *et al.*, 1975).

It is surprising that the occurrence of hydroxy fatty acid glycosides has not been more widely reported. A search for these lipids in other species of plants and microorganisms might well result in the discovery of small quantities that have been missed so far. A more definite explanation of their biological function might then be obtained.

References

Asselineau, C., and Asselineau, J., 1978, Trehalose-containing Glycolipids, in *Progress in the Chemistry of Fats and Other Lipids* (R. T. Holman, ed.), Vol. 16, pp. 59–99, Pergamon, Oxford.

Bergan, T., 1981, Human- and animal-pathogenic members of the genus *Pseudomonas*, in *The Prokaryotes* (M. P. Starr, H. Stolp, H. G. Trüper, A. Balows, and H. G. Schlegal, eds.), Vol. 1, pp. 666–700, Springer-Verlag, Berlin.

Bhattacharjee, S. S., Haskins, R. H., and Gorin, P. A. J., 1970, Location of acyl groups on two partially acylated glycolipids from strains of *Ustilago* (smut fungi), *Carbohydr. Res.* **13**:235.

Boothroyd, B., Thorn, J. A., and Haskins, R. H., 1955, Biochemistry of the Ustilaginales X. The biosynthesis of ustilagic acid, *Can. J. Biochem. Physiol.* **33**:289.

Bouveng, H. O., and Lindberg, B., 1960, Methods in structural polysaccharide chemistry, in *Advances in Carbohydrate Chemistry* (M. L. Wolfrom, ed.), Vol. 15, pp. 53–89, Academic, New York.

Breithaupt, T. B., and Light, R. J., 1982, Affinity chromatography and further characterization of the glucosyltransferases involved in hydroxydocosanoic acid sophoroside production in *Candida bogoriensis, J. Biol. Chem.* **257**:9622.

Bucholtz, M. L., and Light, R. J., 1976a, Acetylation of 13-sophorosyloxydocosanoic acid by an acetyltransferase purified from *Candida borgoriensis, J. Biol. Chem.* **251**:424.

Bucholtz, M. L., and Light, R. J., 1976b, Hydrolysis of 13-sophorosyloxydocosanoic acid esters by acetyl- and carboxylesterases isolated from *Candida bogoriensis, J. Biol. Chem.* **251**:431.

Burger, M. M., Glaser, L., and Burton, R. M., 1963, The enzymatic synthesis of a rhamnose-containing glycolipid by extracts of *Pseudomonas aeruginosa, J. Biol. Chem.* **238**:2595.

Cooper, D. G., and Paddock, D. A., 1984, Production of a biosurfactant from *Torulopsis bombicola*, *Appl. Environ. Microbiol.* **47**:173.

Cutler, A. J., and Light, R. J., 1982, Effect of glucose concentration in the growth medium on the synthesis of fatty acids by the yeast *Candida bogoriensis, Can. J. Microbiol.* **28**:223.

Deinema, M. H., 1961, Intra- and extracellular lipid production by yeasts, Thesis, *Mededel. Landbouwhogeschool Wageningen* **61**:1.

Dutton, G. G. S., 1974, Applications of gas–liquid chromatography to carbohydrates. Part II, in *Advances in Carbohydrate Chemistry and Biochemistry* (R. S. Tipson and D. Horton, eds.), Vol. 30, pp. 9–110, Academic, New York.

Edwards, J. R., and Hayashi, J. A., 1965, Structure of a rhamnolipid from *Pseudomonas aeruginosa*, *Arch. Biochem. Biophys.* **111**:415.

Elbein, A. D., 1981, The structure and biosynthesis of lipopolysaccharides and glycoproteins, in *The Phytochemistry of Cell Recognition and Cell Surface Interactions* (F. A. Loewus and C. A. Ryan, eds.), Vol. 15, pp. 1–24, Plenum, New York.

Esders, T. W., and Light, R. J., 1972a, Glucosyl- and acetyltransferases involved in the biosynthesis of glycolipids from *Candida bogoriensis, J. Biol. Chem.* **247**:1375.

Esders, T. W., and Light, R. J., 1972b, Characterization and *in vivo* production of three glycolipids from *Candida bogoriensis:* 13-Glucopyranosylglucopyranosyloxydocosanoic acid and its mono- and diacetylated derivatives, *J. Lipid Res.* **13**:663.

Gorin, P. A. J., Spencer, J. F. T., and Tulloch, A. P., 1961, Hydroxy fatty acid glycosides of sophorose from *Torulopsis magnoliae, Can. J. Chem.* **39**:846.

Guerra-Santos, L., Kappeli, O., and Fiechter, A., 1984, *Pseudomonas aeruginosa* biosurfactant production in continuous culture with glucose as carbon source, *Appl. Environ. Microbiol.* **48**:301.

Haskins, R. H., 1950, Biochemistry of the Ustilaginales I. Preliminary cultural studies of *Ustilago zeae, Can. J. Res.* **C28**:213.

Haskins, R. H., and Thorn, J. A., 1951, Biochemistry of the Ustilaginales. VII. Antibiotic activity of ustilagic acid, *Can. J. Bot.* **29**:585.

Heinz, E., Tulloch, A. P., and Spencer, J. F. T., 1969, Stereospecific hydroxylation of long-chain compounds by a species of *Torulopsis, J. Biol. Chem.* **244**:882.

Heinz, E., Tulloch, A. P., and Spencer, J. F. T., 1970, Hydroxylation of oleic acid by cell free extracts of a species of *Torulopsis, Biochim. Biophys. Acta* **202**:49.

Hirayama, T., and Kato, I., 1982, Novel methyl rhamnolipids from *Pseudomonas aeruginosa, FEBS Lett.* **139**:81.

Hisatsuka, K., Nakahara, T., Sano, N., and Yamada, K., 1971, Formation of rhamnolipid by *Pseudomonas aeruginosa* and its function in hydrocarbon fermentation, *Agr. Biol. Chem.* **35**:686.

Imoto, M., Yoshimura, H., Sakaguchi, N., Kusumoto, S., and Shiba, T., 1985, Total synthesis of *Escherichia coli* lipid A, *Tetrahedron Lett.* **26**:1545.

Inoue, S., and Ito, S., 1982, Sophorolipids from *Torulopsis bombicola* as microbial surfactants in alkane fermentations, *Biotechnol. Lett.* **4**:3.

Ito, S., and Inoue, S., 1982, Sophorolipids from *Torulopsis bombicola:* Possible relation to alkane uptake, *Appl. Environ. Microbiol.* **43**:1278.

Ito, S., Kinta, M., and Inoue, S., 1980, Growth of yeasts on *n*-alkanes: Inhibition by a lactonic sophorolipid produced by *Torulopsis bombicola, Agr. Biol. Chem.* **44**:2221.

Itoh, S., and Suzuki, T., 1972, Effect of rhamnolipids on growth of *Pseudomonas aeruginosa* mutant deficient in *n*-paraffin-utilizing ability, *Agr. Biol. Chem.* **36**:2233.

Itoh, S., Honda, H., Tomita, F., and Suzuki, T., 1971, Rhamnolipids produced by *Pseudomonas aeruginosa* grown on n-paraffin, *J. Antibiot.* **24**:855.

Jarvis, F. G., and Johnson, M. J., 1949, A glycolipid produced by *Pseudomonas aeruginosa, J. Am. Chem. Soc.* **71**:4124.

Kamibayashi, A., and Matsui, M., 1963, Metabolic products of *Ustilago maydis.* II. Production of ustilagic acid by a mutant of *Ustilago maydis* and degradation products of ustilagic acid, *Kogyo Gijutsuin, Hakko Kenkyusho Kenkyu Hokoku,* **24**:137 (*Chem. Abs.* 1966, 64, 1059a).

Kao Soap Co. Ltd., 1979, Verfahren zur Herstellung eines Hydroxyfettsaureesters, *German patent* 2,834,117, (*Chem. Abs.* 1979, **90**, 168086).

Kao Soap Co. Ltd., 1980a, Kosmetische Zusammensetzung zur Haut- und Haarbehandlung, *German patent* 2,938,383, (*Chem. Abs.* 1980, **93**, 166117).

Kao Soap Co. Ltd., 1980b, Kosmetikpraparat, *German patent* 2,939,519 (*Chem. Abs.* 1981, **94**, 20375).

Kawasaki, T., Okabe, H., and Nakatsuka, I., 1971, Studies on resin glycosides. I. Reinvestigation of the components of pharbitin, a resin glycoside of the seeds of *Pharbitis nil* Choisy, *Chem. Pharm. Bull.* **19**:1144.

Key, B. A., Gray, G. W., and Wilkinson, S. G., 1970, The purification and chemical composition of the lipopolysaccharide of *Pseudomonas alcaligenes, Biochem. J.* **120**:559.

Khanna, S. N., and Gupta, P. C., 1967, Structure of Muricatin, *Phytochemistry* **6**:735.

Kosaric, N., Cairns, W. L., Gray, N. C. C., Stechey, D., and Wood, J., 1984, The role of nitrogen in multiorganism strategies for biosurfactant production, *J. Am. Oil Chem. Soc.* **61**:1735.

Kotowycz, G., and Lemieux, R. U., 1973, Nuclear magnetic resonance in carbohydrate chemistry, *Chem. Rev.* **73**:669.

Lang, S., Gilbon, A., Syldatk, C., and Wagner, F., 1984, Comparison of interfacial active properties of glycolipids from microorganisms, in *Surfactants Solution* (Proceedings of International Symposium) (K. L. Mittal and B. Lindman, eds.), Vol. 2, pp. 1365–1376, Plenum, New York.

Legler, G., 1965, Die bestandteile des giftigen glykosidharzes aus *Ipomoea fistulosa* Mart. et. Chois, *Phytochemistry* **4**:29.

Lemieux, R. U., 1953a, Biochemistry of the Ustilaginales. VIII. The structures and configurations of the ustilic acids, *Can. J. Chem.* **31**:396.

Lemieux, R. U., 1953b, Synthetic musks from a metabolic product of corn smut in artificial culture, *Perfumery Essent. Oil Record* **44**:136.

Lemieux, R. U., and Charanduk, R., 1951, Biochemistry of the Ustilaginales VI. The acyl groups of ustilagic acid, *Can. J. Chem.* **29**:759.

Lemieux, R. U., and Giguere, J., 1951, Biochemistry of the Ustilaginales. IV. The configurations of some β-hydroxyacids and the bioreduction of β-ketoacids, *Can. J. Chem.* **29**:678.

Lemieux, R. U., Thorn, J. A., and Bauer, H. F., 1953, Biochemistry of the Ustilaginales. IX. The β-D-cellobioside units of the ustilagic acids, *Can. J. Chem.* **31**:1054.

Lemieux, R. U., Bundle, D. R., and Baker, D. A., 1975, The properties of a "synthetic" antigen related to the human blood-group Lewis a, *J. Am. Chem. Soc.* **97**:4076.

Linstead, R. P., Owen, L. N., and Webb, R. F., 1953, Elimination reactions of esters. Part I. The formation of αβ-unsaturated acids from β-acyloxy-compounds, *J. Chem. Soc.* **1953**:1211.

Liptak, A., Chari, V. M., Kreil, B., and Wagner, H., 1978, Structure proof of muricatin B: 11-hydroxyhexadecanoic acid dirhamnoside, *Phytochemistry* **17**:997.

Lonngren, J., and Svensson, S., 1974, Mass spectroscopy in structural analysis of natural carbohydrates, in *Advances in Carbohydrate Chemistry and Biochemistry* (R. S. Tipson and D. Horton, eds.), Vol. 29, pp. 41–106, Academic, New York.

Okabe, H., and Kawasaki, T., 1970, Structures of pharbitic acids C and D, *Tetrahedron Lett.* **11**:3123.

Okabe, H., and Kawasaki, T., 1972, Studies on resin glycosides. III. Complete structures of pharbitic acids C and D, *Chem. Pharm. Bull.* **20**:514.

Okabe, H., Koshito, N., Tanaka, K., and Kawasaki, K., 1971, Studies on resin glycosides. II. Unhomogeneity of "pharbitic acid" and isolation and partial structures of pharbitic acids C and D, the major constituents of "pharbitic acid," *Chem. Pharm. Bull.* **19**:2394.

Osol, A., and Farrar, G. E., 1947, *The Dispensatory of the United States of America*, 24th ed., p. 592, 602, J.B. Lippincott, Philadelphia.

Power, F. B., and Rogerson, H., 1910, Chemical examination of Jalap, *J. Am. Chem. Soc.* **32**:80.

Sarin, J. P. S., Garg, H. S., Khanna, N. M., and Dhar, M. M., 1973, Ipolearoside: A new glycoside from *Ipomoea leari* with anti-cancer activity, *Phytochemistry,* **12**:2461.

Shellard, E. J., 1961, The chemistry of some convolvulaceous resins. Part I. Vera Cruz jalap, *Planta Med.* **9**:102.

Singh, S., and Stacey, B. E., 1973, A new β-D-quinovoside from commercial *Ipomoea purga*, *Phytochemistry* **12**:1701.

Smith, C. R., 1970, Optically active long-chain compounds and their absolute configurations, in *Topics in Lipid Chemistry* (F. D. Gunstone, ed.), Vol. 1, pp. 277–368, Wiley, New York.

Smith, C. R., Niece, L. H., Zobel, H. F., and Wolff, I. A., 1964, Glycosidic constituents of *Ipomoea parasitica* seed, *Phytochemistry,* **3**:289.

Spencer, J. F. T., Tulloch, A. P., and Gorin, P. A. J., 1962, Fermentation of long-chain compounds by *Torulopsis magnoliae*. II. Factors influencing production of hydroxy fatty acid glycosides, *Biotechnol. Bioeng.* **4**:271.

Spencer, J. F. T., Gorin, P. A. J., and Tulloch, A. P., 1970, *Torulopsis bombicola* sp. n., *Antonie van Leeuwenhoek* **36**:129.

Spencer, J. F. T., Spencer, D. M., and Tulloch, A. P., 1979, Extracellular glycolipids of yeasts, in *Economic Microbiology*, Vol. 3: *Secondary Products of Metabolism* (A. H. Rose, ed.), pp. 523–540, Academic, New York.

Stodola, F. H., Deinema, M. H., and Spencer, J. F. T., 1967, Extracellular lipids of yeasts, *Bacteriol. Rev.* **31**:194.

Thorn, J. A., and Haskins, R. H., 1951, Biochemistry of the Ustilaginales V. Factors affecting the formation of ustilagic acid by *Ustilago zeae*, *Can. J. Bot.* **29**:403.

Tokyo Yakuruto Co. Ltd., 1970, Fermentive preparation of ustilagic acid, *Japanese patent* 7,017,151 (*Chem. Abs.* 1970, **73**, 129585).

Tulloch, A. P., 1966, Solvent effects on the nuclear magnetic resonance spectra of methyl hydroxystearates, *J. Am. Oil Chem. Soc.* **43**:670.

Tulloch, A. P., 1968, Absolute configurations of 13-hydroxydocosanoic and 17-hydroxyoctadecanoic acids, *Can. J. Chem.* **46**:3727.

Tulloch, A. P., 1971, Diesters of diols in wheat leaf wax, *Lipids* **6**:641.

Tulloch, A. P., 1976, Structures of extracellular lipids produced by yeasts, in *Glycolipid Methodology* (L. A. Witting, ed.), pp. 329–344, American Oil Chemists' Society, Champaign, Illinois.

Tulloch, A. P., and Hill, A., 1968, Synthesis and nuclear magnetic resonance spectra of some partially acylated β-D-glucopyranosides, *Can. J. Chem.* **46**:2485.

Tulloch, A. P., and Spencer, J. F. T., 1964, Extracellular glycolipids of *Rhodotorula* species. The isolation and synthesis of 3-D-hydroxypalmitic and 3-D-hydroxystearic acids, *Can. J. Chem.* **42**:830.

Tulloch, A. P., and Spencer, J. F. T., 1966, Fermentation of long chain compounds by *Torulopsis magnoliae*. III. Preparation of dicarboxylic acids from hydroxy fatty acid sophorosides, *J. Am. Oil Chemists Soc.* **43**:153.

Tulloch, A. P., and Spencer, J. F. T., 1968, Fermentation of long-chain Compounds by *Torulopsis apicola*. IV. Products from esters and hydrocarbons with 14 and 15 carbon atoms and from methyl palmitoleate, *Can. J. Chem.* **46**:1523.

Tulloch, A. P., and Spencer, J. F. T., 1972, Formation of a long-chain alcohol ester of hydroxy fatty acid sophoroside by fermentation of fatty alcohol by a *Torulopsis* species, *J. Org. Chem.* **37**:2868.

Tulloch, A. P., Spencer, J. F. T., and Gorin, P. A. J., 1962, The fermentation of long-chain compounds by *Torulopsis magnoliae*. I. Structures of the hydroxy fatty acids obtained by the fermentation of fatty acids and hydrocarbons, *Can. J. Chem.* **40**:1326.

Tulloch, A. P., Hill, A., and Spencer, J. F. T., 1967, A new type of macrocyclic lactone from *Torulopsis apicola*, *Chem. Commun.* **1967**:584.

Tulloch, A. P., Hill, A., and Spencer, J. F. T., 1968a, Structure and reactions of lactonic and acidic sophorosides of 17-hydroxyoctadecanoic acid, *Can. J. Chem.* **46**:3337.

Tulloch, A. P., Spencer, J. F. T., and Deinema, M. H., 1968b, A new hydroxy fatty acid sophoroside from *Candida bogoriensis*, *Can. J. Chem.* **46**:345.

Wagner, H., and Kazmaier, P., 1971, Struktur der operculinsaure (rhamnoconvolvulinsaure) aus *Ipomoea operculata* Martin, *Tetrahedron Lett.* **12**:3233.

Wagner, H., and Kazmaier, P., 1977, Struktur der operculinsaure aus dem harz von *Ipomoea operculata*, *Phytochemistry* **16**:711.

Wagner, H., and Schwarting, G., 1977, Struktur der microphyllinsaure aus dem harz von *Convolvulus microphyllus*, *Phytochemistry* **16**:715.

Wagner, H., Wenzel, G., and Chari, V. M., 1978, The turpethinic acids of *Ipomoea turpethum* L. Chemical constituents of the Convolvulaceae resins. III, *Planta Med.* **33**:144.

Wagner, H., Schwarting, G., Varljen, J., Bauer, R., Hamdard, M. E., El-Faer, M. Z., and Beal, J., 1983, Chemical constituents of Convolvulaceae resins. IV. The glycosidic acids of *Ipomoea quamoclit, I. lacunosa, I. pandurata* and *Convolvulus al-sirensis, Planta Med.* **49**:154.

Yamaguchi, M., Sato, A., and Yukuyama, A., 1976, Microbial production of sugar-lipids, *Chem. Ind.* **1976**:741.

Chapter 7

Synthesis of Glycoglycerolipids

Jill Gigg and Roy Gigg

7.1. Introduction

Each class of the naturally occurring glycosyldiacylglycerols is usually isolated as a family of molecular species because a spectrum of fatty acids is normally present. Synthetic methods have the advantage not only of making the minor components of natural lipid mixtures more readily available and providing confirmation of structure, but also of allowing the preparation of individual species with a precise fatty acid composition. Synthetic and partially synthetic methods are available for the preparation of these lipids containing saturated and unsaturated fatty acids, which also allow the introduction of fatty acids with isotopic, fluorescence, or spin labeling, providing glycosyldiacylglycerols useful for physical studies on membranes.

Synthetic natural and unnatural glycosyldiacylglycerols, which may contain the oligosaccharide sequences present in naturally occurring glycoproteins, have also been prepared for incorporation into artificial membranes (liposomes) for immunological and other biological studies.

This chapter covers the chemical synthesis of glycoglycerolipids, including mono-, di-, and triglycosyldiacylglycerols, glycosylalkylacylglycerols, and glycosyldialkylglycerols (see Chapters 1, 3, and 4, *this volume*) as well as some phosphatidylglucosyldiacylglycerols (see Chapter 2, *this volume*).

7.2. Monoglycosylglycerolipids

7.2.1. 1,2-trans-Monoglycosyldiacylglycerols

The first synthesis of a β-galactosyldiacylglycerol was reported by Wehrli and Pomeranz (1969) and represented the first approach to the synthesis of the plant monogalactosyldiacylglycerol (Chapter 3, *this volume*). Condensation of racemic isopropylidene glycerol with acetobromogalactose and subsequent removal of the isopropylidene group from the product gave a β-galac-

Jill Gigg and Roy Gigg • Laboratory of Lipid and General Chemistry, National Institute for Medical Research, London NW7 1AA, England.

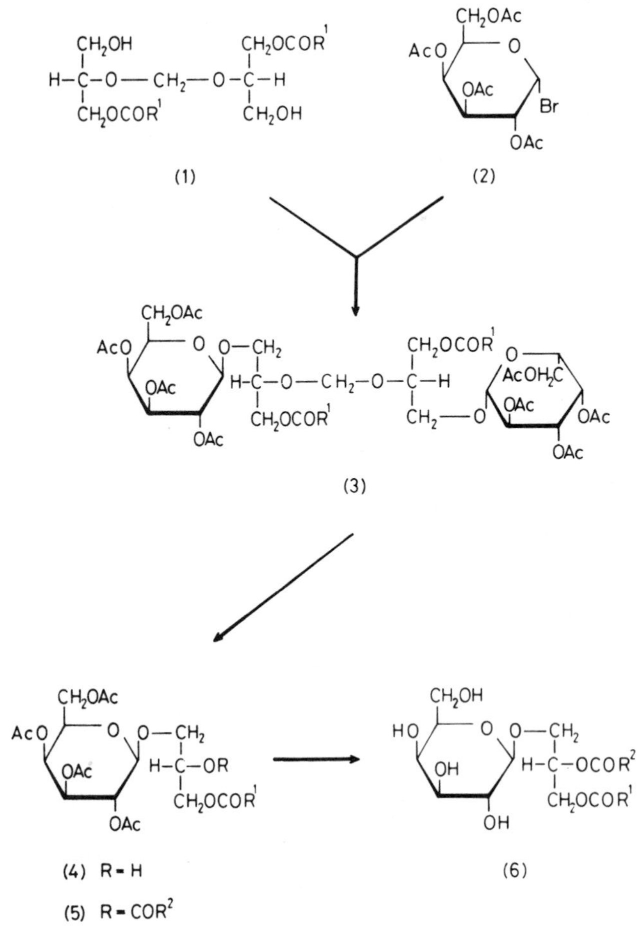

Figure 7-1. Synthesis of a galactosyldiacylglycerol. Ac = CH₃CO-. (From Wehrli and Pomeranz, 1969.)

tosylglycerol, acetylated on the galactose portion. This was acylated with fatty acid chlorides, and the acetyl groups were removed preferentially, by the action of hydrazine, to give a β-galactosyldiacylglycerol.

Wehrli and Pomeranz (1969) also prepared the chiral β-galactosyl di-acylglycerol (**6**, Fig. 7-1) by condensation of acetobromogalactose (**2**) with the glycerol derivative bis(1-acyl-2-methyleneglycerol) (**1**). Subsequent acidic hydrolysis of the product (**3**) gave the acetylated galactosylmonoacylglycerol (**4**), which was acylated with a fatty acid chloride to give the acetylated galactosyldiacylglycerol (**5**). The acetyl groups were again removed by hydrazinolysis to give the galactosyldiacylglycerol (**6**), stereochemically identical to the plant monogalactosyldiacylglycerol (see Chapter 3, *this volume*). This route allowed the preparation of galactosyldiacylglycerols with two different fatty acids: saturated and unsaturated.

Bashkatova *et al.* (1971*a,b,* 1972*a,* 1973) applied the ortho-acetate glycosi-

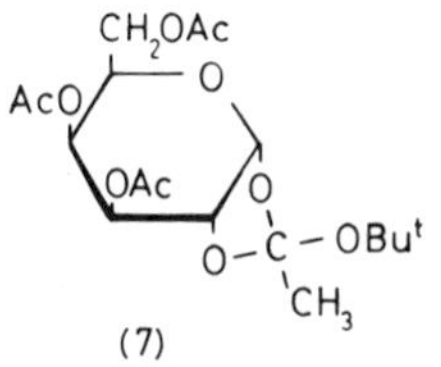

(7)

dation procedure to diacylglycerols for the synthesis of β-glycosyl di-acylglycerols. Thus, condensation of the *t*-butyl (Bu^t) orthoacetate (**7**) with a 1,2-di-*O*-acyl-*sn*-glycerol in the presence of lutidinium perchlorate and subsequent removal of the acetyl groups by hydrazinolysis gave the β-galactosyl diacylglycerol (**6**). The corresponding α-mannosyl and β-glucosyldiacylglyc-erols were also prepared by this method (Bashkatova *et al.*, 1971*a,b*, 1972*a,b*, 1973) and the extensive work of this group has been reviewed (Shvets *et al.*, 1973). The same group of workers (Kaplun *et al.*, 1975, 1977*a,b*, 1978; Vol-kova *et al.*, 1979) has also investigated the stereospecificity of the glycosidation reactions using orthoesters or acetobromo sugars in the synthesis of galac-tosyl, glucosyl, and mannosyldiacylglycerols and have shown that some 1,2-*cis*-glycosyldiacylglycerols are also formed in the reactions.

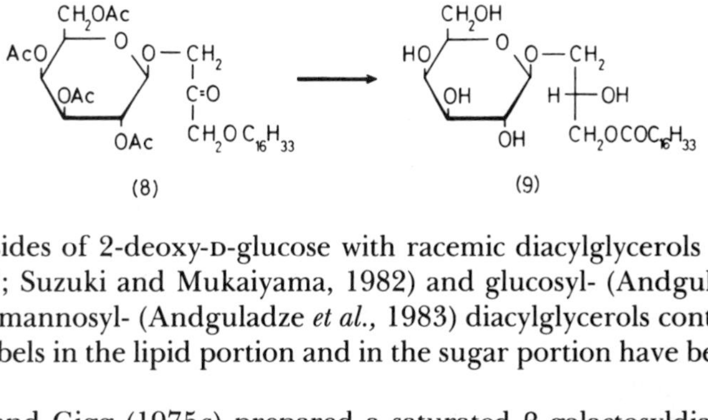

(8)   (9)

Glycosides of 2-deoxy-D-glucose with racemic diacylglycerols (Morozova *et al.*, 1982; Suzuki and Mukaiyama, 1982) and glucosyl- (Andguladze *et al.*, 1981) and mannosyl- (Andguladze *et al.*, 1983) diacylglycerols containing flu-orescent labels in the lipid portion and in the sugar portion have been synthe-sized.

Gent and Gigg (1975*a*) prepared a saturated β-galactosyldiacylglycerol from an intermediate prepared by condensation of acetobromogalactose with 1,2-di-*O*-benzyl-*sn*-glycerol. Subsequent removal of the benzyl groups by hy-drogenolysis and acetonation of the product gave the isopropylidene glycerol derivative (**10**). This was converted into the tribenzyl derivative (**11**), which was required as an intermediate for the preparation of the di- and trigalac-tosyldiacylglycerols of plants (see Sections 7.3 and 7.4) and was also used

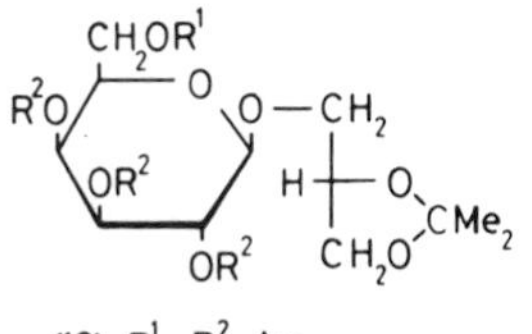

(10) R^1= R^2= Ac
(11) R^1= H ; R^2= Bn
(12) R^1= R^2= Bn

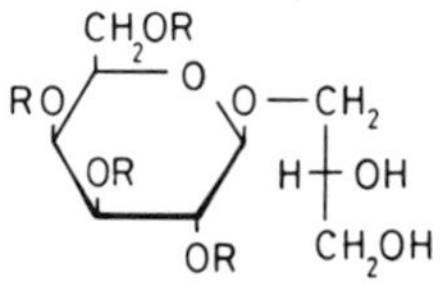

(13) R = CH(OMe)Me

(Gent and Gigg, 1975*a*) for synthesis of a 6-*O*-acyl-β-galactosyldiacylglycerol. A partial synthesis of 6-*O*-acyl β-galactosyldiacylglycerols, starting from natural galactosyldiacylglycerols, has been described by Heinz and Tulloch (1969). The perbenzylated galactosylglycerol derivative (**12**) prepared by benzylation of the alcohol (**11**) was used by Gent and Gigg (1975*a*) for the synthesis of the β-galactosyldiacylglycerol (**6**). This was achieved by hydrolysis of the isopropylidene group from the glycerol, followed by acylation with stearoyl chloride and subsequent removal of the benzyl groups by hydrogenolysis. 1,2-di-*O*-Benzyl-*sn*-glycerol has also been used by Batrakov *et al.* (1976) for the preparation of β-galactosyl-, β-glucosyl-, and α-mannosyldiacylglycerols. The latter two glycolipids are found in bacteria (see Chapter 1, *this volume*). In this preparation the dibenzylglycerol was condensed with the acetobromo-sugars to give the β-glycosides and the benzyl groups were then removed by hydrogenolysis. The glycerol moiety was acylated with fatty acid chlorides and the acetyl groups were then removed from the sugar portion by hydrazinolysis. Several improvements to this procedure, resulting in higher yields (70–75%) and purer products have been reported (Mannock *et al.*, 1987; van Boeckel *et al.*, 1985).

Heinz (1971) described a partial synthesis of β-galactosyldiacylglycerols, containing specified fatty acids, from the galactosyldiacylglycerols isolated from plants. Treatment of the galactosyldiacylglycerol (**6**) with methylvinyl ether in the presence of an acid catalyst and subsequent deacylation with alkali gave the tetramethoxyethylgalactosylglycerol derivative (**13**). This was reacylated with fatty acids and the methoxyethyl groups removed by acidic hydrolysis to give new galactosyldiacylglycerols. Partial acylation of compound (**13**) has also been used (Heisig and Heinz, 1972) to prepare monoacyl derivatives of β-galactosyl glycerol. A similar partially synthetic route has been used (Nishida and Yamada, 1985) for the preparation of galactosyldiacylglycerols containing spin-labeled fatty acids.

7.2.2. Seminolipid

Seminolipids, which occur in testes and brain of animals, are sulfated galactosylacylalkylglycerols (see Chapter 4, *this volume*). Gigg (1978, 1979) has described a synthesis of 3-*O*-(β-D-galactopyranosyl-3-sulfate)-2-*O*-hexadecanoyl-1-*O*-hexadecyl-*sn*-glycerol (**25**, Fig. 7-2) (seminolipid) from an intermediate (**14**) prepared by condensation of acetobromogalactose with 2-*O*-(but-2-enyl)-1-*O*-hexadecyl-*sn*-glycerol. The condensation product was saponified to give (**15**), which was converted into the isopropylidene derivative (**16**), and this was benzylated to give the dibenzyl compound (**17**). Acidic hydrolysis of the isopropylidene group gave (**18**) and preferential allylation of the 3-hydroxyl group of galactose gave (**19**), which on benzylation gave compound (**20**). This was treated with potassium tert-butoxide in dimethylsulfoxide, which removed the but-2-enyl group and isomerised the allyl group to give (**21**) and subsequent acylation with hexadecanoyl chloride in pyridine gave (**22**). Acidic hydrolysis removed the prop-1-enyl group from the 3-position of galactose to give (**23**) which was sulfated with the pyridine–sulfur

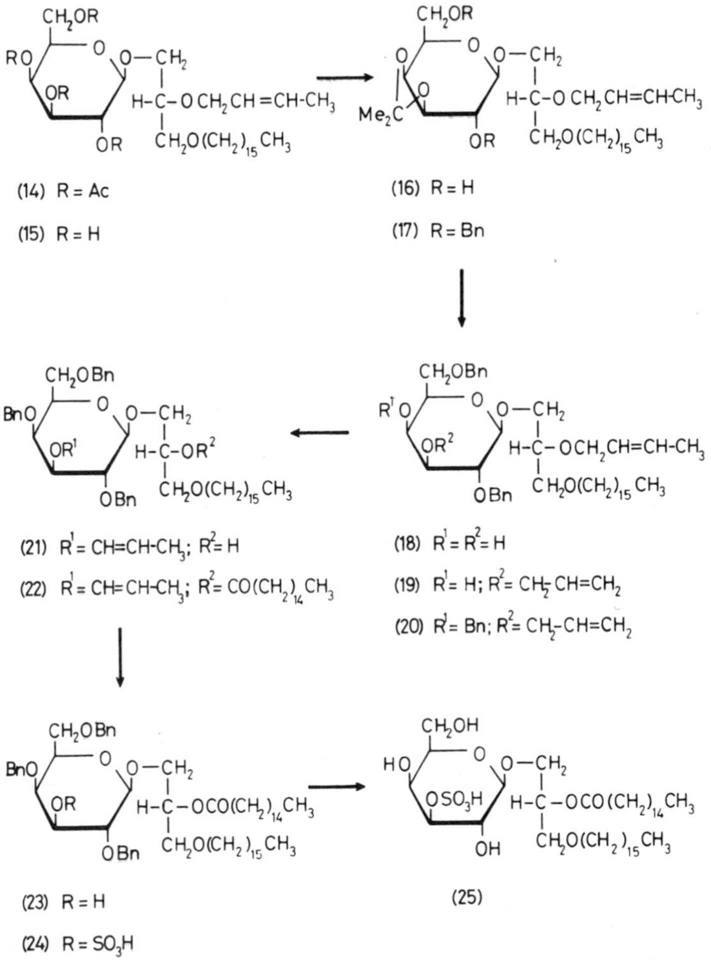

Figure 7-2. Synthesis of seminolipid. Ac = CH_3CO-; Bn = $-CH_2Ph$. (From Gigg, 1979.)

trioxide complex to give (**24**). This was converted into seminolipid (**25**) by removal of the benzyl groups by hydrogenolysis.

7.2.3. 1,2-cis-Glycosyldiacylglycerols

α-Glucosyl diacylglycerols (**26**) occur in bacteria (see Chapter 1, *this volume*) and the 6-deoxy-6-sulfo derivative of α-glucosyldiacylglycerols (sulfoquinovosyldiacylglycerols) (**27**) are major components of plant lipids (see Chapter 3, *this volume*). Phosphorylated derivatives (**28**) of α-glucosyldiacylglycerols also occur in streptococci (see Chapter 1, *this volume*). The syntheses of all these lipids have been reported.

Gigg *et al.* (1977) condensed the glucosyl chloride derivative (**29**) (Fig. 7-3) with 1,2,-di-*O*-(but-2-enyl)-*sn*-glycerol (Gent and Gigg, 1976) to give the

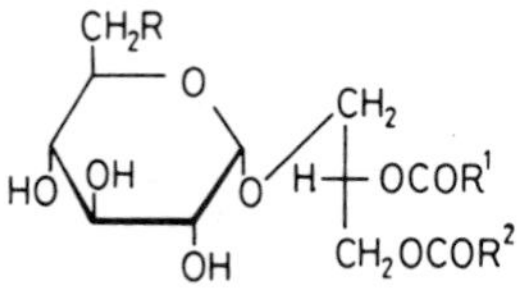

(26) R = OH

(27) R = SO₃H

(28) R = OP(O)OCH₂CH(OCOR)CH₂OCOR
          OH

α-glucoside (**30**). This was treated with potassium tert-butoxide in dimethylsulfoxide, which removed the acetyl and the but-2-enyl groups and the product was converted into the acetonide (**31**). Subsequent benzylation of (**31**), deacetonation, acylation with fatty acids, and removal of the benzyl groups gave the saturated α-glucosyldiacylglycerol (**26**).

The acetonide (**31**) was also used (Gigg, 1978; Gigg *et al.*, 1980) for the

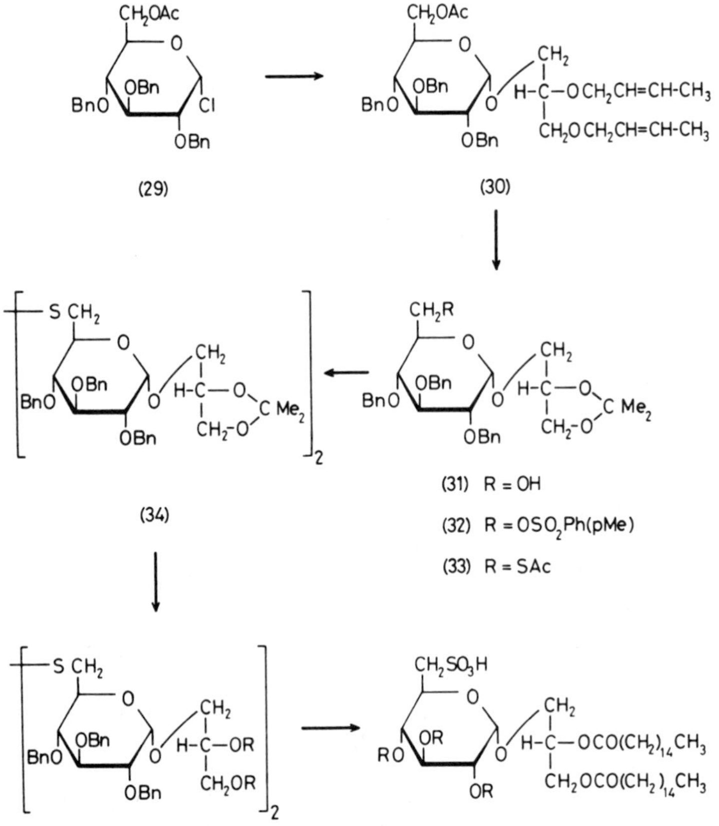

(29)   (30)

(34)

(31) R = OH

(32) R = OSO₂Ph(pMe)

(33) R = SAc

(35) R = H   (37) R = Bn

(36) R = CO(CH₂)₁₄CH₃   (38) R = H

Figure 7-3. Synthesis of a sulfoquinovosyldiacylglycerol. Ac = CH₃CO-; Bn = -CH₂Ph.

synthesis of a sulfoquinovosyldiacylglycerol (Fig. 7-3). The toluene-*p*-sulfonate (**32**) was converted into the thioacetate (**33**), which, after basic hydrolysis and subsequent oxidation with iodine, gave the disulfide (**34**). Acidic hydrolysis of the isopropylidene group from the disulfide (**34**) gave the alcohol (**35**) and acylation of this with hexadecanoyl chloride gave the ester (**36**). Compound (**36**) was oxidised with 3-chloroperbenzoic acid to give the 6-deoxy-6-sulfo derivative (**37**), which on debenzylation gave the sulfoquinovosyldiacylglycerol (**38**).

The 6-*sn*-phosphatidylglucosyl diacylglycerol (**46**) (Fig. 7-4) of *Streptococci* has been synthesised (van Boeckel and van Boom, 1980, 1985*a*) from the glucosylglycerol derivative (**39**). Removal of the but-2-enyl groups with potassium tert-butoxide in dimethyl sulfoxide gave the diol (**40**), which was acylated to give (**41**) and the benzyl groups were removed by hydrogenolysis to give (**42**). Compound (**42**) was treated with 1,3-dichloro-1,1,3,3-tetraisopro-

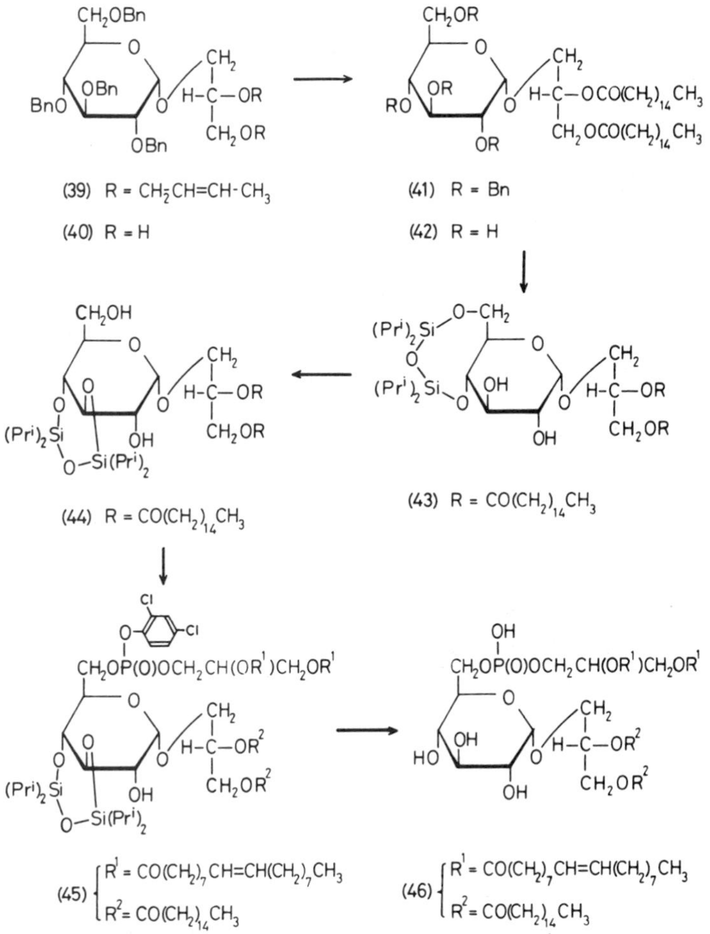

Figure 7-4. Synthesis of phosphatidylglucosyldiacylglycerol. Bn = -CH_2Ph; Pr^i = -$CH(CH_3)_2$. (From van Boeckel and van Boom, 1985*a*.)

pyldisiloxane in pyridine to give the 4,6-protected glucose derivative (**43**) and this was isomerised to the 3,4-protected derivative (**44**) on treatment with an acid catalyst. Compound (**44**) was phosphorylated to give (**45**) and the protecting groups were removed to give the 6-*sn*-phosphatidylglucosyldiacylglycerol (**46**) containing oleic and palmitic acids. van Boeckel *et al.* (1985) also described the synthesis of the related lipid containing a β-linked glucose residue using similar methods.

7.2.4. Monoglycosyldialkylglycerols

α- and β-glycosides of 2-fluoro-2-deoxy-D-mannose with 1,2,-di-*O*-tetradecyl-*sn*-glycerol have been prepared by Ogawa and Takahishi (1983). Earlier, Heinz *et al.* (1979) prepared ether analogues of β-galactosyldiacylglycerols by condensation of acetobromogalactose with 1,2,-di-*O*-octadec-9-enyl-*sn*-glycerol. Ether analogues of β-glucosyldiacylglycerols with a variety of alkyl groups have also been synthesized by Six *et al.* (1983) and β-glucosyl and β-galactosyl derivatives of 1,2-di-*O*-hexadecyl-*sn*-glycerol were prepared by Endo *et al.* (1982). A stereoselective reduction of the β-galactosyl derivative (**8**) with L-Selectride is reported (Mukaiyama *et al.*, 1982) to give a high yield of 3-*O*-(β-D-galactopyranosyl)-1-*O*-hexadecyl-*sn*-glycerol (**9**). Weber and Benning (1986) prepared β-glucosyl and β-maltosyl derivatives of 1-*O*-hexadecyl- and 1-*O*-octadecyl-2-*O*-methyl-*sn*-glycerol as well as racemic and ^{14}C-labeled derivatives.

7.3. Diglycosylglycerolipids

7.3.1. Diglycosyldiacylglycerols

Wehrli and Pomeranz (1969) obtained the first synthetic diglycosyldiacylglycerol, which was derived from cellobiose [β-D-glucopyranosyl(1 → 4)-D-glucose]. Acetobromocellobiose was condensed with racemic isopropylidene glycerol, and subsequent acid hydrolysis of the product, to remove the isopropylidene group, acylation with fatty acids and deacetylation with hydrazine gave the β-cellobiosyldiacylglycerol. The Russian group (Bashkatova *et al.*, 1971*c*, 1972*d*; Shvets *et al.*, 1973) also described a synthesis of a β-cellobiosyldiacylglycerol by condensation of an orthoester derivative of acetylated cellobiose with a diacylglycerol and subsequent deacetylation of the product with hydrazine.

A diglucosyldiacylglycerol (**53**) (Fig. 7-5) derived from kojibiose (α-D-glucopyranosyl-(1 → 2)-D-glucose] was prepared by van Boeckel and van Boom (1981, 1985b) by condensation of the perbenzylated glucosylbromide (**48**) with the α-glucosyldiacylglycerol derivative (**47**) (see also Fig. 7-4), which was protected on the 4- and 6-positions of glucose with the tetraisopropyldisiloxane-1,3-diyl group. The condensation product (**49**) was deprotected to give the α-kojibiosyl diacylglycerol (**53**) such as occurs in streptococci (see Chapters 1 and 2, *this volume*). Isomerisation of the 4,6-*O*-tetraisopropyldisiloxane-1,3-

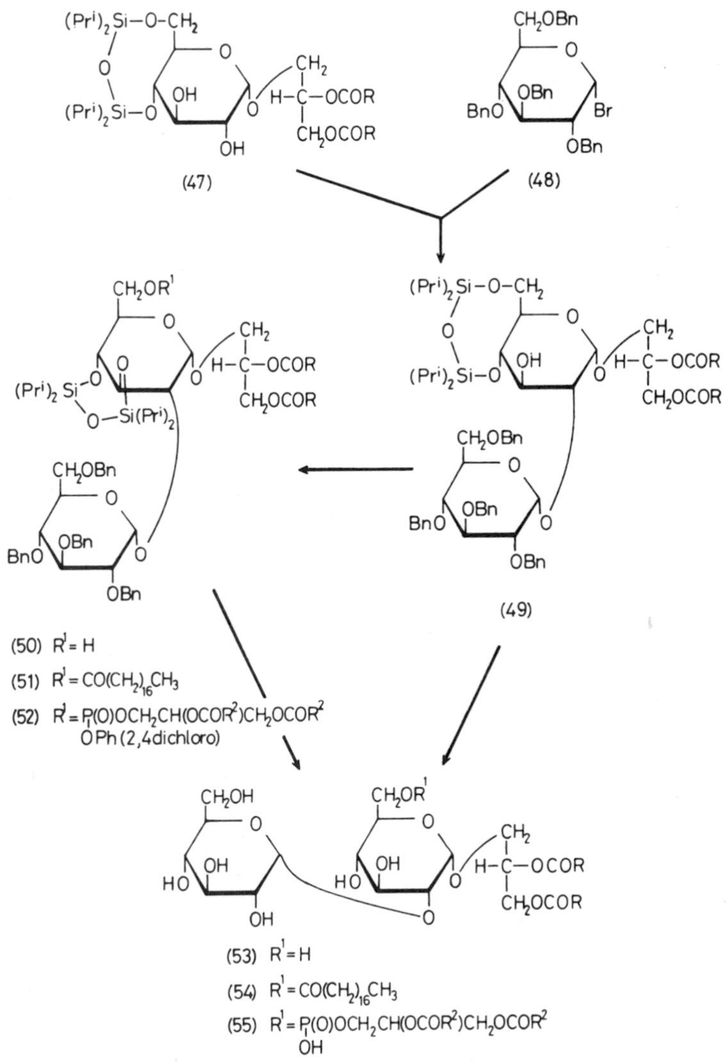

Figure 7-5. Synthesis of diglucosyldiacylglycerol and phosphatidyldiglucosyldiacylglycerol. Bn = -CH₂Ph; Pr^i = -CH(CH$_3$)$_2$. (From van Boeckel and van Boom, 1985*b*.)

diyl group in compound (**49**) with an acid catalyst in *N,N*-dimethylformamide gave the 3,4-*O*-tetraisopropyldisiloxane-1,3-diyl derivative (**50**) and this served as the intermediate for the synthesis (van Boeckel and van Boom, 1981, 1985*b*) of two other glycolipids derived from α-kojibiosyldiacylglycerol. Acylation of (**50**) with hexadecanoyl chloride in pyridine gave (**51**) which was deprotected to give the acylated α-kojibiosyldiacylglycerol (**54**) that has been isolated from *Streptococcus lactis* (see Chapters 1 and 2, *this volume*). Phosphorylation of (**50**) with a protected phosphatidic acid derivative gave (**52**) and subsequent deprotection of the molecule gave the phosphatidyl α-diglucosyldiacylglycerol (**55**) which is found in streptococci (see Chapter 2, *this volume*).

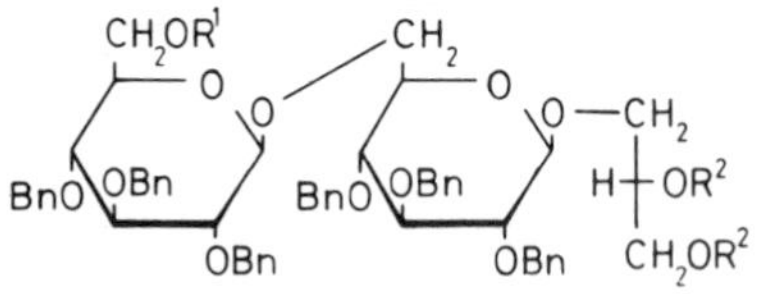

(56) $R^1 = H$; $R^2 = CO(CH_2)_{16}CH_3$

(57) $R^1 = CO(CH_2)_2COCH_3$; $R^2 = CO(CH_2)_{16}CH_3$

(58) $R^1 = P(O)OCH_2CH(OBn)CH_2OBn$
　　　　OH

　　$R^2 = CO(CH_2)_{16}CH_3$

Oltvoort *et al.* (1982) described the synthesis of a derivative (**56**) of a diglucosyl diacylglycerol with the structure β-D-glucopyranosyl(1 → 6)-β-D-glucopyranosyl-(1 → 3)-1,2-di-*O*-acyl-*sn*-glycerol. In this derivative, all the sugar hydroxyl groups, except the 6-position of the terminal glucose, were protected with benzyl groups. This was prepared via the 6-*O*-levulinoyl derivative (**57**) by preferential removal of the levulinoyl group with hydrazine. Phosphorylation of compound (**56**) with a protected 1,2-di-*O*-benzyl-*sn*-glycerol-3-phosphate gave (**58**), which was debenzylated to give the glycerol-phosphate ester of the diglucosyldiacylglycerol, which is a fragment of the membrane teichoic acid of *Staphylococcus aureus* (see Chapter 2, *this volume*). Further work by the same group (Oltvoort *et al.*, 1984) provided a synthetic lipoteichoic acid fragment with three glycerol phosphate residues attached to the diglucosyldiacylglycerol.

Prinz *et al.* (1985) have described the synthesis of 1-*O*-β-D-maltosyl-3-*O*-alkyl-*sn*-glycerols, [i.e., α-D-glucopyranosyl-(1 → 4)-β-D-glucopyranosyl-(1 → 1)-3-*O*-alkyl-*sn*-glycerols] (**59**) by the condensation of acetobromomaltose with 2-*O*-benzyl-3-*O*-alkyl-*sn*-glycerol and subsequent deprotection. The properties of these compounds (critical micelle concentration, hemolytic activity, and antitumor efficiency) were compared with those of analogous compounds containing a phosphorylcholine group instead of the disaccharide.

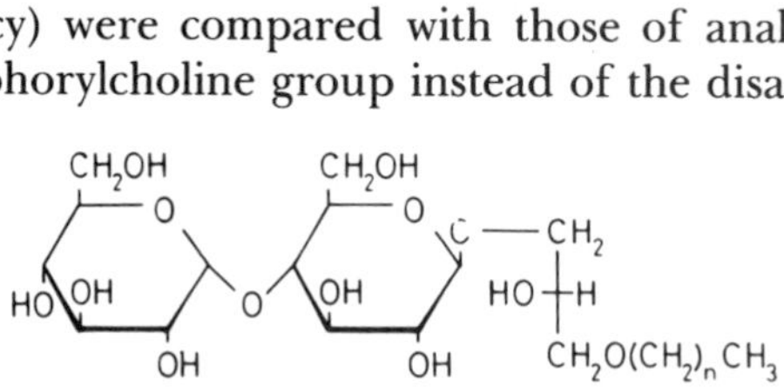

(59) n = 11,13,15 or 17

A digalactosyldiacylglycerol (**68**) (Fig. 7-6) such as occurs in plants (see Chapter 3) [i.e., α-D-galactopyranosyl-(1 → 6)-β-D-galactopyranosyl(1 → 3)-1,2-di-*O*-acylglycerol] was synthesized (Gent and Gigg, 1975b; Manthorpe and Gigg, 1980) from the β-galactosylglycerol derivative (**63**), which was also used in the synthesis of a β-galactosyldiacylglycerol (see Section 7.2.1). Condensation of acetobromogalactose with 1,2-di-*O*-benzyl-*sn*-glycerol gave the β-galac-

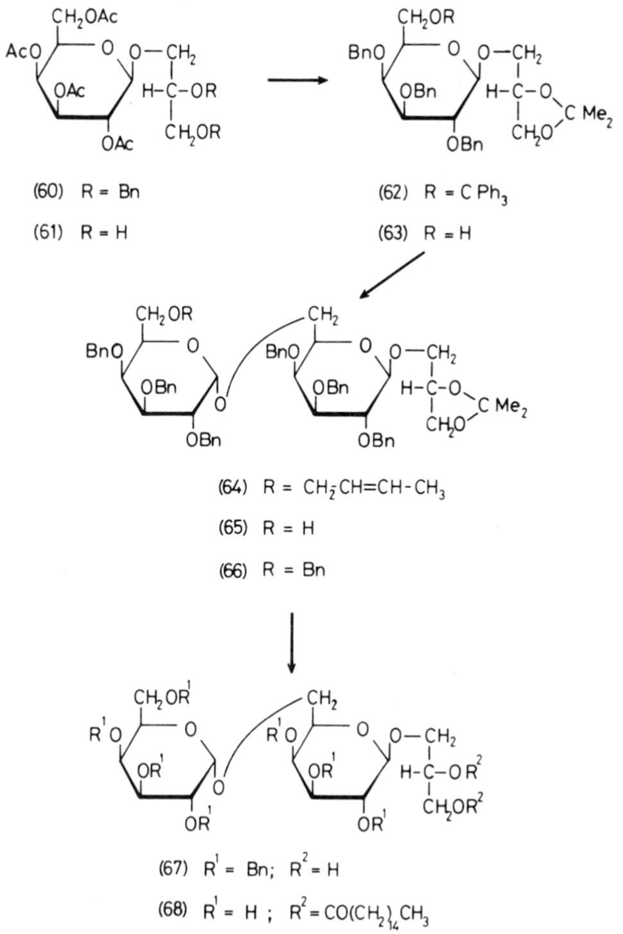

Figure 7-6. Synthesis of digalactosyldiacylglycerol. Ac = CH_3CO-; Bn = $-CH_2Ph$. (From Gent and Gigg, 1975*b*.)

toside (**60**), which on hydrogenolysis gave (**61**); this was converted into (**63**) via the trityl derivative (**62**). Glycosidation of (**63**) with 2,3,4-tri-*O*-ben-zyl-6-*O*-(but-2-enyl)-α-D-galactopyranosyl chloride gave the disaccharide de-rivative (**64**). The but-2-enyl group was removed by treatment with potassium tert-butoxide in dimethyl sulfoxide to give (**65**), and this was benzylated to give (**66**). Acidic hydrolysis removed the isopropylidene group and the product (**67**) was acylated with a fatty acid chloride and the benzyl groups removed to give the plant digalactosyldiacylglycerol (**68**).

Heinz (1971) described the partial synthesis of digalactosyldiacylglycerols from natural plant digalactosyldiacylglycerols by deacylation and reacylation reactions as described in Section 7.2.1 for the partial synthesis of β-galactosyl-diacylglycerols.

Gigg *et al.* (1987) have prepared a synthetic glycosyldiacylglycerol (**75**) (Fig. 7-7) containing the epitope of the major serologically active glycolipid of

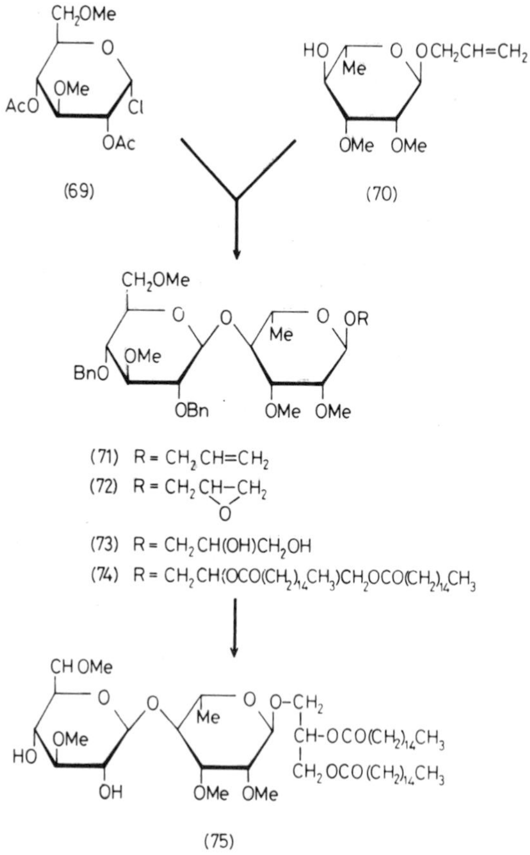

Figure 7-7. Synthesis of glycosyldiacylglycerol containing the epitope of the serologically active glycolipid from *Mycobacterium leprae*. Ac = CH_3CO-; Bn = $-CH_2Ph$.

Mycobacterium leprae for immunological studies. Condensation of the choride (**69**) derived from 3,6-di-*O*-methylglucose with allyl 2,3-di-*O*-methyl-α-L-rhamnopyranoside (**70**), alkaline hydrolysis of the product and subsequent benzylation gave the allyl glycoside of the protected disaccharide (**71**), which was converted into the epoxide (**72**) by the action of 3-chloroperbenzoic acid. Basic hydrolysis of the epoxide (**72**) gave the glycerol glycoside (**73**), which was acylated with hexadecanoyl chloride to give (**74**); removal of the benzyl groups by hydrogenolysis gave the glycosyldiacylglycerol (**75**).

7.3.2. Diglycosyldialkylglycerols

Cellobiosyl-, maltosyl-, and lactosyl-derivatives of 1,2-di-*O*-tetradecyl-*sn*-glycerol and several other monoglycosyl derivatives of this glycerol ether derivative have been prepared by Ogawa and Beppu (1982*a*). Both α- and β-linked derivatives were obtained by condensation of the peracetylated glycosylbromides with the dialkylglycerol. A β-lactosyl derivative [β-D-galactopyranosyl-(1 → 4)-β-D-glucopyranosyl(1 → 3)-1,2-di-*O*-tetradecyl-*sn*-glycerol] was also obtained by Ogawa *et al.* (1981) by condensation of octa-*O*-acetyl-β-lactose with 1,2-di-*O*-tetradecyl-*sn*-glycerol in the presence of trimethylsilyltri-fluoromethane sulfonate.

Ogawa and Beppu (1982*b*) prepared α-D-mannopyranosyl(1 → 2)-α-D-mannopyranosyl(1 → 3)-1,2-di-*O*-tetradecyl-*sn*-glyceorol and α-D-mannopyra-nosyl-(1 → 6)-α-D-mannopyranosyl-(1 → 3)-1,2-di-*O*-tetradecyl-*sn*-glycerol, as glycolipid models of the high mannose type glycans of glycoproteins, for studies in membrane function. β-D-Galactopyranosyl-(1 → 3)-2-acetamido-2-deoxy-β-D-galactopyranosyl(1 → 3)-1,2-di-*O*-tetradecyl-*sn*-glycerol has been prepared (Ogawa and Beppu, 1982*c*), using 3,4,6-tri-*O*-acetyl-2-deoxy-2-pht-halimido-β-D-galactopyranosyl(1 → 3)-1,2-di-*O*-tetradecyl-*sn*-glycerol as a key intermediate. The disaccharide in this diglycosyldialkylglycerol has the termi-nal sequence of the glycosphingolipid, asialo-GM1.

Ogawa and Sugimoto (1984) have also described the preparation of di-glycosyldialkylglycerols containing *N*-acetylneuraminic acid by the condensa-tion of neuraminic acid derivatives with 1,2-di-*O*-tetradecyl-*sn*-glycerol. α-NeuAc-(2 → 9)-α-NeuAc-(2 → 3)-1,2-di-*O*-tetradecyl-*sn*-glycerol and β-NeuAc-(2 → 9)-α-NeuAc-(2 → 3)-1,2-di-*O*-tetradecyl-*sn*-glycerol were prepared. The α-(2 → 9)-linked neuraminic acid derivative was prepared as an analogue of a pro-posed structural unit of the group C meningococcal polysaccharide.

7.4. Triglycosylglycerolipids

7.4.1. Triglycosyldiacylglycerols

A saturated analogue of the trigalactosyldiacylglycerol, α-D-galac-topyranosyl(1 → 6)-α-D-galactopyranosyl(1 → 6)-β-D-galactopyranosyl(1 → 3)-1,2-di-*O*-acylglycerol, which occurs in plants (see Chapter 3, *this volume*) was synthe-sized (Gent and Gigg, 1975*c*) from the digalactosylglycerol derivative (**65**) (Fig. 7-6) by reaction with 2,3,4-tri-*O*-benzyl-6-*O*-(but-2-enyl)-α-D-galactopyranosyl chloride and further processing as described for the synthesis of the digalac-tosyldiacylglycerol (see Section 7.3). It had been reported (Kenny and Newton, 1973) that the trigalactosyldiacylglycerol fraction from spinach glycolipids had similar serological activity to the glycolipids of *Mycoplasma pneumoniae*, but the synthetic trigalactosyldiacylglycerol described above was not active (Gigg, 1977).

Ogawa and Horisaki (1983) described the synthesis of α-D-glucopyranosyl-6-sulfate-(1 → 6)-α-D-glucopyranosyl(1 → 6)-α-D-glucopyrano-syl-(1 → 3)-1-*O*-hexadecyl-2-*O*-hexadecanoyl-*sn*-glycerol, a structure previously proposed for a sulfated glyceroglycolipid from human gastric secretion (see Chapter 4, *this volume*). The natural and synthetic materials were however not identical.

Ogawa and Beppu (1982*b*) prepared α-D-mannopyranosyl(1 → 2)-α-D-mannopyranosyl(1 → 2)-α-D-mannopyranosyl(1 → 3)-1,2-di-*O*-tetradecyl-*sn*-glycerol and α-D-mannopyranosyl(1 → 3)-α-D-mannopyranosyl[(6 → 1)-α-D-mannopyranosyl](1 → 3)-1,2-di-*O*-tetradecyl-*sn*-glycerol using the partially benzylated mannosyl chlorides, 2-*O*-acetyl-3,4,6-tri-*O*-benzyl-α-D-man-nopyranosyl chloride and 3,6-di-*O*-acetyl-2,4-di-*O*-benzyl-α-D-mannopyrano-syl chloride, as intermediates. These triglycosyldialkylglycerols were prepared as analogues of the high-mannose type glycals of glycoproteins for the investigation of biological function in membranes.

7.4.2. *Triglycosyldi-O-Phytanylglycerol*

van Boeckel *et al.* (1981, 1984) described the synthesis of β-D-glucopyranosyl-(1 → 6)-α-D-mannopyranosyl-(1 → 2)-α-D-glucopyranosyl-(1 → 1)-2,3-di-*O*-phytanyl-*sn*-glycerol (**79**) (Fig. 7-8) and β-D-galactopyranosyl-(1 → 6)-α-D-man-nopyranosyl-(1 → 2)-α-D-glucopyranosyl(1 → 1)-2,3-di-*O*-phytanyl-*sn*-glycerol (**82**). These glycolipids are constituents of cells of *Halobacterium marismortui* and of the purple membrane of *Halobacteria*, respectively (see Chapter 1, *this*

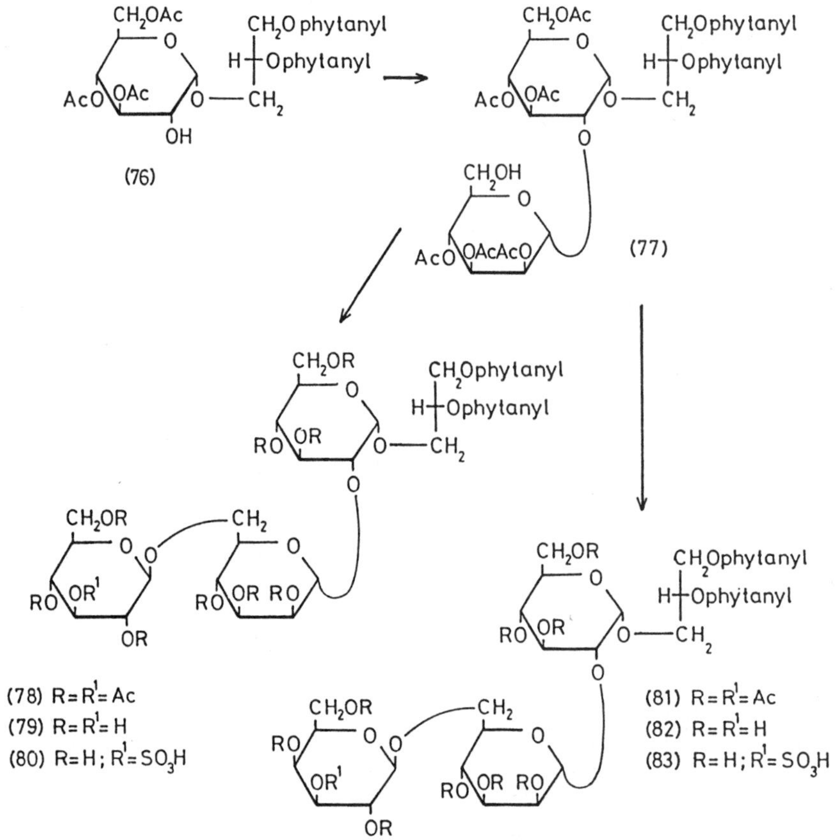

Figure 7-8. Synthesis of di-*O*-phytanylglycerol glycolipids of *Halobacteria*. Ac = CH₃CO-. (From van Boeckel *et al.*, 1981, 1984.)

volume). β-D-Glucopyranosyl-3-sulfate(1 → 6)-α-D-mannopyranosyl(1 → 2)-α-D-glucopyranosyl(1 → 1)-2,3-di-*O*-phytanyl-*sn*-glycerol (**80**) was also synthesized. This is the sulfated glucose analogue of the galactose sulfate-containing glycolipid (**83**), which occurs in *Halobacterium cutirubrum* (see Chapter 1, *this volume*).

These glycolipids were prepared using the intermediate (**76**), which was synthesized from α-D-glucopyranosyl(1 → 1)-2,3-di-*O*-phytanyl-*sn*-glycerol using the 4,6-*O*-tetraisopropyldisiloxane-1,3-diyl and the 2-*O*-(2-dibromomethylbenzoyl)-protecting groups. The α-mannosyl group was added to (**76**) using 2,3,4-tri-*O*-acetyl-6-*O*-(2,2,2-trichloroethoxycarbonyl)-β-D-mannopyranosyl bromide, and the trichloroethoxycarbonyl group was removed to give the disaccharide derivative (**77**). The latter was condensed with acetobromoglucose or acetobromogalactose, to give the acetylated intermediates (**78**) and (**81**), respectively, and these were deacylated to give the glycolipids (**79**) and (**82**). The sulfated triglycosyldiphytanylglycerol (**83**), present in the purple membrane of *Halobacteria* (see Chapter 1, *this volume*), could also be synthesized from (**77**) by condensation with a suitably protected galactose derivative using similar methods to those developed for the synthesis of the sulfolipid (**80**).

References*

Andguladze, M. K., Kaplun, A. P., and Shvets, V. I., 1981, Investigations in the field of glycosyl diglycerides. IX. Synthesis of fluorescence-labelled glycosyl diglycerides, *Zh. Org. Khim.* **17**:2323 (*J. Org. Chem. USSR* **17**:2075).

Andguladze, M. K., Kaplun, A. P., and Shvets, V. I., 1983, Investigations in the field of glycosyl diglycerides. X. Synthesis of mannosyl diglycerides with fluorescent labels, *Zh. Org. Khim.* **19**:961 (*J. Org. Chem. USSR* **19**:852).

Bashkatova, A. I., Smirnova, G. V., Shvets, V. I., and Evstigneeva, R. P., 1971a, Glycosyl diglycerides. I. Synthesis of mannopyranosyl and glucopyranosyl diglycerides, *Zh. Org. Khim.* **7**:1644 (*J. Org. Chem. USSR* **7**:1707).

Bashkatova, A. I., Smirnova, G. V., Volynskaya, V. N., Shvets, V. I., and Evstigneeva, R. P., 1971b, Synthesis of galactosyldiglyceride, *Dokl. Akad. Nauk SSSR (Biochem.)* **201**:1231 (*Proc. Acad. Sci. USSR Biochem.* **201**:357).

Bashkatova, A. I., Shvets, V. I., and Evstigneeva, R. P., 1971c, Synthesis of diglycosyldiglycerides, *Zh. Org. Khim.* **7**:2627 (*J. Org. Chem. USSR* **7**:2729).

Bashkatova, A. I., Volynskaya, V. N., Smirnova, G. V., Shvets, V. I., and Evstigneeva, R. P., 1972a, Glycosyldiglycerides. II. Synthesis of monoglycosyldiglycerides *via* isomeric orthoesters, *Zh. Org. Khim.* **8**:543 (*J. Org. Chem. USSR* **8**:548).

Bashkatova, A. I., Senyushkina, N. N., Shvets, V. I., and Evstigneeva, R. P., 1972b, Directed synthesis of glycoside diglycerides of different fatty acids, *Zh. Org. Khim.* **8**:208 (*J. Org. Chem. USSR* **8**:212).

Bashkatova, A. I., Senyushkina, N. N., Shvets, V. I., and Evstigneeva, R. P., 1972c, Glycosyl diglycerides. III. Synthesis of unsaturated mannosyl diglycerides, *Zh. Org. Khim.* **8**:2271 (*J. Org. Chem. USSR* **8**:2317).

Bashkatova, A. I., Shvets, V. I., and Evstigneeva, R. P., 1972d, Glycosyl diglycerides. IV. Synthesis of diglucosyl diglyceride, *Zh. Org. Khim.* **8**:2277 (*J. Org. Chem. USSR* **8**:2323).

Bashkatova, A. I., Smirnova, G. V., Volynskaya, V. N., Shvets, V. I., and Evstigneeva, R. P., 1973,

*References in parentheses after Russian journals refer to the page numbers of the English translation of these journals together with their English titles.

Investigations into the field of glycosyl diglycerides. V. Synthesis of galactosyl diglyceride, *Zh. Org. Khim.* **9**:1393 (*J. Org. Chem. USSR* **9**:1422).

Batrakov, S. G., Ilina, E. F., and Panosyan, A. G., 1976, Synthesis of monoglycosyl diglycerides and their derivatives, *Izv. Akad. Nauk SSSR Ser. Khim.* **1976**:643 (*Bull. Acad. Sci. USSR Chem.* **1976**:626).

Endo, T., Inoue, K., and Nojima, S., 1982, Physical properties and barrier functions of synthetic glyceroglycolipids, *J. Biochem. (Tokyo)* **92**:953.

Gent, P. A., and Gigg, R., 1975a, Synthesis of 1,2-di-*O*-hexadecanoyl-3-*O*-(beta-D-galactopyrano-syl)-L-glycerol (a "galactosyl diglyceride") and 1,2-di-*O*-octadecanoyl-3-*O*-(6-*O*-octadecanoyl-beta-D-ga lactopyranosyl)-L-glycerol, *J. Chem. Soc. Perkin Trans.* 1 **1975**:364.

Gent, P. A., and Gigg, R., 1975b, Synthesis of 3-*O*-{6-*O*-(alpha-D-galactopyranosyl)-beta-D-galac-topyranosyl}-1,2-di-*O*-stearoyl-L-glycerol, a "digalactosyl diglyceride," *J. Chem. Soc. Perkin Trans.* 1 **1975**:1521.

Gent, P. A., and Gigg, R., 1975c, Synthesis of 3-*O*-[6-*O*-[6-*O*-(alpha-D-galactopyranosyl)-alpha-D-galactopyranosyl]-beta-D-galactopyranosyl]-1,2-di-*O*-stearoyl-L-glycerol, a "trigalactosyl di-glyceride," *J. Chem. Soc. Perkin Trans.* 1 **1975**:1779.

Gent, P. A., and Gigg, R., 1976, Synthesis of 3-0-(3,4-di-*O*-benzyl-alpha-D-glucopyranosyl)-1,2-*O*-isopropylidene-*sn*-glycerol, *Chem. Phys. Lipids* **17**:111.

Gigg, J., Gigg, R., Payne, S., and Conant, R., 1987, The allyl group for protection in carbohydrate chemistry. Part 19. The coupling of allyl 2,3-di-*O*-methyl-4-*O*-(3,6-di-*O*-methyl-beta-D-glucopyranosyl)-alpha-L-rhamnopyrano side to bovine serum albumin. Preparation of a diag-nostic reagent for antibodies to the major glycolipid of *Mycobacterium leprae* (the leprosy bacillus) in human sera, *J. Chem. Soc. Perkin Trans.* 1 **1987**:1165.

Gigg, R., 1977, Studies on the synthesis of serologically active glycolipids, *ACS Symp. Ser.* **39**:253.

Gigg, R., 1978, Studies on the synthesis of sulfur-containing glycolipids ("sulfoglycolipids"), *ACS Symp. Ser.* **77**:44.

Gigg, R., 1979, The allyl ether as a protecting group in carbohydrate chemistry. Part 10. Synthesis of 3-*O*-(beta-D-galactopyranosyl-3-sulphate)-2-*O*-hexadecanoyl -1-*O*-hexadecyl-L-glycerol, "seminolipid," *J. Chem. Soc. Perkin Trans.* 1 **1979**:712.

Gigg, R., Penglis, A. A. E., and Conant, R., 1977, Synthesis of 3-*O*-(alpha-D-glucopyranosyl)-1,2-di-*O*-stearoyl-L-glycerol, a "glucosyl diglyceride," *J. Chem. Soc. Perkin Trans.* 1 **1977**:2014.

Gigg, R., Penglis, A. A. E., and Conant, R., 1980, Synthesis of 3-*O*-(6-deoxy-6-sulpho-alpha-D-glucopyranosyl)-1,2-di-*O*-hexadecanoyl-L-glycerol, "sulphoquinovosyl diglyceride," *J. Chem. Soc. Perkin Trans.* 1 **1980**:2490.

Heinz, E., 1971, Semisynthetic galactolipids of plant origin, *Biochim. Biophys. Acta* **231**:537.

Heinz, E., and Tulloch, A. P., 1969, Reinvestigation of the structure of acyl galactosyl diglyceride from spinach leaves, *Z. Physiol. Chem.* **350**:493.

Heinz, E., Siebertz, H. P., and Linscheid, M., 1979, Synthesis and enzymatic conversion of an ether analogue of monogalactosyl diacylglycerol, *Chem. Phys. Lipids* **24**:265.

Heisig, O. M. R. A., and Heinz, E., 1972, Semisynthetic lysogalactolipids of plant origin, *Phytochemistry* **11**:815.

Kaplun, A. P., Kalugin, V. E., Shvets, V. I., and Evstigneeva, R. P., 1975, Stereospecificity of glycosylation during glucosylglycerol synthesis, *Bioorg. Khim.* **1**:1675 (*Soviet J. Bioorg. Chem.* **1**:1212).

Kaplun, A. P., Shvets, V. I., and Evstigneeva, R. P., 1977a, Studies of glycosyl diglycerides. VI. Study of the glycosylation of glycerol derivatives by orthoesters and D-glucose or D-galactose acylhalogenoses, *Bioorg. Khim.* **3**:222 (*Soviet J. Bioorg. Chem.* **3**:165).

Kaplun, A. P., Shvets, V. I., and Evstigneeva, R. P., 1977b, Investigations in the field of glycosyl diglycerides. VII. Stereochemical direction of the glycosylation of glycerol derivatives by orthoesters and D-mannose alpha-acylhalogenoses, *Zh. Org. Khim.* **13**:1606 (*J. Org. Chem. USSR* **13**:1483).

Kaplun, A. P., Shvets, V. I., and Evstigneeva, R. P., 1978, Studies in the field of glycosyl di-glycerides. VIII. Synthesis of naturally occurring 1,2-*trans*- and 1,2-*cis*-glycosyl diglycerides, *Zh. Org. Khim.* **14**:256 (*J. Org. Chem. USSR* **14**:236).

Kenny, G. E., and Newton, R. M., 1973, Close serological relationship between glycolipids of *Mycoplasma pneumoniae* and glycolipids of spinach, *Ann. NY Acad. Sci.* **225**:54.

Mannock, D. A., Lewis, R. N. A. H., and McElhaney, R. N., 1984, An improved procedure for the preparation of 1,2-di-*O*-acyl-3-*O*-(β-D-glucopyranosyl)-*sn*-glycerols, *Chem. Phys. Lipids* **43**:113.

Manthorpe, P. A., and Gigg, R., 1980, Allyl ethers as protecting groups. Synthesis of 3-*O*-[6-*O*-(alpha-D-galactopyranosyl)-beta-D-galactopyranosyl]-1,2-di-*O*-stearoyl-L-glycerol, a "digalactosyl diglyceride," *Methods Carbohydr. Chem.* **8**:305.

Morozova, N. G., Volkova, L. V., Dmitrieva, O. A., and Evstigneeva, R. P., 1982, Synthesis of deoxyglycosides of 1,2-diglycerides. *Zh. Obshch. Khim.* **52**:1651 (*J. Gen. Chem. USSR* **52**:1460).

Mukaiyama, T., Tanaka, S., and Asami, M., 1982, New method for the synthesis of glycosyl glycerides—the stereoselective reduction of glycosides of 1-alkoxy-3-hydroxyacetone, *Chem. Lett.* **1982**:433.

Nishida, I., and Yamada, M., 1985, Semisynthesis of a spin-labelled monogalactosyl diacylglycerol and its application to the assay for galactolipid-transfer activity in spinach leaves, *Biochim. Biophys. Acta* **813**:298.

Ogawa, T., and Beppu, K., 1982*a*, Synthesis of 3-*O*-glycosyl-1,2-di-*O*-tetradecyl-*sn*-glycerol, *Agric. Biol. Chem.* **46**:255.

Ogawa, T., and Beppu, K., 1982*b*, Synthesis of glycoglycerolipids: 3-*O*-mannooligosyl-1,2-di-*O*-tetradecyl-*sn*-glycerol *Agric. Biol. Chem.* **46**:263.

Ogawa, T., and Beppu, K., 1982*c*, Synthesis of 3-*O*-(2-acetamido-2-deoxy-3-*O*-beta-D-galactopyranosyl-beta-D-galactopyranosyl)-1,2-di-*O*-tetradecyl-*sn*-glycerol, *Carbohydr. Res.* **101**:271.

Ogawa, T., and Horisaki, T., 1983, Synthesis of 2-*O*-hexadecanoyl-1-*O*-hexadecyl-[alpha-Glc-6SO₃Na-(1 → 6)-alpha-Glc-(1 → 6)-alpha-Glc-(1 → 3)-*sn*-glycerol: A proposed structure for the glyceroglucolipids of human gastric secretion and of the mucous barrier of rat-stomach antrum, *Carbohydr. Res.* **123**:C1.

Ogawa, T., and Sugimoto, M., 1984, Synthesis of alpha- and beta-(2 → 9)-linked disialylglycerolipids, *Carbohydr. Res.* **128**:C1.

Ogawa, T., and Takahishi, Y., 1983, Synthesis of and glycosidation by 2-deoxy-2-fluoro-D-mannopyranose, *J. Carbohydr. Chem.* **2**:461.

Ogawa, T., Beppu, K., and Nakabayashi, S., 1981, Trimethylsilyltrifluoromethanesulphonate as an effective catalyst for glycoside synthesis, *Carbohydr. Res.* **93**:C6.

Oltvoort, J. J., van Boeckel, C. A. A., de Koning, J. H., and van Boom, J. H., 1982, A simple approach to the synthesis of a membrane teichoic acid fragment of *Staphylococcus aureus*, *Recl.: J. R. Neth. Chem. Soc.* **101**:87.

Oltvoort, J. J., Kloosterman, M., van Boeckel, C. A. A., and van Boom, J. H., 1984, Synthesis of a lipoteichoic acid-carrier fragment of *Staphylococcus aureus*, *Carbohydr. Res.* **130**:147.

Prinz, H., Six, L., Reuss, K.-P., and Lieflander, M., 1985, Stereoselective synthesis of long-chain 1-*O*-(beta-D-maltosyl)-3-*O*-alkyl-*sn*-glycerols (alkyl glyceryl ether lysoglycolipids), *Liebigs Ann. Chem.* **1985**:217.

Shvets, V. I., Bashkatova, A. I., and Evstigneeva, R. P., 1973, Synthesis of glycosyl diglycerides, *Chem. Phys. Lipids* **10**:267.

Six, L., Ruess, K.-P., and Lieflander, M., 1983, An efficient and stereoselective synthesis of 1,2-*O*-dialkyl-3-*O*-beta-D-glycosyl-*sn*-glycerols, *Tetrahedron Lett.* **24**:1229.

Suzuki, K., and Mukaiyama, T., 1982, Highly stereoselective synthesis of alpha-D-glucopyranosides by the *N*-iodosuccinimide-promoted internal cyclization, *Chem. Lett.* **1982**:1525.

van Boeckel, C. A. A., and van Boom, J. H., 1980, Synthesis of phosphatidyl-alpha-glucosyldiacylglycerol containing palmitic and oleic acid esters, *Tetrahedron Lett.* **21**:3705.

van Boeckel, C. A. A., and van Boom, J. H., 1981, Synthesis of membrane substances: Phosphatidyl-alpha-diglucosyl diglyceride and related glycolipids, *Chem. Lett.* **1981**:581.

van Boeckel, C. A. A., and van Boom, J. H., 1985*a*, Synthesis of phosphatidyl-alpha-glucosyl glycerol containing a dioleoyl phosphatidyl moiety. Application of the tetraisopropyldisiloxane-1,3-diyl (TIPS) protecting group in sugar chemistry. Part III, *Tetrahedron* **41**:4545.

van Boeckel, C. A. A., and van Boom, J. H., 1985*b*, Synthesis of Streptococci phosphatidyl-alpha-diglucosyldiglyceride and related glycolipids. Application of the tetraisopropyldisiloxane-1,3-diyl (TIPS) protecting group in sugar chemistry. Part V, *Tetrahedron* **41**:4567.

van Boeckel, C. A. A., Westerduin, P., and van Boom, J. H., 1981, Synthesis of 2,3-di-*O*-phytanyl-1-*O*[glucosyl (galactosyl)-beta-(1 → 6)-mannosyl-alpha-(1 → 2)-glucosyl-alpha-(1 → 1)]-*sn*-glycerol. Purple membrane glycolipids, *Tetrahedron Lett.* **22**:2819.

van Boeckel, C. A. A., Westerduin, P., and van Boom, J. H., 1984, Synthesis of two purple-membrane glycolipids and the glycolipid sulfate *O*-(beta-ᴅ-glucopyranosyl-3-sulphate)-(1 → 6)-*O*-alpha-ᴅ-mannopyranosyl-(1 → 2)-*O*-alpha-ᴅ-glucopyranosyl-(1 → 1)-2,3-di-*O*-phytanyl-*sn*-glycerol, *Carbohydr. Res.* **133**:219.

van Boeckel, C. A. A., Visser, G. M., and van Boom, J. H., 1985, Synthesis of phosphatidyl-beta-glucosyl glycerol containing a dioleoyl diglyceride moiety. Application of the tetraisopropyldi-siloxane-1,3-diyl (TIPS) protecting group in sugar chemistry. Part IV, *Tetrahedron* **41**:4557.

Volkova, L. V., Suvorova, S. A., Veres, V., Morozova, N. G., and Evstigneeva, R. P., 1979, Synthesis of 1,2-*cis*-substituted glucosylglycerols, *Zh. Obshch. Khim.* **49**:2148 (*J. Gen. Chem. USSR* **49**:1885).

Weber, N., and Benning, H., 1986, Synthesis of ether glyceroglycolipids, *Chem. Phys. Lipids* **41**:93.

Wehrli, H. P., and Pomeranz, Y., 1969, Synthesis of galactosyl glycerides and related lipids, *Chem. Phys. Lipids* **3**:357.

Index

Acetylation of alditols, 18
N-Acetylglucosamine-1-phosphate archaeol, 52, 53
Alanine ester groups in lipoteichoic acids, 128, 129
  distribution along chain, 181–183
  incorporation of D-alanine, 196, 197
  NMR resonances, 185, 186
  pulse-chase, 197
  ratio alanine/P, 181, 183
  stepwise hydrolysis, 181
  turnover and replacement, 197, 198
Alditol acetates
  GLC analysis, 16, 18, 19, 25
  preparation of, 18, 25
Alkyl chain composition, 41, 42
  in anaerobic bacteria, 42, 44
  of dialkylglycerols, 45
  of plasmalogens, 42, 44
Alkyl chlorides or iodides
  from BCl_3 or HI cleavage of ethers, 65, 68
  identification, 65
Analytical procedures
  acid group titration, 13
  aminosugars, 13
  carbazole reaction, 13
  elemental analysis, 9
  Elson–Morgan procedure, 13
  ester groups, 12, 120
  fatty acid methyl esters, 9, 160
  GLC of sugars, 9, 14, 16–18
  glycerol, 9, 13
  glycolipid quantitation, 9, 14, 16–18, 54, 55
  HPLC of glycolipids, 9, 327, 329
  mass spectrometry, 326, 327, 329, 330, 376, 377, 433–439
  mole ratios of constituents, 9, 170
  NMR spectrometry, 11, 27, 28–29, 326, 327, 377, 480, 482
  sugars, 9
  uronic acids, 13
Anomeric carbons, 11, 26
  chromium trioxide oxidation, 27
  configuration by NMR, 11, 27, 28–29
    by specific glycosidases, 27

Antigenicity of glycolipids, 413
  glycopeptidolipid C-mycosides, 413
  a speculation, 414
  surface antigens from mycobacteria, 413
Antitumor activities of cord factor, 404–406
  potentials for improvement, 407–408
Autochthonous tumors, 406
  therapy for, 406–407
Archaebacteria
  glycolipids in, 44, 45, 74–84
Archaebacterial glycolipids
  carbohydrate structure of, 61
  extraction of, 47
  in extreme halophiles, 48, 72–77
  isolation of
    from extreme halophiles, 47
    from methanogens, 51
    from thermoacidophiles, 53
  in methanogens, 52, 78–80
  molecular constituents, 59, 60
  quantitative analysis of, 54, 55, 59, 60
  spectrometric analysis, 57–59
  structure determination, 61
  in thermoacidophiles, 54, 80–85
  TLC of *Halobacterium* lipids, 55, 56, 57
  TLC of methanogen lipids, 58
  TLC mobilities, 49–50
Archaeol; *see also* Diphytanylglycerol
  in archaebacteria, 70, 71, 74–75, 76, 78
  C_{20},C_{25} analogue, 45, 46
  C_{25},C_{25} analogue, 45, 46
  cyc-C_{40} analogue, 45, 46
  cleavage of ether linkages, 65, 68
  definition, 2, 45
  derived glycolipids, 48, 52
  derived phosphoglycolipids, 52
  derived sulfated glycolipids, 48, 74–75
  glycerol configuration, 45, 65, 67
  HPLC separation, 65
  identification, 64, 65
  infrared (IR) spectra, 65
  mass spectra (CI-MS), 67
  NMR spectra, 65, 66
  optical rotation (molecular rotation), 62
  structure, 46
  TLC separation, 64